AF356177

# Human Papillomavirus and Cancers

# Human Papillomavirus and Cancers

Maria Lina Tornesello
Franco M. Buonaguro

Basel • Beijing • Wuhan • Barcelona • Belgrade • Novi Sad • Cluj • Manchester

Maria Lina Tornesello
Department of
Translational Research
IRCCS Fondazione Pascale
Napoli
Italy

Franco M. Buonaguro
Department of
Translational Research
IRCCS Fondazione Pascale
Napoli
Italy

*Editorial Office*
MDPI AG
Grosspeteranlage 5
4052 Basel, Switzerland

This is a reprint of articles from the Special Issue published online in the open access journal *Cancers* (ISSN 2072-6694) (available at: www.mdpi.com/journal/cancers/special_issues/HP_Cancer).

For citation purposes, cite each article independently as indicated on the article page online and using the guide below:

Lastname, A.A.; Lastname, B.B. Article Title. *Journal Name* **Year**, *Volume Number*, Page Range.

**ISBN 978-3-7258-1714-6 (Hbk)**
**ISBN 978-3-7258-1713-9 (PDF)**
**https://doi.org/10.3390/books978-3-7258-1713-9**

# Contents

# About the Editors

**Maria Lina Tornesello**

Dr Maria Lina Tornesello is head of the Progression Markers and Viruses laboratory, and acting Director of Molecular Biology and Viral Oncology, at the Istituto Nazionale Tumori Fondazione Pascale, Napoli, Italy. Dr Tornesello graduated summa cum laude in Biological Science in 1988 and achieved specialty in Microbiology and Virology in 1994 from the University Federico II (Napoli, Italy). She has carried out post-doctoral research programs at the Istituto Nazionale Tumori Fondazione Pascale and the Tumor Biology Program, Fred Hutchinson Cancer Research Center, Seattle, WA, USA. In 1993, she became an assistant member and in 2001 an associate member of the Experimental Oncology at the Istituto Nazionale Tumori Fondazione Pascale. The research activities of Dr Tornesello mainly focus on viral-related tumorigenesis, with a particular focus on the study of viral and host interactions in HPV and genital cancer, HHV8 and Kaposi sarcoma, and HCV/HBV and liver cancer. More recently, her research activities have been extended to the study of genetic alterations and pathway deregulation in the pathogenesis of virus-driven tumors.

**Franco M. Buonaguro**

Franco M. Buonaguro, M.D. Director Emeritus of Molecular Biology and Viral Oncology Unit at Natl Cancer Inst, Naples (IT), graduated cum laude in 1977 at the "Federico II" Medical School, Naples (IT). He is board-certified in Endocrinology (1982) and Microbiology and Virology (1992). He is a postdoctoral fellow at the Department of Cell Biology, Argonne Natl Laboratory, Argonne, IL, USA (1979–81), WHO Fellow and research associate at the Tumor Biology Program, FHCRC Seattle USA (1983–86). Since 1987, he has been a member at the Natl Cancer Institute "Istituto Nazionale Tumori - IRCCS Fondazione Pascale", Naples (IT).

# Preface

The human papillomaviruses (HPVs) are the major cause of nearly all cervical cancers and a significant fraction of several other human malignancies arising from the mucosal squamous epithelia of the anus, vagina, vulva, penis and oropharynx. Despite the promise of HPV vaccination for the global prevention of cervical cancer, HPV-related cancers will remain a major health problem for decades to come.

The long-term persistent infection, the integration of the viral DNA in the human genome and the constitutive expression of HPV oncoproteins cause the accumulation of various molecular changes in the infected cells, leading to cancer development and progression. Many genetic and epigenetic alterations, as well as complex molecular networks, have been identified by "omics" technologies in HPV-related cancers.

The articles of this reprint cover several relevant aspects of the HPV-related cancer topic, including the role of oncoviral proteins in cell transformation, the gene mutational profile of viral and host genomes, expression levels of miRNAs, gene methylation, immune response and new therapeutic opportunities, including cancer vaccines for HPV-related cancers.

**Maria Lina Tornesello and Franco M. Buonaguro**

*Editors*

*Editorial*

# Human Papillomavirus and Cancers

**Maria Lina Tornesello * and Franco M. Buonaguro ***

Molecular Biology and Viral Oncology Unit, Istituto Nazionale Tumori IRCCS "Fondazione G. Pascale", via Mariano Semmola, 80131 Naples, Italy

* Correspondence: m.tornesello@istitutotumori.na.it (M.L.T.); f.buonaguro@istitutotumori.na.it (F.M.B.); Tel.: +39-081-590-3609 (M.L.T.); +39-081-590-3830 (F.M.B.)

Received: 24 November 2020; Accepted: 8 December 2020; Published: 15 December 2020

Persistent infection with oncogenic human papillomaviruses (HPVs) is the main cause of nearly all cervical cancers as well as of a significant proportion of other malignancies arising from the mucosal squamous epithelia of the anogenital tract as well as of the head and neck region [1]. While HPV vaccination programs will have a great impact on the global prevention of cervical neoplasia, other HPV-related cancers will continue to be a serious health problem for some decades to come [2]. Many efforts are still necessary to understand the complex interplay between viral and host factors and to find the best approach for prevention, early diagnosis, and treatment of HPV-related diseases.

In the Special Issue titled "Human papillomavirus (HPV) and cancer", several experts from all over the world have reported and discussed the most recent advances from basic science to clinical management of HPV-related malignancies. A special focus has been given to the oncogenic role of epigenetic factors, viral proteins and immune response in HPV-driven cancers, as well as to the new anti-cancer opportunities including HPV-based therapeutic vaccines.

Novel classes of non-coding RNAs (ncRNAs), such as circular RNAs (circRNAs), pico RNAs (piRNAs) and long noncoding RNAs (lncRNAs), have been found deregulated differently in diverse histotypes of HPV-driven tumours. Casarotto et al. in their review reported that two oncogenic circRNAs are over-expressed and able to sponge specific miRNAs in HPV-positive cervical cancer and derived cell lines [3]. On the other hand, a large number of piRNAs have been found differently expressed in HPV-positive and HPV-negative head and neck squamous cell carcinoma (HNSCC). Among these, the expression of five piRNAs has shown to be associated with worse overall survival in viral-driven HNSCC. PiRNAs, similar only in size to miRNAs, are able to associate specifically with P-element-induced wimpy testis (PIWI) proteins causing epigenetic changes that are important for cell transformation [3].

The role of HPV infection in HNSCC and its significance as a prognostic marker that is indicative of clinical outcome has emerged in the last few decades [4]. The incidence of HPV-driven HNSCC varies in diverse geographical regions [5]. Alsbeih et al. investigated the HPV distribution and its prognostic value in HNSCC of Saudi patients [6]. The results of the study showed that HPV prevalence is significantly lower in Saudi HNSCC than in other parts of the world and, consistent with studies performed in other countries, they observed that patients with HPV/p16 positive tumours had a better overall survival.

The alteration of metabolism in cancer cells is crucial for tumour growth; however, the effect of HPV infection on metabolic pathways has not been well characterized in HNSCC. Prusinkiewicz et al. analysed the RNAseq profiles of HPV-positive and HPV-negative HNSCCs retrieved from the TCGA and identified many differentially expressed metabolic genes between the two cancer groups [7]. Importantly, genes involved in glycolysis were down-regulated while those involved in the tricarboxylic acid cycle, oxidative phosphorylation, and β-oxidation were up-regulated in HPV-positive HNSCC compared to HPV-negative HNSCC. The reduced expression of several genes was predictive of better survival in patients with HPV-positive HNSCC.

Many biomarkers have been identified in HPV-negative HNSCC; however, few of them have been validated for clinical use. Mittal et al. investigated the prognostic role of centrosome amplification in

HPV-negative oropharyngeal SCC (OPSCC) [8]. Centrosome amplification was found more expressed in HPV-negative compared to HPV-positive OPSCC biopsies and associated with poor overall survival. Therefore, centrosome amplification may serve as a novel prognostic biomarker for patients with HPV-negative OPSCC.

The expression of cell cycle regulators such as D-type cyclins has been found frequently deregulated in HNSCC. Novotný et al. performed a comparative analysis of paired tumour/peri-tumour tissues and showed that cyclin D1 was upregulated in 18% of HNSCC and downregulated in 23% of carcinomas, mainly in HPV-positive samples [9]. Moreover, the change in CCND1 expression was observed to be compensated by cyclin D2 expression independently from the HPV status.

The class II major histocompatibility complex (MHC class II) molecules become expressed on the surface of epithelial cells during the inflammation process and they can function as antigen presenting cells (APCs). Gameiro et al. analysed the HNSCC RNA-seq data retrieved from TCGA in order to determine the effect of HPV on the expression of MHC-II genes and other immune related genes [10]. All MHC-II genes along with genes encoding various co-stimulatory molecules involved in T-cell activation were found to be significantly upregulated in HPV-positive tumours compared to HPV-negative HNSCC. Therefore, the highly immunogenic tumour microenvironment observed in HPV-positive HNSCC may be due to the antigen presentation of epithelial cells.

The favourable prognosis of HPV-positive/p16-positive cancer patients, as reported in several clinical trials, demands for de-escalation therapies. However, it is very important to implement HPV testing methods that accurately distinguish HPV-driven OPSCC from HPV-negative tumours. Borena et al. used a sandwich ELISA test to detect the expression of HPV 16, 18 and 45 E7 oncoproteins and compared their results with HPV DNA positivity and p16 immunohistochemistry (IHC) as the reference method [11]. The authors found a significant concordance between E7 oncoprotein detection and the reference method and propose larger studies to confirm the diagnostic value of the assay.

The juvenile-onset recurrent respiratory papillomatosis (JoRRP), caused by the infection with HPV 6 and 11, is a rare and severe respiratory disease that follows an unpredictable clinical course. Lépine et al. analysed viral biomarkers of JoRRP severity by using a chromogenic in situ hybridization (CISH) method detecting the E6 and E7 mRNA of HPV 6 and 11 [12]. They stratified samples in low staining versus high staining and concluded that HPV E6 and E7 CISH might be a biomarker predictive of disease aggressiveness in JoRRP.

Several infectious agents, in addition to oncogenic viruses, may contribute to increase the risk of cancer development. Kofler et al. investigated the role of some sexually transmitted pathogens, including Ureaplasma spp., Chlamydia trachomatis, Mycoplasma hominis, Mycoplasma genitalium and HPV, in oropharyngeal carcinoma [13]. HPV DNA was detected in almost 70% of the oropharyngeal cell exfoliates collected by brushing from OPSCC patients. Conversely, the low prevalence of Ureaplasma spp. and the absence of the other pathogens among patients with oropharyngeal cancer do not support their oncogenic role. Moreover, HPV detection in cell brushing of OPSCC patients following surgery has a prognostic significance. Indeed, seven out of 62 HPV positive patients remained positive at post-treatment follow-up and five had a tumour relapse/progression. Importantly, all HPV-negative patients remained free of disease suggesting the relevance of HPV testing after treatment [14].

Clinical behaviour of HNSCC mainly depends on the site and the HPV status. Numerous studies have shown that virus-related HNSCCs possess peculiar clinical and biological features. Perri et al. described the molecular differences and similarities between HPV-negative and HPV-positive HNSCC including the better prognosis and response to therapies of the latter group of patients [15]. Remarkably, patients with HPV-driven HNSCCs are frequently more responsive to conservative treatments and immunotherapy, opening questions about the use of pre-therapy assessment in order to guide the treatment strategy.

The antiviral agent cidofovir has been previously shown to have an anti-proliferative effect on cervical cancer derived cell lines. Verhees et al. investigated the effect of cidofovir on the growth of HPV-positive and -negative HNSCC-derived cell lines [16]. Cidofovir was able to inhibit the cell proliferation and to cause γ-H2AX accumulation as well as upregulation of DNA repair proteins. The effect was marked in HPV-positive cells and indicative of the occurrence of mitotic catastrophe.

The incidence of HPV-related cancers is significantly higher in HIV-positive subjects compared to the general population and highly active anti-retroviral therapy (HAART) does not seem to have changed the trend. Alam et al. reported that protease inhibitors (PI) caused destabilization of cell-cell junctions within the stratified epithelium of three-dimensional tissues derived from primary human gingiva and cervical epithelial cells [17]. This caused enhanced infectivity of HPV16 to the basal layers and de novo virus biosynthesis suggesting an increased the risk of HPV-driven oral and cervical cancer development in HIV-positive patients under HAART treatment.

Therapeutic targeting of viral oncogenes represents a promising strategy to treat HPV-related cancers. Ehrke-Schulz et al. employed the adenoviral vectors HCAdVs expressing the CRISPR/Cas9 machinery to specifically inactivate the HPV18 or HPV16 E6 genes in HeLa, SiHa and CaSki cervical cancer derived cell lines [18]. Transduced cervical cancer cells showed increased apoptosis and decreased proliferation suggesting that HCAdV can serve as a therapeutic agent when armed with HPV-type-specific CRISPR/Cas9.

Specific antibodies in a single-chain format (scFvs) able to target the E6 and E7 oncoproteins of high-risk HPVs represent new therapeutic molecules against HPV-associated neoplastic lesions. Amici et al. evaluated the antigen-binding capacity of diverse anti-16E7 scFvs by using E7 mutants as well as by performing computational analyses to define length, total net charge, hydrophobicity, polarity, and charge distribution to better adapt the antibodies for clinical use [19]. Hence, scFvs may represent valid candidates for HPV-related cancer immunotherapy.

DNA-based cancer vaccines have a promising application in the prevention as well as in the treatment of diverse types of tumours, including HPV-related cancers. Franconi et al. discussed how the immunogenicity of DNA vaccines can be improved by fusing of HPV antigens with plant gene sequences as well as by using plant-derived immunomodulatory agents [20]. Such molecules have shown to interfere with the processes of carcinogenesis by modulating many different cellular pathways and reducing the drug resistance of tumours.

HPV is the most common sexually transmitted infection in the world and in the large majority of cases, the virus is cleared by the immune system. However, in several cases, HPV is able to escape the immune control and to cause cancer development. Ferreira et al. described several strategies of immune evasion adopted by HPVs and in particular the ability of E6 and E7 to synergistically target almost all cellular innate immune pathways [21].

More than 90% of anal cancers are associated with HPV infection. The study by Carter et al. aimed at identifying cancer-associated circulating cells in peripheral blood of anal cancer patients in order to monitor treatment efficacy and to diagnose relapse [22]. Nucleated cells were isolated from the blood of anal cancer patients and cancer cells identified by using pan-cytokeratin and CD45 antibodies. The successful identification of cancer cells opened new opportunities for the diagnosis and prognosis of anal cancer patients.

The growth of human keratinocytes in three-dimensional (3D) cultures emulating the original stratified epithelium represents an important technical advancement for in vitro studies of HPV-related carcinogenesis. De Gregorio et al., in their review, critically described the diverse in vitro models of HPV-related cancers from the simplified "raft culture" to the complex three-dimensional (3D) organotypic models with a special focus on the artificial tissues containing the tumour microenvironment (TME) components [23]. Novelties in the field of 3D cultures of HPV-associated anogenital and oropharyngeal malignancies have been extensively discussed. Moreover, Maru et al. reported interesting results on the propagation of organoids from the normal squamocolumnar junction (SCJ) cells of the uterine cervix by using Matrigel-based three-dimensional cultures [24]. Such organoids characterized by a dense structure contained mainly squamous cells and in some cases a few mucin-producing endocervix cells. Therefore, such organoids are likely to provide a novel platform to study HPV carcinogenesis in ecto and endocervical cells.

Cervical intraepithelial neoplasia (CIN) may either regress or progress to cervical cancer depending on several factors related to the virus and host interplay. There are no validated biomarkers able to distinguish progressing from regressing neoplastic lesions. Taguchi et al. used a continuous-time multistate Markov model to estimate the bidirectional transition of cervical lesions for designated HPV genotypes [25]. Their model was applied to a retrospective cohort and was able to highlight major differences in the progression patterns between high-risk HPV-related CINs.

Cervical cancer prevention is mainly based on the screening of cytological smears and the treatment of precancerous lesions. The HPV testing in self-sampled cervical exfoliates has been shown to increase the participation of women in oncological screenings particularly in never- or under-screened populations. Hawkes et al., in their review, discussed several studies that have examined the self-collection of cervical samples for HPV-based cervical screening including the collection devices and assays [26]. The authors concluded that self-collection will be crucial for global screening programs and the elimination of cervical cancer as a public health problem.

In conclusion, the HPV and cancer Special Issue covers a substantial portion of the recent knowledge and latest accomplishments in the HPV-related cancer field.

**Author Contributions:** M.L.T. and F.M.B. wrote the editorial. All authors have read and agreed to the published version of the manuscript.

**Funding:** This research received no external funding.

**Acknowledgments:** We acknowledge the authors for their contribution in this Special Issue.

**Conflicts of Interest:** The authors declare no conflict of interest.

## References

1. De Sanjosé, S.; Serrano, B.; Tous, S.; Alejo, M.; Lloveras, B.; Quirós, B.; Clavero, O.; Vidal, A.; Ferrándiz-Pulido, C.; Pavón, M.Á.; et al. RIS HPV TT, VVAP and head and neck study groups. Burden of human papillomavirus (HPV)-related cancers attributable to HPVs 6/11/16/18/31/33/45/52 and 58. *JNCI Cancer Spectr.* **2019**, *2*, pky045. [CrossRef]

2. Van Schalkwyk, C.; Moodley, J.; Welte, A.; Johnson, L.F. Estimated impact of human papillomavirus vaccines on infection burden: The effect of structural assumptions. *Vaccine* **2019**, *37*, 5460–5465. [CrossRef]

3. Casarotto, M.; Fanetti, G.; Guerrieri, R.; Palazzari, E.; Lupato, V.; Steffan, A.; Polesel, J.; Boscolo-Rizzo, P.; Fratta, E. Beyond microRNAs: Emerging role of other non-coding RNAs in HPV-driven cancers. *Cancers* **2020**, *12*, 1246. [CrossRef]

4. Tornesello, M.L.; Perri, F.; Buonaguro, L.; Ionna, F.; Buonaguro, F.M.; Caponigro, F. HPV-related oropharyngeal cancers: From pathogenesis to new therapeutic approaches. *Cancer Lett.* **2014**, *351*, 198–205. [CrossRef]

5. Chaturvedi, A.K.; Anderson, W.F.; Lortet-Tieulent, J.; Curado, M.P.; Ferlay, J.; Franceschi, S.; Rosenberg, P.S.; Bray, F.; Gillison, M.L. Worldwide trends in incidence rates for oral cavity and oropharyngeal cancers. *J. Clin. Oncol.* **2013**, *31*, 4550–4559. [CrossRef]

6. Alsbeih, G.; Al-Harbi, N.; Bin Judia, S.; Al-Qahtani, W.; Khoja, H.; El-Sebaie, M.; Tulbah, A. Prevalence of human papillomavirus (HPV) infection and the association with survival in Saudi patients with head and neck squamous cell carcinoma. *Cancers* **2019**, *11*, 820. [CrossRef]

7. Prusinkiewicz, M.A.; Gameiro, S.F.; Ghasemi, F.; Dodge, M.J.; Zeng, P.Y.F.; Maekebay, H.; Barrett, J.W.; Nichols, A.C.; Mymryk, J.S. Survival-associated metabolic genes in human papillomavirus-positive head and neck cancers. *Cancers* **2020**, *12*, 253. [CrossRef]

8. Mittal, K.; Choi, D.H.; Wei, G.; Kaur, J.; Klimov, S.; Arora, K.; Griffith, C.C.; Kumar, M.; Imhansi-Jacob, P.; Melton, B.D.; et al. Hypoxia-induced centrosome amplification underlies aggressive disease course in HPV-negative oropharyngeal squamous cell carcinomas. *Cancers* **2020**, *12*, 517. [CrossRef]

9. Novotný, J.; Bandúrová, V.; Strnad, H.; Chovanec, M.; Hradilová, M.; Šáchová, J.; Šteffl, M.; Grušanović, J.; Kodet, R.; Pačes, V.; et al. Analysis of HPV-positive and HPV-negative head and neck squamous cell carcinomas and paired normal mucosae reveals cyclin D1 deregulation and compensatory effect of cyclin D2. *Cancers* **2020**, *12*, 792. [CrossRef]

10. Gameiro, S.F.; Ghasemi, F.; Barrett, J.W.; Nichols, A.C.; Mymryk, J.S. High level expression of MHC-II in HPV+ head and neck cancers suggests that tumor epithelial cells serve an important role as accessory antigen presenting cells. *Cancers* **2019**, *11*, 1129. [CrossRef]

11. Borena, W.; Schartinger, V.H.; Dudas, J.; Ingruber, J.; Greier, M.C.; Steinbichler, T.B.; Laimer, J.; Stoiber, H.; Riechelmann, H.; Kofler, B. HPV-induced oropharyngeal cancer and the role of the E7 oncoprotein detection via brush test. *Cancers* **2020**, *12*, 2388. [CrossRef]

12. Lépine, C.; Voron, T.; Berrebi, D.; Mandavit, M.; Nervo, M.; Outh-Gauer, S.; Péré, H.; Tournier, L.; Teissier, N.; Tartour, E.; et al. Juvenile-onset recurrent respiratory papillomatosis aggressiveness: In situ study of the level of transcription of HPV E6 and E7. *Cancers* **2020**, *12*, 2836. [CrossRef]

13. Kofler, B.; Laimer, J.; Bruckmoser, E.; Steinbichler, T.B.; Runge, A.; Schartinger, V.H.; von Laer, D.; Borena, W. The role of HPV and non-HPV sexually transmitted infections in patients with oropharyngeal carcinoma: A case control study. *Cancers* **2020**, *12*, 1192. [CrossRef]

14. Kofler, B.; Borena, W.; Dudas, J.; Innerhofer, V.; Dejaco, D.; Steinbichler, T.B.; Widmann, G.; von Laer, D.; Riechelmann, H. Post-treatment HPV surface brushings and risk of relapse in oropharyngeal carcinoma. *Cancers* **2020**, *12*, 1069. [CrossRef]

15. Perri, F.; Longo, F.; Caponigro, F.; Sandomenico, F.; Guida, A.; della Vittoria Scarpati, G.; Ottaiano, A.; Muto, P.; Ionna, F. Management of HPV-related squamous cell carcinoma of the head and neck: Pitfalls and Caveat. *Cancers* **2020**, *12*, 975. [CrossRef]

16. Verhees, F.; Legemaate, D.; Demers, I.; Jacobs, R.; Haakma, W.E.; Rousch, M.; Kremer, B.; Speel, E.J. The antiviral agent cidofovir induces DNA damage and mitotic catastrophe in HPV-positive and -negative head and neck squamous cell carcinomas in vitro. *Cancers* **2019**, *11*, 919. [CrossRef]

17. Alam, S.; Chatterjee, S.; Kang, S.D.; Milici, J.; Biryukov, J.; Chen, H.; Meyers, C. Anti-retroviral protease inhibitors regulate human papillomavirus 16 infection of primary oral and cervical epithelium. *Cancers* **2020**, *12*, 2664. [CrossRef]

18. Ehrke-Schulz, E.; Heinemann, S.; Schulte, L.; Schiwon, M.; Ehrhardt, A. Adenoviral vectors armed with PAPILLOMAVIRUs oncogene specific CRISPR/Cas9 kill human-papillomavirus-induced cervical cancer cells. *Cancers* **2020**, *12*, 1934. [CrossRef]

19. Amici, C.; Donà, M.G.; Chirullo, B.; Di Bonito, P.; Accardi, L. Epitope mapping and computational analysis of anti-HPV16 E6 and E7 antibodies in single-chain format for clinical development as antitumor drugs. *Cancers* **2020**, *12*, 1803. [CrossRef]

20. Franconi, R.; Massa, S.; Paolini, F.; Vici, P.; Venuti, A. Plant-derived natural compounds in genetic vaccination and therapy for HPV-associated cancers. *Cancers* **2020**, *12*, E3101. [CrossRef]

21. Ferreira, A.R.; Ramalho, A.C.; Marques, M.; Ribeiro, D. The interplay between antiviral signalling and carcinogenesis in human papillomavirus infections. *Cancers* **2020**, *12*, 646. [CrossRef]

22. Carter, T.J.; Jeyaneethi, J.; Kumar, J.; Karteris, E.; Glynne-Jones, R.; Hall, M. Identification of cancer-associated circulating cells in anal cancer patients. *Cancers* **2020**, *12*, 2229. [CrossRef]

23. De Gregorio, V.; Urciuolo, F.; Netti, P.A.; Imparato, G. In vitro organotypic systems to model tumor microenvironment in human papillomavirus (HPV)-related cancers. *Cancers* **2020**, *12*, 1150. [CrossRef]

24. Maru, Y.; Kawata, A.; Taguchi, A.; Ishii, Y.; Baba, S.; Mori, M.; Nagamatsu, T.; Oda, K.; Kukimoto, I.; Osuga, Y.; et al. Establishment and molecular phenotyping of organoids from the squamocolumnar junction region of the uterine cervix. *Cancers* **2020**, *12*, 694. [CrossRef]

25. Taguchi, A.; Hara, K.; Tomio, J.; Kawana, K.; Tanaka, T.; Baba, S.; Kawata, A.; Eguchi, S.; Tsuruga, T.; Mori, M.; et al. Multistate markov model to predict the prognosis of high-risk human papillomavirus-related cervical lesions. *Cancers* **2020**, *12*, 270. [CrossRef] [PubMed]

26. Hawkes, D.; Keung, M.H.T.; Huang, Y.; McDermott, T.L.; Romano, J.; Saville, M.; Brotherton, J.M.L. Self-collection for cervical screening programs: From research to reality. *Cancers* **2020**, *12*, 1053. [CrossRef]

**Publisher's Note:** MDPI stays neutral with regard to jurisdictional claims in published maps and institutional affiliations.

*Review*

# Plant-Derived Natural Compounds in Genetic Vaccination and Therapy for HPV-Associated Cancers

**Rosella Franconi [1,*], Silvia Massa [2,*], Francesca Paolini [3], Patrizia Vici [4] and Aldo Venuti [3]**

[1] Division of Health Protection Technology, Department for Sustainability, Italian National Agency for New Technologies, Energy and Sustainable Economic Development, ENEA, 00123 Rome, Italy

[2] Division of Biotechnology and Agroindustry, Department for Sustainability, ENEA, 00123 Rome, Italy

[3] HPV-UNIT—UOSD Tumor Immunology and Immunotherapy, IRCCS Regina Elena National Cancer Institute, 00144 Rome, Italy; francesca.paolini@ifo.gov.it (F.P.); aldo.venuti@ifo.gov.it (A.V.)

[4] Division of Medical Oncology B, IRCCS Regina Elena National Cancer Institute, 00144 Rome, Italy; patrizia.vici@ifo.gov.it

* Correspondence: rosella.franconi@enea.it (R.F.); silvia.massa@enea.it (S.M.); Tel.: +39-06-3048-4482 (R.F.); +39-06-3048-4052 (S.M.)

Received: 4 September 2020; Accepted: 21 October 2020; Published: 23 October 2020

**Simple Summary:** DNA vaccination represents a useful approach for human papillomavirus (HPV) cancer therapy. The therapeutic potential of plant-based natural compounds for control of HPV-associated cancers has been also widely explored. Genetic vaccines for HPV-associated tumors that include plant protein-encoding gene sequences, used alone or in combinations with plant metabolites, are being investigated but are still in their infancy. Main focus of this paper is to provide an overview of the current state of novel therapeutic strategies employing genetic vaccines along with plant-derived compounds and genes. We highlight the importance of multimodality treatment regimen such as combining immunotherapy with plant-derived agents.

**Abstract:** Antigen-specific immunotherapy and, in particular, DNA vaccination provides an established approach for tackling human papillomavirus (HPV) cancers at different stages. DNA vaccines are stable and have a cost-effective production. Their intrinsic low immunogenicity has been improved by several strategies with some success, including fusion of HPV antigens with plant gene sequences. Another approach for the control of HPV cancers is the use of natural immunomodulatory agents like those derived from plants, that are able to interfere in carcinogenesis by modulating many different cellular pathways and, in some instances, to reduce chemo- and radiotherapy resistance of tumors. Indeed, plant-derived compounds represent, in many cases, an abundantly available, cost-effective source of molecules that can be either harvested directly in nature or obtained from plant cell cultures. In this review, an overview of the most relevant data reported in literature on the use of plant natural compounds and genetic vaccines that include plant-derived sequences against HPV tumors is provided. The purpose is also to highlight the still under-explored potential of multimodal treatments implying DNA vaccination along with plant-derived agents.

**Keywords:** plant molecules; HPV-related tumors; DNA vaccination; multimodal treatments; immunomodulation; immunotherapy; combined DNA vaccine/plant molecule therapy; chimeric vaccine

---

## 1. Introduction

Cancer is considered the leading cause of death in wealthy countries, and 15–20% of all human cancers are associated with viral infections [1,2]. Human papillomavirus (HPV) is the most common sexually transmitted virus and HPV-related cancers account for 8% of all human cancers and their annual incidence is approximately 15 per 100,000 among women and men [3,4]. The incidence of HPV-related cancers remains high despite the introduction of prophylactic vaccines [5]. Currently, commercially available prophylactic HPV vaccines are suggested for use in women up to 45 years of age, but are mainly administered in the 9–15 years' cohort. Since most cancers develop in decades after the initial HPV infection, the impact of this vaccination program will only be seen in the long-term. Therefore, the creation of a therapeutic vaccine able to provide similar results to treatments in use in clinical practice, such as surgery or chemotherapy, represents a challenge for the eradication of HPV-induced tumors. However, no therapeutic vaccines are licensed for clinical use yet. Currently, several types of therapeutic HPV vaccines are being tested [6–9]. In this article, the status of therapeutic, "plant-inspired" HPV genetic vaccines is reviewed, together with the therapeutic potential of plant-based natural compounds. The analysis of published data demonstrated that the power of plant-based molecules/vaccines in the development of therapeutic vaccines against HPV-disease is very strong and that plant molecules may render the immune system more prone to a vaccine response.

In the past, we described that a plant extract of *Nicotiana benthamiana*, containing ectopically expressed HPV 16 E7 protein, induced a cell-mediated immune response able to protect vaccinated mice from tumor challenge, notably without any adjuvant [10,11].

This extract induced maturation of human dendritic cells (DCs) that became able to prime in vitro human blood-derived T lymphocytes from healthy individuals to induce HPV 16 E7-specific cytotoxic response [12]. A similar ability to increase immunogenicity was described for an E7 protein-based vaccine produced in the microalga *Chlamydomonas reinhardtii* [13]. A common and relevant feature of these vaccines was the intrinsic adjuvant activity.

Meanwhile, it was established that enhanced release of HPV antigens from tumor cells pretreated with chemo/radiotherapy can be modulated by plant-derived chemotherapeutic agents. Indeed, chemotherapeutic agents such as Apigenin [14] and Epigallocatechin [15] induced a powerful cell-mediated response when used in combination with DNA vaccines. On the contrary, another chemotherapeutic agent such as Saffron [16] has proven its anticancer effects used alone. Therefore, it is important to explore the mechanisms of action of both plant molecules and DNA vaccination to identify the best combination for HPV-related cancer treatment.

Beside plant molecules, plant viruses and gene sequences encoding plant proteins (including signal sequences) have been employed to improve HPV therapeutic genetic vaccine. Fusion of HPV 16 E7 gene to the Potato Virus X coat protein (PVX-CP) gene increased the rate of proteasomal degradation in transfected cells and, as a consequence, vaccine efficacy [17]. Indeed, peptides produced by the proteasomal degradation of cytosolic proteins bind to MHC I molecules and the rate of antigen degradation by the ubiquitin-proteasome pathway affects antigen presentation by MHC I, that is of pivotal importance for vaccine activity. In addition, we demonstrated that sequences derived from plant proteins improved genetic E7-based vaccine formulations [9,18]. Further, a plant signal sequence fused to synthetic E7 and L2 (i.e., the minor HPV capsid protein) genes of HPV 16 was able to elicit strong specific IgG humoral responses associated to E7 specific T-cell mediated immunity [19,20]. This chimeric vaccine, with preventive and therapeutic effects against HPV infections, offers excellent prospects for the future of DNA vaccine research. These promising efforts to create new therapeutic vaccines will help control HPV-associated malignancies alongside conventional methods of treating HPV [21].

In this review, we focus on the most relevant aspects of plant-derived compounds and genetic vaccines that might be decisive for the future development of cost-effective HPV vaccines.

## 2. HPV Carcinogenesis

The high-risk (HR) HPV types (i.e., HPV 16, 18, 31, 33, 35, 39, 45, 51, 52, 56, 58, 59, 68, 73, 82) are considered to be the main etiological agents of genital tract cancers, such as cervical, vulvar, vaginal, penile, and anal cancers, and of a subset of head and neck cancers. Among the 15 most frequent oncogenic HPV types associated with these cancers, HPV 16 is the most common and associated with the highest risk of progression to cancer [22–24].

The primary evidence of cancer development is HPV integration into the host genome in malignant tumors. HPV integration can take place in multiple non-recurrent regions of amplification and in flanked regions where deletions occur. There is a robust association between HPV insertional breakpoints and genomic structural alterations, which ultimately results in genomic instability, a peculiar sign of HPV-positive tumors. HPV alone is necessary but not sufficient to induce tumors, while it holds an important role in cancer maintenance. However, other genetic and epigenetic events are required for cancer development. This phenomenon may partly explain the latency period that occurs in tumor development and how persistent infection with HR HPVs is necessary for progression to high-grade lesions or cancer [25,26].

Other risk factors for progression to high-grade dysplasia and cancer include age over 30 years, infection with multiple HPV types, immunosuppression, and tobacco use [27,28].

HR HPV E6 and E7 are oncoproteins that bind and promote degradation of tumor suppressor proteins, p53 and retinoblastoma (pRb), respectively. However, HPVs interact with many other cancer-relevant pathways, even in a p53- and/or pRb-independent way. In addition, these oncoproteins may deregulate intracellular microRNA (miRNA) networks, and many HPV-altered miRNAs have been linked to carcinogenesis [29]. E6/E7 oncoproteins represent an ideal set of targets for a therapeutic vaccine against HPV-associated cancer because these proteins not only induce tumorigenesis but also are constitutively expressed in HPV-infected pre-malignant and malignant cells. Since there is evidence that regression of HPV-associated lesions is linked to the presence of a cellular, but not humoral immune response, a therapeutic vaccine able to induce a selective robust E6/E7-specific T-cell response is highly welcome [30].

## 3. HPV Vaccines

Identifying HPV as an etiological factor of cervical cancer and other HPV-associated malignancies helped in the development of immunization strategies to prevent infection and associated diseases caused by HPV. Since 2013, HPV vaccines (bivalent and quadrivalent) have been included in the national immunization programs of at least 66 nations, including North America and Western Europe, primarily [28]. Recombinant HPV virus-like particles (VLPs) are being produced at commercial level via heterologous expression of the major capsid protein L1 in yeast or insect cells [31]. From the morphology viewpoint, VLPs are similar to natural HPV virions with considerable potentialities to induce animal and human type-specific antibody responses [32].

### 3.1. HPV Preventive Vaccines

Preventive HPV vaccines aim to prevent HPV infection by inducing a neutralizing antibody response. Improved understanding of protective humoral immune response against primary HPV infection has led to the development of preventive HPV vaccines targeting L1 and/or L2 viral capsid proteins [33]. Because of the difficulties of in vitro cultivating HR HPVs and their oncogenic nature, live attenuated or inactivated vaccines could not be safely developed for humans. Therefore, studies were focused on alternative methodologies as virus-like particles (VLPs). It was demonstrated that inoculum of VLPs from L1 protein of specific papillomaviruses (PVs) could protect against PV infection [34].

Development of this technology led to the production of current VLP-based preventive vaccines targeting L1 in order to generate neutralizing antibodies against HPV. Commercially available efficacious prophylactic vaccines include the bivalent Cervarix (GlaxoSmithKline, Brentford, UK) [35] as well as multivalent Gardasil-4 and Gardasil-9 (Merck) [36,37]. The recent development of Gardasil-9 has increased preventive coverage from HPV types 6, 11, 16, and 18 to 6, 11, 16, 18, 31, 33, 45, 52, and 58 [34]. Prophylactic HPV vaccines have been shown to effectively prevent vaccinated individuals from contracting HPV infections [38] but these preventive vaccines have not been successful in treating established HPV infections [39].

More than 10 years have elapsed since HPV vaccination was implemented, and a systematic review of HPV vaccination programs in 14 countries that included over 60 million people vaccinated presented evidence of vaccine efficacy [40]. In comparison with the period before vaccine introduction, the prevalence of cervical precancerous lesions decreased by 51% among girls aged 15–19 years and by 31% among women aged 20–24 years at up to 9 years after vaccination began.

However, there is still an urgent need for the development of therapeutic HPV vaccines to tackle existing HPV infections, prevent the development of cancer, and act as immunotherapies for HPV-associated malignancies.

*3.2. Therapeutic Vaccines*

Many different technologies have been utilized to develop therapeutic vaccines and most of them target E6 and/or E7 oncoproteins of HR HPV because they are constantly expressed in HPV-associated cancer [6–8]. It is worth mentioning that the first HPV vaccine administered to women was a live recombinant vaccinia virus expressing the E6/E7 oncoproteins of HPV types 16 and 18 [41]. Live vector vaccines employing viruses (Ankara modified vaccine virus, TG4001 Transgene Inc. France) or bacteria (Listeria monocytogenes, ADXS11-001 Advaxis Inc, Princeton, NJ, USA) are in clinical trials with promising results (NCT03260023 and NCT02853604, respectively). Nevertheless, live vector-based vaccines pose potential safety risks, in particular in immunocompromised people [35]. Furthermore, using the same vector for repeated immunizations leads to a limitation of the immune response [35,38].

Protein- or peptide-based vaccines have been also evaluated and tested in clinical trials with some interesting outcomes for synthetic long peptide-vaccine in early stages of HPV carcinogenesis [42–44]. More challenging approaches such as vaccines based on dendritic cell (DC), tumor cells or adoptive T-cell therapy (ACT) have been developed but they cannot be easily performed and require specialized clinical centers [45–48]. On the contrary, technologies utilizing DNA or RNA vaccines can be easily performed and are in advanced clinical trials, as detailed in the following sections.

3.2.1. DNA Vaccines

Nowadays, nucleic acid therapeutics accounts as promising alternatives to conventional vaccine approaches. Once a DNA vaccine has reached the nucleus of a myocyte, a primary keratinocyte, or a resident antigen presenting cell (APC), the expressed antigen gene is processed by cell machinery. Cross or direct priming of DC produces the presentation of the antigen within the class I or II major histocompatibility complex (MHC) on their surface [49] for immune recognition. However, this process is much more complex and further studies are needed to ensure that DNA vaccines can activate all the complex mechanism of co-stimulatory signals that lead to the activation and expansion of CD4+, CD8+, and naive B-effector cells. In particular, a therapeutic DNA vaccine must be able to generate both a CD8+ response, which directly kills infected or tumor cells, and a CD4+ helper response, which is able to increase and maintain the cytolytic response [50].

In addition, DNA vaccines are characterized by ease of production and high stability. Their safety and use in different administrations without losing their efficacy make them the ideal treatment for the control of HPV infections and associated diseases [51,52]. DNA vaccines also sustain the expression of antigens within cells for longer periods of time when compared with RNA or protein-based vaccines.

Many different HPV DNA vaccines have been constructed and proven to be active in pre-clinical models and few of them are in clinical trials (for review see [21]). In particular, VGX3100 (Inovio Pharmaceuticals Inc., Plymouth Meeting, PA, USA) is close to be used in humans. VGX-3100 is a plasmid DNA-based immunotherapy (HPV 16 E6/E7, HPV 18 E6/E7 DNA delivered intramuscularly by electroporation) under investigation for the treatment of HPV 16 and HPV 18 infection and pre-cancerous lesions of the cervix, vulva, and anus (Phase II/III) (NCT01304524 and NCT03603808) [53].

Two studies that are currently in phase III (NCT03721978 and NCT03185013) using VGX-3100 against cervical cancer show promising results.

VGX-3100 has the potential to be the first approved treatment for HPV infection of the cervix and the first non-surgical treatment for precancerous cervical lesions. VGX-3100 works by stimulating cellular and humoral responses against HPV 16 and HPV 18 E6/E7 oncogenes.

Another DNA vaccine with potential clinical use is the GX-188E (Genexine, Inc., Seongnam, Korea). This vaccine consists of a tissue plasminogen activator (tpa) signal sequence, an FMS-like tyrosine kinase 3 ligand (Flt3L), and shuffled E6 and E7 genes of HPV type 16/18. Flt3L and tpa are included in the fusion gene to promote antigen presentation and trafficking of the fused protein to the secretory pathway, respectively [54]. Recently, GX-188E was described to be highly efficacious in patients with grade 3 cervical intraepithelial neoplasia (CIN3) (NCT02139267) [55].

### 3.2.2. RNA-Based Vaccines

The use of mRNA has several beneficial features: (i) Safety: there is no potential risk of infection or insertional mutagenesis; (ii) efficacy: some adjustments make mRNA stable and highly translatable; (iii) production: mRNA vaccines have the potential for fast, cheap, and scalable production, essentially because of the high yields of in vitro transcription reactions [56]. RNA-based vaccines are created using naked RNA replicons derived from alphaviruses to promote antigen-specific immune response. The replicon-based vectors can replicate in a wide range of cell types, with different expression of antigens. These RNA replicons are less stable than DNA. A combined approach with DNA-launched RNA replicon, termed "suicidal" DNA was developed for HPV vaccine in preclinical models [57]. The anti-tumor properties of some mRNAs expressing oncogene proteins such as E6 and E7 are now known. The therapeutic efficacy of this approach was assessed for TC-1 tumor lesions, demonstrating that the RNA-vaccine induced CD8 T cells to migrate to the tumor tissue [58]. Up-to-date mRNA-based vaccines are developed by CureVac (Tübingen, Germany) and represent a potential new approach in cancer treatment. For the first time, mRNA could be optimized to mobilize the patient's immune system to fight cancer with a specific immune humoral and cellular response elicited by the RNActive® vaccine. CureVac's RNActive® cancer vaccines (CV9103 and CV9104) have successfully completed Phase I/IIa clinical studies in prostate cancer and non-small cell lung cancer [59,60].

Recently, this technology has been used to develop prophylactic vaccines for infectious diseases such as COVID-19 due to the severe acute respiratory syndrome coronavirus 2 (SARS-CoV-2) [61]. Unfortunately, although the strategy of using RNA is so promising and inexpensive, there is still no news on clinical trials for HPV-associated diseases.

### 4. The Role of Adjuvants in Cancer Vaccines

The innate immune system has a key role in triggering an active adaptive T-cell response in the initial phase of in vivo priming [62]. Thus, many molecules that activate innate immunity and support T-cell response have been tested as vaccine adjuvants in clinical use. These adjuvants can also operate as a local depot for antigen protection from degradation [63].

Toll-like receptor (TLR) agonists, (such as CpG, imiquimod and poly I:C, that activate TLR9, and TLR7 and TLR3, respectively [64]), have been employed, as well as cytokines [65,66] and glycolipid ligands [67]. TLRs and agonists of CD40 were able to stimulate tumor specific immunity that in turn elicited cancer regression [68]. Same activity was reported for TLR3 agonists in combination with Freund's incomplete adjuvant (IFA) or anti-PD-1 antibodies [69].

Bacteria- such as IFA and Montanide or virus-derived molecules are the most utilized adjuvants. Bacille Calmette-Guerin (BCG) is utilized as a therapeutic vaccine and its adjuvant activity is mediated mostly by TLRs. On the contrary, TLR independent adjuvant activity can be elicited by cytosolic nucleic acids secreted by bacteria. Cyclic dinucleotides (CDNs) may activate stimulator of interferon genes (STING) that activates TANK-binding kinase 1/interferon regulatory factor 3 (TBK1/IRF3), nuclear factor κB (NF-κB), and signal transducer and activator of transcription 6 (STAT6) signaling pathways causing type 1 Interferon (IFN 1) and proinflammatory cytokine activation in response to deviant host cells (danger associated molecular patterns, DAMPS) or cytosolic double stranded DNA (dsDNA) from pathogens [70,71]. Indeed, cancer vaccines with STING agonists were proven efficacious in different pre-clinical animal models and were shown to induce a marked programmed death ligand 1 (PD-L1) up-regulation, which was associated with tumor-infiltrating CD8(+) IFNγ(+) T cells. A synergistic activity with PD-1 blockade was demonstrated in poorly immunogenic tumors that were no responder to PD-1 blockade alone [72]. Association of anti PD-1/PD-L1 antibodies and STING agonists are now under study in clinical trials (NCT03172936).

Systemic adjuvants are represented by cytokines and monoclonal antibodies. However, conflicting results were reported in human cancer for cytokines [73–75] or antibodies [76] administration and the differences in treatment schedule may account for these results that limit their clinical use. In addition, evidence that a specific cancer vaccine adjuvant is superior to another one is lacking.

Plant extracts are emerging as new adjuvant compounds in cancer vaccines. *Nicotiana benthamiana* plant extracts as well as *Chlamydomonas reinhardtii*, a well-characterized unicellular alga, and hairy root cultures may display adjuvating activity in cancer vaccines significantly eliciting type 1 helper T cells (Th1) and cytotoxic T-lymphocyte (CTL) responses [10,11,13,77–79]. Moreover, plant extracts are able to exert immunomodulatory activity in vitro on DC [12]. Adjuvant activity of plant components has been also reported in studies on Zera® peptide, a self-assembling domain of the maize gamma-zein seed storage protein. This peptide is able to target recombinant proteins to endoplasmic reticulum and to determine their accumulation as protein bodies (PBs). These PBs induce stronger immune responses compared to the soluble recombinant proteins. Zera® peptide has been either fused or combined (i.e., mixed) to a harmless shuffled HPV 16 E7 (16E7SH) synthetic protein. Significantly higher humoral and cellular immune responses to E7 were induced either as ZERA-16E7SH fusion protein or as Zera® PBs mixture with 16E7SH compared to 16E7SH alone. This effect is supposedly determined by a more efficient antigen presentation by PBs and suggests that Zera® may act as an adjuvant [80].

Thus, different plant components may exert common potentiating activity on therapeutic vaccines, which further strengthens the plant-based platforms as useful tools for vaccine preparation.

## 5. Plant Metabolites Targeting HPV Tumors

Even if significant progress was made against HPV disease through preventive vaccination, and despite the success of experimental therapeutic vaccines, early and efficient treatment of HPV cancers is still a challenging issue. For this reason, an active area of research has involved and still considers plants as a source of potential pharmaceutical agents for treatment of HPV-associated tumors. A well-known example is genital warts, that can be treated with plant-based anticancer therapies, such as vincristine, vinblastine, paclitaxel, camptothecin, and podophyllotoxin [81].

Indeed, plant-derived compounds represent about 75% of the whole approved anti-tumor drugs, as either natural products themselves or as molecules mimicking or directly deriving from natural sources [82]. In many cases, plant-derived compounds can be considered an abundantly available, cost-effective source of ingredients. They can be either harvested directly in nature, or obtained from plant cell cultures.

Many plant-derived compounds possess the specificities of ideal chemopreventive agents, with no effect on normal cells, bioavailability, multiple mechanisms of action, easy manner of administration and significant cellular effects in combating oncogenesis, as they may prevent carcinogens from reaching their targets, inhibit malignant cell proliferation, to modulate tumor suppressing agents and immune surveillance [83,84].

Moreover, some of them (like polyphenols) have a prominent role in neutralizing reactive oxygen species, that are well-established messengers in intracellular signaling inducing oncogenesis and as genotoxic damage inducers [85]. As an example, high levels of 8-OhdG (8-hydroxy-2'-deoxyguanosine, one of the predominant free-radicals induced in oxidative lesions) are specific to cervical carcinogenesis and characteristic of the progression from squamous intraepithelial lesions (SIL) to invasive carcinoma [86]. In addition, lipid peroxidation and block of antioxidant functions are seen in patients with several malignant pathologies of the cervix [87,88].

Despite the significant advantages of high specificity and low toxicity of plant-derived compounds as anti-cancer agents, main drawbacks can be the rapid catabolism and the low bioavailability at the target site. Combination with existing drugs, to reach synergistic effects, or the use of nanoformulation of polyphenols, to prevent their degradation, have showed promising results [89].

Phytochemicals, as either purified and characterized entities (i.e., mainly secondary metabolites), or extracts and mixtures composed by different herbal derivatives, were, indeed, the first compounds to be used in the search of tools able to tackle cervical cancer even before its etiology was discovered and since HeLa (HPV 18 positive) cells were developed in 1952 [90,91].

It has been shown that anti-cervical cancer drugs can be found in several ethno-botanical sources and there is a wealth of plant extracts that were described to have HPV-related effects (for a very exhaustive list see [92]). Nevertheless, very often these compounds are not widely distributed in the plant kingdom and not reasonably accessible. In other cases, mixtures have no indication of effective constituents and their roles in HPV-specific cytotoxicity. Also the potential role of traditional Chinese medicine has been evaluated by in vitro and in vivo experiments with studies exploring the mechanisms of action of its active components (reviewed in [93]). All the compounds tested so far need to be screened further and on a really large scale especially in vivo and in clinical settings to finally establish their HPV-specific effects in order to establish them as useful, efficacious, cost-effective, and clinically available therapeutics. Nevertheless, the available studies give a strong idea of the variety of targets and of the potential of plant-derived compounds against HPV lesions.

In the following sections, some studies describing effects of phytochemicals or extracts on HPV cancer cells in vitro and in vivo will be described. A particular mention for those directly affecting HPV E6/E7 activity or displaying adjuvant properties in combination with chemo- or radiation therapies will be deserved.

### 5.1. HPV-Related In Vitro and In Vivo Studies Based on Purified Phytochemicals

Several in vitro studies mainly focused on purified phytochemicals and, in particular, on polyphenols, among which mainly flavonoids, as well as on other chemical species such as alkaloids, polysaccharides and protein-based entities, along with plant extracts (Table 1).

**Table 1.** Plant phytochemicals and extracts in relation with their effects on human papillomavirus (HPV) cancer in vitro and in vivo.

| Compound Type | Phytochemicals | Study Type | Cell Type | Activity | Mechanism of Action | References |
|---|---|---|---|---|---|---|
| Purified Polyphenols (Flavonoids, Anthocyanins) | Black rice anthocyanin and cyanidin 3-glucoside | In vitro | HeLa | Inhibition of proliferation and induction of apoptosis | Dose- and time-dependent by apoptosis induction through Bax/Bcl-2 | [94] |
| Purified Polyphenols (Flavonoids, Cathechins) | Epigallocatechin-3-gallate (EGCG) | In vitro and in vivo | CaSki | Inhibition of proliferation and induction of apoptosis | Dose-dependent apoptosis induction through arrest of cell cycle in the G1 phase. Possible gene regulatory role | [95] |
| | | In vitro | HeLa | Inhibition of proliferation and induction of apoptosis | Inhibition of HPV E6/E7 expression and of estrogen receptor $\alpha$ and aromatase by a time-dependent manner mediated by apoptosis | [96,97] |
| | | In vitro | TCL-1 (HPV-immortalized cervical epithelial cells) Me18 HeLa | Inhibition of proliferation (adenocarcinoma less responsive to EGCG growth inhibition) and apoptosis induction | Dose-dependent increased expression of p53 and p21 | [98] |
| | | In vitro | CaSki HeLa C33A | Suppression of growth | Time- and dose- dependent, possibly via regulating the expression of miRNAs | [99] |
| | | In vitro | HeLa | Repression of hypoxia- and serum-induced HIF-1$\alpha$ and VEGF | Via MAPK and PI3K/AKT pathways | [100] |
| Purified Polyphenols (Flavonoids, Flavanones) | Naringenin | In vitro | SiHa | Inhibition of proliferation and induction of apoptosis | Cell cycle arrest at the G2/M phase and induction of apoptosis | [101] |
| | | In vitro | HeLa | Inhibition of proliferation and induction of apoptosis | Reduced expression of NF-κB p65 subunit, COX-2, and caspase-1 | [102] |
| | Naringenin-loaded nanoparticles | In vitro | HeLa | Inhibition of proliferation and induction of apoptosis and cytotoxicity | Dose-dependent cytotoxicity, apoptosis, reduction of intracellular glutathione levels, alterations in mitochondrial membrane potential, increased intracellular ROS and lipid peroxidation. | [103] |
| | Hesperetin | In vitro | SiHa | Reduction in cell viability and induction of apoptosis | Increased expression of caspases, p53, Bax, and Fas death receptor and its adaptor protein | [104] |

**Table 1.** *Cont.*

| Compound Type | Phytochemicals | Study Type | Cell Type | Activity | Mechanism of Action | References |
|---|---|---|---|---|---|---|
| Purified Polyphenols (Flavonoids, Flavones) | Apigenin | In vitro | CaSki HeLa C33A | Inhibition of proliferation and induction of apoptosis | G1 phase growth arrest through p53-dependent apoptosis and increased expression of p21/WAF1, Fas/APO-1 and caspase-3. Decreased expression of Bcl-2 | [105] |
| | | In vitro | HeLa | Modifications in cell motility, inhibition of translocation | Interference with gap junctions | [106,107] |
| | Daidzein | In vitro | HeLa | Inhibition of proliferation | Cell cycle, cell growth, and telomerase activity alterations | [108] |
| | Jaceosidin | In vitro | SiHa Caski | Inhibition of the function of E6 and E7 oncogenes | Impairment of binding to p53 and pRb | [109] |
| | Luteolin | In vitro | Caski E6/E7 immortalized human foreskin keratinocytes (HFK) primary HFKs | E6 inhibition | Binding at the interface between E6 and E6AP mimicking leucines in the conserved α-helical motif of E6AP | [110] |
| | | In vivo | HeLa | Induction of apoptosis | TRAIL-induced apoptosis by both extrinsic and intrinsic apoptotic pathways | [111] |
| | Wogonin | In vitro | SiHa CasKi | Induction of apoptosis | Suppression of E6 and E7 and increase in p53 and pRb | [112] |
| Purified Polyphenols (Flavonoids, Flavonols) | Quercetin | In vitro | HeLa | Induction of apoptosis Induction of mitochondrial apoptosis | G2/M phase cell cycle arrest, apoptosis, inhibition of anti-apoptotic AKT and Bcl-2 expression | [113] |
| | Kaempferol | In vitro | HeLa | Inhibition of proliferation | G2/M phase growth arrest, decrease of cyclin B1 and CDK1, inhibition of NF-kB nuclear translocation, upregulation of Bax and downregulation of Bcl-2 | [114] |
| | Fisetin | In vivo/in vitro | HeLa | Inhibition of proliferation and reduction of tumor growth by apoptosis | Apoptosis due to activation of the phosphorylation ERK1/2, inhibition of ERK1/2 by PD98059, activation of caspase-8/caspase-3 pathway | [115] |

**Table 1.** *Cont.*

| Compound Type | Phytochemicals | Study Type | Cell Type | Activity | Mechanism of Action | References |
|---|---|---|---|---|---|---|
| Purified Polyphenols (Flavonoids, Phenolic Acids) | Gallic acid | In vitro | HeLa | Induction of apoptosis and/or necrosis | ROS increase and GSH depletion | [116] |
| Purified Polyphenols (Flavonoids, Stilbenes) | Resveratrol | In vitro | SiHa HeLa | Inhibition of proliferation, induction of autophagy and apoptosis | Cathepsin L-mediated mechanism | [117] |
| | | In vitro | HeLa | Suppression of invasion and migration | Generation of ROS | [118] |
| | | In vitro | SiHa HeLa CaSki C33A | Decrease in the angiogenic activity, induction of autophagy | Decreased expression of metalloproteinases. Inhibition of AKT and ERK1/2, destabilization of lysosomes, increased cytosol translocation | [119] |
| | | In vitro | SiHa HeLa C33A | Inhibition of proliferation | Induction of cell apoptosis | [120] |
| | | In vitro | HeLa | Inhibition of proliferation | Sensitization to tumor necrosis factor-related apoptosis-inducing ligand (TRAIL) | [121] |
| Purified Polyphenols (Curcuminoids) | Curcumin (diferuloylmethane) | In vitro | HeLa SiHa C33A | Inhibition of proliferation Induction of apoptosis | Down-regulation of HPV-18 transcription, inhibition of AP-1 binding activity and reversion of the expression of c-fos and fra-1 | [122] |
| | | | | | Downregulation of viral oncogenes E6 and E7, NF-kB and AP-1 COX-2, iNOS and cyclin D1 | [123–125] |
| | | | | | Upregulation of Bax, AIF, release of cytochrome c and downregulation of Bcl-2, Bcl-XL, COX-2, iNOS and cyclin D1 | [126] |
| | MS17 curcumin analogue 1,5-Bis(2-hydroxyphenyl)-1,4-pentadiene-3-one | In vitro | HeLa CaSki | Cytotoxic, anti-proliferative and apoptosis-inducing potential. | Apoptosis through activation of caspase-3 in CaSki cells. Down-regulation of HPV18 and HPV16 E6 and E7 oncogene expression. | [127] |

**Table 1.** *Cont.*

| Compound Type | Phytochemicals | Study Type | Cell Type | Activity | Mechanism of Action | References |
|---|---|---|---|---|---|---|
| Purified Polyphenols (Lignans) | Methylenedioxy lignan | In vitro | HeLa | Inhibition of proliferation and apoptosis | Apoptosis and inhibition of telomerase activity | [128] |
| | Nor-dihydro-guaiaretic acid | In vitro | SiHa | Promotion of apoptosis | Downregulation of HPV E6 and E7 transcription and expression | [129] |
| Purified Diterpenoids | Tanshinone IIA | In vitro | HeLa SiHa CasKi C33A | Inhibition of growth promotion of apoptosis | Downregulation of HPV E6 and E7 expression, cell cycle arrest | [130] |
| Purified Alkaloids | Berberine | In vitro | HeLa | Inhibition of growth | Reduced expression of E6 and E7 with increase in p53 and pRb expression, loss of telomerase protein, hTERT | [131] |
| | | | SiHa | | Alter epigenetic modifications and disrupt microtubule network by targeting p53 | [132] |
| Purified Steroid Lactones | Withaferin A | In vitro | CasKi | Promotion of apoptosis | E6 and E7 repression | [133] |
| Purified Pyranocoumarin compounds | Decursin Decursinol | In vitro | Hela | Promotion of apoptosis | Induction of TRAIL expression | [134] |
| Polysaccharides fractions | From *Solanum nigrum* | In vitro and in vivo | U14 | Promotion of apoptosis and inhibition of tumor growth | CD4$^+$/CD8$^+$ ratio modification | [135] |
| Lectins | From *Astragalus mongholicus* | In vitro | HeLa | Promotion of apoptosis and Antiproliferation | Upregulation of p21 and p27 and reduction of active complex cyclin E/CDK2 kinase | [136] |
| Peptide fractions | From *Triticum aestivum* | In vitro | HeLa | Antiproliferation | Induction of DNA damage and G2 arrest, inactivation of the CDK1-cyclin B1 complex and increase of active chk1 kinase expression | [137] |
| | From *Abrus precatorius* | | | | Induction of apoptosis; generation of ROS, decrease of Bcl-2/Bax ratio, induction of mitochondrial permeability transition | [138] |

**Table 1.** *Cont.*

| Compound Type | Phytochemicals | Study Type | Cell Type | Activity | Mechanism of Action | References |
|---|---|---|---|---|---|---|
| Extracts | Fractionated extract of *Bryophyllum pinnata* (Bryophyllin A) Crude extract of *Phyllanthus emblica* fruits and *Brucea javanica* oil emulsion | In vitro | Hela SiHa Casky | Anti-HPV activities | Inhibitory action on AP-1 and STAT3 and specific downregulation of expression of viral oncogenes E6 and E7 | [139–141] |
| | Lipid-soluble Rhizome extract of *Pinellia pedatisecta* | In vitro | Caski HeLa HBL-100 | Promotion of apoptosis | Increased expression of Caspase-8, Caspase-3, Bax, p53, p21 | [142] |
| | Basant (curcumin, purified saponins, extracts of *Emblica officinalis*, *Mentha citrata* oil, and gel extracts of *Aloe vera*) | In vitro | HeLa | Anti-HPV activities | Inhibition of transduction of HPV16 pseudovirus | [143] |
| | *Cudrania tricuspidata* stem extract | In vitro | SiHa HaCaT human normal keratinocytes | Apoptosis induction and cytotoxic effects in cervical cancer cells with no cytotoxic effect on HaCaT keratinocytes at concentrations of 0.125–0.5 mg/mL. | Dose-dependent mechanism by down-regulation of the E6 and E7 viral oncogenes. Apoptosis induction exclusively based on the increase of mRNA expression of extrinsic factors (i.e., Fas, death receptor 5 and TRAIL) and on activation of caspase-3/caspase-8 and cleavage of poly (ADP-ribose) polymerase (PARP) | [144] |
| | *Ficus carica* fruit latex | In vitro | CaSki HeLa | Inhibition of growth and invasion | Downregulation of the expression of p16 and HPV onco-proteins E6, E7 | [145] |

Polyphenols are a heterogeneous group of chemical substances known to have safe preventive and anticancer effects and ability to specifically target viral oncogenes [146]. It is indubitable that they are of great interest as pharmaceutical agents also against cancers with viral etiology like HPV. As an example of the interest of application of this class of molecules against cancer, the effects of different polyphenols against oral squamous cell carcinoma (both HPV+ and HPV−) has been reported in a wealth of literature (reviewed in [147]).

Anthocyanins (polyphenols, flavonoids) are mainly encountered in berries, currants, eggplant, grapes, and black rice. The feature of anthocyanins, as many tumor cells growth inhibitors, is related to their ability to induce tumor cells apoptosis and neutralize ROS [148]. In particular, anthocyanin from black rice and cyanidin 3-glucoside were found able to block the growth of HeLa cells by apoptosis mediated by Bax/Bcl-2, through a dose- and time-dependent mechanism [94].

(-)-Epigallocatechin-3-gallate (EGCG) (polyphenols, flavonoids, cathechins) extracted from green tea, can contrast tumor cell growth interfering with tumor-related angiogenesis and propensity to metastasize. EGCG was demonstrated also to have a possible gene regulatory role. In the case of HPV, it was shown in vitro that CaSki (HPV 16 positive) cervical cancer cells are guided to apoptosis by EGCG with cell cycle arrest in the G1 phase. Furthermore, in vivo anti-HPV tumor effects of EGCG were also observed [95].

The mechanism found for EGCG suppression of HPV-related cancer cell lines is the inhibition of HPV E6/E7, estrogen receptor $\alpha$, and aromatase expression by apoptosis in a time-dependent manner [96,97]. Indeed, a critical role of estrogens in cervical cancer is known [149]. It was also demonstrated that different responses were found in squamous cell carcinoma and in adenocarcinoma, the latter being less responsive to EGCG inhibition [98]. In HeLa cell line, EGCG was able to repress hypoxia- and serum-induced HIF-1$\alpha$ and VEGF, through MAPK and PI3K/AKT [100].

In HeLa cells, EGCG abrogated HDAC1 activity, also affecting expression of retinoic acid receptor-$\beta$, cadherin 1, and death-associated protein kinase-1 [150], showing participation of these genes in cell processes strongly associated with cancer proliferation.

A genome-wide study reported that EGCG affects also DNA methylation in oral squamous cell carcinoma (CAL-27) [46]. Besides, EGCG was demonstrated to suppresses HeLa, CaSki, and C33A (HPV negative) cell growth via regulating the expression of miRNAs, suggesting as potential therapeutic targets for the control and prevention of cervical cancer [99].

Apigenin (polyphenols, flavonoids, flavones), the main plant-derived bioactive flavone, is abundant in common fruits and vegetables such as parsley and celery, onions, oranges, wheat sprouts, and chamomile [86]. Apigenin has been demonstrated to have anti-carcinogenic effects against CaSki, HeLa, and C33A cervical cancer cell lines [105]. Apigenin treatment arrested HeLa cells growth at G1 phase and consequentially induced p53-dependent apoptosis associated with increased expression of p21/WAF1, Fas/APO-1, and caspase-3 [106]. Apigenin also decreased expression of the antiapoptotic factor Bcl-2 [106]. It was also demonstrated that apigenin-treated HeLa cells can undergo modifications in cell motility and inhibition of translocation (i.e., a reduction of the invasive potential) probably by interference with gap junctions [107].

Jaceosidin (polyphenols, flavonoids, flavones) was first isolated in plants of the *Compositae* family. This compound was demonstrated to inhibit the function of E6 and E7 oncogenes by impairing binding of these oncoproteins with p53 and pRb and making them non-functional [109].

Luteolin (polyphenols, flavonoids, flavones) is most often found in leaves and its sources include celery, broccoli, green pepper, parsley, thyme, perilla, chamomile tea, carrots, olive oil, peppermint, rosemary, oranges, and oregano. It was demonstrated to induce apoptosis in HeLa cells [111]. Luteolin binds to a hydrophobic pocket at the interface between E6 and E6AP and mimics the leucines in the conserved $\alpha$-helical motif of E6AP, displaying an E6 inhibitor activity [110]. Luteolin was also shown to sensitize HeLa cells to TRAIL-induced apoptosis by both extrinsic and intrinsic apoptotic pathways in an in vivo study [111].

Wogonin (polyphenols, flavonoids, flavones), accumulating in the plant *Scutellaria baicalensis*, was found to promote apoptosis through suppression of E6 and E7 and increase in p53 and pRb in SiHa (HPV 16 positive) and CasKi human cervix tumor cells [112].

Curcumin (diferuloylmethane, polyphenols, curcuminoids) is a hydrophobic polyphenol derived from the rhizome of *Curcuma* with a wide spectrum of pharmacological properties among which are anti-inflammatory and antioxidant activities by inhibiting lipo-oxygenase and cyclo-oxygenase [151]. Downregulation of HPV E6 and E7 oncogenes, NF-kB and AP-1, COX-2, iNOS, and cyclin D1 [123–125] represents the main feature of curcumin action affecting HeLa, SiHa, and C33A cells. In addition, growth suppression and apoptosis triggering associated with up-regulation of Bax, release of cytochrome c, and downregulation of Bcl-2 and Bcl-XL, were detected [126]. In HeLa cells, curcumin was also found to down-regulate HPV transcription, to target AP-1 transcription factor affecting expression of E6 and E7, and to block expression of *c*-fos and fra-1 [122].

The cytotoxic potential of diarylpentanoids, curcumin analogues, was evaluated in HeLa and CaSki cervical cancer cell lines as an improved alternative to curcumin. In particular, the MS17 analogue 1,5-Bis(2-hydroxyphenyl)-1,4-pentadiene-3-one exhibited cytotoxic, anti-proliferative, and apoptosis-inducing potential. Apoptosis was sustained by activation of caspase-3 activity in CaSki cells. Quantitative real-time PCR also detected significant down-regulation of HPV 18- and HPV 16-associated E6 and E7 oncogene expression following treatment [127].

Tanshinone IIA (abietane-type diterpenoid) from *Salvia* species, downregulates E6 and E7 and trigger apoptosis, inhibiting growth of HeLa, SiHa, CasKi, and C33A cells [130].

Berberine (benzylisoquinoline alkaloid) mainly accumulates in plants of the *Berberis* species and shows anti-inflammatory and anti-cancer properties with no apparent toxicity. Berberine interferes with of E6, E7, p53, pRb, and c-Fos expression, ultimately leading to the inhibition of cervical cancer cells growth [131]. Furthermore, it determines epigenetic changes and alters microtubules in cervical cancer cells [132].

Withaferin A (Steroid Lactone) from *Withania somnifera* induces apoptosis of CasKi cells through E6 and E7 repression determining inhibition of tumor growth [133].

Other secondary metabolites, together with polysaccharide, lectin, and protein/peptide fractions that are not described in the text are listed with their HPV-related activities in Table 1.

*5.2. HPV-Related In Vitro and In Vivo Studies Based on Plant Extracts or Mixtures*

Compared to purified phytochemicals, few studies are available on the anti-carcinogenic effects of crude or partially fractionated extracts or mixtures composed by different herbal derivatives.

Inhibition of AP-1 and STAT3, known to induce cervical carcinogenesis, and specific down-regulation of viral oncogenes E6 and E7 expression have been demonstrated for rhizome extracts from *Pinellia pedatisecta*, for Bryophyllin A-rich leaf fractionated extracts from *Bryophyllum pinnata*, for fruit extracts from *Phyllanthus emblica* and for oils extracted from *Brucea javanica* [139–142].

Two botanical formulations called, respectively, "Basant", made of purified curcumin and saponins mixed to *Emblica officinalis* and *Aloe vera* extracts and *Mentha citrata* oil [143], and "Praneem", composed by purified saponins, extracts from *Azadirachta indica*, *Emblica officinalis*, and *Aloe vera* mixed to *Mentha citrata* oil, have been shown able to block transfer of HPV16 pseudovirions in HeLa cells [143,152].

The anti-cancer effects of *Cudrania tricuspidata* stem extract were evaluated in HPV-positive cervical cancer cells (CaSki and SiHa cells, $2.5 \times 10^5$ cells/mL) and HaCaT human normal keratinocytes. This extract showed dose-dependent cytotoxic effects in cervical cancer cells with no cytotoxic effect on HaCaT keratinocytes at concentrations of 0.125–0.5 mg/mL. The extract induced apoptosis by down-regulating the E6 and E7 viral oncogenes in SiHa cervical cancer cells. Its mechanism of induction of apoptosis was exclusively based on the increase of mRNA expression of extrinsic factors (i.e., Fas, death receptor 5, and TRAIL) and on activation of caspase-3/caspase-8 and cleavage of polyADP-ribose polymerase. No effects on intrinsic pathway molecules such as Bcl-2, Bcl-xL, Bax, and cytochrome C were observed. These results suggest that this extract can be used as a modulating agent in cervical cancer [144].

The in vitro biological activities of *Ficus carica* fruit latex were explored onto cervical cancer CaSki and HeLa cell lines. Data show that latex inhibits rapid growth and invasion and downregulated the expression of p16 and HPV onco-proteins E6, E7 [145].

*5.3. Evaluation of Plant Compound Adjuvant Activity in Chemo- and Radio-Therapies for HPV-Associated Cancer*

Although prophylactic vaccination represents the most effective method for cervical cancer prevention, chemotherapy is still the primary invasive intervention against HPV cancer lesions. It is urgent to exploit low-toxic natural anticancer drugs on account of high cytotoxicity and side-effects of conventional agents. Moreover, the resistance of cervical tumor cells to chemo- and radiotherapy is one of the crucial problems in the treatment of cervical neoplasia, leading to decreased efficacy or failure of the therapy. In this field, the combination of natural compounds with chemotherapy and radiotherapy in the treatment of cervical cancer was reported to improve in some cases sensitization of HPV cancer cells and to minimize the toxicity of these therapies (Table 2).

**Table 2.** Plant phytochemicals with adjuvant activity in HPV chemo- and radio-therapy.

| Compound Type | Phytochemicals | Study Type | Cell Type | Activity | Mechanism of Action | References |
|---|---|---|---|---|---|---|
| Purified phytochemical with adjuvant activity in HPV chemo-therapies | Curcumin (diferuloylmethane) | In vitro | HeLa | Sensitization to cisplatin, paclitaxel | Induction of apoptosis by down-regulation of NF-kB. | [153] |
| | Tetrahydrocurcuminoids | In vitro | Drug-resistant human cervical carcinoma cell line KB-31 and KB-V-1 | Sensitization to vinblastine, mitoxantrone, and etoposide | Down-regulation of HPV-18 transcription, inhibition of AP-1 binding activity, and reversion of the expression of *c*-fos and fra-1 | [124] |
| | Quercetin | In vitro | HeLa | Sensitization to cisplatin | Enhancement of cancer cells death levels. | [154] |
| | Saikasaponins | In vitro | HeLa SiHa | Sensitization to cisplatin | Reactive oxygen species generation | [155] |
| | Wogonin (5,7-dihydroxy-8-methoxyflavone) | In vitro | A549 HeLa | Sensitization to cisplatin | Reactive oxygen species generation | [156] |
| | Apigenin | In vitro | HeLa SiHa | Sensitization to paclitaxel | Apoptosis through intracellular ROS accumulation | [157] |
| | Formonetin | In vitro | HeLa | Sensitization to epirubicin | Potentiates epirubicin-induced apoptosis via ROS production. | [158] |
| | Tea polyphenols with EGCG | In vitro | SiHa | Sensitization to bleomycin | Activation of caspase-3, -8, -9, and up-regulation of the expression of P53 and Bcl-2 | [159] |
| Purified phytochemical with adjuvant activity in HPV radio-therapies | Resveratrol | In vitro | HeLa SiHa | Increased radiosensitivity and potentiation of apoptosis | Dose-dependent alteration of cell cycle progression and cytotoxic response | [160] |
| | Genistein | In vitro | CaSki Me180 Human esophageal cancer cell lines | Increased radiosensitivity and potentiation of apoptosis | Inhibition of Mcl-1; G(2)M arrest, and activation of the AKT gene | [161–163] |
| | Curcumin | In vitro | HeLa SiHa | Increased radiosensitivity and potentiation of apoptosis | ROS-dependent mechanism | [164] |
| | Ferulic acid | In vitro | HeLa Me180 | Increased radiosensitivity and potentiation of apoptosis | ROS-dependent mechanism | [165] |
| | Quercetin | In vitro and in vivo | DLD1 (human colorectal cancer xenografts) HeLa MCF-7 | Increased radiosensitivity and potentiation of apoptosis | Time- and dose-dependent mechanism and through ROS modulation and downregulation of E6 and E7 expression | [166] |

### 5.3.1. Phytochemicals with Chemosensitizing Effects on Cervical Cancer Cells in Vitro

Curcumin was demonstrated to sensitize cervical cancer cells to taxol- or cisplatin-induced apoptosis by down-regulation of NF-kB [153,167]. Similarly, a class of metabolites of curcumin, tetrahydrocurcuminoids, increased the sensitivity of vinblastine, mitoxantrone, and etoposide in a drug-resistant human cervical carcinoma cell line [124].

Apigenin showed synergistic effects with paclitaxel improving apoptosis rates of HeLa and SiHa cancer cells [157]. Quercetin, saikasaponins (triterpenoid saponins from the plant *Bupleurum falcatum*), wogonin, and apigenin sensitized cervical cancer cells to cisplatin by sensitizing HPV-related cancer cells to paclitaxel-induced apoptosis through intracellular ROS accumulation [154–156]. Also, formononetin (an isoflavone found in a number of plants and herbs such as red clover) was found to sensitize cervical cancer cells to the anthracyclin epirubicin via ROS production [158].

The combination of tea polyphenols with EGCG and bleomycin demonstrated to have therapeutic effects on cervical cancer. Bleomycin is an anti-neoplastic chemotherapeutic used in redox-related cancer, including cervical squamous cell cancer, that cause severe side effects in normal cells such as immune system damage, pneumonitis, and pulmonary fibrosis, which are mediated by redox status disturbances. The combination therapy induced stronger cancer cell apoptosis ability than treated either tea polyphenols or bleomycin alone, by activating caspase-3, -8, -9, and up-regulating the expressions of p53 and Bcl-2 [159].

### 5.3.2. Phytochemicals with Radiosensitizing Effects on HPV-Related Cancer Cells In Vitro and In Vivo

A few phytochemicals have shown radiosensitizing effects. Resveratrol [160], genistein [161–163], curcumin [164], ferulic acid (i.e., a phenolic acid abundant in plant cell walls) [165], and quercetin [166] have shown to increase the pro-apoptotic properties of ionizing radiation on cervical cancer cell lines in vitro and in vivo. A ROS-dependent mechanism has been postulated for these plant compounds. Genistein, a flavone mostly found in legumes, has synergistic radiosensitizing effects against several cancer cell types [168] being able to block the growth of CaSki cells in vitro, probably through cell cycle arrest.

### 5.4. Clinical Evaluation of Plant Compounds

Clinical trials focused on the anti-HPV carcinogenic effects of natural compounds were less frequent than in vitro and in vivo studies (Table 3). EGCG and curcumin were the most investigated compounds.

**Table 3.** Plant phytochemicals and extracts in relation with their effects on HPV cancer evaluated in clinical trial settings.

| Compound Type | Phytochemicals | Disease Stage | Route | Activity | References |
|---|---|---|---|---|---|
| Green tea compounds | 200 mg EGCG+/− "Poly E" (37 mg epigallocatechin + 31 mg epicatechin) | 51 patients with Chronic cervicitis mild dysplasia moderate dysplasia severe dysplasia | Orally (capsule) ± vaginally (ointment) | 20/27 patients (74%) under poly E ointment therapy showed a response. 6/8 patients (75%) under poly E ointment + poly E capsule therapy showed a response, 3/6 patients (50%) under poly E capsule therapy showed a response. 6/10 patients (60%) under EGCG capsule therapy showed a response. Overall, a 69% response rate (35/51) was noted for treatment with green tea extracts, as compared with a 10% response rate (4/39) in untreated controls ($p < 0.05$). | [169] |
| Curcumin-based | Curcumin | Phase I clinical testing 4 cervical intraepithelial neoplasia (CIN) cases | Oral administration of 0.5–12 mg for 3 months | Clinical safety (up to 8 mg/day Histological improvements in 1/4 patients. | [170] |
| | Basant | Phase I/II double-blind clinical trial in women infected with HPV but without high grade CIN | Intra-vaginal application of curcumin-containing capsules or Basant cream | Higher clearance of cervical HPV infection (87.7%) in case of Basant cream 81.3% rate in the case of curcumin capsules compared to the controls (73.3%) with no serious adverse events. | [152] |
| | | 11 women infected with HPV and low grade cervical abnormalities | Intra-vaginal application of Basant capsules | Clearance of HPV16 infection in all the patients (11/11) | [143] |
| Neem-based | Praneem | 20 HPV-infected women +/− early cervical intraepithelial lesions (placebo controlled) | Thirty days intra-vaginal application of Praneem tablets | Clearance of HPV16 infection in 60% of the patients (6/10). Clearance in another 50% after another administration, total 80% HPV clearance rate. | [171] |

The clinical efficacy of EGCG and other green tea compounds (a combination of 200 mg EGCG, 37 mg epigallocatechin, 31 mg epicatechin) delivered orally ± vaginally to patients with HPV cervical lesions was evaluated. In this study, a 69% clearance rate was detected in the treated patients as compared with a 10% response rate in untreated controls, showing that green tea compounds can be effective in the treatment of HPV-related cervical lesions [146].

Green tea extracts components (i.e., polyphenon E, a standardized green tea extract containing 15% green tea polyphenols, and EGCG) were also tested in patients with HPV mild, moderate, and severe dysplasia. One of these studies recruited 90 patients. Components were applied as either ointment or oral administration. In particular, polyphenon E was applied as a topical ointment to 27 patients twice a week. On the other hand, 200 mg of polyphenon E or EGCG were delivered orally on a daily base for eight–twelve weeks.

Despite differences between ointment and capsules, giving indication of a higher efficacy of ointment with respect to capsule administration, overall, 69% of patients (35/51) showed a response to treatment with green tea components, compared to a 10% response rate (4/39) of untreated patients ($p < 0.05$) [169].

Evaluation of curcumin in clinical settings was started with a phase I clinical testing of oral administration of 0.5–12 mg of this compound for 3 months. The main result of this study was to define the safety of oral administration of curcumin as up to 8 mg/day and its bioavailability and efficacy in determining histological improvements in 1 out of the 4 patients [170].

Subsequently, a phase I/II clinical trial demonstrated that intra-vaginal administration of either curcumin capsules or Basant cream in HPV-infected women (without high grade CIN), was able to induce higher infection clearance rate (87.7% for Basant cream, 81.3% for curcumin capsules) than untreated patients (73.3%) [172]. Basant administration cleared HPV16 infection in 100% of patients (i.e., 11 HPV-infected women with low grade cervical abnormalities) recruited in a more recent study [143].

Intra-vaginal administration of Praneem for thirty days to 10 HPV-infected women with low-grade squamous intraepithelial lesions determined the clearance of HPV16 in 60% of patients. Another round of administration was able to induce HPV clearance in 50% of the patients that had not been able to eliminate HPV upon the first treatment [171]. It was postulated that the effect of Praneem administration could be a consequence of its microbicidal activity in the reproductive tract, able to neutralize infections considered a co-factor in HPV carcinogenesis [173].

*5.5. Cytotoxicity of Plant Compounds: Anti-HPV Cancer Efficacy Prediction and Concerns for Healthy Cells*

Previous sections mentioned a non-exhaustive list of plant compounds, extracts, and formulations of different origin and source tested against HPV cancers at various levels. Although many products are in the list, many of these entities need to be further clarified in terms of the role played in cytotoxicity in cancer once translated into clinic and of possible adverse effects in healthy cells. These tasks might be particularly complex to fulfill in the case of multi-component extracts. Indeed, the role of single constituents has to be investigated to evaluate the potential of the plant products as safe and effective anticancer agents.

It is clearly desirable that compounds showing cytotoxic activity in vitro against HPV cancer cell lines have also anti-cancer effects and strong tumor-selective action once tested in clinical trial. Unfortunately, not all the studies demonstrating in vitro efficacy of plant substances on HPV-related cell lines (e.g., HeLa, CaSki, SiHa, C33A) have been translated in in vivo or clinical studies, as already mentioned. The lack of clinical studies is clearly a limitation for the future clinical application of plant compounds. Despite this, it can be said that specifically HeLa and CaSki cells are models commonly used in in vitro cervical cancer research, since they contain the HR HPV types 18 and 16 viral genomes respectively, that are causing seven out of ten cases of invasive cervical cancers. Therefore, cytotoxicity of plant compounds, demonstrated by means of these cell lines in vitro, can be considered truly relevant in view of a clinical application as anti-HPV cancer agents. Moreover, plant products exerted cytotoxic effects on these cell models through specific targeting of E6, E7, and/or other hallmarks of HPV carcinogenesis. Thus, they have even more significance for a possible translation in clinical trial.

In view of clinical administration, special attention should be paid also to the issue of possible unwanted toxicity exerted by some phyto-compounds toward normal cells. Clearly, compounds with selective cytotoxic effects on cancer cells should be preferred. Undoubtedly, literature reporting comparative data on cytotoxicity of plant compounds on cancer and healthy cells should be expanded by further studies. Nevertheless, selective cytotoxic effects and doses on cancer cells have been already demonstrated for some purified compounds and extracts. For dietary compounds such as flavonoids, general safety and selective tumor cytotoxicity have been already well established [146]. Recently, *Ficus carica* latex was reported to inhibit the growth of CaSki and HeLa cells without a cytotoxic effect on human keratinocytes cell line (HaCaT) [145]. For other compounds, enhancement of immunity against cancer was demonstrated. In the case of a lipid-soluble extract of *Pinellia pedatisecta*, an enhancement of antitumor T-cell responses by restoring tumor-associated dendritic cell activation and maturation was shown [174]. Interestingly, in some cases, selective cytotoxic effects on HPV cancer cells have been demonstrated in association with downregulation of E6 and E7 oncogenes. If tumor-specific targeting is demonstrated for a certain compound in vitro, a strong clue for selectivity toward tumor cells in clinical translation is given implicitly. Plant products such as curcumin have shown selective cytotoxicity for HPV 16- and HPV 18-infected cells and induction of apoptosis in cervical cancer cells by downregulation of E6 and E7 oncogenes and of tissue-specific viral gene expression ([96,97,109,110,112,125] Table 1). As another example, the anti-cancer potential of *Cudrania tricuspidata* stem extract was investigated both in HPV-positive cervical cancer cells and in normal HaCaT keratinocytes. The extract showed significant dose-dependent cytotoxic effects in cervical cancer cells and no cytotoxicity on HaCaT. In addition, it was demonstrated that the extract-related apoptosis was induced by down-regulating the E6 and E7 viral oncogenes [144].

A strategy to avoid possible adverse effects in normal cells and to improve bioavailability of plant compounds might be to induce a confined action by tumor-targeting strategies. Nanoparticles of different types, such as multi-functionalized, magnetic or solid lipid particles, dendrimers, liposomes and micelles, some of which are already FDA approved, have been used for targeted delivery of plant compounds without disturbing the physiology of normal cells [175]. Naringenin, generally present in citrus and grapes, inhibits proliferation through cell cycle arrest at the G2/M phase and induction of apoptosis in human cervical SiHa cells. Mainly because of poor bioavailability and instability of the molecule, studies were carried out on naringenin-loaded nanoparticles that demonstrated advantages over free naringenin in HeLa cells through dose-dependent cytotoxicity, apoptosis, reduction of intracellular glutathione levels, alterations in mitochondrial membrane potential, increased intracellular reactive oxygen species (ROS), and lipid peroxidation levels [103] (Table 1). Nanoparticle-targeted delivery would offer also the advantage of multiple product loading (i.e., plant-based drugs along with synthetic drugs) beside possibly contributing to improve efficacy and decreasing unwanted toxicity. In conclusion, efficient targeting strategies accompanied by good toxicology studies could represent the future arena of research in the field of clinical application of plant products for anti-HPV cancer therapies.

## 6. Improving DNA Vaccine Effectiveness by Plant-Derived Solutions

Adjuvant activity is a conceptual issue that, in a broad sense, can be applied to any procedure/compound able to improve vaccine effectiveness. Plants and derived molecule/sequences can offer some solutions based on what we already know about their anti-cancer and apoptotic activity as well as pro-immune properties. A summary of these activities is reported, highlighting the possible adjuvant effect for HPV DNA vaccines.

### 6.1. Improved HPV Genetic Vaccines Including Plant Immune-Modulating Sequences

Although not commercially available yet, genetic vaccination represents a convenient platform with respect to traditional approaches that involve the production and purification of proteins or components of the pathogen. Owing to the ease of preparation, intrinsic safety and stability, DNA vaccination is useful for the quick assay of new synthetic immunogens as well as ensuring the induction of antibodies against conformational epitopes of interest.

Many strategies have been developed to enhance DNA vaccines effectiveness including codon optimization, particular methods of transfection (i.e., electroporation), addition of adjuvants or genetic fusions with immune-stimulating sequences, combination with heterologous boosts, etc.

Several genetic vaccines are in an advanced clinical trial phase for the treatment of HPV-associated malignancies and represent a potential additional weapon to those already available, like chemotherapy and radiotherapy. In this case, improvements of HPV genetic vaccines have focused on (1) increasing DC uptake of HPV antigens, (2) DC processing and presentation of HPV antigens, and (3) enhancing DC and T-cell interactions [21].

The fusion of HPV antigens with immunostimulatory sequences to achieve improvement in their "visibility" to the immune system, has been an approach very often explored in the development of experimental therapeutic vaccines [20]. Nevertheless, the search for innovative immunostimulatory sequences, with increased safety for clinical use, is still an open field. In fact, it is mandatory to avoid possible autoimmune responses induced by proteins of human origin (such as Hsp70 or calreticulin) or weakened immune responses (as in the case of using tetanus toxoids in already vaccinated people).

In an attempt to potentiate a genetic vaccine for HPV, the attenuated E7 gene (E7GGG) of HPV 16 was fused with the gene encoding the coat protein of a plant virus (Potato Virus X, PVX) [16]. This protein had already been used as a carrier able to increase CD4+ T-cell immune response (by "linked T-cell help") and to enhance immunogenicity of silent or poor determinants [176,177].

It was subsequently shown that the inclusion of sequences deriving from plant proteins and peptides with immune-modulatory and anti-cancer activity [178] can potentiate the activity of HPV genetic vaccines. The rationale is that, since in mammals and plants the general mechanisms underlying the innate immune response are highly conserved [179], some plant defense proteins may have an effect also on tumors by modulating innate immune functions and, consequently, also the adaptive response. Some of these plant proteins could also stimulate specific cell-mediated immunity toward tumor antigens, a crucial step for cancer resolution. These plant proteins would therefore behave as adjuvants enhancing the specific immune response toward an antigen.

Among these proteins, some "ribosome inactivating protein" (RIPs) might be included. RIPs show a regulatory and defensive role against pathogens and accumulate in various organs of many plant species [178,180,181]. They are potent inhibitors of protein synthesis (through N-glycosidase activity on rRNAs). This was the first biological feature extensively studied and clinically exploited for the development of selective cytotoxic agents ("immunotoxins") against tumor, immune, or nerve cells. Nevertheless, other biological properties of RIPs, independent of catalytic activity, could prove useful for the design of anticancer vaccines. Among these activities we find the ability to modulate the non-specific and innate immune functions of NK cells [182], CD4+ and CD8+ T cells, and/or cytokine production [183,184] inflammation [185], and apoptosis through multiple pathways [180,186] that have been shown to lead to anti-tumor properties in vivo [184].

Our results indeed support this hypothesis. We demonstrated for the first time that a DNA vaccine involving the fusion of the HPV 16 E7GGG gene with saporin (SAP), a RIP found in *Saponaria officinalis* but rendered no longer catalytically active through mutagenesis (SAP-KQ), potentiates antitumor activity against E7-expressing tumors. The anti-tumor activity was associated with enhanced antibody and cell-mediated immune responses and antigen-specific delayed-type hypersensibility (DTH) [18].

SAP-KQ/E7GGG fusion proteins may undergo rapid degradation via the ubiquitin-proteasome pathway, as postulated also in the case of coat protein of PVX fused to the

C-terminus of E7GGG [17]. Indeed, the A-chain of type II RIPs, once entered the cytosol, is subjected to an efficient protein quality control. Upon interaction with lipid cell membranes, they are possibly recognized as misfolded proteins, ubiquitinated and then targeted to proteasomes for elimination. This is particularly true for RIPs with a high lysine content [187] like saporins, where 10% of the residues is represented by this amino acid. Ultimately, this process could improve the processing of the E7GGG antigen fused to the SAP-KQ which in turn would result in an improvement in the activity of the chimeric vaccine compared to E7GGG alone.

Then, we designed a novel genetic fusion vaccine comprising synthetic genes derived from the E7 and L2 proteins of HPV 16. Here, a signal peptide derived from a plant protein was fused upstream the N-terminal portion of the papillomavirus fusion antigen [19]. Signal sequences (ss, or signal peptides) are short peptides (about 20–30 residues) that influence the targeting pathway of a protein and promote protein secretion or specific post-translational modifications. As a result, ss from highly secreted proteins have been used to enhance protein expression levels of recombinant proteins in different cell systems. Other groups described the possibility to enhance DNA vaccines efficacy by using signal sequences that generally derive from mouse [188,189] or human pathogens [190] and there is no doubt that plant-derived sequences pose less safety concerns.

In a previous work, we had shown that the ss of the polygalacturonase-inhibiting protein (PGIPss) from *Phaseolus vulgaris*, was able to target the HPV 16 E7 protein to the plant secretory compartment, enhancing the E7 protein expression level at least five-fold compared with the unfused version of the antigen [11]. Starting from the observation that ss properties are conserved from bacteria through eukaryotes, we explored the ability of the plant-derived PGIPss in increasing the efficacy of HPV DNA vaccines. We produced a prototype DNA vaccine where the PGIPss was fused upstream of the harmless HPV 16 E7 antigen (named E7GGG or E7*). The PGIPss-E7GGG fusion was able to modify antigen compartmentalization and/or processing in transfected HEK-293cells, promoting E7 protein secretion in the culture medium, and demonstrating its ability to affect the fate of a heterologous protein in mammalian cells [19]. Furthermore, even though secretion was not observed in the culture medium, PGIPss modifies the processing of other constructs like PGIPss-E7GGG-CP, where the E7GGG gene is fused to the PVX coat protein.

With the idea to develop a HPV preventive/therapeutic vaccine, the DNA sequence for PGIPss was fused upstream to a synthetic codon-optimized gene consisting of a cross-reactive epitope of the L2 protein (first 200 aa; $L2_{1-200}$) and E7GGG of HPV 16, and cloned in a mammalian expression vector (pVAX1) (Figure 1). The chimeric DNA vaccine (pVAX-PGIPss-$L2_{1-200}$-E7GGG) was delivered in C57BL/6 mice according to a prime/boost schedule, implying the use of electroporation (EP) after intra-muscular immunization.

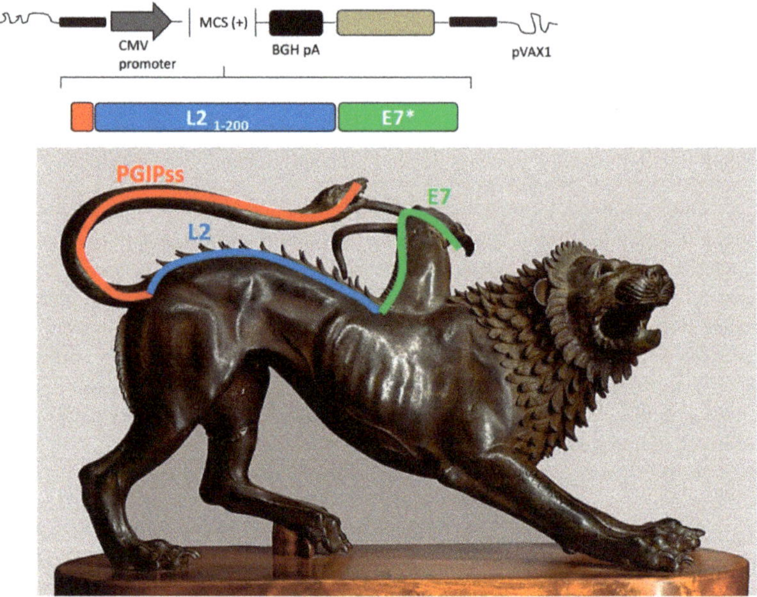



Chimera of Arezzo - National Archaeological Museum,
Florence.

**Figure 1.** Representation of the recombinant gene deriving from the fusion of the sequences encoding HPV antigens (attenuated E7 oncoprotein and part of the L2 structural protein) and a plant signal sequence introduced into the mammalian vector pVAX1. Because of the different origin of the genes and the immunological "dynamics" conferred to the antigens by the fusion, the representation includes an iconographic comparison with the legendary "Chimera" from Greek mythology. Chimera was a hybrid creature, offspring of the giant Thyphon and of the half-woman, half-snake Echidna, incorporating a lion's head with a goat rising from its back and a snaky tail: the different natures indicate the approach of using one nature to obtain a result in the other. CMV: cytomegalovirus promoter; BGH pA: bovine growth hormone poly-adenylation signal; PGIP*ss* (red box/line): signal sequence of the polygalacturonase-inhibiting protein of *Phaseolus vulgaris* gene; E7* (green box/line): attenuated E7 gene of HPV 16 (also named E7GGG); $L2_{1-200}$ (blue box/line): nucleotide sequence corresponding to amino acids 1–200 of the L2 minor capsid protein of HPV type 16.

The immunization protocol, performed together with EP, induced a long-lasting humoral IgG immune response against $L2_{1-200}$ and E7 that persisted for at least six months upon immunization in the mouse model used together with a cell-mediated immune response [19]. Electroporation, indeed, represents an approach, consolidated by recent clinical studies, to increase the effectiveness of DNA vaccines owing to its ability to increase cell permeability with a consequent increase in protein expression level and a better immune response [53,191].

Furthermore, the new DNA vaccine was able to determine an effective tumor regression in vivo in two mouse models: the TC-1 subcutaneous model in C57BL/6 mice and the AT-84 E7 orthotopic oral model in C3H/HeJ mice [192]. The AT-84 E7 cell line was derived from AT-84 cells, that generate a spontaneous oral squamous cell carcinoma in C3H mice [192]. The natural history of the tumor AT-84 and its response to therapy resemble human oral cancer, thus allowing the study of local invasion in a more clinically relevant site. The C3H mice accept AT-84 HPV16 E7 cell grafts without immunosuppression; the derived tumors maintain stably the oncogene expression (as it happens in HPV-related human oral cancer) and grow quickly, allowing fast testing and prediction of therapy effectiveness and of treatment schedules feasibility within few weeks. In addition, we have also developed a rapid and easy way to study in vivo tumor growth by using luciferase reporter gene (bioluminescent AT-84 HPV-16 E7-Luc model) and optical imaging [193]. In vivo imaging studies provide information that cannot be obtained from *post-mortem* analysis alone, representing complementary approaches for monitoring tumor progression and treatment response in orthotopic preclinical studies. In addition, the bioluminescent AT-84 HPV-16 E7-Luc model, generates data that can be compared with those obtained by caliper measurements and allows earlier cancer detection (it has been shown that tumor mass is measurable by luminescence at day 12 when a palpable tumor is still hardly detectable) [192].

The pVAX-PGIPss-L2$_{1-200}$-E7 DNA vaccine was delivered intra-muscle with EP in two different animal models of HP-associated tumor, resulting in a dramatic reduction of tumor growth [19]. Antitumor activity was further investigated showing that PGIPss-L2$_{1-200}$-E7 administration induced a specific cell-mediated immune response against HPV E7 tumor antigen. Further, the ability of the PGIPss to evoke antibody response in mammalian cells was confirmed in this system [20].

As far as we know, this represents the first demonstration that a plant-derived signal sequence has biological activity in mammalian cells, increasing the immunogenicity of an antigen of interest. The chimeric PGIPss-L2$_{1-200}$-E7 genetic vaccine represents a promising candidate against HPV-associated cancers and opens novel perspectives in the design of vaccines for other antigens and/or for different pathogens, in particular against infections where a fast and protective immune response is required (as in the current SARS-CoV-2 pandemic). Moreover, these vaccines, which can be easily produced on an industrial scale, under good manufacturing practices (GMP) conditions, represent safer vaccines as they do not involve the production of chemo/cytokines (which could induce secondary responses) or animal antigens (which could cause autoimmune cross-reactive responses).

*6.2. Combinations of Plant Molecules with HPV DNA Vaccination*

The combination of chemotherapy with immunotherapy (including DNA vaccination) might represent a potential strategy for cancer treatment because certain chemotherapy-based cancer treatments may activate the immune system against the tumor through several molecular and cellular mechanisms and reduce tumor [194].

It has been described that some natural molecules have the ability to increase vaccine-induced immunity against cancer. Furthermore, bulky tumor growth was inhibited by EGCG in combination with a genetic vaccine [15] while a multimodal treatment against progressive tumors based on apigenin and genetic vaccines demonstrated immunotherapeutic effects [14].

Nevertheless, very few works describe the effects of combined used of natural compounds with HPV genetic vaccination.

The immunomodulatory properties of *G. uralensis* polysaccharides were investigated in combination with HPV DNA vaccination [195]. *G. uralensis* polysaccharides injection had no side effect on mice, enhancing the immunity of mice and the antigen-specific cellular and humoral immune responses induced by HPV DNA vaccine.

Saffron (*Crocus sativus L.*) and its components, like monoterpene aldehyde and carotenoids derived from dry stigmas, have been suggested as favorable candidates for cancer prevention. The potential use of saffron derivatives such as extracts or purified components-carotenoids derived from dry stigmas of pure saffron alone or combined with genetic vaccination, was investigated on HPV-related experimental tumor [16]. The in vitro cytotoxic and apoptotic effects of aqueous saffron extract and its components picrocrocin, a monoterpene aldehyde and crocin, a natural carotenoid, were assessed in malignant (TC-1) and non-malignant (COS-7) cell lines. Unlike most carotenoids, that have a limited therapeutic use because of their insolubility in water, glycosylated molecules like crocin and picrocrocin (due to their glycosylated state) are soluble and highly cytotoxic on malignant cells; they therefore represent the most appropriate saffron derivatives for cancer treatment.

The anti-tumor activity of a genetic vaccine candidate and saffron components were evaluated in vivo; mice were challenged subcutaneously with TC-1 tumor cells and on day 3 and 17 they were immunized with E7-NT (gp96) DNA. Crocin, picrocrocin, and saffron extract were given orally at the time of the initial DNA immunization and for the next 16 days. While the multimodal treatment using DNA vaccine along with picrocrocin augmented the anti-tumor effects of picrocrocin, the combination of DNA vaccine with saffron extract and crocin at certain concentrations could not potentiate protective and therapeutic effects compared to mono-therapies for the control of TC-1 tumors. In particular, oral administration of crocin resulted in complete tumor regression, while it did not increase DNA vaccine-mediated anti-tumor effects. These data are apparently in contrast to those from other studies about combination of chemotherapy with EGCG [15], apigenin [14], and cisplatin [196] followed by immunotherapy with DNA vaccination. This highlights that, for proper evaluation of synergist effects of chemo-immunotherapy, the selection of the optimal dose and treatment schedule still represents a critical challenge to overcome [194,197].

## 7. Future Perspectives and Clinical Translation: What Is Needed

Together with evidences of efficacy and safety, the pharmacological use of plant-derived compounds could benefit, in many cases, from their wide availability in nature or by in vitro plant cell/tissue cultures. Certainly, chemically defined and purified entities are more suited and have higher potential as therapeutic compounds against cancer (and HPV-associated pathologies are no exception). All other plant derivatives, such as extracts and mixtures, may suffer from batch to batch variation.

In addition to this, studies to further investigate the encouraging promises of plant-derived compounds and formulations in advanced clinical trial phases, having large numbers of subjects, are needed. These studies should open the way to a deeper evaluation of the synergistic effects of plant compounds with canonical HPV tumor therapy approaches in combinatorial schedules, probably the most feasible and intriguing application of plant-derived compounds in this field.

The use of therapeutic vaccines against HPVs has been shown to induce regression of precancerous lesions of cervix and produce some clinical benefits in cancer patients. The immunosuppressive effects of the cancerous microenvironment can be attenuated by multimodal therapeutic approaches by combining the therapeutic vaccine with radiotherapy, chemotherapy, immunomodulators, and immune checkpoint inhibitors. In fact, experimental clinical evidence has clearly highlighted a synergistic action of these combinatorial approaches [8].

Therapeutic vaccines for pre- and neoplastic lesions of the uterine cervix are becoming a reality and the improvement of all therapeutic strategies associated with a multimodal approach opens a new scenario in the treatment not only of cervical cancer and pre-neoplastic lesions but also of other HPV-associated tumors. Despite to date, no therapeutic vaccine has been approved for clinical use in the treatment of HPV infections and related malignancies, it should be noted that at least some DNA vaccine candidates such as VGX-3100 and GX-188E or ADXS11-001 bacterial vector vaccine are in phase III clinical trials demonstrating that a therapeutic vaccine is in the near future (see NCT03185013, NCT02139267, and NCT02853604, respectively). In addition, therapeutic vaccines could be produced with plant expression systems that make production less expensive, expanding the possibility of their use also in low-middle-income countries where the disease burden is greater. However, these types of vaccines are still in an early stage of experimentation which makes them currently unavailable for clinical translation.

Even for DNA vaccines, their clinical use still requires a better knowledge of the mechanisms through which they are able to induce specific immune responses in vivo.

DNA vaccines can be introduced into the body by intramuscular, intradermal, or mucosal delivery through a variety of technologies, and most of them have been shown to be safe for humans despite causing varying degrees of discomfort. Furthermore, such therapeutic vaccines should not cause the activation of the immunosuppressive population of regulatory T lymphocytes (Treg). To achieve these results, vaccines and in particular DNA vaccines need to be associated to adjuvating procedures, including modification of antigen by fusion with other molecules or modifying the fate of antigen processing or host immune response (i.e., activating DC).

In an effort to make easier the transfer of adjuvated DNA vaccine from pre-clinical studies to clinical trials, it should be conceivable to use compounds and technologies already used in humans for other clinical indications. In this sense, building a DNA genetic vaccine using vectors already used in humans and with few antigen modifications can have a preferential route for their use. However, it is also important to address concerns regarding the potential for oncogenicity associated with administration of oncogenes (E5, E6, or E7) as DNA vaccines into the body. In response to that, all oncogenes in DNA vaccine are harmless version or shuffled epitopes [17,198].

If we can assume that DNA vector and modified antigen sequences are relatively safe, the same cannot be said for adjuvant activities. Such activities often involve molecules interfering with multiple pathways. This makes it difficult to directly use adjuvants in human experimentation without going through long toxicology phases in animal models. In this panorama, plant-based compounds, often derived from common food products (i.e., saffron) offer an undoubted advantage regarding their non-toxicity and, in addition, they often have a millennial use in traditional medicine. Other natural products like apigenin were already tested in several clinical trials and proven to be safe in humans [199,200].

Therefore, the study of plant compounds with adjuvant activity (especially, but not only) for genetic vaccines is a field of research that requires expansion in order to obtain vaccine products that can be used more quickly in humans. Finally, these studies will also be the basis for helping Regulatory Agencies dealing with drugs for human use by providing the necessary knowledge for an assessment of their human use.

## 8. Conclusions

Many efforts have been made to find and produce DNA therapeutic vaccines against HPV-associated lesions and cancers. The most effective vaccines in pre-clinical animal models are in clinical trials and two of them (VGX-3100 DNA and ADXS11-001) are in phase III clinical trials with promising results. Nevertheless, these vaccines are more effective in pre-cancerous lesions (i.e., CIN 2/3) than in cancer indicating that there is need of adjuvants to overcome the immunosuppressive tumor microenvironment. Plant sequences and also other genes from the "green world," such as plant-virus genes and signal sequences could be the answer considering their safety and avoidance of any autoimmune or pathogenic response.

In addition, plant compounds that directly abrogate HPV E6/E7 activity are feasible candidates for a HPV phytotherapy, as they affect the major HPV "oncoplayers." Purified phytochemicals, due to easy extraction and batch to batch consistency, will offer major perspectives to a comparatively safer alternative/combinational approach to HPV current therapy, once their efficacy in clinical trials will be confirmed. A summary of these new therapeutic strategies implying interconnections among DNA vaccines, plant compounds, and plant genes is depicted in Figure 2.

Thus, more studies are to be performed to check the effectiveness of these new therapeutic strategies including combination with existing approaches. Nevertheless, preliminary results open new horizons for the therapy of HPV-associated cancers. In addition, other tumors, where patient-specific neoantigens are detectable as a consequence of tumor-specific mutations, might benefit from these approaches [201].

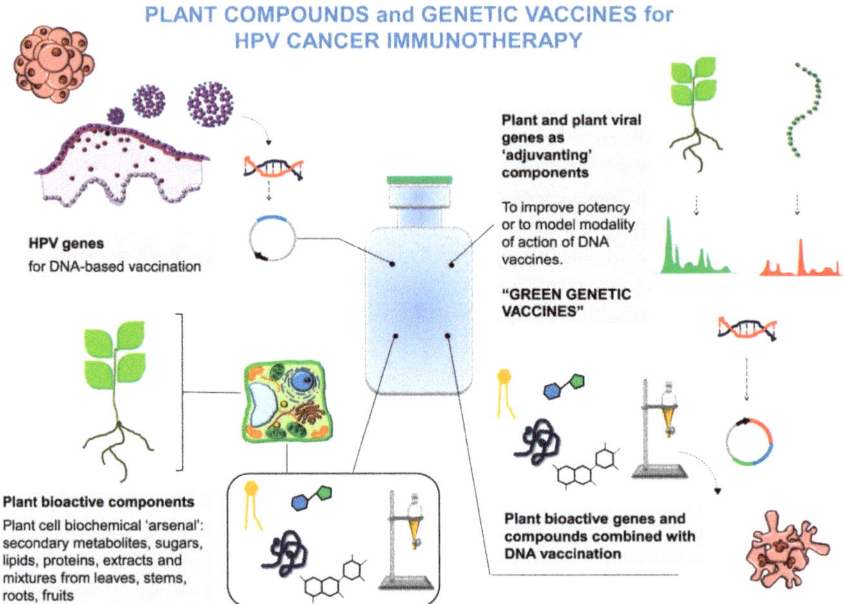


**Figure 2.** Possible new strategies against HPV tumors implying DNA vaccines, plant compounds, and plant gene sequences.

## 9. Patents

Franconi R, Massa S, Venuti A. (2016). Plant protein signal sequence as adjuvant in DNA vaccines. Italian patent 102016000131935, PCT/IT2017/050008, WO 2018/122885 A1. European Patent Application 17826321.6, publication number EP 3,562,504 (EP'504).

**Author Contributions:** Conceptualization, R.F., F.P., S.M., and A.V.; writing—original draft preparation, R.F., F.P., and S.M.; writing—review and editing, F.P., R.F., and A.V.; drawings, A.V., S.M.; supervision, P.V. All authors have read and agreed to the published version of the manuscript.

**Funding:** This research was partially funded by a Research Project "Regione Lazio-Lazio-Innova," funded under the L.R. 13/2008. Prot.# 85-2017-15170.

**Conflicts of Interest:** The authors declare no conflict of interest.

## References


1.  McLaughlin-Drubin, M.E.; Munger, K. Viruses associated with human cancer. *Biochim. Biophys. Acta (BBA) Mol. Basis Dis.* **2008**, *1782*, 127–150. [CrossRef] [PubMed]
2.  Dagenais, G.R.; Leong, D.P.; Rangarajan, S.; Lanas, F.; Lopez-Jaramillo, P.; Gupta, R.; Diaz, R.; Avezum, A.; Oliveira, G.B.F.; Wielgosz, A.; et al. Variations in Common Diseases, Hospital Admissions, and Deaths in Middle-Aged Adults in 21 Countries from Five Continents (PURE): A Prospective Cohort Study. *Lancet* **2020**, *395*, 785–794. [CrossRef]
3.  De Martel, C.; Plummer, M.; Vignat, J.; Franceschi, S. Worldwide burden of cancer attributable to HPV by site, country and HPV type. *Int. J. Cancer* **2017**, *141*, 664–670. [CrossRef] [PubMed]
4.  Cobos, C.; Figueroa, J.A.; Mirandola, L.; Colombo, M.; Summers, G.; Figueroa, A.; Aulakh, A.; Konala, V.M.; Verma, R.; Riaz, J.; et al. The Role of Human Papilloma Virus (HPV) Infection in Non-Anogenital Cancer and the Promise of Immunotherapy: A Review. *Int. Rev. Immunol.* **2014**, *33*, 383–401. [CrossRef]

5. Ferlay, J.; Colombet, M.; Soerjomataram, I.; Mathers, C.; Parkin, D.; Piñeros, M.; Znaor, A.; Bray, F. Estimating the global cancer incidence and mortality in 2018: GLOBOCAN sources and methods. *Int. J. Cancer* **2018**, *144*, 1941–1953. [CrossRef]

6. Lee, S.-J.; Yang, A.; Wu, T.; Hung, C.-F. Immunotherapy for human papillomavirus-associated disease and cervical cancer: Review of clinical and translational research. *J. Gynecol. Oncol.* **2016**, *27*, e51. [CrossRef] [PubMed]

7. Chabeda, A.; Yanez, R.J.; Lamprecht, R.; Meyers, A.E.; Rybicki, E.P.; Hitzeroth, I.I. Therapeutic vaccines for high-risk HPV-associated diseases. *Papillomavirus Res.* **2018**, *5*, 46–58. [CrossRef]

8. Vici, P.; Pizzuti, L.; Mariani, L.; Zampa, G.; Santini, D.; Di Lauro, L.; Gamucci, T.; Natoli, C.; Marchetti, P.; Barba, M.; et al. Targeting immune response with therapeutic vaccines in premalignant lesions and cervical cancer: Hope or reality from clinical studies. *Expert Rev. Vaccines* **2016**, *15*, 1327–1336. [CrossRef]

9. Cordeiro, M.N.; Lima, R.D.C.P.D.; Paolini, F.; Melo, A.R.D.S.; Campos, A.P.F.; Venuti, A.; Freitas, A.C. Current research into novel therapeutic vaccines against cervical cancer. *Expert Rev. Anticancer. Ther.* **2018**, *18*, 365–376. [CrossRef]

10. Franconi, R.; Di Bonito, P.; Dibello, F.; Accardi, L.; Muller, A.; Cirilli, A.; Simeone, P.; Donà, M.G.; Venuti, A.; Giorgi, C. Plant-derived human papillomavirus 16 E7 oncoprotein induces immune response and specific tumor protection. *Cancer Res.* **2002**, *62*, 3654–3658.

11. Franconi, R.; Massa, S.; Illiano, E.; Müller, A.; Cirilli, A.; Accardi, L.; Di Bonito, P.; Giorgi, C.; Venuti, A. Exploiting the Plant Secretory Pathway to Improve the Anticancer Activity of a Plant-Derived HPV16 E7 Vaccine. *Int. J. Immunopathol. Pharmacol.* **2006**, *19*, 187–19719. [CrossRef] [PubMed]

12. Di Bonito, P.; Grasso, F.; Mangino, G.; Massa, S.; Illiano, E.; Franconi, R.; Fanales-Belasio, E.; Falchi, M.; Affabris, E.; Giorgi, C. Immunomodulatory Activity of a Plant Extract Containing Human Papillomavirus 16-E7 Protein in Human Monocyte-Derived Dendritic Cells. *Int. J. Immunopathol. Pharmacol.* **2009**, *22*, 967–978. [CrossRef]

13. Demurtas, O.C.; Massa, S.; Ferrante, P.; Venuti, A.; Franconi, R.; Giuliano, G. A Chlamydomonas-Derived Human Papillomavirus 16 E7 Vaccine Induces Specific Tumor Protection. *PLoS ONE* **2013**, *8*, e61473. [CrossRef]

14. Chuang, C.-M.; Monie, A.; Wu, A.A.; Hung, C.-F. Combination of apigenin treatment with therapeutic HPV DNA vaccination generates enhanced therapeutic antitumor effects. *J. Biomed. Sci.* **2009**, *16*, 49. [CrossRef]

15. Kang, T.H.; Lee, J.H.; Kil Song, C.; Han, H.D.; Shin, B.C.; Pai, S.I.; Hung, C.-F.; Trimble, C.; Lim, J.-S.; Kim, T.W.; et al. Epigallocatechin-3-Gallate Enhances CD8+ T Cell-Mediated Antitumor Immunity Induced by DNA Vaccination. *Cancer Res.* **2007**, *67*, 802–811. [CrossRef] [PubMed]

16. Khavari, A.; Bolhassani, A.; Alizadeh, F.; Bathaie, S.Z.; Balaram, P.; Agi, E.; Vahabpour, R. Chemo-immunotherapy using saffron and its ingredients followed by E7-NT (gp96) DNA vaccine generates different anti-tumor effects against tumors expressing the E7 protein of human papillomavirus. *Arch. Virol.* **2014**, *160*, 499–508. [CrossRef]

17. Massa, S.; Simeone, P.; Muller, A.; Benvenuto, E.; Venuti, A.; Franconi, R. Antitumor Activity of DNA Vaccines Based on the Human Papillomavirus-16 E7 Protein Genetically Fused to a Plant Virus Coat Protein. *Hum. Gene Ther.* **2008**, *19*, 354–364. [CrossRef]

18. Massa, S.; Paolini, F.; Spanò, L.; Franconi, R.; Venuti, A. Mutants of plant genes for developing cancer vaccines. *Hum. Vaccines* **2011**, *7*, 147–155. [CrossRef]

19. Massa, S.; Paolini, F.; Curzio, G.; Cordeiro, M.N.; Illiano, E.; Demurtas, O.C.; Franconi, R.; Venuti, A. A plant protein signal sequence improved humoral immune response to HPV prophylactic and therapeutic DNA vaccines. *Hum. Vaccines Immunother.* **2017**, *13*, 271–282. [CrossRef] [PubMed]

20. Venuti, A.; Franconi, R.; Massa, S.; Mariani, L.; Paolini, F. Therapeutic/Preventive Hpv Vaccines, an Old Idea with Something New: Dna Chimeric Vaccines with Plant-Derived Signal Sequences. In Proceedings of the 32nd International Papillomavirus Conference, Sydney, Australia, 2–6 October 2018.

21. Cheng, M.A.; Farmer, E.; Huang, C.; Lin, J.; Hung, C.-F.; Wu, T.-C. Therapeutic DNA Vaccines for Human Papillomavirus and Associated Diseases. *Hum. Gene Ther.* **2018**, *29*, 971–996. [CrossRef]

22. Burd, E.M. Human Papillomavirus and Cervical Cancer. *Clin. Microbiol. Rev.* **2003**, *16*, 1–17. [CrossRef]

23. Gillison, M.L. Human Papillomavirus-Related Diseases: Oropharynx Cancers and Potential Implications for Adolescent HPV Vaccination. *J. Adolesc. Heal.* **2008**, *43*, S52–S60. [CrossRef]

24. Brianti, P.; De Flammineis, E.; Mercuri, S.R. Review of HPV-related diseases and cancers. *New Microbiol.* **2017**, *40*, 80–85. [PubMed]

25. Pullos, A.; Castilho, R.; Squarize, C. HPV Infection of the Head and Neck Region and Its Stem Cells. *J. Dent. Res.* **2015**, *94*, 1532–1543. [CrossRef] [PubMed]

26. Wu, R.; Paolini, F.; Frank, D.; Kamdar, D.; Curzio, G.; Pichi, B.; Pellini, R.; Spriano, G.; Bonagura, V.; Venuti, A.; et al. Latent human papillomavirus type 16 infection is widespread in patients with oropharyngeal cancers. *Oral Oncol.* **2018**, *78*, 222–224. [CrossRef]

27. Forcier, M.; Musacchio, N. An overview of human papillomavirus infection for the dermatologist: Disease, diagnosis, management, and prevention. *Dermatol. Ther.* **2010**, *23*, 458–476. [CrossRef]

28. Stillo, M.; Santisteve, P.C.; Lopalco, P.L. Safety of human papillomavirus vaccines: A review. *Expert Opin. Drug Saf.* **2015**, *14*, 697–712. [CrossRef]

29. Almeida, A.M.; Queiroz, J.A.; Sousa, F.; Sousa, Â. Cervical cancer and HPV infection: Ongoing therapeutic research to counteract the action of E6 and E7 oncoproteins. *Drug Discov. Today* **2019**, *24*, 2044–2057. [CrossRef]

30. Hoppe-Seyler, K.; Bossler, F.; Braun, J.A.; Herrmann, A.L.; Hoppe-Seyler, F. The HPV E6/E7 Oncogenes: Key Factors for Viral Carcinogenesis and Therapeutic Targets. *Trends Microbiol.* **2018**, *26*, 158–168. [CrossRef]

31. Villa, L.L. Prophylactic HPV vaccines: Reducing the burden of HPV-related diseases. *Vaccine* **2006**, *24*, S23–S28. [CrossRef]

32. Christensen, N.D.; A Reed, C.; Cladel, N.M.; Han, R.; Kreider, J.W. Immunization with viruslike particles induces long-term protection of rabbits against challenge with cottontail rabbit papillomavirus. *J. Virol.* **1996**, *70*, 960–965. [CrossRef] [PubMed]

33. Roden, R.B.; Monie, A.; Wu, T.-C. Opportunities to improve the prevention and treatment of cervical cancer. *Curr. Mol. Med.* **2007**, *7*, 490–503. [CrossRef] [PubMed]

34. Schiller, J.T.; Frazer, I.H.; Lowry, D.R. Chapter Human papillomavirus vaccines. In *Plotkin SA*; Orenstein, W.A., Offit, P.A., Eds.; Vaccines; Saunders/Elsevier: Philadelphia, PA, USA, 2008; pp. 244–257.

35. Schiller, J.T.; Müller, M. Next generation prophylactic human papillomavirus vaccines. *Lancet Oncol.* **2015**, *16*, e217–e225. [CrossRef]

36. Joura, E.A.; Giuliano, A.R.; Iversen, O.-E.; Bouchard, C.; Mao, C.; Mehlsen, J.; Moreira, E.D.; Ngan, Y.; Petersen, L.K.; Lazcano-Ponce, E.; et al. A 9-Valent HPV Vaccine against Infection and Intraepithelial Neoplasia in Women. *N. Engl. J. Med.* **2015**, *372*, 711–723. [CrossRef]

37. Signorelli, C.; Odone, A.; Ciorba, V.; Cella, P.; Audisio, R.A.; Lombardi, A.; Mariani, L.; Mennini, F.S.; Pecorelli, S.; Rezza, G.; et al. Human papillomavirus 9-valent vaccine for cancer prevention: A systematic review of the available evidence. *Epidemiol. Infect.* **2017**, *145*, 1962–1982. [CrossRef]

38. Ma, B.; Maraj, B.; Tran, N.P.; Knoff, J.; Chen, A.; Alvarez, R.D.; Hung, C.-F.; Wu, T.-C. Emerging human papillomavirus vaccines. *Expert Opin. Emerg. Drugs* **2012**, *17*, 469–492. [CrossRef]

39. Yang, A.; Jeang, J.; Cheng, K.; Cheng, T.; Yang, B.; Wu, T.-C.; Hung, C.-F. Current state in the development of candidate therapeutic HPV vaccines. *Expert Rev. Vaccines* **2016**, *15*, 989–1007. [CrossRef]

40. Drolet, M.; Bénard, Élodie; Pérez, N.; Brisson, M.; Ali, H.; Boily, M.-C.; Baldo, V.; Brassard, P.; Brotherton, J.M.L.; Callander, D.; et al. Population-level impact and herd effects following the introduction of human papillomavirus vaccination programmes: Updated systematic review and meta-analysis. *Lancet* **2019**, *394*, 497–509. [CrossRef]

41. Borysiewicz, L.; Fiander, A.; Nimako, M.; Man, S.; Wilkinson, G.; Westmoreland, D.; Evans, A.; Adams, M.; Stacey, S.; Boursnell, M.; et al. A recombinant vaccinia virus encoding human papillomavirus types 16 and 18, E6 and E7 proteins as immunotherapy for cervical cancer. *Lancet* **1996**, *347*, 1523–1527. [CrossRef]

42. Kenter, G.G.; Welters, M.J.P.; Valentijn, A.R.P.M.; Löwik, M.J.G.; Meer, D.M.A.B.-V.D.; Vloon, A.P.G.; Essahsah, F.; Fathers, L.M.; Offringa, R.; Drijfhout, J.W.; et al. Vaccination against HPV-16 Oncoproteins for Vulvar Intraepithelial Neoplasia. *N. Engl. J. Med.* **2009**, *361*, 1838–1847. [CrossRef]

43. Kenter, G.G.; Welters, M.J.; Valentijn, A.P.; Der Meer, D.M.B.-V.; Vloon, A.P.; Drijfhout, J.W.; Wafelman, A.R.; Oostendorp, J.; Fleuren, G.J.; Offringa, R.; et al. Phase I Immunotherapeutic Trial with Long Peptides Spanning the E6 and E7 Sequences of High-Risk Human Papillomavirus 16 in End-Stage Cervical Cancer Patients Shows Low Toxicity and Robust Immunogenicity. *Clin. Cancer Res.* **2008**, *14*, 169–177. [CrossRef] [PubMed]

44. I E Van Poelgeest, M.; Welters, M.J.P.; Van Esch, E.M.G.; Stynenbosch, L.F.M.; Kerpershoek, G.; Meerten, E.L.V.P.V.; Hende, M.V.D.; Löwik, M.J.G.; Meer, D.M.A.B.-V.D.; Fathers, L.M.; et al. HPV16 synthetic long peptide (HPV16-SLP) vaccination therapy of patients with advanced or recurrent HPV16-induced gynecological carcinoma, a phase II trial. *J. Transl. Med.* **2013**, *11*, 88. [CrossRef] [PubMed]

45. Rahma, O.E.; Herrin, V.E.; Ibrahim, R.A.; Toubaji, A.; Bernstein, S.; Dakheel, O.; Steinberg, S.M.; Abu-Eid, R.; Mkrtichyan, M.; Berzofsky, J.A.; et al. Pre-immature dendritic cells (PIDC) pulsed with HPV16 E6 or E7 peptide are capable of eliciting specific immune response in patients with advanced cervical cancer. *J. Transl. Med.* **2014**, *12*, 353. [CrossRef] [PubMed]

46. Chen, B.; Liu, L.; Xu, H.; Yang, Y.; Zhang, L.; Zhang, F. Effectiveness of immune therapy combined with chemotherapy on the immune function and recurrence rate of cervical cancer. *Exp. Ther. Med.* **2015**, *9*, 1063–1067. [CrossRef]

47. Chang, E.Y.; Chen, C.-H.; Ji, H.; Wang, T.-L.; Hung, K.; Lee, B.P.; Huang, A.Y.C.; Kurman, R.J.; Pardoll, D.M.; Wu, T.-C. Antigen-specific cancer immunotherapy using a GM-CSF secreting allogeneic tumor cell-based vaccine. *Int. J. Cancer* **2000**, *86*, 725–730. [CrossRef]

48. Stevanović, S.; Draper, L.M.; Langhan, M.M.; Campbell, T.E.; Kwong, M.L.; Wunderlich, J.R.; Dudley, M.E.; Yang, J.C.; Sherry, R.M.; Kammula, U.S.; et al. Complete Regression of Metastatic Cervical Cancer After Treatment With Human Papillomavirus–Targeted Tumor-Infiltrating T Cells. *J. Clin. Oncol.* **2015**, *33*, 1543–1550. [CrossRef]

49. Caballero, O.L.; Chen, Y.-T. Cancer/testis (CT) antigens: Potential targets for immunotherapy. *Cancer Sci.* **2009**, *100*, 2014–2021. [CrossRef]

50. Parmiani, G.; De Filippo, A.; Novellino, L.; Castelli, C. Unique human tumor antigens: Immunobiology and use in clinical trials. *J. Immunol.* **2007**, *178*, 1975–1979. [CrossRef]

51. Lopes, A.; Vandermeulen, G.; Préat, V. Cancer DNA vaccines: Current preclinical and clinical developments and future perspectives. *J. Exp. Clin. Cancer Res.* **2019**, *38*, 1–24. [CrossRef]

52. Brun, J.; Rajaonarison, J.; Nocart, N.; Hoarau, L.; Brun, S.; Garrigue, I. Targeted immunotherapy of high-grade cervical intra-epithelial neoplasia: Expectations from clinical trials (Review). *Mol. Clin. Oncol.* **2017**, *8*, 227–235. [CrossRef]

53. Trimble, C.L.; Morrow, M.P.; Kraynyak, K.A.; Shen, X.; Dallas, M.; Yan, J.; Edwards, L.; Parker, R.L.; Denny, L.; Giffear, M.; et al. Safety, Efficacy, and Immunogenicity of VGX-3100, a Therapeutic Synthetic DNA Vaccine Targeting Human Papillomavirus 16 and 18 E6 and E7 Proteins for Cervical Intraepithelial Neoplasia 2/3: A Randomised, Double-Blind, Placebo-Controlled Phase 2b Trial. *Lancet* **2015**, *386*, 2078–2088. [CrossRef]

54. Kim, T.J.; Jin, H.-T.; Hur, S.-Y.; Yang, H.G.; Seon, J.M.; Hong, S.R.; Lee, C.-W.; Kim, S.; Chang-Woo, L.; Park, K.S.; et al. Clearance of persistent HPV infection and cervical lesion by therapeutic DNA vaccine in CIN3 patients. *Nat. Commun.* **2014**, *5*, 5317. [CrossRef] [PubMed]

55. Choi, Y.J.; Hur, S.Y.; Kim, T.-J.; Hong, S.R.; Lee, J.-K.; Cho, C.-H.; Park, K.S.; Woo, J.W.; Sung, Y.C.; Suh, Y.S.; et al. A Phase II, Prospective, Randomized, Multicenter, Open-Label Study of GX-188E, an HPV DNA Vaccine, in Patients with Cervical Intraepithelial Neoplasia. *Clin. Cancer Res.* **2019**, *26*, 1616–1623. [CrossRef]

56. Karikó, K.; Muramatsu, H.; A Welsh, F.; Ludwig, J.; Kato, H.; Akira, S.; Weissman, D. Incorporation of Pseudouridine Into mRNA Yields Superior Nonimmunogenic Vector With Increased Translational Capacity and Biological Stability. *Mol. Ther.* **2008**, *16*, 1833–1840. [CrossRef] [PubMed]

57. Kim, T.W.; Hung, C.-F.; Juang, J.; He, L.; Hardwick, J.M.; Wu, T.-C. Enhancement of suicidal DNA vaccine potency by delaying suicidal DNA-induced cell death. *Gene Ther.* **2004**, *11*, 336–342. [CrossRef]

58. Kranz, L.M.; Diken, M.; Haas, H.; Kreiter, S.; Loquai, C.; Reuter, K.C.; Meng, M.; Fritz, D.; Vascotto, F.; Hefesha, H.; et al. Systemic RNA delivery to dendritic cells exploits antiviral defence for cancer immunotherapy. *Nat. Cell Biol.* **2016**, *534*, 396–401. [CrossRef]

59. Kübler, H.; Scheel, B.; Gnad-Vogt, U.; Miller, K.; Schultze-Seemann, W.; Dorp, F.V.; Parmiani, G.; Hampel, C.; Wedel, S.; Trojan, L.; et al. Self-adjuvanted mRNA vaccination in advanced prostate cancer patients: A first-in-man phase I/IIa study. *J. Immunother. Cancer* **2015**, *3*, 26. [CrossRef] [PubMed]

60. Papachristofilou, A.; Hipp, M.M.; Klinkhardt, U.; Früh, M.; Sebastian, M.; Weiss, C.; Pless, M.; Cathomas, R.; Hilbe, W.; Pall, G.; et al. Phase Ib evaluation of a self-adjuvanted protamine formulated mRNA-based active cancer immunotherapy, BI1361849 (CV9202), combined with local radiation treatment in patients with stage IV non-small cell lung cancer. *J. Immunother. Cancer* **2019**, *7*, 38. [CrossRef] [PubMed]

61. Cohen, J. Vaccine designers take first shots at COVID. *Scienceence* **2020**, *368*, 14–16. [CrossRef] [PubMed]

62. Coffman, R.L.; Sher, A.; Seder, R.A. Vaccine Adjuvants: Putting Innate Immunity to Work. *Immunology* **2010**, *33*, 492–503. [CrossRef]

63. Marciani, D.J. Vaccine adjuvants: Role and mechanisms of action in vaccine immunogenicity. *Drug Discov. Today* **2003**, *8*, 934–943. [CrossRef]

64. Pradere, J.-P.; Dapito, D.H.; Schwabe, R.F. The Yin and Yang of Toll-like receptors in cancer. *Oncogene* **2013**, *33*, 3485–3495. [CrossRef]

65. Parmiani, G.; Castelli, C.; Pilla, L.; Santinami, M.; Colombo, M.P.; Rivoltini, L. Opposite immune functions of GM-CSF administered as vaccine adjuvant in cancer patients. *Ann. Oncol.* **2007**, *18*, 226–232. [CrossRef] [PubMed]

66. Del Vecchio, M.; Bajetta, E.; Canova, S.; Lotze, M.T.; Wesa, A.; Parmiani, G.; Anichini, A. Interleukin-12: Biological Properties and Clinical Application. *Clin. Cancer Res.* **2007**, *13*, 4677–4685. [CrossRef] [PubMed]

67. Teng, M.W.L.; A Westwood, J.; Darcy, P.K.; Sharkey, J.; Tsuji, M.; Franck, R.W.; Porcelli, S.A.; Besra, G.S.; Takeda, K.; Yagita, H.; et al. Combined Natural Killer T-Cell Based Immunotherapy Eradicates Established Tumors in Mice. *Cancer Res.* **2007**, *67*, 7495–7504. [CrossRef]

68. Hailemichael, Y.; Dai, Z.; Jaffarzad, N.; Ye, Y.; A Medina, M.; Huang, X.-F.; Dorta-Estremera, S.M.; Greeley, N.R.; Nitti, G.; Peng, W.; et al. Persistent antigen at vaccination sites induces tumor-specific CD8+ T cell sequestration, dysfunction and deletion. *Nat. Med.* **2013**, *19*, 465–472. [CrossRef] [PubMed]

69. Bald, T.; Landsberg, J.; Lopez-Ramos, D.; Renn, M.; Glodde, N.; Jansen, P.; Gaffal, E.; Steitz, J.; Tolba, R.H.; Kalinke, U.; et al. Immune Cell-Poor Melanomas Benefit from PD-1 Blockade after Targeted Type I IFN Activation. *Cancer Discov.* **2014**, *4*, 674–687. [CrossRef]

70. Dubensky, J.T.W.; Kanne, D.B.; Leong, M.L. Rationale, progress and development of vaccines utilizing STING-activating cyclic dinucleotide adjuvants. *Ther. Adv. Vaccines* **2013**, *1*, 131–143. [CrossRef]

71. Barber, G.N. Cytoplasmic DNA innate immune pathways. *Immunol. Rev.* **2011**, *243*, 99–108. [CrossRef] [PubMed]

72. Fu, J.; Kanne, D.B.; Leong, M.; Glickman, L.H.; McWhirter, S.M.; Lemmens, E.; Mechette, K.; Leong, J.J.; Lauer, P.; Liu, W.; et al. STING agonist formulated cancer vaccines can cure established tumors resistant to PD-1 blockade. *Sci. Transl. Med.* **2015**, *7*, 283ra52. [CrossRef]

73. Pilla, L.; Rivoltini, L.; Patuzzo, R.I.; Marrari, A.; Valdagni, R.; Parmiani, G. Multipeptide vaccination in cancer patients. *Expert Opin. Biol. Ther.* **2009**, *9*, 1043–1055. [CrossRef]

74. Kaufman, H.L.; Ruby, C.; Hughes, T.; SlingluffJr, C.L. Current status of granulocyte–macrophage colony-stimulating factor in the immunotherapy of melanoma. *J. Immunother. Cancer* **2014**, *2*, 11. [CrossRef] [PubMed]

75. Lee, P.P.; Wang, F.; Kuniyoshi, J.; Rubio, V.; Stuge, T.B.; Groshen, S.; Gee, C.; Lau, R.; Jeffery, G.; Margolin, K.; et al. Effects of Interleukin-12 on the Immune Response to a Multipeptide Vaccine for Resected Metastatic Melanoma. *J. Clin. Oncol.* **2001**, *19*, 3836–3847. [CrossRef]

76. Barrios, K.; Celis, E. TriVax-HPV: An improved peptide-based therapeutic vaccination strategy against human papillomavirus-induced cancers. *Cancer Immunol. Immunother.* **2012**, *61*, 1307–1317. [CrossRef]

77. Massa, S.; Franconi, R.; Brandi, R.; Muller, A.; Mett, V.; Yusibov, V.; Venuti, A. Anti-cancer activity of plant-produced HPV16 E7 vaccine. *Vaccine* **2007**, *25*, 3018–3021. [CrossRef]

78. Venuti, A.; Massa, S.; Mett, V.; Vedova, L.D.; Paolini, F.; Franconi, R.; Yusibov, V. An E7-based therapeutic vaccine protects mice against HPV16 associated cancer. *Vaccine* **2009**, *27*, 3395–3397. [CrossRef] [PubMed]

79. Massa, S.; Paolini, F.; Marino, C.; Franconi, R.; Venuti, A. Bioproduction of a Therapeutic Vaccine Against Human Papillomavirus in Tomato Hairy Root Cultures. *Front. Plant Sci.* **2019**, *10*, 452. [CrossRef]

80. Whitehead, M.; Öhlschläger, P.; Almajhdi, F.N.; Alloza, L.; Marzábal, P.; E Meyers, A.; Hitzeroth, I.; Rybicki, E.P. Human papillomavirus (HPV) type 16 E7 protein bodies cause tumour regression in mice. *BMC Cancer* **2014**, *14*, 367. [CrossRef]

81. Brower, V. Back to Nature: Extinction of Medicinal Plants Threatens Drug Discovery. *J. Natl. Cancer Inst.* **2008**, *100*, 838–839. [CrossRef] [PubMed]

82. Newman, D.J.; Cragg, G.M. Natural Products As Sources of New Drugs over the 30 Years from 1981 to 2010. *J. Nat. Prod.* **2012**, *75*, 311–335. [CrossRef]

83. Royston, K.J.; Tollefsbol, T.O. The Epigenetic Impact of Cruciferous Vegetables on Cancer Prevention. *Curr. Pharmacol. Rep.* **2015**, *1*, 46–51. [CrossRef] [PubMed]

84. Collins, A.R.; Azqueta, A.; Langie, S. Effects of micronutrients on DNA repair. *Eur. J. Nutr.* **2012**, *51*, 261–279. [CrossRef]

85. Del Rio, D.; Rodriguez-Mateos, A.; Spencer, J.P.; Tognolini, M.; Borges, G.; Crozier, A. Dietary (Poly)phenolics in Human Health: Structures, Bioavailability, and Evidence of Protective Effects Against Chronic Diseases. *Antioxid Redox Sign.* **2013**, *18*, 1818–1892. [CrossRef]

86. Sgambato, A.; Zannoni, G.F.; Faraglia, B.; Camerini, A.; Tarquini, E.; Spada, D.; Cittadini, A. Decreased expression of the CDK inhibitor p27Kip1 and increased oxidative DNA damage in the multistep process of cervical carcinogenesis. *Gynecol. Oncol.* **2004**, *92*, 776–783. [CrossRef]

87. Looi, M.L.; Dali, A.Z.H.M.; Ali, S.A.M.; Ngah, W.Z.W.; Yusof, Y.A.M. Oxidative damage and antioxidant status in patients with cervical intraepithelial neoplasia and carcinoma of the cervix. *Eur. J. Cancer Prev.* **2008**, *17*, 555–560. [CrossRef]

88. Kim, S.Y.; Kim, J.W.; Ko, Y.S.; Koo, J.E.; Chung, H.Y.; Lee-Kim, Y.C. Changes in Lipid Peroxidation and Antioxidant Trace Elements in Serum of Women with Cervical Intraepithelial Neoplasia and Invasive Cancer. *Nutr. Cancer* **2003**, *47*, 126–130. [CrossRef] [PubMed]

89. Španinger, E.; Hrnčič, M.K.; Škerget, M.; Knez, Željko; Bren, U. Polyphenols: Extraction Methods, Antioxidative Action, Bioavailability and Anticarcinogenic Effects. *Molecules* **2016**, *21*, 901. [CrossRef]

90. Scherer, W.F.; Syverton, J.T.; Gey, G.O. Studies on the Propagation in Vitro Of Poliomyelitis Viruses. *J. Exp. Med.* **1953**, *97*, 695–710. [CrossRef]

91. Gey, G.O. Tissue Culture Studies of the Proliferative Capacity of Cervical Carcinoma and Normal Epithelium. *Cancer Res.* **1952**, *12*, 264–265.

92. Wang, S.; Zheng, C.-J.; Peng, C.; Zhang, H.; Jiang, Y.-P.; Han, T.; Qin, L.-P. Plants and cervical cancer: An overview. *Expert Opin. Investig. Drugs* **2013**, *22*, 1133–1156. [CrossRef] [PubMed]

93. Lin, J.; Chen, L.; Qiu, X.; Zhang, N.; Guo, Q.; Wang, Y.; Wang, M.; Gober, H.-J.; Li, D.; Wang, L. Traditional Chinese medicine for human papillomavirus (HPV) infections: A systematic review. *Biosci. Trends* **2017**, *11*, 267–273. [CrossRef]

94. Song, Q.I.N.; Li-qin, J.I.N. The Studies of Cyanidin 3-Glucoside-Induced Apoptosis in Human Cervical Cancer Hela Cells and its Mechanism. *Chin. J. Biochem. Pharm.* **2008**, *6*, 369–373.

95. Ahn, W.S.; Huh, S.W.; Bae, S.-M.; Lee, I.P.; Lee, J.M.; Namkoong, S.E.; Kim, C.K.; Sin, J.-I. A Major Constituent of Green Tea, EGCG, Inhibits the Growth of a Human Cervical Cancer Cell Line, CaSki Cells, through Apoptosis, G1 Arrest, and Regulation of Gene Expression. *DNA Cell Biol.* **2003**, *22*, 217–224. [CrossRef]

96. Sharma, C.; Nusri, Q.E.-A.; Begum, S.; Javed, E.; Rizvi, T.A.; Hussain, A. (-)-Epigallocatechin-3-Gallate Induces Apoptosis and Inhibits Invasion and Migration of Human Cervical Cancer Cells. *Asian Pac. J. Cancer Prev.* **2012**, *13*, 4815–4822. [CrossRef]

97. Qiao, Y.; Cao, J.; Xie, L.; Shi, X. Cell growth inhibition and gene expression regulation by (-)-epigallocatechin-3-gallate in human cervical cancer cells. *Arch. Pharmacal Res.* **2009**, *32*, 1309–1315. [CrossRef] [PubMed]

98. Zou, C.; Liu, H.; Feugang, J.M.; Hao, Z.; Chow, H.-H.S.; Garcia, F. Green Tea Compound in Chemoprevention of Cervical Cancer. *Int. J. Gynecol. Cancer* **2010**, *20*, 617–624. [CrossRef]

99. Zhu, Y.; Huang, Y.; Liu, M.; Yan, Q.; Zhao, W.; Yang, P.; Gao, Q.; Wei, J.; Zhao, W.; Ma, L. Epigallocatechin gallate inhibits cell growth and regulates miRNA expression in cervical carcinoma cell lines infected with different high-risk human papillomavirus subtypes. *Exp. Ther. Med.* **2018**, *17*, 1742–1748. [CrossRef]

100. Zhang, Q.; Tang, X.; Lu, Q.-Y.; Zhang, Z.-F.; Rao, J.; Le, A.D. Green tea extract and (−)-epigallocatechin-3-gallate inhibit hypoxia- and serum-induced HIF-1α protein accumulation and VEGF expression in human cervical carcinoma and hepatoma cells. *Mol. Cancer Ther.* **2006**, *5*, 1227–1238. [CrossRef]

101. Ramesh, E.; Alshatwi, A.A. Naringin induces death receptor and mitochondria-mediated apoptosis in human cervical cancer (SiHa) cells. *Food Chem. Toxicol.* **2013**, *51*, 97–105. [CrossRef]

102. Kim, N.-I.; Lee, S.-J.; Lee, S.-B.; Park, K.; Kim, W.-J.; Moon, S.-K. Requirement for Ras/Raf/ERK pathway in naringin-induced G1-cell-cycle arrest via p21WAF1 expression. *Carcinogenesis* **2008**, *29*, 1701–1709. [CrossRef]

103. Krishnakumar, N.; Sulfikkarali, N.; Rajendraprasad, N.; Karthikeyan, S. Enhanced anticancer activity of naringenin-loaded nanoparticles in human cervical (HeLa) cancer cells. *Biomed. Prev. Nutr.* **2011**, *1*, 223–231. [CrossRef]

104. Alshatwi, A.A.; Ramesh, E.; Periasamy, V.; Subash-Babu, P. The apoptotic effect of hesperetin on human cervical cancer cells is mediated through cell cycle arrest, death receptor, and mitochondrial pathways. *Fundam. Clin. Pharmacol.* **2012**, *27*, 581–592. [CrossRef]

105. Oh, E.K.; Kim, H.J.; Bae, S.M.; Park, M.Y.; Kim, Y.W.; Kim, T.E.; Ahn, W.S. Apigenin-Induced Apoptosis in Cervical Cancer Cell Lines. *Korean J. Obstet. Gynecol.* **2008**, *51*, 874–881.

106. Zheng, P.-W.; Chiang, L.-C.; Lin, C.-C. Apigenin induced apoptosis through p53-dependent pathway in human cervical carcinoma cells. *Life Sci.* **2005**, *76*, 1367–1379. [CrossRef]

107. Czyż, J.; Madeja, Z.; Irmer, U.; Korohoda, W.; Hülser, D.F. Flavonoid apigenin inhibits motility and invasiveness of carcinoma cellsin vitro. *Int. J. Cancer* **2004**, *114*, 12–18. [CrossRef]

108. Guo, J.M.; Kang, G.Z.; Xiao, B.X.; Liu, D.H.; Zhang, S. Effect of Daidzein on Cell Growth, Cell Cycle, and Telomerase Activity of Human Cervical Cancer in Vitro. *Int. J. Gynecol. Cancer* **2004**, *14*, 882–888. [CrossRef]

109. Lee, H.G.; Yu, K.A.; Oh, W.K.; Baeg, T.W.; Oh, H.C.; Ahn, J.S.; Jang, W.C.; Kim, J.W.; Lim, J.S.; Choe, Y.K.; et al. Inhibitory Effect of Jaceosidin Isolated from Artemisiaargyi on the Function of E6 and E7 Oncoproteins of HPV. *J. Ethnopharmacol.* **2005**, *98*, 339–343. [CrossRef]

110. Cherry, J.J.; Rietz, A.; Malinkevich, A.; Liu, Y.; Xie, M.; Bartolowits, M.; Davisson, V.J.; Baleja, J.D.; Androphy, E.J. Structure Based Identification and Characterization of Flavonoids That Disrupt Human Papillomavirus-16 E6 Function. *PLoS ONE* **2013**, *8*, e84506. [CrossRef]

111. Yan, J.; Wang, Q.; Zheng, X.; Sun, H.; Zhou, Y.; Li, D.; Lin-Bo, G.; Wang, X. Luteolin enhances TNF-related apoptosis-inducing ligand's anticancer activity in a lung cancer xenograft mouse model. *Biochem. Biophys. Res. Commun.* **2012**, *417*, 842–846. [CrossRef]

112. Kim, M.S.; Bak, Y.; Park, Y.S.; Lee, D.-H.; Kim, J.H.; Kang, J.W.; Song, H.-H.; Oh, S.-R.; Yoon, D.Y. Wogonin induces apoptosis by suppressing E6 and E7 expressions and activating intrinsic signaling pathways in HPV-16 cervical cancer cells. *Cell Biol. Toxicol.* **2013**, *29*, 259–272. [CrossRef]

113. Priyadarsini, R.V.; Murugan, R.S.; Maitreyi, S.; Ramalingam, K.; Karunagaran, D.; Nagini, S. The flavonoid quercetin induces cell cycle arrest and mitochondria-mediated apoptosis in human cervical cancer (HeLa) cells through p53 induction and NF-κB inhibition. *Eur. J. Pharmacol.* **2010**, *649*, 84–91. [CrossRef] [PubMed]

114. Xu, W.; Liu, J.; Li, C.; Wu, H.-Z.; Liu, Y.-W. Kaempferol-7-O-β-d-glucoside (KG) isolated from Smilax china L. rhizome induces G2/M phase arrest and apoptosis on HeLa cells in a p53-independent manner. *Cancer Lett.* **2008**, *264*, 229–240. [CrossRef] [PubMed]

115. Ying, T.-H.; Yang, S.-F.; Tsai, S.-J.; Hsieh, S.-C.; Huang, Y.-C.; Bau, D.-T.; Hsieh, Y.-H. Fisetin induces apoptosis in human cervical cancer HeLa cells through ERK1/2-mediated activation of caspase-8-/caspase-3-dependent pathway. *Arch. Toxicol.* **2011**, *86*, 263–273. [CrossRef] [PubMed]

116. You, B.R.; Moon, H.J.; Han, Y.H.; Park, W.H. Gallic acid inhibits the growth of HeLa cervical cancer cells via apoptosis and/or necrosis. *Food Chem. Toxicol.* **2010**, *48*, 1334–1340. [CrossRef] [PubMed]

117. Hsu, K.-F.; Wu, C.-L.; Huang, S.-C.; Wu, C.-M.; Hsiao, J.-R.; Yo, Y.-T.; Chen, Y.-H.; Shiau, A.-L.; Chou, C.-Y. Cathepsin L mediates resveratrol-induced autophagy and apoptotic cell death in cervical cancer cells. *Autophagy* **2009**, *5*, 451–460. [CrossRef] [PubMed]

118. Kim, Y.S.; Sull, J.W.; Sung, H.J. Suppressing effect of resveratrol on the migration and invasion of human metastatic lung and cervical cancer cells. *Mol. Biol. Rep.* **2012**, *39*, 8709–8716. [CrossRef]

119. García-Zepeda, S.P.; García-Villa, E.; Díaz-Chávez, J.; Hernández-Pando, R.; Gariglio, P. Resveratrol induces cell death in cervical cancer cells through apoptosis and autophagy. *Eur. J. Cancer Prev.* **2013**, *22*, 577–584. [CrossRef]

120. Xin, S.; Shulan, L.; Jing, Z.; Shengnan, L.; Jiyong, G.; Cheng'En, P. Effects of Res on proliferation and apoptosis of human cervical carcinoma cell lines C33A, SiHa and HeLa. *J. Med Coll. PLA* **2009**, *24*, 148–154. [CrossRef]

121. Szliszka, E.; Czuba, Z.; Jernas, K.; Król, W. Dietary Flavonoids Sensitize HeLa Cells to Tumor Necrosis Factor-Related Apoptosis-Inducing Ligand (TRAIL). *Int. J. Mol. Sci.* **2008**, *9*, 56–64. [CrossRef]

122. Maher, D.M.; Bell, M.C.; O'Donnell, E.A.; Gupta, B.K.; Jaggi, M.; Chauhan, S.C. Curcumin suppresses human papillomavirus oncoproteins, restores p53, rb, and ptpn13 proteins and inhibits benzo[a]pyrene-induced upregulation of HPV. *Mol. Carcinog.* **2010**, *50*, 47–57. [CrossRef]

123. Aggarwal, B.B.; Bhatt, I.D.; Ichikawa, H.; Ahn, K.S.; Sethi, G.; Sandur, S.K.; Natarajan, C.; Seeram, N.; Shishodia, S. *Turmeric the Genus Curcuma*; CRC: New York, NY, USA, 2007; pp. 297–368.

124. Limtrakul, P.; Chearwae, W.; Shukla, S.; Phisalphong, C.; Ambudkar, S.V. Modulation of function of three ABC drug transporters, P-glycoprotein (ABCB1), mitoxantrone resistance protein (ABCG2) and multidrug resistance protein 1 (ABCC1) by tetrahydrocurcumin, a major metabolite of curcumin. *Mol. Cell. Biochem.* **2006**, *296*, 85–95. [CrossRef] [PubMed]

125. Divya, C.S.; Pillai, M.R. Antitumor action of curcumin in human papillomavirus associated cells involves downregulation of viral oncogenes, prevention of NFkB and AP-1 translocation, and modulation of apoptosis. *Mol. Carcinog.* **2006**, *45*, 320–332. [CrossRef] [PubMed]

126. Singh, M.; Singh, N. Molecular mechanism of curcumin induced cytotoxicity in human cervical carcinoma cells. *Mol. Cell. Biochem.* **2009**, *325*, 107–119. [CrossRef]

127. Paulraj, F.; Abas, F.; Lajis, N.H.; Othman, I.; Hassan, S.S.; Naidu, R. The Curcumin Analogue 1,5-Bis(2-hydroxyphenyl)-1,4-pentadiene-3-one Induces Apoptosis and Downregulates E6 and E7 Oncogene Expression in HPV16 and HPV18-Infected Cervical Cancer Cells. *Molecules* **2015**, *20*, 11830–11860. [CrossRef]

128. Giridharan, P.; Somasundaram, S.T.; Perumal, K.; A Vishwakarma, R.; Karthikeyan, N.P.; Velmurugan, R.; Balakrishnan, A. Novel substituted methylenedioxy lignan suppresses proliferation of cancer cells by inhibiting telomerase and activation of c-myc and caspases leading to apoptosis. *Br. J. Cancer* **2002**, *87*, 98–105. [CrossRef]

129. Gao, P.; Zhai, F.; Guan, L.; Zheng, J. Nordihydroguaiaretic acid inhibits growth of cervical cancer SiHa cells by up-regulating p21. *Oncol. Lett.* **2010**, *2*, 123–128. [CrossRef] [PubMed]

130. Munagala, R.; Aqil, F.; Jeyabalan, J.; Gupta, R.C. Tanshinone IIA inhibits viral oncogene expression leading to apoptosis and inhibition of cervical cancer. *Cancer Lett.* **2015**, *356*, 536–546. [CrossRef]

131. Mahata, S.; Bharti, A.C.; Shukla, S.; Tyagi, A.; A Husain, S.; Das, B.C. Berberine modulates AP-1 activity to suppress HPV transcription and downstream signaling to induce growth arrest and apoptosis in cervical cancer cells. *Mol. Cancer* **2011**, *10*, 39. [CrossRef]

132. Saha, S.K.; Khuda-Bukhsh, A.R. Berberine alters epigenetic modifications, disrupts microtubule network, and modulates HPV-18 E6–E7 oncoproteins by targeting p53 in cervical cancer cell HeLa: A mechanistic study including molecular docking. *Eur. J. Pharmacol.* **2014**, *744*, 132–146. [CrossRef]

133. Munagala, R.; Kausar, H.; Munjal, C.; Gupta, R.C. Withaferin A induces p53-dependent apoptosis by repression of HPV oncogenes and upregulation of tumor suppressor proteins in human cervical cancer cells. *Carcinogenesis* **2011**, *32*, 1697–1705. [CrossRef]

134. Yim, N.-H.; Lee, J.H.; Cho, W.-K.; Yang, M.C.; Kwak, D.H.; Ma, J.Y. Decursin and decursinol angelate from Angelica gigas Nakai induce apoptosis via induction of TRAIL expression on cervical cancer cells. *Eur. J. Integr. Med.* **2011**, *3*, e299–e307. [CrossRef]

135. Li, J.; Li, Q.-W.; Gao, D.-W.; Han, Z.-S.; Lu, W.-Z. Antitumor and immunomodulating effects of polysaccharides isolated fromSolanum nigrumLinne. *Phytother. Res.* **2009**, *23*, 1524–1530. [CrossRef] [PubMed]

136. Yan, Q.; Li, Y.; Jiang, Z.; Sun, Y.; Zhu, L.; Ding, Z. Antiproliferation and apoptosis of human tumor cell lines by a lectin (AMML) of Astragalus mongholicus. *Phytomedicine* **2009**, *16*, 586–593. [CrossRef]

137. Mancinelli, L.; De Angelis, P.M.; Annulli, L.; Padovini, V.; Elgjo, K.; Gianfranceschi, G.L. A class of DNA-binding peptides from wheat bud causes growth inhibition, G2 cell cycle arrest and apoptosis induction in HeLa cells. *Mol. Cancer* **2009**, *8*, 55. [CrossRef]

138. Wang, P.; Li, J.-C. Trichosanthin-induced specific changes of cytoskeleton configuration were associated with the decreased expression level of actin and tubulin genes in apoptotic Hela cells. *Life Sci.* **2007**, *81*, 1130–1140. [CrossRef]

139. Mahata, S.; Maru, S.; Shukla, S.; Pandey, A.; Mugesh, G.; Das, B.C.; Bharti, A.C. Anticancer property of Bryophyllum pinnata (Lam.) Oken. leaf on human cervical cancer cells. *BMC Complement. Altern. Med.* **2012**, *12*, 15. [CrossRef] [PubMed]

140. Mahata, S.; Pandey, A.; Shukla, S.; Tyagi, A.; Husain, S.A.; Das, B.C.; Bharti, A.C. Anticancer Activity of Phyllanthus emblicaLinn. (Indian Gooseberry): Inhibition of Transcription Factor AP-1 and HPV Gene Expression in Cervical Cancer Cells. *Nutr. Cancer* **2013**, *65*, 88–97. [CrossRef]

141. Hu, Y.; Wan, X.-J.; Pan, L.-L.; Zhang, S.-H.; Zheng, F.-Y. Effects of Brucea javanica oil emulsion on human papilloma virus type 16 infected cells and mechanisms research. *Chin. J. Integr. Tradit. West. Med.* **2013**, *33*, 1545–1551.

142. Li, G.; Jiang, W.; Xia, Q.; Chen, S.-H.; Ge, X.-R.; Gui, S.-Q.; Xu, C.-J. HPV E6 down-regulation and apoptosis induction of human cervical cancer cells by a novel lipid-soluble extract (PE) from Pinellia pedatisecta Schott in vitro. *J. Ethnopharmacol.* **2010**, *132*, 56–64. [CrossRef]

143. Talwar, G.P.; Sharma, R.; Singh, S.; Das, B.C.; Bharti, A.C.; Sharma, K.; Singh, P.; Atrey, N.; Gupta, J.C. BASANT, a Polyherbal Safe Microbicide Eliminates HPV-16 in Women with Early Cervical Intraepithelial Lesions. *J. Cancer Ther.* **2015**, *6*, 1163–1166. [CrossRef]

144. Kwon, S.-B.; Kim, M.-J.; Yang, J.M.; Lee, H.-P.; Hong, J.T.; Jeong, H.-S.; Kim, E.S.; Yoon, D.-Y. Cudrania tricuspidata Stem Extract Induces Apoptosis via the Extrinsic Pathway in SiHa Cervical Cancer Cells. *PLoS ONE* **2016**, *11*, e0150235. [CrossRef] [PubMed]

145. Ghanbari, A.; Le Gresley, A.; Naughton, D.; Kuhnert, N.; Sirbu, D.; Ashrafi, G.H. Biological activities of Ficus carica latex for potential therapeutics in Human Papillomavirus (HPV) related cervical cancers. *Sci. Rep.* **2019**, *9*, 1–11. [CrossRef] [PubMed]

146. Moga, M.A.; Dimienescu, O.G.; Arvatescu, C.A.; Mironescu, A.; Dracea, L.; Ples, L. The Role of Natural Polyphenols in the Prevention and Treatment of Cervical Cancer—An Overview. *Molecules* **2016**, *21*, 1055. [CrossRef] [PubMed]

147. Gutiérrez-Venegas, G.; Sánchez-Carballido, M.A.; Suárez, C.D.; Gómez-Mora, J.A.; Bonneau, N. Effects of flavonoids on tongue squamous cell carcinoma. *Cell Biol. Int.* **2020**, *44*, 686–720. [CrossRef] [PubMed]

148. Chen, P.-N.; Kuo, W.-H.; Chiang, C.-L.; Chiou, H.-L.; Hsieh, Y.-S.; Chu, S.-C. Black rice anthocyanins inhibit cancer cells invasion via repressions of MMPs and u-PA expression. *Chem. Interact.* **2006**, *163*, 218–229. [CrossRef]

149. Hausen, H.Z. Papillomaviruses and cancer: From basic studies to clinical application. *Nat. Rev. Cancer* **2002**, *2*, 342–350. [CrossRef] [PubMed]

150. Khan, M.A.; Hussain, A.; Sundaram, M.K.; Alalami, U.; Gunasekera, D.; Ramesh, L.; Hamza, A.; Quraishi, U. (−)-Epigallocatechin-3-gallate reverses the expression of various tumor-suppressor genes by inhibiting DNA methyltransferases and histone deacetylases in human cervical cancer cells. *Oncol. Rep.* **2015**, *33*, 1976–1984. [CrossRef]

151. Epelbaum, R.; Schaffer, M. Curcumin as an Anti-Cancer Agent: Review of the Gap Between Basic and Clinical Applications. *Curr. Med. Chem.* **2010**, *17*, 190–197. [CrossRef]

152. Talwar, G.P.; Dar, S.A.; Rai, M.; Reddy, K.; Mitra, D.; Kulkarni, S.V.; Doncel, G.F.; Buck, C.B.; Schiller, J.T.; Muralidhar, S.; et al. A novel polyherbal microbicide with inhibitory effect on bacterial, fungal and viral genital pathogens. *Int. J. Antimicrob. Agents* **2008**, *32*, 180–185. [CrossRef]

153. Bava, S.V.; Sreekanth, C.N.; Thulasidasan, A.K.T.; Anto, N.P.; Cheriyan, V.T.; Puliyappadamba, V.T.; Menon, S.G.; Ravichandran, S.D.; Anto, R.J. Akt is upstream and MAPKs are downstream of NF-κB in paclitaxel-induced survival signaling events, which are down-regulated by curcumin contributing to their synergism. *Int. J. Biochem. Cell Biol.* **2011**, *43*, 331–341. [CrossRef]

154. Jakubowicz-Gil, J.; Paduch, R.; Piersiak, T.; Głowniak, K.; Gawron, A.; Kandefer-Szerszeń, M. The effect of quercetin on pro-apoptotic activity of cisplatin in HeLa cells. *Biochem. Pharmacol.* **2005**, *69*, 1343–1350. [CrossRef] [PubMed]

155. Wang, Q.; Zheng, X.-L.; Yang, L.; Shi, F.; Gao, L.-B.; Zhong, Y.-J.; Sun, H.; He, F.; Lin-Bo, G.; Wang, X. Reactive oxygen species-mediated apoptosis contributes to chemosensitization effect of saikosaponins on cisplatin-induced cytotoxicity in cancer cells. *J. Exp. Clin. Cancer Res.* **2010**, *29*, 159. [CrossRef]

156. He, F.; Wang, Q.; Zheng, X.-L.; Yan, J.-Q.; Yang, L.; Sun, H.; Hu, L.-N.; Lin, Y.; Wang, X. Wogonin potentiates cisplatin-induced cancer cell apoptosis through accumulation of intracellular reactive oxygen species. *Oncol. Rep.* **2012**, *28*, 601–605. [CrossRef] [PubMed]

157. Xu, Y.; Xin, Y.; Diao, Y.; Lu, C.; Fu, J.; Luo, L.; Yin, Z. Synergistic Effects of Apigenin and Paclitaxel on Apoptosis of Cancer Cells. *PLoS ONE* **2011**, *6*, e29169. [CrossRef] [PubMed]

158. Lo, Y.-L.; Wang, W. Formononetin potentiates epirubicin-induced apoptosis via ROS production in HeLa cells in vitro. *Chem. Interact.* **2013**, *205*, 188–197. [CrossRef] [PubMed]

159. Alshatwi, A.A.; Periasamy, V.S.; Athinarayanan, J.; Elango, R. Synergistic anticancer activity of dietary tea polyphenols and bleomycin hydrochloride in human cervical cancer cell: Caspase-dependent and independent apoptotic pathways. *Chem. Interact.* **2016**, *247*, 1–10. [CrossRef]

160. Zoberi, I.; Bradbury, C.; Curry, H.A.; Bisht, K.S.; Goswami, P.C.; Roti, J.L.R.; Gius, D. Radiosensitizing and anti-proliferative effects of resveratrol in two human cervical tumor cell lines. *Cancer Lett.* **2002**, *175*, 165–173. [CrossRef]

161. Yashar, C.M.; Spanos, W.J.; Taylor, U.D.; Gercel-Taylor, C. Potentiation of the radiation effect with genistein in cervical cancer cells. *Gynecol. Oncol.* **2005**, *99*, 199–205. [CrossRef]

162. Zhang, B.; Liu, J.-Y.; Pan, J.; Han, S.-P.; Yin, X.-X.; Wang, B.; Hu, G. Combined treatment of ionizing radiation with genistein on cervical cancer HeLa cells. *J. Pharmacol. Sci.* **2006**, *102*, 129–135. [CrossRef]

163. Shin, J.-I.; Shim, J.-H.; Kim, K.-H.; Choi, H.-S.; Kim, J.-W.; Lee, H.G.; Kim, B.Y.; Park, S.-N.; Park, O.-J.; Yoon, D.-Y. Sensitization of the apoptotic effect of gamma-irradiation in genistein-pretreated CaSki cervical cancer cells. *J. Microbiol. Biotechnol.* **2008**, *18*, 523–5317183.

164. Javvadi, P.; Hertan, L.; Kosoff, R.; Datta, T.; Kolev, J.; Mick, R.; Tuttle, S.W.; Koumenis, C. Thioredoxin Reductase-1 Mediates Curcumin-Induced Radiosensitization of Squamous Carcinoma Cells. *Cancer Res.* **2010**, *70*, 1941–1950. [CrossRef] [PubMed]

165. Karthikeyan, S.; Kanimozhi, G.; Prasad, N.R.; Mahalakshmi, R. Radiosensitizing effect of ferulic acid on human cervical carcinoma cells in vitro. *Toxicol. Vitr.* **2011**, *25*, 1366–1375. [CrossRef] [PubMed]

166. Lin, C.; Yu, Y.; Zhao, H.-G.; Yang, A.; Yan, H.; Cui, Y. Combination of quercetin with radiotherapy enhances tumor radiosensitivity in vitro and in vivo. *Radiother. Oncol.* **2012**, *104*, 395–400. [CrossRef] [PubMed]

167. Prusty, B.K.; Das, B.C. Constitutive activation of transcription factor AP-1 in cervical cancer and suppression of human papillomavirus (HPV) transcription and AP-1 activity in HeLa cells by curcumin. *Int. J. Cancer* **2004**, *113*, 951–960. [CrossRef]

168. Hillman, G.G.; Forman, J.D.; Kucuk, O.; Yudelev, M.; Maughan, R.L.; Rubio, J.; Layer, A.; Tekyi-Mensah, S.; Abrams, J.; Sarkar, F.H. Genistein potentiates the radiation effect on prostate carcinoma cells. *Clin. Cancer Res.* **2001**, *7*, 382–390.

169. Ahn, W.-S.; Yoo, J.; Huh, S.-W.; Kim, C.K.; Lee, J.-M.; Namkoong, S.-E.; Bae, S.-M.; Lee, I.P. Protective effects of green tea extracts (polyphenon E and EGCG) on human cervical lesions. *Eur. J. Cancer Prev.* **2003**, *12*, 383–390. [CrossRef]

170. Cheng, A.L.; Hsu, C.-H.; Lin, J.K.; Hsu, M.; Ho, Y.F.; Shen, T.S.; Ko, J.Y.; Lin, J.T.; Lin, B.R.; Ming-Shiang, W.; et al. Phase I clinical trial of curcumin, a chemopreventive agent, in patients with high-risk or pre-malignant lesions. *Anticancer Res.* **2001**, *21*, 2895–2900.

171. Shukla, S.; Bharti, A.C.; Hussain, S.; Mahata, S.; Hedau, S.; Kailash, U.; Kashyap, V.; Bhambhani, S.; Roy, M.; Batra, S.; et al. Elimination of high-risk human papillomavirus type HPV16 infection by 'Praneem' polyherbal tablet in women with early cervical intraepithelial lesions. *J. Cancer Res. Clin.* **2009**, *135*, 1701–1709. [CrossRef]

172. Basu, P.; Dutta, S.; Begum, R.; Mittal, S.; Das Dutta, P.; Bharti, A.C.; Panda, C.K.; Biswas, J.; Dey, B.; Talwar, G.P.; et al. Clearance of cervical human papillomavirus infection by topical application of curcumin and curcumin containing polyherbal cream: A phase II randomized controlled study. *Asian Pac. J. Cancer Prev.* **2013**, *14*, 5753–5759. [CrossRef]

173. Talwar, G.; Raghuvanshi, P.; Mishra, R.; Banerjee, U.; Rattan, A.; Whaley, K.J.; Achilles, S.L.; Zeitlin, L.; Barré-Sinoussi, F.; David, A.; et al. Polyherbal Formulations with Wide Spectrum Antimicrobial Activity Against Reproductive Tract Infections and Sexually Transmitted Pathogens. *Am. J. Reprod. Immunol.* **2000**, *43*, 144–151. [CrossRef]

174. Wang, Y.; Huang, H.; Yao, S.; Li, G.; Xu, C.; Ye, Y.; Gui, S. A lipid-soluble extract of Pinellia pedatisecta Schott enhances antitumor T cell responses by restoring tumor-associated dendritic cell activation and maturation. *J. Ethnopharmacol.* **2019**, *241*, 111980. [CrossRef] [PubMed]

175. Khan, T.; Gurav, P. PhytoNanotechnology: Enhancing Delivery of Plant Based Anti-cancer Drugs. *Front. Pharmacol.* **2018**, *8*, 1002. [CrossRef] [PubMed]

176. Gerloni, M.; Xiong, S.; Mukerjee, S.; Schoenberger, S.P.; Croft, M.; Zanetti, M. Functional cooperation between T helper cell determinants. *Proc. Natl. Acad. Sci. USA* **2000**, *97*, 13269–13274. [CrossRef] [PubMed]

177. Savelyeva, N.; Zhu, D.; Stevenson, F.K. Engineering DNA vaccines that include plant virus coat proteins. *Biotechnol. Genet. Eng. Rev.* **2003**, *20*, 101–116. [CrossRef]

178. Tan, B.K.H.; Vanitha, J. Immunomodulatory and antimicrobial effects of some traditional chinese medicinal herbs: A review. *Curr. Med. Chem.* **2004**, *11*, 1423–1430. [CrossRef]

179. Ronald, P.C.; Beutler, B. Plant and Animal Sensors of Conserved Microbial Signatures. *Science* **2010**, *330*, 1061–1064. [CrossRef]

180. Girbes, T.; Ferreras, J.M.; Arias, F.J.; Stirpe, F. Description, Distribution, Activity and Phylogenetic Relationship of Ribosome-Inactivating Proteins in Plants, Fungi and Bacteria. *Mini-Rev. Med. Chem.* **2004**, *4*, 461–476. [CrossRef]

181. Fang, E.F.; Ng, T.B.; Shaw, P.C.; Wong, R.N.-S. Recent Progress in Medicinal Investigations on Trichosanthin and other Ribosome Inactivating Proteins from the Plant Genus Trichosanthes. *Curr. Med. Chem.* **2011**, *18*, 4410–4417. [CrossRef]

182. Hajto, T.; Hostanska, K.; Weber, K.; Zinke, H.; Fischer, J.; Mengs, U.; Lentzen, H.; Saller, R. Effect of a recombinant lectin, Viscum album agglutinin on the secretion of interleukin-12 in cultured human peripheral blood mononuclear cells and on NK-cell-mediated cytotoxicity of rat splenocytes in vitro and in vivo. *Nat. Immun.* **1998**, *16*, 34–46. [CrossRef]

183. Hajto, T.; Hostanska, K.; Frei, K.; Rordorf, C.; Gabius, H.J. Increased secretion of tumor necrosis factors alpha, interleukin 1, and interleukin 6 by human mononuclear cells exposed to beta-galactoside-specific lectin from clinically applied mistletoe extract. *Cancer Res.* **1990**, *50*, 3322–3326.

184. Bhutia, S.K.; Mallick, S.K.; Maiti, T.K. In vitro immunostimulatory properties of Abrus lectins derived peptides in tumor bearing mice. *Phytomedicine* **2009**, *16*, 776–782. [CrossRef]

185. Zhao, J.; Ben, L.-H.; Wu, Y.-L.; Hu, W.; Ling, K.; Xin, S.-M.; Nie, H.-L.; Ma, L.; Pei, G. Anti-HIV Agent Trichosanthin Enhances the Capabilities of Chemokines to Stimulate Chemotaxis and G Protein Activation, and This Is Mediated through Interaction of Trichosanthin and Chemokine Receptors. *J. Exp. Med.* **1999**, *190*, 101–112. [CrossRef] [PubMed]

186. Sikriwal, D.; Ghosh, P.; Batra, J.K. Ribosome inactivating protein saporin induces apoptosis through mitochondrial cascade, independent of translation inhibition. *Int. J. Biochem. Cell Biol.* **2008**, *40*, 2880–2888. [CrossRef] [PubMed]

187. Hartley, M.; Lord, J. Cytotoxic ribosome-inactivating lectins from plants. *Biochim. Biophys. Acta (BBA) Proteins Proteom.* **2004**, *1701*, 1–14. [CrossRef]

188. Chandra, J.; Dutton, J.L.; Li, B.; Woo, W.-P.; Xu, Y.; Tolley, L.K.; Yong, M.; Wells, J.W.; Leggatt, G.R.; Finlayson, N.; et al. DNA Vaccine Encoding HPV16 Oncogenes E6 and E7 Induces Potent Cell-mediated and Humoral Immunity Which Protects in Tumor Challenge and Drives E7-expressing Skin Graft Rejection. *J. Immunother.* **2017**, *40*, 62–70. [CrossRef]

189. Fonseca, J.A.; McCaffery, J.N.; Cáceres, J.; Kashentseva, E.; Singh, B.; Dmitriev, I.P.; Curiel, D.T.; Moreno, A. Inclusion of the murine IgGκ signal peptide increases the cellular immunogenicity of a simian adenoviral vectored Plasmodium vivax multistage vaccine. *Vaccine* **2018**, *36*, 2799–2808. [CrossRef]

190. Yolitz, J.; Schwing, C.; Chang, J.; Van Ryk, D.; Nawaz, F.; Wei, D.; Cicala, C.; Arthos, J.; Fauci, A.S. Signal peptide of HIV envelope protein impacts glycosylation and antigenicity of gp. *Proc. Natl. Acad. Sci. USA* **2018**, *115*, 2443–2448. [CrossRef]

191. E Broderick, K.; Humeau, L.M. Enhanced Delivery of DNA or RNA Vaccines by Electroporation. *Recent Res. Cancer* **2016**, *1499*, 193–200. [CrossRef]

192. Paolini, F.; Massa, S.; Manni, I.; Franconi, R.; Venuti, A. Immunotherapy in new pre-clinical models of HPV-associated oral cancers. *Hum. Vaccines Immunother.* **2013**, *9*, 534–543. [CrossRef]

193. Venuti, A.; Curzio, G.; Mariani, L.; Paolini, F. Immunotherapy of HPV-associated cancer: DNA/plant-derived vaccines and new orthotopic mouse models. *Cancer Immunol. Immunother.* **2015**, *64*, 1329–1338. [CrossRef] [PubMed]

194. Weir, G.M.; Liwski, R.S.; Mansour, M. Immune Modulation by Chemotherapy or Immunotherapy to Enhance Cancer Vaccines. *Cancers* **2011**, *3*, 3114–3142. [CrossRef] [PubMed]

195. Cai, S.; Adila, A.; Alimu, A.; Chen, Q.; Li, Y.; Li, J.; Fisch, L. Glycyrrhiza Uralensis Crude Polysaccharides Enhance Mouse Immunity and Immune Responses Induced by HPV-DNA Vaccine. *Chinese J. Microbiol. Immunol.* **2018**, *38*, 774–781.

196. Tseng, C.-W.; Hung, C.-F.; Alvarez, R.D.; Trimble, C.; Huh, W.K.; Kim, D.; Chuang, C.-M.; Lin, C.-T.; Tsai, Y.-C.; He, L.; et al. Pretreatment with Cisplatin Enhances E7-Specific CD8+ T-Cell-Mediated Antitumor Immunity Induced by DNA Vaccination. *Clin. Cancer Res.* **2008**, *14*, 3185–3192. [CrossRef] [PubMed]

197. Meir, H.; Kenter, G.; Burggraaf, J.; Kroep, J.R.; Welters, M.; Melief, C.; Burg, S.; Poelgeest, M. The Need for Improvement of the Treatment of Advanced and Metastatic Cervical Cancer, the Rationale for Combined Chemo-Immunotherapy. *Anti-Cancer Agents Med. Chem.* **2014**, *14*, 190–203. [CrossRef]

198. Cordeiro, M.N.; Paolini, F.; Massa, S.; Curzio, G.; Illiano, E.; Silva, A.J.D.; Franconi, R.; Bissa, M.; Morghen, C.D.G.; De Freitas, A.C.; et al. Anti-tumor effects of genetic vaccines against HPV major oncogenes. *Hum. Vaccines Immunother.* **2014**, *11*, 45–52. [CrossRef]

199. Kanerva, A.; Raki, M.; Ranki, T.; Särkioja, M.; Koponen, J.; Desmond, R.A.; Helin, A.; Stenman, U.-H.; Isoniemi, H.; Höckerstedt, K.; et al. Chlorpromazine and apigenin reduce adenovirus replication and decrease replication associated toxicity. *J. Gene Med.* **2007**, *9*, 3–9. [CrossRef]

200. Gates, M.A.; Tworoger, S.S.; Hecht, J.L.; De Vivo, I.; Rosner, B.; Hankinson, S.E. A prospective study of dietary flavonoid intake and incidence of epithelial ovarian cancer. *Int. J. Cancer* **2007**, *121*, 2225–2232. [CrossRef]
201. Schumacher, T.N.; Schreiber, R.D. Neoantigens in cancer immunotherapy. *Science* **2015**, *348*, 69–74. [CrossRef]

**Publisher's Note:** MDPI stays neutral with regard to jurisdictional claims in published maps and institutional affiliations.

*Article*

# Juvenile-Onset Recurrent Respiratory Papillomatosis Aggressiveness: In Situ Study of the Level of Transcription of HPV E6 and E7

Charles Lépine [1,2], Thibault Voron [2], Dominique Berrebi [3], Marion Mandavit [2], Marine Nervo [1,2], Sophie Outh-Gauer [1,2], Hélène Péré [2,4], Louis Tournier [3], Natacha Teissier [5], Eric Tartour [2,6], Nicolas Leboulanger [7], Louise Galmiche [8] and Cécile Badoual [1,2,*]

[1]  Department of Pathology, European Hospital Georges-Pompidou, Assistance Publique-Hôpitaux de Paris, F-75015 Paris, France; charles.lepine@aphp.fr (C.L.); marine.nervo@gmail.com (M.N.); sophieouthgauer@gmail.com (S.O.-G.)

[2]  Université de Paris, PARCC, INSERM-U970, F-75015 Paris, France; thibault.voron@gmail.com (T.V.); marion.mandavit@gmail.com (M.M.); helene.pere@aphp.fr (H.P.); eric.tartour@aphp.fr (E.T.)

[3]  Department of Pathology, Robert Debré Hospital, Assistance Publique-Hôpitaux de Paris, F-75019 Paris, France; dominique.berrebi2@aphp.fr (D.B.); louis_tournier@yahoo.fr (L.T.)

[4]  Department of Virology, European Hospital Georges-Pompidou, Assistance Publique-Hôpitaux de Paris, F-75015 Paris, France

[5]  Department of Pediatric ENT Surgery, Robert Debré Hospital, Assistance Publique-Hôpitaux de Paris, F-75019 Paris, France; natacha.teissier@aphp.fr

[6]  Department of Immunology, European Hospital Georges-Pompidou, Assistance Publique-Hôpitaux de Paris, F-75015 Paris, France

[7]  Department of Pediatric ENT Surgery, Necker-Enfants Malades Hospital, Assistance Publique-Hôpitaux de Paris, F-75015 Paris, France; nicolas.leboulanger@aphp.fr

[8]  Department of Pathology, Necker-Enfants Malades Hospital, Assistance Publique-Hôpitaux de Paris, F-75015 Paris, France; louise.galmiche@gmail.com

*  Correspondence: cecile.badoual@aphp.fr; Tel.: +33-156-093-888

Received: 14 September 2020; Accepted: 28 September 2020; Published: 1 October 2020

**Simple Summary:** Juvenile-onset recurrent respiratory papillomatosis (JoRRP) is a condition related to HPV 6 and 11 infection which is characterized by the repeated growth of benign exophytic papilloma in the respiratory tract of children. Disease progression is unpredictable leading sometimes to airway compromise and death. The aim of this study was to explore p16$^{INK4a}$ and expression of the RNA of HPV genes *E6* and *E7* with a chromogenic in situ hybridization (CISH) as biomarkers of JoRRP aggressiveness on a bicentric cohort of forty-eight children. CISH was scored semi-quantitatively as high (2+ score) and low (1+ score) levels of transcription of *E6* and *E7*. Patients with a 2+ score had a more aggressive disease compared to those with a 1+ score. These data are a first step towards the use of biomarkers predictive of disease severity in JoRRP, this could improve the disease management, for example, by implementing adjuvant treatment at the early stages.

**Abstract:** Juvenile-onset recurrent respiratory papillomatosis (JoRRP) is a condition related to HPV 6 and 11 infection which is characterized by the repeated growth of benign exophytic papilloma in the respiratory tract. Disease progression is unpredictable: some children experience minor symptoms, while others require multiple interventions due to florid growth. The aim of this study was to explore the biomarkers of JoRRP severity on a bicentric cohort of forty-eight children. We performed a CISH on the most recent sample of papilloma with a probe targeting the mRNA of the *E6* and *E7* genes of HPV 6 and 11 and an immunostaining with p16$^{INK4a}$ antibody. For each patient HPV RNA CISH staining was assessed semi-quantitatively to define two scores: 1+, defined as a low staining extent, and 2+, defined as a high staining extent. This series contained 19 patients with a score of 1+ and 29 with a score of 2+. Patients with a score of 2+ had a median of surgical excision (SE) per year that was twice that of patients with a score of 1+ (respectively 6.1 versus 2.8, *p* = 0.036). We found similar results with the median number of SE the first year. Regarding p16$^{INK4a}$, all patients were negative. To conclude, HPV RNA CISH might be a biomarker which is predictive of disease aggressiveness in JoRRP, and might help in patient care management.

**Keywords:** Juvenile-onset recurrent respiratory papillomatosis; HPV 6; HPV 11; RNA; chromogenic in situ hybridization

## 1. Introduction

Recurrent respiratory papillomatosis (RRP) is characterized by the repeated growth of benign exophytic papilloma in the respiratory tract [1,2], primarily in the larynx [1]. The age distribution of RRP is trimodal, with a first peak in children younger than 5 years of age, a second one in adults between 20 and 40 years old and a third in individuals around the age of 64 [3,4]. This condition is referred to as Juvenile-onset Recurrent respiratory papillomatosis (JoRRP) when it occurs in children. There is a limited number of studies with epidemiologic data: in Denmark, between 1969 and 1984 the incidence was 3.6 case per year per 100,000 children [5], whereas in Canada, based on a national database, the incidences and prevalences from 1994 to 2007 were, respectively, 0.24 and 1.11 per 100,000 children, and the median age at diagnosis was 4.4 years with a sex ratio near 1:1 [6]. In the United States of America, data are similar [7]; however, incidence and prevalence seems to depend on socioeconomic status [8]. JoRRP is caused by an HPV infection, mostly by 6 and 11 genotypes [9]. Three modes of transmission are suggested: vertical transmission at birth (however, HPV type concordance between mother and newborn in different studies are contradictory [10–12]), vertical transmission in utero [13] and horizontal transmission via the child's environment [10]. Several studies have also demonstrated that maternal condyloma at the time of delivery is a major risk factor of developing JoRRP [14,15]. While the prevalence of HPV 6 and 11 infection in pregnant women is around 2%, the prevalence of JoRRP is surprisingly low. Thus, HPV infection alone does not explain the development of the disease, and strong arguments suggests that JoRRP is linked to immunity defects and genetic susceptibility. Patients with RRP are associated with HLA DRB1*0102/0301, DQB1*0201/0202 [16,17] and presents a lack of KIR genes 3DS1 et 2DS1 [18]. Moreover, their immune response presents a Th2 polarization [19,20] which is not suitable for viral infection control.

The management of this disease is challenging, due to its unpredictable course: some children experience minor symptoms with spontaneous remission, while others require multiple interventions due to florid growth. In addition, RRP may lead to airway compromise. Malignant transformation to carcinoma rarely occurs, most often over pulmonary spread [21,22]. The standard treatment of JoRRP is a surgical excision (SE) with cold instruments or microdebriders. Multiple endolaryngeal procedures can lead to glottis synechia and irreversible damage to the vocal cords, as well as an impaired social life [23]. To improve the surgical outcome and extend the symptom-free interval, numerous adjuvant treatments have been tried: interferon $\alpha$ [24], celecoxib [25], bevacizumab [26], cidofovir [27,28], PD-1/PD-L1 immunotherapy [29,30] and the quadrivalent HPV vaccine [31]. Currently, routine use of these treatments is not recommended by the International Pediatric Otolaryngology Group [32]. The most promising approaches are the quadrivalent HPV vaccine, bevacizumab and PD-1/PD-L1 immunotherapies which appear to decrease relapses [29,30,33,34]. Systemic bevacizumab, a monoclonal antibody against VEGF-A, seems to be the most promising treatment for aggressive forms of JoRRP, as several studies found a rapid and sustained partial or complete response to treatment in patients with lung involvement [35,36]. In order to improve the management of these children with JoRRP, it seemed important to us to identify biomarkers associated with the severity of the disease. Many studies have focused on finding clinical severity risk factors, of which only early age of onset of the disease [37–39] is currently recognized as such.

We previously described in p16 positive squamous cell carcinoma of the oropharynx two prognostic groups thanks to HPV RNA chromogenic in situ hybridization (CISH) [40]. In brief, HPV RNA CISH was scored semi-quantitatively as "high" and "low" depending on the extent of the staining. RNA CISH high staining was associated with a better overall survival in both univariate and multivariate analyses ($p = 0.033$ and $p = 0.042$, respectively). Based on these results, we decided to explore HPV E6 and E7 transcription with this technique in our cohort of JoRRP. Thus, we hypothesize that a highly transcriptionally active virus could be associated with more severe disease. When it comes to HPV related cancers, the p16 protein is typically essential. It is a surrogate marker of HPV related cancers in various locations, including oropharyngeal squamous cell carcinoma (OPSCC) [2]. p16 is a CDK (cyclin-dependent kinase) inhibitor. This protein is involved in the pRB pathway, implicated in cell cycle regulation; its overexpression avoids phosphorylation of Rb family members, leading to cell cycle arrest into G1 phase [41]. Low-risk HPV produce E6 and E7 proteins which have lower affinity for p53 and pRb proteins [42], and thus, are not theoretically associated with cell cycle progression, nor with p16 overexpression. Conversely, in lesions related with high risk HPV (cancers or intra-epithelial neoplasia), there is an overexpression of the p16 protein, resulting in intense cytoplasmic and nuclear staining of the majority of tumors cells (>70%); this has mainly been studied in the female genital tract and the anus [43–45]. However, low-risk HPV-related cancers have been described in different locations including the anus and head and neck [46–48]. Data over p16 expression are heterogeneous, with some studies finding p16 positivity in these cancers while others do not.

The aim of this study is to explore p16 expression and HPV RNA CISH as in situ biomarkers of JoRRP severity.

## 2. Results

### 2.1. Population

Forty-eight children were included: twenty-two were boys and twenty-six were girls. The average age at diagnosis was 3.8 years, with a median of 2 years. Twenty-seven percent of patients had HPV 11 infection, 65% had HPV 6 infection and 6% had co-infection with HPV 6 and 11. All patients had glottic involvement; 73% of patients had supraglottic complication, 68.7% had subglottic localization, 25% had tracheal involvement and 8% had pulmonary lesions. Regarding adjuvant treatment, 73% of patients received at least one injection of Cidofovir. Patients received an average of 7.1 injections of Cidofovir with a median of 3.5 injections. Six patients (12.5%) received Cidofovir during an SE prior to the study specimen. The delay between the first and last SE was on average 3.6 years and the median was 2 years. Moreover, 71% of patients had a lesion at the last check-up. In addition, in our cohort, a young patient died at the age of 18 from the malignant transformation of a pulmonary localization of her JoRRP into bronchopulmonary squamous cell carcinoma. Her JoRRP progressed for 17 years: 132 SE were performed, with a mean interval between each endoscopy of 47 days. She also received 67 injections of Cidofovir.

### 2.2. p16 Immunochemistry

All patients had a negative staining with p16$^{INK4a}$ antibody.

### 2.3. HPV RNA Chromogenic In Situ Hybridization

All negative control DaPB probes were negatives. For the PPIB housekeeping gene control probe, we found 11 patients with a score of 1+ and 37 with a score of 2+. No statistical correlation was found between the grading of the HPV E6 and E7 probe and the PPIB probe ($p = 0,063$, Chi2 test).

For the HPV RNA probe, we found 19 patients with a score of 1+ and 29 with a score of 2+. The characteristics of these two groups are described in Table 1. The two populations are comparable in terms of HPV type, gender and location of the lesions. The only patient who died of her disease had a score of 2+. Although patients with a score of 2+ had more lung involvement and tracheostomy, the difference was not statistically significant.

**Table 1.** Characteristics of the score 1+ and score 2+ populations.

| Clinical Characteristics | | Score 1+ ($n = 19$) | Score 2+ ($n = 29$) | $p$ |
|---|---|---|---|---|
| Gender | Boys | 10 (53%) | 12 (41%) | 0.444 |
| | Girls | 9 (47%) | 17 (59%) | 0.444 |
| HPV typing [1] | HPV6 and 11 | 1 (5%) | 2 (7%) | 1 |
| | HPV11 | 5 (26%) | 8 (27%) | 0.923 |
| | HPV6 | 12 (63%) | 19 (66%) | 0.867 |
| Onset of disease at age ≤ 3 years | | 12 (63%) | 14 (48%) | 0.312 |
| Tracheostomy | | 2 (10%) | 6 (21%) | 0.451 |
| At least 1 Cidofovir injection | | 11 (57%) | 24 (82%) | 0.058 |
| Extra-laryngeal involvement | | 5 (26%) | 8 (27%) | 0.923 |
| Sub-glottic involvement | | 12 (63%) | 21 (72%) | 0.499 |
| Tracheal involvement | | 5 (26%) | 7 (24%) | 0.8655 |
| Postoperative morbidity | | 4 (21%) | 6 (21%) | 1 |
| Lesion at last check-up | | 11 (63%) | 23 (79%) | 0.110 |
| Pulmonary involvement | | 1 (5%) | 3 (10%) | 1 |
| Death | | 0 | 1 (3%) | 1 |
| Malignant transformation | | 0 | 1 (3%) | 1 |

[1] one patient couldn't have HPV typing due to sample depletion.

The clinical markers of JoRRP aggressiveness are described in Table 2. Patients with a score of 2+ had a median of SE per year that was two times higher than that of patients with a score of 1+. We found similar results with the median number of SE the first year. Also, for the score of 2+, we found 72% patient with more than 4 SE in one year compared to patients with score of 1+, who comprised only 37%. Although the difference is not statistically different, patient with a score of 2+ had a median of interval between each SE two time shorter compared with 1+ patients. We also performed a second statistical analysis excluding the patient who underwent 132 SE and received 67 injections of Cidofovir; the results were similar. We found that patients with a score of 2+ had a median of SE per year that was higher than patients with a score of 1+ (respectively 5.8 versus 2.8, $p = 0.039$), and patients with a score of 2+ also had a higher median number of SE the first year (respectively 5 versus 2, $p = 0.019$). There were no significant differences between the two groups for the other items.

**Table 2.** Comparison of clinical markers of aggressiveness between score 1+ and 2+ populations; results are medians.

| Clinical Markers of Aggressiveness | Score 1+ ($n = 19$) | Score 2+ ($n = 29$) | $p$ |
|---|---|---|---|
| Age at diagnosis (year) | 2 | 4 | 0.676 |
| Years between 1st and last endoscopy | 2.6 | 1.5 | 0.292 |
| Number of SE per year | 2.8 | 6.1 | 0.036 |
| Total number of SE | 8 | 9 | 0.128 |
| Number of SE first year after diagnosis | 2 | 5 | 0.029 |
| Number of patient with more than 4 SE in one year | 7 | 21 | 0.015 |
| Average interval between each SE (days) | 168 | 80 | 0.067 |
| Number of Cidofovir injections | 1 | 5 | 0.053 |
| Cidofovir injection prior to the specimen | 2 | 4 | 1 |

## 3. Discussion

Juvenile recurrent respiratory papillomatosis is a rare disease. In order to improve the management of these patients, it is necessary to carry out studies to find new biomarkers that could predict disease severity. To our knowledge, our cohort of JoRRP is the largest ever studied in Europe. Our population had characteristics which were comparable to those described in national databases in the U.S. and Canada (covering respectively 603 and 243 children with JoRRP [6,49]). We found a median rate of SE per year of 4.8, comparable to the U.S. cohort, which was of 4.3, higher than the Canadian one, i.e., 1.5. Our median age at diagnosis was slightly lower, i.e., 2 years old versus 3 years old in the U.S. cohort and 4 years old in the Canadian one. Interestingly, the percentage of patients treated with Cidofovir was much higher in our cohort than in the Canadian cohort (respectively 73% vs. 4.7%). The differences in terms of Cidofovir treatment could be explained by variability in local practices. Regarding the distribution of HPV types, our data are comparable to those presented in the literature. We found a low proportion of co-infection with HPV6 and 11 (6%) and a predominance of HPV6 (65%), as described elsewhere [50,51]. Currently, no consensual definition exists in the literature for disease severity. Some authors use composite scores incorporating criteria for disease localization, such as the Derkay-Coltrera score, and intervention-related criteria, such as the number of SE per year [6,38,52]. Others use only intervention-related criteria. A total number of SE greater than or equal to 10 or a number of SE/year greater than 3 or 4 is frequently found as a criterion of severity [37]. We were unable to use Derkay-Coltrera score because one of the two centers involved did not use it systematically. Given the absence of consensus, we remained descriptive and compared known intervention-related criteria between our two groups of JoRRP. Thus, our results with the E6 and E7 RNA CISH can be correlated to the activity of the disease and, by inference, to its severity.

To date, E6 and E7 RNA CISH has not been tested in benign HPV-related tumors. It is a great opportunity to have performed, for the first time, a CISH with an HPV6/11 RNA probe on an JoRRP cohort. As expected, all the patients had a positive CISH for E6 and E7 RNA of HPV6/11, confirming the presence of viral transcription in the papilloma of JoRRP. Both populations were comparable in terms of HPV type, gender and location of the lesions. Our results show an association between patients with a score of 2+ and higher activity markers of the disease. Indeed, patients with score of 2+ had a median of SE per year and a median number of SE the first year, twice as high as those of patients with score of 1+; this difference is statistically significant. We also found that patient with a score of 2+ were associated with more than 4 SE in one year, compared to patient with score of 1+. The difference between the two groups for the mean interval between each SE was also important, but not statistically significant. Another striking result was the higher frequency of Cidofovir treatment and the higher number of injection of Cidofovir in patients with a score of 2+; however, these results were close to statistical significance. Another data pointing in this direction is that the patient who died from a carcinomatous transformation of her JoRRP had a score of 2+. Our statistical analyses were not influenced by this patient (who underwent 132 SE and received 67 injections of Cidofovir), since a second statistical analysis was performed excluding this patient, with very similar results. Considering these results, patients with a high level of E6 and E7 transcription (score of 2+) had a more aggressive disease compared those with a low level of transcription (score of 1+). The opposite was observed in our study with OPSCC, as patients with a score of 2+ had a better overall survival compared to those with a score of 1+ [40]. This is likely explained by the immune response failure context in JoRRP, while in OPSCC the presence of more viral RNA in tumors cells could boost the immune response. Our semi-quantitative results with this technique are reliable, since our negative control probes (DaPB) were all negative and our control probes of a housekeeping gene were all positive with a semi-quantitative score not being correlated to the HPV6/11 probe score. However, injection of Cidofovir prior to the SE is a possible confounding factor. Cidofovir is a nucleotide analogue that blocks the DNA replication of viruses by inhibiting their DNA polymerase [53]. The effect of Cidofovir on E6 and E7 mRNA expression is unknown, although it can theoretically be assumed that Cidofovir decreases viral transcription by limiting the number of replicated viral DNA molecules. Thus, it would be interesting to carry out a study comparing the CISH score on samples before and after Cidofovir injection and to look for a correlation with the response to treatment. It can be hypothesized that the HPV6/11 RNA probe could help predict the response to Cidofovir, and thus avoid giving injections to non-responders. The observation of the patient who died from a malignant transformation of her JoRRP seems to be a first argument for the possibility of predicting the response to Cidofovir. Indeed, this patient had received 67 Cidofovir injections, which did not control the disease; she had a CISH score of 2+, with the HPV6/11 RNA probe on the sample taken after receiving Cidofovir injections.

Regarding immunohistochemistry with the p16$^{INK4a}$ antibody, we found, as expected, that there was no overexpression of p16 in the papilloma of JoRRP, even for the patient who died of a carcinomatous transformation. It should be noted, however, that the p16$^{INK4a}$ antibody was performed on a squamous papilloma collected before the onset of the transformation. In addition, no sample of the transformation was available for this patient. Interestingly, Huebbers et al. described a case of transformation of JoRRP related to HPV6. They did not find a p16 positivity in the carcinomatous sample, or in the papilloma sample, but they revealed a single site integration of the viral DNA in the carcinoma sample, while for the papilloma sample, the viral DNA was episomal [54]. Our results are consistent with the data in the literature. Indeed, in a series of condyloma acuminata (mostly related to HPV6 and 11), none showed homogeneous and diffuse p16 staining [55].

## 4. Materials and Methods

### 4.1. Population

This retrospective study was approved by an ethical committee (notice number: CPP2019-02'-019a/2019-00352-55/19.02.05.67237) and by the "Commission Nationale Informatique et Libertés" (application number: 919150). Patients were selected from two pediatric University Hospitals (CHU) in Paris that treat JoRRP: Necker-Enfants Malades Hospital and Robert Debré Hospital. These two hospitals had similar protocols to treat JoRRP, and a SE was performed when patients were dyspneic or when the lesion growth was exponential. One sample per patient was selected, i.e., the most recent one, offering the best possible RNA quality. The inclusion criteria were:

- at least one available sample of laryngeal squamous papilloma.
- for each patient at least one positive in situ hybridization with an HPV "low risk" DNA probe or positive PCR for HPV 6 and/or 11.
- at least one recurrence after diagnosis.

The clinical data were collected retrospectively in March 2018. The following was collected for each patient: gender, age at diagnosis, dates of each SE performed in the two University Hospitals, number of SE, number of Cidofovir injections received, tracheotomy in relation to the disease, presence of surgical sequelae defined as the appearance of synechia at the glottic stage or even stenosis, location of papillomas, presence of lung involvement proven by at least one chest CT scan, presence of a lesion at the last check-up nasofibroscopy, carcinomatous transformation of JoRRP lesions, death related to the disease and cidofovir injection dates to determine whether the sample studied was taken before or after Cidofovir treatment.

From the dates of the SE, an average interval in days between each SE was calculated. The number of SE per year was calculated by dividing the total number of SE by the time in years between the first and last SE.

### 4.2. HPV RNA Chromogenic In Situ Hybridization

We used RNAscope® kits (Advanced Cell Diagnostics Inc., Newak, CA, USA) from the manufacturer ACD™ on FFPE sections of papilloma. For each patient, we used the most recent sample of papilloma (to avoid RNA degradation):

- An in situ hybridization with a probe targeting the mRNA of the E6 and E7 proteins of HPV 6 and 11.
- A negative control with an in situ hybridization probe targeting the RNA of the Bacillus subtilis dihydrodipicolinate reductase bacterial gene transcript (probe name: DaPB).

- A positive control with an in situ hybridization probe targeting the RNA of the Cyclophilin B housekeeping gene (probe name: PPIB), an ubiquitous housekeeping gene.

The FFPE sections were stored at 4 °C before the technique was carried out. The hybridizations were performed according to our laboratory protocol [56]. For each batch of CISH with the RNA probe of HPV 6 and 11, an external positive control was performed.

Each slide was semi-quantitatively scored by three pathologists together (C.B., C.L. and L.G), each slide was assigned a score of 0, 1+ or 2+, as described in Figure 1:

- Score 0: no staining
- Score 1+: at ×20 magnification, staining less than or equal to 50% of the cells, or staining of more than 80% of the cell surface in less than 30% of the tumor cells.
- Score 2+: at ×20 magnification, staining of more than 50% of the cells, or staining of more than 80% of the cell surface in at least 30% of the tumor cells.

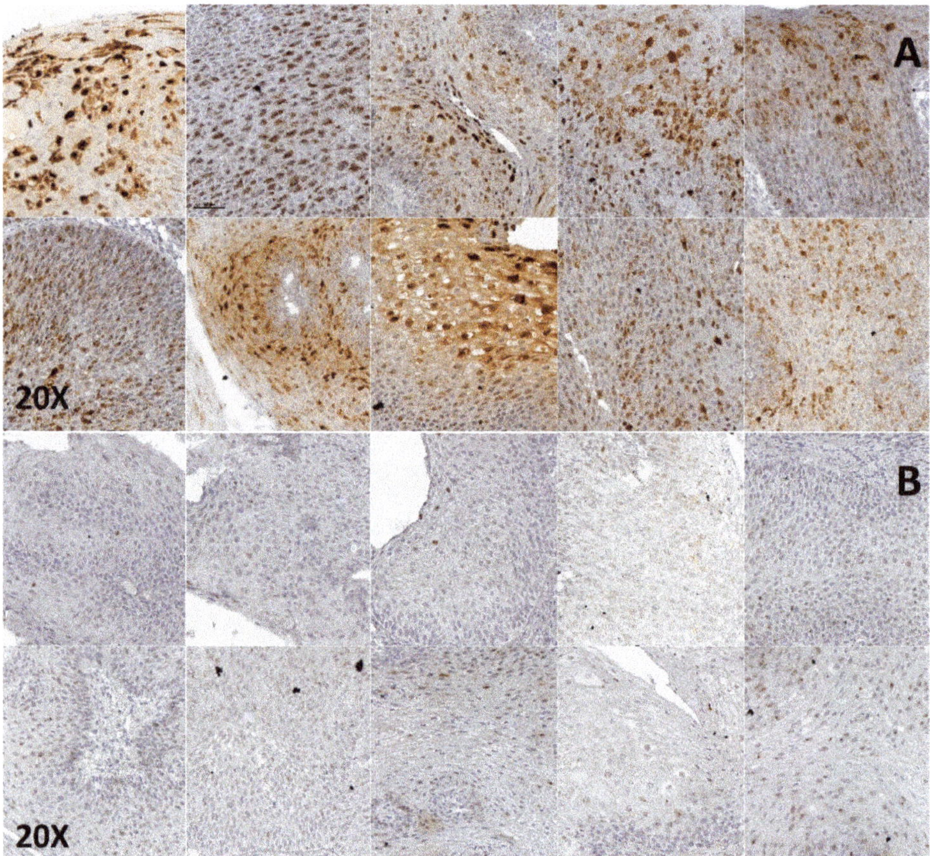


**Figure 1.** E6 and E7 HPV RNA CISH in papillomas of JoRRP: (**A**) 10 patients with score 2+: at ×20 magnification, staining of more than 50% of the cells, or staining of more than 80% of the cell surface in at least 30% of the tumor cells. (**B**) 10 patients with score 1+: at ×20 magnification, staining less than or equal to 50% of the cells, or staining of more than 80% of the cell surface in less than 30% of the tumor cells.

Staining with PPIB probe was also assessed semi-quantitatively following the same rules as previously described.

### 4.3. p16 Immunochemistry

We conducted immunochemistry on FFPE sections of papilloma with $p16^{INK4a}$ antibody (E6H4 clone, manufacturer: Roche, dilution: 1/2), with a Leica™ Bond III® automat (Leica Microsystemes SA, Nanterre, France). The anti-$p16^{INK4a}$ immunohistochemistry was rated positive when more than 70% of the cells displayed clear and homogeneous labeling of the cytoplasm and the nucleus, as shown in the Figure 2.

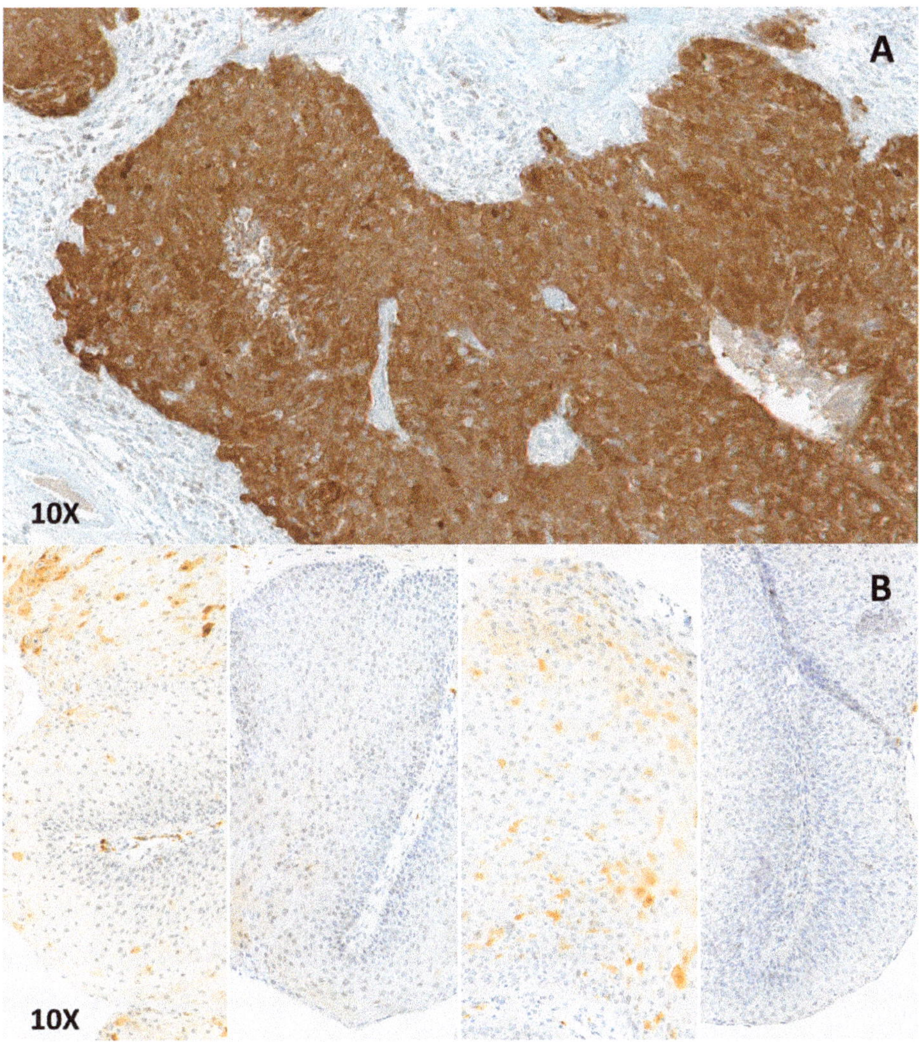


**Figure 2.** Examples of p16 immunostaining: (**A**) Positive immunostaining on an oropharyngeal squamous cell carcinoma related to HPV, nuclear and cytoplasmic staining of more than 70% of the tumoral cells. (**B**) Negative immunostaining: Papilloma in JoRRP from four patients, heterogenous nuclear and cytoplasmic staining of less than 70% of tumoral cells.

*4.4. Statistical Analysis*

Statistical analyses were carried out using the R software. Qualitative variables were analyzed with a Chi2 or Fisher test depending on the sample size. Univariate analyses with quantitative data were performed using a nonparametric Mann-Whitney test. Finally, all tests were bilateral and a $p < 0.05$ was considered significant.

*4.5. Outcome*

Patients were classified into two groups regarding their CISH score: score 1+ and score 2+. The primary outcome was the comparison of clinical markers of disease aggressiveness within these two groups. Secondary outcome was to explore the p16 expression in JoRRP's lesions to find a correlation with a clinical outcome.

## 5. Conclusions

To conclude, we presented here the first results ever described in the JoRRP with a RNA CISH with a probe targeting E6 and E7 from HPV6 and 11. Patients with a high level of E6 and E7 transcription (score 2+) had a more aggressive disease compared to those with a low level of transcription (score 1+). These data are a first step towards the use of biomarkers predictive of disease severity. The use of biomarkers to predict an aggressive disease could improve the management of the disease, for example, by implementing adjuvant treatment at the early stages. It could also be an opportunity to better inform patients and their parents about the potential course of the disease. Our results also point to the potential of HPV6/11 RNA CISH as a predictive test of response to Cidofovir. These data require validation on a larger prospective cohort.

**Author Contributions:** Conceptualization, L.G. and C.B.; methodology, C.B., C.L. and L.G.; validation, C.B., C.L., L.G. and M.M.; formal analysis, C.L. and T.V.; investigation, C.L., M.N., M.M. and H.P.; resources, L.G., D.B., N.T., N.L; data curation, C.L, L.T.; writing—original draft preparation, C.L.; writing—review and editing, L.G., N.T., C.B., S.O.-G., E.T.; visualization, C.L.; supervision, C.B. and L.G.; project administration, C.B. and C.L.; funding acquisition, C.B., C.L., L.G., N.L. and E.T.; All authors have read and agreed to the published version of the manuscript.

**Funding:** This research was funded by LION'S CLUB OF CORRÈZE and the SIRIC CARPEM.

**Conflicts of Interest:** The authors declare no conflict of interest.

## References

1. Kashima, H.; Mounts, P.; Leventhal, B.; Hruban, R.H. Sites of predilection in recurrent respiratory papillomatosis. *Ann. Otol. Rhinol. Laryngol.* **1993**, *102*, 580–583. [CrossRef] [PubMed]
2. El-Naggar, A.K.; Chan, J.K.C.; Grandis, J.R.; Takata, T.; Slootweg, P.J. (Eds.) *WHO Classification of Head and Neck Tumours*, 4th ed.; International Agency for Research on Cancer: Lyon, France, 2017; ISBN 978-92-832-2438-9.
3. Carifi, M.; Napolitano, D.; Morandi, M.; Dall'Olio, D. Recurrent respiratory papillomatosis: Current and future perspectives. *Clin. Risk Manag.* **2015**, *11*, 731–738. [CrossRef] [PubMed]
4. San Giorgi, M.R.M.; van den Heuvel, E.R.; Tjon Pian Gi, R.E.A.; Brunings, J.W.; Chirila, M.; Friedrich, G.; Golusinski, W.; Graupp, M.; Horcasitas Pous, R.A.; Ilmarinen, T.; et al. Age of onset of recurrent respiratory papillomatosis: A distribution analysis. *Clin. Otolaryngol.* **2016**, *41*, 448–453. [CrossRef] [PubMed]
5. Lindeberg, H.; Elbrønd, O. Laryngeal papillomas: The epidemiology in a Danish subpopulation 1965-1984. *Clin. Otolaryngol. Allied Sci.* **1990**, *15*, 125–131. [CrossRef]
6. Campisi, P.; Hawkes, M.; Simpson, K.; Canadian Juvenile Onset Recurrent Respiratory Papillomatosis Working Group. The epidemiology of juvenile onset recurrent respiratory papillomatosis derived from a population level national database. *Laryngoscope* **2010**, *120*, 1233–1245. [CrossRef] [PubMed]
7. Armstrong, L.R.; Preston, E.J.; Reichert, M.; Phillips, D.L.; Nisenbaum, R.; Todd, N.W.; Jacobs, I.N.; Inglis, A.F.; Manning, S.C.; Reeves, W.C. Incidence and prevalence of recurrent respiratory papillomatosis among children in Atlanta and Seattle. *Clin. Infect. Dis.* **2000**, *31*, 107–109. [CrossRef]

8.  Marsico, M.; Mehta, V.; Chastek, B.; Liaw, K.-L.; Derkay, C. Estimating the incidence and prevalence of juvenile-onset recurrent respiratory papillomatosis in publicly and privately insured claims databases in the United States. *Sex. Transm. Dis.* **2014**, *41*, 300–305. [CrossRef]

9.  Donne, A.J.; Hampson, L.; Homer, J.J.; Hampson, I.N. The role of HPV type in Recurrent Respiratory Papillomatosis. *Int. J. Pediatr. Otorhinolaryngol.* **2010**, *74*, 7–14. [CrossRef]

10. Castellsagué, X.; Drudis, T.; Cañadas, M.P.; Goncé, A.; Ros, R.; Pérez, J.M.; Quintana, M.J.; Muñoz, J.; Albero, G.; de Sanjosé, S.; et al. Human Papillomavirus (HPV) infection in pregnant women and mother-to-child transmission of genital HPV genotypes: A prospective study in Spain. *BMC Infect. Dis.* **2009**, *9*, 74. [CrossRef]

11. Smith, E.M.; Parker, M.A.; Rubenstein, L.M.; Haugen, T.H.; Hamsikova, E.; Turek, L.P. Evidence for vertical transmission of HPV from mothers to infants. *Infect. Dis. Obs. Gynecol* **2010**, *2010*, 326369. [CrossRef]

12. Puranen, M.; Yliskoski, M.; Saarikoski, S.; Syrjänen, K.; Syrjänen, S. Vertical transmission of human papillomavirus from infected mothers to their newborn babies and persistence of the virus in childhood. *Am. J. Obstet. Gynecol.* **1996**, *174*, 694–699. [CrossRef]

13. Zouridis, A.; Kalampokas, T.; Panoulis, K.; Salakos, N.; Deligeoroglou, E. Intrauterine HPV transmission: A systematic review of the literature. *Arch. Gynecol. Obstet.* **2018**, *298*, 35–44. [CrossRef] [PubMed]

14. Silverberg, M.J.; Thorsen, P.; Lindeberg, H.; Grant, L.A.; Shah, K.V. Condyloma in pregnancy is strongly predictive of juvenile-onset recurrent respiratory papillomatosis. *Obs. Gynecol.* **2003**, *101*, 645–652.

15. Quick, C.A.; Krzyzek, R.A.; Watts, S.L.; Faras, A.J. Relationship between Condylomata and Laryngeal Papillomata: Clinical and Molecular Virological Evidence. *Ann. Otol. Rhinol. Laryngol.* **1980**, *89*, 467–471. [CrossRef]

16. Bonagura, V.R.; Vambutas, A.; DeVoti, J.A.; Rosenthal, D.W.; Steinberg, B.M.; Abramson, A.L.; Shikowitz, M.J.; Gjertson, D.W.; Reed, E.F. HLA alleles, IFN-gamma responses to HPV-11 E6, and disease severity in patients with recurrent respiratory papillomatosis. *Hum. Immunol.* **2004**, *65*, 773–782. [CrossRef]

17. Gelder, C.M.; Williams, O.M.; Hart, K.W.; Wall, S.; Williams, G.; Ingrams, D.; Bull, P.; Bunce, M.; Welsh, K.; Marshall, S.E.F.; et al. HLA Class II Polymorphisms and Susceptibility to Recurrent Respiratory Papillomatosis. *J. Virol.* **2003**, *77*, 1927–1939. [CrossRef]

18. Bonagura, V.R.; Du, Z.; Ashouri, E.; Luo, L.; Hatam, L.J.; DeVoti, J.A.; Rosenthal, D.W.; Steinberg, B.M.; Abramson, A.L.; Gjertson, D.W.; et al. Activating killer cell immunoglobulin-like receptors 3DS1 and 2DS1 protect against developing the severe form of recurrent respiratory papillomatosis. *Hum. Immunol.* **2010**, *71*, 212–219. [CrossRef]

19. Bonagura, V.R.; Hatam, L.; DeVoti, J.; Zeng, F.; Steinberg, B.M. Recurrent respiratory papillomatosis: Altered CD8(+) T-cell subsets and T(H)1/T(H)2 cytokine imbalance. *Clin. Immunol.* **1999**, *93*, 302–311. [CrossRef]

20. Bonagura, V.R.; Hatam, L.J.; Rosenthal, D.W.; de Voti, J.A.; Lam, F.; Steinberg, B.M.; Abramson, A.L. Recurrent respiratory papillomatosis: A complex defect in immune responsiveness to human papillomavirus-6 and -11. *APMIS* **2010**, *118*, 455–470. [CrossRef]

21. Go, C.; Schwartz, M.R.; Donovan, D.T. Molecular Transformation of Recurrent Respiratory Papillomatosis: Viral Typing and P53 Overexpression. *Ann. Otol. Rhinol. Laryngol.* **2003**, *112*, 298–302. [CrossRef]

22. Gélinas, J.-F.; Manoukian, J.; Côté, A. Lung involvement in juvenile onset recurrent respiratory papillomatosis: A systematic review of the literature. *Int. J. Pediatr. Otorhinolaryngol.* **2008**, *72*, 433–452. [CrossRef] [PubMed]

23. Montaño-Velázquez, B.B.; Nolasco-Renero, J.; Parada-Bañuelos, J.E.; Garcia-Vázquez, F.; Flores-Medina, S.; García-Romero, C.S.; Jáuregui-Renaud, K. Quality of life of young patients with recurrent respiratory papillomatosis. *J. Laryngol. Otol.* **2017**, *131*, 425–428. [CrossRef] [PubMed]

24. Healy, G.B.; Gelber, R.D.; Trowbridge, A.L.; Grundfast, K.M.; Ruben, R.J.; Price, K.N. Treatment of recurrent respiratory papillomatosis with human leukocyte interferon. Results of a multicenter randomized clinical trial. *N. Engl. J. Med.* **1988**, *319*, 401–407. [CrossRef] [PubMed]

25. Limsukon, A.; Susanto, I.; Soo Hoo, G.W.; Dubinett, S.M.; Batra, R.K. Regression of recurrent respiratory papillomatosis with celecoxib and erlotinib combination therapy. *Chest* **2009**, *136*, 924–926. [CrossRef] [PubMed]

26. Rogers, D.J.; Ojha, S.; Maurer, R.; Hartnick, C.J. Use of adjuvant intralesional bevacizumab for aggressive respiratory papillomatosis in children. *JAMA Otolaryngol. Head Neck Surg.* **2013**, *139*, 496–501. [CrossRef] [PubMed]

27. Naiman, A.N.; Ayari, S.; Nicollas, R.; Landry, G.; Colombeau, B.; Froehlich, P. Intermediate-term and long-term results after treatment by cidofovir and excision in juvenile laryngeal papillomatosis. *Ann. Otol. Rhinol. Laryngol.* **2006**, *115*, 667–672. [CrossRef]

28. McMurray, J.S.; Connor, N.; Ford, C.N. Cidofovir efficacy in recurrent respiratory papillomatosis: A randomized, double-blind, placebo-controlled study. *Ann. Otol. Rhinol. Laryngol.* **2008**, *117*, 477–483. [CrossRef]

29. Creelan, B.C.; Ahmad, M.U.; Kaszuba, F.J.; Khalil, F.K.; Welsh, A.W.; Ozdemirli, M.; Grant, N.N.; Subramaniam, D.S. Clinical Activity of Nivolumab for Human Papilloma Virus-Related Juvenile-Onset Recurrent Respiratory Papillomatosis. *Oncologist* **2019**, *24*, 829–835. [CrossRef]

30. Allen, C.T.; Lee, S.; Norberg, S.M.; Kovalovsky, D.; Ye, H.; Clavijo, P.E.; Hu-Lieskovan, S.; Schlegel, R.; Schlom, J.; Strauss, J.; et al. Safety and clinical activity of PD-L1 blockade in patients with aggressive recurrent respiratory papillomatosis. *J. Immunother. Cancer* **2019**, *7*, 119. [CrossRef]

31. Chirilă, M.; Bolboacă, S.D. Clinical efficiency of quadrivalent HPV (types 6/11/16/18) vaccine in patients with recurrent respiratory papillomatosis. *Eur Arch. Otorhinolaryngol.* **2014**, *271*, 1135–1142. [CrossRef]

32. Lawlor, C.; Balakrishnan, K.; Bottero, S.; Boudewyns, A.; Campisi, P.; Carter, J.; Cheng, A.; Cocciaglia, A.; DeAlarcon, A.; Derkay, C.; et al. International Pediatric Otolaryngology Group (IPOG): Juvenile-onset recurrent respiratory papillomatosis consensus recommendations. *Int. J. Pediatr. Otorhinolaryngol.* **2020**, *128*, 109697. [CrossRef] [PubMed]

33. Rosenberg, T.; Philipsen, B.B.; Mehlum, C.S.; Dyrvig, A.-K.; Wehberg, S.; Chirilă, M.; Godballe, C. Therapeutic Use of the Human Papillomavirus Vaccine on Recurrent Respiratory Papillomatosis: A Systematic Review and Meta-Analysis. *J. Infect. Dis.* **2019**, *219*, 1016–1025. [CrossRef] [PubMed]

34. Matsuzaki, H.; Makiyama, K.; Hirai, R.; Suzuki, H.; Asai, R.; Oshima, T. Multi-Year Effect of Human Papillomavirus Vaccination on Recurrent Respiratory Papillomatosis. *Laryngoscope* **2019**. [CrossRef] [PubMed]

35. Bedoya, A.; Glisinski, K.; Clarke, J.; Lind, R.N.; Buckley, C.E.; Shofer, S. Systemic Bevacizumab for Recurrent Respiratory Papillomatosis: A Single Center Experience of Two Cases. *Am. J. Case Rep.* **2017**, *18*, 842–846. [CrossRef]

36. Best, S.R.; Mohr, M.; Zur, K.B. Systemic bevacizumab for recurrent respiratory papillomatosis: A national survey. *Laryngoscope* **2017**, *127*, 2225–2229. [CrossRef]

37. Niyibizi, J.; Rodier, C.; Wassef, M.; Trottier, H. Risk factors for the development and severity of juvenile-onset recurrent respiratory papillomatosis: A systematic review. *Int. J. Pediatr. Otorhinolaryngol.* **2014**, *78*, 186–197. [CrossRef]

38. Buchinsky, F.J.; Valentino, W.L.; Ruszkay, N.; Powell, E.; Derkay, C.S.; Seedat, R.Y.; Uloza, V.; Dikkers, F.G.; Tunkel, D.E.; Choi, S.S.; et al. Age at diagnosis, but not HPV type, is strongly associated with clinical course in recurrent respiratory papillomatosis. *PLoS ONE* **2019**, *14*, e0216697. [CrossRef]

39. Buchinsky, F.J.; Donfack, J.; Derkay, C.S.; Choi, S.S.; Conley, S.F.; Myer, C.M.; McClay, J.E.; Campisi, P.; Wiatrak, B.J.; Sobol, S.E.; et al. Age of child, more than HPV type, is associated with clinical course in recurrent respiratory papillomatosis. *PLoS ONE* **2008**, *3*, e2263. [CrossRef]

40. Augustin, J.; Mandavit, M.; Outh-Gauer, S.; Grard, O.; Gasne, C.; Lépine, C.; Mirghani, H.; Hans, S.; Bonfils, P.; Denize, T.; et al. HPV RNA CISH score identifies two prognostic groups in a p16 positive oropharyngeal squamous cell carcinoma population. *Mod. Pathol.* **2018**, *31*, 1645–1652. [CrossRef]

41. Serra, S.; Chetty, R. p16. *J. Clin. Pathol.* **2018**, *71*, 853–858. [CrossRef]

42. Oh, S.T.; Longworth, M.S.; Laimins, L.A. Roles of the E6 and E7 proteins in the life cycle of low-risk human papillomavirus type 11. *J. Virol.* **2004**, *78*, 2620–2626. [CrossRef] [PubMed]

43. Klaes, R.; Benner, A.; Friedrich, T.; Ridder, R.; Herrington, S.; Jenkins, D.; Kurman, R.J.; Schmidt, D.; Stoler, M.; von Knebel Doeberitz, M. p16INK4a immunohistochemistry improves interobserver agreement in the diagnosis of cervical intraepithelial neoplasia. *Am. J. Surg. Pathol.* **2002**, *26*, 1389–1399. [CrossRef] [PubMed]

44. Albuquerque, A.; Rios, E.; Dias, C.C.; Nathan, M. p16 immunostaining in histological grading of anal squamous intraepithelial lesions: A systematic review and meta-analysis. *Mod. Pathol.* **2018**, *31*, 1026–1035. [CrossRef] [PubMed]

45. O'Neill, C.J.; McCluggage, W.G. p16 expression in the female genital tract and its value in diagnosis. *Adv. Anat. Pathol.* **2006**, *13*, 8–15. [CrossRef]

46. Kreimer, A.R.; Clifford, G.M.; Boyle, P.; Franceschi, S. Human papillomavirus types in head and neck squamous cell carcinomas worldwide: A systematic review. *Cancer Epidemiol. Biomark. Prev.* **2005**, *14*, 467–475. [CrossRef]

47. Guimerà, N.; Lloveras, B.; Lindeman, J.; Alemany, L.; van de Sandt, M.; Alejo, M.; Hernandez-Suarez, G.; Bravo, I.G.; Molijn, A.; Jenkins, D.; et al. The occasional role of low-risk human papillomaviruses 6, 11, 42, 44, and 70 in anogenital carcinoma defined by laser capture microdissection/PCR methodology: Results from a global study. *Am. J. Surg. Pathol.* **2013**, *37*, 1299–1310. [CrossRef]

48. Augustin, J.; Outh-Gauer, S.; Mandavit, M.; Gasne, C.; Grard, O.; Denize, T.; Nervo, M.; Mirghani, H.; Laccourreye, O.; Bonfils, P.; et al. Evaluation of the efficacy of the 4 tests (p16 immunochemistry, polymerase chain reaction, DNA, and RNA in situ hybridization) to evaluate a human papillomavirus infection in head and neck cancers: A cohort of 348 French squamous cell carcinomas. *Hum. Pathol.* **2018**, *78*, 63–71. [CrossRef]

49. Reeves, W.C.; Ruparelia, S.S.; Swanson, K.I.; Derkay, C.S.; Marcus, A.; Unger, E.R. National registry for juvenile-onset recurrent respiratory papillomatosis. *Arch. Otolaryngol. Head Neck Surg.* **2003**, *129*, 976–982. [CrossRef]

50. Wiatrak, B.J.; Wiatrak, D.W.; Broker, T.R.; Lewis, L. Recurrent respiratory papillomatosis: A longitudinal study comparing severity associated with human papilloma viral types 6 and 11 and other risk factors in a large pediatric population. *Laryngoscope* **2004**, *114*, 1–23. [CrossRef]

51. Draganov, P.; Todorov, S.; Todorov, I.; Karchev, T.; Kalvatchev, Z. Identification of HPV DNA in patients with juvenile-onset recurrent respiratory papillomatosis using SYBR Green real-time PCR. *Int. J. Pediatr. Otorhinolaryngol.* **2006**, *70*, 469–473. [CrossRef]

52. Derkay, C.S.; Malis, D.J.; Zalzal, G.; Wiatrak, B.J.; Kashima, H.K.; Coltrera, M.D. A staging system for assessing severity of disease and response to therapy in recurrent respiratory papillomatosis. *Laryngoscope* **1998**, *108*, 935–937. [CrossRef] [PubMed]

53. DE Clercq, E.; Descamps, J.; DE Somer, P.; Holyacute, A. (S)-9-(2,3-Dihydroxypropyl)adenine: An Aliphatic Nucleoside Analog with Broad-Spectrum Antiviral Activity. *Science* **1978**, *200*, 563–565. [CrossRef] [PubMed]

54. Huebbers, C.U.; Preuss, S.F.; Kolligs, J.; Vent, J.; Stenner, M.; Wieland, U.; Silling, S.; Drebber, U.; Speel, E.-J.M.; Klussmann, J.P. Integration of HPV6 and downregulation of AKR1C3 expression mark malignant transformation in a patient with juvenile-onset laryngeal papillomatosis. *PLoS ONE* **2013**, *8*, e57207. [CrossRef] [PubMed]

55. Kazlouskaya, V.; Shustef, E.; Allam, S.H.; Lal, K.; Elston, D. Expression of p16 protein in lesional and perilesional condyloma acuminata and bowenoid papulosis: Clinical significance and diagnostic implications. *J. Am. Acad. Dermatol.* **2013**, *69*, 444–449. [CrossRef] [PubMed]

56. Outh-Gauer, S.; Augustin, J.; Mandavit, M.; Grard, O.; Denize, T.; Nervo, M.; Lépine, C.; Rassy, M.; Tartour, E.; Badoual, C. Chromogenic In Situ Hybridization as a Tool for HPV-Related Head and Neck Cancer Diagnosis. *J. Vis. Exp.* **2019**. [CrossRef] [PubMed]

*Article*

# Anti-Retroviral Protease Inhibitors Regulate Human Papillomavirus 16 Infection of Primary Oral and Cervical Epithelium

**Samina Alam [1], Sreejata Chatterjee [1], Sa Do Kang [1], Janice Milici [1], Jennifer Biryukov [1], Han Chen [2] and Craig Meyers [1,3,***

[1]   Department of Microbiology and Immunology, The Pennsylvania State University College of Medicine, 500 University Drive, Hershey, PA 17033, USA; salam@pennstatehealth.psu.edu (S.A.); sreejatach@gmail.com (S.C.); skang@pennstatehealth.psu.edu (S.D.K.); jmilici@pennstatehealth.psu.edu (J.M.); jenbiryukov@gmail.com (J.B.)

[2]   Section of Research Resources, The Pennsylvania State University College of Medicine, Hershey, PA 17033, USA; hchen3@pennstatehealth.psu.edu

[3]   Department of Obstetrics and Gynecology, The Pennsylvania State University College of Medicine, Hershey, PA 17033, USA

*   Correspondence: cmeyers@pennstatehealth.psu.edu; Tel.: +1-717-531-6240

Received: 27 August 2020; Accepted: 10 September 2020; Published: 18 September 2020

**Simple Summary:** In 2016, globally, 36.7 million people were living with Human Immunodeficiency Virus (HIV), of which 53% had access to anti-retroviral therapy (ART) (UNAIDS 2017 Global HIV Statistics). The risk of Human Papillomavirus (HPV) associated oropharyngeal, cervical and anal cancers are higher among patients infected with HIV in the era of ART. Generally, HPV infections are self-limiting, however, persistent HPV infection is a major risk to carcinogenic progression. Long intervals between initial infection and cancer development imply cofactors are involved. Co-factors that increase infectivity, viral load, and persistence increase risk of cancer. We propose that the ART Protease Inhibitors (PI) class of drugs are novel co-factors that regulate HPV infection in HIV-infected patients. We developed a model system of organotypic epithelium to study impact of PI treatment on HPV16 infection. Our model could be used to study mechanisms of HPV infection in context of ART, and for developing drugs that minimize HPV infections.

**Abstract:** Epidemiology studies suggest that Human Immunodeficiency Virus (HIV)-infected patients on highly active anti-retroviral therapy (HAART) may be at increased risk of acquiring opportunistic Human Papillomavirus (HPV) infections and developing oral and cervical cancers. Effective HAART usage has improved survival but increased the risk for HPV-associated cancers. In this manuscript, we report that Protease Inhibitors (PI) treatment of three-dimensional tissues derived from primary human gingiva and cervical epithelial cells compromised cell-cell junctions within stratified epithelium and enhanced paracellular permeability of HPV16 to the basal layer for infection, culminating in de novo biosynthesis of progeny HPV16 as determined using 5-Bromo-2′-deoxyuridine (BrdU) labeling of newly synthesized genomes. We propose that HAART/PI represent a novel class of co-factors that modulate HPV infection of the target epithelium. Our in vitro tissue culture model is an important tool to study the mechanistic role of anti-retroviral drugs in promoting HPV infections in HAART-naïve primary epithelium. Changes in subsequent viral load could promote new infections, create HPV reservoirs that increase virus persistence, and increase the risk of oral and cervical cancer development in HIV-positive patients undergoing long-term HAART treatment.

**Keywords:** non-AIDS defining cancers (NADC); AIDS defining cancers (ADC); human papillomavirus type 16 (HPV16); opportunistic HPV infections; oropharyngeal cancer; cervical cancer; highly active anti-retroviral therapy (HAART); protease inhibitors; organotypic raft cultures; HPV persistence

## 1. Introduction

In 2016, globally, 36.7 million people were living with Human Immunodeficiency Virus (HIV), of which 53% had access to anti-retroviral therapy (ART) (UNAIDS 2017 Global HIV Statistics). In the United States, approximately 1.2 million people are living with Human Immunodeficiency Virus (HIV)/AIDS [1]. Better tolerated combinations of Highly Active Anti-Retroviral Therapy (HAART) has significantly improved the survival of HIV-positive individuals including reduction of new infections and extended life-span [2]. Declined mortality among HIV-infected individuals has resulted in growth and aging of the HIV-positive populations, which has implications for increased risk of cancer development [3]. Higher cancer risk in HIV/AIDS patients compared to the general population is a result of HIV-related immunosuppression that impairs control of oncogenic viral infections [3], including AIDS-defining cancers (ADC) due to HPV induced cervical cancer, and non-AIDS-defining cancers (NADC) that include HPV associated oropharyngeal and anal cancers [3,4]. Since the introduction of HAART in 1996, rates of ADC, including cervical cancer have decreased, but incident rates of cervical cancer remain elevated in patients undergoing HAART treatment compared to the general population [3]. In contrast, rates of NADCs have increased with respect to both oropharyngeal and anal cancers [3]. Although increasing longevity is the greatest risk factor for NADCs, it is insufficient to explain trends in cancer epidemiology [5].

In 1995, the first generation of Protease Inhibitors (PI) class of anti-retroviral therapy (ART) became commercially available, followed by novel combinations of PIs with other anti-retroviral classes of drugs [5,6]. Multiple epidemiological studies representing diverse cohorts support the finding that oral manifestations of HPV infections increased among HIV-positive patients on long-term HAART compared to patients not taking ART [7–11]. One study analyzed cancer incidence after ART initiation in eight US HIV clinical cohorts who started ART between 1996 and 2011, of which 50% started a PI-containing regimen [6], showed that rates of NADCs rose with longer time on ART [12], and older age was a significant predictor of NADCs, including HPV related malignancies [12]. A retrospective study of patients (1996–1999), analyzed the relationship between exposure to combination HAART therapy and prevalence of oral warts and showed that oral lesions were significantly associated with PI containing regimens compared with another class of HAART [8]. Prevalence of oral warts was 23% of patients on HAART (+PI) and 15% of patients on HAART (−PI) therapy containing HIV non-nucleoside reverse transcriptase inhibitors (NNRTI), when compared to 5% of patients on neither medication [8]. When adjusted for CD4+ count and HIV load, the odds of having oral warts for those on HAART + NNRTI alone showed a non-significant association, but for those on HAART + PI there was a highly significant association, which also suggested that HAART use increased oral warts [8]. Overall, HAART usage has decreased incidents of oral lesions of both viral- and non-viral etiologies and correlates with increased CD4+ T-cell count, but is not statistically significant for decreased HPV infections [13]. Therefore, the burden of NADCs continues to rise, as does the need for cancer detection, prevention and treatment in HIV-positive patients [14].

Studies suggest that patients on HAART are at increased risk of acquiring opportunistic HPV infections and developing oropharyngeal [15,16], anal [17] and cervical cancers [18], compared to the general population. Oncogenic high-risk HPV16 is responsible for more than 60% of oropharyngeal carcinoma, ~90% of tonsillar carcinomas [19], most cases of vulvar carcinoma [20], 90% of anal carcinoma [21] and penile cancers (less than 1% of all male cancers in the United States) with incidence rates that greatly vary across different regions of the world [21]. The oropharyngeal compartment is central to the persistence of HPV, and the virus is more commonly detected in the oral mucosa of HIV-positive patients compared to HIV-negative patients [22]. Up to 56% of HIV-infected adults have detectable HPV DNA, which is significantly higher than the non-HIV infected population [15,23]. In addition, HAART usage is associated with development of adverse oral complications that damage the mucosal epithelium and potentially expose the underlying tissues to HPV infection [24–26]. Thus, viral persistence could determine increased prevalence of oropharyngeal cancer among HIV patients receiving ART compared to the general population [8,16]. HPV16 tends to be persistent and is refractory to clearance in women on HAART [27], and accounts for ~70% of all invasive cervical cancer [28]. Cumulatively, these studies suggest that HAART treatment potentially enhances opportunistic HPV infection, viral persistence and cancer progression.

The molecular mechanism of how HAART exposure sensitizes target epithelium to opportunistic HPV infection, and potentially other viruses, is of significant interest. In the current manuscript, we developed an in vitro model of HPV infection of HAART-naïve primary human gingiva and cervical epithelium, and asked whether treatment with two protease inhibitors, Amprenavir and Kaletra, prime the mucosa for virus infection, and further investigated impact of drug treatments on subsequent viral load.

## 2. Results

### 2.1. Amprenavir Treatment Enhances HPV16 Infection of Primary Oral Tissue

Amprenavir (Agenerase®, GSK) was one of the first PIs in the market that was later removed due to increased viral resistance. The drug binds to the active site of HIV-1 aspartyl protease and prevents processing of viral gag and gag-pol poly-protein precursors resulting in formation of immature non-infectious viral particles [29]. Additionally, prolonged use of Amprenavir was associated with adverse orofacial effects including progressive oral warts that recurred after removal [8]. We previously reported that treatment with Amprenavir impacted growth, differentiation and epithelial repair of gingiva tissues [30]. In the current study, we determined whether PI exposure affected HPV16 infection in an in vitro model of three-dimensional epithelium. Organotypic cultures were derived from primary gingival keratinocytes isolated from mixed pools of human gingiva from patients undergoing dental surgery [30]. On day 8 post-lifting and differentiation at the air-liquid interface, gingiva tissues were treated for 24 h with 7.66 µg/mL Amprenavir (drug $C_{max}$ representing peak blood concentration after drug administration maintained between two dosages for optimal HIV suppression). Drug treatment impacted tissue morphology, compromising cell-cell junctions within the stratified suprabasal as well as the basal layers, altering structural/barrier integrity of desmosome-, tight- and adherens junctions (Figure 1). It is generally thought that HPV infects cells of the basal layer via micro-abrasions, where viral genome amplifies to high copies [31]. Hypothetically, Amprenavir regulated damage of protein complexes at cell-cell contact sites is reminiscent of tissue "wounding" that could provide opportunistic HPV access to basal cells for infection.

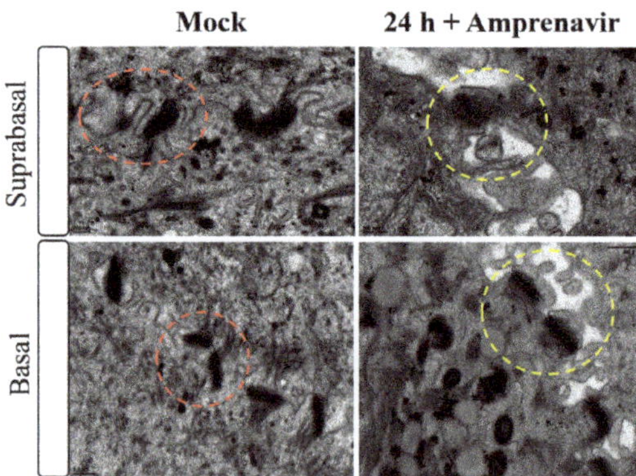



**Figure 1.** Transmission electron microscopy depicting cell-cell junction morphology of day 8 gingiva raft tissue treated with Amprenavir (7.66 µg/mL) for 24 h compared with tissue not drug treated.

We then tested the impact of Amprenavir treatment on HPV16 infection of primary gingiva tissues. Laboratory stocks of HPV16 were prepared from raft tissues derived from cervical cell lines productively infected with HPV16 that were differentiated in culture for 20 d, as described in the Methods sections. Amprenavir concentrations ranging from 7.66–2.5 µg/mL were added to the culture media for 24–72 h, followed by infecting tissues at each time point using increasing doses of HPV16 virus particles as described in this scheme (Figure 2). As controls, untreated tissues were infected with the highest dose of HPV16 virions. Total time spent in culture was 15–18 d, respectively, followed by tissue harvesting and measuring the E1^E4 major spliced transcript, a major hallmark of virus infection. The E1^E4 open reading frame is present in both early and late HPV transcripts, and high level expression of this protein is restricted to differentiated suprabasal cells [32]. The E1^E4 protein has multiple functions and is thought to interact with keratin intermediate filament networks to facilitate network re-organization [33], associate with mitochondria to induce apoptosis [34], bind RNA processing proteins [35], disrupt nuclear dot 10 domains [36] and associate with cellular cyclins to mediate cell cycle arrest in the G1/M phase [37]. In addition, we have reported that the E1^E4 protein may also play a role in HPV capsid assembly, infectivity and virion maturation [38]. In the current manuscript, relative E1^E4 transcript levels in HPV16 infected/Amprenavir treated tissues were compared to HPV16 infected tissues that did not receive drug treatment. Amprenavir treatment renders primary gingiva tissue more favorable to HPV16 infection, compared to untreated tissues that were poorly infected (Figure 3A). Relative fold-change of E1^E4 expression varied relative to drug pre-treatment times across three independent experiments, and is attributed to natural variation in host cell genetics (Figures 3A and S1A,B). However, in all cases, E1^E4 transcript expression in drug treated tissues was significantly increased compared to drug untreated tissues. Throughout the manuscript, unless otherwise stated, for each experiment performed in triplicate, the first data panel is presented in the Results section, and the other two panels in the Supplementary Materials section. Virus infection of tissue was inhibited using antibodies against HPV16 L1 (α-V5) and L2 (α-RG-1) capsid proteins (Figures 3B and S1C,D), thus demonstrating interference with receptor mediated virus entry pathways, as would be expected. Lower doses of Amprenavir (5 µg/mL) also significantly enhanced HPV16 infection only after 24–48 h of treatment (Figure 4A), whereas treatment with 2.5 µg/mL Amprenavir poorly supported infection (Figure 4B). These results are significant as we show for the first time that three-dimensional tissues derived from primary human epithelial cells can be infected with a high-risk HPV in vitro. Importantly, this could relate to the clinically relevant oral HPV16 infections observed in HIV+ patients undergoing HAART treatment.

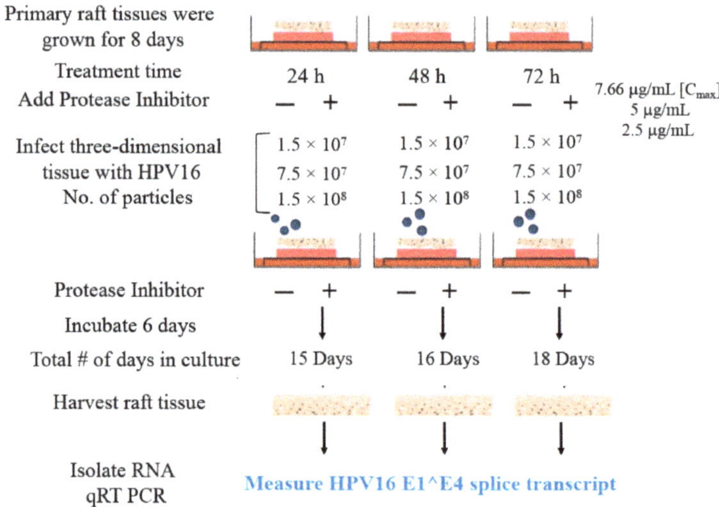



**Figure 2.** Schematic of raft tissue growth, drug treatment and infection using three doses of standard laboratory stocks of HPV16 virions.

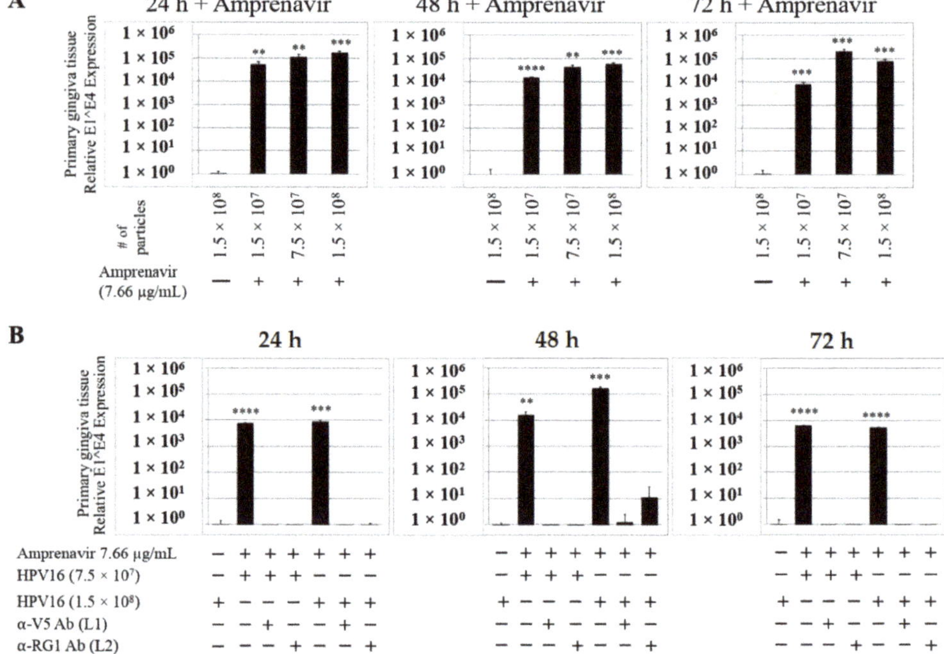



**Figure 3.** Amprenavir (7.66 µg/mL) treatment sensitizes primary gingiva tissue to HPV16 infection (**A**) Comparative expression of HPV16 E1^E4 transcripts in drug treated tissues compared with virus infected tissues not drug treated. (**B**) Inhibition of virus infection of tissues using HPV16 pre-incubated with α-V5 and α-RG1. Data were analyzed as mean ± SD. *p*-values were calculated using two-tailed Student's *t*-tests. Quantitative data are presented as mean ± standard deviation. Significance was based on pairwise Student's *t*-test. Comparisons are indicated as $0.001 < p < 0.01$ by **; $0.0001 < p < 0.001$ by ***; and $p < 0.0001$ by ****.

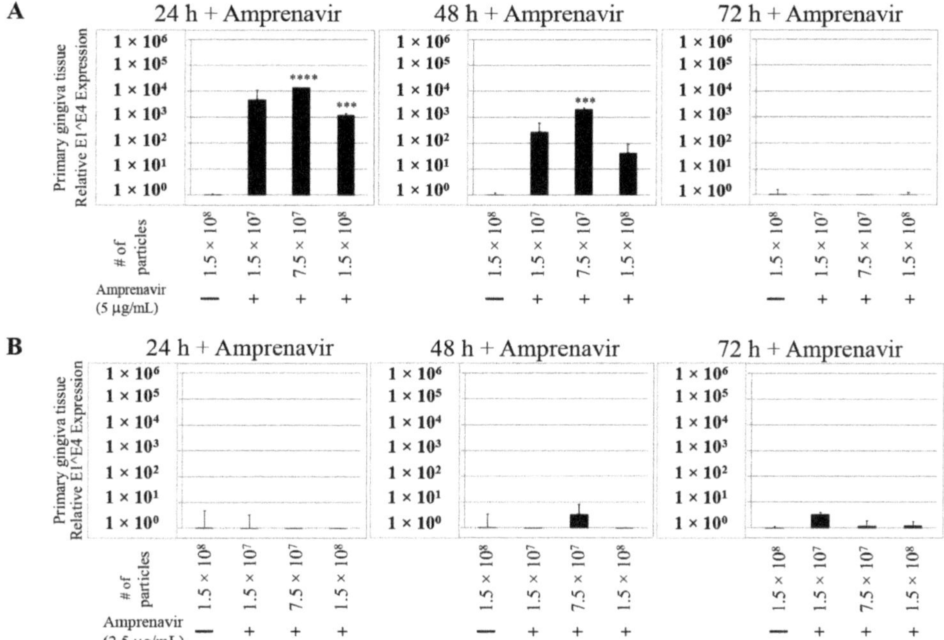

**Figure 4.** Dose-dependent Amprenavir treatment modulates HPV16 infection of gingiva tissues. (**A**) Comparative expression of HPV16 E1^E4 transcripts in tissues treated with 5 µg/mL Amprenavir. Results shown are average of three individual experiments. (**B**) Comparative expression of HPV16 E1^E4 transcripts in tissues treated with 2.5 µg/mL Amprenavir. Data were analyzed and is presented as mean ± SD. *p*-values were calculated using two-tailed Student's *t*-tests. Quantitative data are presented as mean ± standard deviation. Significance was based on pairwise Student's *t*-test. Comparisons are indicated as $0.0001 < p < 0.001$ by ***; and $p < 0.0001$ by ****.

Further analysis showed that virus infection of Amprenavir (7.66 µg/mL) treated tissues correlated with changes in putative progeny viral titers, a milestone in the viral life-cycle (Figure 5, top panel and Figure S1E,F). Such progeny HPV16 virions (*prog*-HPV16) poorly infected monolayer HaCaT cells compared to parental HPV16 (*P*-HPV16) used above for infecting raft tissues (Figure 5, bottom panel), suggesting that extended time in culture may be needed for virus capsid maturation to occur. This concept is based on our previously reported finding that improved HPV16 infectivity over time is a function of capsid maturation with respect to disulfide bond formation that determines virus infectivity [39,40]. Raft tissues treated with a range of Amprenavir concentrations (7.66, 5 and 2.5 µg/mL) and infected with either $7.5 \times 10^7$ and $1.5 \times 10^8$ of *P*-HPV16 virions, were cultured for a further 18–24 d, followed by preparing crude virus (CV) stocks of *prog*-HPV16 from tissues and virus titer determination (Figures 6A and S1G,H). In this experiment, Amprenavir dose-dependent changes in progeny virus titers over time was observed. Progeny HPV16 virus stocks isolated from tissues treated with 7.66 µg/mL Amprenavir were further purified using Optiprep gradient fractionation, where infectious virus particles with fully mature capsids are known to partition within fractions #5–#8 [41]. Infectivity of *prog*-HPV16 in fraction #7 [F#7] samples was measured by infecting HaCaT monolayer cultures, and determining fold-change expression of E1^E4 transcripts compared with *P*-HPV16 [F#7] virus (Figure 6B). Infectivity of *prog*-HPV16 [F#7] improved with increased time in culture and trended towards infectivity of *P*-HPV16 [F#7] stocks. Similar to *P*-HPV16, infection of *prog*-HPV16 [F#7] was inhibited with α-V5 and α-RG1 antibodies, suggesting that capsid structure was conserved between *prog*-HPV16 and *P*-HPV16 (Figure 6B). Virus infected gingiva tissues treated with 5 µg/mL Amprenavir also produced infectious *prog*-HPV16, that was neutralized with anti-capsid antibodies (Figure 7), suggesting that fluctuations of HAART concentrations as would occur in patients undergoing treatment, would not affect biochemical integrity of new HPV particles synthesized in target tissues. In contrast, long-term cultures treated with 2.5 µg/mL Amprenavir produced low titers of *prog*-HPV16 (Figures 6A and S1G,H). Cumulatively, these results suggest that Amprenavir acts as a co-factor to sensitize HPV16 infection and further increases viral load in oral tissues, at least via *prog*-HPV16 biosynthesis.

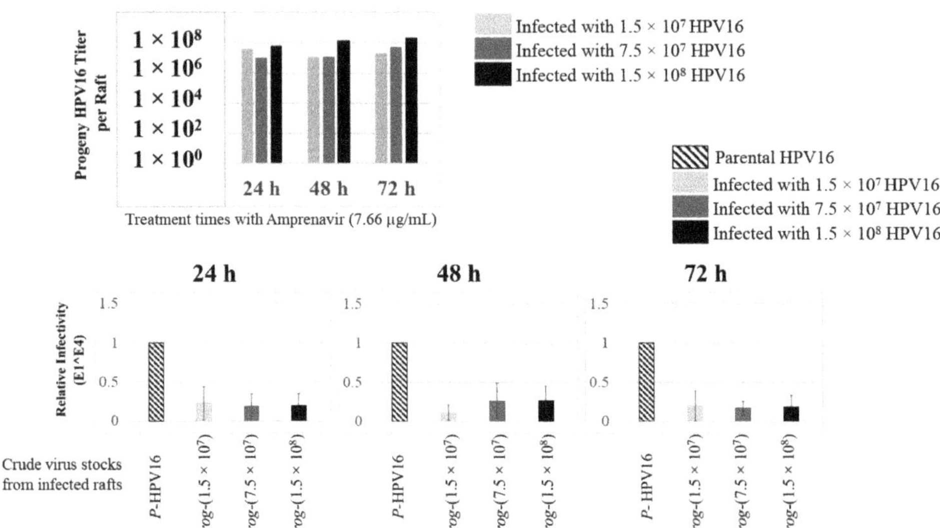


**Figure 5.** Progeny HPV16 stocks isolated from day 18 gingiva tissues treated with Amprenavir are poorly infectious. Top panel: Progeny HPV16 virus stock (day18 harvest) titers isolated from raft tissues infected with three virus doses indicated in light grey bars: $1.5 \times 10^7$ HPV16 virions; Grey bars: $7.5 \times 10^7$ HPV16 virions; Black bars: $1.5 \times 10^8$ HPV16 virions. Bottom panel: Corresponding infectivity of progeny virion stocks (5 MOI) (Multiplicity of Infection) in HaCaT monolayer cells compared with standard (Parental) laboratory stocks. Infection results shown are average of three experiments and is presented as mean ± SD.

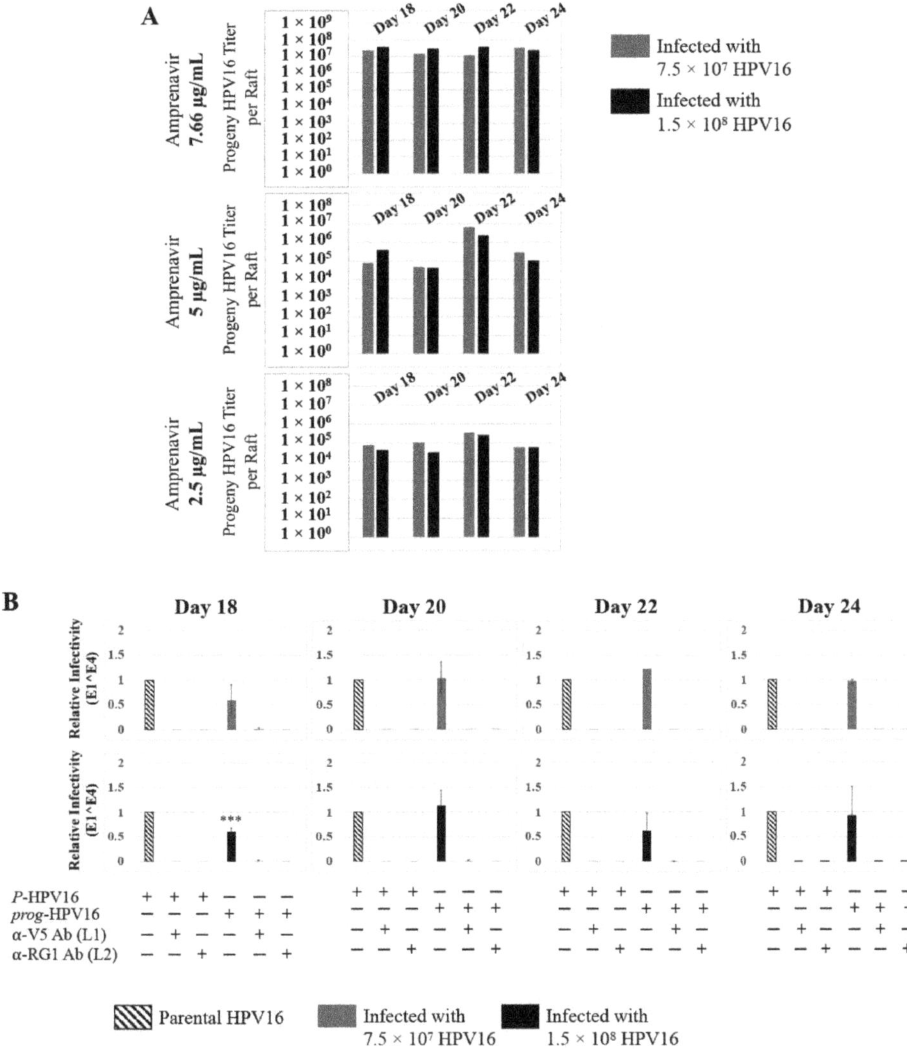

**Figure 6.** Extended culturing of infected gingiva tissues treated with Amprenavir determines progeny virus titers. (**A**) Raft tissues (day 18–24) infected with two virus doses modulate *prog*-HPV16 titers in an Amprenavir concentration dependent manner. Grey bars: infected with $7.5 \times 10^7$ *P*-HPV16 virions; Black bars: infected with $1.5 \times 10^8$ *P*-HPV16 virions. (**B**) Infectivity of *prog*-HPV16 Optiprep [F#7] compared with *P*-HPV16 [F#7] (1 MOI), and infection inhibition using α-V5 and α-RG1 monoclonal antibodies. Infection results shown are average of three experiments and is presented as mean ± SD. *p*-values were calculated using two-tailed Student's *t*-tests. Infectivity of *prog*-HPV16 were not significantly different compared with *P*-HPV16. Comparisons are indicated as $0.0001 < p < 0.001$ by ***.

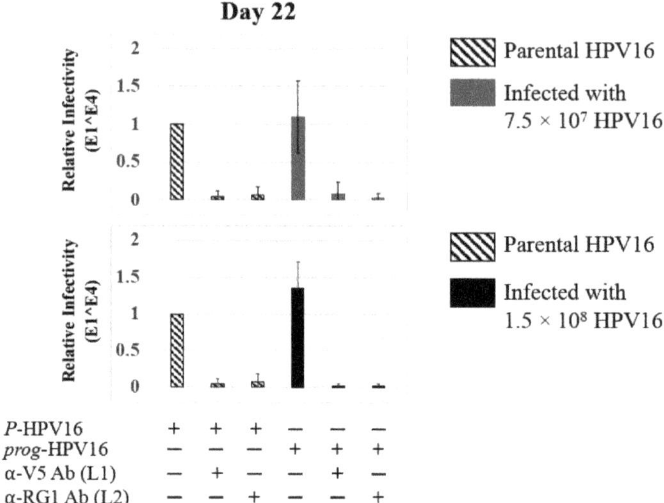


**Figure 7.** Low Amprenavir concentrations determine production and infectivity of *prog*-HPV16. Optiprep [F#7] *prog*-HPV16 from tissues treated with 5 μg/mL Amprenavir compared with *P*-HPV16 Optiprep [F#7] (1 MOI in HaCaT cells), and infection inhibition using α-V5 and α-RG1 monoclonal antibodies. Infection results shown are average of three experiments and is presented as mean ± SD. *p*-values were calculated using two-tailed Student's *t*-tests. Infectivity of *prog*-HPV16 was not significantly different compared with *P*-HPV16.

## 2.2. Using BrdU to Label Newly Synthesized HPV16 Genomes Distinguishes between Parental and Progeny Virions

To confirm that Amprenavir treatment induced de novo virus biosynthesis in gingiva tissues, we developed protocols to distinguish "input" *P*-HPV16 from "output" *prog*-HPV16 using BrdU labeling of newly replicated genomes. First, culture conditions were optimized for BrdU-labeling of replicating viral genomes, utilizing HPV16 positive organotypic cultures routinely used for generating infectious HPV16 standard laboratory stocks. In this system, high-risk HPV-positive cervical cell lines maintaining episomal copies of viral genomes are allowed to grow and differentiate over a period of 20 d [41], during which time tissue stratification occurs synonymously with viral genome amplification, late gene L1 and L2 capsid protein expression, culminating in virion morphogenesis [31]. We reported that HPV genome amplification occurs on day 8 prior to capsid protein expression on day 10, potentially in mechanistic tandem for genome encapsidation and virus assembly [42]. To maximally enable BrdU incorporation into replicating viral genomes with minimal toxicity to host tissues, on day 8 post-lifting of rafts to the air-liquid interface, 25 μM BrdU was added to media and maintained until tissue harvesting (Figure 8A). Infectivity of HPV16 stocks grown in the presence of BrdU (*P*-HPV16-BrdU) was similar to control/unlabeled *P*-HPV16, including the ability of neutralizing anti-capsid antibodies to inhibit virus infection of HaCaT monolayer cultures (Figure 8B).

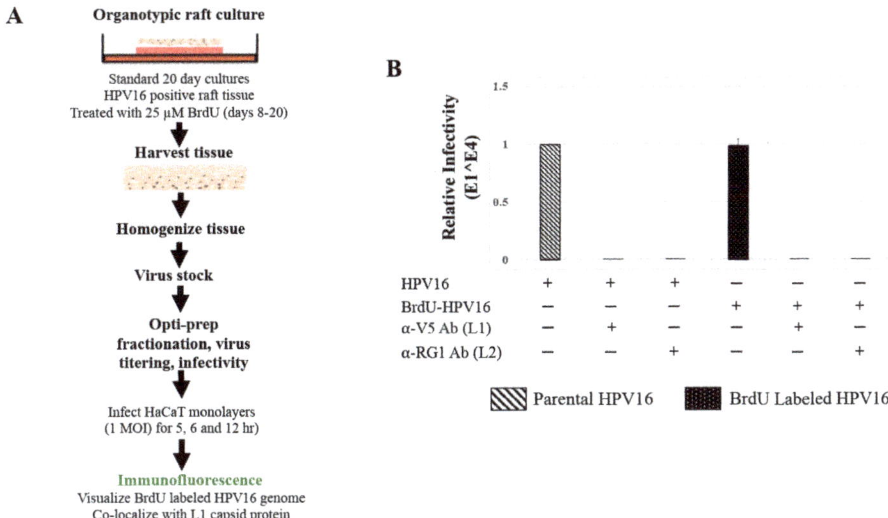



**Figure 8.** BrdU Labeling of 20 d Standard Laboratory HPV16 Stocks. (**A**) Schematic for BrdU labeling of HPV16 virus in 20 d raft tissues derived from human cervical cell line, maintaining episomal HVP16 genomes. (**B**) Infectivity comparison of BrdU-labeled HPV16 versus control unlabeled virus stocks (1 MOI in HaCaT cells). Optiprep gradient purified virus [F#7] stocks was used. Infection was in the presence of α-V5 and α-RG1 monoclonal antibodies. Infection results shown are average of three experiments and is presented as mean ± SD. *p*-values were calculated using two-tailed Student's *t*-tests. Infectivity comparisons were not significantly different.

To visualize HPV16-BrdU labeled genomes, HaCaT monolayer cultures were infected with *P*-HPV16-BrdU virus from [F#7]-stocks purified using Optiprep gradient fractionation, and immunofluorescence staining was performed to detect labeled genomes using a mouse anti-BrdU antibody. Additionally, to confirm that BrdU-labeled genomes were virion encapsidated, we further co-localized BrdU immunofluorescence with HPV16 L1 capsid protein using a rabbit anti-HPV16 L1 antibody. Significant co-localization was determined between viral genome/capsid complexes manifesting as punctate spots in perinuclear regions (Pearson's correlation coefficient for co-localization between 0.560 ± 0.023 and 0.626 ± 0.076) (Figure 9A). Time course experiments were also performed to determine kinetics of HPV16-BrdU cell uptake and entry. The L1 capsid staining alone was weakly visualized at 3 h post-infection, suggesting virus attachment to cells (Figure 9B). However, BrdU immunofluorescence was not detected suggesting the absence of capsid uncoating. In contrast, at 5 h post-infection, significant co-localization was found between BrdU-labeled genomes and L1, suggesting initial stages of virus disassembly. Additionally, HPV16-BrdU staining appeared in polarized clusters, indicating mass virus trafficking within vesicles around perinuclear locations. In contrast, at 6 h post-infection, co-localization between L1 and genomes was decreased (Pearson's coefficient 0.264 ± 0.017), indicating further advancement of viral capsid/genome disassembly. At 12 h post-infection, co-localization of the two fluorophores was not significant (Pearson's coefficient 0.057 ± 0.019). As controls, non-specific binding of anti-BrdU and anti-L1 antibodies was not detected in uninfected HaCaT cells (Figure S2A). Lastly, neutralizing α-V5/L1 capsid protein complexes were visualized at 6h post-infection, where L1 staining appeared in clusters depicting non-uniform spots (Figure 9B, right panel, compare L1 staining pattern with left panel, 6h post-infection without α-V5), that correlate with inhibition of virus infection (Figure 6B). Antibody sequestration or "clumping" of L1 presumably serves as a mechanism that prevents virus infectivity. Our studies present first time data that show biosynthesis of a high-risk papillomavirus in three-dimensional culture, that is amenable to BrdU-labeling and is able to be visualized following infection using confocal imaging.

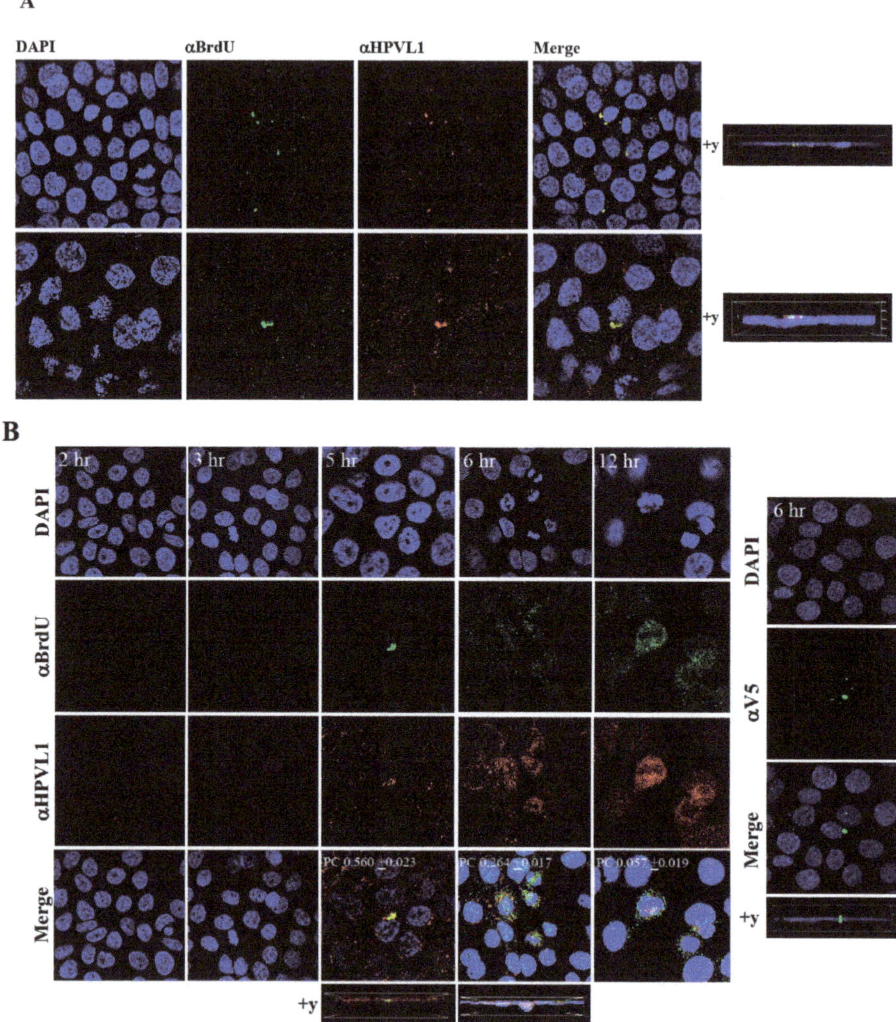



**Figure 9.** Visualization of BrdU-labeled viral genomes. (**A**) Confocal microscopy imaging of Optiprep [F#7] BrdU-labeled HPV16 genomes co-localized with L1 capsid proteins in HaCaT cells 5 h post-infection. Cells were infected using 1 MOI of virus. (**B**) Left panel: Time-dependent co-localization of HPV16-BrdU labeled genomes and L1 capsids in HaCaT monolayers, 2 h, 3 h, 5 h and 6 h post-infection. Optiprep gradient virus in [F#7] stocks (1 MOI) was used. Right panel: Immunofluorescence of α-V5/L1 capsid protein complexes that correlate with inhibition of infection. Optiprep gradient fractionated [F#7] virus stock was incubated with α-V5/L1 followed by infecting HaCaT monolayers (1 MOI) and imaging 6 h post-infection (compare with left panel, 6h post-infection L1 staining pattern without α-V5 incubation). Pearson's coefficients illustrating co-localization of BrdU labeled genome and L1 capsid protein are presented in the merged images. Data represent mean Pearson's coefficient ± SD, calculated from 5 independent images.

### 2.3. BrdU Labels Newly Synthesized Progeny Virus in Amprenavir Treated Gingiva Tissue Infected with P-HPV16

The techniques developed above to label newly synthesized HPV16 genomes packaged into virions provided a foundation for further determining whether Amprenavir treated primary gingiva tissues infected with *P*-HPV16 also induced de novo *prog*-HPV16 biosynthesis. Primary tissues were optimized for BrdU-labeling of replicating genomes incorporated into new virions using a modified protocol (Figure 10). On day 8 post-lifting, primary gingiva tissues were pre-treated with Amprenavir for 72 h, followed by infecting with one of two doses of *P*-HPV16 as indicated. At 24 h post-infection, BrdU was added to the media and tissues were cultured until harvesting on day 22. Then, BrdU-labeled *prog*-HPV16 (*prog*-HPV16-BrdU) CV stocks were purified using Optiprep gradient fractionation. Monolayer HaCaT cultures were infected with [F#7]-derived *prog*-HPV16-BrdU (1 MOI) and visualized using confocal imaging (Figure 11A). Significant viral genome/L1 co-localization was determined 5 h post-infection (Pearson's coefficient 0.657 ± 0.060), that was decreased at 6 h (Pearson's coefficient 0.204 ± 0.109), suggesting that kinetics of virus disassembly was similar to *P*-HPV16-BrdU (Figure 9B). Further, punctate α-V5/L1 capsid-protein complexes were visualized that correlate with inhibition of infection (Figure 11A, right panel, compare L1 staining pattern with left panel, 6 h post-infection without α-V5; and Figure 6B). We present first time data that show Amprenavir treatment and infection of three-dimensional tissue with authentic HPV16 virus results in productive infection and de novo biosynthesis of infectious progeny HPV16 in vitro.

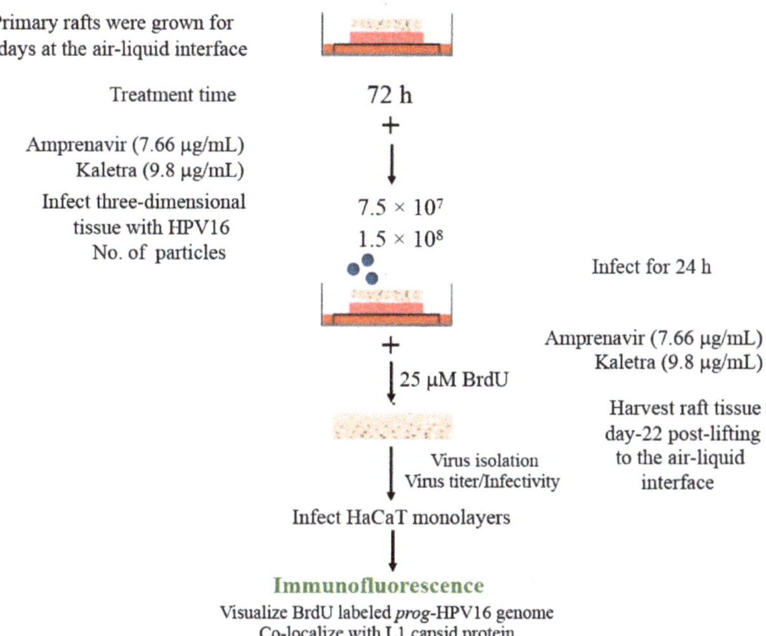


**Figure 10.** BrdU labeling of progeny HPV16 biosynthesized in primary tissues. Schematic for BrdU labeling of *prog*-HPV16 biosynthesized in 22 d PI treated tissues. In all cases, Optiprep gradient purified [F#7] day 22 virions were used for infections (1 MOI).

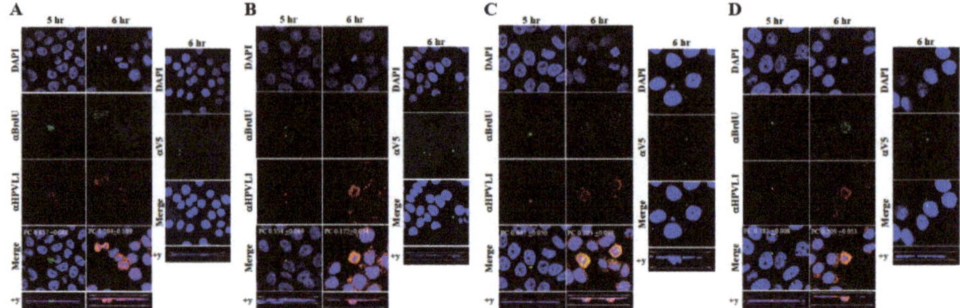

**Figure 11.** Immunofluorescence co-localization of BrdU-labeled genomes and L1 capsids of *prog*-HPV16. (**A**) Progeny HPV16 biosynthesized in Amprenavir treated gingiva tissues. (**B**) *Prog*-HPV16 biosynthesized in Kaletra treated gingiva tissues. (**C**) *Prog*-HPV16 biosynthesized in Amprenavir treated cervical tissues. (**D**) *Prog*-HPV16 biosynthesized in Kaletra treated cervical tissues. Right panels of (**A–D**): Immunofluorescence of α-V5/L1 capsid protein complexes that correlate with inhibition of infection. *Prog*-HPV16 [F#7] stocks biosynthesized in 22 d raft tissues was incubated with α-V5/L1 followed by infecting HaCaT monolayers and then imaging 6 h post-infection (compare with 6 h post-infection L1 staining pattern without α-V5 incubation, left panel). Pearson's coefficients illustrating co-localization of BrdU labeled genome and L1 capsid protein are presented in the merged images. Data represent mean Pearson's coefficient ± SD, calculated from 5 independent images.

## 2.4. Kaletra® Treatment Also Induces De Novo Biosynthesis of Progeny Virions in Gingiva Tissues

To confirm that our findings were not restricted to Amprenavir treatment, we utilized Kaletra (Lopinavir/ritonavir, Abbott Laboratories), a PI currently prescribed for HIV treatment. Use of Kaletra was approved in 2000 by the US FDA for the treatment of HIV infection in adults and children. Similar to Amprenavir, prolonged use of Kaletra is associated with several adverse orofacial effects [43,44]. We reported that Kaletra treatment also affected growth, differentiation and epithelial repair of gingiva tissues [45]. In the current study, we determined the impact of Kaletra treatment ($C_{max}$ 9.8 μg/mL) on *P*-HPV16 infection and subsequent progeny viral load in primary tissues. Like Amprenavir, Kaletra treatment also rendered primary gingiva tissue more favorable to HPV16 infection, compared with untreated tissues (Figure 12A). Again, relative fold-change of E1^E4 expression varied relative to drug pre-treatment times across three independent experiments attributed to natural variation in host cell genetics (Figures 12A and S3A,B). Virus infection of tissue was neutralized using antibodies against HPV16 L1 and L2 capsid proteins (Figures 12B and S3C,D). Raft tissues treated with a range of Kaletra concentrations (9.8, 6 and 3 μg/mL) were infected with one of two doses of *P*-HPV16 and grown in long-term cultures, and putative *prog*-HPV16 CV stocks were titered and were further determined to be drug dose-dependent (Figures 13A and S3E,F). Infectivity of *prog*-HPV16 in [F#7]-stocks were similar to that of *P*-HPV16 [F#7], and were inhibited with α-V5 and α-RG1 antibodies, suggesting that regardless of whether Amprenavir or Kaletra was tested, capsid structure in relation to infectivity was conserved between *prog*-HPV16 and *P*-HPV16 (Figure 13B). In contrast, long-term cultures treated with 6 and 3 μg/mL Kaletra produced low *prog*-HPV16 titers (Figure 13A and Supplementary Figure S3E,F). We also confirmed that Kaletra treatment induced de novo virion biosynthesis. Our studies show that Kaletra treatment and infection of three-dimensional tissue with *P*-HPV16 virus also results in production of BrdU-labeled *prog*-HPV16 in vitro (Figure 11B). Time-line experiments were performed to show significant *prog*-HPV16 BrdU-labeled genome/L1 co-localization after 5 h of infection of HaCaT monolayer cells (Pearson's coefficient 0.554 ± 0.069), that was decreased at 6 h post-infection (Pearson's coefficient 0.172 ± 0.054) (Figure 11B). In addition, α-V5/L1 capsid protein complexes were similarly visualized that correlate with inhibition of infection (Figure 11B right panel, compare L1 staining pattern with left panel, 6 h post-infection without α-V5; and Figure 13B). We previously reported that Kaletra treatment of primary gingiva tissues also mediated disruption of protein complexes that regulate cell-cell junctions [45] that could provide HPVs with efficient access to their target cells in the basal layer. Overall, these results indicate that Kaletra also acts as a co-factor to sensitize HPV16 infection of oral tissue.

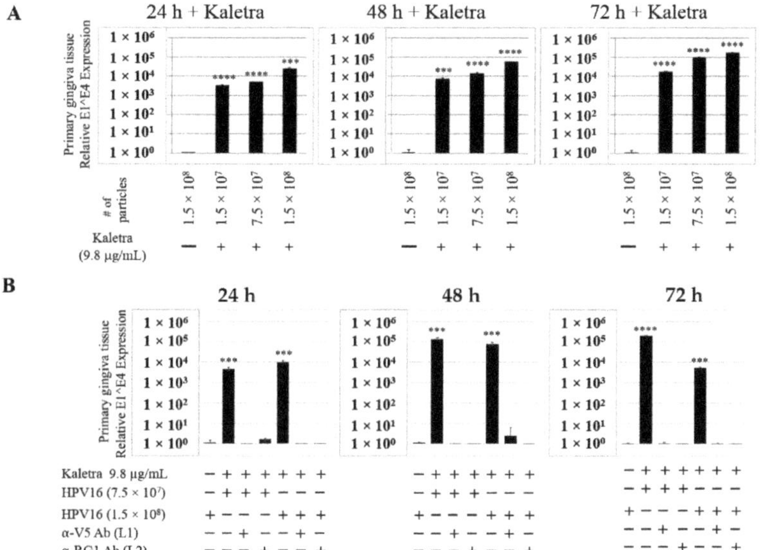

**Figure 12.** Kaletra (9.8 µg/mL) treatment sensitizes primary gingiva tissue to HPV16 infection. (**A**) Comparative expression of HPV16 E1^E4 transcripts in Kaletra treated tissues compared with virus infected tissues not drug treated. Three P-HPV16 doses were used for infecting rafts as indicated. (**B**) Inhibition of virus infection of tissues using HPV16 pre-incubated with α-V5 and α-RG1. Data were analyzed as mean ± SD. *p*-values were calculated using two-tailed Student's *t*-tests. Significance was based on pairwise Student's *t*-test. Comparisons are indicated as $0.0001 < p < 0.001$ by ***; and $p < 0.0001$ by ****.

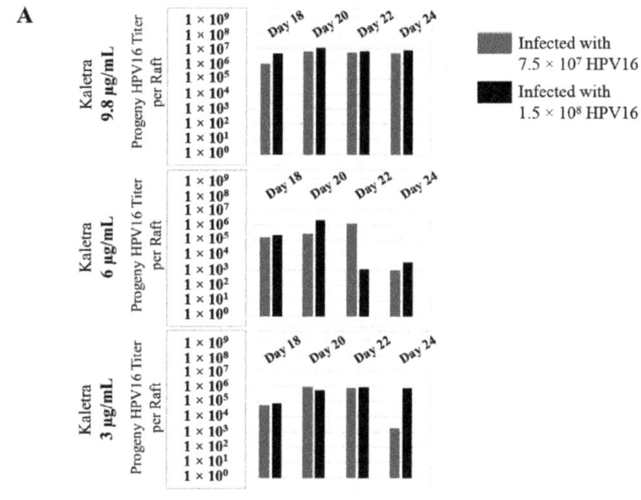

**Figure 13.** *Cont.*

**B**

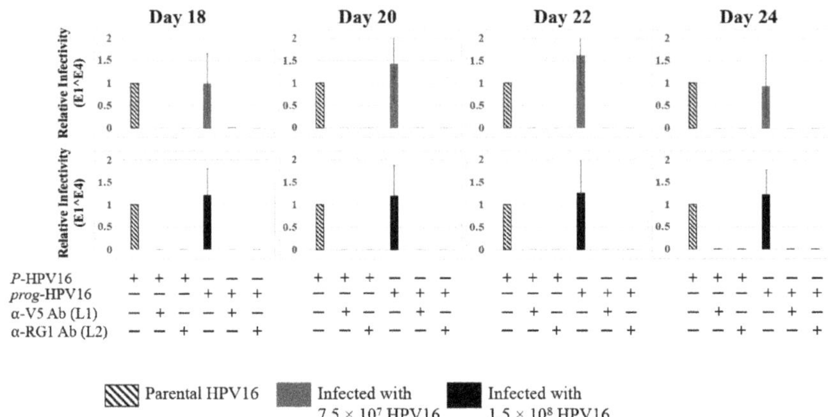



**Figure 13.** Extended culturing of Kaletra infected gingiva tissues determines progeny virus titers. (**A**) Raft tissues (day 18–24) infected with two virus doses modulates *prog*-HPV16 titers in a Kaletra concentration dependent manner. Grey bars: infected with $7.5 \times 10^7$ *P*-HPV16 virions; Black bars: infected with $1.5 \times 10^8$ *P*-HPV16 virions. (**B**) Infectivity of *prog*-HPV16 Optiprep [F#7] compared with *P*-HPV16 [F#7] (1 MOI), and infection inhibition using α-V5 and α-RG1 monoclonal antibodies. Infection results shown are average of three experiments and is presented as mean ± SD. *p*-values were calculated using two-tailed Student's *t*-tests. Infectivity of *prog*-HPV16 were not significantly different compared with *P*-HPV16.

### 2.5. Amprenavir Treatment Allows for Virus Transit through Gingiva Tissue Layers for Infecting Basal Cells

Thus far, we have shown that untreated gingiva tissues were poorly infected with HPV16 compared with tissues treated with PI (Figures 3A and 12A). Amprenavir mediated disruption of protein complexes that regulate cell-cell junctions (Figure 1) could provide HPVs with more efficient access to their target cells in the basal layer. In order to visually correlate HPV16 infection with virus localization within different layers of the stratified epithelium, we used *P*-HPV16-BrdU to infect Amprenavir treated primary gingiva tissues and performed immunofluorescent staining of BrdU-labeled HPV16 virus particles in transit through the tissue over a period of 12–72 h post-layering of virus on top of tissues (Figure 14A,B). At 12 h post-infection, HPV16-BrdU was mostly localized in the cornified and upper portions of the suprabasal layer irrespective of Amprenavir treatment. At increasing times post-addition of virus, in drug-treated tissues, HPV16-BrdU was localized throughout the suprabasal layer as well as cells within the basal layer, whereas in untreated tissues, the virus was restricted in the upper cornified layers of the tissues. Significant co-localization was determined between viral genome/capsid complexes, manifesting as punctate spots in perinuclear regions in cells of the basal layer (Pearson's correlation coefficient for co-localization between $0.878 \pm 0.094$ and $0.644 \pm 0.107$) across 24–72 h post-layering of virus on tissues (Figure 14B). We also performed control experiments to rule out the possibility of "bleed-through" of the fluorescence emission of Alexa Fluor 488 and Alexa Fluor 568, the two fluorophores used in our studies, and found no evidence of crossover fluorescence (Figure S2B). Taken together, our results suggest that Amprenavir mediated disruption of cell–cell barrier integrity likely plays a role in enhancing HPV16 transit through the epithelium to the target basal cells for infection, whereas non-disrupted cell-cell junctions in control tissues impede virus transit through the tissue layers.

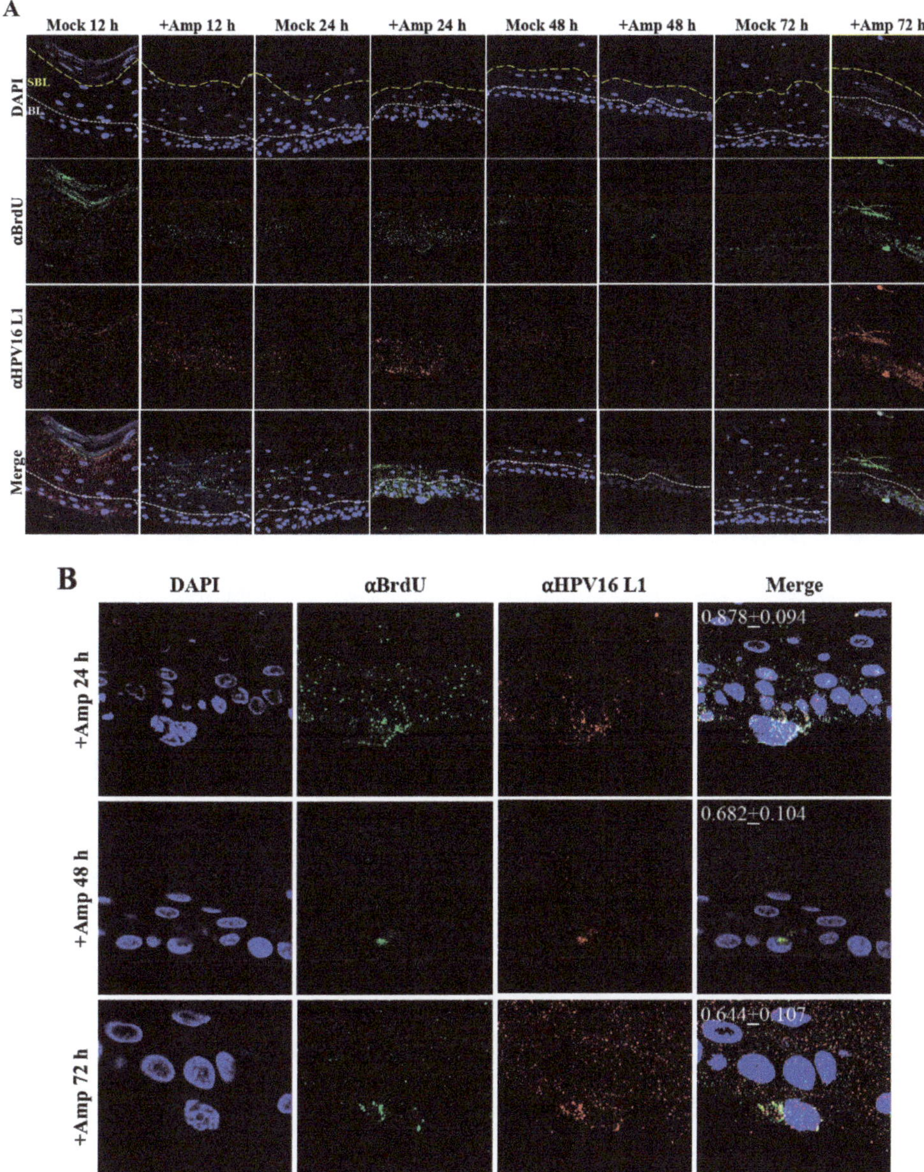

**Figure 14.** Time-course visualization of HPV16-BrdU transit through primary gingiva tissues. (**A**) Amprenavir (7.66 µg/mL) treated gingiva tissues (72 h) were infected with $5 \times 10^6$ HPV16-BrdU virions and tissues harvested and fixed periodically 12–72 h post-layering of virus on top of tissue. Immunofluorescence staining/confocal analysis of tissue sections staining the HPV16-BrdU genomes/L1 capsid complexes within cornified, suprabasal and basal layers. (**B**) 20× magnification of infected basal cells shown in (**A**). Pearson's coefficients illustrating co-localization of BrdU labeled genome and L1 capsid protein in basal layer cells are presented in the merged images. Data represent mean Pearson's coefficient ± SD, calculated from 3 independent images.

### 2.6. Primary Cervix Tissues Differentially Regulate Progeny HPV16 Biosynthesis When Treated with Amprenavir and Kaletra

We also compared PI treatment and HPV16 infection of primary cervical epithelium and measured impact on de novo *prog*-HPV16 biosynthesis. Organotypic cultures were derived from primary cervical keratinocytes isolated from patients undergoing hysterectomy. Amprenavir and Kaletra were added at the $C_{max}$ dosage (7.66 µg/mL and 9.8 µg/mL, respectively) to the culture media for 24–72 h, followed by infecting tissues using increasing doses of *P*-HPV16 (Figures 15A and 16A). Untreated tissues infected with the highest dose of HPV16 virions were used as controls. Expression of the E1^E4 spliced transcript in drug treated tissues was measured and relative levels compared to HPV16 infected, drug untreated controls. Similar to gingiva tissues, treatment with Amprenavir and Kaletra, rendered primary cervical tissue more favorable to HPV16 infection, compared to untreated tissues that were poorly infected (Figures 15A and S4A,B; Figures 16A and S5A,B). Again, relative E1^E4 expression levels were non-linear with regard to drug pre-treatment times. Observed fold-changes of E1^E4 expression varied relative to drug pre-treatment times across three independent experiments, once again suggesting a universal role for host genetics of target tissues. Virus infected Amprenavir and Kaletra treated tissues, respectively, were also inhibited using antibodies against HPV16 L1 and L2 capsid proteins (Figures 15B and S4C,D; Figures 16B and S5C,D), as would be expected. These results are significant as we show for the first time that three-dimensional tissues derived from primary cervical epithelial cells can be infected with a high-risk HPV in vitro, in the context of HIV-positive patients co-infected with HPVs undergoing HAART treatment.

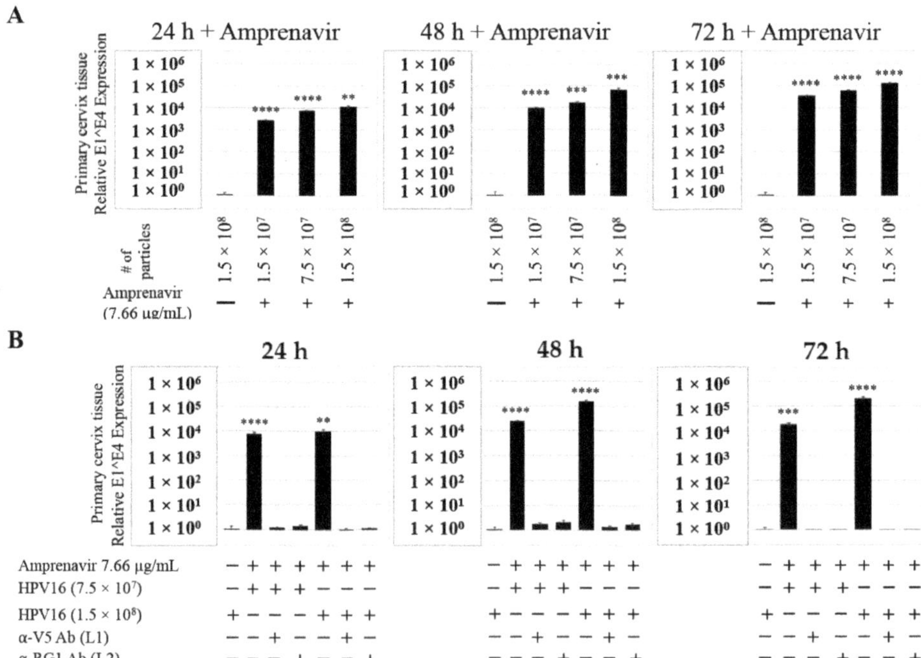

**Figure 15.** Amprenavir (7.66 µg/mL) treatment sensitizes primary cervical tissue to HPV16 infection. (**A**) Comparative expression of HPV16 E1^E4 transcripts in drug treated tissues compared with virus infected tissues not drug treated. (**B**) Inhibition of virus infection of tissues using HPV16 pre-incubated with α-V5 and α-RG1. Data were analyzed as mean ± SD. *p*-values were calculated using two-tailed Student's *t*-tests. Comparisons are indicated as $0.001 < p < 0.01$ by **; $0.0001 < p < 0.001$ by ***; and $p < 0.0001$ by ****.

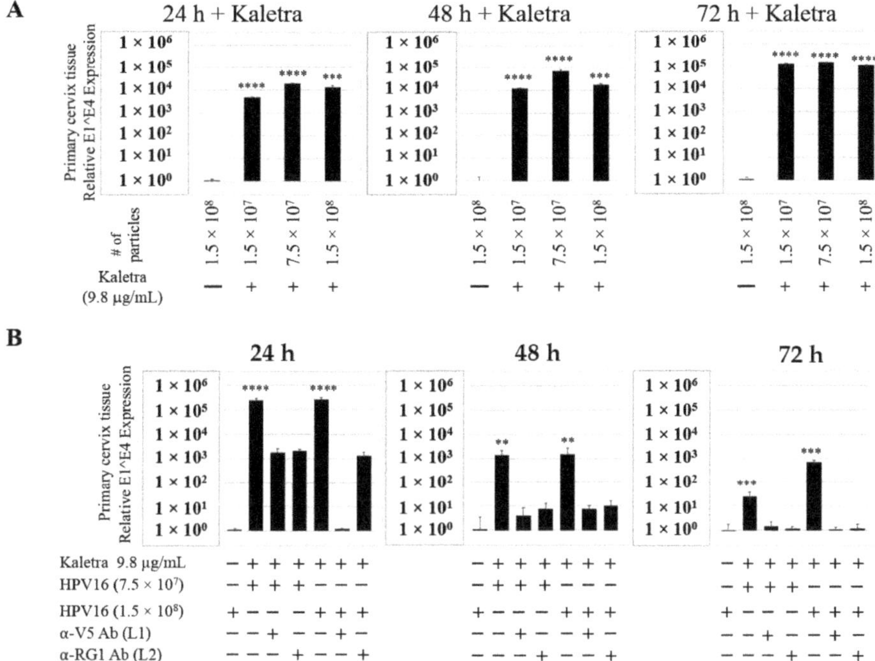

**Figure 16.** Kaletra (9.8 µg/mL) treatment sensitizes primary cervical tissue to HPV16 infection. (**A**) Comparative expression of HPV16 E1^E4 transcripts in Kaletra (9.8 µg/mL) treated tissues compared with virus infected tissues not drug treated. (**B**) Inhibition of HPV16 infection using α-V5 and α-RG1 of tissues treated with Kaletra. Data were analyzed as mean ± SD. *p*-values were calculated using two-tailed Student's *t*-tests. Comparisons are indicated as $0.001 < p < 0.01$ by **; $0.0001 < p < 0.001$ by ***; and $p < 0.0001$ by ****.

Further analysis showed that *P*-HPV16 infected tissues differentially correlated with changes in *prog*-HPV16 titers in the context of PI treatment utilized. Primary cervical tissues treated with Amprenavir (Figures 17A and S4E,F) induced virus titers that were comparable to those of infected gingiva tissues at all concentrations tested (Figures 6A and S1G,H). In contrast, treatment with the $C_{max}$ dose of Kaletra negatively regulated progeny virus biosynthesis in cervical tissues (Figures 18A and S5E,F) compared with gingiva tissues (Figures 13A and S3E,F). Inability to synthesize *prog*-HPV16 in presence of Kaletra treatment was not due to ability of the cervical tissues to be infected, as clear infection, as determined using E1^E4 expression, was observed at all time points post-infection (Figures 16A and S5A,B). In contrast, treating cervical tissues with low concentrations of Kaletra (6–3 µg/mL) resulted in production of low yet measurable *prog*-HPV16 titers in long-term cultures (Figures 18A and S5E,F). Infectivity of *prog*-HPV16 produced in Amprenavir (7.66 µg/mL) or Kaletra (3 µg/mL) treated cervical tissue was similar to that of *P*-HPV16, as was the ability to be inhibited with α-V5 and α-RG1 antibodies, thereby suggesting that biochemical nature of newly synthesized progeny virions was conserved regardless of whether oral or cervical tissue was examined, or specific PI and concentration used for treatments (Figures 17B and 18B). Progeny HPV16 biosynthesized in the presence of the two PIs was also confirmed using BrdU-labeling of genomes (Figure 11C,D). At 5 h post-infection in HaCaT monolayers, significant co-localization was also observed between BrdU labeled genomes and L1 capsids (Amprenavir 0.647 ± 0.050; Kaletra 0.782 ± 0.008), that was decreased at 6 h post-infection (Amprenavir 0.205 ± 0.098; Kaletra 0.208 ± 0.053), a further measure of capsid uncoating within infected cells. Additionally, α-V5/L1 capsid protein complexes correlate with inhibited infection (Figure 11C,D, right panels, compare L1 staining pattern with left panel, 6 h post-infection without α-V5; and Figures 17B and 18B). Cumulatively, these results suggest that both PIs act as co-factors to sensitize HPV infection of cervical tissues. Further, differences in progeny virus titers were noted with regards to Kaletra treatment in tissues isolated from different anatomic sites.

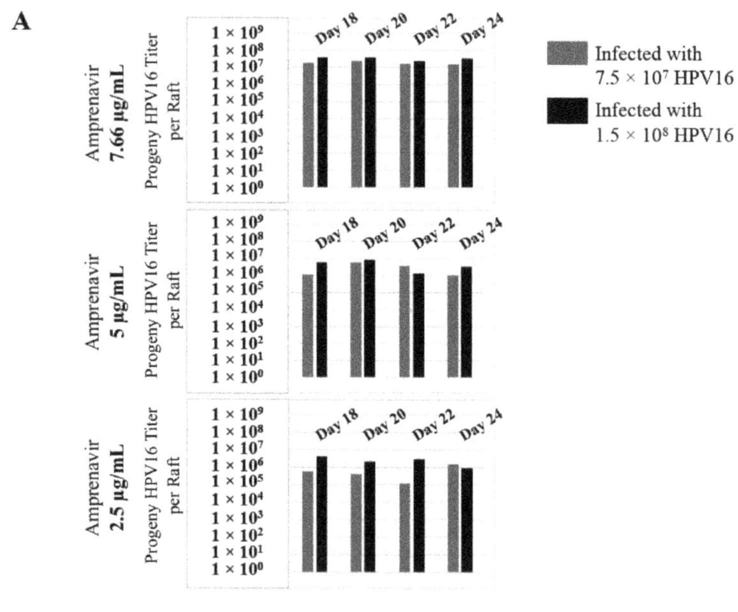


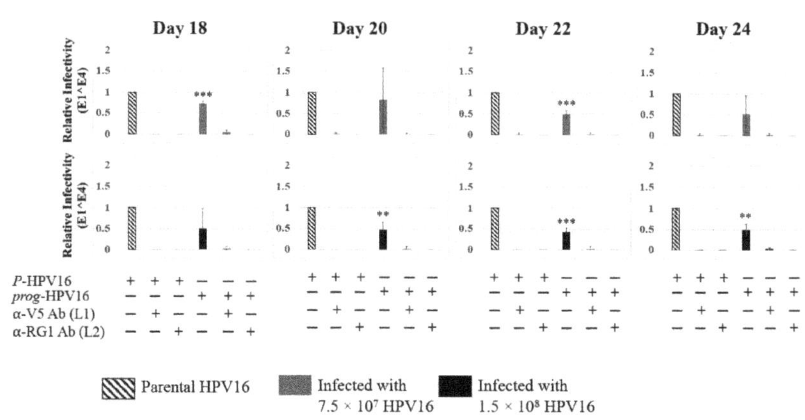


**Figure 17.** Extended culturing of Amprenavir treated HPV16 infected cervical tissues modulates progeny virus titers. (**A**) Raft tissues (day 18–24) infected with two virus doses modulate *prog*-HPV16 titers in an Amprenavir concentration dependent manner. Grey bars: infected with $7.5 \times 10^7$ *P*-HPV16 virions; Black bars: infected with $1.5 \times 10^8$ *P*-HPV16 virions. (**B**) Infectivity of *prog*-HPV16 Optiprep [F#7] compared with *P*-HPV16 [F#7] (1 MOI in HaCaT cells), and infection inhibition using α-V5 and α-RG1 monoclonal antibodies. Infection results shown are average of three experiments and is presented as mean ± SD. *p*-values were calculated using two-tailed Student's *t*-tests. Infectivity of *prog*-HPV16 were not significantly different compared with *P*-HPV16. Comparisons are indicated as $0.001 < p < 0.01$ by **; $0.0001 < p < 0.001$ by ***.

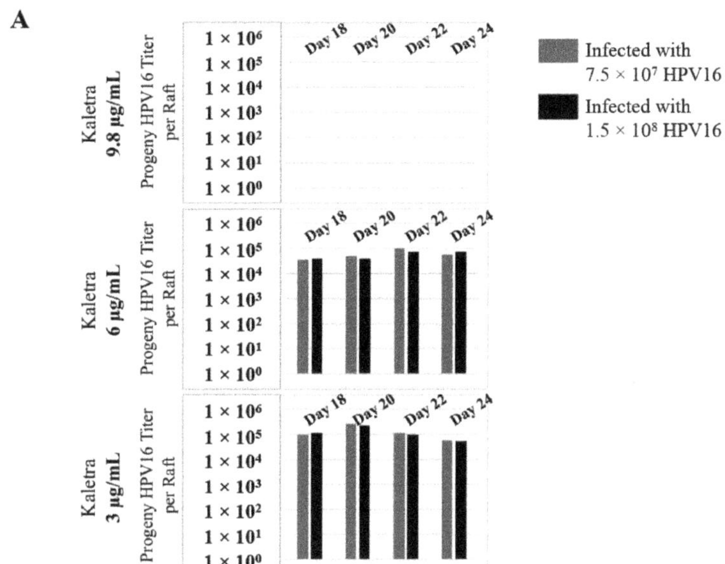

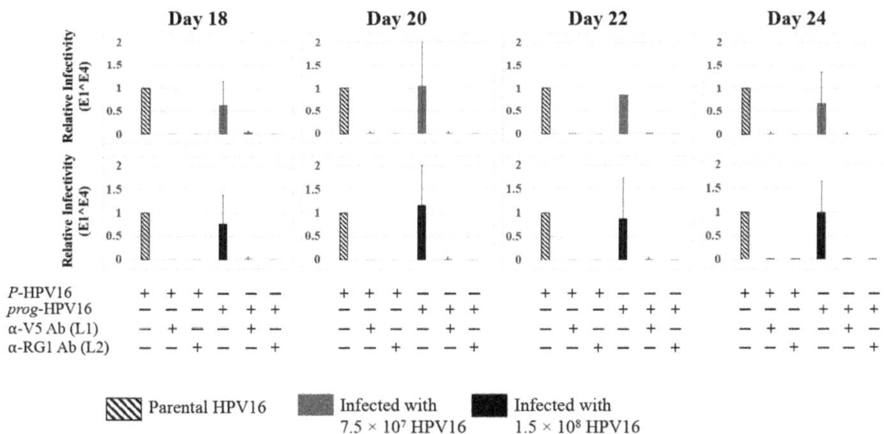

**Figure 18.** Extended culturing of Kaletra treated HPV16 infected cervical tissues modulates progeny virus titers. (**A**) Raft tissues (day 18–24) infected with two virus doses modulates *prog*-HPV16 titers in a Kaletra concentration dependent manner. Grey bars: infected with $7.5 \times 10^7$ *P*-HPV16 virions; Black bars: infected with $1.5 \times 10^8$ *P*-HPV16 virions. (**B**) Infectivity of concentrated virus stocks isolated from raft tissues treated with Kaletra (3 µg/mL) compared with *P*-HPV16 (1 MOI in HaCaT cells), and infection inhibition using α-V5 and α-RG1 monoclonal antibodies. Infection results shown are average of three experiments and is presented as mean ± SD. *p*-values were calculated using two-tailed Student's *t*-tests. Infectivity of *prog*-HPV16 were not significantly different compared with *P*-HPV16.

## 2.7. Amprenavir Treatment of Primary Cervix Tissue Allows for Virus Transit through Layers for Infecting Basal Cells

Similar to primary gingiva tissues, cervical tissues not treated with PIs were poorly infected with HPV16 compared with tissues that were drug treated (Figures 15A and 16A). In order to visually correlate HPV16 infection with virus localization within different layers of the cervical epithelium, we used *P*-HPV16-BrdU to infect Amprenavir treated primary cervical tissues and performed immunofluorescent staining of virus particles in transit through the tissue over a period of 12–72 h post-infection (Figure 19A,B). At 12 h post-infection, HPV16-BrdU was mostly localized in the cornified and suprabasal layer in tissues not treated with Amprenavir, whereas HPV16-BrdU was localized within the suparabasal layers in drug-treated tissues. At increasing times post-layering of virus on top of tissues, in drug treated tissues, HPV16-BrdU was localized throughout the suprabasal layer as well as within the basal cells. In contrast, virus particles were impeded in the upper cornified layers of untreated tissues. Significant co-localization was determined between viral genome/capsid complexes manifesting as punctate spots in perinuclear regions in basal cells (Pearson's correlation coefficient for co-localization between 0.578 ± 0.063 to 0.613 ± 0.072) across 12–48 h, followed by a decrease to non-significant levels at 72 h (Pearson's correlation coefficient for co-localization 0.365 ± 0.0615) post-layering of virus on tissues. These kinetics suggest that at 72 h post-addition of virus, HPV16 in infected basal layer cells of cervical tissues have undergone significant disassembly. This observation is in contrast to gingiva tissues where significant co-localization of viral genome/capsid complexes were observed across all times analyzed including the 72 h time point (Figure 14A,B). These results also suggest that target epithelium from different anatomic sites may regulate kinetics of virus infection of basal cells and downstream establishment of genome replication. Taken together, our results suggest that Amprenavir de-regulation of cell-cell barrier integrity in cervical tissue could play a role in enhancing HPV16 transit through the epithelium to the basal layer for infection.

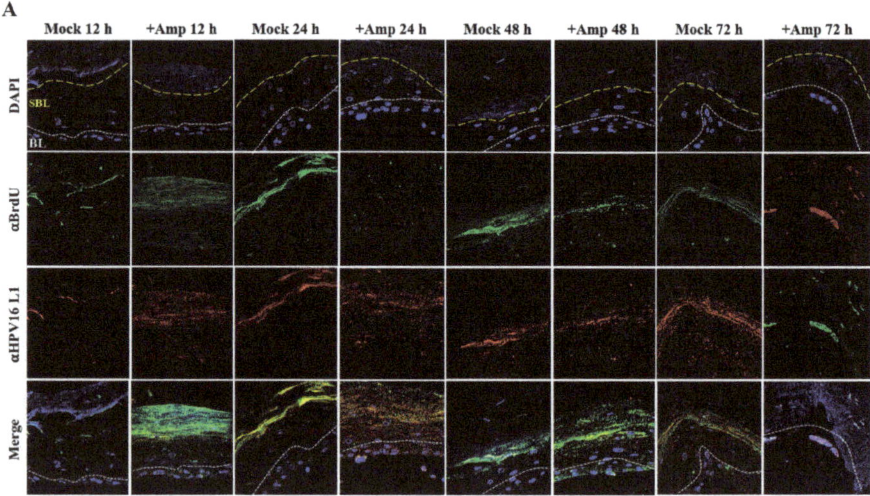



**Figure 19.** *Cont.*

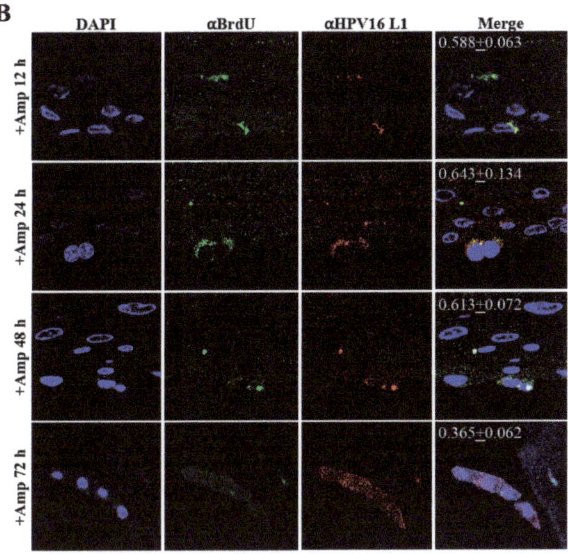

**Figure 19.** Time-course visualization of HPV16-BrdU transit through primary cervix. (**A**) Amprenavir (7.66 µg/mL) treated cervical tissues (72 h) were infected with $5 \times 10^6$ HPV16-BrdU virions and tissues harvested and fixed periodically 12–72 h post-layering of virus on top of tissue. Immunofluorescence staining/confocal analysis of tissue sections staining the HPV16-BrdU genomes/L1 capsid complexes within cornified, suprabasal and basal layers. (**B**) 20× magnification of infected basal cells in (**A**). Pearson's coefficients illustrating co-localization of BrdU labeled genome and L1 capsid protein in basal layer cells are presented in the merged images. Data represent mean Pearson's coefficient ± SD, calculated from 3 independent images.

## 2.8. Progeny HPV16 Can Be Serially Propagated in the Organotypic Epithelium Model

Use of three-dimensional cultures has revolutionized propagation of HPV in the laboratory for conducting detailed studies of the viral life cycle. First, this technique enabled the production of any high-risk HPV type for which the cloned viral genome is available [46–49]. Viral genomes are electroporated into isolated human mucosal or cutaneous keratinocyte of choice, resulting in chronically infected cell-lines that stably maintain viral episomes. Immortalized cell lines are then differentiated in organotypic cultures for producing infectious HPV stocks. Second, high-risk HPV positive cell lines are generated via acute infection of target epithelial cells that can be grown in three-dimensional cultures for virus production (Chatterjee and Meyers, manuscript in preparation). Third, in the current study, we show that PI treated stratified epithelium can be infected with HPV16 resulting in de novo biosynthesis of infectious progeny virus. Our techniques developed in the current study are important tools to study the mechanistic role of anti-retroviral drugs in promoting HPV infections in HAART-naïve primary epithelium. A related long-term goal is to also understand how changes in viral load may promote new infections of surrounding healthy tissue, thereby potentially creating HPV reservoirs that increase virus persistence, and increase the risk of oral and cervical cancer development in HIV-positive patients undergoing long-term HAART treatment. Therefore, we further asked whether Amprenavir-treated cervical epithelium could be used for serial propagation of virus, such that *prog*-HPV16 isolated from one epithelium is used to infect new epithelium. To account for variation in host genetics, organotypic tissues derived from primary cervical epithelial cells isolated from six different hosts, were treated with Amprenavir and infected with HPV16 as described in the scheme presented here (Figure 20A). Resultant progeny CV stocks were concentrated and titered, followed by infecting a new set of tissues derived from another set of six different hosts, that were further treated with Amprevanir in culture, followed by repeating the process twice more. Cumulatively, the end results depict serial infection of 24 individual host tissues. Moreover, variation in progeny viral titers derived in serial infection/passaging is indicative of differences in host genetics (Figure 20B). Our results show that HPV16 can be propagated in the infected epithelial model, albeit in the presence of PI. In comparison, raft tissues in the first set of infections not treated with Amprenavir displayed very low titers and were unable to be used for serial infection. This is significant in terms of defining one mechanism of long-term virus persistence in HIV infected patients undergoing HAART treatment, since *prog*-HPV16 produced in one area of infected epithelium has the potential to infect neighboring uninfected epithelium, that could eventually translate to creation of HPV reservoirs, virus persistence and cancer progression.

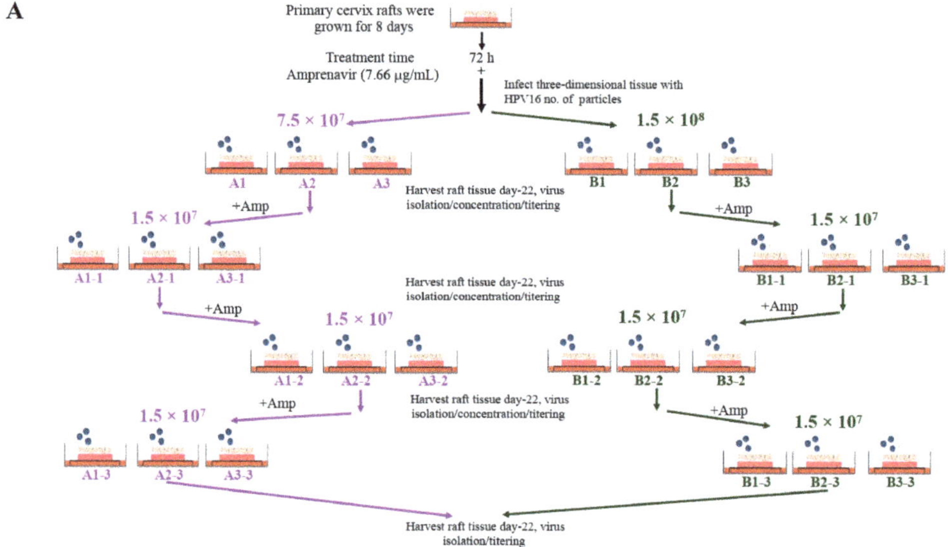


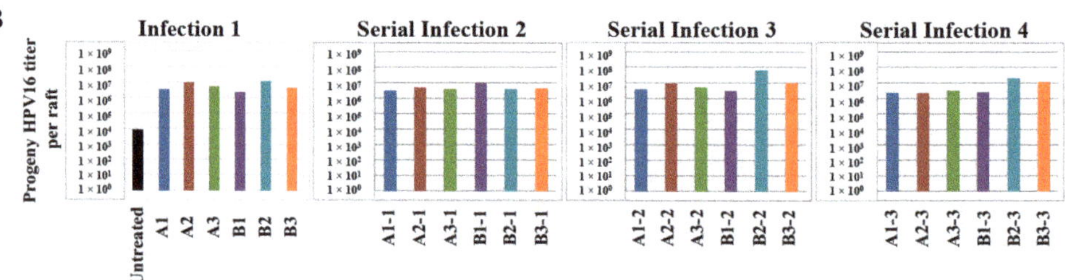


**Figure 20.** Serial propagation of progeny HPV16 in primary cervical tissues treated with Amprenavir. (**A**) Schematic for serial propagation of HPV16 in Amprenavir (7.66 µg/mL) treated primary cervical tissues. (**B**) *Prog*-HPV16 titers in serial infections representing 24 different host lines. Each infected raft tissue designated in (**A**) was titered and presented in graphs. Each bar is color coded to represent serial infections (Serial infections 3–4) from the same initial infection (Infection 1). Virus stocks were concentrated prior to infecting the next set of raft tissues.

## 3. Discussion

In the current study, we report for the first time that three-dimensional tissues derived from primary epithelial oral and cervical cells can be productively infected with authentic HPV16 in vitro.

Our ability to reproduce in vitro human epithelium capable of replicating the complete HPV life cycle provides an opportunity to investigate the effect of HAART/PI as potential co-factors that modulate infection and viral load, factors that determine HPV persistence and cancer of mucosal epithelium. With the advent of HAART, the prevalence of oropharyngeal cancer and persistence of oral warts has increased among HIV patients undergoing anti-retroviral therapy [24]. The underlying cause of increased opportunistic high-risk HPV16 oral infections in patients on HAART treatment is unknown. It is thought that prolonged use of HAART adversely affects turnover rate of the mucosa, which could affect acquisition and establishment of oral disease [24,25,50]. HAART usage is associated with development of adverse oral complications, resulting from oral and perioral manifestations due to oral ulcerations, epithelial hyperplasia and xerostomia [24,25], that damage the mucosal epithelium and potentially expose the underlying tissues to infections due to multiple microorganisms, including HPV infections [24–26]. Damage-induced inflammation in the oral epithelium decrease patient adherence to drug regimens [51], that ultimately correlate with suboptimal drug levels and development of drug resistance that could compromise future therapy. Patients on HAART may develop painful oral lesions that affect chewing and swallowing, further contributing to development of malnutrition and weight loss, and also adds to the increased morbidity [52]. The risk of HPV-associated oropharyngeal, cervical and anal cancer are higher among HIV-infected patients in the ART era compared to the general population [16]. The incidence of HPV-associated anal cancer is 80 times, and cervical cancer is 22 times higher in HIV-infected individuals compared to HIV-uninfected individuals [53]. Additionally, HIV-infected individuals have a six-fold greater risk for oropharyngeal and tonsillar cancers than do HIV-uninfected individuals [26]. In contrast, the relationship between the incidence of penile and vulvar cancers and ART exposure have not been reported. However, some studies do indicate HIV infection and immune-compromise as risk-factors for penile cancer, albeit without analyzing whether such patients were undergoing HAART therapy [54]. Another study looked at the relationship between vulvar and other gynecologic cancers in HIV infected women receiving ART, but limiting their analysis to only determining patient survival [55]. Since penile and vulvar cancers also pose significant public health problems in many parts of the developing world, the epidemiology of such neoplasms, in context of availability of HAART therapy, would be expected to vary among different populations. Overall, availability of HAART has extended life of HIV infected patients, but associated with increased incidence of NADC as the leading cause of morbidity and mortality [1,56]. Increasing age/longevity is the greatest risk factor for NADCs, but not sufficient to explain these trends in cancer epidemiology [5].

With ready access to HAART, survival of the population newly infected with HIV is only marginally shorter than that of the HIV-uninfected population [57–59]. Consequently, the number of people living with HIV/AIDS has increased four-fold [60,61]. Longer life expectancy afforded by HAART treatment is also associated with increased risk of NADCs compared to the general population [62–67]. In 2003, the total number of NADCs exceeded the number of ADCs among people with HIV/AIDS [3]. Cancer deaths were responsible for more than one third of all deaths in HIV-infected patients [17], of which NADCs accounted for 26% of deaths, representing head and neck cancers (8%) and anal cancer (8%), among other malignancies [68]. As a result, NADCs currently comprise the majority of the global burden of cancer in HIV-infected populations, and represent an important public health concern. Our studies have begun to provide a handle on molecular links to epidemiology data that first described the impact of long-term ART on HPV associated oral cancers. How these drugs regulate HPV infection of epithelia, viral load and subsequent risk for cancer progression remains to be determined. In particular, our future studies will reveal novel drivers and pathways related to anatomical site specific impact of PIs on the natural history of HPV in HIV+ patients undergoing HAART treatment.

HIV infection and associated immune suppression is linked with patient susceptibility to opportunistic infections [69]. Immunosuppression may play a role during the early stages of oral HPV carcinogenesis. HPV infections are self-limiting; however, virus persistence is increased in HIV-positive individuals due to immune-dysfunction and reduced HPV clearance, including individuals on long-term ART [70]. The higher prevalence of oral HPV infection among HIV-infected patients could be explained by an increased risk of incident infection due to immunosuppression rather than by reduced clearance [71–73]. Importantly, direct effects of HIV-1 transactivator protein tat and gp120 have been shown to modulate disruption of tight junctions in oral mucosal epithelium, that could facilitate HPV infection and reduce clearance [74,75], thereby suggesting a potential mechanism of HPV entry and infection of oral tissue. On the flip-side, one study suggested that HAART treatment could itself abrogate the barrier function of the oral epithelium, thereby increasing invasiveness, and thus malignancy of the HPV-infection [76]. In support of these observations, one study demonstrated that HIV infected patients with an undetectable HIV load had a six-fold risk of presenting HPV oral lesions [11]. Understanding the mechanisms of HPV infection via the paracellular route in HAART treated tissues would provide future opportunities to identify novel/alternate pathways of epithelial cell entry/infection, and key proteins utilized by HPV in this process, as well as delineate mechanisms of HPV persistence in HIV+ patients undergoing therapy. Clearance rates of oral HPV infections in HIV-positive patients are also determined by such factors as sexual behavior and immunosuppression that increase the risk of oral HPV infections [26]. HPV acquisition is increased by high-risk sexual behavior in populations considered at higher risk of acquiring HIV [74]. Alternatively, the high HPV detection rates could be due to increased HPV replication and/or persistence rather than increased HPV acquisition. If persistence of oral HPV leads to HPV-related disease, similar to the genital tract, then increased persistence of HPV could explain the increased prevalence or oral warts in HAART treated HIV+ individuals [15]. Therefore, treatment of HIV, rather than HIV immunosuppression, potentially plays a role in HPV infections in HIV infection [15].

Persistent HPV infections are a major risk to carcinogenic progression. Long intervals between initial infection and the development of cancer imply cofactors are involved. Co-factors that increase infectivity, viral load, and persistence all increase the risk of cancer. We propose that HAART/PI is a novel class of co-factors that regulates HPV infection, and subsequent viral load that could determine persistence and cancer risk. Our ongoing studies focus on mechanisms of HAART induced molecular changes that favor opportunistic HPV infections and changes in HPV load documented in HIV infected patients undergoing treatment (Alam et al., manuscript in preparation). Our studies utilizing the organotypic tissues provide a foundation for understanding mucosal wound healing and regulation of epithelial barrier integrity that promote HPV infection and provide future opportunities to study cellular mechanisms that control HPV infection of epithelial cells. Additionally, our studies also have the potential for future design of novel HIV therapeutics that could protect integrity of the epithelial cell-cell adhesion that minimize opportunistic HPV infections in HIV+ patients. Our in vitro culture system could be applicable as a screening platform for different classes of HAART drugs and their ability to sensitize mucosal epithelia to promote opportunistic HPV infections and subsequent viral load. Identification of infection pathways via the paracellular route could also be used to design therapeutics that minimize the risk of opportunistic HPV infections of target tissues in HIV+ patients undergoing HAART treatment.

## 4. Materials and Methods

### 4.1. Isolation of Gingival and Cervical Keratinocytes

Gingiva tissue was obtained from patients undergoing dental surgery [30]. Cervix tissue was obtained from patients undergoing hysterectomy. To maintain confidentiality, tissue samples were devoid of any identification, such as name, race and age. Approval to collect patient samples as "discarded tissues" was obtained from the Penn State University College of Medicine Institutional Review Board (IRB# 25284). Mixed pools of epithelial cells were isolated from tissues as previously described [30]. Briefly, the connective tissue and dermis were removed from the epithelium and discarded. The epithelial tissue was washed three times with phosphate-buffered saline (PBS) containing 50 µg/mL Gentamycin sulfate (Gibco BRL, Bethesda, MD, USA) and 2× Nystatin (Sigma

Chemical Co., St Louis, MO, USA). The epithelial tissue was then minced with scissors and trypsinized into a single-cell suspension using a spinner flask. The suspension was removed, 20 mL of E medium containing 5% fetal calf serum (FCS) was added and cells were pelleted using centrifugation. The supernatant was aspirated and the cell pellet was resuspended in 1 mL of 154 Medium (Cascade Biologics Inc., Portland, OR, USA) supplemented with the Human Keratinocyte Growth Supplement Kit (Cascade Biologics, Inc.) followed by adding to a 10-cm tissue culture plate containing an additional 7 mL of 154 medium. To the spinner flask, 20 mL of fresh trypsin was added to remaining tissue to obtain a second and third round of single-cell suspension. When the cultures became ~70% confluent, they were split 1:3. When cells of the first passage were 70% confluent, the cells were used for growing raft cultures.

### 4.2. Growth of Keratinocytes in Organotypic Cultures

Raft cultures were grown as previously described [30]. Briefly, mouse fibroblast 3T3 J2 were trypsinized and resuspended in 10% reconstitution buffer, 10% 10× DMEM (Dulbecco's Modified Eagle Medium) (Life Technologies, Gaithersburg, MD, USA), 2.4 µL/mL of 10M NaOH, and 80% collagen (Dickinson, Franklin Lakes, NJ, USA). Cells were added at a concentration of $2.5 \times 10^5$ cells/mL. The mixture when then aliquoted into 6 well plates at 2.5 mL per well and incubated at 37 °C for 2–4 h to allow solidification of the collages matrices. Two mL E-media was then added to each to well to allow the matrix to equilibrate. Human gingiva and cervical epithelial cells were trypsinized and resuspended at $2 \times 10^6$ cells/mL in E-media and 1 mL of cell suspension was added to each well of the 6 well plate on top of the collagen matrices. Epithelial cells were allowed to attach to the dermal equivalent for 2 h in the presence of 0.005 µg/mL EGF (Epidermal Growth Factor). After removal of the media, the collagen matrices were lifted onto stainless steel grids at the air-liquid interface. The raft cultures were fed by diffusion from below with E-media without EGF for 7 days. On day 8, the rafts were treated with Amprenavir and Kaletra using concentrations and treatment times as described herein. Control raft tissues were fed with E-media and 0.01% ethanol. Raft tissues were fed and treated every other day. Raft tissues were harvested at times as discussed herein.

### 4.3. Protease Inhibitors

Kaletra capsules (200 mg/50 mg) (Lopinavir/ritonavir, Abbott Laboratories) were purchased from the pharmacy at the Milton S Hershey Medical Center, Penn State University College of Medicine. One tablet was crushed into powder and stock solutions were prepared in 70% ethanol. Appropriate dilutions were prepared in E-Media to reach the correct final concentrations prior to feeding the cultures. Amprenavir powder was obtained through the NIH AIDS Reagent Program (Fisher Bioservices, MD, USA) as mentioned in the Acknowledgements sections.

### 4.4. HPV16 Infection of Primary Raft Tissues

Standard laboratory stocks of HPV16 were prepared as described below. Virus stocks were diluted in 200 µL E-Media without serum and gently added drop-wise on top of raft tissues. The beaded droplets were carefully coalesced with a pipette tip to form a uniform layer without disturbing the epithelium.

### 4.5. Transmission Electron Microscopy

Whole raft tissues were harvested without the attached collagen layer and were fixed with 2.5% glutaraldehyde and 2% paraformaldehyde in 0.1 M cacodylate buffer (pH 7.4), and further fixed in 1% osmium tetroxide in 0.1 M cacodylate buffer (pH 7.4) for 1 h. Samples were dehydrated in a graduated ethanol series: pure acetone embedded in LX-112 (Ladd Research, Williston, VT, USA). Thin sections (60 nM) were stained with uranyl acetate and lead citrate and viewed in a JEOL JEM1400 Transmission Electron Microscope (JEOL USA Inc., Peabody, MA, USA).

### 4.6. Production of HPV16 Laboratory Stocks in Organotypic Raft Cultures

Immortalized cervical keratinocytes stably maintaining HPV16 genomes were cultured with J2 3T3 feeder cells and maintained in E-medium and further used for growing the standard 20 day organotypic cultures. Immortalized human cervical keratinocytes persistently infected with HPV16 (cell line HCK16-8) were seeded ($1 \times 10^6$ cells) onto each collagen matrix consisting of rat-tail type 1 collagen and containing J2 3T3 feeder cells. Following cell attachment and growth to confluence, the matrices were lifted onto stainless steel grids and fed with E-medium supplemented with 10 µM 1,2-dioctanoyl-*sn*-glycerol (C8:O; Sigma Chemical Company, St. Louis, MO, USA) via diffusion from below, as previously described [41]. Raft cultures were allowed to stratify and differentiate for 20 d. Raft cultures were fed every other day, until harvesting tissues on day 20. For BrdU labeling of virions, BrdU (5-Bromo-2′-deoxyuridine) (cat# B5002 Sigma) was added to the media starting on day 8 of raft growth at a final concentration of 25 µM, and replenished during feeding every other day, until harvesting on day 20, as described in this manuscript. Virus stocks were further prepared and titers determined as described below.

### 4.7. HPV16 Isolation and Optiprep Purification of Virions

HPV infected raft tissues were harvested as described [41]. For preparing CV stocks, two rafts were Dounce homogenized in 500 µL of phosphate buffer (0.05 M sodium phosphate [pH 8.0], 2 mM $MgCl_2$). Homogenizers were rinsed with an additional 250 µL of phosphate buffer. Non-encapsidated viral genomes were digested by the addition of 1.5 µL (375 U) of benzonase to 750 µL of virus preps, followed by incubation at 37 °C for 1 h. Samples were adjusted to 1 M NaCl by adding 188 µL of ice-cold 5 M NaCl. Samples were further vortexed and centrifuged at 4 °C at 10,500 rpm for 10 min. The supernatants (CV stocks) were stored at −80 °C for further analysis. Optiprep purification: Optiprep purification of CV stocks was performed as previously described [41]. Briefly, Optiprep gradients were prepared by underlaying 27%, 23% and 39% Optiprep. Gradients were allowed to diffuse for 1 h at room temperature. Then 300 µL of clarified, benzonase-treated CV stock was layered on top of the gradient. Tubes were then centrifuged in a SW55 rotor (Beckman, Pasadena, CA, USA) at 234,000× *g* for 3.5 h at 18 °C. After centrifugation, 11–500 µL fractions were carefully collected, top to bottom, from each tube. Virus titers in fractions were determined as described below. Where specified, CV stocks were concentrated using Amicon® Ultra-4 Centrifugal Filters (30 K) (Merck Millipore, Burlington, MA, USA). Samples were centrifuged for 30 min at 3000 rpm, and stored at −80 °C for further analysis.

### 4.8. Titering HPV16 Virus Stocks

HPV16 titers were measured using qPCR-based DNA encapsidation assay as previously described [41]. To detect endonuclease-resistant genomes in CV stocks and Optiprep fractions the following method was used. Briefly, viral genomes were released from 10 µL benzonase-treated CV stock or 20 µL Optiprep fraction by re-suspension in 200 µL HIRT DNA extraction buffer (400 mM NaCl/10 mM Tris-HCl (pH 7.4)/10 mM EDTA (pH 8.0)), 2 µL 20 mg/mL proteinase K, and 10 µL 10% SDS, for 2–4 h at 37 °C. Following digestion, the DNA was extracted twice using phenol-choloroform-isoamyl alcohol (25:24:1), followed by extraction in an equal amount of chloroform. DNA was ethanol precipitated overnight at −20 °C. Samples were centrifuged, and the DNA pellet was washed with 70% ethanol and resuspended in 20 µL of Tris-EDTA overnight. To quantify viral genomes, a Thermo Scientific Maxima SYBR Green qPCR kit was utilized. Amplification of the HPV16 E2 open reading frame (ORF) was performed using 0.3 µM of forward primer HPV16E2-5′ and HPV16E2-3′ (Table S1). Amplification of the E2 ORF of serially diluted pBSHPV16 DNA, ranging from $10^8$–$10^4$ copies/µL, was used to generate a standard curve. A Bio-Rad iQ5 Multicolor Real-Time qPCR machine and software were utilized for PCR amplifications and subsequent data analysis.

*4.9. RT-qPCR Infectivity Assays in HaCaT Monolayer Cultures*

All infectivity studies were performed using HaCaT keratinocytes. HaCaT cells were seeded 50,000 cells/well in 24-well plates and infectivity assays were performed as previously described [41]. Briefly, cells were incubated with virus (CV stocks using MOI of 10, or Optiprep fraction #7 using MOI of 1) samples in cell culture medium for 48 h at 37 °C in 5% CO2 followed by mRNA harvesting using the RNeasy kit (Qiagen, Hilden, Germany). Infections were analyzed using a RT-qPCR based assay detecting levels of the E1^E4 splice transcript (QuantiTect Probe RT-PCR Kit). The HPV16 E1^E4 transcript was detected using 4 μM of the forward primer HPV16E1^E4-5′ and reverse primer HPV16E1^E4-3′ and using 0.2 μM of HPV16 E1^E4 fluorogenic probe (Table S1). The TATA-binding protein (TBP) amplicons were detected using 0.125 μM primers TBP-5′ and TBP-3′, and 0.2 μM of fluorogenic probe (Table S1). For each sample, the E1^E4 transcript abundance was normalized to TBP using infection of standard HPV16 laboratory stocks (either CV stock or Optiprep fraction #7) as controls, arbitrary designated as 1, using MOIs as described herein.

*4.10. Virus Neutralization and Infectivity Assays in HaCaT Monolayer Cultures*

For neutralization assays, virus samples were co-incubated with antibodies in 500 μL culture medium for 1 h at 37 °C prior to infecting HaCat monolayer cultures followed by RNA harvesting as described above. For these experiments, conformation-dependent anti-L1 mouse antibody H16.V5 (1:1000 dilution; a kind gift from Neil Christensen, Penn State College of Medicine) or the anti-L2 mouse antibody RG-1 (1:500 dilution; a kind gift from Richard Roden, John Hopkins) were used. RNA samples were harvested followed by the infectivity assay as described above [41].

*4.11. RNA Isolation from Raft Tissue Samples and RT-qPCR to Determine E1^E4 Transcript Expression in Infectivity Assays*

DNase I treated total RNA was isolated from 30 mg raft tissue using the RNeasy Fibrous Tissue Mini Kit (Qiagen cat. No. 74704), as per instructions provided by the manufacturer. Infections were analyzed using RT-qPCR based assay detecting levels of the E1^E4 splice transcript and TBP amplicons as described above. Untreated tissues infected with the highest dose of HPV16 virions were used as controls. Expression abundance of the E1^E4 spliced transcript in HPV16 infected/drug treated tissues was normalized to TBP and relative levels compared to HPV16 infected/non-drug treated tissues as controls (arbitrary designated as 1).

*4.12. Immunofluorescence Analysis of BrdU-Labeled HPV16 Virions Infecting HaCat Monolayer Cultures*

HaCaT monolayer cultures were plated on glass coverslips and incubated 10–12 h when cells reached 70% confluency. Cells were infected with HPV16-BrdU or *prog*-HPV16-BrdU Optiprep gradient [F#7] stocks using an MOI of 1. Infected cells were incubated for times as described herein. Post-infection, cells were fixed in 4% *w/v* phosphate buffered-paraformaldehyde (pH 7.4) for 10 min, washed three times in PBS, and permeabilized with 0.2% Triton X-100 in PBS for 10 min, followed by washing three times in PBS. Coverslips were blocked with 5% goat serum/1% bovine serum albumin in PBS for 20 min followed by co-incubation with anti-BrdU anti-mouse IgG1 (IIB5, Abcam) (1:1000) and anti-L1 anti-rabbit antibody (1:1000) (a kind gift from Dr. Neil Christensen, Penn State College of Medicine) for 60 min. Coverslips were washed three times with PBS and further co-incubated with secondary antibodies Alexa Fluor 488-conjugated anti-mouse IgG (2 μg/mL; Invitrogen, Carlsbad, CA, USA) (1:1000) and Alexa Fluor 568-conjugated anti-rabbit IgG (1 μg/mL; Invitrogen) (1:1000) in the absence of serum. Coverslips were washed three times with PBS and nuclei were stained with Hoechst (1:5000) for 10 min. Cells were washed twice with PBS, and mounted in ProLong Diamond (Invitrogen). Images were taken on a C2+ confocal microscope system (Nikon, Melville, NY, USA). Images were processed using NIS Elements software. Images also show volume renderings of z-stacks. Pearson's coefficients were determined and statistical analysis was performed on 5 separate images.

*4.13. Immunofluorescence Analysis of Raft Tissue Sections Post-Layering with HPV16-BrdU*

Raft cultures were treated with Amprenavir and followed by layering of HPV16-BrdU on top of tissues for 12, 24, 48 and 72 h, as indicated followed by harvesting. Tissues were fixed in 10% buffered formalin, embedded in paraffin and 4 µM sections were prepared. For immunofluorescence staining, the slides were submerged in xylene for de-paraffinization and then rehydrated. Antigen retrieval was achieved by submerging the slides in Tris-EDTA buffer (pH 9) in a 90 °C water bath for 10 min. The slides were then rinsed with Tris-buffered saline (TBS)-Tween and blocked with Background Sniper blocking reagent (Biocare Medical, Pacheco, CA, USA). The slides were then stained with the primary antibodies (anti-BrdU and anti-HPV16 L1, 1:1000 dilution each antibody) overnight at 4 °C. The slides were then rinsed with TBS-Tween 3 times, 10 min each, and stained with secondary antibody (Alexa Fluor 488 and Alexa Fluor 568; (Life Technologies, Carlsbad, CA, USA) diluted 1:1000 for 1 h at room temperature. The slides were then rinsed once with (TBS)-Tween, followed by staining with Hoechst nuclear stain (1:5000) dilution for 5 min, and then rinsed twice with TBS-Tween. All primary and secondary antibodies were diluted in Da Vinci Green diluent (Biocare Medical). Slides were washed with PBS, and mounted in ProLong Diamond (Invitrogen). Images were taken on a C2+ confocal microscope system (Nikon). Images were processed using NIS Elements software (Nikon, Melville, NY, USA).

*4.14. Statistical Analysis*

Data were analyzed using Prism 8.0 by Graphpad (La Jolla, CA, USA). Quantitative data are presented as mean ± standard deviation. Significance was based on pairwise Student's *t*-test. Comparisons with $p > 0.05$ are indicated by NS (not significant); $0.01 < p < 0.05$ by *; $0.001 < p < 0.01$ by **; $0.0001 < p < 0.001$ by ***; and $p < 0.0001$ by ****.

## 5. Conclusions

The risk of HPV-associated oropharyngeal, cervical and anal cancers are higher among HIV-infected patients in the ART era compared to the general population [16]. Generally, HPV infections are self-limiting; however, persistent HPV infection is a major risk to carcinogenic progression. Prolonged use of HAART adversely affects turnover rate of the oral epithelium, leading to oral complications that could affect acquisition and establishment of HPV infection and oral disease, and the same mechanism could affect turnover of the cervical epithelium. Our results presented in this manuscript indicate that HAART treatment creates favorable cellular conditions for opportunistic HPV infections in target epithelium. The organotypic raft tissues can physiologically model carcinogenic stages from precancerous to cancer [77–79]. Using this system, our ability to reproduce in vitro human epithelium capable of replicating the complete HPV life cycle is an opportunity to investigate how HAART manipulates normal cellular mechanisms and signaling pathways to promote HPV16 infection and de novo virus biosynthesis, which is an important milestone in driving virus persistence. We propose that HAART is a potential co-factor that modulates HPV infection and subsequent changes in viral load that could determine viral persistence and cancer of the oral cavity and cervix, and potentially the anal canal. Our future studies are geared towards mapping the molecular interaction of HAART with the drug-naïve primary epithelium, and how this interaction affects downstream cellular targets that regulate HPV infection, subsequent viral load and cancer progression.

**Supplementary Materials:** The following are available online at http://www.mdpi.com/2072-6694/12/9/2664/s1, Figure S1: Amprenavir (7.66 µg/mL) Treatment Sensitizes Primary Gingiva Tissue to HPV16 Infection, Figure S2: Control staining for confocal imaging, Figure S3: Kaletra (9.8 µg/mL) Treatment Sensitizes Primary Gingiva Tissue to HPV16 Infection, Figure S4: Amprenavir (7.66 µg/mL) Treatment Sensitizes Primary Cervical Tissue to HPV16 Infection, Figure S5: Kaletra (9.8 µg/mL) Treatment Sensitizes Primary Cervical Tissue to HPV16 Infection, Table S1: Primer and Probe Sequences.

**Author Contributions:** C.M. and S.A. conceptualized and designed the research plan; S.A. performed all experiments with assistance from S.C., S.D.K., J.M., and J.B.; H.C. prepared tissue for TEM and data processing;

S.A. and C.M. wrote the manuscript and all authors edited it. All authors have read and agreed to the published version of the manuscript.

**Funding:** Funding for this study was supported by grants R01CA225268 and U01CA179724 to CM.

**Acknowledgments:** The authors thank Neil Christensen and Sarah Brendle for providing the HPV16L1 mouse anti-V5 and rabbit anti-HPV16L1 antibodies, and Richard Roden for providing the HPV16 L2 anti-RG-1 antibody. The authors thank Nick Buchkovich and Nick Streck for their advice with confocal microscopy imaging. The authors also thank Debra Shearer for histological expertise. The following reagent was obtained through the AIDS Reagent Program, Division of AIDS, NIAID, NIH: Amprenavir from NIAID, DAIDS (cat# 8148).

**Conflicts of Interest:** The authors declare no conflict of interest.

## References

1. Shiels, M.S.; Pfeiffer, R.M.; Gail, M.H.; Hall, H.I.; Li, J.; Chaturvedi, A.K.; Bhatia, K.; Uldrick, T.S.; Yarchoan, R.; Goedert, J.J.; et al. Cancer Burden in the HIV-Infected Population in the United States. *J. Natl. Cancer Inst.* **2011**, *103*, 753–762. [CrossRef] [PubMed]

2. Cohen, M.S.; Chen, Y.Q.; McCauley, M.; Gamble, T.; Hosseinipour, M.C.; Kumarasamy, N.; Hakim, J.G.; Kumwenda, J.; Grinsztejn, B.; Pilotto, J.H.; et al. Prevention of HIV-1 Infection with Early Antiretroviral Therapy. *N. Engl. J. Med.* **2011**, *365*, 493–505. [CrossRef]

3. Shiels, M.S.; Engels, E.A. Evolving epidemiology of HIV-associated malignancies. *Curr. Opin. HIV AIDS* **2017**, *12*, 6–11. [CrossRef] [PubMed]

4. Hernández-Ramírez, R.U.; Shiels, M.S.; Dubrow, R.; Engels, E.A. Cancer risk in HIV-infected people in the USA from 1996 to 2012: A population-based, registry-linkage study. *Lancet HIV* **2017**, *4*, e495–e504. [CrossRef]

5. Rubinstein, P.G.; Aboulafia, D.M.; Zloza, A. Malignancies in HIV/AIDS. *AIDS* **2014**, *28*, 453–465. [CrossRef] [PubMed]

6. Yanik, E.L.; Napravnik, S.; Cole, S.R.; Achenbach, C.J.; Gopal, S.; Olshan, A.; Dittmer, D.P.; Kitahata, M.M.; Mugavero, M.J.; Saag, M.; et al. Incidence and Timing of Cancer in HIV-Infected Individuals Following Initiation of Combination Antiretroviral Therapy. *Clin. Infect. Dis.* **2013**, *57*, 756–764. [CrossRef] [PubMed]

7. Patton, L.L.; McKaig, R.; Strauss, R.; Rogers, D.; Eron, J.J. Changing prevalence of oral manifestations of human immuno-deficiency virus in the era of protease inhibitor therapy. *Oral Surg. Oral Med. Oral Pathol. Oral Radiol. Endodontol.* **2000**, *89*, 299–304. [CrossRef]

8. Greenspan, D.; Canchola, A.J.; MacPhail, L.A.; Cheikh, B.; Greenspan, J.S. Effect of highly active antiretroviral therapy on frequency of oral warts. *Lancet* **2001**, *357*, 1411–1412. [CrossRef]

9. Greenwood, I.; Zakrzewska, J.; Robinson, P. Changes in the prevalence of HIV-associated mucosal disease at a dedicated clinic over 7 years. *Oral Dis.* **2002**, *8*, 90–94. [CrossRef]

10. King, M.D.; Reznik, D.A.; O'Daniels, C.M.; Larsen, N.M.; Osterholt, D.; Blumberg, H.M. Human Papillomavirus-Associated Oral Warts among Human Immunodeficiency Virus-Seropositive Patients in the Era of Highly Active Antiretroviral Therapy: An Emerging Infection. *Clin. Infect. Dis.* **2002**, *34*, 641–648. [CrossRef]

11. Anaya-Saavedra, G.; Flores-Moreno, B.; García-Carrancá, A.; Irigoyen-Camacho, E.; Guido-Jiménez, M.; Ramírez-Amador, V. HPV oral lesions in HIV-infected patients: The impact of long-term HAART. *J. Oral Pathol. Med.* **2012**, *42*, 443–449. [CrossRef] [PubMed]

12. Labarga, P. Cancer in HIV patients. *AIDS Rev.* **2013**, *15*, 237–238. [PubMed]

13. Patton, L.L.; Ramirez-Amador, V.; Anaya-Saavedra, G.; Nittayananta, W.; Carrozzo, M.; Ranganathan, K. Urban legends series: Oral manifestations of HIV infection. *Oral Dis.* **2013**, *19*, 533–550. [CrossRef] [PubMed]

14. Wemmert, S.; Linxweiler, M.; Lerner, C.; Bochen, F.; Kulas, P.; Linxweiler, J.; Smola, S.; Urbschat, S.; Wagenpfeil, S.; Schick, B. Combinational chromosomal aneuploidies and HPV status for prediction of head and neck squamous cell carcinoma prognosis in biopsies and cytological preparations. *J. Cancer Res. Clin. Oncol.* **2018**, *144*, 1129–1141. [CrossRef]

15. Cameron, J.E.; Mercante, D.; O'brien, M.; Gaffga, A.M.; Leigh, J.E.; Fidel, P.L.; Hagensee, M.E. The impact of highly active antiretroviral therapy and immunodeficiency on human papillomavirus infection of the oral cavity of human immunodeficiency virus-seropositive adults. *Sex. Transm. Dis.* **2005**, *32*, 703–709. [CrossRef]

16. Shiboski, C.; Lee, A.; Chen, H.; Webster-Cyriaque, J.; Seaman, T.; Landovitz, R.J.; John, M.; Reilly, N.; Naini, L.; Palefsky, J.; et al. Human papillomavirus infection in the oral cavity of HIV patients is not reduced by initiating antiretroviral therapy. *AIDS* **2016**, *30*, 1573–1582. [CrossRef]

17. Wang, C.-C.J.; Sparano, J.A.; Palefsky, J. Human Immunodeficiency Virus/AIDS, Human Papillomavirus, and Anal Cancer. *Surg. Oncol. Clin. N. Am.* **2017**, *26*, 17–31. [CrossRef]

18. Oliver, N.T.; Chiao, E.Y. Malignancies in women with HIV infection. *Curr. Opin. HIV AIDS* **2017**, *12*, 69–76. [CrossRef]

19. Purgina, B.; Pantanowitz, L.; Seethala, R.R. A Review of Carcinomas Arising in the Head and Neck Region in HIV-Positive Patients. *Pathol. Res. Int.* **2011**, *2011*, 1–12. [CrossRef]

20. Preti, M.; Rotondo, J.C.; Holzinger, D.; Micheletti, L.; Gallio, N.; McKay-Chopin, S.; Carreira, C.; Privitera, S.S.; Watanabe, R.; Ridder, R.; et al. Role of human papillomavirus infection in the etiology of vulvar cancer in Italian women. *Infect. Agents Cancer* **2020**, *15*, 20–28. [CrossRef]

21. Moscicki, A.-B.; Palefsky, J.M. Human Papillomavirus in Men. *J. Low. Genit. Tract Dis.* **2011**, *15*, 231–234. [CrossRef] [PubMed]

22. Ammatuna, P.; Campisi, G.; Giovannelli, L.; Giambelluca, D.; Alaimo, C.; Mancuso, S.; Margiotta, V. Presence of Epstein-Barr virus, cytomegalovirus and human papillomavirus in normal oral mucosa of HIV-infected and renal transplant patients. *Oral Dis.* **2001**, *7*, 34–40. [PubMed]

23. D'Souza, G.; Fakhry, C.; Sugar, E.A.; Seaberg, E.C.; Weber, K.; Minkoff, H.L.; Anastos, K.; Palefsky, J.M.; Gillison, M.L. Six-month natural history of oral versus cervical human papillomavirus infection. *Int. J. Cancer* **2007**, *121*, 143–150. [CrossRef] [PubMed]

24. De Almeida, V.; Lima, I.; Ziegelmann, P.; Paranhos, L.R.; De Matos, F.R. Impact of highly active antiretroviral therapy on the prevalence of oral lesions in HIV-positive patients: A systematic review and meta-analysis. *Int. J. Oral Maxillofac. Surg.* **2017**, *46*, 1497–1504. [CrossRef]

25. Toljić, B.; Trbovich, A.M.; Petrović, S.M.; Kannosh, I.Y.; Dragović, G.; Jevtović, D.; De Luka, S.R.; Ristić-Djurović, J.L.; Milašin, J. Ageing with HIV—A periodontal perspective. *New Microbiol.* **2018**, *41*, 61–66.

26. Beachler, D.C.; Sugar, E.A.; Margolick, J.B.; Weber, K.M.; Strickler, H.D.; Wiley, D.; Cranston, R.D.; Burk, R.D.; Minkoff, H.; Reddy, S.; et al. Risk Factors for Acquisition and Clearance of Oral Human Papillomavirus Infection Among HIV-Infected and HIV-Uninfected Adults. *Am. J. Epidemiol.* **2014**, *181*, 40–53. [CrossRef]

27. Blitz, S.; Baxter, J.; Raboud, J.; Walmsley, S.L.; Rachlis, A.; Smaill, F.; Ferenczy, A.; Coutlée, F.; Hankins, C.; Money, D. Evaluation of HIV and Highly Active Antiretroviral Therapy on the Natural History of Human Papillomavirus Infection and Cervical Cytopathologic Findings in HIV-Positive and High-Risk HIV-Negative Women. *J. Infect. Dis.* **2013**, *208*, 454–462. [CrossRef]

28. Muñoz, N.; Bosch, F.X.; De Sanjose, S.; Herrero, R.; Castellsagué, X.; Shah, K.V.; Snijders, P.J.; Meijer, C.J.L.M. Epidemiologic Classification of Human Papillomavirus Types Associated with Cervical Cancer. *N. Engl. J. Med.* **2003**, *348*, 518–527. [CrossRef]

29. Wang, Y.; Lv, Z.; Chu, Y. HIV protease inhibitors: A review of molecular selectivity and toxicity. *HIV/AIDS Res. Palliat. Care* **2015**, *7*, 95–104. [CrossRef]

30. Israr, M.; Mitchell, D.; Alam, S.; Dinello, D.; Kishel, J.J.; Meyers, C. Effect of the HIV protease inhibitor amprenavir on the growth and differentiation of primary gingival epithelium. *Antivir. Ther.* **2010**, *15*, 253–265. [CrossRef]

31. Moody, C.A.; Laimins, L.A. Human papillomavirus oncoproteins: Pathways to transformation. *Nat. Rev. Cancer* **2010**, *10*, 550–560. [CrossRef] [PubMed]

32. Middleton, K.; Peh, W.; Southern, S.; Griffin, H.; Sotlar, K.; Nakahara, T.; El-Sherif, A.; Morris, L.; Seth, R.; Hibma, M.; et al. Organization of Human Papillomavirus Productive Cycle during Neoplastic Progression Provides a Basis for Selection of Diagnostic Markers. *J. Virol.* **2003**, *77*, 10186–10201. [CrossRef] [PubMed]

33. Doorbar, J.; Ely, S.; Sterling, J.; McLean, C.; Crawford, L. Specific interaction between HPV-16 E1–E4 and cytokeratins results in collapse of the epithelial cell intermediate filament network. *Nature* **1991**, *352*, 824–827. [CrossRef] [PubMed]

34. Raj, K.; Berguerand, S.; Southern, S.; Doorbar, J.; Beard, P. E1^E4 Protein of Human Papillomavirus Type 16 Associates with Mitochondria. *J. Virol.* **2004**, *78*, 7199–7207. [CrossRef] [PubMed]

35. Doorbar, J.; Elston, R.C.; Napthine, S.; Raj, K.; Medcalf, E.; Jackson, D.; Coleman, N.; Griffin, H.M.; Masterson, P.; Stacey, S.; et al. The E1^E4 protein of human papillomavirus type 16 associates with a putative rna helicase through sequences in its c terminus. *J. Virol.* **2000**, *74*, 10081–10095. [CrossRef]
36. Roberts, S.; Hillman, M.L.; Knight, G.L.; Gallimore, P.H. The ND10 Component Promyelocytic Leukemia Protein Relocates to Human Papillomavirus Type 1 E4 Intranuclear Inclusion Bodies in Cultured Keratinocytes and in Warts. *J. Virol.* **2003**, *77*, 673–684. [CrossRef]
37. Moody, C.A. Mechanisms by which HPV Induces a Replication Competent Environment in Differentiating Keratinocytes. *Viruses* **2017**, *9*, 261. [CrossRef]
38. Biryukov, J.; Myers, J.C.; McLaughlin-Drubin, M.E.; Griffin, H.M.; Milici, J.; Doorbar, J.; Meyers, C. Mutations in HPV18 E1^E4 Impact Virus Capsid Assembly, Infectivity Competence, and Maturation. *Viruses* **2017**, *9*, 385. [CrossRef]
39. Conway, M.J.; Cruz, L.; Alam, S.; Christensen, N.D.; Meyers, C. Differentiation-Dependent Interpentameric Disulfide Bond Stabilizes Native Human Papillomavirus Type 16. *PLoS ONE* **2011**, *6*, e22427. [CrossRef]
40. Israr, M.; Biryukov, J.; Ryndock, E.J.; Alam, S.; Meyers, C. Comparison of human papillomavirus type 16 replication in tonsil and foreskin epithelia. *Virology* **2016**, *499*, 82–90. [CrossRef]
41. Conway, M.J.; Alam, S.; Ryndock, E.J.; Cruz, L.; Christensen, N.D.; Roden, R.B.S.; Meyers, C. Tissue-Spanning Redox Gradient-Dependent Assembly of Native Human Papillomavirus Type 16 Virions. *J. Virol.* **2009**, *83*, 10515–10526. [CrossRef] [PubMed]
42. Ozbun, M.A.; Meyers, C. Characterization of late gene transcripts expressed during vegetative replication of human papillomavirus type 31b. *J. Virol.* **1997**, *71*, 5161–5172. [CrossRef] [PubMed]
43. Lee, C.A.; Mistry, D.; Sharma, R.; Coatesworth, A.P. Rhinological, laryngological, oropharyngeal and other head and neck side effects of drugs. *J. Laryngol. Otol.* **2006**, *120*, 1–10. [CrossRef] [PubMed]
44. Borrás-Blasco, J.; Navarro-Ruiz, A.; Borrás, C.; Casterá, E. Adverse cutaneous reactions associated with the newest antiretroviral drugs in patients with human immunodeficiency virus infection. *J. Antimicrob. Chemother.* **2008**, *62*, 879–888. [CrossRef]
45. Israr, M.; Mitchell, D.; Alam, S.; Dinello, D.; Kishel, J.J.; Meyers, C. The HIV protease inhibitor lopinavir/ritonavir (Kaletra) alters the growth, differentiation and proliferation of primary gingival epithelium. *HIV Med.* **2010**, *12*, 145–156. [CrossRef]
46. Meyers, C.; Frattini, M.; Hudson, J.; Laimins, L. Biosynthesis of human papillomavirus from a continuous cell line upon epithelial differentiation. *Science* **1992**, *257*, 971–973. [CrossRef]
47. Meyers, C.; Mayer, T.J.; Ozbun, M.A. Synthesis of infectious human papillomavirus type 18 in differentiating epithelium transfected with viral DNA. *J. Virol.* **1997**, *71*, 7381–7386. [CrossRef]
48. E McLaughlin-Drubin, M.; Christensen, N.D.; Meyers, C. Propagation, infection, and neutralization of authentic HPV16 virus. *Virology* **2004**, *322*, 213–219. [CrossRef]
49. E McLaughlin-Drubin, M.; Wilson, S.; Mullikin, B.; Suzich, J.; Meyers, C. Human papillomavirus type 45 propagation, infection, and neutralization. *Virology* **2003**, *312*, 1–7. [CrossRef]
50. Beachler, D.C.; Abraham, A.G.; Silverberg, M.J.; Jing, Y.; Fakhry, C.; Gill, M.J.; Dubrow, R.; Kitahata, M.M.; Klein, M.B.; Burchell, A.N.; et al. Incidence and risk factors of HPV-related and HPV-unrelated Head and Neck Squamous Cell Carcinoma in HIV-infected individuals. *Oral Oncol.* **2014**, *50*, 1169–1176. [CrossRef]
51. Li, H.; Marley, G.; Ma, W.; Wei, C.; Lackey, M.; Ma, Q.; Renaud, F.; Vitoria, M.; Beanland, R.; Doherty, M.; et al. The Role of ARV Associated Adverse Drug Reactions in Influencing Adherence Among HIV-Infected Individuals: A Systematic Review and Qualitative Meta-Synthesis. *AIDS Behav.* **2016**, *21*, 341–351. [CrossRef] [PubMed]
52. Perera, R.A.P.M.; Tsang, C.S.P.; Samaranayake, L.; Lee, M.P.; Li, P. Prevalence of oral mucosal lesions in adults undergoing highly active antiretroviral therapy in Hong Kong. *J. Investig. Clin. Dent.* **2012**, *3*, 208–214. [CrossRef] [PubMed]
53. Brickman, C.; Palefsky, J. Cancer in the HIV-Infected Host: Epidemiology and Pathogenesis in the Antiretroviral Era. *Curr. HIV/AIDS Rep.* **2015**, *12*, 388–396. [CrossRef] [PubMed]
54. Douglawi, A.; Masterson, T.A. Updates on the epidemiology and risk factors for penile cancer. *Transl. Androl. Urol.* **2017**, *6*, 785–790. [CrossRef] [PubMed]
55. Levinson, K.L.; Riedel, D.J.; Ojalvo, L.S.; Chan, W.; Angarita, A.M.; Fader, A.N.; Rositch, A.F. Gynecologic cancer in HIV-infected women: Treatment and outcomes in a multi-institutional cohort. *AIDS* **2018**, *32*, 171–177. [CrossRef]

56. Pettit, A.C.; Giganti, M.J.; Ingle, S.M.; May, M.T.; Shepherd, B.E.; Gill, M.J.; Fätkenheuer, G.; Abgrall, S.; Saag, M.S.; Del Amo, J.; et al. Increased non-AIDS mortality among persons with AIDS-defining events after antiretroviral therapy initiation. *J. Int. AIDS Soc.* **2018**, *21*, e25031. [CrossRef]

57. Lima, V.D.; Hogg, R.S.; Harrigan, P.R.; Moore, D.; Yip, B.; Wood, E.; Montaner, J.S. Continued improvement in survival among HIV-infected individuals with newer forms of highly active antiretroviral therapy. *AIDS* **2007**, *21*, 685–692. [CrossRef]

58. Teeraananchai, S.; Kerr, S.J.; Amin, J.; Ruxrungtham, K.; Law, M. Life expectancy of HIV-positive people after starting combination antiretroviral therapy: A meta-analysis. *HIV Med.* **2016**, *18*, 256–266. [CrossRef]

59. Patterson, S.; Cescon, A.; Samji, H.; Chan, K.; Zhang, W.; Raboud, J.; Burchell, A.N.; Cooper, C.; Klein, M.B.; Rourke, S.B.; et al. Life expectancy of HIV-positive individuals on combination antiretroviral therapy in Canada. *BMC Infect. Dis.* **2015**, *15*, 274. [CrossRef]

60. Palella, F.J.; Delaney, K.M.; Moorman, A.; Loveless, M.O.; Fuhrer, J.; Satten, G.A.; Aschman, D.J.; Holmberg, S.D. Declining Morbidity and Mortality among Patients with Advanced Human Immunodeficiency Virus Infection. *N. Engl. J. Med.* **1998**, *338*, 853–860. [CrossRef]

61. Hall, H.I.; Song, R.; Rhodes, P.; Prejean, J.; An, Q.; Lee, L.M.; Karon, J.; Brookmeyer, R.; Kaplan, E.H.; McKenna, M.T.; et al. Estimation of HIV Incidence in the United States. *JAMA* **2008**, *300*, 520–529. [CrossRef] [PubMed]

62. Clifford, G.M.; Polesel, J.; Rickenbach, M.; Maso, L.D.; Keiser, O.; Kofler, A.; Rapiti, E.; Levi, F.; Jundt, G.; Fisch, T.; et al. Cancer Risk in the Swiss HIV Cohort Study: Associations With Immunodeficiency, Smoking, and Highly Active Antiretroviral Therapy. *J. Natl. Cancer Inst.* **2005**, *97*, 425–432. [CrossRef] [PubMed]

63. Herida, M.; Mary-Krause, M.; Kaphan, R.; Cadranel, J.; Poizot-Martin, I.; Rabaud, C.; Plaisance, N.; Tissot-Dupont, H.; Boué, F.; Lang, J.-M.; et al. Incidence of Non–AIDS-Defining Cancers Before and During the Highly Active Antiretroviral Therapy Era in a Cohort of Human Immunodeficiency Virus–Infected Patients. *J. Clin. Oncol.* **2003**, *21*, 3447–3453. [CrossRef]

64. Patel, P.; Hanson, D.L.; Sullivan, P.S.; Novak, R.M.; Moorman, A.; Tong, T.C.; Holmberg, S.D.; Brooks, J.T.; Adult and Adolescent Spectrum of Disease Project and HIV Outpatient Study Investigators. Incidence of Types of Cancer among HIV-Infected Persons Compared with the General Population in the United States, 1992–2003. *Ann. Intern. Med.* **2008**, *148*, 728–736. [CrossRef] [PubMed]

65. Powles, T.; Robinson, D.; Stebbing, J.; Shamash, J.; Nelson, M.; Gazzard, B.; Mandelia, S.; Møller, H.; Bower, M. Highly Active Antiretroviral Therapy and the Incidence of Non–AIDS-Defining Cancers in People With HIV Infection. *J. Clin. Oncol.* **2009**, *27*, 884–890. [CrossRef] [PubMed]

66. Hoffmann, C.; Horst, H.-A.; Weichenthal, M.; Hauschild, A. Malignant Melanoma and HIV Infection—Aggressive Course despite Immune Reconstitution. *Oncol. Res. Treat.* **2004**, *28*, 35–37. [CrossRef]

67. Lanoy, E.; Dores, G.M.; Madeleine, M.M.; Toro, J.R.; Fraumeni, J.F.; Engels, E.A. Epidemiology of nonkeratinocytic skin cancers among persons with AIDS in the United States. *AIDS* **2009**, *23*, 385–393. [CrossRef]

68. Morlat, P.; Roussillon, C.; Henard, S.; Salmon, D.; Bonnet, F.; Cacoub, P.; Georget, A.; Aouba, A.; Rosenthal, E.; May, T.; et al. Causes of death among HIV-infected patients in France in 2010 (national survey). *AIDS* **2014**, *28*, 1181–1191. [CrossRef]

69. Phillips, A.N.; Lundgren, J. The CD4 lymphocyte count and risk of clinical progression. *Curr. Opin. HIV AIDS* **2006**, *1*, 43–49. [CrossRef]

70. Speicher, D.J.; Ramirez-Amador, V.; Dittmer, D.P.; Webster-Cyriaque, J.; Goodman, M.T.; Moscicki, A.-B. Viral infections associated with oral cancers and diseases in the context of HIV: A workshop report. *Oral Dis.* **2016**, *22*, 181–192. [CrossRef]

71. Beachler, D.C.; Weber, K.M.; Margolick, J.B.; Strickler, H.D.; Cranston, R.D.; Burk, R.D.; Wiley, D.; Minkoff, H.; Reddy, S.; Stammer, E.E.; et al. Risk factors for oral HPV infection among a high prevalence population of HIV-positive and at-risk HIV-negative adults. *Cancer Epidemiol. Biomark. Prev.* **2011**, *21*, 122–133. [CrossRef]

72. Mooij, S.H.; Boot, H.J.; Speksnijder, A.G.C.L.; Stolte, I.G.; Meijer, C.J.L.M.; Snijders, P.J.; Verhagen, D.W.M.; King, A.J.; De Vries, H.J.C.; Quint, W.; et al. Oral human papillomavirus infection in HIV-negative and HIV-infected MSM. *AIDS* **2013**, *27*, 2117–2128. [CrossRef] [PubMed]

73. Kreimer, A.R.; Alberg, A.J.; Daniel, R.; Gravitt, P.E.; Viscidi, R.; Garrett-Mayer, E.; Shah, K.V.; Gillison, M.L. Oral Human Papillomavirus Infection in Adults Is Associated with Sexual Behavior and HIV Serostatus. *J. Infect. Dis.* **2004**, *189*, 686–698. [CrossRef] [PubMed]

74. Tugizov, S.M.; Herrera, R.; Chin-Hong, P.; Veluppillai, P.; Greenspan, D.; Berry, J.M.; Pilcher, C.D.; Shiboski, C.H.; Jay, N.; Rubin, M.; et al. HIV-associated disruption of mucosal epithelium facilitates paracellular penetration by human papillomavirus. *Virology* **2013**, *446*, 378–388. [CrossRef] [PubMed]
75. Kim, R.H.; Yochim, J.M.; Kang, M.; Shin, K.-H.; Christensen, R.; Park, N.-H. HIV-1 Tat enhances replicative potential of human oral keratinocytes harboring HPV-16 genome. *Int. J. Oncol.* **2008**, *33*, 777–782. [PubMed]
76. Tugizov, S.M.; Webster-Cyriaque, J.; Syrianen, S.; Chattopadyay, A.; Sroussi, H.; Zhang, L.; Kaushal, A. Mechanisms of Viral Infections Associated with HIV. *Adv. Dent. Res.* **2011**, *23*, 130–136. [CrossRef]
77. McCance, D.J.; Kopan, R.; Fuchs, E.; Laimins, L.A. Human papillomavirus type 16 alters human epithelial cell differentiation in vitro. *Proc. Natl. Acad. Sci. USA* **1988**, *85*, 7169–7173. [CrossRef]
78. Ozbun, M.A.; Meyers, C. Transforming growth factor beta1 induces differentiation in human papillomavirus-positive keratinocytes. *J. Virol.* **1996**, *70*, 5437–5446. [CrossRef]
79. Rader, J.S.; Golub, T.R.; Hudson, J.B.; Patel, D.; Bedell, M.A.; Laimins, L.A. In vitro differentiation of epithelial cells from cervical neoplasias resembles in vivo lesions. *Oncogene* **1990**, *5*, 571–576.

 *cancers*

*Article*

# HPV-Induced Oropharyngeal Cancer and the Role of the E7 Oncoprotein Detection via Brush Test

Wegene Borena [1], Volker H. Schartinger [2], Jozsef Dudas [2], Julia Ingruber [2], Maria C. Greier [2], Teresa B. Steinbichler [2], Johannes Laimer [3], Heribert Stoiber [1], Herbert Riechelmann [2] and Barbara Kofler [2,*]

[1] Institute of Virology, Department of Hygiene, Microbiology, Social Medicine,
Medical University of Innsbruck, Peter-Mayr-Strasse 4b, 6020 Innsbruck, Austria;
Wegene.Borena@i-med.ac.at (W.B.); heribert.stoiber@i-med.ac.at (H.S.)
[2] Department of Otorhinolaryngology, Medical University of Innsbruck, Anichstrasse 35,
6020 Innsbruck, Austria; volker.schartinger@i-med.ac.at (V.H.S.); jozsef.dudas@i-med.ac.at (J.D.);
julia.ingruber@tirol-kliniken.at (J.I.); maria.greier@tirol-kliniken.at (M.C.G.);
teresa.steinbichler@i-med.ac.at (T.B.S.); herbert.riechelmann@i-med.ac.at (H.R.)
[3] University Hospital of Cranio-Maxillofacial and Oral Surgery, Medical University of Innsbruck,
Anichstrasse 35, 6020 Innsbruck, Austria; johannes.laimer@i-med.ac.at
* Correspondence: ba.kofler@tirol-kliniken.at; Tel.: +43-43-50-504-23141

Received: 19 July 2020; Accepted: 21 August 2020; Published: 23 August 2020

**Abstract:** *Background:* High risk human papillomavirus (hr-HPV)-associated oropharyngeal cancers (OPCs) are characterized by significantly better therapy responses. In order to implement a de-escalated treatment strategy for this tumor entity, it is highly crucial to accurately distinguish HPV-associated OPCs from non-HPV-associated ones. *Methods:* In this prospective study, 56 patients with histologically confirmed OPC were evaluated. A commercially available sandwich ELISA test system was used for the detection of hr-HPV E7 oncoprotein targeting the genotypes 16, 18 and 45. Results were presented as optical density. Positivity for HPV DNA and p16 immunohistochemistry (IHC) was taken as the reference method. *Results:* E7 positivity was significantly associated with the reference method ($p$ = 0.048). The sensitivity, specificity, positive predictive value and negative predictive value for the E7 oncoptotein was 60.9% (95% CI 38.5 to 80.3%), 66.7% (95% CI 46% to 83.5%), 64.2% (95% CI 49.4 to 77.4%) and 63.01% (95% CI 48.9–75.2%), respectively, for the cutoff provided by the manufacturer. *Conclusions:* We found a significant association between E7 oncoprotein detection and the currently used combination. We believe that the use of the ELISA based E7 antigen test could be a valuable addition in cases of ambiguous findings and may be used in combination with other techniques to distinguish between HPV-driven and non-HPV-driven OPCs. However, the low sensitivity of the assay coupled with the small sample size in our study may represent a limitation. We recommend that future larger studies elucidate the diagnostic value of the E7 brush test.

**Keywords:** human papillomavirus; oropharyngeal cancer; E7 oncoprotein; brush test; p16 IHC

---

## 1. Introduction

An increase in oropharyngeal cancer (OPC) was first observed in the United States at the beginning of the 21st century [1,2]. Since then, a steady increase in OPCs in the USA and Europe has been described, caused by high risk human papillomavirus (hr-HPV) [3,4], while the number of smokers and the incidence of tobacco related head and neck cancers have declined [5]. Changes in sexual behavior in the last decade, like high numbers of oral sex partners, seem to play an important etiological role in the rising incidence of HPV-positive OPCs [6]. The HPV-positive OPC is a distinct tumor entity that can be distinguished from HPV-negative OPCs by its etiology, molecular characteristics and clinical presentation [7,8]. Patients with an HPV-positive OPC have a substantially better prognosis [2,9]. In a retrospective European study of 259 OPC patients, the HPV status was the most important parameter for overall survival, regardless of the treatment strategies. HPV association was shown to positively influence the survival more than, for example, the size of the primary tumor or smoking status [7]. Hence, precise distinction between HPV-driven and non-HPV-driven tumors is essential. The detection of HPV DNA in tumor tissue does not necessarily speak of an HPV-driven tumor as only a small proportion of HPV infections lead to a transforming lesion [10].

This highlights the need to incorporate a diagnostic surrogate marker which precisely distinguishes between a transient and a transforming hr-HPV infection. The biomarker currently in use for the diagnosis of HPV-driven OPC is the copresence of HPV DNA and overexpression of the cellular marker p16 protein [11,12].

P16 is highly expressed in tissues undergoing cell cycle deregulation, suggesting that the detected hr-HPV DNA in the tumor tissue may be the cause of OPC. However, being merely a cellular marker, p16 is also overexpressed in lesions with no HPV association. Rasmussen and coworkers, for example, observed in a cohort of 1243 OPC patients a group of p16-positive but HPV DNA-negative patients. These patients experienced a significantly higher hazard ratio (HR) for metastatic recurrence as compared to HPV+/p16+ patients (HR = 2.56) ($p$ = 0.006) [12]. This may translate into lower reliability of this marker in correctly identifying HPV-induced OPC and justifies the need to search for a surrogate marker which is closely linked to hr-HPV oncogenesis. One possible alternative is the detection of upregulated expression of hr-HPV oncoproteins. Hr-HPV oncoproteins (E6 and E7) play a major role in HPV-associated malignant transformation since they are able to inactivate tumor suppressor proteins and consequently inhibit cell cycle control mechanisms [13,14]. The molecular mechanism of the E7 oncoprotein is the inactivation of the retinoblastoma (Rb) protein—a tumor suppressor cellular protein which controls key regulators of S-phase genes. Hr-HPV E7 oncoproteins interact with Rb at a higher efficiency than low-risk HPV E7 oncoproteins. The interaction of E7 with Rb causes disruption of the growth-suppressive Rb-E2F complexes, promoting $G_1$-S cell cycle transition and uncontrolled cellular replication [15,16].

A recently published study evaluated ELISA-based detection of hr-HPV E7 oncoprotein as a screening method in cervical samples of healthy women. Agorastos and coworkers found in this study that E7 oncoprotein detection might be a promising marker for precisely distinguishing transformation-relevant hr-HPV infections from transient ones [17]. The ELISA-based procedure is easy to perform, less time-consuming and requires only a basic laboratory setup. Although this sandwich-based E7 antigen ELISA test has become available commercially within the last couple of years, no previous study ever evaluated this assay among patients with OPC.

Confronted with the increasing need to establish a de-escalated therapy strategy explicitly for patients with HPV-driven OPC, we questioned whether this ELISA-based E7 oncoprotein test could be a reliable option in accurately distinguishing between HPV-driven and non-HPV-driven OPC.

## 2. Results

### 2.1. Study Population

During the study period, 56 patients with OPC were included. Of the study population, 46 (82.1%) patients were male, the mean age at diagnosis was 65.4 (standard deviation ± 10.12) years and 26 (46.4%) patients were positive for the E7 oncoprotein (Table 1). The mean follow-up time was 8.0 months ± 6 months.

In 23 (41.1%) OPC patients, HPV DNA was detected, and the most common genotype was HPV 16 (83%) (Table 2).

**Table 1.** Study population.

| Variables | N (%) |
| --- | --- |
| Male | 46 (82.1%) |
| Female | 10 (17.9%) |
| Mean age | 65.4 years (±10.12) |
| E7 positivity * | 26 (46.4%) |

* E7 oncoprotein for human papillomavirus 16, 18 and/or 45 without differentiation.

**Table 2.** HPV genotypes in oropharyngeal cancer (OPC) patients.

| HPV Subtypes | Number and Percent of HPV + OPC Patients |
| --- | --- |
| HPV 16 | 19 patients (83%) |
| HPV 18 | 2 patients (9%) |
| HPV 33 | 1 patient (4%) |
| HPV 58 | 1 patient (4%) |

The E7 oncoprotein was detected in 26 patients (46.4%) and was associated with the American Society of Anesthesiologists (ASA) classification ($p = 0.04$) and smoking was inversely and significantly associated with E7 oncoprotein positivity ($p = 0.009$). Further clinico-pathological parameters are shown in Table 3.

**Table 3.** Clinico-pathological characteristics of OPC patients.

| Variables | E7 Positive ($n = 26$) | E7 Negative ($n = 30$) | $p$-Value |
| --- | --- | --- | --- |
| **Sex** | | | |
| Male | 22 | 24 | $p = 0.73$ |
| Female | 4 | 6 | |
| **Age** | | | |
| ≤65 years | 10 | 10 | $p = 0.45$ |
| >65 years | 16 | 20 | |
| **ASA score** | | | |
| ASA I/II | 19 | 14 | $p = 0.04$ |
| ASA III/IV | 7 | 16 | |
| **Smoking** | | | |
| Non-smokers | 16 | 8 | $p = 0.009$ |
| Smoker | 10 | 22 | |
| **Alcohol consumption** | | | |
| Daily | 8 | 13 | $p = 0.33$ |
| Not daily | 18 | 17 | |
| **Clinical T-stage** | | | |
| cT1/T2 | 14 | 16 | $p = 1.0$ |
| cT3/4 | 12 | 14 | |
| **UICCC** | | | |
| Stage I | 1 | 3 | |
| Stage II | 4 | 4 | $p = 0.84$ |
| Stage III | 6 | 7 | |
| Stage IV | 15 | 16 | |
| **Subsite oropharynx** | | | |
| Palatine tonsil | 17 | 19 | |
| Base of tongue | 9 | 6 | $p = 0.14$ |
| Uvula | 0 | 3 | |
| Lateral pharyngeal wall | 0 | 2 | |

**Table 3.** *Cont.*

| Variables | E7 Positive ($n$ = 26) | E7 Negative ($n$ = 30) | $p$-Value |
|---|---|---|---|
| **Therapy** | | | |
| Surgery only | 3 | 5 | |
| Surgery and PORT | 3 | 4 | |
| Surgery and RCT/RIT | 1 | 0 | $p$ = 0.72 |
| Primary RCT/RIT | 14 | 13 | |
| Primary RT | 2 | 3 | |
| Chemo only | 2 | 3 | |
| **p16** | | | |
| Positive | 17 | 12 | $p$ = 0.05 |
| Negative | 9 | 18 | |
| **HPV DNA** | | | |
| Positive | 14 | 9 | $p$ = 0.06 |
| Negative | 12 | 21 | |
| **Follow up** | | | |
| No recurrence | 15 | 20 | $p$ = 0.55 |
| Progression | 6 | 9 | |

OPC, oropharyngeal cancer; UICC, Union for International Cancer Control; PORT, postoperative radiation; RCT, radiochemotherapy; RIT, radioimmunotherapy; RT, radiotherapy; ASA, American Society of Anesthesiologists.

## 2.2. Detection of E7 Oncoprotein, HPV DNA and p16

In 26/56 patients, E7 oncoprotein was detected; in 23/56 patients, HPV DNA was detected, and in 29/56 patients, p16 immunohistochemistry (IHC) positivity was detected. There was a high association between p16 and HPV DNA ($p$ < 001) (Table 4). Patients being positive or negative for both test methods, HPV DNA and p16 IHC, served as reference method. The E7 oncoprotein was associated ($p$ = 0.048) with the reference method, including 50 patients with concordant results for p16 and HPV DNA (Table 5). The two patients who were positive for HPV genotypes other than those detectable by the E7 oncoprotein ELISA tests were in one case positive (HPV 33) and in one case negative (HPV 58) for the E7 oncoprotein.

**Table 4.** HPV DNA and p16.

| HPV DNA | p16 | | | |
|---|---|---|---|---|
| | Negative | Positive | Total | $p$-Value |
| Negative | 27 | 6 | 33 | |
| Positive | 0 | 23 | 23 | $p$ < 0.001 |
| Total | 27 | 29 | 56 | |

**Table 5.** E7 oncoprotein and reference method.

| Reference Method | E7 Oncoprotein | | | |
|---|---|---|---|---|
| | Negative | Positive | Total | $p$-Value |
| Negative | 18 | 9 | 27 | |
| Positive | 9 | 14 | 23 | $p$ = 0.048 |
| Total | 27 | 23 | 50 | |

Mean optical density (OD) and 95% CI of the E7 oncoprotein ELISA was shown to be significantly higher among patients who were positive for the reference method (0.18, 95% CI 0.05–0.26) (median OD = 0.15) as compared to those who were negative for the reference method (0.09, 95% CI 0.05–1.13) (median OD = 0.04) ($p$ = 0.031) (Figure 1).

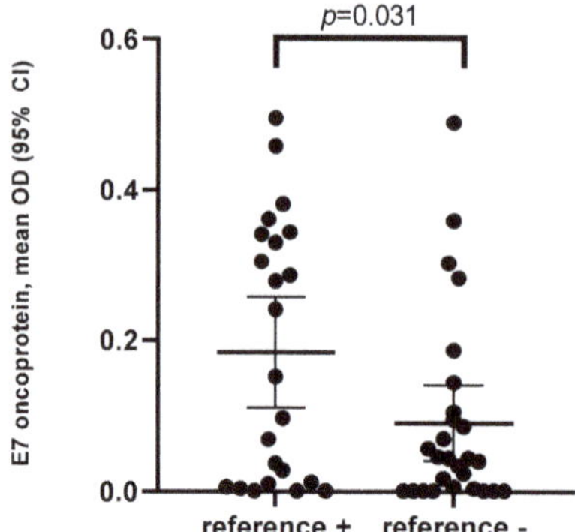

**Figure 1.** Mean optical density (OD) of the E7 oncoprotein level by the reference method (p16 and HPV DNA) among patients with histologically confirmed oropharyngeal cancer (OPC).

In addition to the cutoff value provided by the manufacturer, we also analyzed the performance of E7 oncoprotein using other arbitrarily selected cutoff points. The best agreement with the reference method is achieved when using a cutoff OD value of 0.2. With this cutoff value, the E7 oncoprotein ELISA shows the highest specificity (and the highest % agreement) as compared to the cutoff value provided by the manufacturer (Table S1).

*2.3. Sensitivity, Specificity and Accuracy*

Sensitivity, specificity, positive predictive value and negative predictive value for the E7 oncoptotein was 60.9% (95% CI 38.5 to 80.3%), 66.7% (95% CI 46% to 83.5%), 64.2% (95% CI 49.4 to 77.4%) and 63.01% (95% CI 48.9–75.2%), respectively. The percent agreement between the standard approach and E7 method was 64%. Concordant results of the three test methods are shown in Figure 2.

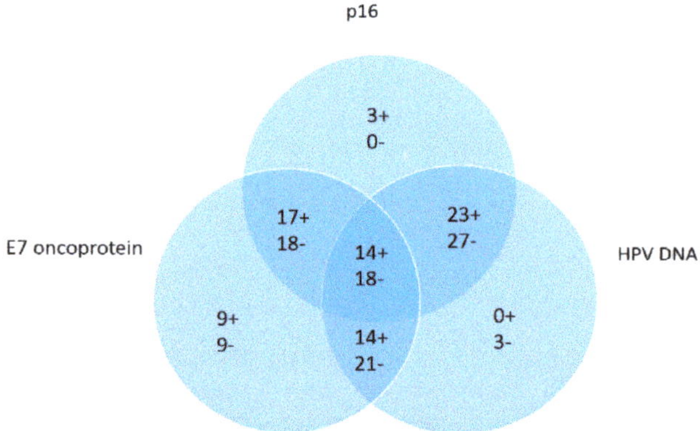


**Figure 2.** Venn diagram for E7 oncoprotein, p16 IHC and HPV DNA, (+) describing positive and (−) describing negative results. Field in the center describes concordant results for all 3 test methods (14 positive and 18 negative). The overlapping fields describe concordant results of E7 and HPV DNA (14 positive and 21 negative), E7 and p16 (17 positive and 18 negative) and HPV DNA and p16 (23 positive and 27 negative). The fields on the outside describe single positivity or negativity for the single test methods.

## 3. Discussion

To our knowledge, this is the first study evaluating the role of E7 oncoprotein detection in patients with OPC for a precise distinction of HPV-driven from non-HPV-driven OPCs with a simple tumor surface brush sample. This test was initially developed to triage HPV-positive women being screened for cervical cancer [17]. A positive result in this assay (values greater than 0.5 pg/well (OD > 0.076)) was defined to be consistent with upregulated E7 function, corresponding to a transcriptionally relevant hr-HPV infection. Compared to p16 IHC, this ELISA-based E7 oncoprotein assay is easy to perform and less time-consuming. As different de-escalation treatment strategies to reduce toxicity in HPV-positive OPC patients are in current development [18–20], the detection of HPV-induced OPCs is essential. As previously mentioned, p16 is not sufficient as a single test method, particularly because it is a cell cycle deregulation marker, which may be expressed also in other tumor setups with no HPV involvement. Rb loss through a non-HPV associated mutation can result likewise in p16 expression [21]. In this case, the presence of a high level of E7 oncoprotein, the only source of which is transcriptionally active hr-HPV infection detected, may be a more specific biomarker.

Despite the low sensitivity and specificity of the assay, E7 oncoprotein detection was significantly associated with the reference method ($p = 0.048$), consisting of concordant HPV DNA and p16 results.

E7 oncoprotein positivity was also associated with the ASA score ($p = 0.04$), indicating the lower rate of comorbidities [22] in HPV-positive patients [23]. This assumes that E7 oncoprotein detection may identify younger and healthier patients benefiting from a de-escalation therapy. Further findings were an association between the E7 oncoprotein and smoking status; the E7 oncoprotein was more often expressed in non-smokers ($p = 0.009$). This is in accordance with previous studies describing HPV-positive patients to be more commonly non-smokers, whereas smoking and alcohol consumption are the pathogenic mechanisms in non-HPV-driven OPCs [24–26]. There was no statistically significant difference regarding the subsite of the OPC; however, despite the low sample size of this subsite, no E7 oncoprotein expression was found in OPCs of the uvula or lateral pharyngeal wall.

Interestingly, in nine patients, a single positive result for the E7 oncoprotein was found, and also, in nine patients, a single negative result was found. The single negative ones may indicate an OPC definitely of non-HPV origin despite the presence of HPV DNA in the lesion, i.e., a non-transforming HPV co-infection.

The single negative result of E7 could indicate episomally latent hr-HPV in the cells of a tobacco smoker and alcohol-induced carcinoma without HPV being involved in the carcinogenesis. A further explanation may be the low sensitivity of ELISA-based antigen tests, with a consequence of failing to detect low levels but yet relevant amounts of E7 oncoprotein.

The single E7-positive result may also encourage us to consider resetting the cutoff value for E7 positivity. Since the test kit was originally approved for triaging in cervical cancer screening, it may be essential to re-evaluate the assay in the context of OPC to optimize an appropriate threshold for this entity. In our study, we observed much better performance characterized by the best percent agreement and high specificity when using a higher cutoff value than provided by the manufacturer (also supported by receiver operating characteristic analysis).

Data from cervical cancer studies show that, in some tumors, although definitely caused by HPV, tests are negative for HPV DNA [27,28]. A plausible explanation is the non-productive nature of the infection in the oncogenic setup, in which viral DNA is integrated in the host. This situation, being transcriptionally active, may be characterized by high expression of transforming proteins, with less virus being released. The fact that we are taking brush samples from the surface of the tumor may support this notion. We cannot explain with certainty the single E7 positive cases. They may be due to the non-productive nature of the infection in tumors. Such phenomena are known among patients with invasive cervical cancers [29].

Moreover, in this study, a certain proportion of patients (6 patients, 10.7%) positive for p16 IHC and negative for HPV DNA was observed. The importance of additional HPV DNA testing to identify HPV-positive OPCs was described in several studies [12,30,31]. As previously described, these patients seem to have a less favorable prognosis than HPV DNA and p16-positive patients [12,32]. Weinberger and coworkers classified 79 OPC patients into class I (HPV DNA negative/p16 low positive), class II (HPV DNA positive/p16 low positive) and class III (HPV DNA positive/ p16 high positive). Class III patients had improved overall survival ($p = 0.0095$) and disease-free survival ($p = 0.03$) in comparison to class I and class II OPC patients [33].

Detection of the HPV oncoproteins in the clinical routine of OPC patients is still challenging. The brush test is very feasible and easily applied; however, further studies based on our results and further developments may be necessary. The brush is designed for the cervix, so, in the oropharyngeal region, a smaller brush would be more applicable in order to ensure that the tumor surface can be brushed more precisely without touching the surrounding tissue. A major limitation of the study is the lack of data on E7 mRNA expression. Taken by some as the gold standard for the classification of transcriptionally active hr-HPV, results of mRNA PCR may have supplemented the low sensitivity of this ELISA-based antigen assay. However, we used the combination of HPV DNA and p16 detection as a reference method, since this combination is the standard method for the diagnosis HPV-driven tumor according to many guidelines [11,12]. The role of routine E7 mRNA detection should, however, be elucidated in a future study. Since the great majority (>95%) of HPV driven tumors are due to HPV 16 and HPV 18 [34], the fact that our ELISA assay was limited to the three genotypes (16, 18 and 45) is a negligible limitation. A possible cross-reaction between other genotypes genetically closely linked to these three genotypes cannot be excluded. A further limitation of the study is the fact that this ELISA assay detects only E7 protein and not E6, another oncogenic protein owned by HPV which is of carcinogenic significance in a different pathway. The additional detection of E6 oncoprotein might have added valuable information to the data already generated by this study. Further studies need to elaborate on the diagnostic value of this method to additionally determine E6 oncoprotein for these hr-HPV types in order to accurately identify HPV-driven OPCs.

## 4. Materials and Methods

This was a prospectively designed study, and patients presenting with OPC between January 2018 and June 2020 at the Department of Otorhinolaryngology, Medical University Innsbruck, Austria were included. Each patient who agreed to participate in this study gave informed consent. The study was conducted in full accordance with the principles expressed in the Declaration of Helsinki and was approved by the ethics committee of the Medical University Innsbruck. The respective reference number was 1147/2018.

Patients were included only when the histology confirmed squamous cell carcinoma of the oropharynx and the patient's age was at least 18 years. Patients with other diagnoses than OPC in the histopathological examination were excluded. With the exception of 3 patients, all patients had a primary tumor in the oropharyngeal regions, and 3 patients presented with a recurrence or second primary in the oropharynx. In all patients, the E7 brush test was performed before treatment.

### 4.1. Specimen Harvest and Handling

Details of specimen harvesting and handling has been described in a previous work [35]. In short, each patient suspected to have OPC received a panendoscopy under anesthesia. In this procedure, two different cytology brush tests for HPV DNA and E7 oncoprotein detection were conducted (digene® HC2 DNA Collection Device, Qiagen, Hilden, Germany and ThinPrep® PreservCyt Solution, Hologic, Manchester, UK) by gently brushing the tumor surface. The brushes were then placed in a sterile container and sent to the Institute of Virology, Medical University Innsbruck. Furthermore, tumor biopsies were obtained and fixed in formalin for routine histopathological examination at the Department of Pathology, Medical University Innsbruck. Additionally, a tumor biopsy for p16 IHC detection was kept in cell culture medium and immediately sent to the Laboratory for Molecular Biology and Oncology, Department of Otorhinolaryngology.

### 4.2. E7 Oncoprotein Detection by Brush Test

Detection of E7 oncoprotein for genotypes of HPV 16, 18 and/or 45 was conducted using a sandwich ELISA test system (recomWell HPV 16/18/45, Mikrogen, Neurid, Germany) developed and validated (CE-labelled) initially to support diagnostic and therapeutic decisions in cervical cancer screening [17]. The ELISA microtiter plates (MTPs) were coated with rabbit monoclonal antibodies (RabMabs) specific for the three genotypes mentioned above. The remaining oropharyngeal swab samples, after cytology and HPV genotyping, were centrifuged and the pellets were incubated with the RabMabs after a couple of lysis steps. Biotinylated polyclonal goat-anti-E7 antibodies were used to detect E7 antigens which remained bound to the monoclonal antibodies after final washing steps. Results were provided as optical density (OD), with a limit of detection of 0.5 pg of protein per well, with the corresponding cutoff value of the OD being 0.076. We also evaluated the association between the reference method and E7 oncoprotein positivity using further arbitrarily selected cutoff values.

### 4.3. DNA Amplification and HPV Genotyping

For HPV DNA detection, real-time PCR was used based on the amplification of the L1 open reading frames (ORF). As internal control for the availability of cellular material, a PCR for the housekeeping gene beta globin was performed. The HPV DNA was considered positive if the fluorescence signal appeared before the fortieth cycle [36]. Further genotyping was performed on all HPV-positive samples using reverse line blot hybridization on nitrocellulose membrane strips containing genotype specific probes (AmpliQuality HPV-TYPE EXPRESS, AB Analitica®, Padova, Italy) [37]. With this genotyping kit, it is possible to identify 40 different HPV types, including all hr and several low-risk HPV types.

### 4.4. Immunohistochemistry

First, five-micrometer thin paraffin sections were dewaxed and then antigens were retrieved in an automated staining system (Ventana, Discovery, Tucson, AZ, USA). For p16 detection, a commercial diagnostic assay was used (CINtec® Histology V-Kit, Roche diagnostics, Basel, Switzerland). The staining was completed by using a universal secondary antibody solution, the DAB MAP Kit and hematoxylin counterstaining (both Ventana products). One experienced observer evaluated the tumor cell areas, and specimens were considered p16-positive if $\geq$66% of the cells in the tumor areas revealed immunohistochemical reaction products.

### 4.5. Data Analysis

Patient clinical data were presented in tabular form. A comparison of HPV DNA, E7 detection and p16 IHC in samples obtained from OPC patients was performed. For each investigated variable, a binary outcome (positive/negative) was obtained. Combined hr-HPV DNA positivity and p16 positivity served as the reference method. Contingency tables were analyzed with Fisher exact test or Pearson chi-square. Diagnostic accuracy parameters including sensitivity and specificity were calculated using the diagnostic test routines of MedCalc. For data analysis, SPSS Statistics 24 software (IBM Corporation, Armonk NY, USA) was used.

## 5. Conclusions

We do not have strong evidence that confirms that ELISA-based E7 oncoprotein replaces the reference method for diagnosis of HPV-associated OPC. However, looking at the significant association with the reference method of diagnosis, it could be a valuable addition in the case of ambiguous findings. We strongly recommend further investigation and optimization of the test in large prospective studies.

**Supplementary Materials:** The following are available online at http://www.mdpi.com/2072-6694/12/9/2388/s1, Table S1: Comparing the performance of hr-HPV E7 oncoprotein ELISA in detecting HPV-driven OPC* (*n* = 50).

**Author Contributions:** Methodology, W.B., J.D.; software, V.H.S.; validation, W.B., B.K., J.L.; formal analysis, V.H.S.; investigation, B.K., W.B., T.B.S.; resources, H.R.; data curation, V.H.S., H.S.; writing—original draft preparation, B.K.; writing—review and editing, W.B., J.D., J.L.; visualization, J.I., M.C.G.; supervision, W.B.; project administration, W.B., B.K.; funding acquisition, H.R. All authors have read and agreed to the published version of the manuscript.

**Funding:** This research received no external funding.

**Acknowledgments:** The authors thank the biomedical analysts at the Institute of Virology, Bettina Hofer and Agnes Mayr, for supporting this work by conducting E7 ELISA.

**Conflicts of Interest:** The authors declare no conflict of interest.

## References

1. Chaturvedi, A.K.; Engels, E.A.; Pfeiffer, R.M.; Hernandez, B.Y.; Xiao, W.; Kim, E.; Jiang, B.; Goodman, M.T.; Sibug-Saber, M.; Cozen, W.; et al. Human papillomavirus and rising oropharyngeal cancer incidence in the United States. *J. Clin. Oncol.* **2011**, *29*, 4294–4301. [CrossRef] [PubMed]
2. Mahal, B.A.; Catalano, P.J.; Haddad, R.I.; Hanna, G.J.; Kass, J.I.; Schoenfeld, J.D.; Tishler, R.B.; Margalit, D.N. Incidence and Demographic Burden of HPV-Associated Oropharyngeal Head and Neck Cancers in the United States. *Cancer Epidemiol. Biomark. Prev.* **2019**, *28*, 1660–1667. [CrossRef] [PubMed]
3. Wittekindt, C.; Wagner, S.; Bushnak, A.; Prigge, E.S.; von Knebel Doeberitz, M.; Wurdemann, N.; Bernhardt, K.; Pons-Kuhnemann, J.; Maulbecker-Armstrong, C.; Klussmann, J.P. Increasing Incidence rates of Oropharyngeal Squamous Cell Carcinoma in Germany and Significance of Disease Burden Attributed to Human Papillomavirus. *Cancer Prev. Res. (Phila)* **2019**, *12*, 375–382. [CrossRef] [PubMed]
4. Dahlstrom, K.R.; Bell, D.; Hanby, D.; Li, G.; Wang, L.E.; Wei, Q.; Williams, M.D.; Sturgis, E.M. Socioeconomic characteristics of patients with oropharyngeal carcinoma according to tumor HPV status, patient smoking status, and sexual behavior. *Oral Oncol.* **2015**, *51*, 832–838. [CrossRef]
5. Sturgis, E.M.; Cinciripini, P.M. Trends in head and neck cancer incidence in relation to smoking prevalence: An emerging epidemic of human papillomavirus-associated cancers? *Cancer* **2007**, *110*, 1429–1435. [CrossRef]
6. Schnelle, C.; Whiteman, D.C.; Porceddu, S.V.; Panizza, B.J.; Antonsson, A. Past sexual behaviors and risks of oropharyngeal squamous cell carcinoma: A case-case comparison. *Int. J. Cancer* **2017**, *140*, 1027–1034. [CrossRef]
7. Wagner, S.; Wittekindt, C.; Sharma, S.J.; Wuerdemann, N.; Juttner, T.; Reuschenbach, M.; Prigge, E.S.; von Knebel Doeberitz, M.; Gattenlohner, S.; Burkhardt, E.; et al. Human papillomavirus association is the most important predictor for surgically treated patients with oropharyngeal cancer. *Br. J. Cancer* **2017**, *116*, 1604–1611. [CrossRef]
8. Kofler, B.; Laban, S.; Busch, C.J.; Lorincz, B.; Knecht, R. New treatment strategies for HPV-positive head and neck cancer. *Eur. Arch. Oto-Rhino-Laryngol.* **2014**, *271*, 1861–1867. [CrossRef]
9. Mirghani, H.; Bellera, C.; Delaye, J.; Dolivet, G.; Fakhry, N.; Bozec, A.; Garrel, R.; Malard, O.; Jegoux, F.; Maingon, P.; et al. Prevalence and characteristics of HPV-driven oropharyngeal cancer in France. *Cancer Epidemiol.* **2019**, *61*, 89–94. [CrossRef]
10. Jung, A.C.; Briolat, J.; Millon, R.; de Reynies, A.; Rickman, D.; Thomas, E.; Abecassis, J.; Clavel, C.; Wasylyk, B. Biological and clinical relevance of transcriptionally active human papillomavirus (HPV) infection in oropharynx squamous cell carcinoma. *Int. J. Cancer* **2010**, *126*, 1882–1894. [CrossRef]
11. Mena, M.; Taberna, M.; Tous, S.; Marquez, S.; Clavero, O.; Quiros, B.; Lloveras, B.; Alejo, M.; Leon, X.; Quer, M.; et al. Double positivity for HPV-DNA/p16(ink4a) is the biomarker with strongest diagnostic accuracy and prognostic value for human papillomavirus related oropharyngeal cancer patients. *Oral Oncol.* **2018**, *78*, 137–144. [CrossRef] [PubMed]
12. Rasmussen, J.H.; Gronhoj, C.; Hakansson, K.; Friborg, J.; Andersen, E.; Lelkaitis, G.; Klussmann, J.P.; Wittekindt, C.; Wagner, S.; Vogelius, I.R.; et al. Risk profiling based on p16 and HPV DNA more accurately predicts location of disease relapse in patients with oropharyngeal squamous cell carcinoma. *Ann. Oncol.* **2019**, *30*, 629–636. [CrossRef] [PubMed]
13. Narisawa-Saito, M.; Kiyono, T. Basic mechanisms of high-risk human papillomavirus-induced carcinogenesis: Roles of E6 and E7 proteins. *Cancer Sci.* **2007**, *98*, 1505–1511. [CrossRef] [PubMed]
14. zur Hausen, H. Papillomaviruses causing cancer: Evasion from host-cell control in early events in carcinogenesis. *J. Natl. Cancer Inst.* **2000**, *92*, 690–698. [CrossRef] [PubMed]

15. Moody, C.A.; Laimins, L.A. Human papillomavirus oncoproteins: Pathways to transformation. *Nat Rev. Cancer* **2010**, *10*, 550–560. [CrossRef]
16. Gonzalez, S.L.; Stremlau, M.; He, X.; Basile, J.R.; Munger, K. Degradation of the retinoblastoma tumor suppressor by the human papillomavirus type 16 E7 oncoprotein is important for functional inactivation and is separable from proteasomal degradation of E7. *J. Virol.* **2001**, *75*, 7583–7591. [CrossRef]
17. Agorastos, T.; Chatzistamatiou, K.; Moysiadis, T.; Kaufmann, A.M.; Skenderi, A.; Lekka, I.; Koch, I.; Soutschek, E.; Boecher, O.; Kilintzis, V.; et al. Human papillomavirus E7 protein detection as a method of triage to colposcopy of HPV positive women, in comparison to genotyping and cytology. Final results of the PIPAVIR study. *Int. J. Cancer* **2017**, *141*, 519–530. [CrossRef]
18. Foster, C.C.; Seiwert, T.Y.; MacCracken, E.; Blair, E.A.; Agrawal, N.; Melotek, J.M.; Portugal, L.; Brisson, R.J.; Gooi, Z.; Spiotto, M.T.; et al. Dose and Volume De-Escalation for Human Papillomavirus-Positive Oropharyngeal Cancer is Associated with Favorable Post-Treatment Functional Outcomes. *Int. J. Radiat. Oncol. Biol. Phys.* **2020**. [CrossRef]
19. Bigelow, E.O.; Seiwert, T.Y.; Fakhry, C. Deintensification of treatment for human papillomavirus-related oropharyngeal cancer: Current state and future directions. *Oral Oncol.* **2020**, *105*, e104652. [CrossRef]
20. Patel, R.R.; Ludmir, E.B.; Augustyn, A.; Zaorsky, N.G.; Lehrer, E.J.; Ryali, R.; Trifiletti, D.M.; Adeberg, S.; Amini, A.; Verma, V. De-intensification of therapy in human papillomavirus associated oropharyngeal cancer: A systematic review of prospective trials. *Oral Oncol.* **2020**, *103*, e104608. [CrossRef]
21. Liang, C.; Marsit, C.J.; McClean, M.D.; Nelson, H.H.; Christensen, B.C.; Haddad, R.I.; Clark, J.R.; Wein, R.O.; Grillone, G.A.; Houseman, E.A.; et al. Biomarkers of HPV in head and neck squamous cell carcinoma. *Cancer Res.* **2012**, *72*, 5004–5013. [CrossRef] [PubMed]
22. Riechelmann, H.; Neagos, A.; Netzer-Yilmaz, U.; Gronau, S.; Scheithauer, M.; Rockemann, M.G. The ASA-score as a comorbidity index in patients with cancer of the oral cavity and oropharynx. *Laryngo-Rhino-Otol.* **2006**, *85*, 99–104. [CrossRef] [PubMed]
23. Grisar, K.; Dok, R.; Schoenaers, J.; Dormaar, T.; Hauben, E.; Jorissen, M.; Nuyts, S.; Politis, C. Differences in human papillomavirus-positive and -negative head and neck cancers in Belgium: An 8-year retrospective, comparative study. *Oral Surg. Oral Med. Oral Pathol. Oral Radiol.* **2016**, *121*, 456–460. [CrossRef] [PubMed]
24. Beynon, R.A.; Lang, S.; Schimansky, S.; Penfold, C.M.; Waylen, A.; Thomas, S.J.; Pawlita, M.; Waterboer, T.; Martin, R.M.; May, M.; et al. Tobacco smoking and alcohol drinking at diagnosis of head and neck cancer and all-cause mortality: Results from head and neck 5000, a prospective observational cohort of people with head and neck cancer. *Int. J. Cancer* **2018**, *143*, 1114–1127. [CrossRef] [PubMed]
25. Deschler, D.G.; Richmon, J.D.; Khariwala, S.S.; Ferris, R.L.; Wang, M.B. The "new" head and neck cancer patient-young, nonsmoker, nondrinker, and HPV positive: Evaluation. *Otolaryngol. Head Neck Surg.* **2014**, *151*, 375–380. [CrossRef]
26. Lassen, P.; Lacas, B.; Pignon, J.P.; Trotti, A.; Zackrisson, B.; Zhang, Q.; Overgaard, J.; Blanchard, P.; Group, M.C. Prognostic impact of HPV-associated p16-expression and smoking status on outcomes following radiotherapy for oropharyngeal cancer: The MARCH-HPV project. *Radiother. Oncol. J. Eur. Soc. Ther. Radiol. Oncol.* **2018**, *126*, 107–115. [CrossRef]
27. Kaliff, M.; Karlsson, M.G.; Sorbe, B.; Bohr Mordhorst, L.; Helenius, G.; Lillsunde-Larsson, G. HPV-negative Tumors in a Swedish Cohort of Cervical Cancer. *Int. J. Gynecol. Pathol.* **2020**, *39*, 279–288. [CrossRef]
28. Arroyo Muhr, L.S.; Lagheden, C.; Eklund, C.; Lei, J.; Nordqvist-Kleppe, S.; Sparen, P.; Sundstrom, K.; Dillner, J. Sequencing detects human papillomavirus in some apparently HPV-negative invasive cervical cancers. *J. Gen. Virol.* **2020**, *101*, 265–270. [CrossRef]
29. Mills, A.M.; Dirks, D.C.; Poulter, M.D.; Mills, S.E.; Stoler, M.H. HR-HPV E6/E7 mRNA In Situ Hybridization: Validation Against PCR, DNA In Situ Hybridization, and p16 Immunohistochemistry in 102 Samples of Cervical, Vulvar, Anal, and Head and Neck Neoplasia. *Am. J. Surg. Pathol.* **2017**, *41*, 607–615. [CrossRef]
30. Rietbergen, M.M.; Snijders, P.J.; Beekzada, D.; Braakhuis, B.J.; Brink, A.; Heideman, D.A.; Hesselink, A.T.; Witte, B.I.; Bloemena, E.; Baatenburg-De Jong, R.J.; et al. Molecular characterization of p16-immunopositive but HPV DNA-negative oropharyngeal carcinomas. *Int. J. Cancer* **2014**, *134*, 2366–2372. [CrossRef]
31. Parfenov, M.; Pedamallu, C.S.; Gehlenborg, N.; Freeman, S.S.; Danilova, L.; Bristow, C.A.; Lee, S.; Hadjipanayis, A.G.; Ivanova, E.V.; Wilkerson, M.D.; et al. Characterization of HPV and host genome interactions in primary head and neck cancers. *Proc. Natl. Acad. Sci. USA* **2014**, *111*, 15544–15549. [CrossRef] [PubMed]

32.  Perrone, F.; Gloghini, A.; Cortelazzi, B.; Bossi, P.; Licitra, L.; Pilotti, S. Isolating p16-positive/HPV-negative oropharyngeal cancer: An effort worth making. *Am J Surg Pathol* **2011**, *35*, 774–777. [CrossRef] [PubMed]

33.  Weinberger, P.M.; Yu, Z.; Haffty, B.G.; Kowalski, D.; Harigopal, M.; Brandsma, J.; Sasaki, C.; Joe, J.; Camp, R.L.; Rimm, D.L.; et al. Molecular classification identifies a subset of human papillomavirus–associated oropharyngeal cancers with favorable prognosis. *J. Clin. Oncol.* **2006**, *24*, 736–747. [CrossRef] [PubMed]

34.  LeConte, B.A.; Szaniszlo, P.; Fennewald, S.M.; Lou, D.I.; Qiu, S.; Chen, N.W.; Lee, J.H.; Resto, V.A. Differences in the viral genome between HPV-positive cervical and oropharyngeal cancer. *PLoS ONE* **2018**, *13*, e0203403. [CrossRef] [PubMed]

35.  Kofler, B.; Borena, W.; Manzl, C.; Dudas, J.; Wegscheider, A.S.; Jansen-Durr, P.; Schartinger, V.; Riechelmann, H. Sensitivity of tumor surface brushings to detect human papilloma virus DNA in head and neck cancer. *Oral Oncol.* **2017**, *67*, 103–108. [CrossRef] [PubMed]

36.  Abreu, A.L.; Souza, R.P.; Gimenes, F.; Consolaro, M.E. A review of methods for detect human Papillomavirus infection. *Virol. J.* **2012**, *9*, e262. [CrossRef] [PubMed]

37.  Mason, A.G.S.; Vettorato, M.; Negri, G.; Mian, C.; Brusauro, F.; Bortolozzo, K. Detection of high-risk HPV genotypes in cervical samples: A comparison study of a novel real time pcr/reverse line blot-based technique and the digene HC2 assay. *Pathologica* **2012**, *6*, 104–146.

*Article*

# Identification of Cancer-Associated Circulating Cells in Anal Cancer Patients

**Thomas J. Carter [1,†], Jeyarooban Jeyaneethi [2,†], Juhi Kumar [2], Emmanouil Karteris [2], Rob Glynne-Jones [1,2] and Marcia Hall [1,2,*]**

[1]  Mount Vernon Cancer Centre, Middlesex HA6 2RN, UK; thomascarter@nhs.net (T.J.C.); rob.glynnejones@nhs.net (R.G.-J.)
[2]  Department of Life Sciences, Brunel University, London UB83PH, UK; jeyarooban.jeyaneethi@brunel.ac.uk (J.J.); juhi.kumar@brunel.ac.uk (J.K.); Emmanouil.karteris@brunel.ac.uk (E.K.)
*  Correspondence: marcia.hall@nhs.net
†  T.J.C. and J.J. contributed equally to this manuscript.

Received: 11 July 2020; Accepted: 6 August 2020; Published: 10 August 2020

**Abstract:** Whilst anal cancer accounts for less than 1% of all new cancer cases, incidence rates have increased by up to 70% in the last 30 years with the majority of cases driven by human papilloma virus (HPV) infection. Standard treatment for localised anal cancer is chemoradiotherapy (CRT). Localised progression is the predominant pattern of relapse but well under 50% of cases are salvaged by surgery, predominantly because confirming recurrence within post-radiation change is very challenging. Identifying cancer-associated circulating cells (CCs) in peripheral blood could offer a corroborative method of monitoring treatment efficacy and identifying relapse early. To study this, nucleated cells were isolated from the blood of patients with anal cancer prior to, during, and after CRT and processed through the Amnis® ImageStream®X Mk II Imaging Flow Cytometer, without prior enrichment, using Pan-cytokeratin (PCK), CD45 antibodies and making use of the DNA dye DRAQ5. Analysis was undertaken using IDEAS software to identify those cells that were PCK-positive and DRAQ5-positive as well as CD45-negative; these were designated as CCs. CCs were identified in 7 of 8 patients; range 60–876 cells per mL of blood. This first report of the successful identification of CCs in anal cancer patients raises the possibility that liquid biopsies will find a future role as a prognostic/diagnostic tool in this patient group.

**Keywords:** anal cancer; HPV; cancer-associated circulating cells; liquid biopsies

---

## 1. Introduction

Anal cancer is a rare cancer, accounting for less than 1% of all cancer diagnoses [1] and around 2% of all cancers of the gastrointestinal tract [2]. Squamous cell carcinoma of the anus (SCCA) is the predominant histological subtype, accounting for over 90% [3,4]. Incidence rates of SCCA have increased by up to 70% over the last 30 years, with the majority of cases occurring as a result of human papilloma virus (HPV) infection [5]. Risk factors for SCCA include immunosuppression, multiple sexual partners, history of anal intercourse and smoking [6,7]. Most cases of SCCA are diagnosed at an early stage (T1/T2); metastatic disease accounts for <5% of new cases [8].

The standard treatment for localised SSCA is chemoradiotherapy (CRT) [8]. CRT consists of a pyrimidine analogue (5-fluorouracil or capecitabine) and mitomycin C (MMC) delivered concurrently with radiation treatment [9]. Long-term response rates to this treatment are favorable at 80–90%; however loco-regional failure can occur, especially in patients with T3/T4 disease [10]. Tumour HPV status is an important predictor of response, with HPV-16 or -18 driven SCCA more likely to respond to CRT, whilst HPV-negative SCCA are less likely to respond [8,11]. After definitive CRT, approximately

10–15% of patients have persistent cancer, and 15–30% develop subsequent local recurrence after initial complete response [12]. Cross-sectional imaging (computed tomography; CT or magnetic resonance; MR) is generally used to assess response, although recurrence can be difficult to distinguish from treatment effect in the first 6 months following treatment completion [13]. Assessment prior to this can over-estimate the necessity for salvage surgery, which is successful in fewer than 50% of patients with loco-regional recurrences [10]. Prognosis for patients in whom salvage surgery is unsuccessful and for those with metastatic disease remains poor, with median overall survival (mOS) varying from 8 to 34 months [12,14]. Identification of biomarkers to confirm/refute imaging changes suggestive of recurrent/residual local disease after CRT could identify those who should be offered salvage surgery earlier versus those who could safely be followed-up, thus avoiding unnecessary extensive pelvic surgery. Assessment of circulating cells from liquid biopsies are an emerging technology which could provide valuable information on treatment failure, before recurrence disease is clearly evident via cross-sectional imaging [15].

Liquid tumour biopsies from blood possess two distinct advantages over conventional tumour biopsies; (i) they are minimally invasive and (ii) they can enable the tracking of cancers in real time, including changes in response to treatment [11]. Blood tests able to identify relapse would be invaluable, especially where conventional tissue biopsy is inadvisable due to inaccessible sites of recurrence or where the patient is too unwell [15,16]. Circulating tumour DNA (ctDNA) has begun to transition into the clinic [17–19], with potential roles including in the selection of cancer patients with advanced disease for appropriate clinical trials [20]. Studies with digital drop PCR (ddPCR) technology recently utilised to detect HPV16 ctDNA in blood are on-going. To date, HPV ctDNA has been detected in patients with early-stage SCCA, and those with advanced disease, with evidence that HPV ctDNA levels are higher in patients with advanced disease [21,22].

Another avenue interrogating liquid biopsies is to isolate and characterize malignant cells which have detached from the primary tumour mass and entered the blood stream; known as cancer-associated circulating cells (CCs) [23,24]. CCs have been observed in a number of cancers including breast, prostate and colorectal cancer [25], and are normally detected via two methods: CellSearch and ISET (Isolation by Size of Tumor cell) assay. CellSearch uses magnetically tagged antibodies in order to enrich the samples before staining with antibodies which target cell surface markers such as epithelial cell adhesion molecule (EpCAM) to identify malignant cells, and to date, is the only FDA approved method to detect CCs in breast, colorectal and prostate cancer. One major limitation to this method is the use of EpCAM alone to enrich samples and identify CCs. EpCAM is down-regulated in many cancers, especially during the process of metastasis, when epithelial to mesenchymal transition (EMT) occurs [26]. In contrast, ISET uses filtration to enrich the liquid biopsy sample based on size differences of CCs. Filtered samples are then immunostained before observation and identification under a light microscope by a trained user. Although not FDA-approved, ISET can detect non-epithelial CCs using a filtration chamber pressurised at 5–9 kPa to capture cells larger than 8 µm onto a membrane as a method of enrichment. However, there are a high number of false positives detected in volunteer patients using ISET alone, a limitation for any filtration-based method of enrichment [27]. Whilst liquid biopsy technologies are increasingly employed in a number of malignancies, in patients with SCCA, a definitive role has not yet been identified [28]. To the best of our knowledge, there are no studies exploring CCs in anal cancer.

Our group has previously demonstrated the utility of liquid biopsies in order to detect CCs in both ovarian [29] and lung cancers [23]. No prior enrichment is undertaken both to mitigate the loss of CCs during sample processing and to be able to examine all sizes of CCs as it is becoming increasingly recognised that many CCs are smaller than 8 µm. Our group has successfully identified CCs by labelling nucleated circulating cells with pan-cytokeratin (PCK) and CD45 markers. This differentiates epithelial cells (PCK+) from haematopoitic (CD45+) non-epithelial (PCK−) cells within the blood stream. Identification as cancer-associated CCs was further confirmed using WT1 antibody in high-grade serous ovarian cancer patient samples, and TTF1 in lung cancer patient samples [23]. Using these techniques,

CCs can be identified and quantified at higher numbers compared to alternative techniques. We have also shown that CC levels fall in response to treatment in ovarian cancer patients [29]. Within the present study, we sought to apply these techniques to peripheral blood samples obtained from SCCA patients before, during and after CRT, to determine whether appropriate PCK+ CCs were present and could be identified. To the best of our knowledge, this is the first study demonstrating the potential presence of cancer-associated non-haemopoietic CCs in this patient group.

## 2. Results

### 2.1. Patient Demographics

Blood samples were collected from eight patients diagnosed with localised SCCA. Patients were staged using the TNM (tumour, nodes, metastasis) staging. All patients had localised disease, the majority were T2 (range T1-4), one patient had local nodal involvement (N1) and no patients had distant metastases (M0). The patients were all female, and they all received radical treatment with chemoradiotherapy (41.4 Gy in 23 fractions with concurrent MMC and capecitabine, as part of the ACT-IV clinical trial to which all eight patients were co-recruited). End of treatment (EOT) MR imaging confirmed a complete radiological response in all patients. To date, no patients have clinical or radiological evidence of disease relapse or recurrence, with three patients (38%) followed up for >12 months from EOT at the time of analysis. In addition, no patients have yet undergone re-biopsy to confirm response due to the risk of post-biopsy necrosis. Baseline blood samples were obtained for all eight patients, with serial blood samples (>3) available for five patients (63%). Full staging and follow-up details can be found in Table 1.

**Table 1.** Patient demographics.

| Patient Number | Age at Diagnosis | Tumour Staging | Number of Samples | Imaging Follow Up |
| --- | --- | --- | --- | --- |
| 1 | 51 | T4N0 | 1 | 12 m |
| 2 | 53 | T2N0 | 4 | 12 m |
| 3 | 64 | T1N0 | 4 | 6 m |
| 4 | 68 | T2N0 | 4 | 6 m |
| 5 | 74 | T1/2N0 | 3 | 6 m |
| 6 | 78 | T2N0 | 3 | 6 m |
| 7 | 53 | T2N1 | 1 | 6 m |
| 8 | 48 | T2N0 | 2 | EOT |

EOT = End of Treatment (done 3 m following end of CRT), m = months. T = Tumour, N = Nodes.

### 2.2. Isolation and Identification of Circulating Tumour Cells

Following sample collection and preparation, cells were stained and processed using the ImageStream®X Mk II; images were analysed using the IDEAS software [30]. Collected images were filtered using the criteria as shown in Figure 1A. In brief, images were initially selected based on size and circularity to demonstrate that they are single cells. From this data, only images which were adequately in focus for analytical purposes were chosen. Focused cells that were double-positive for PCK and DRAQ5 were then selected for full assessment and quantification. The final collection of images for every blood sample were then manually analyzed by two independent parties. Positive CCs were identified as being nucleated cells which stained for PCK and DRAQ5 but were negative for CD45. Hematopoietic cells positive for CD45 were excluded from quantification regardless of PCK expression. Examples of CCs and hematopoietic cells identified using IDEAS software in this patient group are shown in Figure 1B.

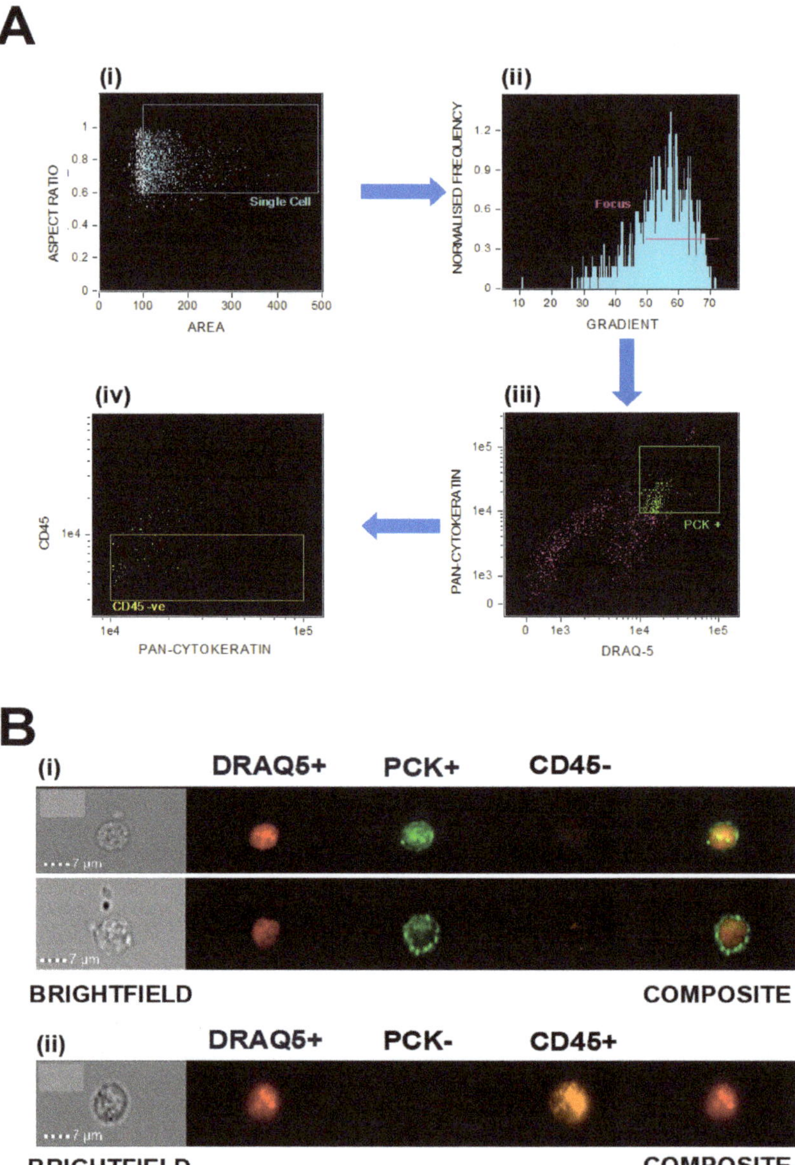



**Figure 1.** (**A**) Graphs illustrating the criteria used to filter the images taken from ImageStream[®]X Mk II for analysis in IDEAS. Each image is displayed by a single dot and the selection criteria appears either as a box (i, iii, iv) or a line (ii); (i) shows the images filtered *via* size and circularity; (ii) represents the single cells filtered by focus of the images; (iii) shows the in-focus cells that are double-positive stained for PCK and DRAQ5; and (iv) represents the double-positively stained cells that are negative for CD45 staining. (**B**) Images showing (i) two PCK+/CD45− CCs and (ii) an example of a PCK−/CD45+ leucocyte. All images are stained positive with DRAQ5 nuclear staining.

### 2.3. Correlation of CC Numbers with Clinical Timepoints

Baseline blood samples were obtained for all eight patients as close to the start of treatment as possible. For four patients, baseline samples were obtained before the first day of CRT, samples for three patients were taken on the first day of CRT, and for one patient, the baseline sample was taken 5 days after starting CRT. With respect to follow-up samples, five patients had a follow-up sample obtained either during CRT or within one week of finishing, five patients had a sample taken between one week and 3 months of finishing CRT and three patients had samples obtained >3 months after finishing CRT (Table 2).

**Table 2.** CC samples obtained (per patient).

| Patient # | Baseline Sample | Sample 2 | Sample 3 | Sample 4 |
|:---:|:---:|:---:|:---:|:---:|
| 1 | On Day 1 | NS | NS | NS |
| 2 | Before Day 1 | <3 m | >3 m | >3 m |
| 3 | On Day 1 | During CRT | <3 m | >3 m |
| 4 | Before Day 1 | During CRT | <3 m | >3 m |
| 5 | On Day 1 | During CRT | <3 m | NS |
| 6 | After Day 1 | During CRT | <3 m | NS |
| 7 | Before Day 1 | NS | NS | NS |
| 8 | Before Day 1 | During CRT | NS | NS |

NS = No Sample.

Non-haematopoietic, cancer-associated CCs were identified in baseline samples from seven of eight patients (88%) with a mean number of 334 CCs identified/mL blood (range 0–876). No reason was identified to differentiate the patient in whom no CCs were identified from the remaining seven patients. In control samples, non-hematopoietic CCs were identified in 13 out of 28 volunteers (46%) with a mean of 19 cells/mL blood (0–330). In patient samples taken during CRT ($n = 5$), the mean number of CCs identified was 299/mL blood (60–433), although in four patients, CC numbers increased during CRT. For samples taken within 3 months of completing CRT ($n = 5$), the mean number of CCs was 224/mL blood (0–602), and for those samples obtained >3 months from treatment completion ($n = 3$), the mean number of CCs was 49/mL of blood (0–84) (Figure 2A). For the one patient for whom there were two samples obtained over 3 months from treatment completion (patient 2), only the first sample was included in this analysis.

Following this, patients for whom there were >3 samples (patients 2, 3, 4, 5, and 6) were analysed individually to evaluate changes in CC numbers over time and following treatment (Figure 2B). In one patient (patient 5), CC numbers appear to fall during treatment from baseline, and fell further following treatment. For the remaining four patients (2, 3, 4, 6), CC numbers appeared to increase during, or in the weeks following treatment, with subsequent drops in numbers following this. For patient 2, CC numbers fell to 62 cells/mL >3 months following treatment, but when levels were measured 8 months after this, an increase to 84 cells/mL was observed, the significance of which is unclear. At time of analysis, all patients remained disease-free on clinical surveillance.

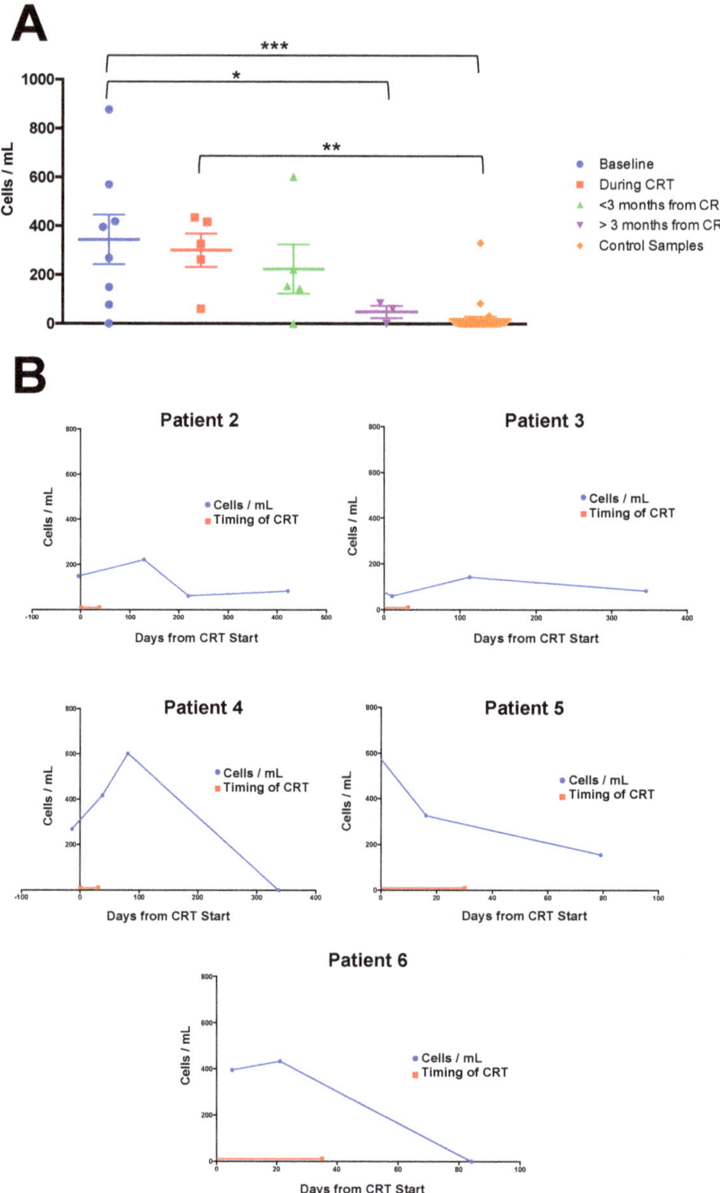

**Figure 2.** (**A**) CC counts were divided into baseline samples, those taken during CRT (and up to one week after EOT), those taken within 3 months of EOT and those taken over 3 months from EOT. Control samples were obtained from 28 volunteers. When analysed using a one-way analysis of variance (ANOVA), a significant difference was observed between baseline levels and control samples ($p < 0.001$), baseline samples and those taken >3 m from treatment ($p < 0.05$) and between samples taken during treatment and controls samples ($p < 0.01$). (**B**) Individual profiles of CC count (cells/mL) over time in the five patients for whom there were 3 or more samples available. Treatment timing is shown is red, whilst blue lines represent CC counts.

## 3. Discussion

We have previously demonstrated the presence of numerous non-hematopoietic CCs in the blood stream of patients with advanced epithelial ovarian cancer and lung cancer, by examining the entire liquid biopsy and avoiding the use of enrichment methods typically employed in CellSearch and ISET technologies [23,29,31,32]. In the present study, we applied these methods to peripheral blood samples obtained from patients with localised SCCA, successfully isolating and identifying cells within the blood of patients which stain positive for the DRAQ5 nuclear marker and the PCK epithelial cell marker, and negative for the pan-hematopoietic maker CD45 for first time (Figure 1). This pattern of expression suggests that these cells represent cancer-associated non-hematopoietic CCs.

Following identification of isolated PCK (+), CD45 (−) CCs, baseline (pre-CRT) CC numbers were compared with samples obtained during CRT, and after CRT and with those from healthy volunteers (Figure 2A). Our results demonstrate significantly higher numbers of CCs in patient samples (7/8 patients (88%), mean of 344 cells/mL, range 0–876, $p < 0.001$), compared to volunteers (13/28 volunteers (46%), mean of 19 cells/mL blood, range 0–330). During treatment, whilst average CC numbers were lower than at baseline, this was statistically non-significant and in four patients, CC numbers actually increased during treatment. However, average CC numbers fell following completion of treatment, with significantly fewer CCs/mL of blood in samples obtained >3 m from the end of treatment (mean 49/mL of blood, range 0–84, $p < 0.05$) compared to baseline; although to date, only 3/8 patients have samples analysed within this time period, and further data collection is ongoing. Taken together, these results suggest a reduction in CC numbers over time from treatment. To provide a clearer impression of the effects of treatment on CC expression, patient samples were then analysed individually (Figure 2B), and this revealed than in 4/5 patients (80%) for whom >3 samples were available, CC numbers appeared to increase during, or in the 3 m following treatment, before subsequently falling. This observation is consistent with previous studies which have shown mobilisation of CCs into the circulation as a result of radiotherapy in patients with lung cancer [33]. Furthermore, an increase in CC numbers in patients during the first three months of treatment for lung cancer was associated with a better progression-free and overall survival [34]. Possible explanations for this include release of CCs prompted by changes in a malignant mass responding to chemotherapy, radiotherapy or damaged during surgical manipulation. Whilst the mechanism is unclear, it is intriguing to consider that the phenotype of the CCs is also important in certain settings. In pancreatic cancer patients, a mesenchymal phenotype for CCs appears to be associated with distant metastases (liver, lung) compared with a pure epithelial CC phenotype which is related to advanced locally recurrent disease [35]. In addition, multicellular, CC microclusters, sometimes including neutrophils, have been identified as a small percentage of the CCs. These microclusters are able to undergo extravasation (via angiopellosis); they have distinct survival and secondary tumour formation abilities, exceeding those of any single CC [36].

During the process of identifying CCs, we too identified various cell doublets including CCs associated with atypical CD45(+), PCK(+) cells (data not shown). The presence of CD45(+), PCK(+) atypical cells associated with CTCs, may represent the presence of atypical cancer-associated macrophage-like cells (CAMLs), which have previously been described in patients with both breast and pancreatic cancer [37]. These CD45(+), PCK(+) cells may either represent macrophages that have engulfed epithelial cellular debris, or CAMLs which have been shown to bind to and migrate in blood attached to CCs [38]. Another possibility is that these cells represent fusions between cancer cells and myeloid cells; atypical cells that may display increased metastatic behavior compared with non-hybrid CCs [39,40]. The presence of these cells demonstrates the complexity of the malignant process and presents an important area of future research to further elucidate the identity of these atypical cells.

Despite these findings, we recognise that there are limitations with the detection of CCs using imaging flow cytometry, as the repertoire of antibody staining is limited for each cell. We have previously demonstrated that using Cell-Free DNA Collection Tubes (Roche), the cell integrity of CCs remains intact for up to 6 days, so we are confident that we are not losing significant numbers of circulating cells or compromising their morphology [29]. We also recognise that CCs may undergo epithelial–mesenchymal transition (EMT) and subsequently loose epithelial markers particularly EpCAM; therefore, it is important in future studies to further characterise the pool of circulating cancer-associated cells using mesenchymal markers such as Vimentin, Fibronectin or N-Cadherin. Furthermore, we and others have also demonstrated the presence of circulating tumor-derived endothelial cells (CTECs) [29], and the properties of endothelial cells within tumours which confer the ability to become CTECs upon entering circulation have recently been the focus of a seminal review article [41]. Further work is on-going to confirm the true prevalence of CTECs in our patient cohorts utilising specific antibodies for endothelial cells (including CD106, CD105, CD34 or CD146). Future work could also exploit the hypoxia driven process of EMT, through the use of emerging technologies to identify abnormal CD31+ CTECS. Identifying these cells in parallel with the PCK+ CCs described in this paper could act as a biomarker not only for treatment response, but also for early metastatic potential [41].

Whilst the results presented showcase convincing data that CCs are present in patients with localised SCCA, further research is needed to fully understand the role of these cells and the implication they have on treatment outcome. After definitive CRT around 10–15% of patients have persistent disease, with a further 15–30% developing subsequent local recurrence after initial complete response. There remains a degree of tension between defining early local recurrence amenable to surgical salvage, and not allowing sufficient time after CRT for a CR to be achieved [12]. Recently, the proportion of patients undergoing salvage surgery seems considerably lower [42] than in older studies [43], reflecting results from the ACT-II clinical trial [12] which demonstrated that delaying surgery is acceptable providing the response trajectory is favourable. Since salvage surgery offers 5-year OS of around 40–45%, serial CC measurements could enable the identification of patients in whom early surgery is necessary versus those who could safely be followed up with imaging; acting as a surrogate marker of response or relapse. Further studies could also investigate whether baseline CC levels are predictive of recurrence. The high levels of CCs in patients undergoing CRT might be indicative of more epithelial cells entering the circulation due to vascular injury. Indeed, increase of CTCs has been documented in patients with advanced esophageal squamous cell carcinoma undergoing CRT [44].

Patient samples continue to be collected for this patient cohort, and future experiments are planned to perform paired ctDNA analysis in an effort to detect and quantify HPV DNA in our patient cohort. Two recent studies have detected HPV ctDNA within the blood of 91.1% of patients with advanced SCCA [45], and 95.6% of patients with early HPV-driven cancers [46]. Combined with knowledge of the immunohistochemical p16 status of our patient cohort, this would provide conclusive evidence that the cells observed and described within this manuscript are cancer-associated CCs. Previously, HPV16 and HPV33 ddPCR assays were used to detect HPV from ctDNA [46]. However, HPV markers can be detected in circulating mononuclear cells from peripheral blood [47]. There is future interest in the application of imaging flow cytometry to our non-enriched cell populations in an attempt to detect HPV-related proteins from liquid biopsies of anal cancer patients.

## 4. Conclusions

In summary, our results confirm the presence of nucleated, non-haematological cells which express epithelial cell surface markers, identified in the bloodstream of patients with clinically localised SCCA. In addition, we have demonstrated that CC numbers fall >3 months following treatment, which is the expected timeframe for response assessment. Data collection is ongoing to evaluate whether CC numbers may provide an early predictor of disease relapse which could be utilised along with serial cross-sectional imaging surveillance to identify those few patients who require radical surgery following CRT for residual or early localised recurrent disease. The identification and quantification of CCs could minimise the mortality among patients with SCCA by providing tailored, personalised, and targeted therapy, whilst also facilitating investigation into new therapeutic targets.

## 5. Materials and Methods

Patient Recruitment: Blood samples were taken from patients enrolled on the CICATRIx clinical study. These patients have advanced cancer and are attending the Mount Vernon Cancer Centre (East and North Hertfordshire NHS Trust). All patients involved in the clinical study provided informed consent for the use of blood samples as well as their participation. The CICATRIx study is a registered and ethics approved study collecting blood samples to explore Circulating tumour cells, cell-free DNA and leucocytes with Imagestream analysis in patients with various cancers. Approved by the West Midlands-South Birmingham Ethics Committee (reference 16/WM/0196).

Blood Sample Preparation: Venepuncture took place using a 21- or 23-G needle, to minimise contamination by skin epithelial cells. Cell-Free DNA Collection Tubes (Roche) were utilized for blood collection to maintain CC membrane integrity for >6 days. A quantity of 1 mL of whole blood from each patient was transferred from Roche tube into a 15 mL Falcon and 9 mL of red blood cell lysis (RBC) buffer was added before gentle rocking and subsequent centrifugation at 2500 rpm for 10 min each to produce a pellet. The process was repeated after removal of supernatant and adding further 3 mL of fresh RBC lysis buffer. Fixation of the pellet was performed by the addition of 1 mL 4% paraformaldehyde in PBS to the pellet and transferring the resultant solution to a 1.5 mL Eppendorf tube. This was left to incubate on ice for 5 min before centrifuging at 4000 rpm for 3 min. After aspiration 1 mL of blocking buffer (10% BSA in PBS) was added and left to incubate on a rotor for an hour followed by addition of antibodies [Pan Cytokeratin (AE1/AE3) and CD45 (Life Technologies)] dissolved in blocking buffer at 1:100 dilution and left to incubate at 4 °C overnight on a rotor. Once the incubation period had elapsed, the sample was centrifuged at 4000 rpm for 3 min and washed with 0.1% Tween in PBS followed by a second centrifugation with the same settings. Finally, 100 µL of Accumax cell detachment solution [StemCell Technologies] was added to the pellet alongside 0.5 µL of the nuclear staining DRAQ5 [BioStatus] ready to be run on the Imagestream Mark II [Luminex].

Running and Analysing Samples: The sample was run using the Imagestream Mark II with each staining expressed in a different channel and subsequently, the files produced were analysed and quantified using the IDEAS software [30]. In order to identify a positive CC, the cell had to express positive staining of Pan Cytokeratin and DRAQ5 with a negative staining of CD45 (in order to exclude lymphocytes). All the samples were analysed by two independent observers (JJ and TC). All statistical tests were performed using GraphPad Prism® Software (GraphPad Software). Statistical analyses were performed using one-way ANOVA with significance determined at the level of $p < 0.05$. $p$ values are indicated in graphs as follows; * $p = 0.01$–$0.05$, ** $p = 0.001$–$0.009$, and *** $p < 0.0009$.

**Author Contributions:** T.J.C. and J.J. contributed to this manuscript equally. Conceptualization, E.K., M.H. and R.G.-J.; methodology, T.J.C., J.J., J.K., E.K.; formal analysis, T.J.C., J.J., J.K., E.K., M.H.; investigation, T.J.C., E.K., M.H., R.G.-J.; resources, M.H. and R.G.-J.; data curation, T.J.C., J.J., J.K.; writing—original draft preparation, T.J.C., J.J., E.K.; writing—review and editing, M.H., R.G.-J.; supervision, E.K., M.H. All authors have read and agreed to the published version of the manuscript.

**Funding:** This research was funded by the Cancer Treatment and Research Trust (CTRT) and Rob-Glynne Jones Charities (PACE #11782).

**Acknowledgments:** We are grateful to the research team at MVCC, and the patients who donated samples to the project.

**Conflicts of Interest:** The authors declare no conflict of interest.

## References

1. Bray, F.; Ferlay, J.; Soerjomataram, I.; Siegel, R.L.; Torre, L.A.; Jemal, A. Global Cancer Statistics 2018: Globocan Estimates of Incidence and Mortality Worldwide for 36 Cancers in 185 Countries. *CA Cancer J. Clin.* **2018**, *68*, 394–424. [CrossRef] [PubMed]
2. Siegel, R.; Ward, E.; Brawley, O.; Jemal, A. Cancer Statistics, 2011: The Impact of Eliminating Socioeconomic and Racial Disparities on Premature Cancer Deaths. *CA Cancer J. Clin.* **2011**, *61*, 212–236. [CrossRef] [PubMed]

3. Kang, Y.-J.; Smith, M.; Canfell, K. Anal Cancer in High-Income Countries: Increasing Burden of Disease. *PLoS ONE* **2018**, *13*, e0205105. [CrossRef] [PubMed]

4. Lewis, G.D.; Haque, W.; Butler, E.B.; Teh, B.S. Survival Outcomes and Patterns of Management for Anal Adenocarcinoma. *Ann. Surg. Oncol.* **2019**, *26*, 1351–1357. [CrossRef] [PubMed]

5. Wilkinson, J.; Morris, E.; Downing, A.; Finan, P.; Aravani, A.; Thomas, J.; Sebag-Montefiore, D. The Rising Incidence of Anal Cancer in England 1990–2010: A Population-Based Study. *Colorectal Dis.* **2014**, *16*, O234–O239. [CrossRef]

6. Patel, J.; Salit, I.E.; Berry, M.J.; de Pokomandy, A.; Nathan, M.; Fishman, F.; Palefsky, J.; Tinmouth, J. Environmental Scan of Anal Cancer Screening Practices: Worldwide Survey Results. *Cancer Med.* **2014**, *3*, 1052–1061. [CrossRef]

7. Ogunbiyi, O.; Scholefield, J.; Smith, J.; Polacarz, S.; Rogers, K.; Sharp, F. Immunohistochemical Analysis of p53 Expression in Anal Squamous Neoplasia. *J. Clin. Pathol.* **1993**, *46*, 507–512. [CrossRef]

8. Glynne-Jones, R.; Saleem, W.; Harrison, M.; Mawdsley, S.; Hall, M. Background and Current Treatment of Squamous Cell Carcinoma of the Anus. *Oncol. Ther.* **2016**, *4*, 135–172. [CrossRef]

9. Northover, J.; Glynne-Jones, R.; Sebag-Montefiore, D.; James, R.; Meadows, H.; Wan, S.; Jitlal, M.; Ledermann, J. Chemoradiation for the Treatment of Epidermoid Anal Cancer: 13-year Follow-Up of the First Randomised UKCCCR Anal Cancer Trial (ACT I). *Br. J. Cancer* **2010**, *102*, 1123–1128. [CrossRef]

10. Glynne-Jones, R.; Nilsson, P.J.; Aschele, C.; Goh, V.; Peiffert, D.; Cervantes, A.; Arnold, D. Anal Cancer: ESMO-ESSO-ESTRO Clinical Practice Guidelines for Diagnosis, Treatment and Follow-Up. *Ann. Oncol.* **2014**, *25*, iii10–iii20. [CrossRef]

11. Moscicki, A.-B.; Darragh, T.M.; Berry-Lawhorn, J.M.; Roberts, J.M.; Khan, M.J.; Boardman, L.A.; Chiao, E.; Einstein, M.H.; Goldstone, S.E.; Jay, N. Screening for Anal Cancer in Women. *J. Low. Genit. Tract Dis.* **2015**, *19*, S26. [CrossRef] [PubMed]

12. Glynne-Jones, R.; Sebag-Montefiore, D.; Meadows, H.M.; Cunningham, D.; Begum, R.; Adab, F.; Benstead, K.; Harte, R.J.; Stewart, J.; Beare, S. Best Time to Assess Complete Clinical Response after Chemoradiotherapy in Squamous Cell Carcinoma of the Anus (Act Ii): A Post-Hoc Analysis of Randomised Controlled Phase 3 Trial. *Lancet Oncol.* **2017**, *18*, 347–356. [CrossRef]

13. Kochhar, R.; Renehan, A.G.; Mullan, D.; Chakrabarty, B.; Saunders, M.P.; Carrington, B.M. The Assessment of Local Response Using Magnetic Resonance Imaging at 3- and 6-month Post Chemoradiotherapy in Patients with Anal Cancer. *Eur. Radiol.* **2017**, *27*, 607–617. [CrossRef] [PubMed]

14. Dewdney, A.; Rao, S. Metastatic Squamous Cell Carcinoma of the anus: Time for a Shift in the Treatment Paradigm? *ISRN Oncol.* **2012**, *2012*, 1–6. [CrossRef]

15. Khoo, B.L.; Grenci, G.; Jing, T.; Lim, Y.B.; Lee, S.C.; Thiery, J.P.; Han, J.; Lim, C.T. Liquid Biopsy and Therapeutic Response: Circulating Tumor Cell Cultures for Evaluation of Anticancer Treatment. *Sci. Adv.* **2016**, *2*, e1600274. [CrossRef]

16. Huang, W.-L.; Chen, Y.-L.; Yang, S.-C.; Ho, C.-L.; Wei, F.; Wong, D.T.; Su, W.-C.; Lin, C.-C. Liquid Biopsy Genotyping in Lung Cancer: Ready for Clinical Utility? *Oncotarget* **2017**, *8*, 18590. [CrossRef]

17. Nawroz, H.; Koch, W.; Anker, P.; Stroun, M.; Sidransky, D. Microsatellite Alterations in Serum Dna of Head and Neck Cancer Patients. *Nat. Med.* **1996**, *2*, 1035–1037. [CrossRef]

18. Chen, X.Q.; Stroun, M.; Magnenat, J.-L.; Nicod, L.P.; Kurt, A.-M.; Lyautey, J.; Lederrey, C.; Anker, P. Microsatellite Alterations in Plasma DNA of Small Cell Lung Cancer Patients. *Nat. Med.* **1996**, *2*, 1033–1035. [CrossRef]

19. Wan, J.C.; Massie, C.; Garcia-Corbacho, J.; Mouliere, F.; Brenton, J.D.; Caldas, C.; Pacey, S.; Baird, R.; Rosenfeld, N. Liquid Biopsies Come of Age: Towards Implementation of Circulating Tumour DNA. *Nat. Rev. Cancer* **2017**, *17*, 223. [CrossRef]

20. Rothwell, D.G.; Ayub, M.; Cook, N.; Thistlethwaite, F.; Carter, L.; Dean, E.; Smith, N.; Villa, S.; Dransfield, J.; Clipson, A. Utility of Ctdna to Support Patient Selection for Early Phase Clinical Trials: The Target Study. *Nat. Med.* **2019**, *25*, 738–743. [CrossRef]

21. Veyer, D.; Pavie, J.; Pernot, S.; Mandavit, M.; Garrigou, S.; Lucas, M.-L.; Gibault, L.; Taly, V.; Weiss, L.; Péré, H. HPV-Circulating Tumoural DNA by Droplet-Based Digital Polymerase Chain Reaction, A New Molecular Tool for Early Detection of HPV Metastatic Anal Cancer? A Case Report. *Eur. J. Cancer* **2019**, *112*, 34–37. [CrossRef]

22. Morris, V.K. Circulating Tumor DNA in Advanced Anal Cancer: A Blood Biomarker Goes Viral. *Clin. Cancer Res.* **2019**, *25*, 2030–2032. [CrossRef]
23. Chudasama, D.; Katopodis, P.; Stone, N.; Haskell, J.; Sheridan, H.; Gardner, B.; Urnovitz, H.; Schuetz, E.; Beck, J.; Hall, M. Liquid Biopsies in Lung Cancer: Four Emerging Technologies and Potential Clinical Applications. *Cancers* **2019**, *11*, 331. [CrossRef]
24. Hou, H.W.; Warkiani, M.E.; Khoo, B.L.; Li, Z.R.; Soo, R.A.; Tan, D.S.-W.; Lim, W.-T.; Han, J.; Bhagat, A.A.S.; Lim, C.T. Isolation and Retrieval of Circulating Tumor Cells Using Centrifugal Forces. *Sci. Rep.* **2013**, *3*, 1259. [CrossRef] [PubMed]
25. Krebs, M.G.; Hou, J.-M.; Ward, T.H.; Blackhall, F.H.; Dive, C. Circulating Tumour Cells: Their Utility in Cancer Management and Predicting Outcomes. *Ther. Adv. Med. Oncol.* **2010**, *2*, 351–365. [CrossRef] [PubMed]
26. Dent, B.M.; Ogle, L.F.; O'Donnell, R.L.; Hayes, N.; Malik, U.; Curtin, N.J.; Boddy, A.V.; Plummer, E.R.; Edmondson, R.J.; Reeves, H.L. High-Resolution Imaging for the Detection and Characterisation of Circulating Tumour Cells from Patients with Oesophageal, Hepatocellular, Thyroid and Ovarian Cancers. *Int. J. Cancer* **2016**, *138*, 206–216. [CrossRef] [PubMed]
27. Castle, J.; Morris, K.; Pritchard, S.; Kirwan, C.C. Challenges in Enumeration of CTCs in Breast Cancer Using Techniques Independent of Cytokeratin Expression. *PLoS ONE* **2017**, *12*, e0175647. [CrossRef] [PubMed]
28. Bernardi, M.-P.; Ngan, S.Y.; Michael, M.; Lynch, A.C.; Heriot, A.G.; Ramsay, R.G.; Phillips, W.A. Molecular Biology of Anal Squamous Cell Carcinoma: Implications for Future Research and Clinical Intervention. *Lancet Oncol.* **2015**, *16*, e611–e621. [CrossRef]
29. Kumar, J.; Chudasama, D.; Roberts, C.; Kubista, M.; Sjöback, R.; Chatterjee, J.; Anikin, V.; Karteris, E.; Hall, M. Detection of Abundant Non-Haematopoietic Circulating Cancer-Related Cells in Patients with Advanced Epithelial Ovarian Cancer. *Cells* **2019**, *8*, 732. [CrossRef]
30. Rogers-Broadway, K.R.; Kumar, J.; Sisu, C.; Wander, G.; Mazey, E.; Jeyaneethi, J.; Pados, G.; Tsolakidis, D.; Klonos, E.; Grunt, T. Differential Expression of mTOR Components in Endometriosis and Ovarian Cancer: Effects of Rapalogues and Dual Kinase Inhibitors on mTORC1 and mTORC2 Stoichiometry. *Int. J. Mol. Med.* **2019**, *43*, 47–56. [CrossRef]
31. Barr, J.; Chudasama, D.; Rice, A.; Karteris, E.; Anikin, V. Lack of Association between Screencell-Detected Circulating Tumour Cells and Long-Term Survival of Patients Undergoing Surgery for Non-Small Cell Lung Cancer: A Pilot Clinical Study. *Mol. Clin. Oncol.* **2020**, *12*, 191–195. [CrossRef] [PubMed]
32. Chudasama, D.; Burnside, N.; Beeson, J.; Karteris, E.; Rice, A.; Anikin, V. Perioperative Detection of Circulating Tumour Cells in Patients with Lung Cancer. *Oncol. Lett.* **2017**, *14*, 1281–1286. [CrossRef] [PubMed]
33. Martin, O.A.; Anderson, R.L.; Russell, P.A.; Cox, R.A.; Ivashkevich, A.; Swierczak, A.; Doherty, J.P.; Jacobs, D.H.; Smith, J.; Siva, S. Mobilization of Viable Tumor Cells into the Circulation during Radiation Therapy. *Int. J. Radiat. Oncol. Biol. Phys.* **2014**, *88*, 395–403. [CrossRef]
34. Shishido, S.N.; Carlsson, A.; Nieva, J.; Bethel, K.; Hicks, J.B.; Bazhenova, L.; Kuhn, P. Circulating Tumor Cells as a Response Monitor in sTage IV Non-Small Cell Lung Cancer. *J. Transl. Med.* **2019**, *17*, 294. [CrossRef] [PubMed]
35. Poruk, K.E.; Valero, V., III; Saunders, T.; Blackford, A.L.; Griffin, J.F.; Poling, J.; Hruban, R.H.; Anders, R.A.; Herman, J.; Zheng, L. Circulating Tumor Cell Phenotype Predicts Recurrence and Survival in Pancreatic Adenocarcinoma. *Ann. Surg.* **2016**, *264*, 1073. [CrossRef] [PubMed]
36. Allen, T.A.; Asad, D.; Amu, E.; Hensley, M.T.; Cores, J.; Vandergriff, A.; Tang, J.; Dinh, P.-U.; Shen, D.; Qiao, L. Circulating Tumor Cells Exit Circulation While Maintaining Multicellularity, Augmenting Metastatic Potential. *J. Cell Sci.* **2019**, *132*, jcs231563. [CrossRef]
37. Adams, D.L.; Martin, S.S.; Alpaugh, R.K.; Charpentier, M.; Tsai, S.; Bergan, R.C.; Ogden, I.M.; Catalona, W.; Chumsri, S.; Tang, C.-M. Circulating Giant Macrophages as a Potential Biomarker of Solid Tumors. *Proc. Natl. Acad. Sci. USA* **2014**, *111*, 3514–3519. [CrossRef]
38. Broncy, L.; Paterlini-Bréchot, P. Cancer-Associated Circulating Atypical Cells with both Epithelial and Macrophage-Specific Markers. *J. Lab. Precis. Med.* **2018**, *3*, 91. [CrossRef]
39. Gast, C.E.; Silk, A.D.; Zarour, L.; Riegler, L.; Burkhart, J.G.; Gustafson, K.T.; Parappilly, M.S.; Roh-Johnson, M.; Goodman, J.R.; Olson, B. Cell Fusion Potentiates Tumor Heterogeneity and Reveals Circulating Hybrid Cells That Correlate With Stage and Survival. *Sci. Adv.* **2018**, *4*, eaat7828. [CrossRef]

40. Szczerba, B.M.; Castro-Giner, F.; Vetter, M.; Krol, I.; Gkountela, S.; Landin, J.; Scheidmann, M.C.; Donato, C.; Scherrer, R.; Singer, J. Neutrophils Escort Circulating Tumour Cells to Enable Cell Cycle Progression. *Nature* **2019**, *566*, 553–557. [CrossRef]

41. Lin, P.P. Aneuploid Circulating Tumor-Derived Endothelial Cell (CTEC): A Novel Versatile Player in Tumor Neovascularization and Cancer Metastasis. *Cells* **2020**, *9*, 1539. [CrossRef] [PubMed]

42. Patel, S.V.; Ko, G.; Raphael, M.J.; Booth, C.M.; Brogly, S.B.; Kalyvas, M.; Li, W.; Hanna, T. Salvage APR for Anal Squamous Cell Carcinoma: Utilization, Risk Factors and Outcomes in a Canadian Population. *Dis. Colon Rectum* **2020**. [CrossRef] [PubMed]

43. Renehan, A.G.; Saunders, M.P.; Schofield, P.F.; O'Dwyer, S.T. Patterns of Local Disease Failure and Outcome after Salvage Surgery in Patients with Anal Cancer. *Br. J. Surg.* **2005**, *92*, 605–614. [CrossRef]

44. Su, P.-J.; Wu, M.-H.; Wang, H.-M.; Lee, C.-L.; Huang, W.-K.; Wu, C.-E.; Chang, H.-K.; Chao, Y.-K.; Tseng, C.-K.; Chiu, T.-K. Circulating Tumour Cells as an Independent Prognostic Factor in Patients with Advanced Oesophageal Squamous Cell Carcinoma Undergoing Chemoradiotherapy. *Sci. Rep.* **2016**, *6*, 31423. [CrossRef] [PubMed]

45. Bernard-Tessier, A.; Jeannot, E.; Guenat, D.; Debernardi, A.; Michel, M.; Proudhon, C.; Vincent-Salomon, A.; Bièche, I.; Pierga, J.-Y.; Buecher, B. Clinical Validity of HPV Circulating Tumor Dna in Advanced Anal Carcinoma: An Ancillary Study to the Epitopes-HPV02 Trial. *Clin. Cancer Res.* **2019**, *25*, 2109–2115. [CrossRef] [PubMed]

46. Damerla, R.R.; Lee, N.Y.; You, D.; Soni, R.; Shah, R.; Reyngold, M.; Katabi, N.; Wu, V.; McBride, S.M.; Tsai, C.J. Detection of Early Human Papillomavirus–Associated Cancers by Liquid Biopsy. *JCO Precis. Oncol.* **2019**, *3*, 1–17. [CrossRef] [PubMed]

47. Foresta, C.; Bertoldo, A.; Garolla, A.; Pizzol, D.; Mason, S.; Lenzi, A.; De Toni, L. Human Papillomavirus Proteins Are Found in Peripheral Blood and Semen Cd20+ and Cd56+ Cells during Hpv-16 Semen Infection. *BMC Infect. Dis.* **2013**, *13*, 593. [CrossRef] [PubMed]

*Article*

# Adenoviral Vectors Armed with PAPILLOMAVIRUs Oncogene Specific CRISPR/Cas9 Kill Human-Papillomavirus-Induced Cervical Cancer Cells

**Eric Ehrke-Schulz, Sonja Heinemann, Lukas Schulte, Maren Schiwon and Anja Ehrhardt ***

Institute for Virology and Microbiology, Department for Human Medicine, Faculty of Health,
Center for Biomedical Education and Research (ZBAF), Witten/Herdecke University, Stockumer Street 10,
58453 Witten, Germany; Eric.Ehrke-Schulz@uni-wh.de (E.E.-S.); Sonja.Heinemann@uni-wh.de (S.H.);
Lukas.Schulte@uni-wh.de (L.S.); m.schiwon@gmx.de (M.S.)
* Correspondence: anja.ehrhardt@uni-wh.de

Received: 15 May 2020; Accepted: 14 July 2020; Published: 17 July 2020

**Abstract:** Human papillomaviruses (HPV) cause malignant epithelial cancers including cervical carcinoma, non-melanoma skin and head and neck cancer. They drive tumor development through the expression of their oncoproteins E6 and E7. Designer nucleases were shown to be efficient to specifically destroy HPV16 and HPV18 oncogenes to induce cell cycle arrest and apoptosis. Here, we used high-capacity adenoviral vectors (HCAdVs) expressing the complete CRISPR/Cas9 machinery specific for HPV18-E6 or HPV16-E6. Cervical cancer cell lines SiHa and CaSki containing HPV16 and HeLa cells containing HPV18 genomes integrated into the cellular genome, as well as HPV-negative cancer cells were transduced with HPV-type-specific CRISPR-HCAdV. Upon adenoviral delivery, the expression of HPV-type-specific CRISPR/Cas9 resulted in decreased cell viability of HPV-positive cervical cancer cell lines, whereas HPV-negative cells were unaffected. Transduced cervical cancer cells showed increased apoptosis induction and decreased proliferation compared to untreated or HPV negative control cells. This suggests that HCAdV can serve as HPV-specific cancer gene therapeutic agents when armed with HPV-type-specific CRISPR/Cas9. Based on the versatility of the CRISPR/Cas9 system, we anticipate that our approach can contribute to personalized treatment options specific for the respective HPV type present in each individual tumor.

**Keywords:** papillomavirus; HPV; CRISPR; gene therapy; viral vector; adenovirus

---

## 1. Introduction

Human papillomaviruses (HPV) are small non-enveloped epitheliotropic DNA viruses with a circular genome comprised of approximately 8000 base pairs (bp). So called high-risk HPV are responsible for the development of malignant cervical carcinoma and their precursor lesions cervical intraepithelial neoplasia (CIN) [1]. Currently over 200 different HPV types are known. Due to their carcinogenic properties, especially HPV16, 18 and 31 (but also 33, 35, 39, 45, 51, 52, 56, 58, 59, 68, 73, and 82) are classified as high-risk HPV types [2]. HPV DNA was also found in oropharyngeal carcinomas summarized as head and neck cancer (HNC) and are therefore regarded as important carcinogens [3]. Furthermore, HPV are a cofactor in the development of dermatologic malignancies, such as Squamous cell carcinoma (SCC) and basal cell carcinoma (BCC) of the skin summarized as non-melanoma skin cancer (NMSC), and their precursor lesions, actinic keratosis (AK) [4]. Despite the availability of a protective vaccination against high-risk HPV, not all girls and young women (as well as boys and young men) are vaccinated. Therefore, it is likely, that HPV-associated tumors represent a continuous health care burden. HPV-associated tumors such as cervical carcinoma, NMSC and HNC can be removed surgically, but a complete detection and removal of tumor tissue including possible precursor lesions in the immediate vicinity is difficult as non-infected regions, HPV infected but non transformed regions, precursor lesions and invasive carcinomas are located in close proximity of an epithelial area [5]. The conventional cancer therapies such as radiation or chemotherapy in concert with surgical resection increase the chances of recovery but are associated with strong side effects. Therefore, the development and testing of alternative treatment strategies are desirable.

In the context of cervical infection, the HPV genome can be integrated into the chromosomal DNA of basal epithelial cells. During this process, the episomal, circular HPV genome is linearized within the HPV E2 gene, leading to a loss of E2 function. As the HPV E2 protein negatively regulates HPV E6 and E7 gene expression, the loss of HPV E2 function leads to unregulated overexpression of the HPV oncogenes E6 and E7, that mediate transformation of the host cell by inducing cell proliferation, bypassing cell cycle control and inhibition of apoptosis. The HPV E6 protein interacts with the tumor suppressor protein p53, causes its degradation and thus prevents cell cycle control and the initiation of apoptosis [6]. The HPV E7 protein interacts with the retinoblastoma protein pRB and causes the release of the pRB-bound E2F transcription factor. E2F subsequently induces the expression of genes that induce cell cycle progression [7,8]. In recent years, several approaches have been described to induce cell cycle arrest and induction of apoptosis in HPV positive cancer cells, either by specifically inhibiting HPV oncoprotein interactions using intracellular antibodies [9] or peptides such as HPV E7 antagonist [10], E6-binding aptamer [11], E6-AP mimetic epitope [12], or RNA molecules like Anti-E6 ribozyme [13]. Moreover, downregulating HPV oncogene expression using RNA interference [14–21] has been shown to be a promising treatment option for HPV induced tumors. However, only the most recent studies tried to translate these in vitro findings toward in vivo applications using non-viral RNA delivery or AAV-mediated viral delivery.

Designer nucleases such as zinc finger nucleases (ZFN), transcription activator-like effector nuclease (TALEN) and especially clustered regularly interspaced short palindromic repeats (CRISPR/Cas9) are highly efficient customizable molecular scissors for sequence-specific induction of in/del mutations at the DNA-target site. This gene disruption strategy has been applied in antiviral approaches against DNA viruses such as Hepatis B virus (HBV) [22–24]. In the context of HPV-related cancer, it was reported that cleavage of the HPV18 origin of replication by artificial zinc-finger proteins fused to a bacterial nuclease inhibited HPV DNA replication [25]. CRISPR/Cas9 mediated inactivation of HPV18- and HPV16 E6 or E7 resulted in the induction of p53 or pRb, leading to cell cycle arrest and cell death [26]. CRISPR/Cas9 mediated disruption of HPV E7 open reading frame (orf) alone induced apoptosis and growth inhibition in HPV16 positive cervical cancer cells [27] and HPV6/11 E7-expressing keratinocytes [28]. Moreover, it was shown that a CRISPR/Cas9 approach to specifically destroy the HPV16 early promoter, HPV16 E6, or HPV16 E7 coding regions lead to decreased cell viability and increased apoptosis induction and decreased growth of SiHa cell-derived tumors implants in BALB/C nude mice [29]. Interestingly cleavage of HPV16 E6 alone was sufficient to induce apoptosis and growth inhibition of HPV16-positive cells [30].

However, delivery approaches to enable comprehensive in preclinical in vivo studies are rare. A recent study showed that non-viral CRISPR/Cas9 delivery using PEGylated liposomes resulted in tumor elimination in vivo [31]. Numerous publications reported the CRISPR/Cas9 delivery using AAV co-transduction approaches. However, these vectors are rather small and do not allow to deliver all CRISPR/Cas9 components including one or several guide RNAs within one vector. Even though CRISPR/Cas9 is particularly suitable for arming conditionally replicating Adenoviruses (CRAdVs)/oncolytic Adenoviruses [32–34], the potential of viral delivery of HPV-specific CRISPR/Cas9 was not fully exploited. Adenoviruses (AdV) infect a great variety of different cells types and tissues and can enter quiescent as well as dividing cells. Another benefit of the AdV vector system arises from the non-integrating, episomal persistence of the viral genome [35]. As AdV do not integrate their genome into host cell chromosomes genotoxicity related to insertion into transcribed genomic loci is circumvented. To translate the non-viral CRISPR approaches of the preceding studies into an all-in-one viral vector delivery approach to enable future in vivo studies (high-capacity adenoviral vectors (HCAdV)) offer several benefits. In HCAdV genomes all viral coding sequences have been removed and only the inverted terminal repeats (ITR) and the packaging signal that is necessary for vector genome replication and efficient genome packaging are still present [36,37]. The HCAdV packaging capacity of up to 35 kb allows transporting the whole CRISPR/Cas9 machinery including several gRNAs [38]. As HCAdV do not express AdV genes, they are regarded as less immunogenic than early generation AdV vectors [39,40]. Nevertheless, production of HCAdV are time and work intensive when compared with Lentivirus- or AAV-vector platforms, hampering their exploration for specific applications.

In this study, we aim to exploit the advantages of the HCAdV platform and constructed HCAdVs expressing the complete CRISPR/Cas9 machinery including Cas9 and a gRNA specific for HPV16-E6 [29] or HPV18E6 [26,38] within a single vector. These non-replicating HCAdV vectors express the HPV E6 specific CRISPR/Cas9 and destroy the respective E6 oncogene. Upon vector transduction of HPV positive cervical carcinoma cell lines HeLa, CaSki, and SiHa and HPV-negative lung carcinoma cell line A549, we examined the potential of the HPV E6 specific CRISPR/Cas9 expressing HCAdV to mediate apoptosis induction, inhibition of tumor cell growth, and increase of tumor cell death.

## 2. Results

### 2.1. HPV Oncogene Specific CRISPR Expressed from HCAdVs Efficiently Disrupts PV E6 Oncogenes

A schematic overview of the HCAdV vector genomes used in this study is presented in Figure 1. As a control virus, we used the E1- and E3-deleted first generation adenoviral vector ΔE1-ΔE3-AdV5, which can replicate in HPV-positive cancer cells. Furthermore, two HCAdVs-encoding CRISPR/Cas9 and gRNA against the HPV16 oncogene E6 (HCAdV-CRISPR-HPV16E6gRNA) and the HPV18 oncogene E6 (HCAdV-GFP-CRISPR-HPV18E6gRNA) were generated (Figure 1).

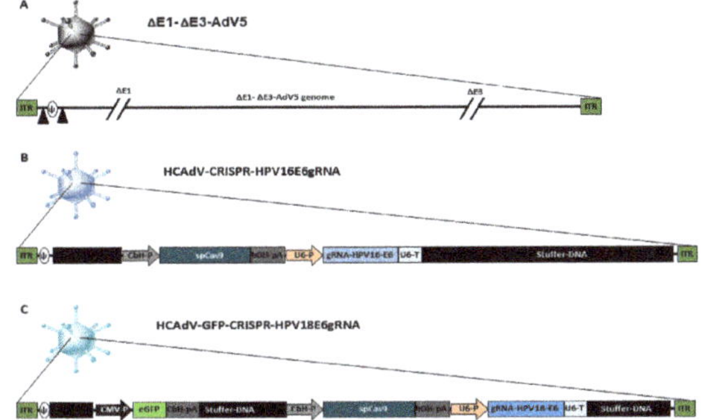

**Figure 1.** Schematic presentation of the genome organization of the vectors used in this study. (**A**) ΔE1-ΔE3-AdV5 contains a wild type AdV5 genome lacking the E1- and E3 gene, the packaging signal is flanked by loxP sites indicated by black triangles. (**B**) HCAdV-CRISPR-HPV16E6gRNA contains a spCas9-gene from Staphylococcus pyogenes controlled by a CbH-promoter and a bGH-pA and a gRNA expression cassette with specificity toward HPV16-E6 controlled by a human U6-promoter and a U6-terminator sequence. (**C**) HCAdV-CRISPR-HPV18E6gRNA contains a spCas9-gene controlled by a CbH-promoter and a bGH-pA and a gRNA expression cassette with specificity towards HPV18-E6 controlled by a human U6-Promoter and a U6-terminator sequence. It also contains an enhanced GFP transgene expression cassette controlled by a CMV promoter and a CbH-pA. CbH-P, chicken β actin hybrid promoter; bGH-pA, bovine growth hormone polyadenylation signal; gRNA, guide RNA; U6-P, human U6- small nuclear RNA promoter; U6-pA, U6- small nuclear RNA polyadenylation signal; CMV-P, Cytomegalovirus promoter, CbH-pA chicken β actin hybrid polyadenylation signal; ITR, inverted terminal repeat; Ψ, AdV5 packaging signal.

To prove CRISPR-mediated mutation induction at the target sites of the respective gRNAs, HeLa and SiHa cells were transduced with HPV18-E6 specific CRISPR-HCAdV or HPV16-E6 specific CRISPR-HCAdV, respectively. Then, 48 h post-transduction, genomic DNA of HPV18-E6-specific CRISPR-HCAdV-treated HeLa cells and HPV16-E6-specific CRISPR-HCAdV-treated SiHa cells was isolated for mutation detection. HPV-E6 loci surrounding the respective gRNA target sites were amplified by PCR. Resulting PCR amplicons were then subjected to hetero-duplex formation and digested with T7 endonuclease 1. Specific cleavage products of the expected size were detected for HPV18-E6 and HPV16-E6 respectively indicating successful mutation induction at the predicted target sites (Figure 2).

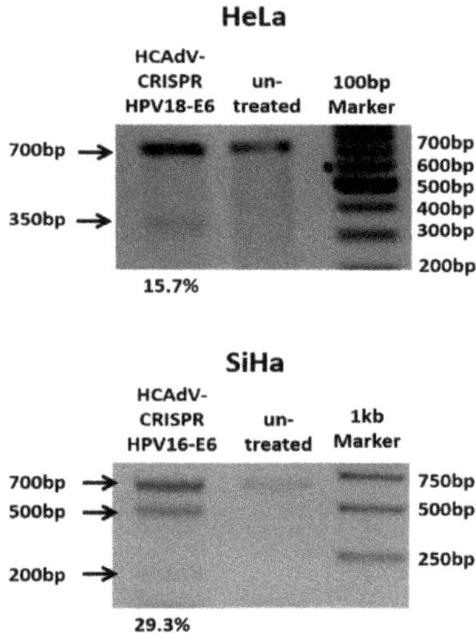


**Figure 2.** Heteroduplex based mutation detection of HPV16-E6 and HPV18-E6 after treatment of SiHa or HeLa cells at multiplicity of infection (MOI) 500 per cell. 48 h post transduction the T7E1-assay showed efficient mutagenesis of HPV-E6-genes of cells treated with HPV specific CRISPR HCAdV compared to cells that were not transduced. Arrows indicate the expected size of HPV-E6 specific PCR-products (~700 bp) and expected size of mutation specific cleavage products (~350 bp for HPV18E6 and 200 bp and 500 bp for HPV16E6). Mutation rates calculated in percentages from the differences in band strength of the original PCR product and cleavage products are depicted below.

## 2.2. Decrease of HPV Tumor Cell Survival

To examine the potential of HCAdV armed with HPV-E6 specific CRISPR/Cas9 to reduce HPV tumor specific cell survival, we transduced HeLa, SiHa and CaSki cervical cancer cells as well as HPV negative A549 lung carcinoma cells with HPV18-E6 specific CRISPR-HCAdV, HPV16-E6 specific CRISPR-HCAdV, ΔE1-ΔE3-AdV5, or AdV storage buffer. Compared to untreated controls or AdV storage buffer treated controls, cervical carcinoma cells treated with HPV-E6-specific CRISPR-HCAdV showed a reduced number of surviving cells (Figure 3). SiHa cells transduced with HPV16-E6-specific CRISPR-HCAdV showed a significant reduction of 85.4% of surviving cells, whereas transduction with ΔE1-ΔE3-AdV5 led to a reduction of 44.3% of viable cells. CaSki cells transduced with HPV16-E6-specific CRISPR-HCAdV showed a reduction of 29.6% of metabolizing cells (Figure 3). In contrast, transduction with ΔE1-ΔE3-AdV5 resulted in a reduction of 30.4% of viable cells. In HeLa cells transduced with HPV18-E6-specific CRISPR-HCAdV showed a viability reduction of 33.7%, whereas transduction with ΔE1-ΔE3-AdV5 decrease viable cells of 15.8% (Figure 3). Even though a trend for reduction of cell viability can be observed CaSki and HeLa cells, values obtained for these cells were not statistically significant. For HPV-negative A549 cells, transduced with CRISPR expressing HCAdV, cell viability was also reduced. Transduction with HPV16-E6 specific CRISPR-HCAdV led to a reduction of 22% of viable cells and for HPV18-E6 specific CRISPR-HCAdV to a reduction of 52.1% of viable cells. ΔE1-ΔE3-AdV5 resulted in a reduction of 36% when measuring viable cells in A549 cells. Compared with untreated or buffer treated controls, values were not statistically significant (Figure 3).

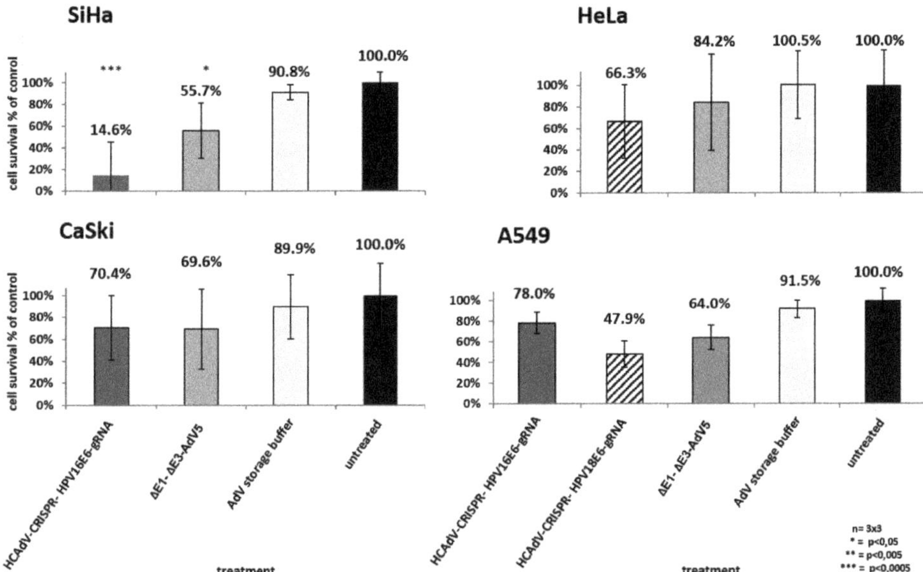



**Figure 3.** Monitoring of cell survival in SiHa, HeLa, CaSki, and A549 cells. Cells were transduced at confluency with HPV-E6-specific CRISPR/Cas9-HCAdV or ΔE1-ΔE3-AdV5 with 1000 infectious vector particles per cell. Cell viability was measured using a CCK-8 assay five days post-transduction. HPV-positive tumor cells showed reduced viability compared to untreated controls. Standard deviations of mean values are shown as error bars. Statistically significant differences compared to untreated controls are shown as one, two, or three stars indicating *p* values < 0.05, <0.005, and <0.0005 respectively.

Following the CCK-8 cell viability screening, the medium was removed, and cells were subjected to methylene blue staining to confirm the previous results using a different methodology that visualizes the healthy attached cells. The results of the methylene blue staining support the results obtained for the CCK-8-based viability assay and showed even stronger effects on the attachment of cells as quantified by the CCK-8 assay. In HeLa, SiHa, and CaSki, a clear decrease of attached cells could be seen after transduction with the respective vector at MOI 1000, whereas untreated controls (MOI 0) or AdV storage-buffer-treated controls were well attached (Supplementary Materials Figure S1). A549 cells showed reduction in cell attachment when treated with HPV18-E6 or HPV16-E6-specific CRISPR-HCAdV or ΔE1-ΔE3-AdV5 (Figure S1).

## 2.3. Cervical Cancer Cell Lines Show Different Susceptibility to AdV5

To find out whether the differences in the effect of the HPVE6 specific CRISPR/Cas9 expressing HCAdV on different cervical cancer cell lines is caused by different transduction efficiencies of the vector, we determined the susceptibility of SiHa, HeLa, and CaSki cells to AdV5. We infected each respective cell line with defined numbers of viral particles of a GFP-luciferase expressing E3 deleted AdV5.

24 h post transduction with 20 viral particles per cell, quantification of luciferase activity of transduced cells showed a significant 100.4-fold increase in luminescence in SiHa cells compared to CaSki cells, whereas HeLa cells revealed a 2.1-fold increase in luciferase expression levels compared to

CaSki cells (Figure 4A). At low virus concentration, SiHa cells seem to be more susceptible to AdV5 infection than HeLa and CaSki cells (Figure 4A).

Due to saturation of the luminescence signal at higher viral particle numbers, we compared susceptibility of the different cell lines to AdV5 by quantifying the fluorescent signal from vector-derived GFP expression. Quantification of the mean fluorescence intensity 48 h post transduction of each respective cell line with 1000 viral particles per cell showed a significant 1.5-fold increased fluorescence signal in SiHa and HeLa cells if directly compared to CaSki cells, respectively. No difference was observed between SiHa and HeLa cells (Figure 4B).

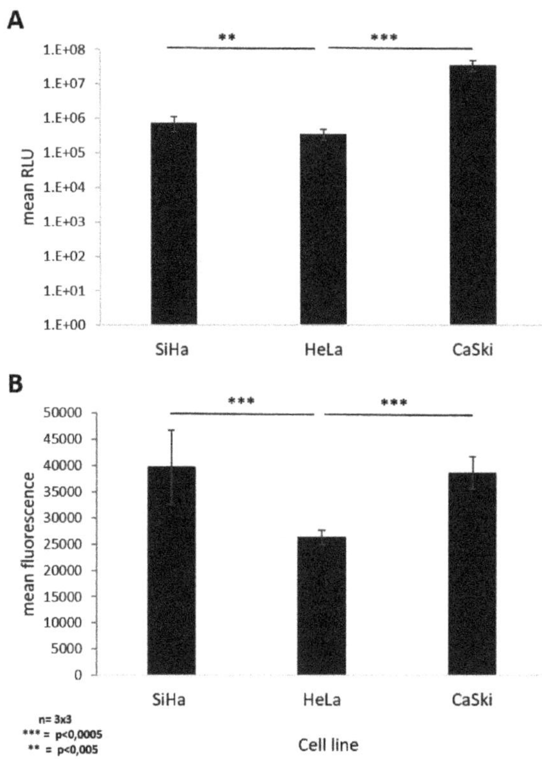


**Figure 4.** Monitoring cell susceptibility of SiHa, HeLa, and CaSki cells to AdV5. Siha, HeLa, and Caski cells were infected with E3-deleted AdV5-expressing GFP and luciferase at different doses. (**A**) AdV5 mediated luminescence 24 h post transduction with 20 viral particles per cell (vpc). (**B**) AdV 5 mediated fluorescence 48 h post transduction with 1000 vpc. Standard deviations of mean values are shown as error bars. The line above the columns indicate which sampled were compared to each other Statistically significant differences of the cell lines compared to each other are shown as two or three stars, indicating *p* values < 0.005, and 0.0005 respectively.

## 2.4. Reduction of Proliferation of HPV Positive Cancer Cell Lines

To investigate whether HPV-E6 specific CRISPR-HCAdV can reduce proliferation of HPV-induced cervical cancer cells, we transduced HPV18 containing HeLa cells, HPV16-positive SiHa and CaSki and SiHa cervical cancer cells as well as HPV-negative A459 lung carcinoma cells. We applied the vectors HPV18-E6 specific CRISPR-HCAdV, HPV16-E6 specific CRISPR-HCAdV or ΔE1-ΔE3-AdV5 at MOI 1000 and monitored the increase of viable cells for eight days. Transduction with HPV16-E6-specific CRISPR-HCAdV inhibited cell proliferation of SiHa cells and the number of viable cells significantly differed from untreated controls already three days post-transduction. In contrast, transduction with ΔE1-ΔE3-AdV5 only led to a significant reduction of cell proliferation that was significantly different from untreated controls after day 6 (Figure 5). Transduction with HPV16-E6 specific CRISPR-HCAdV inhibited cell proliferation of CaSki cells and the number of viable cells was significantly reduced compared to untreated controls already four days post-transduction. Transduction with ΔE1-ΔE3-AdV5 also resulted in a significant reduction of cell proliferation that was significantly different from untreated controls after day 6 (Figure 5). Transduction with HPV18-E6 specific CRISPR-HCAdV strongly inhibited cell proliferation of HeLa cells, which was in sharp contrast to untreated controls already three days post-transduction. Transduction with ΔE1-ΔE3-AdV5 resulted in a less pronounced reduction of cell proliferation that was still significantly different from untreated controls between days 4–6 (Figure 5).

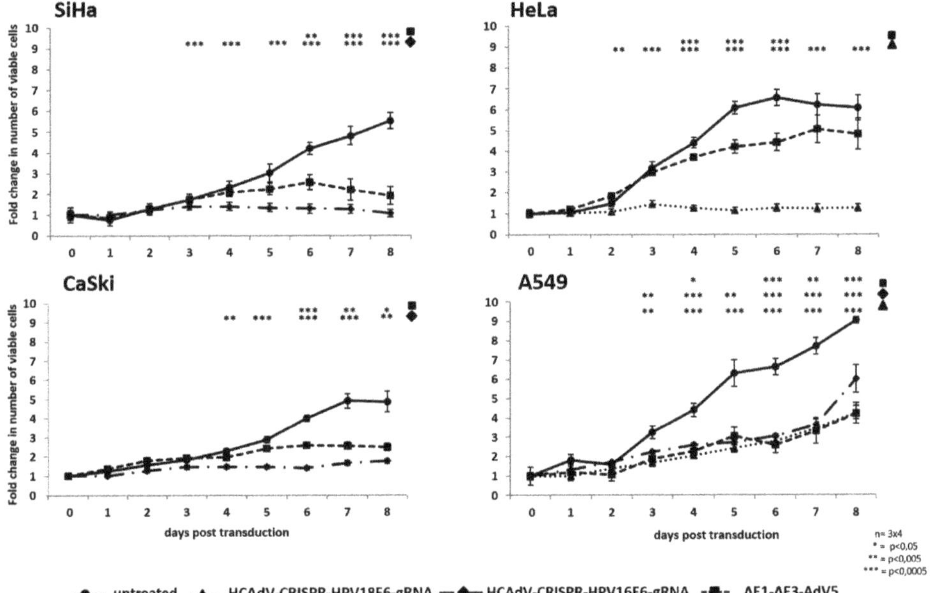


**Figure 5.** Monitoring of cell proliferation in SiHa, HeLa, CaSki, and A549 cells. Following transduction with the respective HPV-E6-specific CRISPR/Cas9-HCAdV or ΔE1-ΔE3-AdV5 with 1000 infectious vector particles per cell, cell proliferation was monitored for eight days. The relative number of viable cells at each time point post-transduction was measured by CCK-8 assay. Standard deviations of mean values are depicted as error bars. Statistically significant differences compared to untreated controls are shown as one, two or three stars, indicating *p* values < 0.05, <0.005, and <0.0005 respectively.

Note that transduction with HPV16-E6-specific CRISPR-HCAdV, HPV18-E6-specific CRISPR-HCAdV, or ΔE1-ΔE3-AdV5 also reduced cell proliferation of A549 cells, which was different from untreated controls. Even though cell proliferation was reduced proliferation continued until the end of the experiment independent of the treatment (Figure 5). Compared to untreated control cells, treatment with HPV-E6 specific CRISPR-HCAdV or ΔE1-ΔE3-AdV5 led to a decreased cell proliferation in HPV-induced tumor cells. In HeLa cells, cell proliferation was strongly inhibited by the HPV18-E6-specific CRISPR-HCAdV and ΔE1-ΔE3-AdV5. In SiHa and CaSki cells, HPV16-E6-specific CRISPR-HCAdV and HPV18-E6-specific CRISPR-HCAdV and ΔE1-ΔE3-AdV5 reduced cell proliferation when compared to untreated cells. In contrast, HPV negative A549 cells were almost unaffected by HPV-E6-specific CRISPR-HCAdV when compared to untreated controls. As seen in HPV-positive cells ΔE1-ΔE3-AdV5 was also able to inhibit cell proliferation in

HPV-negative A549 cells. Taken together, these results suggested that tumor cell proliferation inhibition of specific HPV-E6-specific CRISPR-HCAdV is specific for HPV-positive cervical cancer cells and that ΔE1-ΔE3-AdV5 acts through a different mechanism possibly conditional replication in tumor cells.

### 2.5. Vector Treatment Lead to Increase in p53 Protein Levels

To prove that HCAdV delivery of HPV16E6 or HPV18-specific CRISPR/Cas9 and subsequent E6 mutagenesis increased apoptosis in a p53 dependent manner, we performed cell western analysis to quantify the change in cellular p53 level upon vector transduction. The results show a significant of 1.6-fold and 1.3-fold increase of p53 for SiHa and CaSki cells, respectively, when treated with HCAdV-CRISPR-HPV16E6gRNA. HeLa and A549 cell treated with HCAdV-CRISPR-HPV18E6gRNA showed no change in p53 levels (Figure 6).

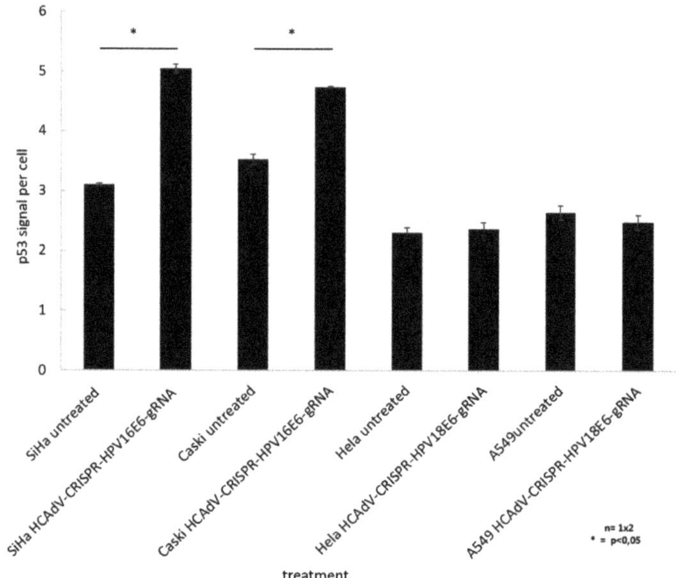


**Figure 6.** Monitoring change in p53 levels in treated SiHa, HeLa, and CaSki cells. In cell western analyses based on p53 quantification of cells treated with MOI 1000 of respective vectors compared to untreated controls 96 h post transduction. For each well, the fluorescent intensity values of the p53 antibody measurement at 800 nm was divided by the mean fluorescent intensity of cell tag700 at 700 nm to normalize for the cell number (p53 signal per cell). Standard deviations of mean values are shown as error bars. Statistically significant differences compared to untreated controls are shown as one star, indicating *p* values < 0.05.

### 2.6. Increase in Apoptosis Induction

As high-risk HPV-E6 proteins interact with the cellular p53 tumor suppressor and mediate its proteasomal degradation, HPV-induced cervical cancer cells can circumvent p53 mediated apoptosis induction. Inactivation of HPV-E6 should inhibit HPV-E6-mediated p53 degradation, leading to a re-accumulation p53 protein and a reactivation of p53 mediated apoptosis induction. To investigate whether HCAdV armed with HPV-E6-specific CRISPR/Cas9 specifically induce cell death in HPV tumor cells by mediating apoptosis induction, we transduced HPV18-containing HeLa cells, SiHa and CaSki cervical cancer cells, and A459 lung carcinoma cells. We applied the vectors HPV18-E6-specific CRISPR-HCAdV, HPV16-E6-specific CRISPR-HCAdV, and ΔE1-ΔE3-AdV5. Two days post-transduction, the increase of apoptosis induction was determined by measuring the activation of effector Caspases 3/7. Compared to untreated controls, transduction of SiHa cells with HPV16-E6 specific CRISPR-HCAdV significantly increased Caspase 3/7 induction (7.1-fold), whereas transduction with ΔE1-ΔE3-AdV5 only increased Caspase 3/7 induction 1.4-fold (Figure 7). Transduction of CaSki cells with HPV16-E6-specific CRISPR-HCAdV significantly increased Caspase 3/7 induction 2.19-fold and transduction with ΔE1-ΔE3-AdV5 increased Caspase 3/7 induction 1.28-fold (Figure 7). Transduction of HeLa cells with HPV18-E6 specific CRISPR-HCAdV increased Caspase 3/7 induction 1.49-fold, whereas transduction with ΔE1-ΔE3-AdV5 significantly increased Caspase 3/7 induction 1.78-fold (Figure 7). Transduction of A549 cells with HPV16-E6-specific CRISPR-HCAdV, HPV18-E6-specific CRISPR-HCAdV, or ΔE1-ΔE3-AdV5 moderately reduced Caspase 3/7 induction but without any statistical significance.

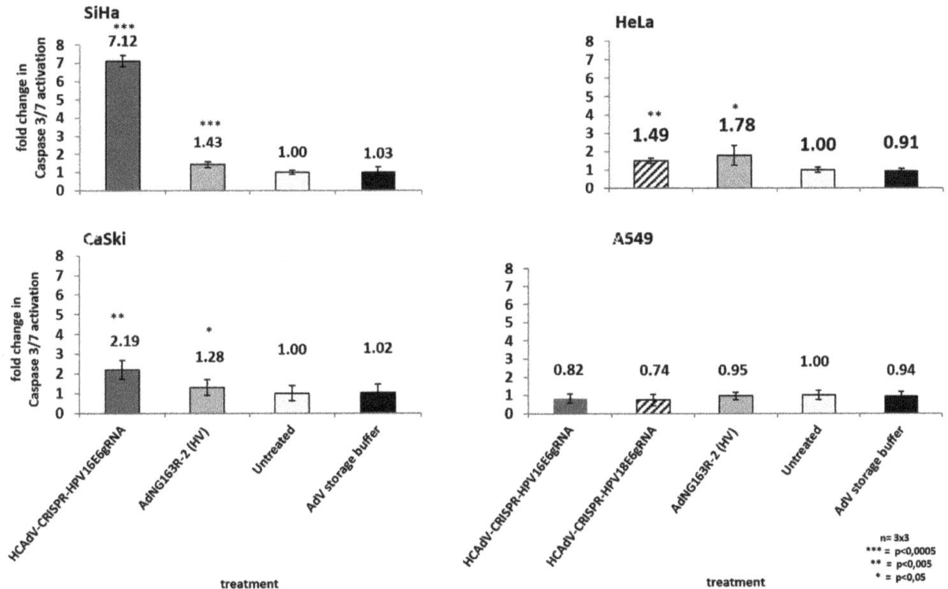

**Figure 7.** Monitoring apoptosis induction in SiHa, HeLa, CaSki, and A549 cells. Here, 48 h post-transduction with the respective HPV-E6-specific CRISPR/Cas9-HCAdV and ΔE1-ΔE3-AdV5 at 1000 infectious vector particles per cell, caspase 3/7 activation was measured. Standard deviations of mean values are shown as error bars. Statistically significant differences compared to untreated controls are shown as one, two, or three stars, indicating *p* values of <0.05, <0.005, and <0.0005 respectively.

## 3. Discussion

Previous studies demonstrated that designer-nuclease-mediated destruction of HPV oncogenes are efficient strategies to combat HPV-induced tumors by inducing apoptosis [26–30,41]. To exploit this strategy as a therapeutic intervention, we aimed at translating the findings based on CRISPR/Cas9-mediated HPV oncogene destruction toward a potential in vivo delivery strategy using an all-in-one CRISPR-HCAdVs containing a Cas9 expression cassette and a HPV E6 specific gRNA expression. We tested the effects of the HPVE6 specific CRISPR/Cas9 expressing HCAdV on viability, proliferation and apoptosis induction of cervical cancer cell lines compared to a ΔE1-ΔE3-AdV5-CRAdV that kills tumor cells but not non-transformed cells due to the cytopathic effect caused by AdV replication [42–47]. The results of this study prove that HCAdV can be used as delivery vehicles for the HPV-E6-specific CRISPR/Cas9 machinery. Transduction of three different HPV-induced cervical cancer cell lines HeLa, CaSki, and SiHa with HPV-oncogene-specific CRISPR-HCAdV inhibited tumor cell proliferation and increased cell death though induction of apoptosis.

Noteworthy, even though the same cell numbers were infected with the same MOI, the antitumor effects in the cell lines investigated in this study differ. In SiHa cells, the antitumor effects of our vector treatment were strong, whereas CaSki cells showed no or at least weak response to the treatment (Figures 3 and 5–7). The response to vector treatment in part reflects the different susceptibility of the studied cell lines to the vector. When treated with low viral particle numbers, SiHa cells are 100-fold more susceptible to AdV5 than CaSki cells (Figure 4A). In contrast, at high viral particle numbers per cell, the difference in vector susceptibility is less pronounced. HeLa and SiHa cells are 1.5-fold more susceptible than CaSki cells with no differences between SiHa and CaSki cells (Figure 4B). Therefore, differences in vector susceptibility only partially explain the differences in response to vector treatment.

Another difference between the cervical cancer cells studied here is the genome copy number of HPV genomes integrated into the chromosomes. Those differences might also contribute to the differences in response to vector treatment. SiHa cells contain 1–2 HPV16 genome copies integrated into their chromosomes, whereas CaSki cells contain more than 600 copies of HPV16 in 11 chromosomal sites, as well as low copy numbers of partial HPV18 genomes [48–51]. The intermediate response in HeLa cells to the vector treatment might also be explained by the intermediate number of HPV18 genomes [52] in HeLa cells. HeLa cells contain at least five copies of the HPV18 genome [53,54]. Therefore, the difference in response to the vector treatment could be explained by a large number of uncut HPV copies remaining within HeLa and even more in CaSki cells leading to ongoing anti-apoptotic E6 effects within these cells. In SiHa cells, CRISPR/Cas9 seemed to have mutated enough E6 sequences on genome level to inhibit E6 effects. HPV copy numbers in cervical tumors depends on the integration frequencies and chromosome duplication events and can vary between different lesions. As HPV integration status and copy number seem to influence the effect of an anti-HPV-CRISPR approach, molecular diagnostics determining the HPV type, integration status, and viral load, can be predictive for the efficiency of such an anti-HPV-CRISPR HCAdV treatment. As we observed differences in the susceptibility of the different cells to the vector and differences in response to the treatment, pretreatment diagnostics could also help predict the optimal vector dose. This strategy enables to identify the optimal dose to achieve the desired response in a cell line-dependent manner and in the future to develop a personalized therapy approach for patients.

It remains unclear why the cell survival of HPV negative A549 cells was affected after transduction with HPV oncogene-specific CRISPR-HCAdVs. All used vectors including the $\Delta$E1-$\Delta$E3-AdV5 non-CRISPR control influenced cell viability in A549 cells. This is probably due to the high transduction efficiencies of adenoviral vectors in A549 cells. It is well established that A549 cells are highly susceptible to adenovirus infection. As all vectors negatively influenced A549 cell survival and proliferation this is probably not due to the genetic cargo, but rather a general reaction to the viral transduction. Zhen et al. [26,29], as well as Kennedy et al. [28], who published the gRNAs used in this study, did not report any off-target effects after plasmid transfection. As cellular toxicity due to Cas9 overexpression has been reported in hematopoietic stem cells [55], the presence of Cas9 protein could have led to decreased cell viability of A549 cells. Before further developing anti-HPV-CRISPR HCAdV treatment, further studies including other HPV-negative control cell lines and especially control vectors containing a non-specific gRNA or vectors containing no gRNA are needed to elucidate whether decreased viability in HPV-negative cells is due to the vector treatment, the presence of Cas9 and potential unspecific action, or an artifact resulting from the assay used to quantify viable cells.

Compared to the CRISPR-expressing HCAdVs, $\Delta$E1-$\Delta$E3-AdV5 showed slightly stronger inhibitory effects on cell viability. However, HPV-specific CRISPR/Cas9-expressing HCAdV led to much stronger Caspase 3/7 induction in HPV-positive cells. However, in A549, no significant induction of apoptosis was observed, indicating that the inhibitory effects on viability and proliferation of HPV positive cells were specific to the CRISPR/Cas9-mediated E6 disruption and induction of apoptosis. Previous studies have shown that transfection of HPV-positive cervical cancer cells with HPV-specific CRISPR/Cas9 expression plasmids lead to increased p53 levels [26,30]. Following adenoviral delivery of HPV specific CRISPR/Cas9 we also observed an increase in p53 levels (Figure 6), indicating that the elimination of HPV oncogenes leads to a re-accumulation or less p53 degradation, which explains the higher number of cells that showed increased Caspase 3/7 activity upon vector treatment (Figure 7).

HCAdV armed with HPV-E6-specific CRISPR/Cas9 drive cell death and inhibition of cell proliferation of HPV-positive cervical cancer cells through apoptosis induction, whereas ΔE1-ΔE3-AdV5 did not significantly increase apoptosis induction in HPV-positive or HPV-negative cell lines. The observed ΔE1-ΔE3-AdV5-mediated tumor cell killing is apoptosis independent and could rather be explained as a result of viral replication. This shows that the effects seen after transduction with HPV-specific CRISPR/Cas9-expressing HCAdV are related to the specific effects of it CRISPR/Cas9 cargo rather than the transduction process or potential replication of remaining helper virus particles within the HCAdV preparations. The versatility of the CRISPR/Cas9 system in combination with HCAdV delivery which can be constructed in a facilitated way through our CRISPR-HCAdV production pipeline [38] offers the possibility to develop personalized treatment depending on the causative HPV type present in a respective patient tumor. Here, precise diagnostics of the HPV type and viral load within the tumor will help to choose the optimal treatment regimen.

Translation of the present results towards preclinical in vivo models is challenging. Xenograft models in immune-deficient mice or transgenic mice expressing HPV oncogenes are a considerable model system that can be exploited. Here, the route of administration will be of particular interest. Systemic injection of AdV vectors based on AdV serotype 5 will efficiently transduce the liver [56], limiting the effectiveness for treating epithelial tumors. Regarding the necessity to treat epithelia with precursor lesions, invasive tumors and infected areas without transformation in close proximity [5], alternative strategies should be considered. As AdVs are able to penetrate the upper layers of epithelium of the mouse skin without inducing cytotoxic effects [57], vector delivery through microporation [58,59] could be a relevant option.

Besides the route of administration modification of the vectors used for this approach can contribute to optimize delivery to relevant cells. Chemical modification of vector capsids [60] or using Darpins as adaptors between target cells and vector [61] could direct the vectors to the desired cell type to increase specificity and efficiency of the approach. Recently, an alternative AdV serotype became available [62–64]. Vectorization of such serotypes could contribute to an improved efficiency of the approach.

However, the setup of transgenes used in the HCAdV-expressing HPV-specific CRISPR/Cas9 used in this study does not exploit the full antitumor potential of what HCAdV could achieve. Additional gRNAs targeting could be included to enhance the anti-HPV effects. Addition of RNA interference directed against HPV oncogene mRNA [14,18] could be added to reach synergistic effects. A combined expression of immune modulators [65] or anti-neoangiogenic factors [66] could also be considered. Arming CRAdV, with HPV oncogene specific CRISPR/Cas9 could enhance efficacy through synergistic effects. Therefore, combining the effects of HPV-oncogene-specific designer nucleases with oncolytic virotherapy is also a reasonable strategy that deserves further commitment.

## 4. Materials and Methods

### 4.1. Viral Vectors

An E1and E3-deleted, first-generation Adenovirus AdNG163R-2 (ΔE1-ΔE3-AdV5) was produced as previously described [37]. HPV16 and HPV18-specific CRISPR/Cas9-expressing HCAdV based on human AdV serotype 5 (AdV5) were produced as previously described [38]. In brief, HCAdV-CRISPR-HPV16E6gRNA and HCAdV-CRISPR-HPV18E6gRNA contained a staphylococcus pyogenes Cas9 gene driven by a CbH promoter and a gRNA expression unit controlled by human U6-RNA promoter. HCAdV-CRISPR-HPV18E6gRNA contained an additional GFP expression cassette driven by CMV promoter (Figure 2). The gRNA sequences for HPV18-E6

(GCGCTTTGAGGATCCAACA) and HPV16-E6 (CAACAGTTACTGCGACGTG) were previously published [26,29].

Quantification of infectious virus particles within the vector preparations was carried out as previously described [67,68]. Briefly, HEK293 cells were transduced with defined volumes of the purified vectors. Three hours post transduction cells were washed with 1x phospahate-buffered saline (PBS) prior to harvesting and harvested cells were centrifuged and resuspended in 1x PBS to remove non-infective particles and centrifuged again. Following isolation of genomic DNA from transduced cells, HCAdV vector genome copy numbers were determined by HCAdV-specific q-PCR.

### 4.2. Cell Lines and Cell Culture

The cervical cancer cell lines SiHa and CaSki containing HPV16 and HeLa-containing HPV18 genomes integrated into their chromosomal DNA as well as HPV-negative A549 lung carcinoma cells were cultured in DMEM medium supplemented with 10% fetal bovine serum (FBS, PAN Biotec, Aidenbach, Germany), 100 U/mL Penicillin and 100 µg/mL Streptomycin (PAN Biotec) unless stated otherwise. Cells were grown at 37 °C and 5% $CO_2$.

### 4.3. Mutation Induction and Detection

Determination of genome targeting efficiency using T7 Endonuclease I was performed using a protocol adapted from previously published protocols [69]. Briefly, $2 \times 10^6$ cells were seeded into 24-well plates and subsequently infected with the respective HPV type-specific CRISPR-HCAdVs. Then, 48 h post-infection, genomic DNA was extracted from the cells for subsequent PCR using a Nucleo Spin Tissue kit (Macherey-Nagel, Düren, Deutschland). PCR amplicons covering the respective HPV-E6 region surrounding the respective gRNA binding sites was amplified using Phusion high fidelity DNA polymerase kit (New England Biolabs, Frankfurt, Germany) according to manufacturer's instructions. The HVP18-E6 amplicon was generated using the primer-set HPV18T7E1fwd (5′ CTTGCATAACTATATCCACTCCC 3′) and HPV18E6rev (5′ ATTCAACGGTTTCTGGCAC 3′) yielding a PCR product of 656 bp and HPV16E6 region was amplified using the primer-set HPV16E6fwd (5′ TGAACCGAAACCGGTTAGTA 3′) and HPV16E6rev (5′ TGAACCGAAACCGGTTAGTA 3′) (30) yielding a PCR product of 660 bp. PCR products were purified by ethanol precipitation by filling the PCR volume up to 100 µL with $H_2O$, adding of 10 µL NaAc (3M, pH 8.0) and 250 µL ice cold EtOH (100%) followed by centrifugation for 10 min, high speed. The DNA-pellet was washed with 500 µL EtOH (70%) and centrifuged for 5 min at high speed. After centrifugation, the supernatant was discarded, and the pellet was air dried and resuspended in 17.5 µL $H_2O$. Purified amplicons were supplemented with 2µL of Buffer NEB2 (New England Biolabs, Ipswich, MA, USA) and subjected to heteroduplex formation by heating them to 95 °C and ramping down to room temperature using a cooling rate of 0.1 °C/s. After heteroduplex formation, 0.5 µL of T7E1 enzyme (New England Biolabs) was added and the mixture was incubated for 15 min at 37 °C. The reaction was stopped by adding purple loading dye (New England Biolabs, containing 10 mM EDTA, 0.08% SDS) and separated on a 2% agarose gel and at ~90 V for >45 min. Mutation rate was calculated according to the formula: % gene modification = 100 × (1 − (1 − fraction cleaved)1/2) [69].

### 4.4. Cell Viability Assay and Crystal Violet Staining

We seeded $1.5 \times 10^6$ HeLa, CaSki, SiHa, and A549 cells into 24-well plates. Five hours later, cells were transduced with 0, 0.1, 1.0, 10, 100, or 1000 infectious particles of the respective HPV16 or HPV18-specific CRISPR/Cas9-expressing HCAdV or AdNG163R-2. To rule out potential cytotoxic effects of the solvent used to dilute the vectors, negative controls cells were treated with serial dilution of AdV storage buffer (50 mM Hepes pH 7.4, 100 mM NaCl, 10% Glycerol), equaling the highest volume of buffer that was applied for MOI 1000 of the HPV16-specific CRISPR/Cas9-expressing HCAdV. Five days post-transduction, cell-viability was determined using the Cell Counting Kit-8 (CCK-8; Sigma, St. Louis, MO, USA) following manufacturer's instructions. Briefly, 10 µL of CCK8 substrate were applied to each well containing 100 µL of cell culture medium and incubated for 1 h to allow viable cells to convert the CCK-8 substrate into a colored dye. Adsorption at 450 nm was measured using a GENios multi-plate reader (TECAN, Crailsheim, Germany). After CCK-8 readout, the medium was removed by three consecutive washing steps with $1 \times$ PBS and fixed in 10% formaldehyde in PBS for 10 min at room temperature, followed by three consecutive washing steps with $1 \times$ PBS. Adherent cells were stained with crystal violet staining solution for 10 min. Excess staining solution was removed by gently washing the plates in tap water. Subsequently plates were air dried and kept for photo documentation. All experiments were performed three times with triplicates for each sample.

### 4.5. Determination of Cellular Susceptibility to Human AdV Serotype 5 (AdV5)

To determine the susceptibility of the cell lines used in this study by means of GFP reporter gene expression, 40,000 cells of HeLa, SiHa, or CaSki cells were seeded in a 96-well plate. Then, 5 h post-seeding, cells were infected with 1000 viral particles of an E3 deleted AdV5-expressing GFP and luciferase that was produced as previously described [62]. Forty-eight hours post-transduction, virus-mediated GFP fluorescence within transduced cells was quantified using an Infinity 200 pro multi-plate reader (TECAN). To determine the susceptibility of the cell lines used in this study by means of luciferase reporter gene expression, 40,000 cells of HeLa, SiHa, or CaSki cells were seeded per well in a 96-well plate. Five hours post-seeding, cells were infected with 20 viral particles of an E3 deleted AdV5 expressing GFP and luciferase. Then, 24 h post-transduction, quantification of virus-mediated luciferase activity of transduced cells was carried out using the dual-luciferase reporter assay (Promega, Walldorf, Germany) following the manufacturer's instructions. Luminescence measurement was performed using an Infinity 200 pro multi-plate reader (TECAN). Experiments were repeated three times in triplicates.

### 4.6. Proliferation Assay

We seeded $1 \times 10^3$ HeLa, CaSki, SiHa or A549 cells per well into 96-well plates, respectively, and transduced with 1000 infectious vector particles per cell of the HPV16 or HPV18 specific CRISPR/Cas9 expressing HCAdV or AdNG163R-2, respectively. The number of viable cells for each treatment was determined using the Cell Counting Kit-8 (CCK-8; Sigma, St. Louis, MO, USA) in 24 h intervals in quadruplicates for seven days. All experiments were performed three times with quadruplicates for each sample.

### 4.7. In Cell Western Analysis of Cellular p53

To analyze changes in cellular p53 protein levels in response to HPV-specific CRISPR/Cas9-expressing HCAdV treatment, we performed in cell western analysis. Here, 25,000 cells of the respective cell line were seeded to each well of a 96 well plate. SiHa and CaSki cells were treated with HAdV-CRISPR-HPV16E6gRNA and Hela, and A549 cells were treated with HCAdV-CRISPR-HPV18E6gRNA at MOI of 1000, respectively. Controls were left untreated. Then, 96 h post-transduction, the medium was removed, and cells were fixed using $1\times$ PBS, 4% paraformaldehyde for 20 min at room temperature. Fixed cells were washed three times using $1\times$ PBS, 0.1% triton X100 for five minutes to permeabilize the cells. Following cell permeabilization, wells were incubated with blocking solution (LI-COR, Lincoln, NE, USA) for 90 min at room temperature. After blocking, wells were incubated with AlexaFluor790 coupled anti-p53 antibody (p53 (Do-1): sc-126; Santa Cruz, Dallas, Germany) and CellTag 700 Stain (LI-COR, Lincoln, NE, USA) diluted 1:50 and 1:500, respectively, in blocking solution for 150 min at room temperature. Background controls only received anti p53 antibody or cell tag 700, respectively. Finally, plates were washed with 1x PBS, 0,1% tween 20 for five times at room temperature before plates were air dryed and the 700 and 800 nm fluorescent signals of all wells were quantified using the Odyssey CLx scanner (LI-COR) and image studio 5.2 software (LI-COR). Measurements were carried out at a resolution of 169 µm using a focus depth of 3 mm. For each well, the fluorescent intensity values of the p53 antibody (800 nm) was divided by the mean fluorescent intensity of cell tag700 (700 nm) to normalize for the cell number. Data were analyzed in duplicates.

### 4.8. Apoptosis Assay

We seeded $2.5 \times 10^4$ HeLa, CaSki, SiHa, or A549 cells into black-walled 96-well plates with a transparent bottom. Between experimental groups, one row was left empty to prevent the luminescent signal to influence neighboring wells. Five hours post-seeding, cells were transduced with HPV16 or HPV18-specific CRISPR/Cas9-expressing HCAdV or the E1-deleted, first generation Adenovirus (AdNG163R-2), respectively. To rule out potential apoptotic effects of the solvent used to dilute the vectors, untransfected cells were treated with AdV storage buffer (50 mM HEPES, pH 7.4, 100 mM NaCl, 10% Glycerol), equaling the highest volume of virus that was applied for MOI 1000 of the HPV16-specific CRISPR/Cas9-expressing HCAdV. Untreated cells were incubated with cell culture medium alone. Then, 48 h post-transduction, apoptosis induction was monitored by measuring caspase 3/7 activation using luminescent caspase 3/7 glow assay (Promega) following the manufacturer's instructions. Luciferase signal indicating Caspase 3/7–mediated substrate cleavage was detected using a GENios multi-plate reader (TECAN). Cell culture medium alone or medium supplemented with the AdV storage buffer was measured on the same 96-well plate to determine background signal. These background values were subtracted from the sample values. All experiments were performed three times with quadruplicates for each sample.

### 4.9. Statistical Analysis

Statistical significance of differences between experimental groups and untreated controls were analyzed using Students t-test using the software GraphPad Prism 8 (San Diego, CA, USA). Significant differences with $p$ values of $p < 0.0005$, $p < 0.005$, or $p < 0.05$ are depicted as ***, ** or * respectively.

## 5. Conclusions

In summary, this study shows a proof-of-concept for using the CRISPR/Cas9 technology delivered by the most advanced adenoviral vector (HCAdV) to treat HPV-derived tumors. We conclude that HCAdV can serve as HPV-specific cancer gene therapeutic agents when armed with HPV-type-specific CRISPR/Cas9. We believe that our approach can contribute to treatment options specific for the respective HPV type present in each individual tumor. The next step is the translation of this approach into relevant in vivo animal models and, in the long-term run, this approach may lead to a novel concept in personalized tumor therapy.

**Supplementary Materials:** The following are available online at http://www.mdpi.com/2072-6694/12/7/1934/s1, Figure S1. Oncolysis assay to monitor cell viability.

**Author Contributions:** Conceptualization, E.E.-S.; Methodology and experiments, E.E.-S., S.H., L.S. and M.S.; E.E.-S.; Data Analysis, S.H.; Writing—original draft preparation, E.E.-S. and A.E.; Writing—review and editing, E.E.-S. and A.E.; Project administration, E.E.-S. and A.E.; funding acquisition, E.E.-S. and A.E. All authors have read and agreed to the published version of the manuscript.

**Funding:** This research was funded by the foundation "Stiftung Tumorforschung Kopf-Hals."

**Acknowledgments:** We thank the research group of B.A. for sharing CaSki and SiHa cells and their helpful advice.



## References

1. Zur Hausen, H. Papillomaviruses and cancer: From basic studies to clinical application. *Nat. Rev. Cancer* **2002**, *2*, 342–350. [CrossRef]

2. Munoz, N.; Bosch, F.X.; de Sanjose, S.; Herrero, R.; Castellsague, X.; Shah, K.V.; Snijders, J.F.; Meijer, C.J.L.M.L. Epidemiologic classification of human papillomavirus types associated with cervical cancer. *N. Engl. J. Med.* **2003**, *348*, 518–527. [CrossRef] [PubMed]

3. Hubbers, C.U.; Akgul, B. HPV and cancer of the oral cavity. *Virulence* **2015**, *6*, 244–248. [CrossRef]

4. Nindl, I.; Gottschling, M.; Stockfleth, E. Human papillomaviruses and non-melanoma skin cancer: Basic virology and clinical manifestations. *Dis. Markers* **2007**, *23*, 247–259. [CrossRef] [PubMed]

5. Stockfleth, E. The importance of treating the field in actinic keratosis. *J. Eur. Acad Dermatol. Venereol.* **2017**, *31* (Suppl. 2), 8–11. [CrossRef] [PubMed]

6. Lechner, M.S.; Laimins, L.A. Inhibition of p53 DNA binding by human papillomavirus E6 proteins. *J. Virol.* **1994**, *68*, 4262–4273. [CrossRef]

7. Munger, K.; Werness, B.A.; Dyson, N.; Phelps, W.C.; Harlow, E.; Howley, P.M. Complex formation of human papillomavirus E7 proteins with the retinoblastoma tumor suppressor gene product. *EMBO J.* **1989**, *8*, 4099–4105. [CrossRef]

8. Dyson, N.; Howley, P.M.; Munger, K.; Harlow, E. The human papilloma virus-16 E7 oncoprotein is able to bind to the retinoblastoma gene product. *Science* **1989**, *243*, 934–937. [CrossRef]

9. Amici, C.; Visintin, M.; Verachi, F.; Paolini, F.; Percario, Z.; Di Bonito, P.; Mandarino, A.; Affabris, E.; Venuti, A.; Accardi, L. A novel intracellular antibody against the E6 oncoprotein impairs growth of human papillomavirus 16-positive tumor cells in mouse models. *Oncotarget* **2016**, *7*, 15539–15553. [CrossRef]

10. Guo, C.P.; Liu, K.W.; Luo, H.B.; Chen, H.B.; Zheng, Y.; Sun, S.N.; Zhang, Q.; Huang, L. Potent anti-tumor effect generated by a novel human papillomavirus (HPV) antagonist peptide reactivating the pRb/E2F pathway. *PLoS ONE* **2011**, *6*, e17734. [CrossRef]

11. Butz, K.; Denk, C.; Ullmann, A.; Scheffner, M.; Hoppe-Seyler, F. Induction of apoptosis in human papillomaviruspositive cancer cells by peptide aptamers targeting the viral E6 oncoprotein. *Proc. Natl. Acad. Sci. USA* **2000**, *97*, 6693–6697. [CrossRef] [PubMed]

12. Liu, Y.; Liu, Z.; Androphy, E.; Chen, J.; Baleja, J.D. Design and characterization of helical peptides that inhibit the E6 protein of papillomavirus. *Biochemistry* **2004**, *43*, 7421–7431. [CrossRef] [PubMed]

13. Zheng, Y.F.; Rao, Z.G.; Zhang, J.R. Effects of anti-HPV16 E6-ribozyme on the proliferation and apoptosis of human cervical cancer cell line CaSKi. *Di Yi Jun Yi Da Xue Xue Bao* **2002**, *22*, 496–498. [PubMed]

14. Butz, K.; Ristriani, T.; Hengstermann, A.; Denk, C.; Scheffner, M.; Hoppe-Seyler, F. siRNA targeting of the viral E6 oncogene efficiently kills human papillomavirus-positive cancer cells. *Oncogene* **2003**, *22*, 5938–5945. [CrossRef]

15. Jiang, M.; Milner, J. Selective silencing of viral gene expression in HPV-Positive human cervical carcinoma cells treated with siRNA, a primer of RNA interference. *Oncogene* **2002**, *21*, 6041–6048. [CrossRef] [PubMed]

16. Yamato, K.; Yamada, T.; Kizaki, M.; Ui-Tei, K.; Natori, Y.; Fujino, M.; Nishihara, T.; Ikeda, Y.; Nasu, Y.; Saigo, K.; et al. New highly potent and specific E6 and E7 siRNAs for treatment of HPV16 positive cervical cancer. *Cancer Gene Ther.* **2008**, *15*, 140–153. [CrossRef] [PubMed]

17. Nishida, H.; Matsumoto, Y.; Kawana, K.; Christie, R.J.; Naito, M.; Kim, B.S.; Miyata, K.; Taguchi, A.; Tomio, K.; Yamashita, A.; et al. Systemic delivery of siRNA by actively targeted polyion complex micelles for silencing the E6 and E7 human papillomavirus oncogenes. *J. Control. Release* **2016**, *231*, 29–37. [CrossRef] [PubMed]

18. Sato, N.; Saga, Y.; Uchibori, R.; Tsukahara, T.; Urabe, M.; Kume, A.; Fujiwara, H.; Suzuki, M.; Ozawa, K.; Mizukami, H.; et al. Eradication of cervical cancer in vivo by an AAV vector that encodes shRNA targeting human papillomavirus type 16 E6/E7. *Int. J. Oncol.* **2018**. [CrossRef]

19. Niu, X.Y.; Peng, Z.L.; Duan, W.Q.; Wang, H.; Wang, P. Inhibition of HPV 16 E6 oncogene expression by RNA interference in vitro and in vivo. *Int. J. Gynecol. Cancer* **2006**, *16*, 743–751. [CrossRef]

20. Chang, J.T.C.; Kuo, T.F.; Chen, Y.J.; Chiu, C.C.; Lu, Y.C.; Li, H.F.; Shen, C.-R.; Cheng, A.-J. Highly potent and specific siRNAs against E6 or E7 genes of HPV16- or HPV18-infected cervical cancers. *Cancer Gene Ther.* **2010**, *17*, 827. [CrossRef]

21. Lea, J.S.; Sunaga, N.; Sato, M.; Kalahasti, G.; Miller, D.S.; Minna, J.D.; Muller, C.Y. Silencing of HPV 18 oncoproteins With RNA interference causes growth inhibition of cervical cancer cells. *Reprod. Sci.* **2007**, *14*, 20–28. [CrossRef] [PubMed]

22. Schiwon, M.; Ehrke-Schulz, E.; Oswald, A.; Bergmann, T.; Michler, T.; Protzer, U.; Ehrhardt, A. One-Vector System for Multiplexed CRISPR/Cas9 against Hepatitis B Virus cccDNA Utilizing High-Capacity Adenoviral Vectors. *Mol. Ther.-Nucl. Acids* **2018**, *12*, 242–253. [CrossRef] [PubMed]

23. Weber, N.D.; Stone, D.; Jerome, K.R. TALENs targeting HBV: Designer endonuclease therapies for viral infections. *Mol. Ther. J. Am. Soc. Gene Ther.* **2013**, *21*, 1819–1820. [CrossRef] [PubMed]

24. Bloom, K.; Ely, A.; Mussolino, C.; Cathomen, T.; Arbuthnot, P. Inactivation of hepatitis B virus replication in cultured cells and in vivo with engineered transcription activator-like effector nucleases. *Mol. Ther. J. Am. Soc. Gene Ther.* **2013**, *21*, 1889–1897. [CrossRef]

25. Mino, T.; Mori, T.; Aoyama, Y.; Sera, T. Gene- and protein-delivered zinc finger-staphylococcal nuclease hybrid for inhibition of DNA replication of human papillomavirus. *PLoS ONE* **2013**, *8*, e56633. [CrossRef]

26. Kennedy, E.M.; Kornepati, A.V.; Goldstein, M.; Bogerd, H.P.; Poling, B.C.; Whisnant, A.W.; Kastan, M.B.; Cullen, B.R. Inactivation of the human papillomavirus E6 or E7 gene in cervical carcinoma cells by using a bacterial CRISPR/Cas RNA-guided endonuclease. *J. Virol.* **2014**, *88*, 11965–11972. [CrossRef]

27. Hu, Z.; Yu, L.; Zhu, D.; Ding, W.; Wang, X.; Zhang, C.; Wang, L.; Jiang, X.; Shen, H.; He, D.; et al. Disruption of HPV16-E7 by CRISPR/Cas system induces apoptosis and growth inhibition in HPV16 positive human cervical cancer cells. *Biomed. Res. Int.* **2014**, *2014*, 612823. [CrossRef]

28. Liu, Y.C.; Cai, Z.M.; Zhang, X.J. Reprogrammed CRISPR-Cas9 targeting the conserved regions of HPV6/11 E7 genes inhibits proliferation and induces apoptosis in E7-transformed keratinocytes. *Asian J. Androl.* **2016**, *18*, 475–479. [CrossRef]

29. Zhen, S.; Hua, L.; Takahashi, Y.; Narita, S.; Liu, Y.H.; Li, Y. In vitro and in vivo growth suppression of human papillomavirus 16-positive cervical cancer cells by CRISPR/Cas9. *Bioch. Biophys. Res. Commun.* **2014**, *450*, 1422–1426. [CrossRef]

30. Yu, L.; Wang, X.; Zhu, D.; Ding, W.; Wang, L.; Zhang, C.; Jiang, X.; Shen, H.; Liao, S.; Ma, D.; et al. Disruption of human papillomavirus 16 E6 gene by clustered regularly interspaced short palindromic repeat/Cas system in human cervical cancer cells. *Onco Targets Ther.* **2015**, *8*, 37–44. [CrossRef]

31. Jubair, L.; Fallaha, S.; McMillan, N.A.J. Systemic Delivery of CRISPR/Cas9 Targeting HPV Oncogenes Is Effective at Eliminating Established Tumors. *Mol. Ther. J. Am. Soc. Gene Ther.* **2019**. [CrossRef] [PubMed]

32. Heise, C.; Kirn, D.H. Replication-Selective adenoviruses as oncolytic agents. *J. Clin. Investig.* **2000**, *105*, 847–851. [CrossRef] [PubMed]

33. Russell, S.J.; Peng, K.W.; Bell, J.C. Oncolytic virotherapy. *Nat. Biotechnol.* **2012**, *30*, 658–670. [CrossRef] [PubMed]

34. Yuan, M.; Webb, E.; Lemoine, N.R.; Wang, Y. CRISPR-Cas9 as a Powerful Tool for Efficient Creation of Oncolytic Viruses. *Viruses* **2016**, *8*, 72. [CrossRef]

35. Jager, L.; Ehrhardt, A. Persistence of high-capacity adenoviral vectors as replication-defective monomeric genomes in vitro and in murine liver. *Hum. Gene Ther.* **2009**, *20*, 883–896. [CrossRef]

36. Parks, R.J.; Chen, L.; Anton, M.; Sankar, U.; Rudnicki, M.A.; Graham, F.L. A helper-dependent adenovirus vector system: Removal of helper virus by Cre-Mediated excision of the viral packaging signal. *Proc. Natl. Acad. Sci. USA* **1996**, *93*, 13565–13570. [CrossRef]

37. Palmer, D.; Ng, P. Improved system for helper-dependent adenoviral vector production. *Mol. Ther. J. Am. Soc. Gene Ther.* **2003**, *8*, 846–852. [CrossRef] [PubMed]

38. Ehrke-Schulz, E.; Schiwon, M.; Leitner, T.; David, S.; Bergmann, T.; Liu, J.; Ehrhardt, A. CRISPR/Cas9 delivery with one single adenoviral vector devoid of all viral genes. *Sci. Rep.* **2017**, *7*, 17113. [CrossRef] [PubMed]

39. Schiedner, G.; Morral, N.; Parks, R.J.; Wu, Y.; Koopmans, S.C.; Langston, C.; Graham, F.L.; Beaudet, A.L.; Kochanek, S. Genomic DNA transfer with a high-capacity adenovirus vector results in improved in vivo gene expression and decreased toxicity. *Nat. Genet.* **1998**, *18*, 180–183. [CrossRef]

40. Ehrhardt, A.; Kay, M.A. A new adenoviral helper-dependent vector results in long-term therapeutic levels of human coagulation factor IX at low doses in vivo. *Blood* **2002**, *99*, 3923–3930. [CrossRef] [PubMed]

41. Yoshiba, T.; Saga, Y.; Urabe, M.; Uchibori, R.; Matsubara, S.; Fujiwara, H.; Mizukami, H. CRISPR/Cas9-mediated cervical cancer treatment targeting human papillomavirus E6. *Oncol. Lett.* **2019**, *17*, 2197–2206. [CrossRef] [PubMed]

42. Chellappan, S.; Kraus, V.B.; Kroger, B.; Munger, K.; Howley, P.M.; Phelps, W.C.; Nevins, J.R. Adenovirus E1A, simian virus 40 tumor antigen, and human papillomavirus E7 protein share the capacity to disrupt the interaction between transcription factor E2F and the retinoblastoma gene product. *Proc. Natl. Acad. Sci. USA* **1992**, *89*, 4549–4553. [CrossRef] [PubMed]

43. Sang, N.; Avantaggiati, M.L.; Giordano, A. Roles of p300, pocket proteins, and hTBP in E1A-mediated transcriptional regulation and inhibition of p53 transactivation activity. *J. Cell. Biochem.* **1997**, *66*, 277–285. [CrossRef]

44. Whyte, P.; Williamson, N.M.; Harlow, E. Cellular targets for transformation by the adenovirus E1A proteins. *Cell* **1989**, *56*, 67–75. [CrossRef]

45. Steinwaerder, D.S.; Carlson, C.A.; Lieber, A. DNA Replication of First-Generation Adenovirus Vectors in Tumor Cells. *Hum. Gene Ther.* **2000**, *11*, 1933–1948. [CrossRef]

46. Steinwaerder, D.S.; Carlson, C.A.; Lieber, A. Human papilloma virus E6 and E7 proteins support DNA replication of adenoviruses deleted for the E1A and E1B genes. *Mol. Ther.* **2001**, *4*, 211–216. [CrossRef]

47. Abou, E.L.; Hassan, M.A.; van der Meulen-Muileman, I.; Abbas, S.; Kruyt, F.A. Conditionally replicating adenoviruses kill tumor cells via a basic apoptotic machinery-independent mechanism that resembles necrosis-like programmed cell death. *J. Virol.* **2004**, *78*, 12243–12251. [CrossRef]

48. Raybould, R.; Fiander, A.; Wilkinson, G.W.; Hibbitts, S. HPV integration detection in CaSki and SiHa using detection of integrated papillomavirus sequences and restriction-site PCR. *J. Virolog. Methods* **2014**, *206*, 51–54. [CrossRef]

49. Yee, C.; Krishnan-Hewlett, I.; Baker, C.C.; Schlegel, R.; Howley, P.M. Presence and expression of human papillomavirus sequences in human cervical carcinoma cell lines. *Am. J. Pathol.* **1985**, *119*, 361–366.

50. Baker, C.C.; Phelps, W.C.; Lindgren, V.; Braun, M.J.; Gonda, M.A.; Howley, P.M. Structural and transcriptional analysis of human papillomavirus type 16 sequences in cervical carcinoma cell lines. *J. Virol.* **1987**, *61*, 962–971. [CrossRef] [PubMed]

51. Mincheva, A.; Gissmann, L.; zur Hausen, H. Chromosomal integration sites of human papillomavirus DNA in three cervical cancer cell lines mapped by in situ hybridization. *Med. Microbiol. Immunol.* **1987**, *176*, 245–256. [CrossRef] [PubMed]

52. Maglennon, G.A.; McIntosh, P.; Doorbar, J. Persistence of viral DNA in the epithelial basal layer suggests a model for papillomavirus latency following immune regression. *Virology* **2011**, *414*, 153–163. [CrossRef] [PubMed]

53. Adey, A.; Burton, J.N.; Kitzman, J.O.; Hiatt, J.B.; Lewis, A.P.; Martin, B.K.; Qiu, R.; Lee, C.; Shendure, J. The haplotype-resolved genome and epigenome of the aneuploid HeLa cancer cell line. *Nature* **2013**, *500*, 207–211. [CrossRef] [PubMed]

54. Macville, M.S.E.; Padilla-Nash, H.; Keck, C.; Ghadimi, B.M.; Zimonjic, D.; Popescu, N.; Ried, T. Comprehensive and definitive molecular cytogenetic characterization of HeLa cells by spectral karyotyping. *Cancer Res.* **1999**, *59*, 141–150.

55. Li, C.; Psatha, N.; Gil, S.; Wang, H.; Papayannopoulou, T.; Lieber, A. HDAd5/35(++) Adenovirus Vector Expressing Anti-CRISPR Peptides Decreases CRISPR/Cas9 Toxicity in Human Hematopoietic Stem Cells. *Mol. Ther. Methods Clin. Dev.* **2018**, *9*, 390–401. [CrossRef]

56. Alemany, R.; Suzuki, K.; Curiel, D.T. Blood clearance rates of adenovirus type 5 in mice. *J. Gen. Virol.* **2000**, *81 Pt 11*, 2605–2609. [CrossRef]

57. Lu, B.; Federoff, H.J.; Wang, Y.; Goldsmith, L.A.; Scott, G. Topical Application of Viral Vectors for Epidermal Gene Transfer. *J. Investig. Dermatol.* **1997**, *108*, 803–808. [CrossRef]

58. Bramson, J.; Dayball, K.; Evelegh, C.; Wan, Y.H.; Page, D.; Smith, A. Enabling topical immunization via microporation: A novel method for pain-free and needle-free delivery of adenovirus-based vaccines. *Gene Ther.* **2003**, *10*, 251–260. [CrossRef]

59. Weiss, R.; Hessenberger, M.; Kitzmuller, S.; Bach, D.; Weinberger, E.E.; Krautgartner, W.D.; Hauser-Kronberger, C.; Malissen, B.; Boehler, C.; Kalia, Y.N.; et al. Transcutaneous vaccination via laser microporation. *J. Control. Release* **2012**, *162*, 391–399. [CrossRef]

60. Yoon, A.R.; Hong, J.; Kim, S.W.; Yun, C.O. Redirecting adenovirus tropism by genetic, chemical, and mechanical modification of the adenovirus surface for cancer gene therapy. *Exp. Opin. Drug Deliv.* **2016**, *13*, 843–858. [CrossRef]

61. Dreier, B.; Honegger, A.; Hess, C.; Nagy-Davidescu, G.; Mittl, P.R.; Grutter, M.G.; Belousova, N.; Mikheeva, G.; Krasnykh, V.; Plückthun, A. Development of a generic adenovirus delivery system based on structure-guided design of bispecific trimeric DARPin adapters. *Proc. Natl. Acad. Sci. USA* **2013**, *110*, E869–E877. [CrossRef] [PubMed]

62. Zhang, W.; Fu, J.; Liu, J.; Wang, H.; Schiwon, M.; Janz, S.; Schaffarcyzk, L.; von de Golta, L.; Ehrke-Schulz, E.; Solanki, M.; et al. An engineered Virus Library as a Resource for the Spectrum-Wide Exploration of Virus and Vector Diversity. *Cell Rep.* **2017**, *19*, 1698–1709. [CrossRef] [PubMed]
63. Zhang, W.; Fu, J.; Ehrhardt, A. Novel Vector Construction Based on Alternative Adenovirus Types via Homologous Recombination. *Hum. Gene Ther. Methods* **2018**, *29*, 124–134. [CrossRef]
64. Zhang, W.; Ehrhardt, A. Getting genetic access to natural adenovirus genomes to explore vector diversity. *Virus Genes* **2017**. [CrossRef]
65. Cerullo, V.; Capasso, C.; Vaha-Koskela, M.; Hemminki, O.; Hemminki, A. Cancer-Targeted Oncolytic Adenoviruses for Modulation of the Immune System. *Curr. Cancer Drug Targets* **2018**, *18*, 124–138. [CrossRef] [PubMed]
66. Angarita, F.A.; Acuna, S.A.; Ottolino-Perry, K.; Zerhouni, S.; McCart, J.A. Mounting a strategic offense: Fighting tumor vasculature with oncolytic viruses. *Trends Mol. Med.* **2013**, *19*, 378–392. [CrossRef] [PubMed]
67. Ehrke-Schulz, E.; Zhang, W.; Schiwon, M.; Bergmann, T.; Solanki, M.; Liu, J.; Boehme, P.; Leitner, T.; Ehrhardt, A. Cloning and Large-Scale Production of High-Capacity Adenoviral Vectors Based on the Human Adenovirus Type 5. *J. Vis Exp.* **2016**, e52894. [CrossRef]
68. Jager, L.; Hausl, M.A.; Rauschhuber, C.; Wolf, N.M.; Kay, M.A.; Ehrhardt, A. A rapid protocol for construction and production of high-capacity adenoviral vectors. *Nat. Protoc.* **2009**, *4*, 547–564. [CrossRef]
69. Guschin, D.Y.; Waite, A.J.; Katibah, G.E.; Miller, J.C.; Holmes, M.C.; Rebar, E.J. A rapid and general assay for monitoring endogenous gene modification. *Methods Mol. Biol.* **2010**, *649*, 247–256. [CrossRef]

*Article*

# Epitope Mapping and Computational Analysis of Anti-HPV16 E6 and E7 Antibodies in Single-Chain Format for Clinical Development as Antitumor Drugs

**Carla Amici [1], Maria Gabriella Donà [2], Barbara Chirullo [3], Paola Di Bonito [4] and Luisa Accardi [4,***

[1]  Department of Biology, University of Rome Tor Vergata, 00133 Rome, Italy; carami371@gmail.com
[2]  STI/HIV Unit, Istituto Dermatologico San Gallicano IRCCS, 00144 Rome, Italy; mariagabriella.dona@ifo.gov.it
[3]  Department of Food Safety, Nutrition and Veterinary Public Health, Istituto Superiore di Sanità, 00161 Rome, Italy; barbara.chirullo@iss.it
[4]  Department of Infectious Diseases, Istituto Superiore di Sanità, 00161 Rome, Italy; paola.dibonito@iss.it
*  Correspondence: luisa.accardi@iss.it

Received: 10 June 2020; Accepted: 3 July 2020; Published: 6 July 2020

**Abstract:** Human Papillomavirus 16-associated cancer, affecting primarily the uterine cervix but, increasingly, other body districts, including the head–neck area, will long be a public health problem, despite there being a vaccine. Since the virus oncogenic activity is fully ascribed to the viral E6 and E7 oncoproteins, one of the therapeutic approaches for HPV16 cancer is based on specific antibodies in single-chain format targeting the E6/E7 activity. We analyzed the Complementarity Determining Regions, repositories of antigen-binding activity, of four anti-HPV16 E6 and -HPV16 E7 scFvs, to highlight possible conformity to biophysical properties, recognized to be advantageous for therapeutic use. By epitope mapping, using E7 mutants with amino acid deletions or variations, we investigated differences among the anti-16E7 scFvs in terms of antigen-binding capacity. We also performed computational analyses to determine whether length, total net charge, surface hydrophobicity, polarity and charge distribution conformed well to those of the antibodies that had already reached clinical use, through the application of developability guidelines derived from recent literature on clinical-stage antibodies, and the Therapeutic Antibodies Profiler software. Overall, our findings show that the scFvs investigated may represent valid candidates to be developed as therapeutic molecules for clinical use, and highlight characteristics that could be improved by molecular engineering.

**Keywords:** therapeutic antibodies; single-chain antibody fragments; HPV-associated cancer; clinical stage antibodies; Human Papillomavirus 16 E6 and E7 oncoproteins

---

## 1. Introduction

Recombinant monoclonal antibodies (mAbs) are among the classes of therapeutic drugs that convey most of the large funds from the biotechnology industry. Since 1986 up until to May 2020, the European Medicines Agency and the US Food and Drug Administration have approved ninety-four antibody therapies for the European or US market, while sixteen are under review [1] (Antibody Society. Approved antibodies. Available at https://www.antibodysociety.org/resources/approved-antibodies/).

Among the different formats of recombinant antibodies, single-chain variable antibody fragments (scFvs), consisting of the variable domains of the heavy (VH) and light (VL) immunoglobulin chains joined by a flexible linker, have probably the primacy of versatility. Indeed, they can be easily engineered by molecular biology techniques according to the purpose; e.g., grafted to different scaffolds, expressed as intracellular antibodies (intrabodies) by eukaryotic viral or non-viral vectors for delivery to tumor cells and tissues, or even administered as purified proteins directly to target cells [2,3]. A number of scFvs were utilized to specifically inhibit different protein functions and showed effective anti-tumor activity both in vitro and in vivo [3–5].

Among the over 200 Human Papillomavirus (HPV) genotypes discovered in humans (PaVe Database) [6], only twelve to fourteen are defined as high risk (HR) and are causally related to virtually all tumor lesions of the cervix, a high proportion of squamous cell carcinomas (SCC) in the ano-genital region and an increasing number of those in the head and neck area (HNSCC) [7,8]. Among the HR HPVs, HPV16 is the most represented in all body districts, and almost the only genotype present in the HPV-related oropharyngeal SCC, which comprise 30% of the total HNSCC [9].

The whole pro-oncogenic activity of the HR HPVs is in charge of the E6 and E7 oncoproteins, which are the first (Early) viral proteins transcribed from the same mRNA during infection. E7 is a 98 amino acid (aa) phosphoprotein comprising three conserved regions: CR1 (aa 2 to 15), CR2 (aa 16 to 37) and CR3 (aa 38 to 98) [10,11]. The CR3 region at the C-terminus is highly structured and comprises a zinc finger domain (aa 58–61 and 91–94) proposed to be involved in protein oligomerization [12]. The E7 N-terminus, which includes residues 1–40 (CR1/CR2), is instead unstructured and represents an intrinsically disordered region determining the protein plasticity, where the capacity to assume different conformations determines the E7 stability and correlates with its capacity to bind to different targets, involved or not in the transforming activity [13]. In particular, aa residues 21–26 in the CR2 region are responsible for the binding to the pRb tumor suppressor, crucial for the transforming activity. Furthermore, cysteine (C) at position 24 is a redox center that, under oxidative stress in HPV-transformed cells, undergoes glutathionilation, hindering pRb binding. This region also binds to TMEM173/STING, causing the inhibition of the antiviral response [14]. Instead CR1 has a role in transformation unrelated to the interaction with pRb, where the deletion of positions 6–10 (PTLHE) inactivates the E7 transforming activity but does not affect its transactivating capacity [15].

E6 is a 158 aa protein characterized by the presence of two Zinc finger domains (residues 37–73 and 110–146) linked by a LXXLL binding motif responsible for the proteasome-mediated degradation of a number of cellular substrates, including the p53 tumor suppressor [16]. The PDZ sequences (residues 156–158) at the C-terminus also participate in the E6 oncogenicity by the interaction with proteins implicated in cellular adhesion and polarity control [17].

The continuous expression of E6 and E7 during a persistent infection with HR-HPV can induce, over the years, the development of tumors in which the two oncoproteins represent the Tumor Associate Antigens (TAA) [18,19]. Therefore, targeting E6 and E7 offers the possibility of counteracting the development and/or progression of pre-tumor and tumor lesions for which the current HPV vaccine, effective in preventing infection, is not useful, since it has not been designed for therapeutic purposes. That early, HPV-induced pre-neoplastic lesions are localized in a limited area, which is an additional advantage for therapy, as it allows for local treatment [20]. This approach may also represent a therapeutic option for HPV-positive oropharyngeal SCC, which presently has poor disease outcome [21]. At present, the available prophylactic vaccines are not approved for the prevention of this tumor.

In the past few years, we isolated four scFvs against the E7 (scFv9, scFv32, scFv43 and scFv51) and one scFv against the E6 (scFvI7) oncoprotein of HPV16. The scFv43 was modified by site-directed mutagenesis to increase its stability [22], and the resulting scFv43M2 used thereafter. ScFv43M2 fused with the SEKDEL signal for retention in the endoplasmic reticulum, and scFvI7 fused with the nuclear localization signal (NLS), expressed as intrabodies, demonstrating a strong antiproliferative activity in HPV16–positive cells in vitro as well as antitumor efficacy in preclinical models in vivo [23–25].

Here, we compared the reactivity of the anti-HPV16 E6 (16E6) and -HPV16 E7 (16E7) scFvs against their targets, and identified the E7 regions recognized by the different anti-16E7 scFvs.

It is well known that the Complementarity Determining Regions (CDRs) present in the antibody VH and VL domains are responsible for various antibody properties in addition to the ability of antigen-binding, including solubility and stability.

Many studies analyzed the CDRs charge of therapeutic antibodies in an attempt to decipher a common signature and predict the proper functioning of new antibody candidates. One of these studies analyzed more than 100 clinical-stage antibodies and found a strong correlation between the negative net charge of CDRs on the one hand, and a high binding specificity, good solubility, efficient refolding and the low self-association and aggregation of the antibody molecule on the other hand [26]. In particular, it was reported that negative charged residues, such as aspartate (D) and glutamine (E) positively correlate with the probability of favorable scFv properties, whereas positive charged residues, such as arginine (R) and lysine (K), show a negative correlation. An exception to this rule is represented by the hydrophobic leucine (L) residues, a high number of which correlates with a lower specificity of binding and poor biophysical properties.

A different study analyzed a large set of post-phase I clinical-stage antibody therapeutics (CSTs) and identified shared CDR characteristics indicative of a possible development as antibodies for clinical use, leading to the implementation of the Therapeutic Antibody Profiler (TAP) software [27]. TAP allows for the comparative analysis of new antibody candidates with CSTs by analyzing five key parameters: (i) total CDR length; (ii) extent and magnitude of surface hydrophobicity (Patches of Surface Hydrophobicity Metric, PSH); (iii) Patches of Positive Charge Metric (PPC); (iv) Patches of Negative Charge Metric (PNC); (v) asymmetry in the net heavy- and light-chain surface changes (Structural Fv Charge Symmetry Parameter, SFvCSP), responsible for high viscosity.

Bearing in mind the possible use of our scFvs for the treatment of HPV16 lesions, we evaluated the scFv specificity based on the charge of the aa residues that form their CDRs, and investigated by TAP the properties potentially favoring their therapeutic development.

## 2. Results

### 2.1. CDRs aa Sequences of the Anti-16E6 and -16E7 ScFvs

Sequence analysis of the four anti-16E7 scFv9, scFv32, scFv43 and scFv51 confirmed the VH origin from the DP47 germline gene. As far as it concerns the VL origin, scFv43 derives from the DPK22, while scFv9, scFv32 and scFv51 derive from the DPL16 germline gene [28]. The diversity of the ETH-2 library ($10^8$ clones) used to select these antibodies is entirely up to the VH and VL CDR3 that have been modified by the random mutagenesis of selected nucleotide positions [28].

The CDRs of the VL and VH domains of the anti-16E6 scFvI7 were compared by sequence alignment to the IgG database using the IgG blast tool. Of note, due to the library construction, the VH and VL in the anti-16E6 scFv are in reverse order compared to the anti-16E7 scFvs, with the VL deriving from murine IgKV6-23 germline gene located at the N-terminus upstream of the linker, and the VH deriving from IGHV1S81 located at the C-terminus, downstream thereof.

The aa compositions of the CDRs are shown in Table 1. ScFv43M2 is indicated in place of the originally selected scFv43. The two scFvs have the same specificity but scFv43M2 carries an aa variation from T to M at position VH34 (numbering according to Kabat) [29]. The change, obtained by site-directed mutagenesis, re-established the original sequence of the germline gene according to the VH consensus, and successfully improved the scFv stability [22].

**Table 1.** Complementarity-determining Regions (CDRs) of the Anti-16E7 and -16E6 ScFvs.

| scFv | CDR1–IMGT (27–38) | CDR2–IMGT (56–65) | CDR3–IMGT (105–117) | CDR1–IMGT (27–38) | CDR2–IMGT (56–65) | CDR3–IMGT (105–117) |
|---|---|---|---|---|---|---|
| anti-16E7 | | VH (aa) | | | VL (aa) | |
| scFv9 | GFTF … SSYA | ISGS..GGST | ARGVGAFRPYRKHE | SLR … … SYY | GK … … .N | NSSPFE..HNLVV |
| scFv32 | GFTF … .SSYA | ISGS..GGST | AKQLHK … TLFDY | SLR … … SYY | GK … … .N | NSSPNK..ANPVV |
| scFv43M2 | GFTF … .SSYA | ISGS..GGST | AKVRR … .RFDY | QSVS … ..SSY | GA … … .S | QQRHG … .NPAT |
| scFv51 | GFTF … .SSYA | ISGS..GGST | AKHLK … .GFDY | SLR … … SYY | GK … … .N | NSSLQH..PPRVV |
| anti-16E6 | | VL (aa) | | | VH (aa) | |
| scFvI7 | QDV … … GTA | WA … … .S | QQYSS … .YPYT | GYTF … .TSHW | INPS..NGRT | ARYDG … .YFDY |

The aa sequences of the VH and VL CDRs of the anti-16E7 and -16E6 scFvs are shown. The aa positions indicated are according to the IMGT unique numbering, which provides standardized limits for the CDRs (CDR1-IMGT: 27 to 38, CDR2-IMGT: 56 to 65 and CDR3-IMGT: 105 to 117). Gaps, indicated by dots represent unoccupied positions. Note that the 14 aa length of the scFv9 VH-CDR3 exceeds the most common length for this region, which is up to 13 aa; therefore, the additional position 112.1 was created between positions 111 and 112.

## 2.2. Analysis of the Anti-16E6 and Anti-16E7 ScFvs Binding Activity

The reactivity of all the anti-16 E7 scFvs was confirmed by a comparative ELISA using the purified scFvs proteins against the recombinant antigen, as reported in the Materials and Methods. As shown in Figure 1A, all the anti-16E7 scFvs were able to recognize their antigen. Since reactivity against multiple nonspecific antigens was observed for scFv9 (personal communication), the features of this antibody were not further investigated. ScFvI7 and an anti-16E6 polyclonal Ab were included in the analysis as internal controls, since their anti-E6 binding activity had been previously demonstrated, as shown in Figure 1B [25].

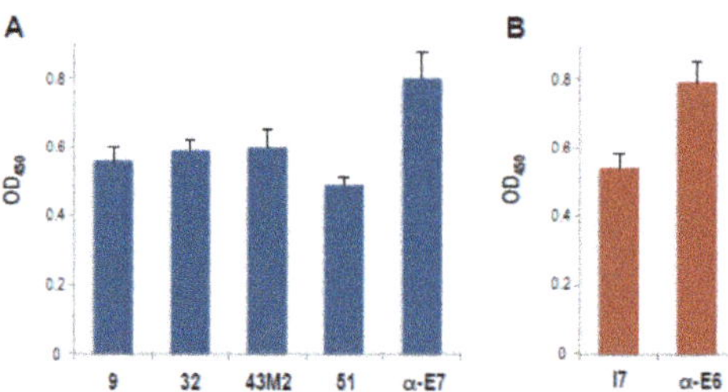


**Figure 1.** Characterization of the anti-16E7 and anti-16E6 scFvs reactivity against the respective recombinant oncoproteins by ELISA. Data represent the mean ± SD of samples in quadruplicate from a representative experiment of three with similar results. (**A**) the anti-16E7 scFv9 (9), scFv32 (32), scFv43M2 (43M2), scFv51 (51) and (**B**) the anti-16E6 scFvI7 (I7) are indicated. Anti-16E7 polyclonal Ab and anti-16E6 mAb were used as positive controls.

We previously demonstrated that scFv43M2 and scFv51 specifically bind to the E7 N-terminus (residues 1–54) region that retains important functions, including the transforming activity. From the same studies, we also know that scFv43M2 and scFv51 bind to different E7 epitopes [30]. Instead, we had no indication of the scFv32 binding regions on E7. To map the binding sites of scFv32, scFv43M2 and scFv51 on E7 in detail, as antigens we used a number of GST-tagged E7 proteins carrying either deletions or single aa variations previously designed to map the contribution of specific domains of the E7 gene product in the transcriptional trans-activation and cellular transformation functions [31]. The E7 mutants with aa deletions or variations are represented in Figure 2A,B, respectively.

The ELISA results demonstrated that scFv32 and scFv51 did not recognize $E7_{\Delta 2-15}$ and $E7_{\Delta 10-20}$, while they were still able to recognize $E7_{\Delta 21-35}$. This means that the E7 region, comprised between aa at positions 2 and 20, which includes CR1 and part of CR2, is critical for the binding of scFv32 and scFv51 to E7, while the adjacent region 21–35 is dispensable. Therefore, the two scFvs have a similar valence with regard to the binding of their target E7. On the other hand, scFv43M2 binding was greatly reduced using E7$_{\Delta21-35}$ and absent using E7$_{\Delta2-15}$ and E7$_{\Delta10-20}$, as shown in Figure 2A, indicating that this scFv recognizes CR1 and almost the whole CR2. Western blot analysis confirmed the ELISA results (personal communication).

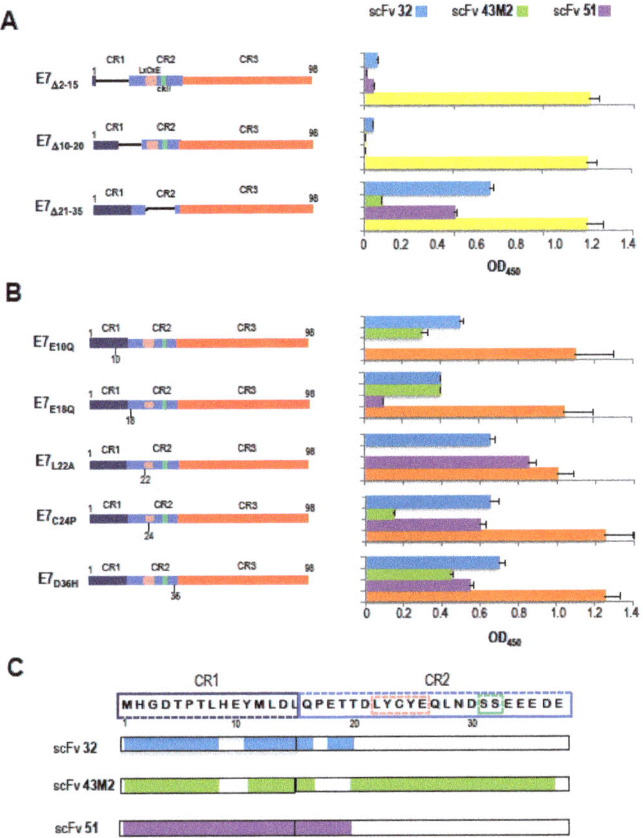

**Figure 2.** Epitope mapping of anti-E7 scFvs binding site on the 16E7 oncoprotein. (**A**) Schematic representation of 16E7 deletion mutants with the deleted amino acids, indicated by Δ, is shown on the left. The conserved CR1, CR2 and CR3 regions, the conserved LxCxE motif responsible for pRb binding and the casein kinase II (CKII) phosphorylation site in the CR2, are also indicated. The bars on the right represent the reactivity in ELISA, of scFv32 (light blue), scFv43M2 (green), scFv51 (violet) and in-house-made anti-E7 mouse polyclonal antibody (yellow), against each mutant. Data are expressed as the mean ± SD of three independent experiments, each performed in triplicate. (**B**) Schematic representation of the 16E7 mutants carrying single aa variations is shown on the left. E7$_{E10Q}$ has a variation from glutamic acid (E) to glutamine (Q) at position 10, and the same change is present in E7$_{EQ18}$ at position 18, while E7$_{L22A}$ has a variation from leucine (L) to alanine (A) at position 22. E7$_{C24P}$ has proline (P) instead of cysteine (C) at position 24, while E7$_{D36H}$ has histidine (H) instead of aspartic acid (D) at position 36. The results of the ELISA performed with the same mutants are shown on the right. The color code for scFvs is the same as in panel A. The reactivity of the anti-E7 mAb is indicated in orange. Values are the mean ± SD of four independent experiments, each performed in triplicate. (**C**) The aa sequence of the E7 N-terminal region. CR1 and CR2 sequences, as well as the pRb binding and the CKII-phosphorylation sites in the CR2, are delineated by dashed line boxes with color codes, as in panels A and B. The E7 regions, bound by the scFvs indicated on the left, are highlighted by the respective color codes.

The E7 proteins carrying single aa variations allowed to better define the epitope binding sites on E7. As shown in Figure 2B, scFv32 and scFv43M2 retained the binding to $E7_{E10Q}$ and $E7_{E18Q}$ mutants while both variations abrogated the binding of scFv51. The A22L and C24P variations abolished the scFv43M2 binding to E7, while they were irrelevant for the scFv32 and scFv51 binding. Lastly, all the scFvs could bind to the $E7_{D36H}$ mutant. Again, the ELISA results were confirmed by Western blot analysis (personal communication).

Taken together, these results indicate that scFv32 is able to recognize an E7 domain, comprised between positions 1 and 20, which includes CR1 and part of CR2, as well as scFv51, which, however, seems to be affected by the E18Q variation. On the other side, scFv43M2, not recognizing the E7 with A22L and C24P variations, nor the $E7_{\Delta21-35}$ mutant, binds to a wider region, comprising CR1 and CR2, and confirms the potential ability to interfere directly with the pRb-E7 interaction. A schematic representation of the scFvs binding sites on the E7 region, including the first N-terminal 37 aa, based on the results above described, is shown in Figure 2C.

To better define the binding differences among the three anti-16E7 scFvs, we analyzed, by Surface Plasmon Resonance (SPR), the ability of the purified scFv proteins to bind to the recombinant wild type 16E7, covalently immobilized on the surface of a CM5 sensor chip. This was accomplished in a competition test by studying the binding interference among the three scFvs. As shown in Figure 3, the saturation of E7 with scFv32 was able to partially hamper the scFv51 binding, indicating that the E7-binding regions of the two scFvs were at least partly overlapped. ScFv43M2 injected after scFv51 was still able to bind to the sensor chip, showing that its ability to bind the E7 is not hampered by the previously injected scFvs. From this result we can also deduce that scFv43M2 has binding epitopes different from both scFv32 and scFv51.

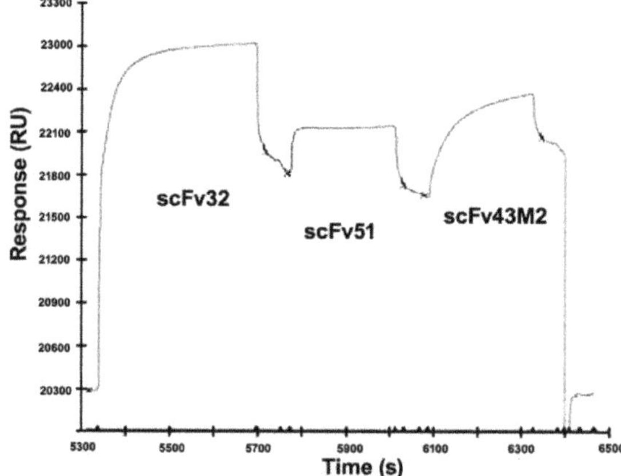


**Figure 3.** Surface Plasmon Resonance (SPR) analysis of the anti-16E7 scFv32, scFv43M2 and scFv51 binding to the 16E7 protein. The sensorgram shows the relative binding of the three scFvs to the recombinant 16E7 immobilized on the sensor chip. ScFv32 at saturating concentration was injected, followed by scFv51 and then by scFv43M2.

### 2.3. Computational Analyses of The Anti-16E7 and -16E6 Scfvs aa Sequences

To evaluate the potential therapeutic efficacy of our anti-16E6 and -16E7 scFvs, we first calculated the CDR3 net charge at neutral pH, based on the presence of positively and negatively charged aa residues as predictors of antibody specificity [25]. As reported in Table 2, both the VH and VL CDR3s were in the range of −1 to +3.1. By adding the net charges of the VH and VL CDR3, most anti-16E7 scFvs have a slightly positive net charge in the range of +1 to +2, with the exception of scFv43M2 (+4.1), while the anti-16E6 scFv has slightly negative CDR3 net charge (−1).

**Table 2.** Analysis of the ScFv CDRs Net Charge.

| Anti-16E7 ScFvs | VH CDR3 Net Charge | VL CDR3 Net Charge | Total CDR Net Charge |
|---|---|---|---|
| scFv9 | +3.1 (+3.0) | −0.9 (−1.0) | +4.2 (+4.0) |
| scFv32 | +1.1 (+1.0) | +1.0 (+1.0) | +4.1 (+3.9) |
| scFv43M2 | +3.0 (+3.0) | +1.1 (+1.0) | +4.1 (+4.0) |
| scFv51 | +1.1 (+1.0) | +1.1 (+1.0) | +4.2 (3.9) |
| **Anti-16E6 scFvs** | **VL CDR3 Net Charge** | **VH CDR3 Net Charge** | **Total CDR Net Charge** |
| scFvI7 | 0.0 (0.0) | −1.0 (−1.0) | −0.9 (−1.0) |

The theoretical net charges of the scFv VH and VL CDR3 were considered separately, and the entire set of six CDRs were calculated by adding the negative charges of glutammate (E, −1) and aspartate (D, −1), and the positive charges of arginine (R, +1), lysine (K, +1) and histidine (H, + 0.1) or using the free tool by Kozlowski LP (2016), IPC-Isoelectric point Calculator with Biology Direct 11: 55 (in brackets, the values calculated at pH 7.4) [32].

Since it has been reported that the net charge of the entire set of six CDRs is a better predictor of antibody specificity than the CDR3 charge, or even of the entire scFv, we then calculated the theoretical total CDR net charge at neutral pH, as shown in Table 2. All the anti-16E7 scFvs showed a positive total net charge of around +4.0, different from the anti-16E6 scFvI7 that had a slightly negative total CDRs net charge (−0.9). Furthermore, since non polar or polar non charged aa residues in CDRs were also reported to contribute positively or negatively to specificity [26], we calculated their number in our scFvs and found that the CDR aa residues positively related to specificity are 1.4 to 1.8 fold those negatively related.

We then analyzed our scFvs by TAP software, and the results, reported in Table 3, show the average values obtained for the properties analyzed. We did not include scFv9 in this analysis because of its abovementioned nonspecific reactivity, nor did we include scFvI7. Indeed, due to the murine scaffold, in the case of the therapeutic development of scFvI7, it is necessary to graft the CDRs into a scaffold of human origin.

The anti-16E7 scFv32 and scFv51 display values in full accordance with CSTs, whereas the anti-16E7 scFv43M2 has suboptimal PPC, with this value falling in the extreme 5% of the CSTs distribution.

**Table 3.** Therapeutic Antibody Profiling of ScFv Candidates.

| TAP Metrics | scFv32 | scFv43M2 | scFv51 |
|---|---|---|---|
| CDR length | 47 | 44 | 45 |
| CDR vicinity PSH score | 133.6982 | 119.791 | 134.2631 |
| CDR vicinity PPC score | 0.4761 | 3.1258 | 0.4023 |
| CDR vicinity PNC score | 0.0 | 0.0 | 0.0 |
| sFvCSP score | 2.1 | 12.4 | 2.31 |

Comparison among the anti-16E7 scFvs using the TAP computational tool by Raybould et al. [27] for evaluation of possible therapeutic development. Green boxes indicate a good agreement of the scores with those of post-phase I clinical-stage antibody therapeutics (CSTs); the amber box indicates a score that is less represented in the metric distribution of the 242 antibody therapeutics, considered for setting the parameters. PSH = Patches of Surface Hydrophobicity Metric; PPC = Patches of Positive Charge Metric; PNC = Patches of Negative Charge Metric; SFvCSP = Structural Fv Charge Symmetry Parameter. Results are available at the following links: http://opig.stats.ox.ac.uk/webapps/newsabdab/sabpred/tap_results/20200502_0570079 for scFv32; http://opig.stats.ox. ac.uk/webapps/newsabdab/sabpred/tap_results/20200502_0199339 for scFv43M2; http://opig.stats.ox.ac.uk/webapps/ newsabdab/sabpred/tap_results/20200502_0278604 for scFv51 (accessed on 1 June 2020).

## 3. Discussion

Targeting the HPV oncoprotein activity has been explored in a number of therapeutic approaches for HPV-associated tumors, mainly because it warrants high precision and specificity due to the oncoprotein key role in HPV tumor onset and progression [33]. Because of the dominant role played by HPV16 in HPV-associated cancers, most of the studies have focused on the oncoproteins encoded by this genotype.

Our approach aimed at hampering the activity of the 16E6 and 16E7 oncoproteins through antibodies in single-chain format (scFvs), which, due to their flexible format, can be tailored and improved according to their needs.

In our laboratory, we isolated a number of anti-16E7 scFvs, namely scFv9, scFv32, scFv43M2 and scFv51, and the anti-16E6 scFvI7. All the scFvs exhibit a good reactivity for their respective antigens, as shown in Figure 1.

It is well known that the 16E7 is a multi-functional protein that possesses a mosaic of oncogenic activities whose partial or total ablation has therapeutic implications. Since the different scFvs selected, all recognizing the E7 N-terminus [30], can bind to different epitopes and potentially interfere with different functions of this protein, to map in detail such epitopes, we took advantage of a series of E7 mutants carrying either aa deletions or single aa variations, that we used in immunoassays. The observation that neither scFv32 nor scFv51 were able to recognize the $E7_{\Delta2-15}$ and $E7_{\Delta10-20}$ deletion mutants made it clear that both scFvs recognize the E7 N-terminal region, including the CR1 domain and some amino acids of CR2, critical for the trans-activating and anti-apoptotic activity of the oncoprotein [10]. In the meantime, they both recognize $E7_{\Delta21-35}$, and their binding is not affected by variations at positions 22 and 24, indicating that their function is not directly involved in hampering the E7 binding to pRb. On the other hand, scFv32 was able to bind to the $E7_{E10Q}$ and $E7_{E18Q}$ mutants, suggesting that scFv32 binding is not affected by variation from a negatively charged residue (E) to a non-charged residue (Q). ScFv51 behaves differently, since both variations at positions 10 and 18 abolish its binding to E7. The scFv43M2 binding to E7 is instead hampered by variations at positions 22 and 24, confirming the already demonstrated interference of this antibody with the E7 binding to the pRb tumor suppressor [30]. Finally, all the scFvs can bind to $E7_{D36H}$, indicating that the CKII serine-specific kinase region [34,35] is not involved in the anti-E7 scFv activities. The results obtained with Surface Plasmon Resonance confirmed the partial overlapping of the scFv32 and scFv51 binding sites, and also that they differ from the epitopes bound by scFv43M2.

Studies on antibody optimization for therapeutic use have long found that the CDR charge is relevant for the antibody effectiveness, in influencing properties, such as folding stability, solubility and pharmacokinetics [36]. Recently, the theoretical CDR net charge was demonstrated to be a strong predictor of the antibody specificity, where negatively charged CDR3 and total CDRs were found to confer antibodies a higher specificity with respect to the positively charged CDRs [26]. When analyzing our anti-E6 and -E7 scFvs in terms of CDR3 and total CDR net charge, to identify which of them have promising characteristics for therapeutic development, we found values of CDR net charge of around +4.0, except for the anti-16E6 scFv, which exhibits a slightly negative charge and values ranging from −1.0 to + 3.0 for the VH and VL CDR3 charges.

In their paper, Rabia et al. analyzed 137 antibodies in clinical stage to identify values of CDR charges optimal for antibody specificity and biophysical properties. Of note, even though our scFvs have values of CDR charge not fully matching the optimal values, 20–35% (depending on the number of parameters considered) of the antibodies analyzed by Rabia with CDR net charge $\geq + 2.0$ had favorable biophysical properties [26].

Furthermore, when we considered other aa residues—uncharged polar and non-polar—that can contribute negatively or positively to CDR specificity [26], we found that the number of residues with positive correlation far exceeds that of residues with negative correlation. Additionally, the low number of L residues in the CDRs of scFv32 and scFv51, and the absence of L in scFv43M2 and scFvI7, are not in contrast with a specific interaction between the scFvs and their targets. We also compared our scFvs to accredited therapeutic antibodies by using the recently developed Therapeutic Antibodies Profiler (TAP) [27]. TAP considers both the sequence and structural properties of a large set of CSTs and establishes threshold values for five parameters examined so that new candidate molecules can be scored to highlight potential weaknesses, discouraging therapeutic development. TAP analysis of our scFvs showed a good agreement in general with the optimal values for all the parameters, with only few values falling in the extreme 5% of the CSTs distribution. However, it is important to note that, as far as it concerns the properties considered, all our scFvs show values falling within the range of values belonging to the CSTs analyzed by the TAP software [27]. Additionally, scFv43M2, which a displays suboptimal PPC score, has already shown an effective antiproliferative and antitumor activity [24,25]. In this regard, it should be remembered that TAP analysis allows a theoretical prediction based on the assumption that antibodies that have reached phase I clinical trials have characteristics favorable to therapeutic development, but it does not consider antibody features, such as immunogenicity or stability, which are equally important for therapeutic applications. In fact, few antibodies in clinical use have parameters which do conform to the desirable values according to TAP.

In summary, our anti-E7 scFvs scFv32 and 51, although differing in the CDR3 sequences, bind to an almost identical region of E7, have overlapping characteristics with regard to their E7 binding capacity and exhibit similar values of CDR3, total CDR net charge and TAP analyses. Hence, from the point of view of therapeutic development, these two scFvs can be considered equivalent, with an advantage for scFv51 because of its anti-proliferative activity, already demonstrated. ScFv43M2 binds instead to a different region of E7, including the pRb-binding region, which is critical for the protein transforming activity, and was characterized in previous studies showing its effectiveness as an antitumor molecule. Therefore, it is worth pursuing future research regarding the clinical application of these antibodies, alone or in combination.

We believe that the information obtained by these analyses could represent a starting point to optimize the properties of the scFvs and to support decisions on their therapeutic development. Indeed, small variations in the aa sequence can have an impact on the antibody biophysical characteristics, as demonstrated by the increase in scFv43 half-life obtained by site-directed mutagenesis [22].

Once particularly favorable CDRs have been identified in an antibody framework, the cassettes, including the VH and VL CDRs, can be grafted onto different antibody scaffolds, as needed [37].

In addition, the information acquired could help identify scFvs to be used in combination or for the construction of bi-specific scFvs, in order to enhance the antitumor effect by targeting different epitopes of the same oncoprotein, or both the E6 and E7 proteins at the same time.

## 4. Materials and Methods

### 4.1. ETH-2 and SPLINT Libraries of Recombinant Single-Chain Antibodies

ETH-2 is a phage display library of human recombinant antibodies in single-chain format, consisting of a single polypeptide chain that includes an antibody VH joined by a flexible polypeptide linker to a VL domain [28]. Only DPK-22 and DPL-16 for the light chain, and DP-47 for the heavy chain, were used as germ-line genes for the library construction [28]. To create a large repertoire of antibodies, random loops of 4-5-6 amino acids were appended to position 95 of the VH CDR3, whereas six aa positions were altered in the VL CDR3. The library was cloned in the NcoI and NotI restriction sites of the pDN332 phagemid vector, containing the M13 origin of replication, the *E. coli* origin of replication, an Amp resistance, a peptide leader and a lac-Z promotor [28]. All scFvs expressed using this vector were fused with a FLAG-tag and a 6xHistidine-tag (His-tag) at the C-terminus.

The Single Pot Library of Intracellular antibodies (SPLINT) is a murine naïve library of scFv fragments expressed in the yeast cytoplasm. SPLINT construction was detailed elsewhere [38].

### 4.2. ScFv Selection

The selection of the anti-16E7 scFvs from the ETH-2 library was described in Accardi et al. [23]. Three rounds of panning in solution were carried out against the biotinylated recombinant His-E7 protein, according to Pini et al. [28]. The selected anti-E7 scFvs are cloned in the pDN332 phagermid vector under the lac z-promoter control.

The selection of the anti-16E6 scFvI7 from the murine SPLINT is described elsewhere [25]. Briefly, the SPLINT was transformed in the L40 yeast containing 16E6-expressing bait. After two yeast screenings for LacZ activity and histidine prototrophy, one positive clone, specifically interacting with

16E6 bait and not interacting with lamin bait used as an irrelevant antigen, was identified among transformants by Intracellular Antibody Capture Technology (IACT) [38].

### 4.3. Sequencing

For the sequence analysis of the CDR3 regions responsible for the diversity of the anti-16E7 antibodies, two primers were used, specifically:

DP47CDR2back (priming in the VH germline gene, before the VH CDR3)
5′-TAC TAC GCA GAC TCC GTC AAC-3′;
fdseq1 (priming at the beginning of the phage gene III, which is located downstream of the scFv sequence);
5′-GAA TTT TCT GTA TGA GG-3′.
Other primers designed to cover the whole scFv sequences were used, specifically:
PelBback (priming on the PelB leader, which is located upstream of the scFv sequence);
5′-AGC CGC TGG ATT GTT ATT AC-3′
C3 (closer to the VH CDR3):
5′-TACTACGCAGACTCCGTGAAG-3′
GVL (closer to the VLCDR3):
5′-CTCTCCTGCAGGGCCAG-3′.
The scFvI7 cloned in scFvExpress was sequenced using the following primers to cover both strands:
ScFvExRev
5′-GAG GGG CAA ACA ACA GAT GG-3′;
antiE6seqDir
5′-GTC CCT GAT CGC TTC ACA GG-3′;
antiE6seqRev
5′-CCC AGA ACC GCT GGT CGA CC-3′.
Sequence alignments to the NCBI database were carried out using Immunoglobulin BLAST.

### 4.4. Plasmids for Protein Expression

The anti-16E7 scFvs were all inserted in the pDN332 phagemid, also allowing expression in the prokaryotic systems [28]. Interestingly, the coding sequences of the anti-16E6 scFvI7, which had been selected as an intrabody, were subcloned into the scFvExCyto-SV5 eukaryotic vector for expression in cell cytoplasm, and into the pQE30 prokaryotic vector for protein expression [25,39].

Full-length 16E6 and 16E7, fused to a 6-His Tag tail, were constructed by cloning in the pQE-30 vectors (Qiagen, Chatsworth, Ca), as described in Di Bonito et al. 2006 [40] and Accardi et al. 2005 [23]. The JM109 strain of *E. coli* was transformed with the recombinant pQE-30 plasmids.

The recombinant plasmids expressing mutant E7 proteins carrying specific deletions or aa variations (kindly provided by David Pim, ICGEB, Trieste, Italy) were cloned in the pGEX-2T vector (Sigma-Aldrich, Italy) and expressed as Glutathione-transferase (GST)-fusion proteins in the *E. coli* DH5a strain.

### 4.5. Protein Purification

The extraction of scFvs and of the E6 and E7 proteins was performed from the respective transformed bacteria, and the proteins were purified using protein A-Sepharose CL-4B agarose beads (Amersham Biosciences) for the scFvs, and Ni-NTA agarose beads (Qiagen) for the oncoproteins, as previously reported [22,23,39,40].

The purity of the proteins was evaluated by Coomassie Blue Staining after SDS-PAGE, and the protein concentration was determined by Bradford assay (BioRad, Italy).

The E7 mutants were purified by affinity chromatography using GST-Sepharose (Invitrogen-ThermoFisher Scientific, Waltham, MA, USA) as described in Accardi et al. [30].

### 4.6. ScFv Reactivity

The reactivity of the purified scFvs (3 µg/mL) towards the recombinant His-E7 or His-E6 proteins (0.3 µg/well), immobilized onto microtiter 96-well plates in carbonate buffer (pH 9.4) at 4° O/N, was tested in ELISA. The experimental conditions used in this assay are different from those used for scFv selection by Phage Display, where the biotinylated His-E7 was employed for panning in solution. The scFv binding was detected by incubation with mouse anti-Flag M2 monoclonal antibody (mAb) (2 µg/mL) (Sigma, St. Louis, MO, USA), which recognizes the FLAG-tag at the scFv C terminus. Anti-E7 polyclonal Ab produced in mice in our laboratory and anti-E6 mAb (Invitrogen-ThermoFisher Scientific, Europe), diluted 1:500, were used as positive controls [40]. After extensive washing, the immune complexes were revealed by goat anti-mouse Horseradish peroxidase-conjugated (GAM-HRP) IgG (Amresco, Krackeler Scientific, Albany, NY, USA) and by using the TMB substrate kit for peroxidase (Vector Laboratories, Inc., Burlingame, CA, USA) for colorimetric evaluation. OD was measured at 450 nm in a microtiter plate reader (iMark, Bio-Rad, CA).

### 4.7. Epitope Mapping of the Anti-16E7 ScFvs

Analysis by ELISA was performed using a number of GST-tagged E7 proteins carrying either a deletion of aa stretches or a single aa variation as coating antigens (kindly provided by Dr David Pim). In particular, the $E7_{\Delta2-15}$ and $E7_{\Delta10-20}$ deletion mutants lack two different but overlapping regions, the first one belonging to CR1 and the second one also including few amino acids of CR2, whereas $E7_{\Delta21-35}$ maps in CR2 include the aa 20–29 involved in the binding of pRb and TMEM173/STING. As far as regards the single aa variations, both $E7_{E10Q}$ and $E7_{E18Q}$ have a variation from the negative charged glutamic acid (E) to the polar non-charged glutamine (Q); $E7_{L22A}$ has a variation from the hydrophobic leucine (L) to a small, less hydrophobic alanine (A); $E7_{C24P}$ has a variation from the reactive, sulfur-containing cysteine (C) to the cyclic non-polar proline (P); $E7_{D36H}$ has a variation from the negative charged aspartic acid (D) to the positively charged histidine (H). The scFvs in immune-complexes were detected as described above and according to the protocol reported elsewhere [30]. The non-specific signal of the scFvs bound to the recombinant GST protein (Kerafast, Boston, USA) was used as a cut-off value.

Western blotting analysis was performed to confirm the ELISA results. Each mutant E7 protein was separated on 15% SDS-PAGE and electro-blotted with semidry apparatus (Bio-Rad, CA) onto PVDF Immobilon-P membrane (Millipore). The membrane was then cut in strips (1 µg of protein/strip) and incubated with the purified scFvs at 2 µg/mL in 2% non-fat dry milk (NFDM, BioRad, Italy) followed by mouse anti-Flag M2 mAb (Sigma, St. Louis, MO, USA) and GAM-HRP (Amresco, Krackeler Scientific, Albany, NY, USA). In-house-produced mouse anti-E7 polyclonal Ab and anti-E7 mAb (Zymed, ThermoFisher Scientific, Europe), diluted 1:500, were used as positive controls. Immune-complexes were revealed by ECL with Super Signal West Pico Chemiluminescent substrate (ThermoFisher Scientific, Europe).

### 4.8. Surface Plasmon Resonance (SPR)

SPR was carried out as described in Accardi et al. [33]. Briefly, the E7 recombinant protein was immobilized to the Biacore CM5 sensor chip using conventional amine coupling. The reaction was performed by injecting E7 at a flow rate of 5 µL/min for 7 min in HBS-P (10 mM Hepes pH 7.4, 0.15 M NaCl, 0.005% (v/v) surfactant P20) at 25 °C. For the epitope mapping, the E7 surface was saturated by an injection of the purified scFv32 at a flow rate of 10 µL/min. Then, 40 µL of the purified scFv51 were injected at 500 nM, followed by the scFv43M2 at the same concentration.

### 4.9. Computational Analyses

Theoretical CDR3 net charge and total CDR charge were calculated, according to Rabia et al. [26] and using the free tool by Kozlowski [32].

TAP analysis was performed using the tool described in Raybould et al. [27] and which is freely available at http://opig.stats.ox.ac.uk/webapps/sabdab-sabpred/TAP.php.

## 5. Conclusions

We performed a theoretical validation of several scFvs against the 16E6 and 16E7 oncoproteins to be used as intrabodies for the treatment of HPV16-associated lesions, and highlighted some characteristics supporting their therapeutic development. Molecular engineering could help improve and tailor specific scFv properties according to the therapeutic needs, even in the direction of a personalized medicine.

The opportunity of using purified scFv proteins directly as therapeutic tools is intriguing as it embodies a very safe delivery system, free from the critical issue of exogenous DNA being internalized and expressed by the cells. However, protein delivery may be less efficient because self-limiting and requiring the continuous administration of the therapeutic molecules; in turn, involves the production of large amounts of the molecule in Good Manufacturing Practice conditions.

**Author Contributions:** Conceptualization, L.A. and C.A.; methodology, L.A., M.G.D. and B.C.; validation, L.A. and M.G.D.; investigation, L.A., B.C. and M.G.D.; resources, L.A. and P.D.B.; writing—original draft preparation, L.A.; writing—review and editing, L.A., C.A., M.G.D., B.C. and P.D.B.; visualization, L.A. and C.A.; supervision, L.A. and C.A.; project administration, L.A.; funding acquisition, L.A. and P.D.B. All authors have read and agreed to the published version of the manuscript.

**Funding:** This research was funded by intramural funds of the Istituto Superiore di Sanità.

**Conflicts of Interest:** The authors declare no conflict of interest.

## References

1. Antibody Society. Approved Antibodies. Available online: https://www.antibodysociety.org/resources/approved-antibodies/ (accessed on 1 June 2020).
2. Holliger, P.; Hudson, P.J. Engineered antibody fragments and the rise of single domains. *Nat. Biotechnol.* **2005**, *23*, 1126–1136. [CrossRef] [PubMed]
3. Accardi, L.; Di Bonito, P. Antibodies in single-chain format against tumour-associated antigens: Present and future applications. *Curr. Med. Chem.* **2010**, *17*, 1730–1755. [CrossRef]
4. Frenzel, A.; Schirrmann, T.; Hust, M. Phage display-derived human antibodies in clinical development and therapy. *mAbs* **2016**, *8*, 1177–1194. [CrossRef] [PubMed]
5. Lin, Y.; Chen, Z.; Hu, C.; Chen, Z.S.; Zhang, L. Recent progress in antitumor functions of the intracellular antibodies. *Drug Discov. Today* **2020**, *26*, 1359–6446. [CrossRef] [PubMed]
6. PaVE: Papilloma Virus Genome Database. Available online: https://pave.niaid.nih.gov/ (accessed on 1 June 2020).
7. Zur Hausen, H. Papillomaviruses and cancer: From basic studies to clinical application. *Nat. Rev. Cancer* **2002**, *2*, 342–350. [CrossRef]
8. De Martel, C.; Plummer, M.; Vignat, J.; Franceschi, S. Worldwide burden of cancer attributable to HPV by site, country and HPV type. *Int. J. Cancer* **2017**, *141*, 664–670. [CrossRef]
9. Castellsagué, X.; Alemany, L.; Quer, M.; Halec, G.; Quirós, B.; Tous, S.; Clavero, O.; Alòs, L.; Biegner, T.; Szafarowski, T.; et al. HPV involvement in head and neck cancers: comprehensive assessment of biomarkers in 3680 patients. *J. Natl. Cancer Inst.* **2016**, *108*, djv403. [CrossRef]
10. Donà, M.G. E7 oncoprotein of human papillomarvirus: Functions and strategies of inactivation for the treatment of HPV-associated cancer. In *Oncogene Proteins: New Research*; Arthur, H., Malloy, E., Carson, C., Eds.; Nova Science Publishers, Inc.: Hauppauge, NY, USA, 2008; pp. 113–155.
11. Todorovic, B.; Massimi, P.; Hung, K.; Shaw, G.S.; Banks, L.; Mymryk, J.S. Systematic analysis of the amino acid residues of human papillomavirus type 16 E7 conserved region 3 involved in dimerization and transformation. *J. Virol.* **2011**, *85*, 10048–10057. [CrossRef]
12. Smal, C.; Alonso, L.G.; Wetzler, D.E.; Heer, A.; de Prat Gay, G. Ordered self-assembly mechanism of a spherical oncoprotein oligomer triggered by zinc removal and stabilized by an intrinsically disordered domain. *PLoS ONE* **2012**, *7*, e36457. [CrossRef]

13. García-Alai, M.M.; Alonso, L.G.; de Prat-Gay, G. The N-terminal module of HPV16 E7 is an intrinsically disordered domain that confers conformational and recognition plasticity to the oncoprotein. *Biochemistry* **2007**, *46*, 10405–10412. [CrossRef]

14. Lau, L.; Gray, E.E.; Brunette, R.L. Stetson DB. DNA tumor virus oncogenes antagonize the cGAS-STING DNA-sensing pathway. *Science* **2015**, *350*, 568–571. [CrossRef]

15. Gulliver, G.A.; Herber, R.L.; Liem, A.; Lambert, P.F. Both conserved region 1 (CR1) and CR2 of the human papillomavirus type 16 E7 oncogene are required for induction of epidermal hyperplasia and tumor formation in transgenic mice. *J. Virol.* **1997**, *71*, 5905–5914. [CrossRef]

16. Nominé, Y.; Masson, M.; Charbonnier, S.; Zanier, K.; Ristriani, T.; Deryckère, F.; Sibler, A.P.; Desplancq, D.; Atkinson, R.A.; Weiss, E.; et al. Structural and functional analysis of E6 oncoprotein: Insights in the molecular pathways of human papillomavirus-mediated pathogenesis. *Mol. Cell* **2006**, *21*, 665–678. [CrossRef] [PubMed]

17. Massimi, P.; Gammoh, N.; Thomas, M.; Banks, L. HPV E6 specifically targets different cellular pools of its PDZ domain-containing tumour suppressor substrates for proteasome-mediated degradation. *Oncogene* **2004**, *23*, 8033–8039. [CrossRef] [PubMed]

18. Graham, S.V. The human papillomavirus replication cycle, and its links to cancer progression: A comprehensive review. *Clin. Sci.* **2017**, *131*, 2201–2222. [CrossRef]

19. Cohen, P.A.; Jhingran, A.; Oaknin, A.; Denny, L. Cervical cancer. *Lancet* **2019**, *393*, 169–182. [CrossRef]

20. Doorbar, J. Molecular biology of human papillomavirus infection and cervical cancer. *Clin. Sci.* **2006**, *110*, 525–541. [CrossRef] [PubMed]

21. Alsahafi, E.; Begg, K.; Amelio, I.; Raulf, N.; Philippe Lucarelli, P.; Sauter, T.; Tavassoli, M. Clinical update on head and neck cancer: Molecular biology and ongoing challenges. *Cell Death Dis.* **2019**, *10*, 540. [CrossRef]

22. Donà, M.G.; Giorgi, C.; Accardi, L. Characterization of antibodies in single-chain format against the E7 oncoprotein of the human papillomavirus type 16 and their improvement by mutagenesis. *BMC Cancer* **2007**, *7*, 25. [CrossRef]

23. Accardi, L.; Donà, M.G.; Di Bonito, P.; Giorgi, C. Intracellular anti-E7 human antibodies in single-chain format inhibit proliferation of HPV16-positive cervical carcinoma cells. *Int. J. Cancer* **2005**, *116*, 564–570. [CrossRef]

24. Accardi, L.; Paolini, F.; Mandarino, A.; Percario, Z.; Di Bonito, P.; Di Carlo, V.; Affabris, E.; Giorgi, C.; Amici, C.; Venuti, A. In vivo antitumor effect of an intracellular single-chain antibody fragment against the E7 oncoprotein of human papillomavirus 16. *Int. J. Cancer* **2014**, *134*, 2742–2747. [CrossRef]

25. Amici, C.; Visintin, M.; Verachi, F.; Paolini, F.; Percario, Z.; Di Bonito, P.; Mandarino, A.; Affabris, E.; Venuti, A.; Accardi, L. A novel intracellular antibody against the E6 oncoprotein impairs growth of human papillomavirus 16-positive tumor cells in mouse models. *Oncotarget* **2016**, *7*, 15539–15553. [CrossRef] [PubMed]

26. Rabia, L.A.; Zhang, Y.; Ludwig, S.D.; Julian, M.C.; Tessier, P.M. Net charge of antibody complementarydetermining regions is a key predictor of specificity. *Protein Eng. Des. Sel.* **2019**, *31*, 1–10. [CrossRef]

27. Raybould, M.I.J.; Marks, C.; Krawczyk, K.; Taddese, B.; Nowak, J.; Lewis, A.P.; Bujotzek, A.; Shi, J.; Deane, C.M. Five computational developability guidelines for therapeutic antibody profiling. *Proc. Natl. Acad. Sci. USA* **2019**, *116*, 4025–4030. [CrossRef] [PubMed]

28. Pini, A.; Viti, F.; Santucci, A.; Carnemolla, B.; Zardi, L.; Neri, P.; Neri, D. Design and use of a phage display library. Human antibodies with subnanomolar affinity against a marker of angiogenesis eluted from a two-dimensional gel. *J. Biol Chem.* **1998**, *273*, 21769–21776. [CrossRef]

29. Kabat, E.A.; Wu, T.T. Identical V region aa sequences and segments of sequences in antibodies of different specificities. Relative contributions of VH and VL genes, minigenes, and complementarity-determining regions to binding of antibody-combining sites. *J. Immunol.* **1991**, *147*, 1709–1719. [PubMed]

30. Accardi, L.; Donà, M.G.; Mileo, A.M.; Paggi, M.G.; Federico, A.; Torreri, P.; Petrucci, T.C.; Accardi, R.; Pim, D.; Tommasino, M.; et al. Retinoblastoma-independent antiproliferative activity of novel intracellular antibodies against the E7 oncoprotein in HPV 16-positive cells. *BMC Cancer* **2011**, *11*, 17. [CrossRef]

31. Phelps, W.C.; Münger, K.; Yee, C.L.; Barnes, J.A.; Howley, P.M. Structure-function analysis of the human papillomavirus type 16 E7 oncoprotein. *J. Virol.* **1992**, *66*, 2418–2427. [CrossRef]

32. Kozlowski, L.P. IPC—Isoelectric Point Calculator. *Biol. Direct.* **2016**, *11*, 55. [CrossRef]

33. Hoppe-Seyler, K.; Bossler, F.; Braun, J.A.; Herrmann, A.L.; Hoppe-Seyler, F. The HPV E6/E7 oncogenes: Key factors for viral carcinogenesis and therapeutic targets. *Trends Microbiol.* **2018**, *26*, 158–168. [CrossRef]

34. Edmonds, C.; Vousden, K.H. A point mutational analysis of human papillomavirus type 16 E7 protein. *J. Virol.* **1989**, *63*, 2650–2656. [CrossRef] [PubMed]

35. Barbosa, M.S.; Edmonds, C.; Fisher, C.; Schiller, J.T.; Lowy, D.R.; Vousden, K.H. The region of the HPV E7 oncoprotein homologous to adenovirus E1a and Sv40 large T antigen contains separate domains for Rb binding and casein kinase II phosphorylation. *EMBO J.* **1990**, *9*, 153. [CrossRef] [PubMed]

36. Datta-Mannan, A.; Thangaraju, A.; Leung, D.; Tang, Y.; Witcher, D.R.; Lu, J.; Wroblewski, V.J. Balancing charge in the complementarity-determining regions of humanized mAbs without affecting pl reduces non-specific binding and improves the pharmacokinetics. *MAbs* **2015**, *7*, 483–493. [CrossRef] [PubMed]

37. Aubrey, N.; Billiald, P. Antibody fragments humanization: Beginning with the end in mind. *Methods Mol. Biol.* **2019**, *1904*, 231–252. [CrossRef] [PubMed]

38. Visintin, M.; Quondam, M.; Cattaneo, A. The intracellular antibody capture technology: Towards the high-throughput selection of functional intracellular antibodies for target validation. *Methods* **2004**, *34*, 200–214. [CrossRef] [PubMed]

39. Verachi, F.; Percario, Z.; Di Bonito, P.; Affabris, E.; Amici, C.; Accardi, L. Purification and characterization of antibodies in single-chain format against the E6 oncoprotein of human papillomavirus type 16. *BioMed Res. Int.* **2018**, *20*, 6583852. [CrossRef]

40. Di Bonito, P.; Grasso, F.; Mochi, S.; Accardi, L.; Donà, M.G.; Branca, M.; Costa, S.; Mariani, L.; Agarossi, A.; Ciotti, M.; et al. Serum antibody response to Human papillomavirus (HPV) infections detected by a novel ELISA technique based on denatured recombinant HPV16 L1, L2, E4, E6 and E7 proteins. *Infect. Agents Cancer* **2006**, *1*. [CrossRef]

*Review*

# Beyond MicroRNAs: Emerging Role of Other Non-Coding RNAs in HPV-Driven Cancers

Mariateresa Casarotto [1], Giuseppe Fanetti [2], Roberto Guerrieri [1], Elisa Palazzari [2], Valentina Lupato [3], Agostino Steffan [1], Jerry Polesel [4], Paolo Boscolo-Rizzo [5] and Elisabetta Fratta [1,*]

[1] Division of Immunopathology and Cancer Biomarkers, Centro di Riferimento Oncologico di Aviano (CRO), IRCCS, National Cancer Institute, 33081 Aviano (PN), Italy; mtcasarotto@cro.it (M.C.); roberto.guerrieri@cro.it (R.G.); asteffan@cro.it (A.S.)

[2] Division of Radiotherapy, Centro di Riferimento Oncologico di Aviano (CRO), IRCCS, National Cancer Institute, 33081 Aviano (PN), Italy; giuseppe.fanetti@cro.it (G.F.); elisa.palazzari@cro.it (E.P.)

[3] Division of Otolaryngology, General Hospital "Santa Maria degli Angeli", 33170 Pordenone, Italy; valentinalupato@gmail.com

[4] Division of Cancer Epidemiology, Centro di Riferimento Oncologico di Aviano (CRO), IRCCS, National Cancer Institute, 33081 Aviano (PN), Italy; polesel@cro.it

[5] Section of Otolaryngology, Department of Neurosciences, University of Padova, 31100 Treviso, Italy; paolo.boscolorizzo@unipd.it

* Correspondence: efratta@cro.it; Tel.: +390434659569

Received: 27 March 2020; Accepted: 12 May 2020; Published: 15 May 2020

**Abstract:** Persistent infection with high-risk Human Papilloma Virus (HPV) leads to the development of several tumors, including cervical, oropharyngeal, and anogenital squamous cell carcinoma. In the last years, the use of high-throughput sequencing technologies has revealed a number of non-coding RNA (ncRNAs), distinct from micro RNAs (miRNAs), that are deregulated in HPV-driven cancers, thus suggesting that HPV infection may affect their expression. However, since the knowledge of ncRNAs is still limited, a better understanding of ncRNAs biology, biogenesis, and function may be challenging for improving the diagnosis of HPV infection or progression, and for monitoring the response to therapy of patients affected by HPV-driven tumors. In addition, to establish a ncRNAs expression profile may be instrumental for developing more effective therapeutic strategies for the treatment of HPV-associated lesions and cancers. Therefore, this review will address novel classes of ncRNAs that have recently started to draw increasing attention in HPV-driven tumors, with a particular focus on ncRNAs that have been identified as a direct target of HPV oncoproteins.

**Keywords:** HPV; squamous cell carcinoma; non-coding RNAs; circular RNAs; PIWI-interacting RNAs; long non-coding RNAs

---

## 1. Introduction

Worldwide, 4.5% of all cancers (630,000 new cancer cases per year) are attributable to Human Papilloma Virus (HPV) infection [1]. HPVs are a heterogeneous group of small non-envelope double-stranded circular DNA viruses targeting the basal cells of stratified epithelia [2,3]. The IARC Working Group has classified alpha-HPV types 16, 18, 31, 33, 35, 39, 45, 51, 52, 56, 58, 59 as carcinogenic to humans; these high-risk (HR)-HPVs are responsible for virtually all carcinomas of the cervix and different proportions of carcinomas of the anus, vagina, penis, vulva, and oropharynx (Table 1) [4]. Among the HR-HPV types, HPV16 is responsible for the majority of HPV-driven cancers. In addition, some HPV types of the beta genus showing cutaneous tropism have been proposed to cooperate with ultraviolet radiation in the development of non-melanoma skin cancer [5].

**Table 1.** Worldwide burden of cancer attributable to Human Papilloma Virus (HPV) by site.

| Tumor Site | Predominant HPV Types * | HPV Attributable Fraction (%) | New Cases Attributable to HPV | Prognostic Significance of HPV-Positivity | References |
|---|---|---|---|---|---|
| Head and neck cancer | | | | | |
| Oropharynx | HPV16; HPV33; HPV35 | 30.1 | 42,000 | Better survival | [1,6–8] |
| Oral cavity | HPV16; HPV52; HPV35 | 2.2 | 5900 | Inconclusive | [1,6–8] |
| Larynx | HPV16; HPV31; HPV33 | 2.4 | 4100 | Inconclusive | [1,6–8] |
| Cervical cancer | HPV16; HPV18; HPV45 | 100 | 570,000 | - | [1,6,9] |
| Anal cancer | HPV16; HPV18 | 88.0 | 29,000 | Better prognosis in men | [1,6,10] |
| Penile cancer | HPV16; HPV6; HPV18 | 50.0 | 18,000 | Inconclusive | [1,6,11,12] |
| Vulval cancer | HPV16; HPV33 | 24.9 | 11,000 | Better survival | [1,6,13,14] |
| Vaginal cancer | HPV16; HPV18; HPV73 | 78.0 | 14,000 | - | [1,6,15] |

* HPV16 is by far the most predominant type in all HPV-driven cancer.

Cervical squamous cell carcinoma (CSCC) is the fourth most common cancer in women worldwide [16]. Despite the spread of screening programs has significantly reduced mortality, nearly 50% of patients worldwide are still diagnosed with locally advanced stages. Concurrent platinum based chemoradiation is the current standard treatment of locally advanced CSCC [17]. Several studies have shown improved local control and survival with the use of concurrent chemoradiation with respect to radiotherapy alone but in these patients, recurrence rate and mortality remain still high [18,19]. Infection with HR-HPV is the most significant risk factor for CSCC. Several studies shown that the sustained expression of the oncogenic genes E6 and E7 of HPV is involved in CSCC progression [20–24] but the prognostic role of HPV expression genes is not fully elucidated yet. In clinical practice there are not available prognostic factors that can guide therapeutic choice in CSCC patients, and several studies are needed to improve our knowledge, especially on the role of HPV and other molecular and genomic factors.

The role of HPV in head and neck squamous cell carcinoma (HNSCC) has emerged in the last decades, with relevant etiological and clinical aspects. Nowadays, approximately 30% of oropharyngeal squamous cell carcinoma (OPSCC) is attributable to HPV worldwide [1], but this proportion is expected to increase in the close future. Therefore, HPV has been included as one of the strongest prognostic factors of OPSCC alongside the already well-defined stage, smoking, performance status, and quality of treating facilities [25]. Compared to HPV-negative counterparts, HPV-positive OPSCC patients show peculiar clinico-pathological features and improved prognosis [26]. On this basis, a different TNM staging has been proposed for HPV-positive OPSCC [27]. Notably, a gender-specific trend has also emerged for HPV-driven OPSCC. In fact, mirroring the downward trend of CSCC due to HPV vaccination programs, the HPV-driven OPSCC incidence is expected to decline in women, whereas the incidence among men has been increasing over the last years [28]. One possible explanation could lie in the profound differences observed in male versus female immune responses in cancer since it has become increasingly evident that the major susceptibility of women to a variety of autoimmune diseases might contribute to enhanced immune surveillance against various tumor types [29]. Sex hormones can also affect the immune system since high estrogen levels have been shown to promote antibody production, whereas androgens have been reported to suppress immune function [30]. Consistent with this evidence, only a small proportion of seroconversions occur in men following HPV infection [31], and HPV seroprevalence in men is significantly lower than that reported among women [32]. The combination of clinical stage, HPV status, and smoking history lead to the definition of three different OPSCC risk groups with different prognosis [33]. Despite a more precise risk assessment, the therapeutic options remain unchanged and include chemo-radiation or surgery with or without adjuvant (chemo-) radiotherapy in the radical setting [34]. Due to the evidence of a better prognosis in HPV-driven OPSCC, several strategies in treatment de-intensification are under evaluation with the purpose to maintain efficacy and reduce short- and long-term treatment related side effects [35–42]. These studies are only part of a growing literature in the field of reduction of aggressiveness of treatments for HPV-driven OPSCC, and mainly focus on the low risk OPSCC patients. Hopefully, other trials that are still ongoing may also help clinicians in the choice of the optimal strategy to offer to HPV-driven OPSCC (in particular results from PATHOS, HN002, HN005, and KEYCHAIN trials). Although clinical trials move in the direction of reduced treatment intensity, 20% of HPV-positive OPSCC patients relapse and even die of the disease [43]. For that reason, the identification of novel prognostic factors is urgently needed not only to select HPV-positive OPSCC patients that may benefit from de-intensified treatments, but also to identify those patients at higher risk of relapse.

Squamous cell carcinoma of the anal canal (SCCA) is a rare cancer [44] associated with HPV infection in 80–85% of patients (usually HPV16 or HPV18 genotypes in Europe) [45]. Other important risk factors for SCCA include human immunodeficiency virus, immune suppression in transplant recipients, and the use of immunosuppressant drugs. Definitive chemoradiotherapy (CRT) is the current standard of care for patients with locally advanced SCCA, whereas surgery as a salvage treatment is indicated for patients with persistent disease after CRT or local relapse [46]. Prognostic factors for survival in SCCA include male sex, positive lymph nodes, and primary tumor size greater than 5 cm [46]. Despite advances in the understanding of biology and pathogenesis of SCCA, there is considerable heterogeneity in terms of outcome, particularly for more advanced stages. Only a limited number of biomarkers have been investigated and at present there are not current available factors to guide prognosis or select treatment. In a systematic review, Lampejo et al. [47] examined 29 different biomarkers, but the tumor suppressor genes p53 and p21 were the only significantly related to prognosis. Therefore, since there are no current biomarkers that strongly predict response to CRT and prognosis in SCCA patients, the investigation of HPV-related biomarkers would be an interesting objective.

## 2. HPV-Driven Cancerogenesis

The HPV genome is organized into three regions: a non-coding region, termed the long control region, which contains the early promoter and regulatory element involved in viral replication and transcription, and two protein-coding regions, the early (E) region coding proteins regulating viral transcription (E2), viral DNA replication (E1, E2), cell proliferation (E5, E6, E7), and viral particle release (E4), and the late (L) region which encodes the structural proteins (L1 and L2). E5, E6, and E7 are viral oncogenes; several studies on mucosal HR HPVs have established that E6 and E7 play a pivotal role in altering host immune response and promoting cell proliferation and transformation [3].

The best characterized interactions, whose maintenance is considered fundamental for the neoplastic phenotype, are those between E6 and E7 with p53 and pRb, respectively. By suppressing p53 activity, HPV is able to bypass cellular senescence. On the other hand, the release of E2F transcription factors allows for unscheduled cell proliferation [48]. The E6 oncoprotein of HR-HPV binds the E6-associate protein (E6AP), an E3 ubiquitin ligase that ubiquitinates target proteins for subsequent proteasome degradation. P53, a transcription factor that induces cell cycle arrest or apoptosis in response to cellular stress or DNA damage, is the best characterized target of E6/E6AP heterodimer-induced degradation leading to the loss of tumor suppression activity, accumulation of

DNA mutations, and to genomic instability [49]. E7 of HR-HPV types interacts and inactivates pRb and related pocket proteins (p107 and p130), which is in control of the G1-S phase transition by binding the transcription factor E2F [50]. As a consequence, E2F is released, with consequent promotion of cell G1-S phase transition, and transcription of genes, such as cyclin E and cyclin A, which are required for cell cycle progression [51]. Furthermore, by recruiting p300/CBP and pRb, E7 brings the histone acetyltransferase domain of p300/CBP into proximity to pRb and promotes its acetylation, leading to cell cycle deregulation [52]. In addition, cells harboring transforming HR-HPV infection acquire the capability to replicate indefinitely through the ability of E6 to reactivate the expression of telomerase (a ribonucleoprotein complex containing an internal RNA component and a catalytic protein, TERT, with telomere specific reverse transcriptase activity) by significantly upregulating TERT promoter activity [53].

Although the main mechanism of the malignant transformation induced by HPV is orchestrated by the abovementioned transforming activity of the viral E6 and E7 oncoproteins, the control of gene expression by specific non-coding RNAs (ncRNAs) may give a significant contribution in the process of transformation. Regarding the relationship between transforming HPV infection and the expression pattern of host ncRNAs, numerous studies, mainly focused on the study of micro RNAs (miRNAs), have shown a different miRNAs expression in HPV-positive tumor cells compared to the negative counterpart [54–56]. However, different investigations give conflicting results with a significant proportion of miRNAs being upregulated in one study but downregulated in another study [57]. Interestingly, both by standard sequencing and next generation sequencing, it has been successfully demonstrated that HPVs are able to generate both their own miRNAs and circular RNAs (circRNAs) [58,59]. Although the levels of expression are rather low, the frequent identification of viral miRNAs in cell lines and their higher expression in high-grade lesions suggest that they probably have a role in viral replication and malignant transformation [58].

Overall, although the transforming activity of HPV is mainly based on the degradation of p53 and pRb induced, respectively, by the viral oncoproteins E6 and E7, numerous other mechanisms including the contribution of ncRNAs generated both by the host cell and by the virus seem to participate in the process of carcinogenesis and tumor progression of HPV-induced tumors.

## 3. Non-Coding RNAs

In the last years, ncRNAs have emerged as key players in regulating the expression levels of the coding RNAs and other cellular processes [60]. Generally, ncRNAs with lengths exceeding 200 nucleotides are known as long non-coding RNAs (lncRNAs) or circRNAs, whereas all smaller transcripts are defined as small ncRNAs (sncRNAs); among sncRNAs small interfering RNAs (siRNAs), miRNAs, and P-element-induced wimpy testis (PIWI)-interacting RNAs (piRNAs) have been extensively studied so far (Figure 1) [61]. With the development of high-throughput sequencing technology and bioinformatics, an increasing number of ncRNAs are gradually being discovered. To date, multiple functional tumor-associated ncRNAs have been described, and several studies have shown they have either oncogenic or tumor-suppressive properties in cancer (for review see Diamantopoulos et al. [62]). Increasing evidence has revealed that ncRNAs play key roles not only in tumor progression and metastasis, but also in chemoresistance [63–66]. In particular, ncRNAs have been found to act as mediators of drug-resistance mechanisms through their ability to impair cell cycle arrest and apoptosis [67], but also to induce and modulate epithelial–mesenchymal transition (EMT) and cell adhesion-associated signaling pathways [68–70].

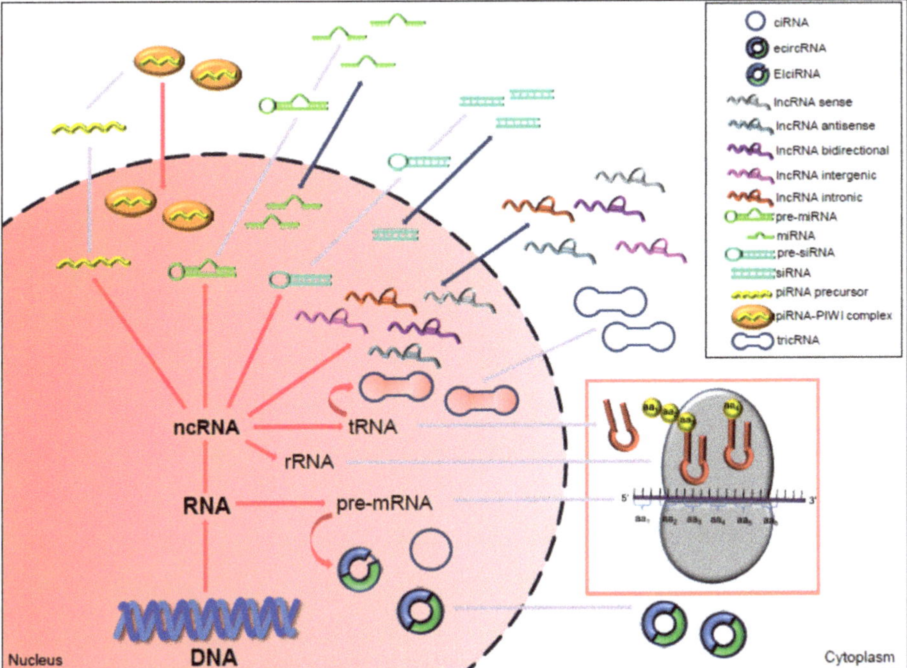



**Figure 1.** Coding and non-coding classes of RNA. Precursor messenger RNA (pre-mRNA) gives rise to mRNA, which is further translated into protein. Non-coding RNAs (ncRNAs) comprise transfer RNA (tRNA), ribosomal RNA (rRNA), and a large variety of regulatory ncRNAs, including micro RNAs (miRNAs), P-element-induced wimpy testis (PIWI)-interacting RNAs (piRNAs), small interfering RNAs (siRNAs), long non-coding RNAs (lncRNAs) and circular RNAs (circRNAs). Mature miRNAs and siRNAs are transcribed as precursors that undergo a series of nuclear and cytoplasmic processing events, and act in both nucleus and cytoplasm. Similarly, piRNAs are generated from long single-stranded piRNA precursors that are exported in the cytoplasm where they are processed; mature piRNAs are then transported into the nucleus in complex with PIWI. Most circRNAs that derive from linear pre-mRNAs, and are classified in exon-derived circRNAs (ecircRNAs), containing only exons and completely lacking introns, circular intronic RNAs (ciRNAs), which consists of only introns, exon-intron cirRNAs (ElciRNAs), in which one intron is inserted between two exons. RNA circularization can also occur through tRNA intron splicing thus generating tRNA intronic circRNAs (tricRNAs). CiRNAs and ElciRNAs are mainly nuclear, whereas ecircRNAs and tricRNAs are synthesized in the nuclear compartment and then exported to the cytosol. The lncRNAs biogenesis is mostly similar to mRNA, but they are located in the nucleus or cytoplasm, and rarely encode proteins. LncRNAs are classified as sense, antisense, bidirectional, intronic, or intergenic with respect to nearby protein-coding genes.

Besides miRNAs, that have been extensively studied in the last years [20,24,71], other ncRNAs (i.e., circRNAs, piRNAs, and lncRNAs) are drawing increasing attention nowadays since they have been found to play a role in HPV-driven tumors, suggesting that they could function as predictive biomarkers and therapeutic targets. In fact, many circRNAs, piRNAs, and lncRNAs involved in HPV-driven tumors have recently been characterized and several models of action have also been proposed; in some cases, deregulation of a specific ncRNA has come from two or more different studies (Table A1). Given these considerations, in this review we mainly focus our attention on these ncRNAs, classifying them according to the different HPV-driven tumor types.

## 4. CircRNAs Expression in HPV-Driven Cancers

Even though circRNAs are derived from linear pre-mRNAs, they are generally presented as covalently linked circles lacking both 5′ cap and 3′ poly(A) tails [72]. Over 80% of the identified circRNAs is exon-derived circRNA, containing only exons and completely lacking introns [73]. However, the splicing mechanism of circRNAs is complicated and the same position of a gene can produce different types of circRNAs. Consistently, three other types of circRNAs have been also identified by high-throughput sequencing: circular intronic RNAs, which consists of only introns, exon-intron cirRNAs, in which one intron is inserted between two exons, and tRNA intronic circRNAs, which can form stable circRNA via pre-tRNA splicing [73]. One of the most widely studied functions of circRNAs is their role as miRNA sponges and as modulators of splicing or transcription. In addition, circRNAs interact with RNA-binding proteins, and transport substances and information. Furthermore, the presence of short sequences containing N6-methyladenosine (m6A) as internal ribosomal entry site [74] allows to a small number of circRNAs to be translated into peptides or proteins that are functionally different from their linear transcripts (for review see [75,76]. There is increasing evidence that circRNAs play important roles in the development of several cancers [75,77,78]; however, information regarding their involvement in HPV-driven cancer and their potential prognostic role still remains significantly limited to CSCC (Table 2). In fact, although the role of circRNAs in HNSCC has been recently reviewed [79], little is known about circRNA expression in HPV-driven OPSCC [59], probably because HPV status was not fully reported in all studies.

**Table 2.** Non-coding RNA and PIWI-like proteins expression in Human Papilloma Virus (HPV)-driven cancers.

| HPV-Driven Cancer Type | CircRNAs ID | Sample Source | Expression Change | Function/Effect | Targets | Prognostic Value | Notes | References |
|---|---|---|---|---|---|---|---|---|
| CSCC | CircRNA8924 | Tissues, Cell lines | Up | Promote proliferation, cell cycle progression, migration, and invasion | MiR-518-d-5p/miR-519-5p | | | [80] |
| | Circ_0005576 | Tissues, Cell lines | Up | Promote tumor progression | MiR-153 | Correlated with advanced FIGO stage, lymph node metastasis and poor prognosis | | [81] |

| HPV-Driven Cancer Type | PiRNAs ID | Samples Source | Expression Change | Function/Effect | Targets | Prognostic Value | Notes | References |
|---|---|---|---|---|---|---|---|---|
| HNSCC | FR018916, FR140858, FR197104, FR237180, FR298757 | TCGA | Down | | | PiRNA expression signature can predict OS in HPV positive patients | Downregulated in HPV16/18 HNSCC samples compared to cases harboring other HPV types | [82] |
| | PiR-36742 | TCGA | Up | | | | Deregulated in smoking HPV-positive patients | [83] |
| | PiR-33519 | | Up | | | | | |
| | PiR-36743 | | Up | | | | | |
| | PiR-34291 | | Up | | | | | |
| | PiR-36340 | | Down | | | | | |
| | PiR-62011 | | Down | | | | | |
| | PiR-30652 | TCGA | Up | | | Significantly predictive of patient outcome | Significantly associated with higher histologic grade | |
| | PiR-33686 | | Up | | | | | |
| | PiR-36340 | | Down | | | | | |
| | PiR-45029 | | Down | | | | | |

**Table 2.** *Cont.*

| HPV-Driven Cancer Type | PIWI-Like Proteins ID | Samples Source | Expression Change | Function/Effect | Targets | Prognostic Value | Notes | References |
|---|---|---|---|---|---|---|---|---|
| CSCC | PIWIL1 | Tissues, Cell lines | Up | Promote tumorigenesis and tumor progression, suppress chemotherapy sensitivity | | | | [84] |
| | PIWIL2 | Tissues, Cell lines | Up | Promote tumorigenesis, induce H3K9 acetylation and reduce H3K9 trimethylation | | | PIWIL2 activation in CSCC appears to depend on the integration of HR-HPV DNA | [85] |
| | PIWIL4 | Tissues, Cell lines | Up | Promote proliferation, inhibit apoptosis | P14ARF/p53 pathway | | | [86] |
| HNSCC | PIWIL4 | TCGA consortium | Up | | | | Upregulated in smoking HPV-positive patients | [83] |

| HPV-Driven Cancer Type | LncRNAs ID | Samples Source | Expression Change | Function/Effect | Targets | Prognostic Value | Notes | References |
|---|---|---|---|---|---|---|---|---|
| CSCC | ENST00000503812 | Cell lines | Up | Impair DNA repair, induce immune response | | | Negatively correlated with RAD51B and IL-28A expression | [87,88] |
| | ENST00000420168, ENST00000564977 | Cell lines | Up | Promote proliferation, inhibit apoptosis | | | Correlation with the increased expression of FOX Q1 and the reduced expression of caspase-3 | [89] |
| | TCONS_00010232 | Cell lines | Down | | | | | |
| | OIS1 | Tissues, Serum Cell lines | Down | Suppress cell proliferation | MTK-1 | Significant association between OIS1 serum levels and tumor size | | [90] |
| | UCA1 | Cell lines | Up | Induce cisplatin resistance and inhibit apoptosis | | | | [91] |
| | SNHG8 | Cell lines | Up | Promote proliferation and migration, inhibit apoptosis | EZH2 | | | [92] |
| | HOST2 | Tissues, Cell lines | Up | Promote proliferation, migration and invasion, inhibit apoptosis | MiRNA let-7b | | | [93] |
| | MEG3 | Tissues, Serum, Cell lines | Down | Suppress proliferation, promote apoptosis | MiR-21-5p | | Correlated with increased tumor size, advanced FIGO stage, lymph node metastasis, HPV infection, recurrence-free survival and OS | [94–96] |

**Table 2.** *Cont.*

| HPV-Driven Cancer Type | LncRNAs ID | Samples Source | Expression Change | Function/Effect | Targets | Prognostic Value | Notes | References |
|---|---|---|---|---|---|---|---|---|
| HNSCC | LINC01089 | TCGA | Up | | | | | [97] |
| | PTOV1-AS1 | TCGA | Up | | | | | [97] |
| | IL17RA-11 | TCGA Cell lines | Up | Induce radiotherapy sensitivity | | Correlated with better prognosis | HPV infection stimulates ERα to increase lnc-IL17RA-11 expression; this finding suggests why HPV-positive HNSCC are more sensitive to radiotherapy | [98] |
| | EGOT | TCGA data | Up | Promote tumor progression | | | EGOT expression levels vary according to age, N-stage, grade, location, lymph node dissection, and HPV16 status | [99] |

CircRNAs, circular RNAs; CSCC, cervical squamous cell carcinoma; EGOT, eosinophil granule ontogeny transcript; ERα, estrogen receptor α; EZH2, enhancer of zeste homolog 2; FIGO, International Federation of Gynecology and Obstetrics; FOXQ1, oncogene forkhead box Q1; HNSCC, head and neck squamous cell carcinoma; HOST2, human ovarian cancer-specific transcript 2; HR-HPV, high risk HPV; H3K9, Histone H3 Lysine 9; IL17RA-11, Interleukin 17 receptor A; IL-28A, interleukin 28A; LncRNAs, long non-coding RNAs; MEG3, maternally expressed gene 3; MTK-1, mitogen-activated protein kinase kinase kinase 4; OIS1, oncogene-induced senescence 1; OS, overall survival; PiRNAs, PIWI-interacting RNAs; PIWIL, P-element-induced wimpy testis-like protein; PTOV1-AS1, PTOV1 antisense RNA 1; RAD51B, RAD51 paralog B; SNHG8, small nucleolar RNA host gene 8; TCGA, The Cancer Genome Atlas; UCA1, Urothelial Cancer Associated 1.

*CSCC*

By using RNA-seq, Wang et al. [100] explored the expression profiles of several ncRNAs in HPV16-induced CSCC and matched adjacent non-tumor tissues from three patients. Authors identified 99 circRNAs that were differentially expressed in CSCC patients, and 44 circRNAs have not been reported before. In a subsequent study, circRNA microarray demonstrated a significant increase of circRNA8924 expression in CSCC [80]. CircRNA8924 was found to adsorb miR-518-d-5p/miR-519-5p and to promote the expression of the polycomb protein chromobox 8, which has been shown to be a key regulator of several cancers, including CSCC. In fact, circRNA8924 knockdown significantly inhibited the proliferation, migration, and invasion of HPV-positive HeLa and SiHA cell lines both in vitro and in vivo [80]. Similarly, knockdown of circ_0005576 in the HPV-positive HeLa and SiHA cells significantly reduced CSCC aggressiveness. Mechanistically, circ_0005576 facilitated CSCC progression by binding miR-153 and thereby upregulating the kinesin family member 20A [81]. In the last years, other circRNAs have been identified to participate in CSCC tumorigenesis. However, in these studies the HPV-status of CSCC tissues was not defined and/or circRNAs appeared to be deregulated also in HPV-negative CSCC cell lines, indicating their expression might not be limited to HPV infection [101–105].

## 5. PiRNAs and PIWI-Like Proteins Expression in HPV-Driven Cancers

PiRNAs are very similar in size to miRNAs since they are 26–30 nucleotides in length, but far exceed the total number of miRNAs; in fact, about 23,439 piRNAs have been discovered so far [76]. PiRNAs specifically associate with the PIWI proteins, a subfamily of Argonaute proteins, to exert their regulatory functions (for review see Rojas-Ríos et al. [106]). Unlike miRNAs, the piRNA/PIWI complex principally acts through epigenetic silencing rather than mRNA targeting [107]. PiRNAs guide PIWI proteins to the genomic region where they share complementarities, and regulate the epigenetic status of the target sequence by recruiting epigenetic factors required for DNA methylation and/or histone modifications [108]. Besides, the piRNA/PIWI complex can also regulate gene expression at the post-transcriptional level, via alternative splicing or regulating mRNA stability, or at the post-translational level through the binding of the coding protein [109]. Although piRNAs and PIWI proteins have not been extensively studied in cancer, a limited number of published data suggest their expression is altered in HPV-driven tumors, and associated with prognosis (Table 2).

*5.1. CSCC*

Due to restricted expression during embryonic development and in several tumor types, PIWI proteins have been suggested to act as oncogenes and/or to represent a marker of cancer stem cells [110]. Interestingly, the expression of both PIWI-like protein 1 (PIWIL1) and PIWI-like protein 2 (PIWIL2) has been observed in tissues from patients with HPV16-positive CSCC [111,112]. High PIWIL1 expression was significantly associated with CSCC invasion [111], thus supporting the interaction between HPV16 and host cells during CSCC carcinogenesis. According to these data, both in vitro and in vivo studies demonstrated that PIWIL1 expression increased tumorigenesis, resistance to chemotherapeutic drugs, and self-renewal abilities of the HPV-positive HeLa and SiHa cell lines [84]. Feng et al. reported high levels of PIWIL2 in high-grade cervical intraepithelial neoplasia (CIN) and in CSCC, whereas in healthy tissue and low-grade CIN PIWIL2 was weakly expressed [85]. Additionally, the HPV-positive HeLa, SiHA and CaSki cell lines constitutively expressed PIWIL2; in contrast, PIWIL2 expression was undetectable in the HPV-negative C33A cell line [85], indicating that PIWIL2 activation in CSCC might depend on the integration of HR-HPV DNA into the host cell genome. Consistent with this hypothesis, authors demonstrated that PIWIL2 expression was restored in human keratinocyte cells following transfection with lentivirus containing complete HPV16 E6 and E7 sequences. Furthermore, PIWIL2 overexpression significantly induced histone H3 lysine 9 (H3K9) acetylation and decreased H3K9 trimethylation, thus reprogramming human keratinocyte cells into tumor-initiating cells. On the other hand, PIWIL2 knockdown led to an upregulation of p53 and p21, and reduced the tumorigenic potential of the HPV-positive HeLa and SiHa cells both in vitro and in vivo [85]. With regard to other members of the PIWI protein family, the role of PIWI-like protein 4 (PIWIL4) has also been investigated in CSCC, and results showed that PIWIL4 expression promoted a significant increase in cell growth and proliferation, and prevented apoptosis, by inhibiting p14ARF/p53 pathway in HPV-positive HeLa cell line [86].

*5.2. HNSCC*

Aberrant expression of piRNAs has been recently observed in HNSCC samples when compared to normal tissue. In particular, Firmino et al. identified 41 piRNAs that were differently expressed between HPV-positive and -negative HNSCC. Interestingly, 11 piRNAs were deregulated in tumors positive for HPV16 or HPV18 infection and, among them, the expression of piRNAs FR018916, FR140858, FR197104, FR237180, and FR298757 was associated with worse overall survival (OS), thus highlighting their potential clinical utility in HPV-positive HNSCC [82]. Subsequently, Krishnan et al. used RNA-sequencing datasets from The Cancer Genome Atlas (TCGA) to identify 30 piRNAs that were deregulated in HPV-driven HNSCC. Among them, six piRNAs (piR-36742, piR-33519, piR-36743, piR-34291, piR-36340, piR-62011) were aberrantly expressed in smoking versus never smoking HPV-positive HNSCC patients [83], suggesting that some piRNAs may be commonly implicated in smoking-related and HPV-driven HNSCC. Similarly, PIWIL4 was aberrantly expressed in smoking with respect to never smoking HPV-positive HNSCC patients as well [83]. Interestingly, piR-36743 was previously identified to be implicated in breast cancer [113], indicating the ability of the same piRNA to modulate other malignancies. Starting from the 30 HPV-deregulated piRNAs, authors further verified that the expression level of piR-30652, piR-33686, piR-36340, and piR-45029 was significantly associated with higher histologic grade, with piR-30652 being significantly predictive of patient outcome in both univariate and multivariate regression analyses.

*5.3. SCCA*

To date, no information on piRNAs/PIWIs involvement in HPV-driven SCCA is available, and the existing data concerning other sncRNAs (i.e., miRNAs) is also extremely scarce [114,115]. Therefore, there are many unknown questions about piRNAs/PIWIs that need to be explored in HPV-related tumors, especially in HPV-positive SCCA.

## 6. LncRNAs Expression in HPV-Driven Cancers

LncRNAs were initially identified as mRNA-like transcripts that do not code for proteins since they are in many ways very similar to mRNAs, including their biogenesis. However, a further characterization of lncRNAs has allowed to distinguish them from other major classes of RNA transcripts (for review see Karapetyan et al. [116]). LncRNAs can be subdivided according to their biogenesis loci in sense, antisense, bidirectional, intergenic, and intronic lncRNAs (for review see Rinn et al. [117]). Sense lncRNAs are transcribed in the same direction of exons, and they may overlap with introns and part or the entire sequence of protein-coding genes [118]. Besides representing functional RNA molecules able to regulate gene expression, sense lncRNAs can translate into protein [118]. Antisense lncRNAs are transcribed from the antisense strand of protein-coding genes, whereas bidirectional lncRNAs are expressed from the promoter of a protein-coding gene, but in the opposite direction [118]. Intergenic lncRNAs (lincRNAs) originate from the region between two protein-coding genes, and have been found to associate with chromatin modifying proteins [119]. Finally, intronic lncRNAs can be transcribed from an intronic region of a protein-coding gene in the sense or antisense direction [118]. LncRNAs play important roles in various cellular processes since their functions are highly pleiotropic; in fact, lncRNAs can regulate gene expression at many levels, such as epigenetic, transcriptional, post-transcriptional, translational, and post-translational [120]. Therefore, it is not surprising that upon viral infections most modifications occur in lncRNAs expression [121]. Consistently, an increasing number of studies have revealed a large amount of lncRNAs whose expression is deregulated in HPV-driven cancers, with most of them mainly focused on CSCC (Table 2). LncRNAs exhibit a more cell type-specific restricted expression pattern than protein-coding genes [122–124]. In addition, lncRNAs are stable in a broad range of specimen types (FFPE, plasma, and other body fluids), and are easily accessible for analysis using non-invasive methods [125,126], thus resulting appealing as prognostic/predictive biomarkers. So far, a small number of studies has highlighted the significant implication of lncRNAs as prognostic biomarkers in HPV-driven cancers (Table 2).

*6.1. CSCC*

A microarray analysis revealed that thousands of host lncRNAs had differential expression in oncogenic HPV-positive cells compared to the HPV-negative C33A cell line. In particular, 4750 lncRNAs were differentially expressed in the HPV16 positive SiHa cells compared with C33A cell line, including 2127 upregulated and 2623 downregulated lncRNAs. Similarly, 5026 lncRNAs were differentially expressed in the HPV18 positive HeLa cells respect to C33A cell line and, among these deregulated lncRNAs, 2218 were upregulated whereas 2808 were downregulated. In this study, the authors further demonstrated that HPV could exert effects on the development and progression of CSCC via altering the expression of lncRNAs and their downstream mRNAs targets [89]. In fact, in the HPV-positive SiHa cell line, the lncRNAs ENST00000503812 was upregulated whereas the expression of its target genes RAD51 paralog B (RAD51B), which is a component of the DNA double-strand break repair pathway [88], and interleukin-28A, which plays a role in immune defense against viruses [87], was decreased [89]. Interestingly, HPV integration was previously shown to disrupt RAD51B expression in CSCC [127]. Therefore, ENST00000503812 upregulation may impair DNA repair pathway and immune responses in HPV16 positive CSCC cells. In addition, ENST00000420168, ENST00000564977 upregulation and TCONS_00010232 downregulation showed a significant correlation with the increased expression of the oncogene forkhead box Q1 and the reduced expression of the apoptosis-related gene caspase-3 in the HPV-positive HeLa cell line [89]. These results indicate that HPV18 might alter ENST00000420168, ENST00000564977, and TCONS_00010232 expression in order to promote cell proliferation and to prevent apoptosis during CSCC progression. As shown by Zhou et al. [90], the lncRNA oncogene-induced senescence 1 (OIS1) was significantly downregulated in the majority of tumor tissues from HPV-positive CSCC compared with adjacent non-tumor tissues, but not in HPV-negative CSCC patients. Serum levels of OIS1 were also significantly lower in HPV-positive CSCC [90], indicating that OIS1 downregulation was specifically involved in the pathogenesis of HPV-driven CSCC. Accordingly, OIS1 overexpression markedly reduced the proliferation of the HPV-positive SiHa cells, potentially by inhibiting the expression of the mitogen-activated protein kinase kinase kinase 4 (MTK-1) [90]. MTK-1 expression was also reduced following GATA binding protein 6 antisense (GATA6-AS) overexpression in CSCC cell lines; however, GATA6-AS expression levels were revealed to be significantly reduced in both HPV-positive and -negative CSCC patients, suggesting GATA6-AS might play a role in CSCC through an HPV-independent pathway [128]. The study of Wang et al., that was already discussed above, reported 19 lncRNAs that were differentially expressed between HPV-positive CSCC and adjacent non-tumor tissues, and the co-expression network and function prediction suggested that all of them could play a role in HPV-driven CSCC [100]. The majority of these lncRNAs were intergenic, and three lncRNAs have not been described before. Among the differentially expressed lncRNAs, the authors identified urothelial cancer associated 1 (UCA1) which has gained much attention in recent years due to its aberrant expression in several cancers [100]. UCA1 has also been shown to promote cisplatin resistance, suggesting its potential use as a target for a novel therapeutic strategy in CSCC [91]. Unfortunately, the regulatory mechanism between UCA1 expression and cisplatin resistance is still unknown. Small nucleolar RNA host gene 8 (SNHG8) was recently clarified as a critical driving force for the development of HPV-positive CSCC [92]. Enhanced SNHG8 expression was found in HPV-positive CSCC cell lines but not in the HPV-negative C33A cells, thus indicating that HPV infection led to SNHG8 deregulation. In addition, SNHG8 silencing in

HPV-positive HeLa and SiHa cells reduced cell proliferation and migration, and promoted apoptosis. Functional studies revealed that SNHG8 could bind to the enhancer of zeste homolog 2, thus inhibiting the transcription of the tumor suppressor reversion inducing cysteine-rich protein with kazal motifs in CSCC cells [92]. Human ovarian cancer-specific transcript 2 has also been reported to be upregulated in HPV-positive CSCC tissues and cell lines, and to act as sponge for the miRNA let-7b, thus promoting CSCC cell proliferation, migration, and invasion along with reduced apoptosis [93]. Differently, maternally expressed gene 3 (MEG3) was shown to function as a tumor suppressor in CSCC; in fact, MEG3 expression inhibited cell proliferation and promoted apoptosis in CSCC cells through modulating the level of miR-21-5p [96]. In addition, MEG3 expression was negatively correlated not only with CSCC grade and survival but also with HR-HPV infection [96]. Surprisingly, low MEG3 expression was associated with favorable prognosis in HPV-negative HNSCC [129]. However, the clinical significance of this finding is still unclear.

## 6.2. HNSCC

Using TCGA RNA-seq data from 426 HNSCC and 42 adjacent normal tissues, Nohata et al. [97] found 140 lncRNA transcripts that were significantly differentially expressed between HPV-positive and -negative tumors. Several lncRNAs were also deregulated in a panel of HPV-positive cell lines [97], and some of them have been already characterized, such as LINC01089 [130] and PTOV1-AS1 [131]. Interestingly, 19 lncRNAs were specifically expressed in HPV-positive and p53 wild type HNSCC, suggesting they might represent potential key molecules in HPV-driven oncogenesis [97]. A similar study identified eight lncRNAs that were associated with better prognoses in HPV-driven HNSCC, including lnc-IL17RA-11 whose expression promoted HNSCC cell sensitivity to radiotherapy [98]. The regulatory mechanism of lnc-IL17RA-11 upregulation has also been illustrated. HPV infection could stimulate estrogen receptor α to increase lnc-IL17RA-11 expression in order to upregulate the activity of genes involved in processes that enhance sensitivity to radiation therapy [98]. These findings might explain why HPV-positive HNSCC are more sensitive to radiotherapy. By analyzing lncRNAs profiling data and the corresponding clinic-pathologic variables of 371 HNSCC patients from TANRIC and cBioPortal, Cui et al. [132] defined a signature of 15 lncRNAs with prognostic significance for recurrence-free survival. Importantly, when HNSCC patients were stratified according to their HPV status, the 15 lncRNAs signature remained a clinically and statistically significant prognostic model. Similarly, by referring to the TCGA data available from the cBioportal and UALCAN databases, Kolenda et al. [99] provided experimental support on the association of the eosinophil granule ontogeny transcript (EGOT) lncRNA upregulation with the progression of HPV-positive HNSCC, but the exact mechanism for its involvement in HPV infection was not reported.

## 7. NcRNAs Modulated by E6/E7 Oncoproteins

Although the ability of HPV E6 and E7 oncoproteins to modulate the expression of many protein-coding or miRNA-coding genes have been well documented, their role in the regulation of ncRNA in HPV-driven cancers is still largely obscure. So far, the majority of studies have focused specifically on CSCC where both E6 and E7 HPV oncoproteins were shown to modulate the expression of several lncRNA (Figure 2). Furthermore, some of the HPV E6 and/or E7 deregulated lncRNAs have been suggested as potential prognostic biomarkers (Table 3). For instance, E6 was recently proposed to increase the expression of the cervical carcinoma expressed PCNA regulatory (CCEPR) lncRNA in CSCC [133]. A study of Yang et al. reported that CCEPR regulated CSCC cell proliferation by binding and stabilizing PCNA mRNA [134]. Accordingly, high levels of CCEPR indicated poor prognosis in HPV-positive CSCC patients [135]. However, in the study of Sharma et al. perturbation of CCEPR expression did not alter PCNA mRNA levels in CSCC cell lines [133], indicating that PCNA mRNA stabilization might not be the primary mechanism by which CCEPR modulates CSCC proliferation. Besides CCEPR, FAM83H-AS1 was established to be upregulated by HPV16 E6 oncoprotein both in primary keratinocytes and in CSCC tumor samples with the expression being involved in cellular proliferation and migration, and associated with worse OS in CSCC patients [136]. As reported by Barr et al. [136], HPV16 oncogene E6 mediated FAM83H-AS1 upregulation in a p300-dependent manner. Of note, FAM83H-AS1 expression was decreased in HPV18 positive CSCC cell lines, probably because HPV18 E6 does not have the ability to interact with p300 with high efficiency [137]. Other lncRNAs are modulated by HPV16 E6, including H19 and the growth arrest-specific transcript 5 (GAS5) [136]. H19 was found to be upregulated in CSCC tissues compared with adjacent normal tissues, and to promote sirtuin 1 overexpression by sponging miR-138-5p in HPV-positive CSCC HeLa and SiHa cell lines [138]. GAS5 has showed tumor suppressor activity in several tumors, including CSCC where its decreased expression was associated with poor prognosis [139,140], and increased proliferation, invasion, and migration of CSCC cells [139]. More recently, GAS5 overexpression was demonstrated to enhance cisplatin sensitivity in HPV-positive CSCC SiHa cells by directly targeting miR-21 and regulating Akt phosphorylation [141], and to improve the radio-sensitivity of HPV-positive CSCC SiHa cells via inducing immediate early response 3 expression by sponging miR-106b both in vitro and in vivo [142]. Interestingly, GAS5 expression might depend on another lncRNA called GAS5-AS1, but the regulatory mechanism between GAS5 and GAS5-AS1 has yet to be elucidated [143].

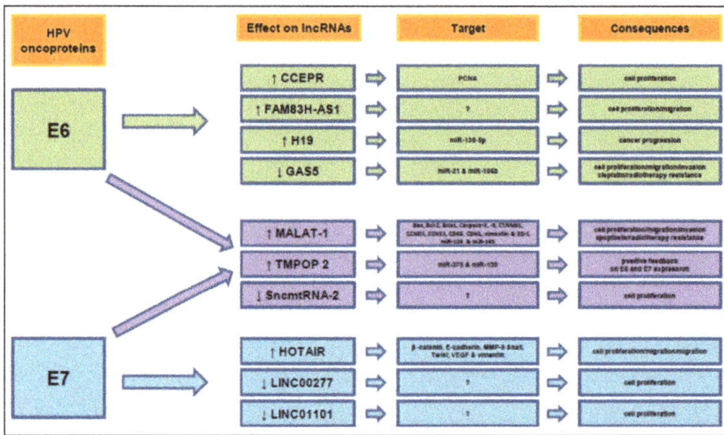


**Figure 2.** Schematic diagram of Human Papilloma Virus (HPV) E6/E7 oncoproteins affecting long non-coding RNAs (lncRNAs) expression in cervical squamous cell carcinoma. Bax, bcl-2-like protein 4; BclxL, B-cell lymphoma-extra large; Bcl-2, B-cell lymphoma 2; CCEPR, cervical carcinoma expressed PCNA regulatory; CDH1, E-cadherin; CDK6, cell division protein kinase 6; CCND1, cyclinD1; CCNE1, cyclinE; CTNNB1, b-catenin; FAM83H-AS1, FAM83H Antisense RNA 1; GAS5, growth arrest-specific transcript 5; HOTAIR, HOX transcript antisense RNA; MALAT-1, metastasis associated lung adenocarcinoma transcript 1; MMP-9, metalloproteinase-9; PCNA, proliferating cell nuclear antigen; TMPOP2, Thymopoietin pseudogene 2; VEGF, vascular endothelial growth factor; ZO-1, zonula occludens-1.

On the other hand, the expression of a number of lncRNAs is exclusively modulated by the HPV E7 oncoprotein. Along this line, Sharma et al. [144] reported that E7 could be involved in CSCC proliferation and metastasis through regulating the expression and function of Hox transcript antisense intergenic RNA (HOTAIR), which might represent a new marker of CSCC recurrence and poor prognosis [145]. HOTAIR possesses distinct binding domains for chromatin-modifying complexes and histone demethylases [146]. Thanks to that, HOTAIR might partially regulate the expression of several genes involved in cell proliferation, migration, invasion, and EMT in CSCC [145]. Interestingly, genetic variations within HOTAIR appeared to modify the risk of CSCC. In particular, the HOTAIR rs2366152C polymorphism was frequently reported in low HOTAIR expressing HPV-positive CSCC where it allowed miR-22 to directly bind to HOTAIR [147]. The gain of the miR-22 binding site in HOTAIR was found to be concordant with miR-22 overexpression, which led to reduced E7 expression in low HOTAIR HPV-positive CSCC cells [147]. HOTAIR expression not only characterized HPV-driven CSCC, but also negatively correlated with the proportion of myeloid-derived suppressor cells in the blood samples of patients with HPV-positive HNSCC [148]; unfortunately, a causal relationship has not yet been established. In a subsequent study, loss-of-function assays allowed to identify EWSAT1-Ewing sarcoma-associated transcript 1 (LINC00277) and LINC01101 as the most upregulated lncRNAs in HPV-positive CSCC HeLa cells transfected with siRNA-HPV18 E7 [149]. Consistent with these data, HPV-positive CSCC tissues exhibited significantly reduced expression of both lncRNAs, and low expression of LINC00277 and LINC01101 could predict poor prognostic features [149].

**Table 3.** Prognostic value of long non-coding RNAs (lncRNAs) modulated by Human Papilloma Virus E6/E7 oncoproteins in cervical squamous cell carcinoma.

| LncRNAs ID | Sample Description | Expression Change | Prognostic Value | References |
|---|---|---|---|---|
| CCEPR | Cell lines, Tissues | Up | Positively correlated with advanced FIGO stage, lymph node metastasis, HPV infection, and poor prognosis | [133–135] |
| FAM83H-AS1 | Cell lines, Tissues | Up | Poor prognosis | [136] |
| GAS5 | Cell lines, Tissues | Down | Poor prognosis | [139,140] |
| HOTAIR | Cell lines, Tissues | Up | Disease recurrence and poor prognosis | [145] |
| | Cell lines, Tissues | | Rs2366152C polymorphism associates to reduced HOTAIR expression and CSCC metastatic molecular signatures | [147] |
| LINC00277, LINC01101 | Cell lines, Tissues | Down | Poor prognosis | [149] |

CCEPR, cervical carcinoma expressed PCNA regulatory; CSCC, cervical squamous cell carcinoma; FAM83H-AS1, FAM83H Antisense RNA 1; FIGO, International Federation of Gynecology and Obstetrics; GAS5, growth arrest-specific transcript 5; HOTAIR, HOX transcript antisense RNA.

Additionally, several lncRNAs are specifically regulated by both HPV E6 and E7 oncoproteins, including the metastasis associated lung adenocarcinoma transcript 1 (MALAT-1) which is one of the most extensively characterized lncRNA so far. By using loss of function assays Guo et al. revealed, for the first time, the tumor promoting role of MALAT-1 in the HPV-positive CSCC CaSki cells. In fact, MALAT-1 silencing impaired the migration ability of CaSki cells, increased caspase-8, caspase-3, and Bax levels, and reduced Bcl-2 and Bcl-xL expression [150]. Subsequently, other research groups studies confirmed the oncogenic functions of MALAT-1 in HPV-driven CSCC since they found that MALAT-1 positively regulated the expression of genes involved in cell cycle regulation [151], and in cell migration [152]. Therefore, MALAT-1 knockdown in HPV-positive CSCC CaSki cells led to G1 arrest [151], and reduced invasion and metastasis both in vitro and in vivo [152]. In a study of Jang et al. [151], MALAT-1 expression was detected in 6/18 cases of HPV-positive cervical normal cells and 14/22 cases of HPV-positive cervical lesions, suggesting that HPV infection might lead to MALAT-1 activation in CSCC. Consistent with this hypothesis, MALAT-1 expression was found to augment in oral keratinocytes transfected with HPV E6/E7 oncoproteins [151]. MALAT-1 was demonstrated to act as a miRNA sponge as well. For instance, MALAT-1 appeared to contribute to CSCC progression by promoting the growth factor receptor bound protein 2 overexpression through binding and sequestering its major negative regulator miR-124 [153]. Interestingly, MALAT-1 has also been implicated in the mechanism of radio-resistance in HR-HPV-driven CSCC via sponging miR-145 [154].

Besides MALAT-1, He et al. [155] has recently reported that overexpression of HPV16/18 E6 or E7 enhanced the expression of Thymopoietin pseudogene 2 (TMPOP2) lncRNA in CSCC cells, whereas depletion of both HPV16/18 oncoproteins significantly downregulated TMPOP2. Similarly, TMPOP2 was found to regulate the expression of HPV16/18 E6 and E7, thus creating a positive feedback that synergistically sustained CSCC. As shown by authors, the mechanism by which HPV16/18 E6 or E7 enhanced TMPOP2 expression was predominately governed by their capacity to promote the degradation of p53; in fact, p53 was demonstrated to bind TMPOP2 promoter and to repress its transcription. Once expressed, TMPOP2 was able to sequester HPV E6/E7-targeting miR-375 and miR-139, allowing the expression of HPV oncoproteins [155]. Besides genomic ncRNAs, human cells express a unique family of mitochondrial long noncoding RNAs (ncmtRNAs) which comprises sense (SncmtRNA) and two antisense (ASncmtRNA-1 and ASncmtRNA-2) transcripts containing long inverted repeats linked to the 5′ end of the 16S mitochondrial rRNA [156,157]. SncmtRNA and ASncmtRNAs exit the mitochondria and localize to the cytosol and to the nucleus, where they associate with chromatin and nucleoli [158]. SncmtRNA represents a marker of cell proliferation since it is expressed in normal proliferating and in tumor cells but not in resting cells [156,157]. Similarly, ASncmtRNAs are expressed in normal proliferating cells, but they are downregulated in several tumors [157], thus suggesting they might function as tumor suppressor. Immortalization of keratinocytes with the complete genome of HPV16 and HPV18 downregulated ASncmtRNAs and induced a novel sense ncmtRNAs called SncmtRNA-2. Interestingly, although ASncmtRNAs expression was shown to depend on HPV E2 oncoprotein both ASncmtRNA-1 and -2 were downregulated in the HPV-positive Hela and SiHa cell lines which did not express E2 [159]. On the other hand, SncmtRNA-2, whose expression was promoted by E6 and E7 oncoproteins, was not upregulated in Hela and SiHa cells [159], thus suggesting that other cellular factors may be involved in the regulation of ASncmtRNAs and SncmtRNA-2 after HPV transformation.

In addition to lncRNAs, it has been recently reported that HPV16 E7 oncoprotein altered the expression profiles of circRNAs in CSCC cells. In this study, HPV E7 expression altered the expression of 526 circRNAs; among them, 352 were upregulated whereas 174 were downregulated. Subsequent bioinformatic analyses indicated that differently expressed circRNAs were likely to be involved in the mTOR signaling pathway, proline metabolism, and glutathione metabolism [160].

## 8. HR-HPV-Derived NcRNAs

Besides to promote aberrant ncRNAs expression, HPV has been shown to encode its own ncRNAs. For instance, a recent study revealed that HR-HPV produced circE7, a circRNA m6A modified, preferentially localized to the cytoplasm, and associated with polysomes [59]. Zhao et al. [59] demonstrated that circE7 represented only 1–3% of total E7 transcripts but, although its weak expression, it was critically involved in HPV-induced carcinogenesis since it was translated to produce E7 oncoprotein. Accordingly, the disruption of circE7 in HPV-positive CSCC CaSki cells reduced E7 protein levels and inhibited cancer cell growth both in vitro and in vivo. CircE7 could be detected in TCGA RNA-Seq data from HPV-positive HNSCC and CSCC [59], thus suggesting it might be used as a molecular biomarker for the presence of HR-HPV and/or as a potential prognostic indicator of clinical outcome in these patients. CircE7 expression was also found in HPV-driven SCCA [161]. Of note, HPV-positive SCCA with high levels of circE7 showed a trend towards improved survival respect to those with low or absent circE7 [161] that could be probably due to a strong E7-specific immune response.

## 9. NcRNAs as Potential Diagnostic Biomarkers in HPV-Driven Cancers

Consistent with increasing role of ncRNAs in HPV-driven cancers, a number of studies have reported their potential value as diagnostic biomarkers (Table 4). For instance, MEG3 emerged as a powerful tool for prediction of tumor size and lymph node metastasis in patients with CSCC [94]. Of note, low MEG3 expression correlated with MEG3 promoter hypermethylation in both tissues [94] and plasma [95] from CSCC patients. Starting from this evidence, Zhang et al. investigated the diagnostic power of plasma MEG3 methylation with favorable results; in fact, plasma MEG3 methylation had high power to discriminate high-grade CIN patients from healthy controls, and to predict HR-HPV infection and lymph node metastasis [95]. Serum OIS1 was also proven to be an effective diagnostic biomarker in patients with HPV-positive CSCC since it effectively distinguished them from healthy controls [90].

**Table 4.** Diagnostic value of long non-coding RNAs (lncRNAs) in cervical squamous cell carcinoma.

| LncRNAs ID | Cohort Size | | Source of LncRNAs | Sensitivity | Specificity | AUC | Diagnostic Value | References |
|---|---|---|---|---|---|---|---|---|
| MEG3 | 72 cases and 72 normal tissues | | Tissues | 56.1% | 80.6% | 0.745 | Tumor-size <4 cm or ≥ 4 cm | [94] |
| | | | | 70.5% | 67.9% | 0.716 | Lymph node metastasis | |
| | 108 cases | | | 54.8% | 84.8% | 0.753 | Tumor-size <4 cm or ≥ 4 cm | |
| | | | | 76.1% | 85.4% | 0.862 | Lymph node metastasis | |
| MEG3 methylation | 160 CIN I-III, 168 cases, and 328 healthy patients divided into training set and test set randomly and averagely | Training set | Plasma | 73.7% | 94.7% | 0.831 | CIN III | [95] |
| | | | | 75.8% | 88.9% | 0.815 | HR-HPV infection | |
| | | | | 93.3% | 51.9% | 0.741 | Lymph node metastasis | |
| | | | | 84.2% | 52.6% | 0.788 | CIN III | |
| | | Test set | | 78.1% | 70.0% | 0.730 | HR-HPV infection | |
| | | | | 82.4% | 72.0% | 0.804 | Lymph node metastasis | |
| OIS1 | 22 HPV-negative patients, 70 HPV-positive patients, and 40 healthy patients | | Plasma | n.a. * | n.a. * | 0.9207 | Serum OIS1 may be used to diagnose HPV-positive, but not HPV-negative CSCC | [90] |

AUC, area under the curve; CIN, cervical intraepithelial neoplasia; HR-HPV, high risk Human Papilloma Virus; MEG3, maternally expressed gene 3; OIS1, oncogene-induced senescence 1. * Data are not available.

## 10. Therapeutic Targeting of NcRNAs

Given their stability and distinct cytoplasmatic localization, ncRNAs can be used as novel therapeutic molecular tools for the treatment of HPV-driven cancers. At present, a number of RNA-based approaches have been developed to target ncRNAs, including antisense oligonucleotides (ASOs) or siRNAs (for review see Bajan et al. [162]). In this context, treatment with MALAT-1 specific ASO decreased the size and the number of tumor nodules in a pulmonary metastasis model of human lung cancer [163]. Similarly, ASO-mediated knockdown of MALAT-1 expression resulted in slower tumor growth and metastasis reduction in a mouse mammary carcinoma model [164]. More recently, Kim et al. [165] developed nanocomplexes carrying siRNAs against MALAT-1 that efficiently enhanced sensitivity of glioblastoma tumor cells to temozolomide both in vitro and in vivo. In addition to ASO and siRNAs, circRNAs targeting HPV-related RNAs and/or RNA-binding proteins may represent another promising therapeutic approach. For instance, Jost et al. [166] produced an artificial circRNA that efficiently sequestered miRNA-122 in in vitro experiments, thereby inhibiting the propagation of Hepatitis C Virus. These results suggest that RNA-based strategies may improve prognosis and therapeutic response in patients affected by HPV-driven tumors. However, since many ncRNAs are located in the nucleus [167], it should be difficult to achieve their knockdown by using RNA-based approaches. In this case, the clustered regularly interspaced short palindromic repeats (CRISPR)-associated nuclease 9 (CRISPR/Cas9) technology would provide the best option to achieve ncRNA-related genome editing since it directly targets the genomic DNA (for review see Yang et al. [168]). Given these considerations, it is expected that future studies focused on the CRISPR/Cas9 system for editing ncRNAs will receive increased interest.

Although high-throughput technologies have recently enabled the development of small molecular compounds that may potentially inhibit ncRNAs in the coming future [169,170], several studies have demonstrated that existing drugs may also modulate ncRNAs expression. Along this line, Xia et al. revealed that metformin treatment decreased tumor growth and angiogenesis of HPV-positive CSCC cell lines, that was likely to depend on the reduced binding of MALAT-1 to the tumor suppressor miR-142-3p [171]. Besides metformin, a novel chemotherapeutic compound namely Casiopeina II-gly (Cas-II-gly) modulated MALAT-1 expression in HPV-positive CSCC cell lines. By acting on MALAT-1, Cas-II-gly inactivated Wnt pathway, thus inhibiting cell proliferation and promoting apoptosis in HPV-positive HeLa and CaSki CSCC cell lines [172]. Conversely, demethylation of MEG3 promoter by using 5-aza-2-deoxycytidine upregulated MEG3 expression and reduced proliferation of HPV-positive HeLa and CaSki cells, indicating the potential use of epigenetic drugs in HPV-driven cancers [94]. At present, other therapeutic agents have been exploited against MEG3, GAS5, and HOTAIR [173–176], but their effect in HPV-driven cancers still remain to be defined.

## 11. Conclusions

In addition to miRNAs, a huge list of ncRNAs has been identified in HPV-driven cancers so far. In particular, the use of high-throughput sequencing technologies, along with loss- and gain-of-function assays, has demonstrated that the expression of circRNAs, piRNAs, and lncRNAs promoted tumorigenesis and progression of HPV-positive cancers, thus suggesting they may be partially responsible for the clinical behavior of these tumors. Despite these findings, the functional relevance of these ncRNAs in HPV-driven cancers remains rather incomplete, in particular in SCCA where the field of ncRNAs is still at its infancy. Based on these considerations, more efforts will be necessary to profile ncRNAs in each HPV-driven cancer type. Furthermore, it will be crucial to better define molecular mechanisms underlying the association between aberrant ncRNAs expression and HPV infection, and to fully explore ncRNAs that are directly generated from HPV. In this context, in vivo experiments that more closely recapitulate the tumor microenvironment will be fundamental.

Numerous studies have documented that ncRNAs expression is tissue and cancer-specific, suggesting that ncRNAs that are linked to HPV infection could be useful in the early detection of HPV-driven cancers. More importantly, ncRNAs that show aberrant expression in both HPV-positive cancer tissues and biological fluids (i.e., plasma and/or saliva) may have a clinical utility in the non-invasive liquid biopsy approach for monitoring cancer progression and its treatment response. Despite these findings, the applicability of ncRNAs as diagnostic and prognostic biomarkers will require additional studies with larger sample sizes.

HPV-related ncRNAs have been found to be involved in tumor resistance to chemotherapy and/or radiotherapy, thus indicating they may also provide an important step towards personalized treatment, in particular for HPV-driven cancers at high risk of recurrence. As mentioned above, ncRNAs can be directly targeted by RNA-based approaches, but reliable methods for their delivery to tumor cells are needed. CRISPR/Cas9-genome editing or small drug inhibitors will also offer an exceptional opportunity to explore ncRNAs as druggable molecules. However, off-target effects and toxicities should be carefully evaluated before their clinical application. Therefore, although ncRNAs seem to be therapeutically promising, additional in vitro and in vivo preclinical studies are mandatory to design novel and more effective targeted therapies for the treatment of HPV-driven cancers.

**Funding:** This work was supported by 5 × 1000 Ministero della Salute Ricerca Corrente and 5 × 1000 Intramural Grant from CRO.

**Acknowledgments:** We thank E. Vaccher, A. De Paoli, V. Giacomarra, and G. Franchin for their helpful suggestions and scientific support.

**Conflicts of Interest:** The authors declare no conflict of interest.

## Appendix A

**Table A1.** Long non-coding RNAs (lncRNAs) and PIWI-like proteins deregulated in two or more different studies.

| LncRNAs ID | HPV-Driven Cancer Type | References |
|---|---|---|
| CCEPR | CSCC | [133–135] |
| GAS5 | CSCC | [139–142] |
| HOTAIR | CSCC | [144,145,147] |
| MALAT-1 | CSCC | [150–154] |
| MEG3 | CSCC | [94–96] |
| **PIWI-Like Proteins ID** | **HPV-Driven Cancer Type** | **References** |
| PIWIL4 | CSCC, HNSCC | [83,86] |

CCEPR, cervical carcinoma expressed PCNA regulatory; CSCC, cervical squamous cell carcinoma; GAS5, growth arrest-specific transcript 5; HNSCC, head and neck squamous cell carcinoma; HOTAIR, HOX transcript antisense RNA; HPV, Human Papilloma Virus; MALAT-1, metastasis associated lung adenocarcinoma transcript 1; MEG3, maternally expressed gene 3; PIWIL4, PIWI-like protein 4.

## References

1. De Martel, C.; Plummer, M.; Vignat, J.; Franceschi, S. Worldwide burden of cancer attributable to HPV by site, country and HPV type. *Int. J. Cancer* **2017**, *141*, 664–670. [CrossRef] [PubMed]
2. Tommasino, M. The human papillomavirus family and its role in carcinogenesis. *Semin. Cancer Biol.* **2014**, *26*, 13–21. [CrossRef] [PubMed]
3. Gheit, T. Mucosal and Cutaneous Human Papillomavirus Infections and Cancer Biology. *Front. Oncol.* **2019**, *9*, 355. [CrossRef] [PubMed]
4. Bouvard, V.; Baan, R.; Straif, K.; Grosse, Y.; Secretan, B.; El Ghissassi, F.; Benbrahim-Tallaa, L.; Guha, N.; Freeman, C.; Galichet, L.; et al. A review of human carcinogens–Part B: Biological agents. *Lancet Oncol.* **2009**, *10*, 321–322. [CrossRef]
5. Bandolin, L.; Borsetto, D.; Fussey, J.; Da Mosto, M.C.; Nicolai, P.; Menegaldo, A.; Calabrese, L.; Tommasino, M.; Boscolo-Rizzo, P. Beta human papillomaviruses infection and skin carcinogenesis. *Rev. Med. Virol.* **2020**, e2104. [CrossRef]

6. de Martel, C.; Georges, D.; Bray, F.; Ferlay, J.; Clifford, G.M. Global burden of cancer attributable to infections in 2018: A worldwide incidence analysis. *Lancet Glob. Health* **2020**, *8*, e180–e190. [CrossRef]

7. D'Souza, G.; Anantharaman, D.; Gheit, T.; Abedi-Ardekani, B.; Beachler, D.C.; Conway, D.I.; Olshan, A.F.; Wunsch-Filho, V.; Toporcov, T.N.; Ahrens, W.; et al. Effect of HPV on head and neck cancer patient survival, by region and tumor site: A comparison of 1362 cases across three continents. *Oral Oncol.* **2016**, *62*, 20–27. [CrossRef]

8. Castellsagué, X.; Alemany, L.; Quer, M.; Halec, G.; Quirós, B.; Tous, S.; Clavero, O.; Alòs, L.; Biegner, T.; Szafarowski, T.; et al. HPV Involvement in Head and Neck Cancers: Comprehensive Assessment of Biomarkers in 3680 Patients. *J. Natl. Cancer Inst.* **2016**, *108*, djv403. [CrossRef]

9. Guan, P.; Howell-Jones, R.; Li, N.; Bruni, L.; de Sanjosé, S.; Franceschi, S.; Clifford, G.M. Human papillomavirus types in 115,789 HPV-positive women: A meta-analysis from cervical infection to cancer. *Int. J. Cancer* **2012**, *131*, 2349–2359. [CrossRef]

10. Jhaveri, J.; Rayfield, L.; Liu, Y.; Chowdhary, M.; Cassidy, R.J.; Madden, N.A.; Tanenbaum, D.G.; Gillespie, T.W.; Patel, P.R.; Patel, K.R.; et al. Prognostic relevance of human papillomavirus infection in anal squamous cell carcinoma: Analysis of the national cancer data base. *J. Gastrointest. Oncol.* **2017**, *8*, 998–1008. [CrossRef]

11. Kidd, L.C.; Chaing, S.; Chipollini, J.; Giuliano, A.R.; Spiess, P.E.; Sharma, P. Relationship between human papillomavirus and penile cancer-implications for prevention and treatment. *Transl. Androl. Urol.* **2017**, *6*, 791–802. [CrossRef] [PubMed]

12. Backes, D.M.; Kurman, R.J.; Pimenta, J.M.; Smith, J.S. Systematic review of human papillomavirus prevalence in invasive penile cancer. *Cancer Causes Control* **2009**, *20*, 449–457. [CrossRef] [PubMed]

13. Gargano, J.W.; Wilkinson, E.J.; Unger, E.R.; Steinau, M.; Watson, M.; Huang, Y.; Copeland, G.; Cozen, W.; Goodman, M.T.; Hopenhayn, C.; et al. Prevalence of human papillomavirus types in invasive vulvar cancers and vulvar intraepithelial neoplasia 3 in the United States before vaccine introduction. *J. Low. Genit. Tract Dis.* **2012**, *16*, 471–479. [CrossRef] [PubMed]

14. Zhang, J.; Zhang, Y.; Zhang, Z. Prevalence of human papillomavirus and its prognostic value in vulvar cancer: A systematic review and meta-analysis. *PLoS ONE* **2018**, *13*, e0204162. [CrossRef] [PubMed]

15. Insinga, R.P.; Liaw, K.L.; Johnson, L.G.; Madeleine, M.M. A systematic review of the prevalence and attribution of human papillomavirus types among cervical, vaginal, and vulvar precancers and cancers in the United States. *Cancer Epidemiol. Biomark. Prev.* **2008**, *17*, 1611–1622. [CrossRef] [PubMed]

16. zur Hausen, H. Papillomavirus infections—A major cause of human cancers. *Biochim. Biophys. Acta* **1996**, *1288*, F55–F78. [CrossRef]

17. Colombo, N.; Carinelli, S.; Colombo, A.; Marini, C.; Rollo, D.; Sessa, C. Cervical cancer: ESMO Clinical Practice Guidelines for diagnosis, treatment and follow-up. *Ann. Oncol.* **2012**, *23*, vii27–vii32. [CrossRef]

18. Rose, P.G.; Bundy, B.N.; Watkins, E.B.; Thigpen, J.T.; Deppe, G.; Maiman, M.A.; Clarke-Pearson, D.L.; Insalaco, S. Concurrent cisplatin-based radiotherapy and chemotherapy for locally advanced cervical cancer. *N. Engl. J. Med.* **1999**, *340*, 1144–1153. [CrossRef]

19. Chemoradiotherapy for Cervical Cancer Meta-Analysis Collaboration. Reducing uncertainties about the effects of chemoradiotherapy for cervical cancer: A systematic review and meta-analysis of individual patient data from 18 randomized trials. *J. Clin. Oncol.* **2008**, *26*, 5802–5812. [CrossRef]

20. Wang, X.; Wang, H.K.; Li, Y.; Hafner, M.; Banerjee, N.S.; Tang, S.; Briskin, D.; Meyers, C.; Chow, L.T.; Xie, X.; et al. microRNAs are biomarkers of oncogenic human papillomavirus infections. *Proc. Natl. Acad. Sci. USA* **2014**, *111*, 4262–4267. [CrossRef]

21. zur Hausen, H. Papillomaviruses causing cancer: Evasion from host-cell control in early events in carcinogenesis. *J. Natl. Cancer Inst.* **2000**, *92*, 690–698. [CrossRef]

22. McLaughlin-Drubin, M.E.; Munger, K. The human papillomavirus E7 oncoprotein. *Virology* **2009**, *384*, 335–344. [CrossRef] [PubMed]

23. Howie, H.L.; Katzenellbogen, R.A.; Galloway, D.A. Papillomavirus E6 proteins. *Virology* **2009**, *384*, 324–334. [CrossRef] [PubMed]

24. Lajer, C.B.; Garnæs, E.; Friis-Hansen, L.; Norrild, B.; Therkildsen, M.H.; Glud, M.; Rossing, M.; Lajer, H.; Svane, D.; Skotte, L.; et al. The role of miRNAs in human papilloma virus (HPV)-associated cancers: Bridging between HPV-related head and neck cancer and cervical cancer. *Br. J. Cancer* **2012**, *106*, 1526–1534. [CrossRef] [PubMed]

25. Yin, L.X.; D'Souza, G.; Westra, W.H.; Wang, S.J.; van Zante, A.; Zhang, Y.; Rettig, E.M.; Ryan, W.R.; Ha, P.K.; Wentz, A.; et al. Prognostic factors for human papillomavirus-positive and negative oropharyngeal carcinomas. *Laryngoscope* **2018**, *128*, E287–E295. [CrossRef] [PubMed]
26. Gillison, M.L.; Koch, W.M.; Capone, R.B.; Spafford, M.; Westra, W.H.; Wu, L.; Zahurak, M.L.; Daniel, R.W.; Viglione, M.; Symer, D.E.; et al. Evidence for a causal association between human papillomavirus and a subset of head and neck cancers. *J. Natl. Cancer Inst.* **2000**, *92*, 709–720. [CrossRef] [PubMed]
27. Hoffmann, M.; Tribius, S. HPV and Oropharyngeal Cancer in the Eighth Edition of the TNM Classification: Pitfalls in Practice. *Transl. Oncol.* **2019**, *12*, 1108–1112. [CrossRef] [PubMed]
28. Carlander, A.F.; Gronhoj Larsen, C.; Jensen, D.H.; Garnaes, E.; Kiss, K.; Andersen, L.; Olsen, C.H.; Franzmann, M.; Hogdall, E.; Kjaer, S.K.; et al. Continuing rise in oropharyngeal cancer in a high HPV prevalence area: A Danish population-based study from 2011 to 2014. *Eur. J. Cancer* **2017**, *70*, 75–82. [CrossRef]
29. Clocchiatti, A.; Cora, E.; Zhang, Y.; Dotto, G.P. Sexual dimorphism in cancer. *Nat. Rev. Cancer* **2016**, *16*, 330–339. [CrossRef]
30. Fish, E.N. The X-files in immunity: Sex-based differences predispose immune responses. *Nat. Rev. Immunol.* **2008**, *8*, 737–744. [CrossRef]
31. Giuliano, A.R.; Viscidi, R.; Torres, B.N.; Ingles, D.J.; Sudenga, S.L.; Villa, L.L.; Baggio, M.L.; Abrahamsen, M.; Quiterio, M.; Salmeron, J.; et al. Seroconversion Following Anal and Genital HPV Infection in Men: The HIM Study. *Papillomavirus Res.* **2015**, *1*, 109–115. [CrossRef] [PubMed]
32. Giuliano, A.R.; Nyitray, A.G.; Kreimer, A.R.; Pierce Campbell, C.M.; Goodman, M.T.; Sudenga, S.L.; Monsonego, J.; Franceschi, S. EUROGIN 2014 roadmap: Differences in human papillomavirus infection natural history, transmission and human papillomavirus-related cancer incidence by gender and anatomic site of infection. *Int. J. Cancer* **2015**, *136*, 2752–2760. [CrossRef] [PubMed]
33. Ang, K.K.; Harris, J.; Wheeler, R.; Weber, R.; Rosenthal, D.I.; Nguyen-Tan, P.F.; Westra, W.H.; Chung, C.H.; Jordan, R.C.; Lu, C.; et al. Human papillomavirus and survival of patients with oropharyngeal cancer. *N. Engl. J. Med.* **2010**, *363*, 24–35. [CrossRef] [PubMed]
34. De Felice, F.; Tombolini, V.; Valentini, V.; de Vincentiis, M.; Mezi, S.; Brugnoletti, O.; Polimeni, A. Advances in the Management of HPV-Related Oropharyngeal Cancer. *J. Oncol.* **2019**, *2019*, 9173729. [CrossRef]
35. Nichols, A.C.; Theurer, J.; Prisman, E.; Read, N.; Berthelet, E.; Tran, E.; Fung, K.; de Almeida, J.R.; Bayley, A.; Goldstein, D.P.; et al. Radiotherapy versus transoral robotic surgery and neck dissection for oropharyngeal squamous cell carcinoma (ORATOR): An open-label, phase 2, randomised trial. *Lancet Oncol.* **2019**, *20*, 1349–1359. [CrossRef]
36. Chera, B.S.; Amdur, R.J.; Green, R.; Shen, C.; Gupta, G.; Tan, X.; Knowles, M.; Fried, D.; Hayes, N.; Weiss, J.; et al. Phase II Trial of De-Intensified Chemoradiotherapy for Human Papillomavirus-Associated Oropharyngeal Squamous Cell Carcinoma. *J. Clin. Oncol.* **2019**, *37*, 2661–2669. [CrossRef]
37. Mehanna, H.; Robinson, M.; Hartley, A.; Kong, A.; Foran, B.; Fulton-Lieuw, T.; Dalby, M.; Mistry, P.; Sen, M.; O'Toole, L.; et al. Radiotherapy plus cisplatin or cetuximab in low-risk human papillomavirus-positive oropharyngeal cancer (De-ESCALaTE HPV): An open-label randomised controlled phase 3 trial. *Lancet* **2019**, *393*, 51–60. [CrossRef]
38. Marur, S.; Li, S.; Cmelak, A.J.; Gillison, M.L.; Zhao, W.J.; Ferris, R.L.; Westra, W.H.; Gilbert, J.; Bauman, J.E.; Wagner, L.I.; et al. E1308: Phase II Trial of Induction Chemotherapy Followed by Reduced-Dose Radiation and Weekly Cetuximab in Patients With HPV-Associated Resectable Squamous Cell Carcinoma of the Oropharynx- ECOG-ACRIN Cancer Research Group. *J. Clin. Oncol.* **2017**, *35*, 490–497. [CrossRef]
39. Seiwert, T.Y.; Foster, C.C.; Blair, E.A.; Karrison, T.G.; Agrawal, N.; Melotek, J.M.; Portugal, L.; Brisson, R.J.; Dekker, A.; Kochanny, S.; et al. OPTIMA: A phase II dose and volume de-escalation trial for human papillomavirus-positive oropharyngeal cancer. *Ann. Oncol.* **2019**, *30*, 1673. [CrossRef]
40. Swisher-McClure, S.; Lukens, J.N.; Aggarwal, C.; Ahn, P.; Basu, D.; Bauml, J.M.; Brody, R.; Chalian, A.; Cohen, R.B.; Fotouhi-Ghiam, A.; et al. A Phase 2 Trial of Alternative Volumes of Oropharyngeal Irradiation for De-intensification (AVOID): Omission of the Resected Primary Tumor Bed After Transoral Robotic Surgery for Human Papilloma Virus-Related Squamous Cell Carcinoma of the Oropharynx. *Int. J. Radiat. Oncol. Biol. Phys.* **2020**, *106*, 725–732. [CrossRef]

41. Ma, D.J.; Price, K.A.; Moore, E.J.; Patel, S.H.; Hinni, M.L.; Garcia, J.J.; Graner, D.E.; Foster, N.R.; Ginos, B.; Neben-Wittich, M.; et al. Phase II Evaluation of Aggressive Dose De-Escalation for Adjuvant Chemoradiotherapy in Human Papillomavirus-Associated Oropharynx Squamous Cell Carcinoma. *J. Clin. Oncol.* **2019**, *37*, 1909–1918. [CrossRef] [PubMed]

42. Gillison, M.L.; Trotti, A.M.; Harris, J.; Eisbruch, A.; Harari, P.M.; Adelstein, D.J.; Jordan, R.C.K.; Zhao, W.; Sturgis, E.M.; Burtness, B.; et al. Radiotherapy plus cetuximab or cisplatin in human papillomavirus-positive oropharyngeal cancer (NRG Oncology RTOG 1016): A randomised, multicentre, non-inferiority trial. *Lancet* **2019**, *393*, 40–50. [CrossRef]

43. Mirghani, H.; Blanchard, P. Treatment de-escalation for HPV-driven oropharyngeal cancer: Where do we stand? *Clin. Transl. Radiat. Oncol.* **2017**, *8*, 4–11. [CrossRef] [PubMed]

44. Amirian, E.S.; Fickey, P.A., Jr.; Scheurer, M.E.; Chiao, E.Y. Anal cancer incidence and survival: Comparing the greater San-Francisco bay area to other SEER cancer registries. *PLoS ONE* **2013**, *8*, e58919. [CrossRef]

45. Arbyn, M.; de Sanjose, S.; Saraiya, M.; Sideri, M.; Palefsky, J.; Lacey, C.; Gillison, M.; Bruni, L.; Ronco, G.; Wentzensen, N.; et al. EUROGIN 2011 roadmap on prevention and treatment of HPV-related disease. *Int. J. Cancer* **2012**, *131*, 1969–1982. [CrossRef]

46. Glynne-Jones, R.; Nilsson, P.J.; Aschele, C.; Goh, V.; Peiffert, D.; Cervantes, A.; Arnold, D. Anal cancer: ESMO-ESSO-ESTRO Clinical Practice Guidelines for diagnosis, treatment and follow-up †. *Ann. Oncol.* **2014**, *25*, iii10–iii20. [CrossRef]

47. Lampejo, T.; Kavanagh, D.; Clark, J.; Goldin, R.; Osborn, M.; Ziprin, P.; Cleator, S. Prognostic biomarkers in squamous cell carcinoma of the anus: A systematic review. *Br. J. Cancer* **2010**, *103*, 1858–1869. [CrossRef]

48. Horner, S.M.; DeFilippis, R.A.; Manuelidis, L.; DiMaio, D. Repression of the human papillomavirus E6 gene initiates p53-dependent, telomerase-independent senescence and apoptosis in HeLa cervical carcinoma cells. *J. Virol.* **2004**, *78*, 4063–4073. [CrossRef] [PubMed]

49. Martinez-Zapien, D.; Ruiz, F.X.; Poirson, J.; Mitschler, A.; Ramirez, J.; Forster, A.; Cousido-Siah, A.; Masson, M.; Vande Pol, S.; Podjarny, A.; et al. Structure of the E6/E6AP/p53 complex required for HPV-mediated degradation of p53. *Nature* **2016**, *529*, 541–545. [CrossRef] [PubMed]

50. Todorovic, B.; Hung, K.; Massimi, P.; Avvakumov, N.; Dick, F.A.; Shaw, G.S.; Banks, L.; Mymryk, J.S. Conserved region 3 of human papillomavirus 16 E7 contributes to deregulation of the retinoblastoma tumor suppressor. *J. Virol.* **2012**, *86*, 13313–13323. [CrossRef] [PubMed]

51. McLaughlin-Drubin, M.E.; Huh, K.W.; Munger, K. Human papillomavirus type 16 E7 oncoprotein associates with E2F6. *J. Virol.* **2008**, *82*, 8695–8705. [CrossRef] [PubMed]

52. Jansma, A.L.; Martinez-Yamout, M.A.; Liao, R.; Sun, P.; Dyson, H.J.; Wright, P.E. The high-risk HPV16 E7 oncoprotein mediates interaction between the transcriptional coactivator CBP and the retinoblastoma protein pRb. *J. Mol. Biol.* **2014**, *426*, 4030–4048. [CrossRef] [PubMed]

53. Van Doorslaer, K.; Burk, R.D. Association between hTERT activation by HPV E6 proteins and oncogenic risk. *Virology* **2012**, *433*, 216–219. [CrossRef] [PubMed]

54. Chiantore, M.V.; Mangino, G.; Iuliano, M.; Capriotti, L.; Di Bonito, P.; Fiorucci, G.; Romeo, G. Human Papillomavirus and carcinogenesis: Novel mechanisms of cell communication involving extracellular vesicles. *Cytokine Growth Factor Rev.* **2020**, *51*, 92–98. [CrossRef] [PubMed]

55. Martinez, I.; Gardiner, A.S.; Board, K.F.; Monzon, F.A.; Edwards, R.P.; Khan, S.A. Human papillomavirus type 16 reduces the expression of microRNA-218 in cervical carcinoma cells. *Oncogene* **2008**, *27*, 2575–2582. [CrossRef]

56. Wang, X.; Tang, S.; Le, S.Y.; Lu, R.; Rader, J.S.; Meyers, C.; Zheng, Z.M. Aberrant expression of oncogenic and tumor-suppressive microRNAs in cervical cancer is required for cancer cell growth. *PLoS ONE* **2008**, *3*, e2557. [CrossRef] [PubMed]

57. Brakenhoff, R.H.; Wagner, S.; Klussmann, J.P. Molecular Patterns and Biology of HPV-Associated HNSCC. *Recent Results Cancer Res.* **2017**, *206*, 37–56. [CrossRef]

58. Virtanen, E.; Pietilä, T.; Nieminen, P.; Qian, K.; Auvinen, E. Low expression levels of putative HPV encoded microRNAs in cervical samples. *SpringerPlus* **2016**, *5*, 1856. [CrossRef]

59. Zhao, J.; Lee, E.E.; Kim, J.; Yang, R.; Chamseddin, B.; Ni, C.; Gusho, E.; Xie, Y.; Chiang, C.M.; Buszczak, M.; et al. Transforming activity of an oncoprotein-encoding circular RNA from human papillomavirus. *Nat. Commun.* **2019**, *10*, 2300. [CrossRef]

60. Fu, X.-D. Non-coding RNA: A new frontier in regulatory biology. *Natl. Sci. Rev.* **2014**, *1*, 190–204. [CrossRef]

61. Wei, J.W.; Huang, K.; Yang, C.; Kang, C.S. Non-coding RNAs as regulators in epigenetics (Review). *Oncol. Rep.* **2017**, *37*, 3–9. [CrossRef] [PubMed]

62. Diamantopoulos, M.A.; Tsiakanikas, P.; Scorilas, A. Non-coding RNAs: The riddle of the transcriptome and their perspectives in cancer. *Ann. Transl. Med.* **2018**, *6*, 241. [CrossRef] [PubMed]

63. Jia, L.; Yang, A. Noncoding RNAs in Therapeutic Resistance of Cancer. *Adv. Exp. Med. Biol.* **2016**, *927*, 265–295. [CrossRef] [PubMed]

64. Yang, Z.; Li, X.; Yang, Y.; He, Z.; Qu, X.; Zhang, Y. Long noncoding RNAs in the progression, metastasis, and prognosis of osteosarcoma. *Cell Death Amp Dis.* **2016**, *7*, e2389. [CrossRef]

65. Deng, H.; Zhang, J.; Shi, J.; Guo, Z.; He, C.; Ding, L.; Hai Tang, J.; Hou, Y. Role of Long Non-Coding RNA in Tumor Drug Resistance. *Tumour Biol.* **2016**, *37*, 11623–11631. [CrossRef]

66. Corrà, F.; Agnoletto, C.; Minotti, L.; Baldassari, F.; Volinia, S. The Network of Non-coding RNAs in Cancer Drug Resistance. *Front. Oncol.* **2018**, *8*. [CrossRef]

67. Malek, E.; Jagannathan, S.; Driscoll, J.J. Correlation of long non-coding RNA expression with metastasis, drug resistance and clinical outcome in cancer. *Oncotarget* **2014**, *5*, 8027–8038. [CrossRef]

68. Dhamija, S.; Diederichs, S. From junk to master regulators of invasion: LncRNA functions in migration, EMT and metastasis. *Int. J. Cancer* **2016**, *139*, 269–280. [CrossRef]

69. Lin, C.-W.; Lin, P.-Y.; Yang, P.-C. Noncoding RNAs in Tumor Epithelial-to-Mesenchymal Transition. *Stem Cells Int.* **2016**, *2016*, 2732705. [CrossRef]

70. Xu, Y.; Brenn, T.; Brown, E.R.; Doherty, V.; Melton, D.W. Differential expression of microRNAs during melanoma progression: MiR-200c, miR-205 and miR-211 are downregulated in melanoma and act as tumour suppressors. *Br. J Cancer* **2012**, *106*, 553–561. [CrossRef]

71. Snoek, B.C.; Babion, I.; Koppers-Lalic, D.; Pegtel, D.M.; Steenbergen, R.D.M. Altered microRNA processing proteins in HPV-induced cancers. *Curr. Opin. Virol.* **2019**, *39*, 23–32. [CrossRef] [PubMed]

72. Quan, G.; Li, J. Circular RNAs: Biogenesis, expression and their potential roles in reproduction. *J. Ovarian Res.* **2018**, *11*, 9. [CrossRef] [PubMed]

73. Huang, G.; Li, S.; Yang, N.; Zou, Y.; Zheng, D.; Xiao, T. Recent progress in circular RNAs in human cancers. *Cancer Lett.* **2017**, *404*, 8–18. [CrossRef]

74. Yang, Y.; Fan, X.; Mao, M.; Song, X.; Wu, P.; Zhang, Y.; Jin, Y.; Yang, Y.; Chen, L.-L.; Wang, Y.; et al. Extensive translation of circular RNAs driven by N(6)-methyladenosine. *Cell Res.* **2017**, *27*, 626–641. [CrossRef] [PubMed]

75. Bach, D.-H.; Lee, S.K.; Sood, A.K. Circular RNAs in Cancer. *Mol. Ther. Nucleic Acids* **2019**, *16*, 118–129. [CrossRef] [PubMed]

76. Wu, J.; Qi, X.; Liu, L.; Hu, X.; Liu, J.; Yang, J.; Yang, J.; Lu, L.; Zhang, Z.; Ma, S.; et al. Emerging Epigenetic Regulation of Circular RNAs in Human Cancer. *Mol. Ther. Nucleic Acids* **2019**, *16*, 589–596. [CrossRef] [PubMed]

77. Su, M.; Xiao, Y.; Ma, J.; Tang, Y.; Tian, B.; Zhang, Y.; Li, X.; Wu, Z.; Yang, D.; Zhou, Y.; et al. Circular RNAs in Cancer: Emerging functions in hallmarks, stemness, resistance and roles as potential biomarkers. *Mol. Cancer* **2019**, *18*, 90. [CrossRef]

78. Vo, J.N.; Cieslik, M.; Zhang, Y.; Shukla, S.; Xiao, L.; Wu, Y.M.; Dhanasekaran, S.M.; Engelke, C.G.; Cao, X.; Robinson, D.R.; et al. The Landscape of Circular RNA in Cancer. *Cell* **2019**, *176*, 869–881. [CrossRef]

79. Guo, Y.; Yang, J.; Huang, Q.; Hsueh, C.; Zheng, J.; Wu, C.; Chen, H.; Zhou, L. Circular RNAs and their roles in head and neck cancers. *Mol. Cancer* **2019**, *18*, 44. [CrossRef]

80. Liu, J.; Wang, D.; Long, Z.; Li, W. CircRNA8924 Promotes Cervical Cancer Cell Proliferation, Migration and Invasion by Competitively Binding to MiR-518d-5p /519-5p Family and Modulating the Expression of CBX8. *Cell. Physiol. Biochem.* **2018**, *48*, 173–184. [CrossRef]

81. Ma, H.; Tian, T.; Liu, X.; Xia, M.; Chen, C.; Mai, L.; Xie, S.; Yu, L. Upregulated circ_0005576 facilitates cervical cancer progression via the miR-153/KIF20A axis. *Biomed. Pharmacother.* **2019**, *118*, 109311. [CrossRef] [PubMed]

82. Firmino, N.; Martinez, V.D.; Rowbotham, D.A.; Enfield, K.S.S.; Bennewith, K.L.; Lam, W.L. HPV status is associated with altered PIWI-interacting RNA expression pattern in head and neck cancer. *Oral Oncol.* **2016**, *55*, 43–48. [CrossRef] [PubMed]

83. Krishnan, A.R.; Qu, Y.; Li, P.X.; Zou, A.E.; Califano, J.A.; Wang-Rodriguez, J.; Ongkeko, W.M. Computational methods reveal novel functionalities of PIWI-interacting RNAs in human papillomavirus-induced head and neck squamous cell carcinoma. *Oncotarget* **2017**, *9*, 4614–4624. [CrossRef] [PubMed]

84. Liu, W.; Gao, Q.; Chen, K.; Xue, X.; Li, M.; Chen, Q.; Zhu, G.; Gao, Y. Hiwi facilitates chemoresistance as a cancer stem cell marker in cervical cancer. *Oncol. Rep.* **2014**, *32*, 1853–1860. [CrossRef]

85. Feng, D.; Yan, K.; Zhou, Y.; Liang, H.; Liang, J.; Zhao, W.; Dong, Z.; Ling, B. Piwil2 is reactivated by HPV oncoproteins and initiates cell reprogramming via epigenetic regulation during cervical cancer tumorigenesis. *Oncotarget* **2016**, *7*, 64575–64588. [CrossRef]

86. Su, C.; Ren, Z.-J.; Wang, F.; Liu, M.; Li, X.; Tang, H. PIWIL4 regulates cervical cancer cell line growth and is involved in down-regulating the expression of p14ARF and p53. *FEBS Lett.* **2012**, *586*, 1356–1362. [CrossRef]

87. Braciale, T.J.; Hahn, Y.S. Immunity to viruses. *Immunol. Rev.* **2013**, *255*, 5–12. [CrossRef]

88. Takata, M.; Sasaki, M.S.; Sonoda, E.; Fukushima, T.; Morrison, C.; Albala, J.S.; Swagemakers, S.M.A.; Kanaar, R.; Thompson, L.H.; Takeda, S. The Rad51 Paralog Rad51B Promotes Homologous Recombinational Repair. *Mol. Cell. Biol.* **2000**, *20*, 6476–6482. [CrossRef]

89. Yang, L.; Yi, K.; Wang, H.; Zhao, Y.; Xi, M. Comprehensive analysis of lncRNAs microarray profile and mRNA-lncRNA co-expression in oncogenic HPV-positive cervical cancer cell lines. *Oncotarget* **2016**, *7*, 49917–49929. [CrossRef]

90. Zhou, D.; Wu, F.; Cui, Y.; Wei, F.; Meng, Q.; Lv, Q. Long non-coding RNA-OIS1 inhibits HPV-positive, but not HPV-negative cervical squamous cell carcinoma by upregulating MTK-1. *Oncol. Lett.* **2019**, *17*, 2923–2930. [CrossRef]

91. Wang, B.; Huang, Z.; Gao, R.; Zeng, Z.; Yang, W.; Sun, Y.; Wei, W.; Wu, Z.; Yu, L.; Li, Q.; et al. Expression of Long Noncoding RNA Urothelial Cancer Associated 1 Promotes Cisplatin Resistance in Cervical Cancer. *Cancer Biother. Radiopharm.* **2017**, *32*, 101–110. [CrossRef] [PubMed]

92. Qu, X.; Li, Y.; Wang, L.; Yuan, N.; Ma, M.; Chen, Y. LncRNA SNHG8 accelerates proliferation and inhibits apoptosis in HPV-induced cervical cancer through recruiting EZH2 to epigenetically silence RECK expression. *J. Cell. Biochem.* **2020**. [CrossRef] [PubMed]

93. Zhang, Y.; Jia, L.G.; Wang, P.; Li, J.; Tian, F.; Chu, Z.P.; Kang, S. The expression and significance of lncRNA HOST2 and microRNA let-7b in HPV-positive cervical cancer tissues and cell lines. *Eur. Rev. Med. Pharmacol. Sci.* **2019**, *23*, 2380–2390. [CrossRef] [PubMed]

94. Zhang, J.; Lin, Z.; Gao, Y.; Yao, T. Downregulation of long noncoding RNA MEG3 is associated with poor prognosis and promoter hypermethylation in cervical cancer. *J. Exp. Clin. Cancer Res.* **2017**, *36*, 5. [CrossRef]

95. Zhang, J.; Yao, T.; Lin, Z.; Gao, Y. Aberrant Methylation of MEG3 Functions as a Potential Plasma-Based Biomarker for Cervical Cancer. *Sci. Rep.* **2017**, *7*, 6271. [CrossRef]

96. Zhang, J.; Yao, T.; Wang, Y.; Yu, J.; Liu, Y.; Lin, Z. Long noncoding RNA MEG3 is downregulated in cervical cancer and affects cell proliferation and apoptosis by regulating miR-21. *Cancer Biol. Ther.* **2016**, *17*, 104–113. [CrossRef]

97. Nohata, N.; Abba, M.C.; Gutkind, J.S. Unraveling the oral cancer lncRNAome: Identification of novel lncRNAs associated with malignant progression and HPV infection. *Oral Oncol.* **2016**, *59*, 58–66. [CrossRef]

98. Song, L.; Xie, H.; Tong, F.; Yan, B.; Zhang, S.; Fu, E.; Jing, Q.; Wei, L. Association of lnc-IL17RA-11 with increased radiation sensitivity and improved prognosis of HPV-positive HNSCC. *J. Cell. Biochem.* **2019**, *120*, 17438–17448. [CrossRef]

99. Kolenda, T.; Kopczyńska, M.; Guglas, K.; Teresiak, A.; Bliźniak, R.; Łasińska, I.; Mackiewicz, J.; Lamperska, K. EGOT lncRNA in head and neck squamous cell carcinomas. *Pol. J. Pathol.* **2018**, *69*, 356–365. [CrossRef]

100. Wang, H.; Zhao, Y.; Chen, M.; Cui, J. Identification of Novel Long Non-coding and Circular RNAs in Human Papillomavirus-Mediated Cervical Cancer. *Front. Microbiol.* **2017**, *8*. [CrossRef]

101. Gao, Y.-L.; Zhang, M.-Y.; Xu, B.; Han, L.-J.; Lan, S.-F.; Chen, J.; Dong, Y.-J.; Cao, L.-L. Circular RNA expression profiles reveal that hsa_circ_0018289 is up-regulated in cervical cancer and promotes the tumorigenesis. *Oncotarget* **2017**, *8*, 86625–86633. [CrossRef] [PubMed]

102. Zhang, J.; Zhao, X.; Zhang, J.; Zheng, X.; Li, F. Circular RNA hsa_circ_0023404 exerts an oncogenic role in cervical cancer through regulating miR-136/TFCP2/YAP pathway. *Biochem. Biophys. Res. Commun.* **2018**, *501*, 428–433. [CrossRef]

103. Cai, H.; Zhang, P.; Xu, M.; Yan, L.; Liu, N.; Wu, X. Circular RNA hsa_circ_0000263 participates in cervical cancer development by regulating target gene of miR-150-5p. *J. Cell. Physiol.* **2019**, *234*, 11391–11400. [CrossRef] [PubMed]

104. Hu, C.; Wang, Y.; Li, A.; Zhang, J.; Xue, F.; Zhu, L. Overexpressed circ_0067934 acts as an oncogene to facilitate cervical cancer progression via the miR-545/EIF3C axis. *J. Cell. Physiol.* **2019**, *234*, 9225–9232. [CrossRef] [PubMed]

105. Ji, F.; Du, R.; Chen, T.; Zhang, M.; Zhu, Y.; Luo, X.; Ding, Y. Circular RNA circSLC26A4 Accelerates Cervical Cancer Progression via miR-1287-5p/HOXA7 Axis. *Mol. Ther. Nucleic Acids* **2020**, *19*, 413–420. [CrossRef] [PubMed]

106. Rojas-Ríos, P.; Simonelig, M. piRNAs and PIWI proteins: Regulators of gene expression in development and stem cells. *Development* **2018**, *145*, dev161786. [CrossRef]

107. Ng, K.W.; Anderson, C.; Marshall, E.A.; Minatel, B.C.; Enfield, K.S.S.; Saprunoff, H.L.; Lam, W.L.; Martinez, V.D. Piwi-interacting RNAs in cancer: Emerging functions and clinical utility. *Mol. Cancer* **2016**, *15*, 5. [CrossRef]

108. Ross, R.J.; Weiner, M.M.; Lin, H. PIWI proteins and PIWI-interacting RNAs in the soma. *Nature* **2014**, *505*, 353–359. [CrossRef]

109. Yu, Y.; Xiao, J.; Hann, S.S. The emerging roles of PIWI-interacting RNA in human cancers. *Cancer Manag. Res.* **2019**, *11*, 5895–5909. [CrossRef]

110. Litwin, M.; Szczepańska-Buda, A.; Piotrowska, A.; Dzięgiel, P.; Witkiewicz, W. The meaning of PIWI proteins in cancer development. *Oncol. Lett.* **2017**, *13*, 3354–3362. [CrossRef]

111. Liu, W.-K.; Jiang, X.-Y.; Zhang, Z.-X. Expression of PSCA, PIWIL1 and TBX2 and its correlation with HPV16 infection in formalin-fixed, paraffin-embedded cervical squamous cell carcinoma specimens. *Arch. Virol.* **2010**, *155*, 657–663. [CrossRef] [PubMed]

112. He, G.; Chen, L.; Ye, Y.; Xiao, Y.; Hua, K.; Jarjoura, D.; Nakano, T.; Barsky, S.H.; Shen, R.; Gao, J.-X. Piwil2 expressed in various stages of cervical neoplasia is a potential complementary marker for p16. *Am. J. Transl. Res.* **2010**, *2*, 156–169. [PubMed]

113. Hashim, A.; Rizzo, F.; Marchese, G.; Ravo, M.; Tarallo, R.; Nassa, G.; Giurato, G.; Santamaria, G.; Cordella, A.; Cantarella, C.; et al. RNA sequencing identifies specific PIWI-interacting small non-coding RNA expression patterns in breast cancer. *Oncotarget* **2014**, *5*, 9901–9910. [CrossRef] [PubMed]

114. Albuquerque, A.; Fernandes, M.; Stirrup, O.; Teixeira, A.L.; Santos, J.; Rodrigues, M.; Rios, E.; Macedo, G.; Medeiros, R. Expression of microRNAs 16, 20a, 150 and 155 in anal squamous intraepithelial lesions from high-risk groups. *Sci. Rep.* **2019**, *9*, 1523. [CrossRef] [PubMed]

115. Myklebust, M.P.; Bruland, O.; Fluge, O.; Skarstein, A.; Balteskard, L.; Dahl, O. MicroRNA-15b is induced with E2F-controlled genes in HPV-related cancer. *Br. J. Cancer* **2011**, *105*, 1719–1725. [CrossRef]

116. Karapetyan, A.R.; Buiting, C.; Kuiper, R.A.; Coolen, M.W. Regulatory Roles for Long ncRNA and mRNA. *Cancers* **2013**, *5*, 462–490. [CrossRef]

117. Rinn, J.L.; Chang, H.Y. Genome regulation by long noncoding RNAs. *Annu. Rev. Biochem.* **2012**, *81*, 145–166. [CrossRef]

118. Ma, L.; Bajic, V.B.; Zhang, Z. On the classification of long non-coding RNAs. *RNA Biol.* **2013**, *10*, 925–933. [CrossRef]

119. Khalil, A.M.; Guttman, M.; Huarte, M.; Garber, M.; Raj, A.; Rivea Morales, D.; Thomas, K.; Presser, A.; Bernstein, B.E.; van Oudenaarden, A.; et al. Many human large intergenic noncoding RNAs associate with chromatin-modifying complexes and affect gene expression. *Proc. Natl. Acad. Sci. USA* **2009**, *106*, 11667–11672. [CrossRef]

120. Zhang, X.; Wang, W.; Zhu, W.; Dong, J.; Cheng, Y.; Yin, Z.; Shen, F. Mechanisms and Functions of Long Non-Coding RNAs at Multiple Regulatory Levels. *Int. J. Mol. Sci.* **2019**, *20*, 5573. [CrossRef]

121. Wang, P. The Opening of Pandora's Box: An Emerging Role of Long Noncoding RNA in Viral Infections. *Front. Immunol.* **2019**, *9*. [CrossRef] [PubMed]

122. Sartori, D.A.; Chan, D.W. Biomarkers in prostate cancer: What's new? *Curr. Opin. Oncol.* **2014**, *26*, 259–264. [CrossRef] [PubMed]

123. Yang, K.; Hou, Y.; Li, A.; Li, Z.; Wang, W.; Xie, H.; Rong, Z.; Lou, G.; Li, K. Identification of a six-lncRNA signature associated with recurrence of ovarian cancer. *Sci. Rep.* **2017**, *7*, 752. [CrossRef] [PubMed]

124. Bolha, L.; Ravnik-Glavač, M.; Glavač, D. Long Noncoding RNAs as Biomarkers in Cancer. *Dis. Markers* **2017**, *2017*, 7243968. [CrossRef] [PubMed]

125. Huarte, M. The emerging role of lncRNAs in cancer. *Nat. Med.* **2015**, *21*, 1253–1261. [CrossRef] [PubMed]

126. Chandra Gupta, S.; Nandan Tripathi, Y. Potential of long non-coding RNAs in cancer patients: From biomarkers to therapeutic targets. *Int. J. Cancer* **2017**, *140*, 1955–1967. [CrossRef] [PubMed]

127. Ojesina, A.I.; Lichtenstein, L.; Freeman, S.S.; Pedamallu, C.S.; Imaz-Rosshandler, I.; Pugh, T.J.; Cherniack, A.D.; Ambrogio, L.; Cibulskis, K.; Bertelsen, B.; et al. Landscape of genomic alterations in cervical carcinomas. *Nature* **2014**, *506*, 371–375. [CrossRef]

128. Chen, L.; Wang, X.; Song, L.; Yao, D.; Tang, Q.; Zhou, J. Upregulation of lncRNA GATA6-AS suppresses the migration and invasion of cervical squamous cell carcinoma by downregulating MTK-1. *Oncol. Lett.* **2019**, *18*, 2605–2611. [CrossRef]

129. Haque, S.-U.; Niu, L.; Kuhnell, D.; Hendershot, J.; Biesiada, J.; Niu, W.; Hagan, M.C.; Kelsey, K.T.; Casper, K.A.; Wise-Draper, T.M.; et al. Differential expression and prognostic value of long non-coding RNA in HPV-negative head and neck squamous cell carcinoma. *Head Neck* **2018**, *40*, 1555–1564. [CrossRef]

130. Yuan, H.; Qin, Y.; Zeng, B.; Feng, Y.; Li, Y.; Xiang, T.; Ren, G. Long noncoding RNA LINC01089 predicts clinical prognosis and inhibits cell proliferation and invasion through the Wnt/beta-catenin signaling pathway in breast cancer. *OncoTargets Ther.* **2019**, *12*, 4883–4895. [CrossRef]

131. Shin, C.H.; Ryu, S.; Kim, H.H. hnRNPK-regulated PTOV1-AS1 modulates heme oxygenase-1 expression via miR-1207-5p. *BMB Rep.* **2017**, *50*, 220–225. [CrossRef] [PubMed]

132. Cui, J.; Wen, Q.; Tan, X.; Piao, J.; Zhang, Q.; Wang, Q.; He, L.; Wang, Y.; Chen, Z.; Liu, G. An integrated nomogram combining lncRNAs classifier and clinicopathologic factors to predict the recurrence of head and neck squamous cell carcinoma. *Sci. Rep.* **2019**, *9*, 17460. [CrossRef] [PubMed]

133. Sharma, S.; Munger, K. Expression of the cervical carcinoma expressed PCNA regulatory (CCEPR) long noncoding RNA is driven by the human papillomavirus E6 protein and modulates cell proliferation independent of PCNA. *Virology* **2018**, *518*, 8–13. [CrossRef] [PubMed]

134. Yang, M.; Zhai, X.; Xia, B.; Wang, Y.; Lou, G. Long noncoding RNA CCHE1 promotes cervical cancer cell proliferation via upregulating PCNA. *Tumour Biol.* **2015**, *36*, 7615–7622. [CrossRef]

135. Chen, Y.; Wang, C.X.; Sun, X.X.; Wang, C.; Liu, T.F.; Wang, D.J. Long non-coding RNA CCHE1 overexpression predicts a poor prognosis for cervical cancer. *Eur. Rev. Med. Pharmacol. Sci.* **2017**, *21*, 479–483.

136. Barr, J.A.; Hayes, K.E.; Brownmiller, T.; Harold, A.D.; Jagannathan, R.; Lockman, P.R.; Khan, S.; Martinez, I. Long non-coding RNA FAM83H-AS1 is regulated by human papillomavirus 16 E6 independently of p53 in cervical cancer cells. *Sci. Rep.* **2019**, *9*, 3662. [CrossRef]

137. White, E.A.; Kramer, R.E.; Tan, M.J.A.; Hayes, S.D.; Harper, J.W.; Howley, P.M. Comprehensive Analysis of Host Cellular Interactions with Human Papillomavirus E6 Proteins Identifies New E6 Binding Partners and Reflects Viral Diversity. *J. Virol.* **2012**, *86*, 13174–13186. [CrossRef]

138. Ou, L.; Wang, D.; Zhang, H.; Yu, Q.; Hua, F. Decreased Expression of miR-138-5p by lncRNA H19 in Cervical Cancer Promotes Tumor Proliferation. *Oncol. Res. Featur. Preclin. Clin. Cancer Ther.* **2018**, *26*, 401–410. [CrossRef]

139. Cao, S.; Liu, W.; Li, F.; Zhao, W.; Qin, C. Decreased expression of lncRNA GAS5 predicts a poor prognosis in cervical cancer. *Int. J. Clin. Exp. Pathol.* **2014**, *7*, 6776–6783.

140. Li, Y.; Wan, Y.P.; Bai, Y. Correlation between long strand non-coding RNA GASS expression and prognosis of cervical cancer patients. *Eur. Rev. Med. Pharmacol. Sci.* **2018**, *22*, 943–949. [CrossRef]

141. Wen, Q.; Liu, Y.; Lyu, H.; Xu, X.; Wu, Q.; Liu, N.; Yin, Q.; Li, J.; Sheng, X. Long Noncoding RNA GAS5, Which Acts as a Tumor Suppressor via microRNA 21, Regulates Cisplatin Resistance Expression in Cervical Cancer. *Int. J. Gynecol. Cancer* **2017**, *27*, 1096–1108. [CrossRef] [PubMed]

142. Gao, J.; Liu, L.; Li, G.; Cai, M.; Tan, C.; Han, X.; Han, L. LncRNA GAS5 confers the radio sensitivity of cervical cancer cells via regulating miR-106b/IER3 axis. *Int. J. Biol. Macromol.* **2019**, *126*, 994–1001. [CrossRef] [PubMed]

143. Wang, X.; Zhang, J.; Wang, Y. Long noncoding RNA GAS5-AS1 suppresses growth and metastasis of cervical cancer by increasing GAS5 stability. *Am. J. Transl. Res.* **2019**, *11*, 4909–4921. [PubMed]

144. Sharma, S.; Mandal, P.; Sadhukhan, T.; Roy Chowdhury, R.; Ranjan Mondal, N.; Chakravarty, B.; Chatterjee, T.; Roy, S.; Sengupta, S. Bridging Links between Long Noncoding RNA HOTAIR and HPV Oncoprotein E7 in Cervical Cancer Pathogenesis. *Sci. Rep.* **2015**, *5*, 11724. [CrossRef]

145. Kim, H.J.; Lee, D.W.; Yim, G.W.; Nam, E.J.; Kim, S.; Kim, S.W.; Kim, Y.T. Long non-coding RNA HOTAIR is associated with human cervical cancer progression. *Int. J. Oncol.* **2015**, *46*, 521–530. [CrossRef]

146. Bhan, A.; Mandal, S.S. LncRNA HOTAIR: A master regulator of chromatin dynamics and cancer. *Biochim. Biophys. Acta* **2015**, *1856*, 151–164. [CrossRef]

147. Sharma Saha, S.; Roy Chowdhury, R.; Mondal, N.R.; Chakravarty, B.; Chatterjee, T.; Roy, S.; Sengupta, S. Identification of genetic variation in the lncRNA HOTAIR associated with HPV16-related cervical cancer pathogenesis. *Cell. Oncol.* **2016**, *39*, 559–572. [CrossRef]

148. Ma, X.; Sheng, S.; Wu, J.; Jiang, Y.; Gao, X.; Cen, X.; Wu, J.; Wang, S.; Tang, Y.; Tang, Y.; et al. LncRNAs as an intermediate in HPV16 promoting myeloid-derived suppressor cell recruitment of head and neck squamous cell carcinoma. *Oncotarget* **2017**, *8*, 42061–42075. [CrossRef]

149. Iancu, I.V.; Anton, G.; Botezatu, A.; Huica, I.; Nastase, A.; Socolov, D.G.; Stanescu, A.D.; Dima, S.O.; Bacalbasa, N.; Plesa, A. LINC01101 and LINC00277 expression levels as novel factors in HPV-induced cervical neoplasia. *J. Cell. Mol. Med.* **2017**, *21*, 3787–3794. [CrossRef]

150. Guo, F.; Li, Y.; Liu, Y.; Wang, J.; Li, G. Inhibition of metastasis-associated lung adenocarcinoma transcript 1 in CaSki human cervical cancer cells suppresses cell proliferation and invasion. *Acta Biochim. Biophys. Sin. (Shanghai)* **2010**, *42*, 224–229. [CrossRef]

151. Jiang, Y.; Li, Y.; Fang, S.; Jiang, B.; Qin, C.; Xie, P.; Zhou, G.; Li, G. The role of MALAT1 correlates with HPV in cervical cancer. *Oncol. Lett.* **2014**, *7*, 2135–2141. [CrossRef] [PubMed]

152. Sun, R.; Qin, C.; Jiang, B.; Fang, S.; Pan, X.; Peng, L.; Liu, Z.; Li, W.; Li, Y.; Li, G. Down-regulation of MALAT1 inhibits cervical cancer cell invasion and metastasis by inhibition of epithelial-mesenchymal transition. *Mol. Biosyst.* **2016**, *12*, 952–962. [CrossRef] [PubMed]

153. Liu, S.; Song, L.; Zeng, S.; Zhang, L. MALAT1-miR-124-RBG2 axis is involved in growth and invasion of HR-HPV-positive cervical cancer cells. *Tumor Biol.* **2016**, *37*, 633–640. [CrossRef] [PubMed]

154. Lu, H.; He, Y.; Lin, L.; Qi, Z.; Ma, L.; Li, L.; Su, Y. Long non-coding RNA MALAT1 modulates radiosensitivity of HR-HPV+ cervical cancer via sponging miR-145. *Tumour. Biol.* **2016**, *37*, 1683–1691. [CrossRef]

155. He, H.; Liu, X.; Liu, Y.; Zhang, M.; Lai, Y.; Hao, Y.; Wang, Q.; Shi, D.; Wang, N.; Luo, X.-G.; et al. Human Papillomavirus E6/E7 and Long Noncoding RNA TMPOP2 Mutually Upregulated Gene Expression in Cervical Cancer Cells. *J. Virol.* **2019**, *93*, e01808-01818. [CrossRef] [PubMed]

156. Villegas, J.; Burzio, V.; Villota, C.; Landerer, E.; Martinez, R.; Santander, M.; Pinto, R.; Vera, M.I.; Boccardo, E.; Villa, L.L.; et al. Expression of a novel non-coding mitochondrial RNA in human proliferating cells. *Nucleic Acids Res.* **2007**, *35*, 7336–7347. [CrossRef]

157. Burzio, V.A.; Villota, C.; Villegas, J.; Landerer, E.; Boccardo, E.; Villa, L.L.; Martínez, R.; Lopez, C.; Gaete, F.; Toro, V.; et al. Expression of a family of noncoding mitochondrial RNAs distinguishes normal from cancer cells. *Proc. Natl. Acad. Sci. USA* **2009**, *106*, 9430–9434. [CrossRef]

158. Landerer, E.; Villegas, J.; Burzio, V.A.; Oliveira, L.; Villota, C.; Lopez, C.; Restovic, F.; Martinez, R.; Castillo, O.; Burzio, L.O. Nuclear localization of the mitochondrial ncRNAs in normal and cancer cells. *Cell. Oncol.* **2011**, *34*, 297–305. [CrossRef]

159. Villota, C.; Campos, A.; Vidaurre, S.; Oliveira-Cruz, L.; Boccardo, E.; Burzio, V.A.; Varas, M.; Villegas, J.; Villa, L.L.; Valenzuela, P.D.; et al. Expression of mitochondrial non-coding RNAs (ncRNAs) is modulated by high risk human papillomavirus (HPV) oncogenes. *J. Biol. Chem.* **2012**, *287*, 21303–21315. [CrossRef]

160. Zheng, S.-R.; Zhang, H.-R.; Zhang, Z.-F.; Lai, S.-Y.; Huang, L.-J.; Liu, J.; Bai, X.; Ding, K.; Zhou, J.-Y. Human papillomavirus 16 E7 oncoprotein alters the expression profiles of circular RNAs in Caski cells. *J. Cancer* **2018**, *9*, 3755–3764. [CrossRef]

161. Chamseddin, B.H.; Lee, E.E.; Kim, J.; Zhan, X.; Yang, R.; Murphy, K.M.; Lewis, C.; Hosler, G.A.; Hammer, S.T.; Wang, R.C. Assessment of circularized E7 RNA, GLUT1, and PD-L1 in anal squamous cell carcinoma. *Oncotarget* **2019**, *10*, 5958–5969. [CrossRef] [PubMed]

162. Bajan, S.; Hutvagner, G. RNA-Based Therapeutics: From Antisense Oligonucleotides to miRNAs. *Cells* **2020**, *9*, 137. [CrossRef] [PubMed]

163. Gutschner, T.; Hammerle, M.; Eissmann, M.; Hsu, J.; Kim, Y.; Hung, G.; Revenko, A.; Arun, G.; Stentrup, M.; Gross, M.; et al. The noncoding RNA MALAT1 is a critical regulator of the metastasis phenotype of lung cancer cells. *Cancer Res.* **2013**, *73*, 1180–1189. [CrossRef] [PubMed]

164. Arun, G.; Diermeier, S.; Akerman, M.; Chang, K.C.; Wilkinson, J.E.; Hearn, S.; Kim, Y.; MacLeod, A.R.; Krainer, A.R.; Norton, L.; et al. Differentiation of mammary tumors and reduction in metastasis upon Malat1 lncRNA loss. *Genes Dev.* **2016**, *30*, 34–51. [CrossRef] [PubMed]

165. Kim, S.-S.; Harford, J.B.; Moghe, M.; Rait, A.; Pirollo, K.F.; Chang, E.H. Targeted nanocomplex carrying siRNA against MALAT1 sensitizes glioblastoma to temozolomide. *Nucleic Acids Res.* **2018**, *46*, 1424–1440. [CrossRef]

166. Jost, I.; Shalamova, L.A.; Gerresheim, G.K.; Niepmann, M.; Bindereif, A.; Rossbach, O. Functional sequestration of microRNA-122 from Hepatitis C Virus by circular RNA sponges. *RNA Biol.* **2018**, *15*, 1032–1039. [CrossRef]

167. Derrien, T.; Johnson, R.; Bussotti, G.; Tanzer, A.; Djebali, S.; Tilgner, H.; Guernec, G.; Martin, D.; Merkel, A.; Knowles, D.G.; et al. The GENCODE v7 catalog of human long noncoding RNAs: Analysis of their gene structure, evolution, and expression. *Genome Res.* **2012**, *22*, 1775–1789. [CrossRef]

168. Yang, J.; Meng, X.; Pan, J.; Jiang, N.; Zhou, C.; Wu, Z.; Gong, Z. CRISPR/Cas9-mediated noncoding RNA editing in human cancers. *RNA Biol.* **2018**, *15*, 35–43. [CrossRef]

169. Connelly Colleen, M.; Moon Michelle, H.; Schneekloth John, S., Jr. The Emerging Role of RNA as a Therapeutic Target for Small Molecules. *Cell Chem. Biol.* **2016**, *23*, 1077–1090. [CrossRef]

170. Di Giorgio, A.; Duca, M. Synthetic small-molecule RNA ligands: Future prospects as therapeutic agents. *MedChemComm* **2019**, *10*, 1242–1255. [CrossRef]

171. Xia, C.; Liang, S.; He, Z.; Zhu, X.; Chen, R.; Chen, J. Metformin, a first-line drug for type 2 diabetes mellitus, disrupts the MALAT1/miR-142-3p sponge to decrease invasion and migration in cervical cancer cells. *Eur. J. Pharmacol.* **2018**, *830*, 59–67. [CrossRef] [PubMed]

172. Xu, Y.; Zhang, Q.; Lin, F.; Zhu, L.; Huang, F.; Zhao, L.; Ou, R. Casiopeina II-gly acts on lncRNA MALAT1 by miR-17-5p to inhibit FZD2 expression via the Wnt signaling pathway during the treatment of cervical carcinoma. *Oncol. Rep.* **2019**, *42*, 1365–1379. [CrossRef] [PubMed]

173. Hu, D.; Su, C.; Jiang, M.; Shen, Y.; Shi, A.; Zhao, F.; Chen, R.; Shen, Z.; Bao, J.; Tang, W. Fenofibrate inhibited pancreatic cancer cells proliferation via activation of p53 mediated by upregulation of LncRNA MEG3. *Biochem. Biophys. Res. Commun.* **2016**, *471*, 290–295. [CrossRef] [PubMed]

174. Wang, T.-H.; Chan, C.-W.; Fang, J.-Y.; Shih, Y.-M.; Liu, Y.-W.; Wang, T.-C.V.; Chen, C.-Y. 2-O-Methylmagnolol upregulates the long non-coding RNA, GAS5, and enhances apoptosis in skin cancer cells. *Cell Death Dis.* **2017**, *8*, e2638. [CrossRef] [PubMed]

175. Özeş, A.R.; Wang, Y.; Zong, X.; Fang, F.; Pilrose, J.; Nephew, K.P. Therapeutic targeting using tumor specific peptides inhibits long non-coding RNA HOTAIR activity in ovarian and breast cancer. *Sci. Rep.* **2017**, *7*, 894. [CrossRef]

176. Ma, T.; Wang, R.-P.; Zou, X. Dioscin inhibits gastric tumor growth through regulating the expression level of lncRNA HOTAIR. *BMC Complement. Altern. Med.* **2016**, *16*, 383. [CrossRef] [PubMed]

*Article*

# The Role of HPV and Non-HPV Sexually Transmitted Infections in Patients with Oropharyngeal Carcinoma: A Case Control Study

Barbara Kofler [1], Johannes Laimer [2], Emanuel Bruckmoser [3], Teresa B. Steinbichler [1], Annette Runge [1], Volker H. Schartinger [1], Dorothee von Laer [4] and Wegene Borena [4,*]

[1]  Department of Otorhinolaryngology, Medical University of Innsbruck, Anichstrasse 35, 6020 Innsbruck, Austria; ba.kofler@tirol-kliniken.at (B.K.); teresa.steinbichler@i-med.ac.at (T.B.S.); annette.runge@tirol-kliniken.at (A.R.); volker.schartinger@i-med.ac.at (V.H.S.)

[2]  University Hospital of Cranio-Maxillofacial and Oral Surgery, Medical University of Innsbruck, 6020 Innsbruck, Austria; johannes.laimer@i-med.ac.at

[3]  Independent Researcher, 5020 Salzburg, Austria; research@bruckmoser.info

[4]  Institute of Virology, Department of Hygiene, Microbiology, Social Medicine, Medical University of Innsbruck, Peter-Mayr-Strasse 4b, 6020 Innsbruck, Austria; dorothee.von-Laer@i-med.ac.at

*  Correspondence: wegene.borena@i-med.ac.at; Tel.: +43-512-9003-71737; Fax: +43-0512-9003-73701

Received: 18 April 2020; Accepted: 5 May 2020; Published: 8 May 2020

**Abstract:** *Background*: Certain high-risk (hr) types of human papillomavirus (HPV) can cause cervical cancer in women and penile cancer in men. Hr-HPV can also cause cancers of the oropharynx and anus in both sexes. In the anal and cervical region, a contribution of co-infections with Ureaplasma spp. on the persistence of the hr-HPV infection by a profound inflammatory state is suggested. Here, we investigated if non-HPV sexually transmitted infections are associated with oropharyngeal carcinoma (OPC). *Materials and Methods*: In this case-control study, a brush test directly from the tumor surface of OPC patients (study group) and from the oropharynx of healthy volunteers (control group), both groups matching in age and sex, was performed. HPV subtypes were detected using a commercially available test kit. For non-HPV sexually transmitted infections (Ureaplasma spp., Chlamydia trachomatis, Mycoplasma hominis, and Mycoplasma genitalium), a multiplex nucleic acid amplification approach was performed. *Results*: In the study group, 96 patients (23 female/73 male), with histologically confirmed OPC and in the control group 112 patients (19 female/93 male), were included. Oropharyngeal hr-HPV-positivity was detected in 68% (65/96 patients) of the study group and 1.8% (2/112 patients) of the control group ($p < 0.001$). In three patients in the study group, Ureaplasma spp. was detected, whereas no patient was Ureaplasma spp. positive in the control group ($p = 0.097$). Chlamydia trachomatis, Mycoplasma hominis, and Mycoplasma genitalium were negative in both groups. *Conclusion*: Based on the current study, the prevalence of oropharyngeal Ureaplasma spp. among patients with OPC is low and does not support a role in oropharyngeal cancer. However, the detection of the pathogen only among OPC patients but not in the healthy individuals might indicate a potential role and needs further elucidation.

**Keywords:** Ureaplasma spp.; sexually transmitted infections; human papillomavirus; oropharyngeal cancer; brush test

---

## 1. Introduction

Changes in sexual behavior in the last decades, like a high number of oral sex partners, seem to play an important etiological role in the development of human papillomavirus (HPV) positive oropharyngeal carcinoma (OPC) by HPV transmission [1]. Once an oropharyngeal HPV infection has occurred, the infection is usually asymptomatic and eliminated by the immune system within a few months. Only a small proportion of these infections persist and progress to dysplastic lesions or cancer, and only the persistence of the virus is oncogenic [2].

Therefore, it is of high importance to understand which factors and mechanisms may contribute to the persistence of an oropharyngeal HPV-infection and lead to the development of OPC.

Co-factors like smoking and immunosuppression were reported to play a meaningful role in the process of HPV-persistence [3,4]. Studies in the past reported that a co-existence of HPV with sexually transmitted infections (STIs) facilitates the persistence and dysplastic transformations of HPV-associated lesions in the cervix or the anus [5–14].

Mycoplasma and Ureaplasma, a distinguished form of bacteria characterized by their minute size and total lack of a cell wall, were reported to play an important role in HPV infections, abnormal cervical cytopathology, and cervical cancer. Ureaplasma urealyticum and Mycoplasma genitalium may increase the risk of a high-risk (hr) HPV infection, while Ureaplasma urealyticum, Ureaplasma parvum, and Mycoplasma hominis may increase the risk of abnormal cervical cytopathology [15].

The bacteria are widespread in nature and well known as human parasites, often causing chronic asymptomatic infections. Mycoplasmas usually exhibit a rather strict host and tissue specificity, and the primary habitats are the mucous surfaces of the respiratory and urogenital tracts, the eyes, alimentary canal, mammary glands, and joints [16]. Because mycoplasmas possess the smallest genome known for free-living organisms, the autonomy of the bacteria is limited and makes them susceptible to changes in the host organism. Many mycoplasmas themselves cause pathological changes in the host organism, often complicated by immune disorders. Furthermore, mycoplasmas can inhibit the p53-mediated checkpoint control of the cell cycle and apoptosis. This indicates that mycoplasmas might act as a cancer-promoting factors [17]. In vitro studies described the potential of a Mycoplasma species to a malignant transformation and chromosomal instability in long term Mycoplasma infected cell cultures [18–20]. Moreover, in epidemiological studies, the detection of Mycoplasma strains in cancer samples or antibodies against these microorganisms in cancer patients has been documented [21]. Idhal and coworkers evaluated the presence of anti-M genitalium antibody in 291 women with ovarian cancer. Their results were suggestive of a potential association between M genitalium and ovarian cancer ($p = 0.01$) [22]. Barykova and coworkers described M. hominis to be three times more frequent in patients with prostate cancer than in those with benign prostatic hyperplasia and suggested that M. hominis infections may be involved in prostate cancer development [23]. Mizuki and coworkers described a correlation between oral leukoplakia and M. salivarium using an immunohistochemical analysis. The authors observed small granular fluorescing structures in the cytoplasm of oral leukoplakia cells, which they identified, based on its morphology and size, as Mycoplasma species [24]. The substantial increase in the presence of Mycoplasma within the cytoplasm of oral leukoplakia as compared to the control group with normal oral mucosa cells further strengthened the potential role of such bacterial pathogen in the development of malignant lesions [25]. Other common sexually transmitted pathogens like C. trachomatis are considered to be possible co-factors facilitating HPV associated oncogenesis. An association between a C. trachomatis infection and cervical cancer or its precursor lesions has been described in several previous studies. Potential mechanisms include alteration of the epithelial tissue due to local inflammatory response making the region susceptible to HPV infection [26–28].

Colleagues of our study group collected recent anal brush samples in 222 HIV-positive men who have sex with men (MSM) for the detection of Chlamydia trachomatis, Neisseria gonorrhea, Ureaplasma spp., Mycoplasma, and HPV genotypes. Out of these participants, 73% were hr HPV-positive, and 19.4% harbored Ureaplasma spp. Hr HPV-infection was significantly associated with the co-presence of Ureaplasma spp. (OR 2.59, 95% CI: 1.03–6.54) [14].

Given the similar routes of transmission, the co-presence of HPV with other STIs is not a surprise even in the oropharyngeal region. Although previous studies reported on the frequency of Mycoplasma and Ureaplasma spp. detection in the oral cavity [29], data on HPV and non-HPV sexually transmitted infections (STI) are inexistent among patients with OPC.

The aim of the study was to evaluate the prevalence of HPV and non-HPV sexually transmitted infections among patients with a pathologically confirmed malignant tumor of the oropharynx. To further validate the results, outcomes were compared to a control group of volunteers without head and neck squamous cell carcinoma (HNSCC). The null hypothesis was that there is no difference in the prevalence of non-HPV sexually transmitted infections among patients with and without OPC.

## 2. Results

### 2.1. Study Population

A total of 208 adult men and women ($n_{OPC}$ = 96 and $n_{healthy}$ = 112) were included. The mean age (SD) of the study participants was 61.9 (10.1) years. There was no significant difference in the age distribution between OPC patients and healthy controls, 61.9 (9.7) versus 62.1 (10.5) years, respectively. In the study group, the proportion of female patients (23.9%) was slightly higher than in the control group (17%) (Table 1). Further clincal-pathological parameters of the study group revealed that 68% of the OPC patients were HPV-positive, radiochemotherapy was the most common treatment modality, and most patients were classified as UICC stage IV. HPV-positive OPC patients were obviously younger than the control group ($p$ = 0.006), had lower T-stage ($p < 0.001$), and lower ASA scores ($p < 0.001$). P16 was in 88.5% of the HPV-associated OPC patients positive ($p < 0.001$) (Table 2).

**Table 1.** Study population.

| Variables | Study Group $n = 96$ | Control Group $n = 112$ | $p$-Value |
|---|---|---|---|
| **Diagnosis** | OPC | Healthy volunteers | |
| **Mean age at diagnosis** | 61.9 years | 62.1 years | |
| **Sex** | 23 female (23.9%) | 19 female (17%) | |
| | 73 male (76.1%) | 93 male (83%) | |
| **HPV-positivity** | 65 patients (68%) | 2 patients (1.8%) | $p < 0.001$ |
| **Ureaplasma spp.** | 3 patients (3.1%) | 0 patients | $p = 0.097$ |

OPC, oropharyngeal cancer; HPV, human papillomavirus

**Table 2.** Clinical characteristics of the OPC patients.

| Variables | HPV-Positive OSC ($n = 65$) | HPV-Negative OSC ($n = 31$) | $p$-Value |
|---|---|---|---|
| **Sex** | | | |
| Male | 48 (73.8%) | 26 (83.9%) | $p = 0.204$ |
| Female | 17 (26.2%) | 5 (16.1%) | |
| **Age** | | | |
| ≤65 years | 48 | 14 | $p = 0.006$ |
| >65 years | 17 | 17 | |
| **Clinical T-stage** | | | |
| cT1/T2 | 48 | 7 | $p < 0.001$ |
| cT3/4 | 17 | 24 | |
| **UICCC** | | | |
| Stage I | 0 | 2 (6.5%) | |
| Stage II | 3 (4.6%) | 0 | $p = 0.143$ |
| Stage III | 17 (26.2%) | 4 (12.9%) | |
| Stage IV | 45 (69.2%) | 25 (80.6%) | |

**Table 2.** *Cont.*

| Variables | HPV-Positive OSC ($n = 65$) | HPV-Negative OSC ($n = 31$) | $p$-Value |
|---|---|---|---|
| **Therapy** | | | |
| Surgery only | 4 (6.3%) | 5 (17.9%) | |
| Surgery and PORT | 13 (20.6%) | 6 (21.4%) | |
| Surgery and RCT/RIT | 6 (9.5%) | 3 (10.7%) | $p = 0.183$ |
| Primary RCT/RIT | 38 (60.3%) | 11 (39.3%) | |
| Primary RT | 2 (3.2%) | 3 (10.7%) | |
| **p16** | | | |
| Positive | 54 (88.5%) | 3 (12%) | |
| Negative | 7 (11.5%) | 22 (88%) | $p < 0.001$ |
| **ASA score** | | | |
| ASA I/II | 55 (87.3%) | 13 (50%) | |
| ASA III/IV | 8 (12.7%) | 13 (50%) | $p < 0.001$ |

OPC, oropharyngeal cancer; *UICC*, Union for International Cancer Control; *PORT*, postoperative radiation; RCT, radiochemotherapy; RIT, radioimmunotherapy; RT, radiotherapy; *ASA*, American Society of Anesthesiologist.

## 2.2. HPV-DNA Detection

A total of 65 (67%) of the OPC patients were positive for HPV DNA. The most common genotype was HPV 16 ($n_{OPC}$ = 52), followed by HPV 18 ($n_{OPC}$ =4), HPV 33 ($n_{OPC}$ =3) and HPV 35 ($n_{OPC}$ =3). Moreover, the HPV types HPV 31, HPV 40, HPV 56, HPV 58, HPV 61, HPV 62, HPV 66, HPV 70, HPV 73 and HPV 82 were detected as a single infection or additionally to HPV 16 as a multiple infection (Table 3). In the control group, two individuals (1.8%) harbored low-risk HPV DNA in the oropharynx. The detected HPV types were HPV 70 and HPV 64 (Table 3).

**Table 3.** HPV subtypes in OPC patients.

| HPV Subtypes | Number and Percent of HPV+ OPC Patients |
|---|---|
| HPV 16 * | 52 patients (80%) |
| HPV 18 * | 4 patients (6.2%) |
| HPV 33 * | 3 patients (4.6%) |
| HPV 35 * | 3 patients (4.6%) |
| HPV 31 *, HPV 40 **, HPV 56 *, HPV 58 *, HPV 61 **, HPV 62 *, HPV 66 *, HPV 70 **, HPV 73 *, HPV 82 * | each in 1 patient or additionally to HPV 16 as multiple infection |

HPV, Human Papillomavirus; OPC, oropharyngeal cancer; * high-risk (hr) and probably hr-HPV infection; ** low-risk HPV infection

## 2.3. Detection of Non-HPV Sexually Transmitted Infections

Three OPC patients (3.1%) were tested positive for Ureaplasma spp., which was detected in two patients as a co-infection with HPV 16, and in one patient with an HPV-negative OPC. All other non-HPV sexually transmitted infections (Chlamydia trachomatis, Mycoplasma hominis, and Mycoplasma genitalium) were negative in the study group. In the control group, all investigated non-HPV sexually transmitted infections were tested negative. The difference between the frequency of the Ureaplasma spp. infections in the study group (three patients) versus the control group (zero patients) did not reach statistical significance ($p = 0.097$) (Table 1).

## 3. Discussion

To the best of our knowledge, this is the first study to examine the presence of HPV in combination with non-HPV STIs in oropharyngeal samples of patients with histologically confirmed OPCs.

### 3.1. HPV and Oropharyngeal Carcinoma

In our study, we found that the prevalence of HPV-positivity was high (68%) in the 96 OPC patients. HPV-positive OPC patients were younger and had a lower ASA score, which indicates a lower rate of comorbidities [30] as compared to HPV-negative OPC patients. This is in line with other studies, which described how HPV positive patients tend to be young and otherwise healthy [31–33]. The most common genotype detected in our study group was HPV 16 (52 patients, 80%), followed by HPV 18 (four patients, 6.2%), HPV 33 (three patients, 4.6%) and HPV 35 (three patients, 4.6%). This is in line with previous investigations. Moreover, Fossum and coworkers reported on 166 OPC patients HPV 16 to be the predominating genotype in 65% of the patients, followed by HPV 33 (17%), HPV 18 (2%), and HPV 31/35/56/59 in one patient each [34].

In the control group, only two patients (1.8%) were HPV-positive. The lack of hr HPV among the control group in our study may be of high diagnostic value. Although hr HPV negativity may not exclude the presence of OPC, based on our study, older adults with hr HPV in the oropharynx may be suspicious of having a head and neck malignancy. These patients may probably be referred to the department of otorhinolaryngology for clinical examination. There is currently no evidence-based implementation of an OPC screening [35]. Additional studies are necessary in order to establish an OPC risk algorithm and to implement this approach in the associated guidelines.

### 3.2. Non-HPV STI and Oropharyngeal Carcinoma

The use of a multiplex nucleic acid amplification approach enabled the search for the co-existence of bacterial infections, namely, Chlamydia trachomatis, Ureaplasma spp., Mycoplasma hominis, and Mycoplasma genitalium, out of a single brush sample. We used a highly sensitive and well-validated extraction and amplification assays, which make the finding in this study a robust one. Our data show that Ureaplasma spp. was detected in two HPV-positive OPC patients and in one HPV-negative OPC patient. Unfortunately, this low prevalence rate of Ureaplasma spp. makes it impossible to assess the potential role of non-HPV STI in HPV associated malignancy in the oropharynx. With only three OPC patients in the study group and none of the control group testing positive for Ureaplasma spp., as well as none testing positive for the other bacterial pathogens, the prevalence of non-HPV STIs was low in our study ($p = 0.097$).

There is substantial molecular evidence that a non-HPV STI co-infection might be associated with cervical or anal carcinogenesis through the induction of a profound inflammatory state [36–39]. Drago and coworkers detected a cervical HPV infection in 31 (68%) and a Ureaplasma parvum (UP) infection in 21 (46%) out of 64 women. Eighteen patients positive for UP were co-infected with HPV (86%), of which only two patients (11%) had a normal cytology, whereas 16 (89%) had an abnormal cytology, showing a cervical intraepithelial neoplasia. The authors suggested that UP may be involved in HPV persistence and promotes the development of cytological abnormalities [40]. Moreover, Biernat-Sudolska and coworkers described in 387 women an association between the presence of urogenital mycoplasmas and HPV infections in cervical smears of women diagnosed with abnormal cervical cytology. Their statistical analysis demonstrated that the risk of an HPV co-infection increased two-fold with a concomitant Mycoplasma infection and 4.7-fold with a concomitant Ureaplasma urealyticum infection [12].

Given the increasing incidence of HPV-associated OPC [41], the observed Ureaplasma spp. positivity exclusively among OPC patients in our study may be a non-negligible finding, particularly considering the fact that previous studies had found an association between Ureaplasma positivity and anogenital dysplasias [14].

There is very scarce literature on non-HPV STIs in the oropharynx. Naksahima and coworkers investigated the prevalence of STIs in oral gargles in 213 men without OPC attending the sexually transmitted disease (STD) clinic. The authors detected HPV in 18.8%, N. gonorrhoeae in 15.6%, C. trachomatis in 4.2%, M. genitalium in 5.2%, and Ureaplasma spp. in 16.0% of the oral samples [29]. In contrast, Deguchi and coworkers described a low prevalence of genital mycoplasmas and ureaplasmas in the pharynges of Japanese female sex workers. The prevalence in the pharynx of N. gonorrhoeae, C. trachomatis, M. genitalium, M. hominis, U. parvum, and U. urealyticum was 4.0%, 2.0%, 0%, 1.2%, 0.2%, and 0.7%, respectively [42]. Moreover, a Dutch study which aimed to assess spontaneous clearance of chlamydia among high-risk female patients and men who have sex with men found a relatively low prevalence of pharyngeal chlamydia [43].

Although a previous study went further to analyze the role of non-HPV STI in oral swab samples among patients with oral dysplastic lesions, similar data among OPC patients is nonexistent. Mosmann and coworkers examined benign oral lesions or potentially malignant disorders, including oral cancer patients. They found a prevalence of 34% for HPV, 16% for C. trachomatis, and 3% for HSV. They failed to test for Ureaplasma and Mycoplasma spp. [44]. By going further into examining OPC patients, we believe that our study contributes to the pool of data known to date. The search for not only HPV but also for several other non-HPV STIs among these patients makes this prevalence study a useful addition to the current understanding.

Epidemiological surveys confirm that oral sexual practice significantly declines with increasing age [45,46]. Since the median age of our study population was about 60 years, and clearly older than study participants from previous studies, the low prevalence of non-HPV STIs in the oropharyngeal region in this study may be partly explained by this epidemiological factor. The fact that our study participants come from the department of otorhinolaryngology, head and neck surgery, and not from an STD clinic might be an important factor playing a role regarding the low prevalence of oropharyngeal STIs. Furthermore, we used a pharyngeal brush with proven high sensitivity and specificity for HPV detection on OPC patients [47] and not an oral gargle solution. With this test method, we were able to collect samples precisely from the tumor surface of the oropharyngeal cancer. In contrast to the gargle solution, with the tumor surface brushing approach, there is no further contact with other regions of the oropharynx or the oral cavity. A further strength of the study is the fact that the oropharyngeal brush was exclusively performed by otorhinolaryngologists in the study group and by maxillofacial surgeons in the control group.

Prospective studies need to elaborate on the clinical and epidemiological significance of this colonization. If other studies confirm a differential predominance of Ureaplasma spp. among patients with OPC, the issue of early detection and appropriate antibiotic therapy or eradication might be a topic. Due to the low prevalence, there was no clear association between Ureaplasma spp. with HPV-positivity among these patients in our study.

## 4. Materials and Methods

In this prospective case-control study, consecutive patients scheduled for oropharyngeal tumor diagnosis at the outpatient unit of the Department of Otorhinolaryngology, Medical University Innsbruck, Austria, were invited to participate. The admission period was between May 2014 and April 2019. The control group of healthy volunteers was recruited from patients presenting for dentoalveolar procedures or control examinations at the outpatient clinic of Oral and Maxillofacial Surgery, Medical University Innsbruck, Austria, between October and December 2019.

This project was conducted in accordance with STROBE guidelines for observational studies [48]. Written informed consent was obtained from each patient who agreed to participate in this study following a detailed explanation of the procedural workflow. Prior to any patient enrolment, the study had been approved by the institutional board in charge, i.e., the ethics committee of the Medical University Innsbruck, Austria. The respective reference number was 1147/2018. The study was conducted in full accordance with the principles expressed in the declaration of Helsinki.

Inclusion criteria for participation in the study group comprised of a minimum age of at least 18 years and a histologically confirmed OPC. For the control group, healthy volunteers with an age of over 45 years and without previous or current HNSCC were included.

### 4.1. Sample Collection

For sample collection, a validated brush test was used as described in a previous publication [47]. For OPC patients, tumor surface lesions were brushed several times using a cytology brush (digene® HC2 DNA Collection Device, Qiagen, Hilden, Germany) and rinsed in digene Specimen Transport Medium. Particular attention was paid to perform the brush precisely from the tumor surface without the brush coming into contact with the mucosa of the oral cavity. Oropharyngeal brush samples from control patients were collected by the same method.

### 4.2. Nucleic Acid Extraction

Nucleic acid extraction was conducted in a fully automated manner (NucliSens® easyMAG®, Biomerieux, Marcy-l'Étoile, France) according to the manufacturer's instructions. In brief, out of 500 µL of the original sample, 110 µL of purified nucleic acid was extracted using magnetic silica particles. This method is a highly sensitive approach since it allows the capture of all nucleic acid available in a sample, and the separation from silica leads to no loss of extracted material.

### 4.3. HPV DNA Detection and Genotyping

After real-time amplification of the HPV-L1 genome using a one single-step PCR (5 µL of purified nucleic acid), genotyping followed using allele-specific reverse line-blot hybridization of the PCR products. These biotinylated PCR products then bind with target-specific probes that are bound to a nylon membrane (strip), permitting the differentiation of 40 high-risk (hr-HPV) and low-risk HPV genotypes. HPV genotypes detectable in this assay include types 6, 11, 16, 18, 26, 31, 33, 35, 39, 40, 42, 43, 44, 45, 51, 52, 53, 54, 55, 56, 58, 59, 61, 62, 64, 66, 67, 68a/b, 69, 70, 71, 72, 73, 81, 82, 83, 84, 87, 89, and 90. Beta-globin was used as an internal control assuring the validity of the test by controlling for sample quality, the performance of the extraction, and the validity of nucleic acid amplification. Moreover, the presence of the dUTP/UNG system within the reagents of the kit help prevent carry-over contaminations. (Ampliquality Type Express, AB ANALITICA, Padua, Italy).

### 4.4. Detection of Non-HPV STIs

Ten µl of DNA extract (NucliSens® easyMAG®, Biomerieux) was used for the amplification of Ureaplasma spp., Mycoplasma genitalium, Mycoplasma hominis, and Chlamydia trachomatis DNA using a validated multiplex nucleic acid amplification kit (AmpliSens® multiprime-FRT, Moscow, Russia) according to the thermocycling conditions recommended by the manufacturer. The kit is equipped with negative and positive controls, as well as an internal control system verifying the extraction and the amplification steps.

### 4.5. Data Analysis

Data were analyzed using SPSS® (IBM Corp. Released 2016. IBM SPSS Statistics for Windows, Version 24.0. Armonk, NY, USA). Frequency data were tabulated and analyzed with Fisher's exact test or with the Kruskal-Wallis test for ordered alternatives. No statistics were computed for Chlamydia trachomatis, Mycoplasma hominis, and Mycoplasma genitalium because these are constants in both groups [49]. For continuous data, means and standard deviations (SD) are provided unless stated otherwise [50]. A $p$-value of $< 0.05$ was considered statistically significant.

## 5. Conclusions

Our study shows that the prevalence of non-HPV sexually transmitted infections among patients with OPC is low. Although Ureaplasma spp. has been implicated as a potential HPV-associated carcinogen in anogenital cancers, the low frequency in this study (both in the HPV-positive and HPV-negative patients) does not support a role in oropharyngeal cancer. However, beyond the obvious predominance of hr-HPV, the detection of Ureaplasma spp. only among OPC patients but not in the age and sex-matched control group represents an interesting finding. Given the fact that Ureaplasma spp. were previously shown to be associated with hr-HPV infection or HPV-associated dysplasia, further evaluation of a potential link between a HPV and Ureaplasma spp. co-infection in the pathogenesis of oropharyngeal carcinoma should be evaluated in larger studies.

**Author Contributions:** Conceptualization, W.B., B.K., J.L.; Methodology, W.B., D.v.L.; Software, E.B.; Validation, B.K., J.L.; Formal Analysis, W.B.; Investigation, A.R., T.B.S., B.K., J.L.; Resources, W.B., D.v.L.; Data Curation, W.B., B.K.; Writing—Original Draft Preparation, B.K., J.L., E.B.; Writing—Review and Editing, W.B.; Visualization, V.H.S.; Supervision, W.B., D.v.L.; Project Administration, W.B., B.K., V.H.S. All authors have read and agreed to the published version of the manuscript.

**Funding:** This research received no external funding.

**Conflicts of Interest:** The final content of the manuscript has been seen and approved by all authors. They declare that the manuscript is original and has not been published previously nor is it under review by another journal. Furthermore, all authors declare that neither financial interests nor financial support by companies exist. The authors declare no conflict of interest.

## References

1. Schnelle, C.; Whiteman, D.C.; Porceddu, S.V.; Panizza, B.J.; Antonsson, A. Past sexual behaviors and risks of oropharyngeal squamous cell carcinoma: A case-case comparison. *Int. J. Cancer* **2017**, *140*, 1027–1034. [CrossRef] [PubMed]

2. Tam, S.; Fu, S.; Xu, L.; Krause, K.J.; Lairson, D.R.; Miao, H.; Sturgis, E.M.; Dahlstrom, K.R. The epidemiology of oral human papillomavirus infection in healthy populations: A systematic review and meta-analysis. *Oral Oncol.* **2018**, *82*, 91–99. [CrossRef] [PubMed]

3. Kero, K.; Rautava, J.; Syrjanen, K.; Willberg, J.; Grenman, S.; Syrjanen, S. Smoking increases oral HPV persistence among men: 7-year follow-up study. *Eur. J. Clin. Microbiol. Infect. Dis.* **2014**, *33*, 123–133. [CrossRef] [PubMed]

4. Castillejos-Garcia, I.; Ramirez-Amador, V.A.; Carrillo-Garcia, A.; Garcia-Carranca, A.; Lizano, M.; Anaya-Saavedra, G. Type-specific persistence and clearance rates of HPV genotypes in the oral and oropharyngeal mucosa in an HIV/AIDS cohort. *J. Oral Pathol. Med.* **2018**, *47*, 396–402. [CrossRef] [PubMed]

5. Bellaminutti, S.; Seraceni, S.; De Seta, F.; Gheit, T.; Tommasino, M.; Comar, M. HPV and Chlamydia trachomatis co-detection in young asymptomatic women from high incidence area for cervical cancer. *J. Med. Virol.* **2014**, *86*, 1920–1925. [CrossRef]

6. Camporiondo, M.P.; Farchi, F.; Ciccozzi, M.; Denaro, A.; Gallone, D.; Maracchioni, F.; Favalli, C.; Ciotti, M. Detection of HPV and co-infecting pathogens in healthy Italian women by multiplex real-time PCR. *Infez. Med.* **2016**, *24*, 12–17.

7. Liu, H.F.; Liu, M.; Xu, Y.L. Analysis of high-risk HPV infection and cervical cytologic screening in HIV positive women. *Zhonghua Fu Chan Ke Za Zhi* **2016**, *51*, 734–738. [CrossRef]

8. Verteramo, R.; Pierangeli, A.; Mancini, E.; Calzolari, E.; Bucci, M.; Osborn, J.; Nicosia, R.; Chiarini, F.; Antonelli, G.; Degener, A.M. Human Papillomaviruses and genital co-infections in gynaecological outpatients. *BMC Infect. Dis.* **2009**, *9*, 16. [CrossRef]

9. Kim, H.S.; Kim, T.J.; Lee, I.H.; Hong, S.R. Associations between sexually transmitted infections, high-risk human papillomavirus infection, and abnormal cervical Pap smear results in OB/GYN outpatients. *J. Gynecol. Oncol.* **2016**, *27*, e49. [CrossRef]

10. Parthenis, C.; Panagopoulos, P.; Margari, N.; Kottaridi, C.; Spathis, A.; Pouliakis, A.; Konstantoudakis, S.; Chrelias, G.; Chrelias, C.; Papantoniou, N.; et al. The association between sexually transmitted infections, human papillomavirus, and cervical cytology abnormalities among women in Greece. *Int. J. Infect. Dis.* **2018**, *73*, 72–77. [CrossRef]

11. Biernat-Sudolska, M.; Szostek, S.; Rojek-Zakrzewska, D.; Kopec, J.; Zawilinska, B. May ureaplasmas in genital tract of HPV-positive women influence abnormal cytology of cervix? *Prz. Epidemiol.* **2008**, *62*, 447–452.

12. Biernat-Sudolska, M.; Szostek, S.; Rojek-Zakrzewska, D.; Klimek, M.; Kosz-Vnenchak, M. Concomitant infections with human papillomavirus and various mycoplasma and ureaplasma species in women with abnormal cervical cytology. *Adv. Med. Sci.* **2011**, *56*, 299–303. [CrossRef] [PubMed]

13. Lukic, A.; Canzio, C.; Patella, A.; Giovagnoli, M.; Cipriani, P.; Frega, A.; Moscarini, M. Determination of cervicovaginal microorganisms in women with abnormal cervical cytology: The role of Ureaplasma urealyticum. *Anticancer Res.* **2006**, *26*, 4843–4849. [PubMed]

14. Borena, W.; Kruis, S.; Kitchen, M.; Taylor, N.; Gisinger, M.; Oberkofler, H.; Stoiber, H.; Zangerle, R.; von Laer, D.; Sarcletti, M. Anal Ureaplasma spp. positivity among HIV positive men who have sex with men may be associated with high-risk-type HPV infections. *Int. J. Infect. Dis.* **2019**, *84*, 75–79. [CrossRef]

15. Ye, H.; Song, T.; Zeng, X.; Li, L.; Hou, M.; Xi, M. Association between genital mycoplasmas infection and human papillomavirus infection, abnormal cervical cytopathology, and cervical cancer: A systematic review and meta-analysis. *Arch. Gynecol. Obstet.* **2018**, *297*, 1377–1387. [CrossRef]

16. Razin, S.; Yogev, D.; Naot, Y. Molecular biology and pathogenicity of mycoplasmas. *Microbiol. Mol. Biol. Rev.* **1998**, *62*, 1094–1156. [CrossRef]

17. Borchsenius, S.N.; Daks, A.; Fedorova, O.; Chernova, O.; Barlev, N.A. Effects of mycoplasma infection on the host organism response via p53/NF-kappaB signaling. *J. Cell. Physiol.* **2018**, *234*, 171–180. [CrossRef]

18. Zhang, S.; Tsai, S.; Lo, S.C. Alteration of gene expression profiles during mycoplasma-induced malignant cell transformation. *BMC Cancer* **2006**, *6*, 116. [CrossRef]

19. Zhang, S.; Wear, D.J.; Lo, S. Mycoplasmal infections alter gene expression in cultured human prostatic and cervical epithelial cells. *FEMS Immunol. Med. Microbiol.* **2000**, *27*, 43–50. [CrossRef]

20. Zhang, S.; Tsai, S.; Wu, T.T.; Li, B.; Shih, J.W.; Lo, S.C. Mycoplasma fermentans infection promotes immortalization of human peripheral blood mononuclear cells in culture. *Blood* **2004**, *104*, 4252–4259. [CrossRef]

21. Erturhan, S.M.; Bayrak, O.; Pehlivan, S.; Ozgul, H.; Seckiner, I.; Sever, T.; Karakok, M. Can mycoplasma contribute to formation of prostate cancer? *Int. Urol. Nephrol.* **2013**, *45*, 33–38. [CrossRef] [PubMed]

22. Idahl, A.; Lundin, E.; Jurstrand, M.; Kumlin, U.; Elgh, F.; Ohlson, N.; Ottander, U. Chlamydia trachomatis and Mycoplasma genitalium plasma antibodies in relation to epithelial ovarian tumors. *Infect. Dis. Obstet. Gynecol.* **2011**, *2011*, 824627. [CrossRef] [PubMed]

23. Barykova, Y.A.; Logunov, D.Y.; Shmarov, M.M.; Vinarov, A.Z.; Fiev, D.N.; Vinarova, N.A.; Rakovskaya, I.V.; Baker, P.S.; Shyshynova, I.; Stephenson, A.J.; et al. Association of Mycoplasma hominis infection with prostate cancer. *Oncotarget* **2011**, *2*, 289–297. [CrossRef] [PubMed]

24. Mizuki, H. In situ staining with DNA-binding fluorescent dye, Hoechst 33258, to detect microorganisms in the epithelial cells of oral leukoplakia. *Oral Oncol.* **2001**, *37*, 521–526. [CrossRef]

25. Mizuki, H.; Kawamura, T.; Nagasawa, D. In situ immunohistochemical detection of intracellular Mycoplasma salivarium in the epithelial cells of oral leukoplakia. *J. Oral Pathol. Med.* **2015**, *44*, 134–144. [CrossRef]

26. Smith, J.S.; Bosetti, C.; Munoz, N.; Herrero, R.; Bosch, F.X.; Eluf-Neto, J.; Meijer, C.J.; Van Den Brule, A.J.; Franceschi, S.; Peeling, R.W.; et al. Chlamydia trachomatis and invasive cervical cancer: A pooled analysis of the IARC multicentric case-control study. *Int. J. Cancer* **2004**, *111*, 431–439. [CrossRef]

27. Madeleine, M.M.; Anttila, T.; Schwartz, S.M.; Saikku, P.; Leinonen, M.; Carter, J.J.; Wurscher, M.; Johnson, L.G.; Galloway, D.A.; Daling, J.R. Risk of cervical cancer associated with Chlamydia trachomatis antibodies by histology, HPV type and HPV cofactors. *Int. J. Cancer* **2007**, *120*, 650–655. [CrossRef]

28. Vriend, H.J.; Bogaards, J.A.; van Bergen, J.E.; Brink, A.A.; van den Broek, I.V.; Hoebe, C.J.; King, A.J.; van der Sande, M.A.; Wolffs, P.F.; de Melker, H.E.; et al. Incidence and persistence of carcinogenic genital human papillomavirus infections in young women with or without Chlamydia trachomatis co-infection. *Cancer Med.* **2015**, *4*, 1589–1598. [CrossRef]

29. Nakashima, K.; Shigehara, K.; Kawaguchi, S.; Wakatsuki, A.; Kobori, Y.; Nakashima, K.; Ishii, Y.; Shimamura, M.; Sasagawa, T.; Kitagawa, Y.; et al. Prevalence of human papillomavirus infection in the oropharynx and urine among sexually active men: A comparative study of infection by papillomavirus and other organisms, including Neisseria gonorrhoeae, Chlamydia trachomatis, Mycoplasma spp., and Ureaplasma spp. *BMC Infect. Dis.* **2014**, *14*, 43. [CrossRef]

30. Riechelmann, H.; Neagos, A.; Netzer-Yilmaz, U.; Gronau, S.; Scheithauer, M.; Rockemann, M.G. The ASA-score as a comorbidity index in patients with cancer of the oral cavity and oropharynx. *Laryngo-Rhino-Otologie* **2006**, *85*, 99–104. [CrossRef]

31. Mahal, B.A.; Catalano, P.J.; Haddad, R.I.; Hanna, G.J.; Kass, J.I.; Schoenfeld, J.D.; Tishler, R.B.; Margalit, D.N. Incidence and Demographic Burden of HPV-Associated Oropharyngeal Head and Neck Cancers in the United States. *Cancer Epidemiol. Biomark. Prev.* **2019**, *28*, 1660–1667. [CrossRef] [PubMed]

32. Kofler, B.; Laban, S.; Busch, C.J.; Lorincz, B.; Knecht, R. New treatment strategies for HPV-positive head and neck cancer. *Eur. Arch. Oto-Rhino-Laryngol.* **2014**, *271*, 1861–1867. [CrossRef] [PubMed]

33. Grisar, K.; Dok, R.; Schoenaers, J.; Dormaar, T.; Hauben, E.; Jorissen, M.; Nuyts, S.; Politis, C. Differences in human papillomavirus-positive and -negative head and neck cancers in Belgium: An 8-year retrospective, comparative study. *Oral Surg. Oral Med. Oral Pathol. Oral Radiol.* **2016**, *121*, 456–460. [CrossRef] [PubMed]

34. Fossum, G.H.; Lie, A.K.; Jebsen, P.; Sandlie, L.E.; Mork, J. Human papillomavirus in oropharyngeal squamous cell carcinoma in South-Eastern Norway: Prevalence, genotype, and survival. *Eur. Arch. Oto-Rhino-Laryngol.* **2017**, *274*, 4003–4010. [CrossRef]

35. Kreimer, A.R.; Shiels, M.S.; Fakhry, C.; Johansson, M.; Pawlita, M.; Brennan, P.; Hildesheim, A.; Waterboer, T. Screening for human papillomavirus-driven oropharyngeal cancer: Considerations for feasibility and strategies for research. *Cancer* **2018**, *124*, 1859–1866. [CrossRef]

36. Anttila, T.; Saikku, P.; Koskela, P.; Bloigu, A.; Dillner, J.; Ikaheimo, I.; Jellum, E.; Lehtinen, M.; Lenner, P.; Hakulinen, T.; et al. Serotypes of Chlamydia trachomatis and risk for development of cervical squamous cell carcinoma. *JAMA* **2001**, *285*, 47–51. [CrossRef]

37. Castle, P.E.; Giuliano, A.R. Chapter 4: Genital tract infections, cervical inflammation, and antioxidant nutrients–assessing their roles as human papillomavirus cofactors. *J. Natl. Cancer Inst. Monogr.* **2003**, *2003*, 29–34. [CrossRef]

38. Rasmussen, S.J.; Eckmann, L.; Quayle, A.J.; Shen, L.; Zhang, Y.X.; Anderson, D.J.; Fierer, J.; Stephens, R.S.; Kagnoff, M.F. Secretion of proinflammatory cytokines by epithelial cells in response to Chlamydia infection suggests a central role for epithelial cells in chlamydial pathogenesis. *J. Clin. Investig.* **1997**, *99*, 77–87. [CrossRef]

39. Kulkarni, S.; Rader, J.S.; Zhang, F.; Liapis, H.; Koki, A.T.; Masferrer, J.L.; Subbaramaiah, K.; Dannenberg, A.J. Cyclooxygenase-2 is overexpressed in human cervical cancer. *Clin. Cancer Res.* **2001**, *7*, 429–434.

40. Drago, F.; Herzum, A.; Ciccarese, G.; Dezzana, M.; Casazza, S.; Pastorino, A.; Bandelloni, R.; Parodi, A. Ureaplasma parvum as a possible enhancer agent of HPV-induced cervical intraepithelial neoplasia: Preliminary results. *J. Med. Virol.* **2016**, *88*, 2023–2024. [CrossRef]

41. Wittekindt, C.; Wagner, S.; Bushnak, A.; Prigge, E.S.; von Knebel Doeberitz, M.; Wurdemann, N.; Bernhardt, K.; Pons-Kuhnemann, J.; Maulbecker-Armstrong, C.; Klussmann, J.P. Increasing Incidence rates of Oropharyngeal Squamous Cell Carcinoma in Germany and Significance of Disease Burden Attributed to Human Papillomavirus. *Cancer Prev. Res.* **2019**, *12*, 375–382. [CrossRef]

42. Deguchi, T.; Yasuda, M.; Yokoi, S.; Nakano, M.; Ito, S.; Ohkusu, K.; Ezaki, T.; Hoshina, S. Failure to detect Mycoplasma genitalium in the pharynges of female sex workers in Japan. *J. Infect. Chemother.* **2009**, *15*, 410–413. [CrossRef] [PubMed]

43. van Rooijen, M.S.; van der Loeff, M.F.; Morre, S.A.; van Dam, A.P.; Speksnijder, A.G.; de Vries, H.J. Spontaneous pharyngeal Chlamydia trachomatis RNA clearance. A cross-sectional study followed by a cohort study of untreated STI clinic patients in Amsterdam, The Netherlands. *Sex. Transm. Infect.* **2015**, *91*, 157–164. [CrossRef] [PubMed]

44. Mosmann, J.P.; Talavera, A.D.; Criscuolo, M.I.; Venezuela, R.F.; Kiguen, A.X.; Panico, R.; Ferreyra De Prato, R.; Lopez De Blanc, S.A.; Re, V.; Cuffini, C.G. Sexually transmitted infections in oral cavity lesions: Human papillomavirus, Chlamydia trachomatis, and Herpes simplex virus. *J. Oral Microbiol.* **2019**, *11*, 1632129. [CrossRef] [PubMed]

45. D'Souza, G.; Cullen, K.; Bowie, J.; Thorpe, R.; Fakhry, C. Differences in oral sexual behaviors by gender, age, and race explain observed differences in prevalence of oral human papillomavirus infection. *PLoS ONE* **2014**, *9*, e86023. [CrossRef]

46. Bajos, N.; Bozon, M.; Beltzer, N.; Laborde, C.; Andro, A.; Ferrand, M.; Goulet, V.; Laporte, A.; Le Van, C.; Leridon, H.; et al. Changes in sexual behaviours: From secular trends to public health policies. *AIDS* **2010**, *24*, 1185–1191. [CrossRef]

47. Kofler, B.; Borena, W.; Manzl, C.; Dudas, J.; Wegscheider, A.S.; Jansen-Durr, P.; Schartinger, V.; Riechelmann, H. Sensitivity of tumor surface brushings to detect human papilloma virus DNA in head and neck cancer. *Oral Oncol.* **2017**, *67*, 103–108. [CrossRef]

48. Ebrahim, S.; Clarke, M. STROBE: New standards for reporting observational epidemiology, a chance to improve. *Int. J. Epidemiol.* **2007**, *36*, 946–948. [CrossRef]

49. Mehta, C.R. The exact analysis of contingency tables in medical research. *Stat. Methods Med. Res.* **1994**, *3*, 135–156. [CrossRef]
50. Schemper, M.; Smith, T.L. A note on quantifying follow-up in studies of failure time. *Control. Clin. Trials* **1996**, *17*, 343–346. [CrossRef]

 *cancers*

*Review*

# In Vitro Organotypic Systems to Model Tumor Microenvironment in Human Papillomavirus (HPV)-Related Cancers

Vincenza De Gregorio [1,2,*], Francesco Urciuolo [1], Paolo Antonio Netti [1,2,3] and Giorgia Imparato [2,*]

[1] Interdisciplinary Research Centre on Biomaterials (CRIB), University of Naples Federico II, 80125 Naples, Italy; urciuolo@unina.it (F.U.); nettipa@unina.it (P.A.N.)
[2] Center for Advanced Biomaterials for HealthCare@CRIB, Istituto Italiano di Tecnologia, 80125 Naples, Italy
[3] Department of Chemical, Materials and Industrial Production Engineering (DICMAPI), University of Naples Federico II, 80125 Naples, Italy
* Correspondence: vincenza.degregorio@unina.it (V.D.G.); giorgia.imparato@iit.it (G.I.); Tel.: +39-08-1199-33100 (V.D.G.); +39-08-1199-33100 (G.I.); Fax: +39-08-1199-3314 (V.D.G.); +39-08-1199-3314 (G.I.)

Received: 27 March 2020; Accepted: 1 May 2020; Published: 3 May 2020

**Abstract:** Despite the well-known role of chronic human papillomavirus (HPV) infections in causing tumors (i.e., all cervical cancers and other human malignancies from the mucosal squamous epithelia, including anogenital and oropharyngeal cavity), its persistence is not sufficient for cancer development. Other co-factors contribute to the carcinogenesis process. Recently, the critical role of the underlying stroma during the HPV life cycle and HPV-induced disease have been investigated. The tumor stroma is a key component of the tumor microenvironment (TME), which is a specialized entity. The TME is dynamic, interactive, and constantly changing—able to trigger, support, and drive tumor initiation, progression, and metastasis. In previous years, in vitro organotypic raft cultures and in vivo genetically engineered mouse models have provided researchers with important information on the interactions between HPVs and the epithelium. Further development for an in-depth understanding of the interaction between HPV-infected tissue and the surrounding microenvironment is strongly required. In this review, we critically describe the HPV-related cancers modeled in vitro from the simplified 'raft culture' to complex three-dimensional (3D) organotypic models, focusing on HPV-associated cervical cancer disease platforms. In addition, we review the latest knowledge in the field of in vitro culture systems of HPV-associated malignancies of other mucosal squamous epithelia (anogenital and oropharynx), as well as rare cutaneous non-melanoma associated cancer.

**Keywords:** human papillomaviruses (HPVs)-related cancers; tumor microenvironment (TME); 3D organotypic models; cervical cancers; anogenital cancers; oropharynx cancers

---

## 1. Introduction

Human papillomaviruses (HPVs) are double-stranded circular DNA epitheliotropic tumor viruses that are causally associated with all cervical cancers, as well as a significant fraction of several other human malignancies arising from the mucosal squamous epithelia of the anogenital tract (vaginal, vulvar, anal, penile), oral, and oropharyngeal cavity (mouth, throat, nasal sinuses, larynx, pharynx, salivary glands, and neck lymph nodes) and, infrequently, from the cutaneous epithelium (skin) [1–3]. According to 2018 data from the Global Cancer Observatory (GLOBOCAN), it is recognized that HPV contributes to more than 90% of cervical and anal cancers, approximately 78% of vaginal, and 25% of vulvar cancers, almost 53% of penile cancers, and 30% of head and neck cancers (HNCs), including oropharyngeal, oral cavity, and laryngeal cancers (30%, 2.1%, and 2.4%, respectively) (Figure 1) [4].

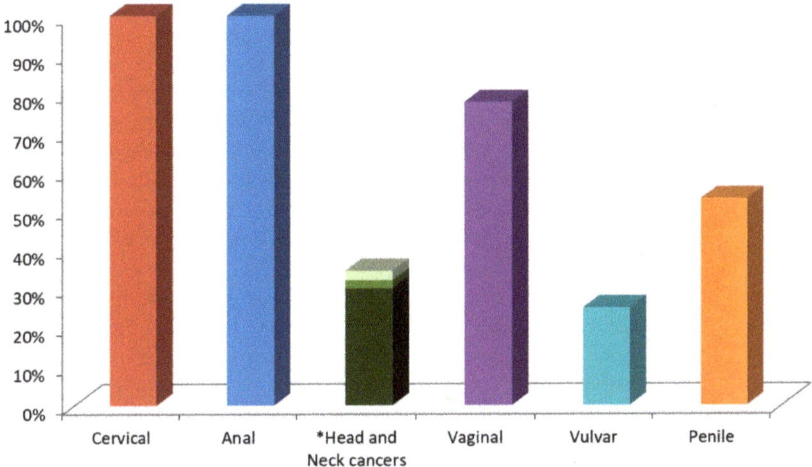

**Figure 1.** Representation of the estimated number percentage of human papillomavirus (HPV)-associated cancers vs. different cancer subsites. *Head and Neck cancers include oropharyngeal, oral cavity and laryngeal cancers.

Recent findings indicate that about 60%–70% of the oropharyngeal cancers may be linked to HPV, although traditionally they were considered to be caused by tobacco and alcohol, or by a combination of these [5,6]. Based on their oncogenic potential, HPVs are classified as low-risk (LR) or high-risk (HR) viruses [7]. LR-HPVs can cause benign genital warts or laryngeal papilloma, whereas HR-HPV types are considered causally associated with nearly all cases of cervical cancer, other cancers of the lower female reproductive system and anus, as well as a high number of oropharyngeal cancers. The HR types include HPV16, 18, 31, 33, 35, 39, 45, 51, 52, 56, 58, and 59. Others are considered as potential HR including HPV 53, 66, 70, 73, and 82. The most virulent HR-HPV genotypes (HPV16 and HPV18) are major contributors to cervical cancer (50% of cervical squamous cell carcinoma are HPV16-positive and 35% of cervical adenocarcinomas are positive for HPV16 and HPV18), with 30% being caused by other HR-HPV types. Of note, most cervical intraepithelial neoplasia (CIN) lesions are HPV-related (HPV6/11/16/18 contribute to 23%–25% of CIN1, 38.4–39% to CIN2, and 58% to CIN3) [8]; HPV-negative CIN has also been reported. HPV16 is most commonly involved in the other HPV-induced cancers (e.g., oropharyngeal cancers ~25%) [9–12]. Recurrent respiratory papillomatosis is highly associated with LR-HPV6/11. HPVs are also responsible for a significant proportion of precancerous lesions of the vulva (vulvar intraepithelial neoplasia grades 2 and 3, VIN2/3), vagina (vaginal intraepithelial neoplasia grades 2 and 3, VaIN2/3), anus (anal intraepithelial neoplasia grades 2 and 3, AIN2/3), penis (penile high-grade squamous intraepithelial lesions), head and neck, as well as genital warts [13].

Although HPV infection is usually solved by the immune system and the vast majority of the virus infections are transient and asymptomatic, persistent HPV infections have an increased chance to induce epithelial cell abnormalities that can ultimately cause cancer [14,15]. HR-HPVs infect a wide range of epithelial sites, but cause cancer at these sites at different frequencies [16]. In detail, HPV initial infection can occur on monolayer squamocolumnar cells, or on squamocolumnar junction cells at the transformation zone regions of the cervix and anus, as well as on the reticulated epithelium of the palatine tonsil. However, HPV infection may also arise upon micro-abrasions of multi-layered epithelium [16,17]. HPV affinity to junctional tissues is due to the fact that the basal cells of the squamocolumnar transformation zone are particularly accessible and are thought to be more receptive toward HPV-mediated transformation [18]. On the other hand, HPV life cycle is closely linked to the differentiation program of the pluristratified epithelial host cell [19,20]. HPV may access dividing basal epithelial cells by falling down micro-abrasions in the epithelium and attaching to cells using common cell surface molecules [21]. The initial HR-HPV type infection determines low grade disease (low-grade squamous intraepithelial lesions, LSIL), due to inhibition of the normal differentiation in the lower third of the epithelium. The lesion may remain low-grade, regress, or progress to severe dysplasia or high-grade squamous intraepithelial lesions (HSIL). This latter stage may persist for several years or may progress from premalignant disease (CIN2/3) to invasive cancer, which, in some cases, leads to metastatic [22]. The relationship between HPV and the host genome may change during progression from premalignant to malignant phases of the disease [23]. The HPV genome integration in the host cell genome often occurs in HSIL, but episomal DNA is found in some cancers [24]. As a consequence, investigating the involvement of the HPV infection in the development of cancer is clinically and scientifically relevant. The progression of HPV replication and the early viral gene expression (E6 and E7) requires a highly-regulated differentiation program of stratified epithelia. In particular, HPV E6 and E7 oncoproteins partially inhibit and/or delay epithelial differentiation in the host cells via a variety of mechanisms, some of which involve the inactivation of pro-proliferative targets, such as retinoblastoma protein (pRb) and p53, promoting the epithelial cell proliferation and evading the apoptosis process [25,26]. Moreover, upon HPV infection, the stratified epithelium starts communication with the underlying stroma. HPVs interact predominantly with extracellular matrix (ECM) components during keratinocytes infection through the link with membrane-associated heparan sulfate proteoglycans, determining the HPV-infected epithelial cells invasion across the stromal barrier [27–29]. More recently, important and emerging roles during the HPV life cycle and HPV-induced disease of the matricellular proteins constituting the underlying stroma, or tumor microenvironment (TME), have become clearer [30]. For several years, the role of the TME in carcinogenesis has been underestimated, considering the stroma as a merely supportive scaffold upon which epithelial cells adhere, but it is now recognized that the stromal-to-HPV-infected-epithelial communication events play a key role in carcinogenesis [31]. Furthermore, stromal fibroblasts, a major cellular component of the connective tissues, also provide important signals in the development and progression of cancer, and it would be interesting to study their role in regulating the HPV life cycle and their presence in the neighborhood of HPV-induced lesions [32]. Of note, there is evidence that cancer-associated fibroblasts (CAFs), fibroblasts activated by paracrine mediators produced by cancer cells, may facilitate HPV-mediated carcinogenesis through a variety of mechanisms involving stromal-to-epithelial crosstalk [33]. Furthermore, the bidirectional communication between epithelial cells and the TME has been reported to affect tumor initiation and neoplastic progression to metastasis and drug resistance [34]. Researchers assume that, in response to this communication, the microenvironment interchanges contact through various stromal-to-epithelial signaling events involved in HPV-positive epithelial cell growth, disease initiation, and maintenance [35]. In vitro model systems able to replicate the interactions between HPV-infected tissue and the surrounding TME are required to an in-depth understanding of these phenomena (Figure 2). The aim of this review article is to summarize the most recent advancement in the field of tissue engineering regarding the development of the 3D organotypic model of HPV-associated disease, focusing firstly on cervical cancer disease, and then to other human mucosal malignancies-derived, spanning from the anogenital tract, oropharynx to cutaneous epithelium. This review also addresses the issue of the cross-talk between the stroma and its microenvironment on HPV-infected epithelia, emphasizing the need for most relevant human in vitro models to study the host-pathogen, as well as HPV-infected-TME interactions in cancer development.

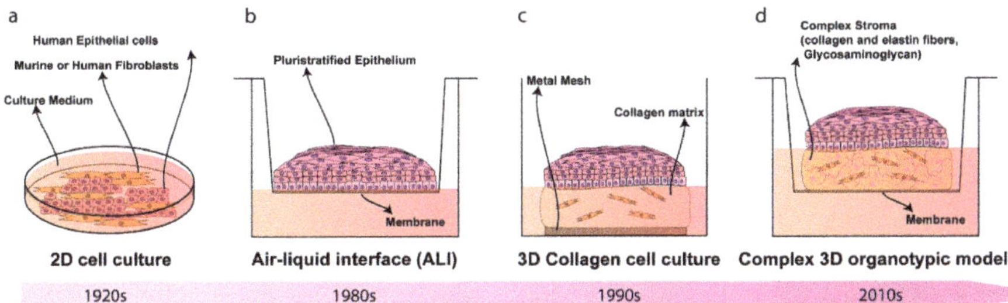



**Figure 2.** Timeline representing the evolution of in vitro culture systems for the study of HPV-related diseases. (**a**) Two-dimensional (2D) cell mono- or co-culture models have been used since the early 1900s; (**b**) in the 1980s, air-liquid interface (ALI) cultures were developed; these models consist of the epithelial cells seeding onto a semi-permeable membrane, allowing the formation of the epithelial strata; (**c**) three-dimensional (3D) collagen cell culture represented the first 3D model that reconstructs both epithelium and stroma and were developed in 1990s: primary or immortalized cells are cultured for 2–3 weeks onto the fibroblasts-feeder-layer or on collagen matrix populated with fibroblasts, intending to mimic the lamina propria. The epithelial cells, then, are induced to stratify and differentiate; (**d**) complex 3D organotypic models built up from tissue explant-repopulated-matrix, de-epidermized culture or endogenously produced extracellular matrix (ECM) was developed in the 2010s. These models provide a complex ECM with a well-differentiated epithelium that physiologically resembles the native tissue.

## 2. HPV-Related Cervical Diseases

In developing countries, cervical cancer is still the second leading cause of cancer death, while in the western world the number of deaths continues to decrease thanks above all to the introduction of an early diagnostic examination [36]. One of the main risk factors for cervical cancer is HPV infection that is sexually transmitted. However, not all HPV infections cause cervical cancer. Most women who come into contact with such viruses are able to resolve the infection, with their immune systems, without subsequent consequences in their health. Finally, it has now been ascertained that only some of the over 100 types of HPV are dangerous from an oncological point of view, while the majority remain silent, or only give rise to small benign tumors called papillomas (also known as genital warts) [17]. Other factors can increase the risk of cervical cancer, including cigarette smoking, familiarity and genetic features, diet and obesity, as well as bacterial infections (e.g., Chlamydia trachomatis) [37–39]. Cervical cancers are classified according to the cells from which they originate in two types: squamous cell carcinoma, which accounts for 80% of cervical cancers, and adenocarcinoma, which accounts for about 15% percent of cervical cancers. Squamous cell carcinoma is a tumor that arises from the cells covering the surface of the exocervix while adenocarcinoma is a cancer that starts from the glandular cells of the endocervix. Adenosquamous carcinomas are known to be uncommon cervical tumors with a mixed origin [40]. The precursor lesions of cervical cancer are classified as a squamous intraepithelial lesion (SIL) or, an alternative term, cervical intraepithelial neoplasia (CIN), and are graded into three categories (I to III) depending on the extent of abnormal maturation: from low-grade lesions (CIN I) to dysplastic (CIN II) and severely dysplastic (CIN III). Viral genome integration has been proposed as an activation mechanism for progression from low- to high-grade lesions [41]. HPVs firstly infect basal epithelial cells, amplifying their genomes as low-copy-number, autonomously replicating in episomal form (establishment phase). Whereas, in the maintenance phase, the viral genomes are stably maintained at an almost constant copy number. The replication of the viral genomes occur during the S-phase, in synchrony with the host DNA replication. Moreover, in the productive or vegetative phase in vivo, which occurs only in highly differentiated cells, a high copy of numbers of the viral genomes are produced, and the packaging of viral capsids occurs only on the most highly differentiated layers of the epithelium [42–44].

### 2.1. In Vitro Model to Reproduce HPV Life Cycle in HPV-Related Cervical Cancers

The first models produced for investigating the HPV replication mechanisms were developed by transplanting infected foreskin explants into immune-compromised mice with the aim to propagate virion stocks from HR-HPVs (HPV16 and HPV18) [45,46]. However, although the animal model can provide considerable basic information on cervical lesion formation and regression, they have a number of limitations linked to the species specificity, different histological appearance, and epithelial tropisms of the HPVs [19,47]. In fact, HPVs are able to reproduce their entire life cycle only in human stratified epithelial cells and cause cancers at discrete epithelial sites for which straightforward in vivo models are missing [48]. The first attempts to produce the HPVs life cycle in vitro have met with little success. Researchers have hardly worked on the development of in vitro model systems that accurately mimic the mechanism of HPV infection in humans. Due to the correlation between the HPV life cycle and squamous epithelial differentiation, the most powerful in vitro models are represented by three-dimensional (3D)-organotypic epithelial tissues, known as "raft" culture [49,50]. Raft cultures have allowed normal keratinocytes to stratify and fully differentiate in air-liquid interface culture to produce a squamous full-thickness epithelial tissue when seeded on the top of a dermal equivalent, consisting of the fibroblast layer or fibroblast-populated collagen gel [51]. These 3D models have provided an environment permissive for recapitulating and modulating the infection program of cancer-associated HR-HPVs [52,53], which cause the majority of cervical cancers. Raft cultures were first developed by Asslineau and Pruniéras [54] and then improved by Kopan et al. [55], who have reproduced the entire HPV life cycle, including virus production, and have developed a dysplastic lesion similar to those observed upon in vivo HPV infection. In subsequent years, the HPV particles assembly has been studied with the use of in vitro-derived particles, such as virus-like particles (VLPs), pseudovirions (PsV), and quasivirions (QV) [56–58]. However, these techniques suffer from technical constraints for the production of "native" HPV virions in 3D organotypic culture. More recently, Conway et al. showed the "native" HPV virion production in a 3D organotypic model, obtained by immortalizing in vitro human foreskin keratinocytes (HFK) on 3T3 feeder layer, which allowed to deep investigate the HPVs' assembly and maturation [59]. Infectious HPV progenies were isolated from such 3D organotypic models initiated with cells that maintain the HPV genome in the extrachromosomal replicative form [60]. In recent years, an increasing number of publications have reported the use of raft cultures produced with in vitro integrated human papillomavirus sequences in the cervical cancer cell line (Table 1). The latter are cervical carcinoma derived cell lines, such as SiHa cells, which contain an integrated HPV16 genome (HPV16, 1–2 copies/cell) [61], CaSki cells that contain an integrated HPV16 genome (600 copies/cell) as well as sequences related to HPV18 [62], and HeLa cells that contain HPV18 sequences cultured on collagen-populated-fibroblast gel [63,64]. C-33a, a pseudodiploid human cell line, was usually used as a control since these cell types are negative for human papillomavirus DNA and RNA [65]. Additionally, established squamous carcinoma cell line were seeded onto collagen plug to reproduce the dysplastic morphologies mimicking the pre-neoplastic lesions seen in vivo [52,66,67]. Other researchers reported the use of HPV-16 episome, containing normal immortalized human keratinocyte line (NIKS) that has been extensively cultured to study some aspects of HPV biology and transformation, particularly on raft culture [68]. As it is known, the physical state of the virus changes in the host cells from 'episomal' to 'integrated' in the polyclonal premalignant lesion, potentially promoting the disruption of human gene loci that are relevant to cancer pathogenesis [69]. In this perspective, Stanley et al. produced a non-tumorigenic human cervical keratinocyte cell line, i.e., W12, from a polyclonal culture of cervical squamous epithelial keratinocytes naturally infected with HPV16, which were derived by explant culture of a low grade-SIL, histologically diagnosed as CIN I [70]. They demonstrated that, at early passages, these 'parental' W12 cells are able to maintain a stable genome and phenotype. In 3D organotypic culture, W12 cells maintain the HPV16 genome as episomes at about 100 copies per cell, recapitulating a low grade-SIL. In long-term culture, W12 cells lose these properties and phenotypically progress to high grade-SIL and then to squamous cervical carcinoma (SCC) with HPV16 DNA integration, recapitulating the host events associated with cervical carcinogenesis in vivo [71].

On the other hand, primary HFK isolated from clinical circumcision at a low passage, that stably maintaining episomal HPV genome, have been frequently used in the laboratory due to the ease of isolation and the high cell yield [72–77]. Notwithstanding, an important aspect to take into account when the researcher decides to build a 3D organotypic model is the use of cells derived from the organs that they intend to study. The epithelia of different organs have a dissimilar differentiation program with or without epithelial keratinization and with different cytokeratins expression at the epithelial layers [78,79]. Moreover, HR-HPVs differently re-program the keratinocytes to express or delay the epithelial differentiation by altering the expression pattern of specific differentiation markers [80]. These phenomena appear to be unique in every anatomical site. For this reason, it is, however, incorrect to use HFK to reproduce cervix uterine models in vitro, although they are more available than cervical tissues. Such models do not reproduce the histological features of the cervical tissue but display the distinctive cutaneous skin morphology. In addition, HPV-infected epithelial mucosa of cervix uterine showed differences from skin particularly in terms of hormone responsiveness and immunological activators [81]. Preferably, 3D organotypic models have to be developed from primary cervical keratinocytes derived from healthy biopsies, and infected by introducing, experimentally, the HPVs or by propagating cervical keratinocytes isolated from naturally HPV-infected biopsies from cervical lesions [17,82]. In this perspective, some scientific reports have established a patient-derived cell culture system by using fresh cervical biopsies as a more accurate alternative to traditional cervical cancer-derived cell lines [83,84]. In detail, De Gregorio et al. developed an organotypic cervix model by using primary cervical cells, obtaining a complete ectocervical epithelium that showed all the characteristic epithelial differentiation markers when seeded onto a complex and auto-produced ECM [83]. Recently, Villa et al. devised a protocol to isolate and resuscitate (after freezing) cervical keratinocytes to model organoid culture [84]. These models closely resemble the in vivo stratified epithelium and may be useful for investigating the complex molecular mechanism of cervical neoplastic transformation related to persistent HPV infection.

## 2.2. In Vitro Organotypic Systems to Model Tumor Microenvironment in HPV-Related Cervical Cancer

It is now established that the stromal microenvironment contributes to tumorigenesis in HPV-related cervical cancer and that the TME sends signals that guide growth, tumor progression, and the formation of metastases, as well as resistance to anticancer drugs [85,86]. CAFs are known to have an active role in tumorigenesis providing relevant signals in the development and progression of the pathology through the release of growth factors that guide ECM remodeling and angiogenesis [87,88]. In addition, alterations in the production of ECM proteins, as well as ECM remodeling enzymes, can influence the stroma by modulating the carcinogenic potential of adjacent epithelial cells [87]. Emerging evidences suggest that a bidirectional crosstalk between HPV-positive epithelia and the underlying stroma occur during cancer progression [88,89]. In recent years, cervix models have been developed by culturing epithelial cells onto human foreskin or mouse 3T3-J2 fibroblasts (Table 1) to generate feeder layers as stromal equivalent, emphasizing the crucial role of the fibroblasts in the epithelial cell culture and propagation [90]. Some researchers have also highlighted the fibroblasts feeder layers involvement in promoting homeostasis and proliferation of the cervical epithelial cells by direct contact-dependent and/or indirect paracrine signaling, including soluble factors, such as growth factors and cytokines, as well as ECM components [90,91]. In this direction, researchers demonstrated that fibroblasts enhance, specifically, the HPV16- and HPV18-positive cervical epithelial cells growth and inhibit the normal epithelial cells growth by a 'double-paracrine' epithelial-stromal signaling mechanism involving, e.g., interleukin-1 proteins production [92,93]. In addition, fibroblasts feeder layers are also required for the HPV genome maintenance as extrachromosomal episomal form in HPV-infected cells, such as W12 [94]. More complex 3D cervix-like constructs, consisting of foreskin fibroblasts embedded within some matrices, such as rat tail fibrillar collagen, polymeric scaffolds, or de-epidermalized matrix, were developed in order to promote the epithelial cell stratification and provide support comparable to the extracellular matrix [49]. Recent studies pointed out that 'human' stromal fibroblasts promoted much more epithelial invasion than 'mouse' fibroblasts in HPV-positive epithelia grown on raft culture, demonstrating the need for associating tissue-specific cells with specific tissue to better reproduce the native microenvironment for in vitro tumor modeling [95,96]. Moreover, 3D organotypic cultures built up from cervical tissue derived from healthy or tumor/cancer-associated keratinocytes and fibroblasts, may be used to correctly resemble, in vitro, the native morphologic and histologic cervical features, and also to better elucidate the interactions between stromal and epithelial compartments in the carcinogenetic process [97,98]. In a pioneering work, a 3D organotypic cervical model was developed by using tissue-specific cells from different organs, among this uterine cervix, to replicate more faithfully the native tissue, showing the epithelial invasion through the basement membrane during tumor progression [99]. Other studies highlighted the crucial role of the stromal-derived factors in promoting epithelial invasion in cervical cancer [100]. Furthermore, another important aspect to take into account when fabricating the engineered tissue equivalents is organ complexity. Reconstructed models should provide the same structure and composition as the cervical mucosa, including, for example, the presence of both collagen and non-collagenous proteins (glycosaminoglycans, etc.) that, in vivo, are involved in the ECM remodeling occurring during epithelial mesenchymal transition (EMT) process [101,102]. Although the ECM represents a non-cellular component within tissues, to which, generally, a supporting role is ascribed, it has been recently recognized that ECM has a fundamental functional repository role for several factors that dynamically modulate the TME [103]. Some molecules are deposited into the ECM and remain latent until activated. Among these, the Matrix Metalloproteinases (MMPs) hollow out space in the matrix allowing cells to migrate by degrading ECM components [104]. MMPs also regulate the epithelial cell migration and interactions within the stroma [105]. To demonstrate this, Fullar et al. investigated the action of the HPV16 on the MMPs produced by epithelial cells and fibroblasts during EMT and carcinogenesis processes [106]. Other researchers highlighted the induction of the ECM remodeling in HR-HPV positive models [107]. It is well known that, in vivo, the cervical stroma undergoes a controlled remodeling by quantitative/qualitative protein changes mediated by specific enzymes and the dysregulation of the ECM composition, structure, stiffness, and abundance affects the pathophysiological tissue status, contributing to several pathological conditions, such as invasive cancer [103,108]. Recently, cervical microtissues have been used as functional units for the manufacture of an endogenous cervical stroma, with stromal characteristics comparable to those of the native cervix. This complex 3D cervix tissue equivalent was provided by a specialized ECM microenvironment featured by an autologous and responsive ECM and an auto-produced basement membrane. Indeed, the 3D completely scaffold-free ECM was able to guide the formation of fully differentiated and stratified epithelium, establishing the correct cross-talk between stroma and epithelium [83]. Organotypic tissues that recapitulate in vitro the composition and structural organization (epithelial stratification, functional basement membrane, fibroblast populated stroma, complex ECM) of their native counterpart represent a new model for studying tumor progression and evaluating combined therapies in non-animal models (Table 2). Consistent with this finding, unpublished work from our laboratory indicates the importance of stromal-to-epithelial communication in guiding the mesenchymalization in HPV-positive epithelia. The crucial role of the cervical cancer-associated fibroblasts, as well as the stromal microenvironment in the biochemical changes that enable the epithelial cells to assume a mesenchymal cell phenotype, were evidenced by analyzing EMT markers expression. Altered expression of the adhesion molecules, the collagenous and non-collagenous proteins as well as the ECM remodeling have been further highlighted. Finally, an up-regulation of the gene expression of late viral proteins on HPV-positive epithelial cells cultured on a diseased cervical model was also found. The epithelium-to-stroma and stroma-to-epithelium crosstalk may be the mechanism through which the viruses manipulate their environment and vice-versa, in the case of tumor-associated viruses, contributing to carcinogenesis [27]. It's also noted that the carcinogenesis of cervical carcinoma implies an intricate interplay of neoplastic, HPV-infected epithelial cells, and stromal tissue including non-tumoral cell types [109]. The HPV-positive epithelium and stromal cells (CAFs, endothelial cells, immune cells and neuronal cells) communicate with HPV-infected epithelium through the exchange of growth factors (Transforming growth factor beta (TGF-β), Vascular endothelial growth factor (VEGF), Heparin-binding EGF-like growth factor (HB-EGF), Epidermal growth factor (EGF)), cytokines (Interleukins (ILs), chemokine (C-X-C motif) ligand 1 (CXCLs) and CC chemokines (CCLs)), neurotransmitters, ECM molecules (MMPs) and other molecules (macrophage colony-stimulating factor (M-CSF), Granulocyte-colony stimulating factor (G-CSF)), leading stromal remodeling, cancer cells proliferation and angiogenesis processes [27]. A thorough mapping of the non-tumoral cell types that populate the TME is critical to understand their unique roles in tumor biology. Interestingly, tumor innervation is associated with worse clinical outcomes in several solid cancers [110,111], emphasizing nerves as microenvironmental factors that may contribute to tumor progression. Scientific reports suggest that carcinogenesis alters cervical innervation, demonstrating the role of the HPV-positive cervical cancer cell lines in effectively stimulate neurite outgrowth [112]. Furthermore, an essential component of the tumor-associated stroma is the vasculature, composed of blood and lymph vessels [109]. The induction of angiogenesis is an early event in cervical carcinogenesis [113]. In details, in low-grade lesions, there is an increase in the number of capillaries in the cervical stroma underlying the dysplastic epithelium. In high-grade lesions there is an additional increase in the number of vessels that appear to be organized into a dense micro-vascular array in close relation to the overlying neoplastic epithelium. Furthermore, in some high-grade lesions, stromal vascular papillae are formed that reach towards the surface of the epithelium [114]. Over the past years, the endothelial cells have been used in vitro as feeder for keratinocytes to support epithelial cell differentiation [32]. Other studies reported the HPV-dependent angiogenic response in terms of proliferation and migration of the endothelial cells when cultured with conditioned media from HPV positive cells [115]. In this perspective, a further implementation of the complexity of the 3D organotypic cervix models may be to insert non-tumor cells (endothelial vascular cells and/or neuronal cells) at the stromal level to study the interaction between stroma and their adjacent complex TME [116,117]. Furthermore, immune cells infiltrate, such as macrophages, and T-cells as potential anti-cancer cellular weaponry may also greatly implement the 3D models' complexity [118,119] (Figure 3). Finally, the reproduction of the microbial species of the cervical microbiota may be a step forward for the reproduction of models that faithfully mimic the native tissue [120,121].

**Table 1.** Summary of different cell types used to develop tissue engineered cervical mucosa.

| Cell Lines | HPV Types | Physical State | References |
| --- | --- | --- | --- |
| Epithelial Cells | | | |
| SiHa | HPV16 | Int. | [61,64] |
| CaSki | HPV16/18 | Int. | [62,64] |
| HeLa | HPV18 | Int. | [63,64] |
| W12 | HPV16 | Int./Epi. | [70,71] |
| C-33a | - | - | [65] |
| NIKS | HPV16 | Epi. | [68] |
| Primary HFK | HPVs | Int. or Epi. | [72–77] |
| Primary HCK | HPVs | Int. or Epi. | [82–84] |
| Stromal cells | | | |
| 3t3J2 fibroblasts | - | - | [90] |
| Primary HFF | - | - | [91–93] |
| Primary HCF | - | - | [83,84,95,106] |

Int. = integrated HPV genome; Epi. = episomal HPV genome; HFK = human foreskin keratinocytes; HCK = human cervical keratinocytes; HFF = human foreskin fibroblasts; HCF = human cervical fibroblasts; NIKS = normal immortalized human keratinocyte line.

**Table 2.** Summary of organotypic cervical models pointing out their advantages and limitations.

| Cell Culture Systems | Advantages | Limitations | References |
|---|---|---|---|
| 2D cell culture 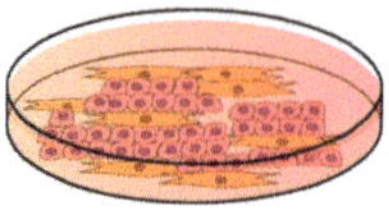 | Simplified model/ Possibility of making co-culture/ Cost effective/ Easy to use | Inability to reproduce HPV life cycle/ Short time culture/ Low biological relevance/ | [56–58,61–63,65,66,70,106] |
| Air-liquid interface culture 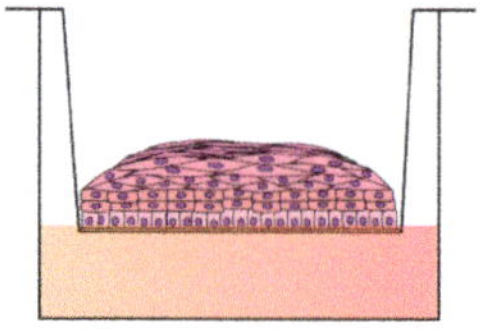 | Recapitulate pluristratified epithelium/ Cell-to-cell-interactions/ HPV genetic studies/ Convenient/ Easy to use | Lack of connective tissue/ Genetic manipulation epithelial cell-dependent/ | [122,123] |
| 3D collagen cell culture 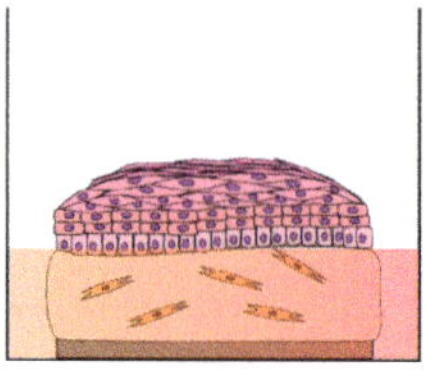 | Resemble the epithelial architecture and differentiation/ Cell-to-ECM signaling/ More complex culture system/ | Hydrogel composition differs to real ECM/ HPV genetic studies epithelial cells-dependent/ Added expensive/ | [49–52,54,55,59,60,64,67,68, 71,74–77,82] |
| Complex 3D organotypic models 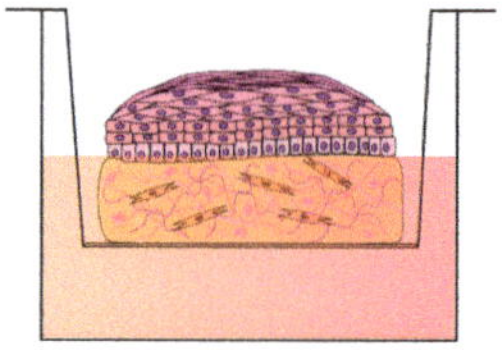 | More accurate culture systems/ Cell-virus-ECM interactions/ ECM complexity/ Patient-specific models/ Long-time culture (3–4 weeks)/ Useful drug testing platform/ Useful for Invasion studies/ | Differences between specimens/ Endpoint assays (genomic, proteomic, metabolomics) dependent of cell culture used/ More expensive/ | [83,84,95,99] |

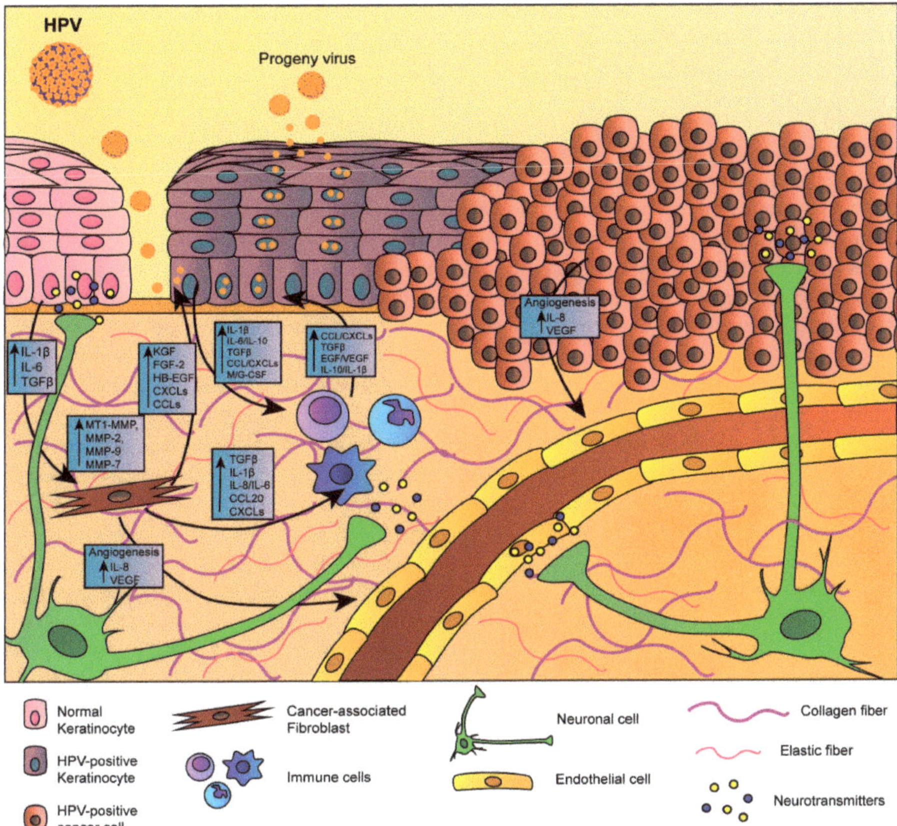



**Figure 3.** Diagrammatic representation of the bidirectional cross-talk between HPV-positive epithelium and tumor microenvironment (TME). HPV-positive epithelium and stromal cells (CAF, endothelial cells, immune cells and neuronal cells) communicate with HPV-infected epithelium through the exchange of growth factors (TGF-β, VEGF, HB-EGF, EGF), cytokines (IL, CXCLs and CCLs), Neurotransmitters and ECM molecules (MMPs) and other molecules (M-CSF, G-CSF). Large black arrows represent the bidirectional communication between HPV-positive epithelium and stromal cells. Small black arrows in the squares represent an increase (up arrow) of specific factor. CAF = Cancer-Associated Fibroblast; IL = Interleukins; CXCs/CCLs = Chemokines; TGF-β = Transforming Growth Factor-beta; EGF = Epidermal Growth Factor; HB-EGF = Heparin-binding EGF-like growth factor; FGF-2 = Fibroblast Growth Factor-2; VEGF = Vascular Endothelial Growth Factor; M-CSF = Macrophage Colony-Stimulating Factor; G-CSF = Granulocyte-Colony Stimulating Factor; MMPs = Metalloproteinases.

## 3. HPV-Related Human Malignancies Arising from Mucosal Squamous Epithelia: Head and Neck and Anogenital-Tract Cancers

### 3.1. Head and Neck Cancers

Head and Neck cancers (HNCs) represent a heterogeneous group of tumors that include cancer of squamous epithelial cells of the oral cavity (mouth and throat), nasal cavity (sinuses), larynx and pharynx, salivary glands and lymph nodes in the neck [3]. Connective tissue cells such as CAFs promote squamous cell carcinoma proliferation, invasion, and metastasis, on the contrary lymphocytes are rarely involved in HNCs (about 10%) [124–126]. A large number of HPV-associated

HNCs are caused by HPV16 and, although are hard to treat, they have favorable prognosis compared to HPV-negative HNCs. Moreover, 3D organotypic models provide a useful tool to evaluate the sensitivity of HPV-positive and HPV-negative cancer cells exposed to different therapeutic strategies to identify potential druggable targets for tailored therapy [127]. Among the heterogeneous group of Head-Neck cancers, HPV-related oropharyngeal cancers, that have a favorable prognosis compared to HPV-unrelated tumors, were deeply studied to have a better understanding of the HPV role in exacerbating malignant phenotypes [126]. In this direction, organotypic raft cultures, which included immortalized oral/oropharyngeal squamous carcinoma cell lines seeded on submucosal equivalents consisting of type I collagen and normal human oral fibroblasts, was used to study key characteristics of cancer [128,129]. Other researchers have developed an oral cancer equivalents system prepared with decellularized human dermal tissue that allowed to study the epithelial stratification and invasion beyond the basement membrane into underlying connective tissue [130]. In another study, Dalley et al. reported the cancer stem cells involvement in the progression of oral dysplasia to squamous cell carcinoma by developing 2D monolayer or 3D organotypic culture of normal, dysplastic, and squamous cell carcinoma-derived oral cell lines. The researchers demonstrated the usefulness of the 3D human oral mucosa equivalent on the detection of the hierarchical expression of cancer stem cell-associated markers (CD44, p75NTR, CD24, and ALDH), information that cannot be revealed in 2D monolayer [131].

Tonsillar carcinomas are among the most frequent HNCs, arising in the reticulated epithelial cells of the crypts with immune cell infiltrations. HR-HPV16 persistent infections are related to oropharyngeal carcinogenesis [80]. However, compared to cervical cancer, little is known about how HPV drives the pathogenesis of oropharyngeal cancer. HPV could establish a productive or abortive infection in keratinocytes of the tonsil crypt, or progresses to cancer through a neoplastic phase, as occurs in cervical HPV infection [132]. In oropharyngeal cancer, the HPV DNA is more frequently found un-integrated and may include novel HPV-human DNA hybrids episomes [127]. Meyers' group recently reported the usefulness of the organotypic raft culture (composed of immortalized primary human tonsil and HFK cell lines persistently infected with HPV16 seeded onto collagen matrices consisting of rat-tail type 1 collagen and containing J23T3 feeder cells) to investigate the life cycle of HPV16 in oral (tonsil) epithelial tissue vs. genital (foreskin) tissue focusing on the titers, infectivity, and maturation of HPV16. They demonstrated that, although some aspects of the HPV16 replication are overlapped in foreskin and tonsil tissue, there are significant differences related to the maturation and final structure of the virions when grown in the two different tissues [133]. In addition, other works reported more complex 3D co-culture systems of tonsil keratinocytes, with immune components providing a more realistic in vitro model [134], since the tonsil has significant infiltration of lymphoid cells into the crypt that potentially explain its improved prognosis [135].

### 3.2. Anogenital HPV-Associated Cancers in Males and Females

To reproduce in vitro anogenital HPV-associated cancers in males and females, the main models developed are the 3D organotypic culture that allows to accurately mimic the life cycle of several cancer-associated HR-HPV types. In the previous sections, we focused on the culture system used to study HPV-related cervical cancer; here, we revised the in vitro model mimicking other organs of the human anogenital apparatus, both in male and female.

### 3.2.1. Vulva and Vagina

Lower genital cancers are frequent malignancies in women prevalently associated with HR-HPVs infection. To date, some studies have reported that vulvar squamous cell carcinoma (VSCC) have different pathological pathways related, or not, to HPV infections. HPV-related VSCC with profound cellular atypia and basaloid/warty histology and VSCC non-dependent to HPV infection with basal atypia and keratinized histology [136–138]. In addition, VSCCs are also associated with chronic inflammatory dermatitis, such as vulvar lichen sclerosus [139]. Limited in vitro models have been developed for better understanding the biology and development of VSCCs. Some researchers have isolated and characterized primary cells from VSCC and normal vulvar tissue adjacent to the tumor in order to develop 3D organotypic and/or in vivo xenografts models [140–142]. Vaginal cancers are uncommon diseases that are also related to HPV infection. Most organotypic models have been developed for the ex vivo study of HIV infection [143]. Few works have reported the development of raft cultures obtained by seeding vaginal cells able to study microRNA biomarkers of oncogenic HPVs infections [144].

### 3.2.2. Anus

Anal cancer is closely related to high-risk HPV persistent infection that occurs in the anal transformation zone, similar to what happens in the uterine cervix [145]. HPV infection leads to the development of an HSIL lesion that can then evolve to invasive carcinoma. Only a few models have been developed due to the lack of immortalized HPV-positive anal epithelial cell lines [10,146]. Among this, Wechsler et al. reported a novel in vitro model of anal cancer pathogenesis using the first HPV-16-transformed anal epithelial cell line (AKC2 cells) that have a poorly differentiated and invasive phenotype in three-dimensional raft culture. In this 3D contest, AKC2 cells express all three HPV-16 oncogenes (E5, E6, and E7) involved in anal cancer progression [147].

### 3.2.3. Penis and Penile Urethra

For several years, the foreskin cell line derived from circumcision has been used to study various HPV-related cancer due to the high availability of this skin fold. The penile urethra, that consists of pluri-stratified squamous cells without keratinization process alike the foreskin or the glans, is also routinely targeted by sexually transmitted viral pathogens such as HIV infection in the mouse model [148]. Limited studies reported the organotypic raft culture of prenatal genital tubercle to investigate the direct effects of the hyperestrogenic state on fetal mouse penile and urethral development [149].

## 4. Other HPV-Related Cancers (Non-Melanoma Skin Cancer)

Among cutaneous malignancies, the non-melanoma skin cancer (NMSC) involves fair-skinned populations and are mainly correlated to solar ultraviolet irradiation [150]. However, this malignancy appears to be also related to HPV infection [151]. Current findings report studies on the critical role of cutaneous HPV infection as a co-factor in association with genomic and mitochondrial mutations induced by ultraviolet irradiation in NMSC development in simplified in vitro models [152]. Other researchers have focused on the study of the HPV transformation mechanisms, as well as the epithelial invasion in the NMSC, by using an in vitro skin-equivalent organotypic model [153]. These models better reproduce the terminal differentiation of the epithelial cells and also the migration and invasion through the underlying dermis after HPV infections.

## 5. Conclusions

In this review, we outlined the current knowledge on the HPV-related cancers modeled in vitro from the simplified 'raft culture' to complex 3D organotypic models focusing on HPV-associated cervical cancer disease platforms and we also reviewed the in vitro culture systems of human HPV-associated mucosal malignancies from the anogenital tract, oropharynx, and cutaneous epithelium. We highlighted the importance of using multicellular models, which involve the use of compartment-specific cell types at both the epithelial and stromal level, and a complex ECM capable of remodeling to fully reproduce the histomorphological features of the tissue in vivo. This review also addresses the issue of the cross-talk between the stroma and its microenvironment on HPV-infected epithelial, emphasizing the need for most promising in vitro models to study host-pathogen, as well as HPV-infected-TME interactions in cancer development in humans.

**Author Contributions:** Writing-original draft preparation, V.D.G.; defining the structure of the review, V.D.G., G.I.; review and editing, V.D.G., G.I., F.U.; supervision, G.I., P.A.N. All authors have read and agreed to the published version of the manuscript.

**Funding:** This research received no external funding.

**Acknowledgments:** The authors thank La Rocca Alessia for her contribution in the graphical design.

**Conflicts of Interest:** The authors declare no conflict of interest.

## References


1. Doorbar, J. Molecular biology of human papillomavirus infection and cervical cancer. *Clin. Sci. (Lond.)* **2006**, *110*, 525–541. [CrossRef]
2. Cubie, H.A. Diseases associated with human papillomavirus infection. *Virology* **2013**, *445*, 21–34. [CrossRef]
3. Syrjanen, S. Human papillomavirus (HPV) in head and neck cancer. *J. Clin. Virol.* **2005**, *32*, S59–S66. [CrossRef] [PubMed]
4. De Martel, C.; Georges, D.; Bray, F.; Ferlay, J.; Clifford, G.M. Global burden of cancer attributable to infections in 2018: A worldwide incidence analysis. *Lancet Glob. Health* **2020**, *8*, e180–e190. [CrossRef]
5. Bansal, A.; Singh, M.P.; Rai, B. Human papillomavirus-associated cancers: A growing global problem. *Int. J. Appl. Basic. Med. Res.* **2016**, *6*, 84–89. [PubMed]
6. Gillison, M.L.; Chaturvedi, A.K.; Lowy, D.R. HPV prophylactic vaccines and the potential prevention of noncervical cancers in both men and women. *Cancer* **2008**, *113*, 3036–3046. [CrossRef]
7. Chan, P.K.; Chang, A.R.; Cheung, J.L.; Chan, D.P.; Xu, L.Y.; Tang, N.L.; Cheng, A.F. Determinants of cervical human papillomavirus infection: Differences between high– and low–oncogenic risk types. *J. Infect. Dis.* **2002**, *185*, 28–35. [CrossRef]
8. Hartwig, S.; Baldauf, J.-J.; Dominiak-Felden, G.; Simondon, F.; Alemany, L.; de Sanjosé, S.; Castellsagué, X. Estimation of the epidemiological burden of HPV-related anogenital cancers, precancerous lesions, and genital warts in women and men in Europe: Potential additional benefit of a nine-valent second generation HPV vaccine compared to first generation HPV vaccines. *Papillomavirus Res.* **2015**, *1*, 90–100.
9. Bosch, F.X.; Broker, T.R.; Forman, D.; Moscicki, A.B.; Gillison, M.L.; Doorbar, J.; Stern, P.L.; Stanley, M.; Arbyn, M.; Poljak, M.; et al. Comprehensive control of human papillomavirus infections and related diseases. *Vaccine* **2013**, *31*, 1–31. [CrossRef]
10. Hoots, B.E.; Palefsky, J.M.; Pimenta, J.M.; Smith, J.S. Human papillomavirus type distribution in anal cancer and anal intraepithelial lesions. *Int. J. Cancer* **2009**, *124*, 2375–2383. [CrossRef]
11. Munoz, N.; Bosch, F.X.; de Sanjose, S.; Herrero, R.; Castellsague, X.; Shah, K.V.; Snijders, P.J.; Meijer, C.J.; International Agency for Research on Cancer Multicenter Cervical Cancer Study Group. Epidemiologic classification of human papillomavirus types associated with cervical cancer. *N. Engl. J. Med.* **2003**, *348*, 518–527. [CrossRef] [PubMed]
12. Ghittoni, R.; Accardi, R.; Chiocca, S.; Tommasino, M. Role of human papillomaviruses in carcinogenesis. *Ecancermedicalscience* **2015**, *9*, 526. [CrossRef] [PubMed]
13. Hartwig, S.; St Guily, J.L.; Dominiak-Felden, G.; Alemany, L.; de Sanjose, S. Estimation of the overall burden of cancers, precancerous lesions, and genital warts attributable to 9-valent HPV vaccine types in women and men in Europe. *Infect. Agent. Cancer* **2017**, *12*, 19. [CrossRef] [PubMed]
14. Steben, M.; Duarte-Franco, E. Human papillomavirus infection: Epidemiology and pathophysiology. *Gynecol. Oncol.* **2007**, *107*, S2–S5. [CrossRef] [PubMed]
15. Shanmugasundaram, S.; You, J. Targeting Persistent Human Papillomavirus Infection. *Viruses* **2017**, *9*, 229. [CrossRef]
16. Doorbar, J.; Griffin, H. Refining our understanding of cervical neoplasia and its cellular origins. *Papillomavirus Res.* **2019**, *7*, 176–179. [CrossRef]
17. Burd, E.M. Human papillomavirus and cervical cancer. *Clin. Microbiol. Rev.* **2003**, *16*, 1–17. [CrossRef]
18. Rajendra, S.; Sharma, P. Transforming human papillomavirus infection and the esophageal transformation zone: Prime time for total excision/ablative therapy? *Dis. Esophagus.* **2019**, *32*. [CrossRef]
19. Egawa, N.; Egawa, K.; Griffin, H.; Doorbar, J. Human Papillomaviruses; Epithelial Tropisms, and the Development of Neoplasia. *Viruses* **2015**, *7*, 3863–3890. [CrossRef]

20. Stanley, M.A. Epithelial cell responses to infection with human papillomavirus. *Clin. Microbiol. Rev.* **2012**, *25*, 215–222. [CrossRef]
21. Graham, S.V. Human papillomavirus: Gene expression, regulation and prospects for novel diagnostic methods and antiviral therapies. *Future Microbiol.* **2010**, *5*, 1493–1506. [CrossRef] [PubMed]
22. Schlecht, N.F.; Platt, R.W.; Duarte-Franco, E.; Costa, M.C.; Sobrinho, J.P.; Prado, J.C.; Ferenczy, A.; Rohan, T.E.; Villa, L.L.; Franco, E.L. Human papillomavirus infection and time to progression and regression of cervical intraepithelial neoplasia. *J. Natl. Cancer Inst.* **2003**, *95*, 1336–1343. [CrossRef] [PubMed]
23. McMurray, H.R.; Nguyen, D.; Westbrook, T.F.; McAnce, D.J. Biology of human papillomaviruses. *Int. J. Exp. Pathol.* **2001**, *82*, 15–33. [CrossRef] [PubMed]
24. McBride, A.A. Replication and partitioning of papillomavirus genomes. *Adv. Virus. Res.* **2008**, *72*, 155–205. [PubMed]
25. Tornesello, M.L.; Buonaguro, L.; Giorgi-Rossi, P.; Buonaguro, F.M. Viral and cellular biomarkers in the diagnosis of cervical intraepithelial neoplasia and cancer. *Biomed. Res. Int.* **2013**, *2013*, 519619. [CrossRef]
26. Howley, P.M.; Livingston, D.M. Small DNA tumor viruses: Large contributors to biomedical sciences. *Virology* **2009**, *384*, 256–259. [CrossRef]
27. Spurgeon, M.E.; Lambert, P.F. Human Papillomavirus and the Stroma: Bidirectional Crosstalk during the Virus Life Cycle and Carcinogenesis. *Viruses* **2017**, *9*, 219. [CrossRef]
28. Surviladze, Z.; Sterkand, R.T.; Ozbun, M.A. Interaction of human papillomavirus type 16 particles with heparan sulfate and syndecan-1 molecules in the keratinocyte extracellular matrix plays an active role in infection. *J. Gen. Virol.* **2015**, *96*, 2232–2241. [CrossRef]
29. Cerqueira, C.; Liu, Y.; Kuhling, L.; Chai, W.; Hafezi, W.; van Kuppevelt, T.H.; Kuhn, J.E.; Feizi, T.; Schelhaas, M. Heparin increases the infectivity of Human Papillomavirus type 16 independent of cell surface proteoglycans and induces L1 epitope exposure. *Cell Microbiol.* **2013**, *15*, 1818–1836. [CrossRef]
30. Maeda, N.; Maenaka, K. The Roles of Matricellular Proteins in Oncogenic Virus-Induced Cancers and Their Potential Utilities as Therapeutic Targets. *Int. J. Mol. Sci.* **2017**, *18*, 2198. [CrossRef]
31. Xu, C.; Langenheim, J.F.; Chen, W.Y. Stromal–epithelial interactions modulate cross-talk between prolactin receptor and HER2/Neu in breast cancer. *Breast Cancer Res. Treat.* **2012**, *134*, 157–169. [CrossRef] [PubMed]
32. Woodby, B.; Scott, M.; Bodily, J. The Interaction Between Human Papillomaviruses and the Stromal Microenvironment. *Prog. Mol. Biol. Transl. Sci.* **2016**, *144*, 169–238. [PubMed]
33. Fiori, M.E.; Di Franco, S.; Villanova, L.; Bianca, P.; Stassi, G.; De Maria, R. Cancer–associated fibroblasts as abettors of tumor progression at the crossroads of EMT and therapy resistance. *Mol. Cancer* **2019**, *18*, 70. [CrossRef] [PubMed]
34. Quail, D.F.; Joyce, J.A. Microenvironmental regulation of tumor progression and metastasis. *Nat. Med.* **2013**, *19*, 1423–1437. [CrossRef] [PubMed]
35. Hemmings, C. Is carcinoma a mesenchymal disease? The role of the stromal microenvironment in carcinogenesis. *Pathology* **2013**, *45*, 371–381. [CrossRef] [PubMed]
36. Ginsburg, O.; Bray, F.; Coleman, M.P.; Vanderpuye, V.; Eniu, A.; Kotha, S.R.; Sarker, M.; Huong, T.T.; Allemani, C.; Dvaladze, A.; et al. The global burden of women's cancers: A grand challenge in global health. *Lancet* **2017**, *389*, 847–860. [CrossRef]
37. Arbyn, M.; Weiderpass, E.; Bruni, L.; de Sanjose, S.; Saraiya, M.; Ferlay, J.; Bray, F. Estimates of incidence and mortality of cervical cancer in 2018: A worldwide analysis. *Lancet Glob. Health* **2020**, *8*, e191–e203. [CrossRef]
38. Lacey, J.V., Jr.; Swanson, C.A.; Brinton, L.A.; Altekruse, S.F.; Barnes, W.A.; Gravitt, P.E.; Greenberg, M.D.; Hadjimichael, O.C.; McGowan, L.; Mortel, R.; et al. Obesity as a potential risk factor for adenocarcinomas and squamous cell carcinomas of the uterine cervix. *Cancer* **2003**, *98*, 814–821. [CrossRef]
39. Koskela, P.; Anttila, T.; Bjorge, T.; Brunsvig, A.; Dillner, J.; Hakama, M.; Hakulinen, T.; Jellum, E.; Lehtinen, M.; Lenner, P.; et al. Chlamydia trachomatis infection as a risk factor for invasive cervical cancer. *Int. J. Cancer* **2000**, *85*, 35–39. [CrossRef]
40. Waggoner, S.E. Cervical cancer. *Lancet* **2003**, *361*, 2217–2225. [CrossRef]
41. Muntean, M.; Simionescu, C.; Taslica, R.; Gruia, C.; Comanescu, A.; Patrana, N.; Fota, G. Cytological and histopathological aspects concerning preinvasive squamous cervical lesions. *Curr. Health Sci. J.* **2010**, *36*, 26–32. [PubMed]
42. Moody, C. Mechanisms by which HPV Induces a Replication Competent Environment in Differentiating Keratinocytes. *Viruses* **2017**, *9*, 261. [CrossRef] [PubMed]

43. McBride, A.A. Mechanisms and strategies of papillomavirus replication. *Biol. Chem.* **2017**, *398*, 919–927. [CrossRef] [PubMed]

44. Hoffmann, R.; Hirt, B.; Bechtold, V.; Beard, P.; Raj, K. Different modes of human papillomavirus DNA replication during maintenance. *J. Virol.* **2006**, *80*, 4431–4439. [CrossRef]

45. Bonnez, W.; Rose, R.C.; Da Rin, C.; Borkhuis, C.; de Mesy Jensen, K.L.; Reichman, R.C. Propagation of human papillomavirus type 11 in human xenografts using the severe combined immunodeficiency (SCID) mouse and comparison to the nude mouse model. *Virology* **1993**, *197*, 455–458. [CrossRef]

46. Kreider, J.W.; Howett, M.K.; Leure-Dupree, A.E.; Zaino, R.J.; Weber, J.A. Laboratory production in vivo of infectious human papillomavirus type 11. *J. Virol.* **1987**, *61*, 590–593. [CrossRef]

47. Doorbar, J. Model systems of human papillomavirus–associated disease. *J. Pathol.* **2016**, *238*, 166–179. [CrossRef]

48. Doorbar, J.; Egawa, N.; Griffin, H.; Kranjec, C.; Murakami, I. Human papillomavirus molecular biology and disease association. *Rev. Med. Virol.* **2015**, *25*, 2–23. [CrossRef]

49. Delvenne, P.; Hubert, P.; Jacobs, N.; Giannini, S.L.; Havard, L.; Renard, I.; Saboulard, D.; Boniver, J. The organotypic culture of HPV–transformed keratinocytes: An effective in vitro model for the development of new immunotherapeutic approaches for mucosal (pre)neoplastic lesions. *Vaccine* **2001**, *19*, 2557–2564. [CrossRef]

50. McLaughlin-Drubin, M.E.; Meyers, C. Propagation of infectious, high-risk HPV in organotypic "raft" culture. *Methods Mol. Med.* **2005**, *119*, 171–186.

51. Arnette, C.; Koetsier, J.L.; Hoover, P.; Getsios, S.; Green, K.J. In Vitro Model of the Epidermis: Connecting Protein Function to 3D Structure. *Methods Enzymol.* **2016**, *569*, 287–308. [PubMed]

52. Dollard, S.C.; Wilson, J.L.; Demeter, L.M.; Bonnez, W.; Reichman, R.C.; Broker, T.R.; Chow, L.T. Production of human papillomavirus and modulation of the infectious program in epithelial raft cultures. OFF. *Genes Dev.* **1992**, *6*, 1131–1142. [CrossRef] [PubMed]

53. Bubb, V.; McCance, D.J.; Schlegel, R. DNA sequence of the HPV–16 E5 ORF and the structural conservation of its encoded protein. *Virology* **1988**, *163*, 243–246. [CrossRef]

54. Asselineau, D.; Prunieras, M. Reconstruction of 'simplified' skin: Control of fabrication. *Br. J. Dermatol.* **1984**, *111*, 219–222. [CrossRef] [PubMed]

55. Kopan, R.; Traska, G.; Fuchs, E. Retinoids as important regulators of terminal differentiation: Examining keratin expression in individual epidermal cells at various stages of keratinization. *J. Cell Biol.* **1987**, *105*, 427–440. [CrossRef]

56. Kirnbauer, R.; Booy, F.; Cheng, N.; Lowy, D.R.; Schiller, J.T. Papillomavirus L1 major capsid protein self–assembles into virus-like particles that are highly immunogenic. *Proc. Natl. Acad. Sci. USA* **1992**, *89*, 12180–12184. [CrossRef] [PubMed]

57. Buck, C.B.; Pastrana, D.V.; Lowy, D.R.; Schiller, J.T. Efficient intracellular assembly of papillomaviral vectors. *J. Virol.* **2004**, *78*, 751–757. [CrossRef]

58. Culp, T.D.; Christensen, N.D. Kinetics of in vitro adsorption and entry of papillomavirus virions. *Virology* **2004**, *319*, 152–161. [CrossRef]

59. Conway, M.J.; Alam, S.; Ryndock, E.J.; Cruz, L.; Christensen, N.D.; Roden, R.B.; Meyers, C. Tissue-spanning redox gradient-dependent assembly of native human papillomavirus type 16 virions. *J. Virol.* **2009**, *83*, 10515–10526. [CrossRef]

60. Ozbun, M.A.; Patterson, N.A. Using organotypic (raft) epithelial tissue cultures for the biosynthesis and isolation of infectious human papillomaviruses. *Curr. Protoc. Microbiol.* **2014**, *34*, 14B.3.1–14B.3.18.

61. El Awady, M.K.; Kaplan, J.B.; O'Brien, S.J.; Burk, R.D. Molecular analysis of integrated human papillomavirus 16 sequences in the cervical cancer cell line SiHa. *Virology* **1987**, *159*, 389–398. [CrossRef]

62. Pattillo, R.A.; Hussa, R.O.; Story, M.T.; Ruckert, A.C.; Shalaby, M.R.; Mattingly, R.F. Tumor antigen and human chorionic gonadotropin in CaSki cells: A new epidermoid cervical cancer cell line. *Science* **1977**, *196*, 1456–1458. [CrossRef] [PubMed]

63. Badal, S.; Badal, V.; Calleja-Macias, I.E.; Kalantari, M.; Chuang, L.S.; Li, B.F.; Bernard, H.U. The human papillomavirus–18 genome is efficiently targeted by cellular DNA methylation. *Virology* **2004**, *324*, 483–492. [CrossRef] [PubMed]

64. Veeraraghavalu, K.; Pett, M.; Kumar, R.V.; Nair, P.; Rangarajan, A.; Stanley, M.A.; Krishna, S. Papillomavirus-mediated neoplastic progression is associated with reciprocal changes in JAGGED1 and manic fringe expression linked to notch activation. *J. Virol.* **2004**, *78*, 8687–8700. [CrossRef] [PubMed]

65. Scheffner, M.; Munger, K.; Byrne, J.C.; Howley, P.M. The state of the p53 and retinoblastoma genes in human cervical carcinoma cell lines. *Proc. Natl. Acad. Sci. USA* **1991**, *88*, 5523–5527. [CrossRef] [PubMed]

66. Meyers, C.; Frattini, M.G.; Hudson, J.B.; Laimins, L.A. Biosynthesis of human papillomavirus from a continuous cell line upon epithelial differentiation. *Science* **1992**, *257*, 971–973. [CrossRef] [PubMed]

67. McLaughlin-Drubin, M.E.; Wilson, S.; Mullikin, B.; Suzich, J.; Meyers, C. Human papillomavirus type 45 propagation, infection, and neutralization. *Virology* **2003**, *312*, 1–7. [CrossRef]

68. Isaacson Wechsler, E.; Wang, Q.; Roberts, I.; Pagliarulo, E.; Jackson, D.; Untersperger, C.; Coleman, N.; Griffin, H.; Doorbar, J. Reconstruction of human papillomavirus type 16-mediated early-stage neoplasia implicates E6/E7 deregulation and the loss of contact inhibition in neoplastic progression. *J. Virol.* **2012**, *86*, 6358–6364. [CrossRef]

69. Duenas-Gonzalez, A.; Lizano, M.; Candelaria, M.; Cetina, L.; Arce, C.; Cervera, E. Epigenetics of cervical cancer. An overview and therapeutic perspectives. *Mol. Cancer* **2005**, *4*, 38. [CrossRef]

70. Stanley, M.A.; Browne, H.M.; Appleby, M.; Minson, A.C. Properties of a non–tumorigenic human cervical keratinocyte cell line. *Int. J. Cancer* **1989**, *43*, 672–676. [CrossRef]

71. Gray, E.; Pett, M.R.; Ward, D.; Winder, D.M.; Stanley, M.A.; Roberts, I.; Scarpini, C.G.; Coleman, N. In vitro progression of human papillomavirus 16 episome-associated cervical neoplasia displays fundamental similarities to integrant–associated carcinogenesis. *Cancer Res.* **2010**, *70*, 4081–4091. [CrossRef] [PubMed]

72. Stoppler, H.; Hartmann, D.P.; Sherman, L.; Schlegel, R. The human papillomavirus type 16 E6 and E7 oncoproteins dissociate cellular telomerase activity from the maintenance of telomere length. *J. Biol. Chem.* **1997**, *272*, 13332–13337. [CrossRef] [PubMed]

73. Kiyono, T.; Foster, S.A.; Koop, J.I.; McDougall, J.K.; Galloway, D.A.; Klingelhutz, A.J. Both Rb/p16INK4a inactivation and telomerase activity are required to immortalize human epithelial cells. *Nature* **1998**, *396*, 84–88. [CrossRef] [PubMed]

74. Blanton, R.A.; Perez-Reyes, N.; Merrick, D.T.; McDougall, J.K. Epithelial cells immortalized by human papillomaviruses have premalignant characteristics in organotypic culture. *Am. J. Pathol.* **1991**, *138*, 673–685.

75. Farwell, D.G.; Shera, K.A.; Koop, J.I.; Bonnet, G.A.; Matthews, C.P.; Reuther, G.W.; Coltrera, M.D.; McDougall, J.K.; Klingelhutz, A.J. Genetic and epigenetic changes in human epithelial cells immortalized by telomerase. *Am. J. Pathol.* **2000**, *156*, 1537–1547. [CrossRef]

76. Schutze, D.M.; Snijders, P.J.; Bosch, L.; Kramer, D.; Meijer, C.J.; Steenbergen, R.D. Differential in vitro immortalization capacity of eleven (probable) [corrected] high-risk human papillomavirus types. *J. Virol.* **2014**, *88*, 1714–1724. [CrossRef]

77. Steenbergen, R.D.; Walboomers, J.M.; Meijer, C.J.; van der Raaij-Helmer, E.M.; Parker, J.N.; Chow, L.T.; Broker, T.R.; Snijders, P.J. Transition of human papillomavirus type 16 and 18 transfected human foreskin keratinocytes towards immortality: Activation of telomerase and allele losses at 3p, 10p, 11q and/or 18q. *Oncogene* **1996**, *13*, 1249–1257.

78. Bosch, F.X.; Leube, R.E.; Achtstatter, T.; Moll, R.; Franke, W.W. Expression of simple epithelial type cytokeratins in stratified epithelia as detected by immunolocalization and hybridization in situ. *J. Cell Biol.* **1988**, *106*, 1635–1648. [CrossRef]

79. Gasparoni, A.; Squier, C.A.; Fonzi, L. Cytokeratin changes in cell culture systems of epithelial cells isolated from oral mucosa: A short review. *Ital. J. Anat. Embryol.* **2005**, *110*, 75–82.

80. Roberts, S.; Evans, D.; Mehanna, H.; Parish, J.L. Modelling human papillomavirus biology in oropharyngeal keratinocytes. *Philos. Trans. R. Soc. Lond. B Biol. Sci.* **2019**, *374*, 20180289. [CrossRef]

81. Wira, C.R.; Rodriguez-Garcia, M.; Patel, M.V. The role of sex hormones in immune protection of the female reproductive tract. *Nat. Rev. Immunol.* **2015**, *15*, 217–230. [CrossRef] [PubMed]

82. Liu, Y.Z.; Wang, T.T.; Zhang, Y.Z. A modified method for the culture of naturally HPV–infected high–grade cervical intraepithelial neoplasia keratinocytes from human neoplastic cervical biopsies. *Oncol. Lett.* **2016**, *11*, 1457–1462. [CrossRef] [PubMed]

83. De Gregorio, V.; Imparato, G.; Urciuolo, F.; Tornesello, M.L.; Annunziata, C.; Buonaguro, F.M.; Netti, P.A. An Engineered Cell-Instructive Stroma for the Fabrication of a Novel Full Thickness Human Cervix Equivalent In Vitro. *Adv. Healthc. Mater.* **2017**, *6*. [CrossRef] [PubMed]

84. Villa, P.L.; Jackson, R.; Eade, S.; Escott, N.; Zehbe, I. Isolation of Biopsy–Derived, Human Cervical Keratinocytes Propagated as Monolayer and Organoid Cultures. *Sci. Rep.* **2018**, *8*, 17869. [CrossRef] [PubMed]

85. Bissell, M.J.; Radisky, D. Putting tumours in context. *Nat. Rev. Cancer* **2001**, *1*, 46–54. [CrossRef] [PubMed]

86. Jiang, E.; Yan, T.; Xu, Z.; Shang, Z. Tumor Microenvironment and Cell Fusion. *Biomed. Res. Int.* **2019**, *2019*, 5013592. [CrossRef] [PubMed]

87. Bonnans, C.; Chou, J.; Werb, Z. Remodelling the extracellular matrix in development and disease. *Nat. Rev. Mol. Cell Biol.* **2014**, *15*, 786–801. [CrossRef]

88. Poltavets, V.; Kochetkova, M.; Pitson, S.M.; Samuel, M.S. The Role of the Extracellular Matrix and Its Molecular and Cellular Regulators in Cancer Cell Plasticity. *Front. Oncol.* **2018**, *8*, 431. [CrossRef] [PubMed]

89. McBride, A.A.; Munger, K. Expert Views on HPV Infection. *Viruses* **2018**, *10*, 94. [CrossRef]

90. Llames, S.; Garcia-Perez, E.; Meana, A.; Larcher, F.; del Rio, M. Feeder Layer Cell Actions and Applications. *Tissue Eng. Part B. Rev.* **2015**, *21*, 345–353. [CrossRef]

91. Schumacher, M.; Schuster, C.; Rogon, Z.M.; Bauer, T.; Caushaj, N.; Baars, S.; Szabowski, S.; Bauer, C.; Schorpp-Kistner, M.; Hess, J.; et al. Efficient keratinocyte differentiation strictly depends on JNK–induced soluble factors in fibroblasts. *J. Invest. Dermatol.* **2014**, *134*, 1332–1341. [CrossRef] [PubMed]

92. Maas-Szabowski, N.; Shimotoyodome, A.; Fusenig, N.E. Keratinocyte growth regulation in fibroblast cocultures via a double paracrine mechanism. *J. Cell Sci.* **1999**, *112*, 1843–1853. [PubMed]

93. Maas-Szabowski, N.; Stark, H.J.; Fusenig, N.E. Keratinocyte growth regulation in defined organotypic cultures through IL-1-induced keratinocyte growth factor expression in resting fibroblasts. *J. Invest. Dermatol.* **2000**, *114*, 1075–1084. [CrossRef] [PubMed]

94. Dall, K.L.; Scarpini, C.G.; Roberts, I.; Winder, D.M.; Stanley, M.A.; Muralidhar, B.; Herdman, M.T.; Pett, M.R.; Coleman, N. Characterization of naturally occurring HPV16 integration sites isolated from cervical keratinocytes under noncompetitive conditions. *Cancer Res.* **2008**, *68*, 8249–8259. [CrossRef]

95. Deng, H.; Hillpot, E.; Mondal, S.; Khurana, K.K.; Woodworth, C.D. HPV16–Immortalized Cells from Human Transformation Zone and Endocervix are More Dysplastic than Ectocervical Cells in Organotypic Culture. *Sci. Rep.* **2018**, *8*, 15402. [CrossRef]

96. Rijal, G.; Li, W. Native-mimicking in vitro microenvironment: An elusive and seductive future for tumor modeling and tissue engineering. *J. Biol. Eng.* **2018**, *12*, 20. [CrossRef]

97. Maufort, J.P.; Shai, A.; Pitot, H.C.; Lambert, P.F. A role for HPV16 E5 in cervical carcinogenesis. *Cancer Res.* **2010**, *70*, 2924–2931. [CrossRef]

98. Hogervorst, M.; Rietveld, M.; de Gruijl, F.; El Ghalbzouri, A. A shift from papillary to reticular fibroblasts enables tumour–stroma interaction and invasion. *Br. J. Cancer* **2018**, *118*, 1089–1097. [CrossRef]

99. Ridky, T.W.; Chow, J.M.; Wong, D.J.; Khavari, P.A. Invasive three–dimensional organotypic neoplasia from multiple normal human epithelia. *Nat. Med.* **2010**, *16*, 1450–1455. [CrossRef]

100. Pickard, A.; Cichon, A.C.; Barry, A.; Kieran, D.; Patel, D.; Hamilton, P.; Salto-Tellez, M.; James, J.; McCance, D.J. Inactivation of Rb in stromal fibroblasts promotes epithelial cell invasion. *EMBO J.* **2012**, *31*, 3092–3103. [CrossRef]

101. Huang, G.; Li, F.; Zhao, X.; Ma, Y.; Li, Y.; Lin, M.; Jin, G.; Lu, T.J.; Genin, G.M.; Xu, F. Functional and Biomimetic Materials for Engineering of the Three–Dimensional Cell Microenvironment. *Chem. Rev.* **2017**, *117*, 12764–12850. [CrossRef] [PubMed]

102. Oxford, J.T.; Reeck, J.C.; Hardy, M.J. Extracellular Matrix in Development and Disease. *Int. J. Mol. Sci.* **2019**, *20*, 205. [CrossRef] [PubMed]

103. Frantz, C.; Stewart, K.M.; Weaver, V.M. The extracellular matrix at a glance. *J. Cell Sci.* **2010**, *123*, 4195–4200. [CrossRef] [PubMed]

104. Amar, S.; Smith, L.; Fields, G.B. Matrix metalloproteinase collagenolysis in health and disease. *Biochim. Biophys. Acta Mol. Cell Res.* **2017**, *1864*, 1940–1951. [CrossRef]

105. Kessenbrock, K.; Plaks, V.; Werb, Z. Matrix metalloproteinases: Regulators of the tumor microenvironment. *Cell* **2010**, *141*, 52–67. [CrossRef]

106. Fullar, A.; Dudas, J.; Olah, L.; Hollosi, P.; Papp, Z.; Sobel, G.; Karaszi, K.; Paku, S.; Baghy, K.; Kovalszky, I. Remodeling of extracellular matrix by normal and tumor–associated fibroblasts promotes cervical cancer progression. *BMC Cancer* **2015**, *15*, 256. [CrossRef]

107. Bergmann, S.; Schoenen, H.; Hammerschmidt, S. The interaction between bacterial enolase and plasminogen promotes adherence of Streptococcus pneumoniae to epithelial and endothelial cells. *Int. J. Med. Microbiol.* **2013**, *303*, 452–462. [CrossRef]

108. Zhai, Y.; Hotary, K.B.; Nan, B.; Bosch, F.X.; Munoz, N.; Weiss, S.J.; Cho, K.R. Expression of membrane type 1 matrix metalloproteinase is associated with cervical carcinoma progression and invasion. *Cancer Res.* **2005**, *65*, 6543–6550. [CrossRef]

109. Sahebali, S.; Van den Eynden, G.; Murta, E.F.; Michelin, M.A.; Cusumano, P.; Petignat, P.; Bogers, J.J. Stromal issues in cervical cancer: A review of the role and function of basement membrane, stroma, immune response and angiogenesis in cervical cancer development. *Eur. J. Cancer Prev.* **2010**, *19*, 204–215. [CrossRef]

110. Ayala, G.E.; Dai, H.; Powell, M.; Li, R.; Ding, Y.; Wheeler, T.M.; Shine, D.; Kadmon, D.; Thompson, T.; Miles, B.J.; et al. Cancer-related axonogenesis and neurogenesis in prostate cancer. *Clin. Cancer Res.* **2008**, *14*, 7593–7603. [CrossRef]

111. Magnon, C.; Hall, S.J.; Lin, J.; Xue, X.; Gerber, L.; Freedland, S.J.; Frenette, P.S. Autonomic nerve development contributes to prostate cancer progression. *Science* **2013**, *341*, 1236361. [CrossRef] [PubMed]

112. Lucido, C.T.; Wynja, E.; Madeo, M.; Williamson, C.S.; Schwartz, L.E.; Imblum, B.A.; Drapkin, R.; Vermeer, P.D. Innervation of cervical carcinoma is mediated by cancer–derived exosomes. *Gynecol. Oncol.* **2019**, *154*, 228–235. [CrossRef] [PubMed]

113. Lee, J.S.; Kim, H.S.; Park, J.T.; Lee, M.C.; Park, C.S. Expression of vascular endothelial growth factor in the progression of cervical neoplasia and its relation to angiogenesis and p53 status. *Anal. Quant. Cytol. Histol.* **2003**, *25*, 303–311. [PubMed]

114. Smith-McCune, K.K.; Weidner, N. Demonstration and characterization of the angiogenic properties of cervical dysplasia. *Cancer Res.* **1994**, *54*, 800–804.

115. Chen, W.; Li, F.; Mead, L.; White, H.; Walker, J.; Ingram, D.A.; Roman, A. Human papillomavirus causes an angiogenic switch in keratinocytes which is sufficient to alter endothelial cell behavior. *Virology* **2007**, *367*, 168–174. [CrossRef]

116. Mazio, C.; Casale, C.; Imparato, G.; Urciuolo, F.; Attanasio, C.; De Gregorio, M.; Rescigno, F.; Netti, P.A. Pre–vascularized dermis model for fast and functional anastomosis with host vasculature. *Biomaterials* **2019**, *192*, 159–170. [CrossRef]

117. D'Anna, R.; Le Buanec, H.; Alessandri, G.; Caruso, A.; Burny, A.; Gallo, R.; Zagury, J.F.; Zagury, D.; D'Alessio, P. Selective activation of cervical microvascular endothelial cells by human papillomavirus 16-e7 oncoprotein. *J. Natl. Cancer Inst.* **2001**, *93*, 1843–1851. [CrossRef]

118. Mazibrada, J.; Ritta, M.; Mondini, M.; De Andrea, M.; Azzimonti, B.; Borgogna, C.; Ciotti, M.; Orlando, A.; Surico, N.; Chiusa, L.; et al. Interaction between inflammation and angiogenesis during different stages of cervical carcinogenesis. *Gynecol. Oncol.* **2008**, *108*, 112–120. [CrossRef]

119. Gun, S.Y.; Lee, S.W.L.; Sieow, J.L.; Wong, S.C. Targeting immune cells for cancer therapy. *Redox Biol.* **2019**, *25*, 101174. [CrossRef]

120. Kyrgiou, M.; Mitra, A.; Moscicki, A.B. Does the vaginal microbiota play a role in the development of cervical cancer? *Transl. Res.* **2017**, *179*, 168–182. [CrossRef]

121. Belkaid, Y.; Harrison, O.J. Homeostatic Immunity and the Microbiota. *Immunity* **2017**, *46*, 562–576. [CrossRef] [PubMed]

122. Ayehunie, S.; Islam, A.; Cannon, C.; Landry, T.; Pudney, J.; Klausner, M.; Anderson, D.J. Characterization of a Hormone-Responsive Organotypic Human Vaginal Tissue Model: Morphologic and Immunologic Effects. *Reprod. Sci.* **2015**, *22*, 980–990. [CrossRef] [PubMed]

123. Gorodeski, G.I.; Romero, M.F.; Hopfer, U.; Rorke, E.; Utian, W.H.; Eckert, R.L. Human uterine cervical epithelial cells grown on permeable support—A new model for the study of differentiation. *Differentiation* **1994**, *56*, 107–118. [PubMed]

124. Heroiu Cataloiu, A.D.; Danciu, C.E.; Popescu, C.R. Multiple cancers of the head and neck. *Maedica (Buchar.)* **2013**, *8*, 80–85.

125. Peltanova, B.; Raudenska, M.; Masarik, M. Effect of tumor microenvironment on pathogenesis of the head and neck squamous cell carcinoma: A systematic review. *Mol. Cancer* **2019**, *18*, 63. [CrossRef]

126. Tumban, E. A Current Update on Human Papillomavirus-Associated Head and Neck Cancers. *Viruses* **2019**, *11*, 992. [CrossRef]

127. Pan, C.; Issaeva, N.; Yarbrough, W.G. HPV-driven oropharyngeal cancer: Current knowledge of molecular biology and mechanisms of carcinogenesis. *Cancers Head Neck* **2018**, *3*, 12. [CrossRef]

128. Lee, S.H.; Lee, C.R.; Rigas, N.K.; Kim, R.H.; Kang, M.K.; Park, N.H.; Shin, K.H. Human papillomavirus 16 (HPV16) enhances tumor growth and cancer stemness of HPV-negative oral/oropharyngeal squamous cell carcinoma cells via miR-181 regulation. *Papillomavirus Res.* **2015**, *1*, 116–125. [CrossRef]

129. Reddout, N.; Christensen, T.; Bunnell, A.; Jensen, D.; Johnson, D.; O'Malley, S.; Kingsley, K. High risk HPV types 18 and 16 are potent modulators of oral squamous cell carcinoma phenotypes in vitro. *Infect. Agent Cancer* **2007**, *2*, 21. [CrossRef]

130. Scanlon, C.S.; Van Tubergen, E.A.; Chen, L.C.; Elahi, S.F.; Kuo, S.; Feinberg, S.; Mycek, M.A.; D'Silva, N.J. Characterization of squamous cell carcinoma in an organotypic culture via subsurface non–linear optical molecular imaging. *Exp. Biol. Med. (Maywood)* **2013**, *238*, 1233–1241. [CrossRef]

131. Dalley, A.J.; AbdulMajeed, A.A.; Upton, Z.; Farah, C.S. Organotypic culture of normal, dysplastic and squamous cell carcinoma–derived oral cell lines reveals loss of spatial regulation of CD44 and p75 NTR in malignancy. *J. Oral. Pathol. Med.* **2013**, *42*, 37–46. [CrossRef] [PubMed]

132. Herfs, M.; Soong, T.R.; Delvenne, P.; Crum, C.P. Deciphering the Multifactorial Susceptibility of Mucosal Junction Cells to HPV Infection and Related Carcinogenesis. *Viruses* **2017**, *9*, 85. [CrossRef] [PubMed]

133. Israr, M.; Biryukov, J.; Ryndock, E.J.; Alam, S.; Meyers, C. Comparison of human papillomavirus type 16 replication in tonsil and foreskin epithelia. *Virology* **2016**, *499*, 82–90. [CrossRef] [PubMed]

134. Lyford-Pike, S.; Peng, S.; Young, G.D.; Taube, J.M.; Westra, W.H.; Akpeng, B.; Bruno, T.C.; Richmon, J.D.; Wang, H.; Bishop, J.A.; et al. Evidence for a role of the PD-1:PD-L1 pathway in immune resistance of HPV-associated head and neck squamous cell carcinoma. *Cancer Res.* **2013**, *73*, 1733–1741. [CrossRef]

135. Ward, M.J.; Thirdborough, S.M.; Mellows, T.; Riley, C.; Harris, S.; Suchak, K.; Webb, A.; Hampton, C.; Patel, N.N.; Randall, C.J.; et al. Tumour-infiltrating lymphocytes predict for outcome in HPV-positive oropharyngeal cancer. *Br. J. Cancer* **2014**, *110*, 489–500. [CrossRef]

136. Hoang, L.N.; Park, K.J.; Soslow, R.A.; Murali, R. Squamous precursor lesions of the vulva: Current classification and diagnostic challenges. *Pathology* **2016**, *48*, 291–302. [CrossRef]

137. Alkatout, I.; Schubert, M.; Garbrecht, N.; Weigel, M.T.; Jonat, W.; Mundhenke, C.; Gunther, V. Vulvar cancer: Epidemiology, clinical presentation, and management options. *Int. J. Womens Health* **2015**, *7*, 305–313. [CrossRef]

138. Del Pino, M.; Rodriguez-Carunchio, L.; Ordi, J. Pathways of vulvar intraepithelial neoplasia and squamous cell carcinoma. *Histopathology* **2013**, *62*, 161–175. [CrossRef]

139. Bleeker, M.C.; Visser, P.J.; Overbeek, L.I.; van Beurden, M.; Berkhof, J. Lichen Sclerosus: Incidence and Risk of Vulvar Squamous Cell Carcinoma. *Cancer Epidemiol. Biomarkers Prev.* **2016**, *25*, 1224–1230. [CrossRef]

140. Dongre, H.; Rana, N.; Fromreide, S.; Rajthala, S.; Boe Engelsen, I.; Paradis, J.; Gutkind, J.S.; Vintermyr, O.K.; Johannessen, A.C.; Bjorge, L.; et al. Establishment of a novel cancer cell line derived from vulvar carcinoma associated with lichen sclerosus exhibiting a fibroblast–dependent tumorigenic potential. *Exp. Cell Res.* **2020**, *386*, 111684. [CrossRef]

141. Van Staveren, W.C.; Solis, D.Y.; Hebrant, A.; Detours, V.; Dumont, J.E.; Maenhaut, C. Human cancer cell lines: Experimental models for cancer cells in situ? For cancer stem cells? *Biochim. Biophys. Acta* **2009**, *1795*, 92–103. [CrossRef] [PubMed]

142. Jung, J. Human tumor xenograft models for preclinical assessment of anticancer drug development. *Toxicol. Res.* **2014**, *30*, 1–5. [CrossRef] [PubMed]

143. Merbah, M.; Introini, A.; Fitzgerald, W.; Grivel, J.C.; Lisco, A.; Vanpouille, C.; Margolis, L. Cervico-vaginal tissue ex vivo as a model to study early events in HIV-1 infection. *Am. J. Reprod. Immunol.* **2011**, *65*, 268–278. [CrossRef] [PubMed]

144. Cheng, Y.; Geng, L.; Zhao, L.; Zuo, P.; Wang, J. Human papillomavirus E6-regulated microRNA-20b promotes invasion in cervical cancer by targeting tissue inhibitor of metalloproteinase 2. *Mol. Med. Rep.* **2017**, *16*, 5464–5470. [CrossRef]

145. Yang, E.J.; Quick, M.C.; Hanamornroongruang, S.; Lai, K.; Doyle, L.A.; McKeon, F.D.; Xian, W.; Crum, C.P.; Herfs, M. Microanatomy of the cervical and anorectal squamocolumnar junctions: A proposed model for anatomical differences in HPV-related cancer risk. *Mod. Pathol.* **2015**, *28*, 994–1000. [CrossRef] [PubMed]

146. Genther, S.M.; Sterling, S.; Duensing, S.; Munger, K.; Sattler, C.; Lambert, P.F. Quantitative role of the human papillomavirus type 16 E5 gene during the productive stage of the viral life cycle. *J. Virol.* **2003**, *77*, 2832–2842. [CrossRef]

147. Wechsler, E.I.; Tugizov, S.; Herrera, R.; Da Costa, M.; Palefsky, J.M. E5 can be expressed in anal cancer and leads to epidermal growth factor receptor–induced invasion in a human papillomavirus 16-transformed anal epithelial cell line. *J. Gen. Virol.* **2018**, *99*, 631–644. [CrossRef]

148. Ganor, Y.; Zhou, Z.; Bodo, J.; Tudor, D.; Leibowitch, J.; Mathez, D.; Schmitt, A.; Vacher–Lavenu, M.C.; Revol, M.; Bomsel, M. The adult penile urethra is a novel entry site for HIV–1 that preferentially targets resident urethral macrophages. *Mucosal. Immunol.* **2013**, *6*, 776–786. [CrossRef]

149. Ma, L.M.; Wang, Z.; Wang, H.; Li, R.S.; Zhou, J.; Liu, B.C.; Baskin, L.S. Estrogen effects on fetal penile and urethral development in organotypic mouse genital tubercle culture. *J. Urol.* **2009**, *182*, 2511–2517. [CrossRef]

150. Pfister, H. Chapter 8: Human papillomavirus and skin cancer. *J. Natl. Cancer Inst. Monogr.* **2003**, *31*, 52–56. [CrossRef]

151. Harwood, C.A.; McGregor, J.M.; Proby, C.M.; Breuer, J. Human papillomavirus and the development of non-melanoma skin cancer. *J. Clin. Pathol.* **1999**, *52*, 249–253. [CrossRef] [PubMed]

152. Akgul, B.; Cooke, J.C.; Storey, A. HPV-associated skin disease. *J. Pathol.* **2006**, *208*, 165–175. [CrossRef] [PubMed]

153. Akgul, B.; Garcia-Escudero, R.; Ghali, L.; Pfister, H.J.; Fuchs, P.G.; Navsaria, H.; Storey, A. The E7 protein of cutaneous human papillomavirus type 8 causes invasion of human keratinocytes into the dermis in organotypic cultures of skin. *Cancer Res.* **2005**, *65*, 2216–2223. [CrossRef] [PubMed]

 *cancers*

*Article*

# Post-Treatment HPV Surface Brushings and Risk of Relapse in Oropharyngeal Carcinoma


Barbara Kofler [1,*], Wegene Borena [2], Jozsef Dudas [1], Veronika Innerhofer [1], Daniel Dejaco [1], Teresa B Steinbichler [1], Gerlig Widmann [3], Dorothee von Laer [2] and Herbert Riechelmann [1]

[1]  Department of Otorhinolaryngology, Medical University of Innsbruck, Anichstrasse 35, 6020 Innsbruck, Austria; jozsef.dudas@i-med.ac.at (J.D.); veronika.innerhofer@tirol-kliniken.at (V.I.); daniel.dejaco@i-med.ac.at (D.D.); teresa.steinbichler@i-med.ac.at (T.B.S.); herbert.riechelmann@i-med.ac.at (H.R.)

[2]  Division of Virology, Department of Hygiene, Microbiology, Social Medicine, Medical University of Innsbruck, Peter-Mayr-Strasse 4b, 6020 Innsbruck, Austria; Wegene.Borena@i-med.ac.at (W.B.); dorothee.von-Laer@i-med.ac.at (D.v.L.)

[3]  Department of Radiology, Medical University of Innsbruck, Anichstrasse 35, 6020 Innsbruck, Austria; gerlig.widmann@i-med.ac.at

*  Correspondence: ba.kofler@tirol-kliniken.at; Tel.: +0043-43-50-504-23141; Fax: 0043-43-50-504-23144


Received: 27 March 2020; Accepted: 23 April 2020; Published: 25 April 2020


**Abstract:** Human papillomavirus (HPV)-positive oropharyngeal squamous cell carcinoma (OPSCC) is a distinct subtype of head and neck cancer. Here, we investigated how frequently brushing remained high-risk (hr)-HPV positive after treatment and whether patients with positive post-treatment brushings have a higher recurrence rate. Following the end of treatment of patients with initially hr-HPV positive OPSCC, surface brushings from the previous tumor site were performed and tested for hr-HPV DNA. Of 62 patients with initially hr-HPV DNA-positive OPSCC, seven patients remained hr-HPV-DNA positive at post-treatment follow-up. Of the seven hr-HPV-positive patients at follow-up, five had a tumor relapse or tumor progression, of whom three died. The majority of patients (55/62) was HPV-negative following treatment. All HPV-negative patients remained free of disease ($p = 0.0007$). In this study, all patients with recurrence were hr-HPV-positive with the same genotype as that before treatment. In patients who were hr-HPV negative after treatment, no recurrence was observed.

**Keywords:** oropharyngeal squamous cell carcinoma; human papillomavirus; recurrence; surface brushing; EGFR


---

## 1. Introduction

Approximately 85% of adults acquire a human papillomavirus (HPV) infection in their life. Most HPV infections are transient, asymptomatic, and eliminated by the immune system [1,2]. However, HPV viral infection can persist latently in a subset of the population. Individuals with persistent high-risk HPV (hr-HPV) infection may acquire epithelial cell abnormalities and subsequently develop cancers at the site of infection [3]. Persistent hr-HPV infection is particularly associated with cervical, anogenital and oropharyngeal cancers [4,5]. Hr-HPV positive oropharyngeal squamous cell carcinoma (OPSCC) is a distinct subtype of head and neck carcinoma. Risk factors that may prevent the natural clearance of oropharyngeal hr-HPV infection are genetic and lifestyle factors like smoking and alcohol consumption [6]. The prevalence of cancer of the oropharynx due to hr-HPV infection has increased, particularly in North America and Europe [7]. It differs from HPV-negative head and neck squamous cell carcinoma (HNSCC) by its risk factor profile, clinical behavior, and molecular biology. Compared to HPV-negative HNSCC, hr-HPV positive OPSCC better responds to treatment and has a significantly better prognosis [8].

It is unclear whether hr-HPV persists in oropharyngeal tissues in patients with hr-HPV-positive OPSCC following cancer treatment and which consequences this might have. Zhang and coworkers collected blood at diagnosis and post-treatment in 64 patients with p16-positive OPSCC to test for serum antibodies to E6 and E7 proteins of HPV 16. At diagnosis, most patients were seropositive to HPV 16 E6 (85%). In the post therapeutic samples, HPV 16 antibody levels decreased slowly over time, but only three patients became seronegative [9]. In another study, salivary and serum immunoglobulin G (IgG) antibodies targeting E2, E6, and E7 were measured in 44 patients with OPSCC at the beginning and 6–7 weeks following the completion of treatment. In this study, E7-directed antibodies were detected in saliva in most of the patients and were associated with the HPV status. The median of salivary E7 antibody levels decreased significantly post-treatment [10].

Fakhry and coauthors used oral rinse samples for HPV detection in 396 patients with oral and oropharyngeal cancer, of which 51% were HPV-positive before therapy. After treatment the HPV prevalence decreased. In patients who received surgical resection, the HPV prevalence decreased from 69.2% to 13.7%. In a subset of patients who required postoperative radiotherapy, the HPV prevalence decreased from 70% to 38% after surgery and then to 1% after radiotherapy. HPV detection in oral rinses was performed several times for patients who received radiotherapy. The median time to clearance was 42 days (95%CI, 37–49 days). The only factor significantly associated with reduced clearance was current smoking. HPV-positivity with the same genotype was detected after treatment in 14.3% of initially HPV-positive patients and among these patients, the cumulative incidence of recurrence was 45.3%. HPV DNA detection after completion of therapy was significantly associated with increased risk of recurrence and death [11].

In cervical dysplasia, hr-HPV infection can persist following treatment and promote disease recurrence. Söderlund-Strand et al. performed a long-term follow-up study obtaining cervical samples for HPV DNA testing and cytological analysis from 178 women with abnormal smears referred for conization. Three years after treatment 3.1% of women were persistently HPV-positive with the same HPV genotype as before treatment. Recurrent or residual cervical intraepithelial neoplasia (CIN) in histopathology was found among 9 (5.1%) women during follow-up. All these women had a type-specific HPV persistence. The authors concluded that only type-specific HPV persistence predicted recurrent or residual disease [12].

In a previous publication, we reported that surface brushings from oropharyngeal cancer reliably detect HPV-DNA. In 53 patients with OPSCC, sensitivity and specificity of the brush test was 86% (95%CI: 70–95%) and 89% (95%CI: 65–99%) [13]. Also, Broglie et al. reported liquid-based brush cytology specimens from oropharyngeal lesions to be a reliable method to identify patients with hr-HPV OPSCC. The authors collected brush cytology specimens prospectively from 50 patients with OPSCC. The accuracy, sensitivity, specificity, positive predictive value, and negative predictive value of brush cytology to identify hr-HPV-DNA-positive and p16-positive OPSCC samples were 88%, 83%, 94%, 95%, and 81%, respectively [14].

In this study, we examined whether hr-HPV is still detectable in surface brushings after treatment in patients with initially hr-HPV-positive OPSCC. Moreover, we compared the course of disease in patients with and without post-treatment hr-HPV in oropharyngeal brushings.

## 2. Results

### 2.1. Study Population and Treatment Outcome

During the study period, 74 patients with hr-HPV DNA-positive OPSCC before treatment were included. From 12 patients, no post-treatment brushings were available because they missed the follow-up or received follow-up in another hospital. The mean age of the remaining 62 patients with post-treatment brushings was 61 years. Forty-eight patients were male, and 14 were female. Most patients had UICC stage IVa and were treated with radiochemotherapy (Table 1). After treatment, all patients (except three) were in complete remission. One patient developed early pulmonal metastasis with complete remission at the primary site, another patient was in partial remission, and the third patient had already disseminated disease during initial diagnosis with pulmonal and bone metastases. This patient received a palliative systemic therapy. The median follow-up time was 29.3 months (95% CI: 25–34 months).

**Table 1.** Study population.

| Variation | Patients ($n = 62$) |
|---|---|
| Gender | |
| Male | 48 |
| Female | 14 |
| Age during Initial Diagnosis | |
| $\leq 50$ | 10 |
| 51–60 | 21 |
| 61–70 | 20 |
| 71–80 | 7 |
| >80 | 4 |
| UICC Stage | |
| stage I | 0 |
| stage II | 5 |
| Stage III | 18 |
| stage IVa | 32 |
| stage IVb | 4 |
| Stage IVc | 2 |
| ASA Score | |
| ASA I/II | 52 |
| ASA III/IV | 9 |
| Therapy | |
| Surgery only | 3 |
| Surgery and PORT | 12 |
| Surgery and RCT/RIT | 5 |
| Primary RCT/RIT | 38 |
| Primary RT | 2 |
| Chemotherapy only | 1 |
| P16 | |
| Negative | 5 |
| Positive | 55 |

ASA, American Society of Anesthesiologists; UICC, Union internationale contre le cancer; PORT, postoperative radiatiotherapy; RCT, radiochemotherapy; RIT, radioimmunotherapy; RT, radiotherapy.

## 2.2. HPV DNA Detection before and after Treatment

Before treatment, the most common HPV subtype was HPV 16 (Table 2). In 60/62 pre-HPV+ patients, p16 immunohistochemistry was available. In 55/60 patients, p16 was positive, and in five patients, it was negative. In the five patients with p16-negative IHC, the HPV genotypes 16, 16, 18, 16 and 40 (multiple infection), and 62 and 82 (multiple infection) were detected by pre-treatment brushing. At follow-up, 7/62 (11.3%) brushings from the previous primary tumor site were hr-HPV

DNA-positive. In five patients, the post-treatment HPV brush test showed the same genotype as before therapy, namely HPV 16 in three patients and HPV 33 and HPV 18 each in one patient. All patients with HPV positivity after therapy with the same genotype as before developed a recurrence or progressive disease.

**Table 2.** Human papillomavirus (HPV) genotypes before and after therapy.

| Patient Number | HPV Genotype before Therapy | HPV Genotype after Therapy | Recurrent Tumor/Tumor Progression |
|---|---|---|---|
| 1–50 | single hr-HPV type * | HPV negative | No |
| 51–55 | multiple hr-HPV types ** | HPV negative | No |
| 56–58 | hr-HPV 16 | hr-HPV 16 | Yes |
| 59 | hr-HPV 33 | hr-HPV 33 | Yes |
| 60 | hr-HP 18 | Hr-HPV 18 | Yes |
| 61 | hr-HPV 33 | hr-HPV 16 | No |
| 62 | hr-HPV 18 | hr-HPV 68b | No |

HPV, human papilloma virus; hr, high risk; * single hr-HPV infection with one of the genotypes HPV 16, 33, 35, or 58; **multiple hr-HPV infections, including almost 1 hr-HPV genotype.

The post-treatment brushing was taken during complete response in one of the five patients positive with the same genotype. This patient with initially cT4cN2cM0 oropharyngeal cancer (Figure 1A) was disease-free after primary radiochemotherapy (Figure 1B) for 17 months and got a recurrence involving the hypopharynx three months after the post-treatment brush was obtained (Figure 1C). Another patient had a complete remission at the primary site but developed new pulmonary metastasis. The positive brushing was taken from normal oropharyngeal mucosa. The third patient had an early recurrence, and the post-treatment brushing was taken from the recurrent tumor surface and the fourth patient was in a palliative setting after diagnosis with a cT3cN2cM1 oropharyngeal cancer, he received four cycles of Carboplatin and Cetuximab and then Nivolumab 240mg/every two weeks. The fifth patient was diagnosed with an initially cT4cN2cM0 oropharyngeal cancer and was treated with primary radiochemotherapy. In this patient, partial remission 12 weeks and 18 weeks after therapy was found. The positive brush was taken from the surface of residual tumor.

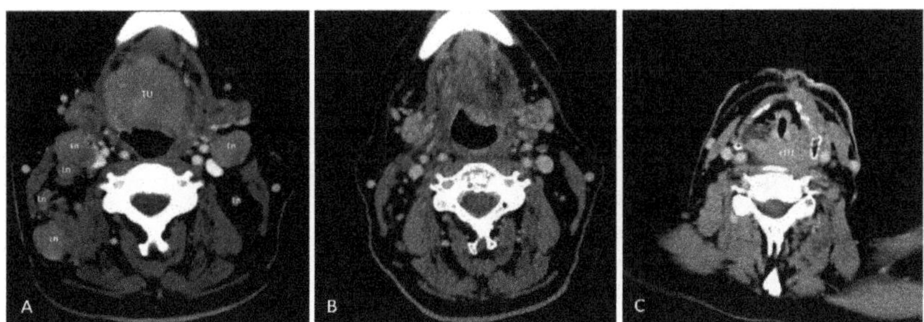

**Figure 1.** Patient presentation in radiological series. Ln, lymph nodes; TU, tumor; rTU, recurrent tumor. (**A**) Oropharyngeal cancer cT4cN2cM0. (**B**) Full remission after primary radiochemotherapy. (**C**) Recurrence involving the hypopharynx.

In two patients, post-treatment brushing was positive for another hr-HPV genotype than before therapy. The genotypes were HPV 16 and 33 before therapy and, in the same order of patients, HPV 68b and 16 thereafter. These two patients were in full remission after treatment, no recurrence or tumor persistence was stated in these patients. After treatment, one hr-negative patient was positive with the low-risk type HPV 6, and 54 patients were negative for all investigated HPV strains.

### 2.3. Post-Treatment HPV-Positivity, Recurrent Disease, and Survival

In 4/7 patients with positive post-treatment hr-HPV DNA detection, a recurrence was observed, and in 1/7, tumor progression was observed (Table 3). In contrast, persistence or recurrence were observed in 0/55 patients who were post-treatment hr-HPV DNA-negative ($p = 0.0007$; OR 244.2; 95% CI: 10.4 to 5757.7). All post-treatment hr-HPV-positive patients were also p16-positive at initial diagnosis. Interestingly, all patients with recurrent or progressive tumor (5/62) were hr-HPV positive with the same genotype than before therapy. Three of these five patients died because of the recurrence or tumor progression 10, 12, and 35 months after diagnosis. In contrast, in the two hr-HPV positive patients after treatment without recurrence, another hr-HPV genotype was detected than before treatment.

**Table 3.** Patients with post-treatment HPV-positivity for the same genotype.

| Patient 1–5 | UICC | Age at Diagnosis | ASA Score | HPV Pre-Treatment | Therapy | HPV Post-Treatment | Course of Disease |
|---|---|---|---|---|---|---|---|
| Patient 1 | Stage III | 55 | 2 | HPV 16 | RCT | HPV 16 | Recurrence |
| Patient 2 | Stage I | 54 | 2 | HPV 16 | Surgery and PORT | HPV 16 | Pulmonal metastasis Locoregional control |
| Patient 3 | Stage III | 73 | 3 | HPV 16 | RCT | HPV 16 | Recurrence |
| Patient 4 | Stage IV | 86 | 2 | HPV 33 | CT only | HPV 33 | Tumor progression |
| Patient 5 | Stage III | 70 | 2 | HPV 18 | RCT | HPV 18 | Recurrence after partial response |

HPV, human papilloma virus; ASA, American Society of Anesthesiologists; UICC, Union internationale contre le cancer; PORT postoperative radiotherapy; RCT radiochemotherapy; CT, chemotherapy.

### 2.4. Factors Associated with Post-Treatment HPV-Positivity

Post-treatment hr-HPV positivity was more frequent in patients whose primary tumor expressed epidermal growth factor receptor (EGFR) (Figure 2). EGFR expression was observed in 5/7 patients with and in 14/44 patients without post-treatment hr-HPV-positivity ($p = 0.025$). Another associated factor was advanced primary tumor T-stage ($p = 0.031$). No association was observed between post-treatment hr-HPV-positivity and UICC stage, treatment modality and radiotherapy dose. Post-treatment hr-HPV positivity was also not associated with PD-L1 expression, patient-reported smoking status, and consumption of alcoholic beverages at initial diagnosis.

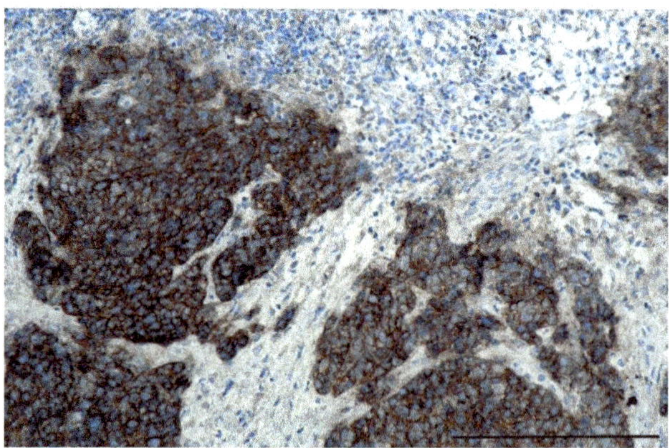

**Figure 2.** Epidermal growth factor receptor (EGFR) expression in a high-risk HPV (hr-HPV)-positive oropharyngeal squamous cell carcinoma (OPSCC) specimen. EGFR, epidermal growth factor receptor; hr, high risk; HPV, human papilloma virus, bar is 1 μm.

*2.5. Post-Treatment HPV-Positivity Virus Persistence or New Infection?*

Before therapy HPV 16 was the most common genotype (50 patients, 80.6%) and HPV 18 the second most common (4 patients, 2.5%). Other genotypes were HPV 33 (three patients, 1.9%) and HPV 35 (three patients, 1.9%). HPV 70, 66, 58, 82, 31, and 40 were detected each in a single patient or additionally to HPV 16 as a multiple infection. As mentioned before, after therapy five patients were hr-HPV-positive with the same genotype than before therapy. Only in these patients a recurrence or tumor progression was stated. The genotypes HPV16 were detected in three patients, and HPV 33 and HPV 18 were each detected in a single patient after therapy. As HPV 18 and HPV 33 are rare in the population of HPV-positive OPSCC patients, we suspect a virus persistence instead of a new infection.

## 3. Discussion

In this study, we questioned whether patients with hr-HPV DNA-positive OPSCC remain hr-HPV DNA-positive after treatment and if post-treatment hr-HPV DNA at the initial tumor site is associated with the rate of disease persistence or recurrence. Before and after treatment, brushings were taken from the oropharynx, including the surface of the previous tumor site and tested for hr-HPV-DNA. Post-treatment brushings were available in 62 patients. Overall, 88.7% of hr-HPV-positive patients were hr-HPV negative at follow-up. In seven patients, hr-HPV after treatment was detected, and all patients hr-HPV-positive for the same genotype developed a recurrence or tumor persistence. Detection of hr-HPV at follow-up was associated with a substantially increased risk for persistent or recurrent disease (OR 244.2; 95% CI: 10.4 to 5757.7).

Post-treatment hr-HPV positivity and persistent or recurrent disease are rare events in hr-HPV-related oropharyngeal carcinoma. Accordingly, the results on potential influencing factors are based on a low number of patients and should be considered with caution. However, our results are in line with previous data.

Also, Hanna and coworkers described a significant decrease in post-treatment E7 antibody levels in the salivary glands of patients with OPSCC [10]. Rettig and coworkers investigated hr-HPV DNA in oral rinses in 157 patients with OPSCC. At initial diagnosis, HPV type 16 was detected in 67/124 patients. After therapy, oral HPV 16 DNA was detected in six patients (9%). All five patients with persistent oral HPV 16 DNA developed a recurrent disease. Of these patients, three died. Persistent HPV 16 DNA detection in oral rinses was associated with a greater than 20-fold increased risk of recurrence (hazard ratio [HR], 29.7 [95% CI, 9.0–98.2]) and death (HR, 23.5 [95% CI, 4.7–116.9]) [15]. In a similarly designed study on 93 patients with OPSCC and HPV 16-positive cancer of unknown primary, pre- and post-treatment serum or saliva samples were taken to detect HPV 16 E6. The authors reported hr-HPV-positive post-treatment saliva to be associated with higher risk of recurrence (hazard ratio [HR], 10.7; 95% CI, 2.36–48.50) and reduced overall survival (HR, 25.9; 95% CI: 3.23–208.00). The combined saliva and plasma post-treatment HPV 16 DNA status was 90.7% specific and 69.5% sensitive in predicting recurrence within three years [16]. Fakhry and coauthors used oral rinse samples for HPV detection and described a significant decrease in HPV DNA after therapy, about 14.3% patients remained HPV-positive compared to 11.3% in our study. Also in this study, the authors described a significantly lower two-year overall survival among HPV-positive patients with persistent HPV detection for the same genotype (tumor-type) after therapy than among those without detectable tumor-type DNA after therapy (68% vs. 95%; adjusted HR, 6.61; 95% CI, 1.86–23.44; $p$ = 0.003), as was recurrence-free survival (55% vs. 88%; adjusted HR, 3.72; 95% CI, 1.71–8.09; $p$ < 0.001) [11].

Although, only few studies analyzed HPV-positivity after treatment in OPSCC patients and although the number of patients in our study is low, post-treatment HPV-positivity seems to be a strong predictor for overall survival. An inclusion of this observation in the clinical management in patients diagnosed with hr-HPV-positive OPSCC should be discussed. Post-treatment HPV detection in the early follow-up period can be a possible new standard to evaluate and predict the clinical course of those patients. As the brush test does not require biopsies and is easy to perform in awake patients without anesthesia, this is a very suitable and simple test method for HPV-DNA detection.

Post-treatment hr-HPV-positive patients may require a very close clinical and radiological follow-up, these patients can be at high risk for a poor overall survival. One of our patients with post-treatment HPV-positivity received an HPV vaccination with Gardasil®9 (Recombinant Human Papillomavirus 9-valent Vaccine, Merck Sharp & Dohme BV, Haarlem, Netherlands) during full remission, hoping to reduce virus activity. However, he developed a locoregional recurrence. Interestingly, 2/7 patients were HPV-positive after therapy with a different hr-HPV genotype, these two patients were in full remission and experienced no recurrence. It seems that a new hr-HPV infection occurred and the risk in these patients for recurrence after therapy is as low as in post-treatment HPV-negative patients. However, higher numbers of patient are needed for further evaluation.

Also, in patients with cervical cancer, persistent HPV infection is associated with an increased risk for recurrence [17]. In a study on 72 women with CIN, persistence or clearance of hr-HPV DNA was described as an early valid prognostic marker of failure or cure after treatment, more accurate than cytology or section margin status at the time of conisation. The absence of hr-HPV DNA had a 100% negative predictive value [18]. Söderlund-Strand et al. described a type-specific HPV-persistence in women with a residual CIN. No recurrent or residual disease was detected in women with any other patterns of HPV positivity, e.g., type change or fluctuating positivity [12].

In our study HPV 16 was the most common genotype before therapy, other genotypes like HPV 18, 33, and 35 were rare. Two patients were pre- and post-treatment positive for comparatively rare HPV genotypes 18 and 33. As reinfection with these rare types is unlikely, we assume tumor virus persistence rather than new HPV infection. Our findings about pretreatment HPV genotypes are in line with other studies. Chatfield-Reed and coauthors reported that of 99 hr-HPV positive HNSCC patients, 75.6% were positive for HPV 16 and 3% for HPV 18. In this study, 16.2% were positive for the genotype HPV 33, which we detected in only 1.9% [19]. Fossum and coworkers reported in 166 OPSCC patients (77% hr-HPV positive), HPV 16 to be the predominating genotype (65%), followed by HPV 33 (17%), HPV 18 (2%), and HPV 31/35/56/59 in one patient each [20].

In this study, post-treatment hr-HPV positivity was associated with clinical T-stage at diagnosis and tumor EGFR expression. An association of post-treatment hr-HPV positivity and primary tumor EGFR expression has not yet been reported. However, this observation is based on very few patients. EGFR protein over-expression has been reported in 70–100% of HNSCC, but 46–72% of OPSCC [21–23]. Although the reason why HPV-positive tumors express less EGFR expression is currently unknown, smoking has been hypothesized to be a contributory factor [24].

We did not observe an association of smoking and post-treatment hr-HPV positivity. This might be due to the low number of patients. Active smoking was admitted by 8/45 patients. Kero et al. reported a correlation between persistent oral HPV infection and smoking in 131 men who were sampled by serial oral scrapings. Genotype-specific HPV persistence was detected in 18/129 men. The mean persistence time ranged from 6 months to 30.7 months. The authors concluded that most of the persisting oral infections in males were caused by HPV 16, and smoking increased the oral hr-HPV persistence [25]. Another assumed risk factor for HPV persistence is immunosuppression. In a cohort of 97 HIV/AIDS patients, a genotype-specific oral and oropharyngeal HPV persistence was described in 33.3% of patients, of which 13.3% were hr-HPV positive [26].

However, the most significant observations of this study are that the majority of patients are hr-HPV-negative after therapy and that post-treatment hr-HPV positive patients seem to have an increased risk of tumor persistence or recurrence.

Limitations of the study are that not all OPSCC patients were enrolled consecutively in the study. From 12 patients, no post-treatment brushings were available because they missed the follow-up or received follow-up in another hospital. This includes also patients who may have had a relapse. Moreover, several otorhinolaryngologists were involved in sampling, but all were instructed how to collect samples. In this study, the brushings were obtained in some patients under general anesthesia during restaging panendoscopy and in some patients awake during clinical follow-up.

The brush test is a simple test method and does not require tumor biopsies for HPV-DNA detection, future clinical usefulness of this test includes an oropharyngeal HPV screening for populations at risk, e.g., immunosuppressed elderly population. Partners of HPV-positive OPSCC patients, who are often concerned about oral HPV transmission, can also be tested by oropharyngeal brushings.

## 4. Material and Methods

The aim of the study was to determine how many patients with hr-HPV-DNA-positive oropharyngeal carcinoma before treatment have positive hr-HPV-DNA detection in the former tumor region after completing therapy. We were also interested in whether post-treatment hr-HPV-DNA detection is related to disease recurrence.

This prospective study was conducted in accordance with the GCP guidelines. Written informed consent was obtained from each patient who agreed to participate in this study following detailed explanation of the procedural workflow. Prior to any patient enrolment, the study had been approved by the institutional board in charge, i.e., the ethics committee of the Medical University Innsbruck, Austria. The respective reference number was 1147/2018. The study was conducted in full accordance with the principles expressed in the Declaration of Helsinki.

### 4.1. Inclusion and Exclusion Criteria

Patients with incident oropharyngeal cancer positive for HPV-DNA before treatment were included. Virus DNA detection was performed by tumor surface brushings and/or by HPV-DNA isolation from formalin-fixed and paraffin-embedded (FFPE) tumor biopsies [13]. All patients were treated at the Department of Otorhinolaryngology—Head and Neck Surgery, Medical University of Innsbruck between April 2014 and October 2019 according to the recommendations of the interdisciplinary tumor board. Patients were invited to undergo a post-treatment surface brushing from the initial tumor site at least 12 weeks following end of treatment. Patients who were not able or willing to undergo a post-treatment brushing were excluded from the analysis. The follow-up for HNSCC patients is standardized, the restaging is performed 8–12 weeks after the treatment and includes a CT scan and panendoscopy with sampling from the former tumor region. In the case of full remission, further clinical and radiological controls are arranged every six weeks for three times, every three months for three times and then every six months. The follow-up ends five years after diagnosis.

### 4.2. Specimen Harvest and Handling

Before the start of the treatment, all patients underwent routine endoscopy under anesthesia, including tracheobronchoscopy, esophagoscopy, and laryngopharyngoscopy. During this examination, a cytology brush (Digene® HC2 DNA Collection Device, Qiagen, Hilden, Germany) of the tumor surface was taken. The brush test reliably detects HPV-DNA in patients with OPSCC, sensitivity and specificity of this method is 86% (95%CI: 70–95%) and 89% (95%CI: 65–99%) [13]. The brush was then kept in a sterile container in 0.05% sodium azide and sent to the Division of Virology of the Medical University of Innsbruck. Biopsies of the tumor were then obtained, fixed in formalin, and sent to the Department of Pathology for routine histopathological investigation. Moreover, immunohistochemistry (IHC) of p16, epidermal growth factor receptor (EGFR), and programmed death-ligand 1 (PD-L1) was performed.

After completion of therapy, a second surface brushing was taken from all subsites of the oropharynx, including the initial tumor site. This second surface brush was obtained during follow-up examinations at least 12 weeks after therapy during a restaging endoscopy or during clinical follow-up.

### 4.3. Processing of Brush Specimens

Nucleic acid extraction was performed within two days of arrival of the samples at the laboratory. The DNA extraction was performed in a fully automated manner (NucliSens® easyMAG®, Biomerieux, Marcy-l'Étoile, France) [27,28]. A total of 500 µL of the sample was pipetted into a disposable well.

After an initial cell lysis, the nucleic acid components were isolated from the mixture using magnetic silica particles, resulting in a total of 110 μL purified DNA extract. Approximately 5–10 μL of this extract was used for HPV-DNA detection and genotyping.

### 4.4. DNA Amplification and HPV Genotyping

For the detection of HPV DNA, real-time polymerase chain reaction (PCR) was used based on the amplification of the L1 open reading frames (ORF). As an internal control for the availability of cell material, a PCR for the household beta-globin gene was performed in parallel. The HPV DNA was considered positive if the fluorescence signal appeared before the 40th cycle [29]. In the next step, all HPV-positive samples were further genotyped after an amplification step using reverse line blot hybridization on nitrocellulose membrane strips containing genotype-specific probes (AmpliQuality HPV-TYPE EXPRESS, AB Analitica®, Padova, Italy) [30]. The genotyping kit is able to identify 40 different HPV types, namely 6, 11, 16, 18, 26, 31, 33, 35, 39, 40, 42, 43, 44, 45, 51, 52, 53, 54, 55, 56, 58, 59, 61, 62, 64, 66, 67, 68a/b, 69, 70, 71, 72, 73, 81, 82, 83, 84, 87, 89, and 90 (AB Analitica®, Padova, Italy). Using the 2′-deoxyuridine 5′-triphosphate/uracil-DNA N-glycosylase dUTP/UNG system, which degrades non-specific residual RNAs, contamination due to carry-over was minimized. The detection limit of the test kit is 1000 viral copies per ml for HPV 16 and HPV 6, which is in accordance with the required detection limit by the World Health Organization (WHO) and the international HPV reference center [31]. Based on the International Agency for Research on Cancer classification, the HPV genotypes were classified as established high-risk, probably high-risk, established low-risk, and uncharacterized [32,33].

### 4.5. IHC

Five-micrometer-thick paraffin sections were dewaxed, and antigens were retrieved in an automated staining system (Ventana, Discovery, Tucson, AZ, USA). A commercial in vitro diagnostic (IVD) certified assay containing ready-to-use prediluted mouse monoclonal antibody was used for p16 detection (CINtec® Histology V-Kit, Roche Diagnostics, Basel, Switzerland). EGFR was detected by a rabbit monoclonal IVD certified antibody (clone: A20-E) of Diagnostic Biosystems (Kosice, Slovakia), PD-L1 was detected by a rabbit monoclonal antibody (clone:E1L3N®) of Cell Signaling Technology (Frankfurt am Main, Germany), both of these primary rabbit monoclonal antibodies were used in a 1:200 final dilution in a Ventana Discovery staining system. Staining was completed using universal secondary antibody solution, hematoxylin counterstaining, and the DAB MAP Kit (all Ventana products, Ventana, Discovery, Tucson, AZ, USA). The tumor cell areas were evaluated by one experienced observer. Specimens were judged p16-positive if ≥60% of the cells in tumor areas revealed immunohistochemical reaction products. Staining of fibroblastic stroma cells was not counted. EGFR staining was scored based on previous publications as a positive membrane reaction in tumor cell nests: "0": no staining, "1": 1–9%, "2": 10–49%, "3": over 50% of the cells positive [34,35]. PD-L1 was estimated in cells located in tumor cell areas as follows: "0": no staining, "1": 1–5%, "2": 5–10%, and "3": at least 10% of the cell stained, as published by Ferris et al. in 2016 [36].

### 4.6. Data Analysis

Frequency data are tabulated and analyzed with Fisher's exact test or with the Kruskal–Wallis test for ordered alternatives [37]. To calculate 95% CIs of proportions, the Wilson method with continuity correction was used [38]. For continuous data, means, and standard deviations (SD) are provided unless stated otherwise. Median follow-up was calculated by with the reverse Kaplan–Meier method [39].

## 5. Conclusions

In this study, most patients with initially hr-HPV-positive OPSCC were HPV-negative after treatment, none of these patients experienced a recurrence. The five patients with recurrence or tumor progression were all post-treatment hr-HPV-positive with the same genotype. No recurrence was observed in two post-treatment hr-HPV-positive patients with a different genotype than at diagnosis.

**Author Contributions:** Conceptualization, B.K. and H.R.; Methodology, W.B., D.v.L. and J.D.; Software, H.R.; Validation, B.K. and H.R.; Formal Analysis, J.D.; Investigation, D.D., V.I., and T.B.S.; Resources, H.R.; Data Curation, B.K. and V.I.; Writing—Original Draft Preparation, B.K.; Writing—Review and Editing, H.R.; Visualization, G.W.; Supervision, H.R.; Project Administration, B.K., D.D., and T.B.S. All authors have read and agreed to the published version of the manuscript.

**Funding:** This research received no external funding.

**Conflicts of Interest:** The authors declare no conflict of interest.

## References

1. Jung, A.C.; Briolat, J.; Millon, R.; De Reynies, A.; Rickman, D.; Thomas, E.; Abecassis, J.; Clavel, C.; Wasylyk, B. Biological and clinical relevance of transcriptionally active human papillomavirus (HPV) infection in oropharynx squamous cell carcinoma. *Int. J. Cancer* **2010**, *126*, 1882–1894. [CrossRef]

2. Zou, J.; Cao, Z.; Zhang, J.; Chen, T.; Yang, S.; Huang, Y.; Hong, D.; Li, Y.; Chen, X.; Wang, X.; et al. Variants in human papillomavirus receptor and associated genes are associated with type-specific HPV infection and lesion progression of the cervix. *Oncotarget* **2016**, *7*, 40135–40147. [CrossRef] [PubMed]

3. Shanmugasundaram, S.; You, J. Targeting Persistent Human Papillomavirus Infection. *Viruses* **2017**, *9*, 229. [CrossRef] [PubMed]

4. Galloway, D.A.; Laimins, L.A. Human papillomaviruses: Shared and distinct pathways for pathogenesis. *Curr. Opin. Virol.* **2015**, *14*, 87–92. [CrossRef] [PubMed]

5. Ducatman, B.S. The Role of Human Papillomavirus in Oropharyngeal Squamous Cell Carcinoma. *Arch. Pathol. Lab. Med.* **2018**, *142*, 715–718. [CrossRef] [PubMed]

6. Haukioja, A.; Asunta, M.; Soderling, E.; Syrjanen, S. Persistent oral human papillomavirus infection is associated with smoking and elevated salivary immunoglobulin G concentration. *J. Clin. Virol.* **2014**, *61*, 101–106. [CrossRef]

7. Mehanna, H.; Beech, T.; Nicholson, T.; El Hariry, I.; McConkey, C.; Paleri, V.; Roberts, S. Prevalence of human papillomavirus in oropharyngeal and nonoropharyngeal head and neck cancer—Systematic review and meta-analysis of trends by time and region. *Head Neck* **2013**, *35*, 747–755. [CrossRef]

8. Fung, N.; Faraji, F.; Kang, H.; Fakhry, C. The role of human papillomavirus on the prognosis and treatment of oropharyngeal carcinoma. *Cancer Metastasis Rev.* **2017**. [CrossRef]

9. Zhang, Y.; Waterboer, T.; Haddad, R.I.; Miles, B.A.; Wentz, A.; Gross, N.D.; Fakhry, C.; Quon, H.; Lorch, J.H.; Gourin, C.G.; et al. Human papillomavirus (HPV) 16 antibodies at diagnosis of HPV-related oropharyngeal cancer and antibody trajectories after treatment. *Oral. Oncol.* **2017**, *67*, 77–82. [CrossRef]

10. Hanna, G.J.; Sridharan, V.; Margalit, D.N.; La Follette, S.K.; Chau, N.G.; Rabinowits, G.; Lorch, J.H.; Haddad, R.I.; Tishler, R.B.; Anderson, K.S.; et al. Salivary and serum HPV antibody levels before and after definitive treatment in patients with oropharyngeal squamous cell carcinoma. *Cancer Biomark* **2017**, *19*, 129–136. [CrossRef]

11. Fakhry, C.; Blackford, A.L.; Neuner, G.; Xiao, W.; Jiang, B.; Agrawal, A.; Gillison, M.L. Association of Oral Human Papillomavirus DNA Persistence With Cancer Progression After Primary Treatment for Oral Cavity and Oropharyngeal Squamous Cell Carcinoma. *JAMA Oncol.* **2019**, *5*, 985–992. [CrossRef] [PubMed]

12. Soderlund-Strand, A.; Kjellberg, L.; Dillner, J. Human papillomavirus type-specific persistence and recurrence after treatment for cervical dysplasia. *J. Med. Virol.* **2014**, *86*, 634–641. [CrossRef] [PubMed]

13. Kofler, B.; Borena, W.; Manzl, C.; Dudas, J.; Wegscheider, A.S.; Jansen-Durr, P.; Schartinger, V.; Riechelmann, H. Sensitivity of tumor surface brushings to detect human papilloma virus DNA in head and neck cancer. *Oral. Oncol.* **2017**, *67*, 103–108. [CrossRef] [PubMed]

14. Broglie, M.A.; Jochum, W.; Forbs, D.; Schonegg, R.; Stoeckli, S.J. Brush cytology for the detection of high-risk HPV infection in oropharyngeal squamous cell carcinoma. *Cancer Cytopathol.* **2015**, *123*, 732–738. [CrossRef]

15. Rettig, E.M.; Wentz, A.; Posner, M.R.; Gross, N.D.; Haddad, R.I.; Gillison, M.L.; Fakhry, C.; Quon, H.; Sikora, A.G.; Stott, W.J.; et al. Prognostic Implication of Persistent Human Papillomavirus Type 16 DNA Detection in Oral Rinses for Human Papillomavirus-Related Oropharyngeal Carcinoma. *JAMA Oncol.* **2015**, *1*, 907–915. [CrossRef]

16. Ahn, S.M.; Chan, J.Y.; Zhang, Z.; Wang, H.; Khan, Z.; Bishop, J.A.; Westra, W.; Koch, W.M.; Califano, J.A. Saliva and plasma quantitative polymerase chain reaction-based detection and surveillance of human papillomavirus-related head and neck cancer. *JAMA Otolaryngol. Head Neck Surg.* **2014**, *140*, 846–854. [CrossRef]

17. Song, D.; Kong, W.M.; Zhang, T.Q.; Jiao, S.M.; Chen, J.; Han, C.; Liu, T.T. The negative conversion of high-risk human papillomavirus and its performance in surveillance of cervical cancer after treatment: A retrospective study. *Arch. Gynecol. Obstet.* **2017**, *295*, 197–203. [CrossRef]

18. Verguts, J.; Bronselaer, B.; Donders, G.; Arbyn, M.; Van Eldere, J.; Drijkoningen, M.; Poppe, W. Prediction of recurrence after treatment for high-grade cervical intraepithelial neoplasia: The role of human papillomavirus testing and age at conisation. *BJOG* **2006**, *113*, 1303–1307. [CrossRef]

19. Chatfield-Reed, K.; Gui, S.; O'Neill, W.Q.; Teknos, T.N.; Pan, Q. HPV33+ HNSCC is associated with poor prognosis and has unique genomic and immunologic landscapes. *Oral. Oncol.* **2020**, *100*, 104488. [CrossRef]

20. Fossum, G.H.; Lie, A.K.; Jebsen, P.; Sandlie, L.E.; Mork, J. Human papillomavirus in oropharyngeal squamous cell carcinoma in South-Eastern Norway: Prevalence, genotype, and survival. *Eur. Arch. Oto Rhino Laryngol.* **2017**, *274*, 4003–4010. [CrossRef]

21. Cohen, E.R.; Reis, I.M.; Gomez, C.; Pereira, L.; Freiser, M.E.; Hoosien, G.; Franzmann, E.J. Immunohistochemistry Analysis of CD44, EGFR, and p16 in Oral Cavity and Oropharyngeal Squamous Cell Carcinoma. *Otolaryngol. Head Neck Surg.* **2017**, *157*, 239–251. [CrossRef] [PubMed]

22. Temam, S.; Kawaguchi, H.; El-Naggar, A.K.; Jelinek, J.; Tang, H.; Liu, D.D.; Lang, W.; Issa, J.P.; Lee, J.J.; Mao, L. Epidermal growth factor receptor copy number alterations correlate with poor clinical outcome in patients with head and neck squamous cancer. *J. Clin. Oncol.* **2007**, *25*, 2164–2170. [CrossRef] [PubMed]

23. Chung, C.H.; Ely, K.; McGavran, L.; Varella-Garcia, M.; Parker, J.; Parker, N.; Jarrett, C.; Carter, J.; Murphy, B.A.; Netterville, J.; et al. Increased epidermal growth factor receptor gene copy number is associated with poor prognosis in head and neck squamous cell carcinomas. *J. Clin. Oncol.* **2006**, *24*, 4170–4176. [CrossRef] [PubMed]

24. Sivarajah, S.; Kostiuk, M.; Lindsay, C.; Puttagunta, L.; O'Connell, D.A.; Harris, J.; Seikaly, H.; Biron, V.L. EGFR as a biomarker of smoking status and survival in oropharyngeal squamous cell carcinoma. *J. Otolaryngol. Head Neck Surg.* **2019**, *48*, 1. [CrossRef] [PubMed]

25. Kero, K.; Rautava, J.; Syrjanen, K.; Willberg, J.; Grenman, S.; Syrjanen, S. Smoking increases oral HPV persistence among men: 7-year follow-up study. *Eur. J. Clin. Microbiol. Infect. Dis.* **2014**, *33*, 123–133. [CrossRef]

26. Castillejos-Garcia, I.; Ramirez-Amador, V.A.; Carrillo-Garcia, A.; Garcia-Carranca, A.; Lizano, M.; Anaya-Saavedra, G. Type-specific persistence and clearance rates of HPV genotypes in the oral and oropharyngeal mucosa in an HIV/AIDS cohort. *J. Oral. Pathol. Med.* **2018**, *47*, 396–402. [CrossRef]

27. Rutjes, S.A.; Italiaander, R.; Van den Berg, H.H.; Lodder, W.J.; De Roda Husman, A.M. Isolation and detection of enterovirus RNA from large-volume water samples by using the NucliSens miniMAG system and real-time nucleic acid sequence-based amplification. *Appl. Environ. Microbiol.* **2005**, *71*, 3734–3740. [CrossRef]

28. Tang, Y.W.; Sefers, S.E.; Li, H.; Kohn, D.J.; Procop, G.W. Comparative evaluation of three commercial systems for nucleic acid extraction from urine specimens. *J. Clin. Microbiol.* **2005**, *43*, 4830–4833. [CrossRef]

29. Abreu, A.L.; Souza, R.P.; Gimenes, F.; Consolaro, M.E. A review of methods for detect human Papillomavirus infection. *Virol. J.* **2012**, *9*, 262. [CrossRef]

30. Mason, S.; Gani, A.; Vettorato, M.; Negri, G.; Mian, C.; Brusauro, F. Detection of high-risk HPV genotypes in cervical samples: A comparison study of a novel real time pcr/reverse line blot-based technique and the digene HC2 assay. *Pathologica* **2012**, *6*, 104.

31. Eklund, C.; Forslund, O.; Wallin, K.L.; Dillner, J. Global improvement in genotyping of human papillomavirus DNA: The 2011 HPV LabNet International Proficiency Study. *J. Clin. Microbiol.* **2014**, *52*, 449–459. [CrossRef] [PubMed]

32. Zur Hausen, H.; De Villiers, E.M. Human papillomaviruses. *IARC Monogr. Eval. Carcinog. Risks Hum.* **2007**, *90*, 1–636.

33. Munoz, N.; Castellsague, X.; De Gonzalez, A.B.; Gissmann, L. Chapter 1: HPV in the etiology of human cancer. *Vaccine* **2006**, *24* (Suppl. 3), 1–10. [CrossRef] [PubMed]

34. Reimers, N.; Kasper, H.U.; Weissenborn, S.J.; Stutzer, H.; Preuss, S.F.; Hoffmann, T.K.; Speel, E.J.; Dienes, H.P.; Pfister, H.J.; Guntinas-Lichius, O.; et al. Combined analysis of HPV-DNA, p16 and EGFR expression to predict prognosis in oropharyngeal cancer. *Int. J. Cancer* **2007**, *120*, 1731–1738. [CrossRef] [PubMed]

35. Bentzen, S.M.; Atasoy, B.M.; Daley, F.M.; Dische, S.; Richman, P.I.; Saunders, M.I.; Trott, K.R.; Wilson, G.D. Epidermal growth factor receptor expression in pretreatment biopsies from head and neck squamous cell carcinoma as a predictive factor for a benefit from accelerated radiation therapy in a randomized controlled trial. *J. Clin. Oncol.* **2005**, *23*, 5560–5567. [CrossRef] [PubMed]

36. Ferris, R.L.; Blumenschein, G., Jr.; Fayette, J.; Guigay, J.; Colevas, A.D.; Licitra, L.; Harrington, K.; Kasper, S.; Vokes, E.E.; Even, C.; et al. Nivolumab for Recurrent Squamous-Cell Carcinoma of the Head and Neck. *N. Engl. J. Med.* **2016**, *375*, 1856–1867. [CrossRef]

37. Mehta, C.R. The exact analysis of contingency tables in medical research. *Stat. Methods Med. Res.* **1994**, *3*, 135–156. [CrossRef]

38. Newcombe, R.G. Improved confidence intervals for the difference between binomial proportions based on paired data. *Stat. Med.* **1998**, *17*, 2635–2650. [CrossRef]

39. Schemper, M.; Smith, T.L. A note on quantifying follow-up in studies of failure time. *Control. Clin. Trials* **1996**, *17*, 343–346. [CrossRef]

*Review*

# Self-Collection for Cervical Screening Programs: From Research to Reality

David Hawkes [1,2,3,*], Marco H. T. Keung [1], Yanping Huang [1], Tracey L. McDermott [1], Joanne Romano [1], Marion Saville [1,4,5] and Julia M. L. Brotherton [1,5,6]

[1]  VCS Foundation, Carlton South, VIC 3053, Australia; hkeung@vcs.org.au (M.H.T.K.); whuang@vcs.org.au (Y.H.); tmcdermo@vcs.org.au (T.L.M.); jromano@vcs.org.au (J.R.); msaville@vcs.org.au (M.S.); jbrother@vcs.org.au (J.M.L.B.)
[2]  Department of Pharmacology and Therapeutics, University of Melbourne, Parkville, VIC 3010, Australia
[3]  Department of Pathology, University of Malaya, Kuala Lumpur 50603, Malaysia
[4]  Department of Obstetrics and Gynaecology, University of Melbourne, Parkville, VIC 3010, Australia
[5]  Department of Obstetrics and Gynaecology, University of Malaya, Kuala Lumpur 50603, Malaysia
[6]  Melbourne School of Population and Global Health, University of Melbourne, Parkville, VIC 3010, Australia
*  Correspondence: dhawkes@vcs.org.au

Received: 11 February 2020; Accepted: 16 April 2020; Published: 24 April 2020

**Abstract:** In 2018, there were an estimated 570,000 new cases of cervical cancer globally, with most of them occurring in women who either had no access to cervical screening, or had not participated in screening in regions where programs are available. Where programs are in place, a major barrier for women across many cultures has been the requirement to undergo a speculum examination. With the emergence of HPV-based primary screening, the option of self-collection (where the woman takes the sample from the vagina herself) may overcome this barrier, given that such samples when tested using a PCR-based HPV assay have similar sensitivity for the detection of cervical pre-cancers as practitioner-collected cervical specimens. Other advantages of HPV-based screening using self-collection, beyond the increase in acceptability to women, include scalability, efficiency, and high negative predictive value, allowing for long intervals between negative tests. Self-collection will be a key strategy for the successful scale up of cervical screening programs globally in response to the WHO call for all countries to work towards the elimination of cervical cancer as a public health problem. This review will examine self-collection for HPV-based cervical screening including the collection devices, assays and possible routine laboratory processes considering how they can be utilized in cervical screening programs.

**Keywords:** human papillomavirus; cervical screening; diagnostic testing; self-collection

## 1. Introduction

Cervical cancer is a major public health problem with approximately 570,000 new cases, and 311,000 deaths globally in 2018, the majority of these in low- and middle-income countries [1]. Even within high income countries, most cervical cancers occur in women who are never or under-screened [2–5].

Participation in a screening program involves a woman knowing the reasons for screening, believing that screening is relevant to her, and that being screened provides benefits that outweigh any potential risks or costs. Being able to access a trusted service provider and the ability to undergo a speculum examination are also contributors to women's decisions about participating in a screening program [6]. Women who live in settings where screening programs are available may be under- or never-screened for a variety of reasons including practical barriers such as work and parenting commitments, financial costs related to attending an appointment for screening, lack of access to appropriate health care services or providers, or other barriers such as a previous negative experience with undergoing a screening test, history of sexual trauma, female genital mutilation, or cultural beliefs [6,7].

Traditional cervical cytology relied on healthcare practitioners visualizing the cervix and sampling cervical cells from the transformation zone of the cervix [8]. Human papillomavirus (HPV)-based cervical screening programs generally use the same sampling approach because HPV positive samples are often then reflexly examined using liquid-based cytology from the same specimen [9]. An advantage of HPV-based testing is that it detects viral nucleic acid, rather than morphological changes in the cell, and as such there is no need to sample from the transformation zone of the cervix because viral nucleic acid is shed from the cervix into the vaginal canal. [10]. Using self-collection, a woman can take the sampling device, insert it into her own vagina to collect a specimen, and then return it to a healthcare provider (either directly or through the mail) without the physical intervention of a healthcare practitioner [11]. This sample can then be processed using a nucleic acid test for the presence of HPV DNA or RNA. In contrast, at present there is not a body of evidence that supports self-collection for morphological analysis (e.g., Pap-stained cells) for the purpose of cervical screening.

Self-sampling has been shown to increase participation in never- or under-screened populations [10]. Several methods for engaging currently non-screening women in cervical screening by utilizing self-collection have been trialed, with varied uptake levels. Pooled analysis of 25 self-collection participation trials showed community campaigns that included community outreach and media support, or door-to-door campaigns, achieved higher participation in the self-collection arms than control arms with practitioner collected cytology or HPV samples, or visual inspection of the cervix with acetic acid by a practitioner (community campaign: 15.6% vs. 6.0%, door-to-door campaign 94.2% vs. 53.0%). Mailing out self-collection kits has also achieved higher participation in the participant arm offered self-collection than the practitioner collection arm (19.2% vs. 11.5%); however, 'opt in' trials did not (7.8% vs. 13.4%) [10]. There is some evidence to suggest that face-to-face interactions with a trained healthcare practitioner have the highest likelihood of the largest increases in participation [12].

Self-sampling facilitated cervical screening can be performed in a variety of settings including health care facilities, homes or workplaces facilitated by community health outreach teams [10], with HPV testing of the samples carried out centrally in a laboratory [13–18], or on location in the case of Point of Care HPV testing assays [19,20].

This flexibility of self-collection based cervical screening can substantially increase participation in cervical screening in rural, remote or low resource settings [12], and therefore has the potential to increase equity in cervical screening. Modelling by the WHO Cervical Cancer Elimination Modelling Consortium predicts that reducing cervical screening inequity, particularly in low- and middle-income countries, can reduce the burden of cervical cancer at a global level [21]. With the World Health Organization's call to eliminate cervical cancer as a public health problem by 2120 [22], and the draft elimination strategy target of 70% of the world's women being screened with a high-performance HPV test by 35 and 45 years of age by 2030 [23], self-collection is likely the only feasible way to scale up and realize this target, given both health workforce constraints and costs, and the limited acceptability to women of speculum-based sample collection in many settings.

In this review, informed by key papers in the literature and our emerging experience in the HPV-based screening program in Australia, we examine a range of different assays and collection devices and consider other pragmatic issues for introducing self-collection into the laboratory as part of an organized HPV-based cervical screening program. We focus not only on emerging research findings but also on how this evidence can be utilized in the paradigm of clinical testing undertaken at pathology laboratories with the sample volumes that could be expected from an organized screening program.

## 2. Materials and Methods

### 2.1. PCR-Based Technologies for HPV Testing of Self-Collected Specimens

There are a large number of HPV testing technologies available but some of the most common are polymerase chain reaction (PCR)-based assays. Major manufacturers such as Abbott, BD, Cepheid, Qiagen, Roche, and Seegene all produce medium- to high-volume, PCR-based HPV assays for use in cervical screening programs. The meta-analysis by Arbyn et al. [10] clearly demonstrated that when self-collected specimens were processed on PCR-based HPV assays, they were as sensitive as clinician-collected specimens for cervical intraepithelial grade 2 or above (CIN2+) across 17 studies, and for CIN3+ across eight studies. There were small but significant reductions in specificity (0.98 (95% CI 0.97–0.99) compared with clinician-collected specimens observed for both CIN2+ and CIN3+. Further analyses [10] also revealed no significant reductions in sensitivity based on the device used for self-collection nor the storage medium. Currently the wealth of evidence supports PCR-based assays for use in self-collection protocols, and this is explicitly stated in the Australian technical requirements for HPV-based cervical screening [24]. It should also be noted that most of the currently utilised medium- to high-volume HPV assays test for the same 14 HPV types; 16, 18, 31, 33, 35, 39, 45, 51, 52, 56, 58, 59, 66, and 68. It is important to note that each different combination of device, buffer and assay/system requires validation, either by the manufacturer or by individual laboratories. A recent study examined dry flocked swabs collected then eluted in ThinPrep media and then tested on six different PCR-based HPV assays, and this was used as the basis for accreditation of this protocol as part of the National Cervical Screening Program in Australia. [25]. As the evidence-base for different combinations of devices and assays grows, there may be refinements to these requirements for component validation, but these are more likely to relate to evidence regarding how stability affects the sensitivity of the assays over time [15].

### 2.2. Non-PCR-Based Technologies for HPV Testing of Self-Collected Specimens

### 2.2.1. Hologic Aptima

There is a wealth of data, including a recent meta-analysis [10], that demonstrate that when self-collected specimens are tested on polymerase chain reaction (PCR)-based diagnostic HPV assays they produce results which are non-inferior to clinician-collected specimens for the detection of underlying high grade cervical lesions (CIN2+). Signal amplification assays such as the Hybrid Capture 2 or Cervista have also been examined but analyses have demonstrated that these types of assays have lower sensitivity for CIN2+ using self-collected specimens than clinician-collected samples [10,26]. The Hologic Aptima assay targets the mRNA of the HPVE6/E7 region of the same 14 types as many of the other clinically validated and automated HPV assays, with a reflex test for partial genotyping of HPV16 and a combined HPV18/45. In a recent meta-analysis, Arbyn et al. [10] examined three studies [27–29] which utilised the Aptima HPV assay and determined that, whilst results from self-collected specimens indicated good specificity for cervical intraepithelial neoplasia grade 2 or above (CIN2+), they were significantly less sensitive than clinician-collected samples. Histological results are generally used as the gold standard for assessing cervical screening tests performance with most studies using either CIN2(+) or CIN3(+) as their marker for cervical disease, a precursor to cervical cancer. In the three studies examined by Arbyn and colleagues for the detection of CIN2+ by self-collected specimens tested for HPV on the Aptima, the results indicated that 4/30 [27], 10/69 [28], and 6/16 [29] of histologically-confirmed CIN2+ cases were missed. All 20 of these cases were detected by the paired clinician-collected specimens.

A recent methodological paper trialled a different protocol for diagnostic testing of self-collected vaginal samples on the Aptima. Borgfeldt and Forslund [30] used an additional heating step (90 °C for 1 h) on specimens stored in the Aptima media. This heating step caused 11/20 previously negative samples to return a positive result on the Aptima assay, and these results were independently confirmed using different assays. When the full cohort of samples (172 specimens) was retested after the novel heating step, the HPV positivity rate increased from 5.3% to 15.9%. This novel protocol raises interesting questions about future optimisation of the Aptima assay for self-collection. It may be worthwhile for the manufacturers to consider further studies, investigating whether the performance of Aptima on self-collected specimens could potentially be improved by the addition of such a heating step.

Another aspect of the Aptima assay is that it lacks a control for cellularity (sample adequacy), and as a result there is no way of knowing how many of the HPV negative specimens actually contained cellular material. In a cervical screening environment where co-testing (HPV and cytology) are always undertaken, there is no need for a cellularity control as a lack of cellular material is easily identified on cytology. However, with self-collection there is an inherent difficulty in knowing the validity of a negative result unless evidence is available to demonstrate that a sample was actually taken by the woman. Even in clinician-collected cervical screening there is a failure of the cellularity control of between 0.1–0.3% [31–33], although inhibition (e.g., by blood or lubricant) can cause the same control failure. There appears to be a growing body of evidence suggesting an increased percentage of invalid specimens (likely due to a lack of endogenous material) in self-collected specimens compared to practitioner-collected samples. [7,25,32–34] suggesting some women may agree to self-collect but actually not undertake the test. The presence of common inhibiting reagents, such as lubricants, is likely to be less frequent than when a speculum examination is undertaken. It should be noted that the presence of the endogenous cellularity control does not indicate that specimens have been taken from the cervix or vagina.

The current data appears to suggest that the Aptima assay may not be optimised for self-collection using the current manufacturer's instructions for use. Further studies using the endpoint of histologically confirmed CIN2/3+ would be useful to confirm the current findings of the Arbyn meta-analysis relating to Aptima [10]. There remains a possibility of further developments to increase the sensitivity of self-collected specimens by altering the extraction protocol. Additionally, the other issue of assay design—the lack of a sample adequacy control—may be resolved by the use of a Hologic-manufactured, research use only cellularity control as is currently in use in Australia.

### 2.2.2. careHPV

careHPV is an adapted form of the Hybrid Capture 2 assay and was developed as an affordable HPV-DNA detection technology [35,36]. It is specifically designed for low-resource settings, and for potential use as a point of care test, and has been utilised in a variety of geographical locations such as Ghana [36], Bhutan [37], Western Kenya [38] and rural China [35], with diverse demographics including amongst the general population, unscreened women, and women living with HIV [35–37]. It should be noted that the suggested cost per test of US$5 may not be achievable [39].

Therefore, is careHPV an appropriate assay for self-collection, when meta-analyses have clearly demonstrated that the Hybrid Capture 2 it was developed from is not [10]? We acknowledge the tension between having an assay that is not cost-prohibitive and able to screen otherwise never-screened women, and the use of an assay that is less sensitive for HPV than a clinician-collected specimen. In general, minor variations in specificity are likely less critical in under- or never-screened populations than reductions in sensitivity.

There is certainly evidence to suggest that self-collection tested on careHPV facilitates screening by being more acceptable than traditional, speculum-assisted, cervical cytology [37]. A number of studies have been undertaken where there is a lack of a comparator test, either clinician-collected specimen tested on careHPV, or testing of the self-collected specimen on a different HPV assay [37,38], which makes interpretation of the value of the results difficult. A study undertaken in Ghana [36] produced highly concordant results between self- and clinician-collected samples tested on careHPV but did not collect further data to allow assessment of the sensitivity of either test for either cytological or histological markers of disease.

The careHPV assay lacks a control for sample adequacy, or for inhibition (such as by blood). The meta-analysis by Arbyn [10] determined that careHPV was significantly less sensitive for HPV associated with CIN2+ lesions when a self-collected sample was used rather than a clinician-collected sample. The current state of data indicates that careHPV, whilst being cheaper and therefore more accessible, has significantly lower sensitivity for HPV from self-collected specimens and is therefore unlikely to be appropriate for cervical screening of these samples because of the risk of cases of disease being missed. This is a significant issue in settings where women may only receive a once- or twice-in-a-lifetime screening.

### 2.3. Self-Collection Devices for HPV Testing

### 2.3.1. Evalyn Brush

Of the current devices in use for self-collection, the Evalyn brush (Rovers Medical Devices, Lekstraat, The Netherlands) is possibly the most established and evidence-based device. It is currently used in a number of jurisdictions, including as the device used in the self-collection pathway of The Netherlands national cervical screening program. The Evalyn brush is marketed as a customized device designed to be used for self-collection for HPV-based cervical screening.

There are a wide range of studies examining the use of Evalyn brush in detecting 14 oncogenic HPV genotypes on multiple clinically validated PCR-based HPV assays, including the Roche cobas 4800 [13–15], Abbott RealTime [16,17], BD Onclarity [18], Cepheid Xpert [15], Seegene Anyplex II [15], GP5+/6+ PCR enzyme immunoassay [40–42], and SPF10 PCR-DEIA-LiPA25 [43,44]. These studies were undertaken across a range of developed countries including China [17], Germany [16], Netherlands [13,40,41], Denmark [14], Norway [15], and Spain [42].

In the Arbyn meta-analysis [10] there were four studies using the Evalyn brush, one tested on the HC2, and three on validated, PCR-based tests. When an Evalyn brush self-collected specimen was tested on the HC2 [45], it had a relative sensitivity compared with clinician-collected specimens of 0.66. When the Evalyn brush self-collected samples were tested on PCR-based assays, they had a relative sensitivity of 0.99. These data suggest that, in most cases, it is likely to be the type of assay, rather than the collection device, that will predict the clinical quality of the result.

An interesting study published by a Norwegian group [17] examined the sensitivity and specificity of the Evalyn brush and a Copan FLOQswab (Copan, Brescia, Italy) compared to clinician-collected samples tested on three PCR-based assays: Cepheid Xpert, Seegene Anyplex II, and the Roche cobas 4800 HPV test. Relative sensitivity, compared to clinician-collected specimens, for CIN3+ demonstrated that the Copan swab was significantly less sensitive than the clinician-collected across all three assays, but that the Evalyn brush was not. The authors presented a further analysis which segregated the results based on the time between collection and specimen preparation (stabilization). If the time between specimen collection and preparation was 28 days or less, there was no longer a significant drop in sensitivity for the Copan swab for CIN3+. There was also variation in which the self-collection device was more sensitive depending on the assay used. Another study [15], comparing the sensitivity for CIN2+ of the Evalyn brush with the Qvintip device (Aprovix, Stockholm, Sweden), found no significant differences between either of the self-collection devices or the clinician-collected sample. For longer intervals between collection and preparation, the Evalyn brush showed similar levels of sensitivity which suggests that this device may be superior when extended delays between collection and processing cannot be avoided [15]. The major concern with using the Evalyn brush, particularly in low- or middle-income countries, is the cost. Prices obviously vary depending on local conditions but the cost of the Evalyn brush appears to be between 3–5 times the price of other self-collection devices, such as the Copan FLOQswab.

### 2.3.2. Viba-Brush

As highlighted above, the one major drawback of the Evalyn brush is its cost. Rovers Medical Devices have sought to address this issue by introducing a more affordable version of the Evalyn collection device called the Viba-brush. The Viba-brush utilises the same collection head but mounts it more simply on a straight handle and is reported to have a price equivalent to other low-cost devices.

There are two paired sample studies [43,44] which examined the relative sensitivity of HPV tests performed on self-collected samples using the Viba-brush compared with clinician-collected specimens using CIN2+ as the clinical endpoint.

In the Dijkstra study [44] undertaken in the Netherlands, 43/135 recruited women had biopsy-confirmed CIN2+. The self-collection had a sensitivity of 93.0% (40/43) and clinician-collection had a sensitivity of 90.7% (39/43) using a PCR-based assay. Two CIN 2 lesions were hrHPV negative in both sample types.

In the study undertaken by Geraets et al. [43] biopsies were undertaken from 49 women (out of a cohort of 182) after evidence of an abnormality was discovered during colposcopy. Two HPV assays were used to test all clinician- and self-collected specimens. The clinician-collected specimens were more sensitive for CIN2+ than the self-collected samples on both assays (Table 1.)

**Table 1.** Sensitivity for CIN2+ of clinician-collected specimens and self-collection using Viba-brush specimens.

| Specimen Type | GP5+/6+-EIA PCR Assay [43] | SPF10 PCR-DEIALiPA25 PCR Assay [43] | GP5+/6+-EIA PCR Assay [44] |
|---|---|---|---|
| Clinician-collected | 48/49 (98.0%) | 49/49 (100%) | 39/43 (90.7%) |
| Self-collected | 43/49 (87.8%) | 47/49 (95.9%) | 40/43 (93.0%) |

### 2.3.3. Qvintip

There are only a few studies in the past ten years which examine the use of the Qvintip device for HPV testing. Four of these studies [46–49] are from the same research group where a senior member was a minority shareholder in the company that produces Qvintip. These studies do provide good information on the acceptability of self-collection using the Qvintip device but lack controls with none comparing paired samples, either two self-collected or one self-collected and one clinician-collected specimens. Follow-up (cytology, histology) was only undertaken on HPV positive participants so sensitivity is not calculable.

There is a direct comparison study by Jentschke and colleagues [16] which provide data that suggest that the Qvintip may be easier to use and less likely to cause discomfort than the Evalyn brush. These data also suggest, although this is not a statistically significant result, that the Qvintip is less sensitive for CIN2+ (83.7%) compared with either the Evalyn brush (89.8%) or a clinician-collected specimen (89.8%). The differences in both ease of use and sensitivity are small and it is possible that these differences may stem from the different instructions for use; the Evalyn brush was inserted into the vagina and rotated five times before being removed, whereas the Qvintip was simply inserted then removed without any rotation. Interestingly, the Qvintip instructions for use [50] now indicate that it should be rotated after insertion into the vagina.

### 2.3.4. Copan FLOQSwab

Another option for self-collection being increasingly used in cervical screening programs is the flocked swab, generally the FLOQswab made by Copan (Brescia, Italy). There are a few studies that examined self-sampling using flocked swabs, but it is difficult to assess these studies together as the specific type of FLOQswab used is not always described. A small study [51] examining the Copan self-collection swab compared with a cotton swab determined that the flocked swab was more sensitive for HPV16, HPV18 and for other high-risk HPV types. Although numbers of histologically confirmed CIN2+ were low, there was a tendency for better detection by the flocked sample (9 CIN2+) compared with the cotton swab (5 CIN2+). Another small study [52] used the Copan ESwab™ for self-collection and compared it to a self-collection sample stabilized on an FTA®-cartridge, and a clinician-collected specimen. The flocked swab had equal or greater sensitivity than both the FTA®-cartridge and the clinician-collected specimen. The flocked swab also matched the clinician–collected sample for detection of cytological high-grade squamous intraepithelial (HSIL) lesions and cervical cancers and was more sensitive than the FTA®-cartridge method. Both of these studies used the clinically-validated Seegene Anyplex HPV assay.

The Copan flocked swab (product number #552) has also be used in the national cervical screening programs in both Malaysia and Australia after successful pilot implementation studies [7,20].

*2.4. Other Considerations for Clinical Testing of Self-Collection Specimens for HPV*

2.4.1. Validated Assays

It is emerging best practice for the use of HPV assays in screening to require the assay to have evidence of being clinically validated, with the two major mechanisms for assessing clinical sensitivity and specificity being the Meijer Criteria [53] or VALGENT [54]. The Meijer Criteria [53] examines the assay's clinical sensitivity and specificity (for histologically confirmed CIN2+) compared with a reference assay, with an additional group of HPV positive enriched specimens for examining intra- and inter-laboratory reproducibility. VALGENT (VALidation of HPV GENotyping Tests) [54] uses a slightly different protocol for assessing specificity and sensitivity compared to a reference assay. It requires 1300 consecutive screening specimens supplemented with 300 abnormal cytology specimens and HPV results, along with follow-up histology where appropriate to determine relative sensitivity and specificity for CIN2+.

Both The Netherlands and Australia have self-collection pathways as part of a HPV-based national cervical screening program, with The Netherlands using the Evalyn brush and laboratories in Australia who are accredited currently all utilizing the Copan FLOQSwab. Currently only Roche cobas HPV assays (cobas or cobas 4800) are being used in these national programs for self-collected specimens. In Australia the availability of self-collection was delayed by lack of any suitable clinically validated HPV assays having manufacturer validated claims for the use of self-collected specimens (on label use), which required laboratories to independently validate their self-collection protocols against paired clinician-collected specimens ([25], other studies not published).

Whilst a number of assays present data from the scientific literature indicating that their assays perform well with self-collected specimens, at the time of writing, there does not appear to be any clinically validated assays with widely accepted (e.g., CE mark or FDA-approved) instructions for use covering self-collected specimens.

2.4.2. Sample Preparation

One of the less commonly addressed issues with self-collection is the actual processing of the sample within an accredited pathology laboratory, as opposed to a research laboratory. Many of the research studies investigating self-collection involve the self-collected specimen being transported to the laboratory as a dry sample where it is then resuspended in a liquid media [7,13,16]. The mechanism by which these dry samples are resuspended is highly varied. The most common buffers used are liquid-based cytology media, such as Hologic PreservCyt or BD SurePath, as these solutions are validated for use with HPV assays. Different volumes are used in different studies and often the mechanism by which the dry sample is eluted into these solutions is poorly described. At present, there is no consensus protocol for sample preparation, but this is likely to change if any of the major assay manufacturers add a self-collection claim to their assays. Use of such a manufacturer's protocol would simplify pathology laboratory accreditation for self-collection.

Another important consideration in the resuspension of dry samples is the availability of media. Many countries do not have stable supplies of liquid-based cytology media and even transporting these media can be difficult, as they are alcohol-based and flammable. This means that they must be shipped as a dangerous good (in volumes > 300 mL), increasing costs which may be prohibitive for low- and middle-income countries. Some manufacturers are looking to validate non-alcohol-based media to reduce these issues, although this may in turn result in reduced stability which could restrict use to point of care testing.

### 2.4.3. Pre-Analytic Considerations

In addition to a lack of validated HPV assays for self-collection and a consensus elution protocol, another issue for future scale up of its use is that self-collection is currently dependent on a number of manual steps. If self-collection is going to become widely available, and laboratories are going to process hundreds of samples a day, there will be a need for pre-analytic automation—both for workflow and for the reduction in inter-operator error. Current automated pre-analytic instruments, such as the Roche p480, do not have the ability to process self-collected devices.

Another possible mechanism would be to elute dry swabs directly into a solution already used for the processing of specimens, such as the BD diluent or the Roche cobas PCR media. There are currently two studies [55,56] which utilised the Roche cobas PCR media as the buffer into which a dry swab was eluted. This methodology (sample in Roche cobas PCR media) could be suitable using the Roche p480 pre-analytic instrument. Both studies transferred the swab directly into the cobas PCR media at the point of collection, and as such there are currently no data on the stability of a swab transported dry to a laboratory before being resuspended into this media. BD diluent has been used as buffer, but only in an ad hoc method rather than using the commercially available diluent tubes [57].

## 3. Discussion

The updated meta-analysis on the validity of using self-collection methods for HPV testing in 2018 by Arbyn et al. [10] was of critical importance as it demonstrated that the body of evidence had grown over the four years since the initial meta-analysis to a point where it could be clearly demonstrated that PCR-based assays offered similar levels of sensitivity for HPV in self-collected specimens compared with clinician-collected specimens. All 11 assays included in the analysis had self-collection sensitivities for HPV that were not significantly lower than the clinician-collected samples.

This review also aimed to summarize the current evidence on the utility of other technologies in national screening programs. careHPV presents as a viable test because of the very low cost, although the exact cost per test appears to be in question [39]. The data presented in the review supports the analysis by Arbyn et al. [10], that the lack of sensitivity of careHPV is likely to result in false negative results for specimens from women with CIN2+.

The Hologic Aptima HPV test uses RNA as its template, which may potentially give better specificity which could be beneficial for managing limited follow-up resources such as colposcopy more effectively. There are only a relatively small number of studies using Aptima and, because of the different mechanism of detection of HPV results compared to other assays, results cannot be pooled with PCR-based tests. All studies comparing self-collection on Aptima indicate that it is a less sensitive test for CIN2+ which suggests that, if there is higher specificity, it is to the detriment of detecting disease.

The evidence supporting non-PCR-based HPV tests for use with self-collected samples is not as favourable or as comprehensive as that supporting PCR-based testing. Further evidence on the application and performance of non-PCR-based technologies for self-sampling are likely to emerge and any new information will be critical for any re-assessment of use of these technologies for self-collection.

In contrast to the importance of correct HPV assay selection for the testing of self-collected specimens, a wide range of vastly different sampling devices appear to produce good results. The Arbyn meta-analysis [10] covered 10 different self-sampling devices and none produced significantly lower sensitivity than the clinician-collected specimen. This review examined a number of these devices in detail. The information seems to suggest that factors other than clinical sensitivity may be the deciding factor for the use of a device within a specific cervical screening program. The cost of self-collection devices can be prohibitive, with the FTA®-cartridge costing more than US$5 whereas a Copan flocked swab is under US$1 [52]. Environmental stability is also a variable requirement depending on the situation—temperature and humidity vary greatly from region to region. Another aspect of a self-collection device is whether it can be modified to high volume throughput. Few of the current device/assay combinations are designed for medium to high throughput (>200, and >1000 clinical specimens a day respectively), the exception being the cobas swab shipped in cobas PCR media. As we move towards self-collection as a primary cervical screening pathway, throughput will have to be considered, and incorporated into the design of the next general of pre-analytic automation.

Another emerging methodology for self-collection is the use of urine for HPV-based cervical screening. This particular method was not covered in this review as there is a paucity of data on the predictive value of a urine HPV positive result as it relates to clinical outcomes such as histologically—confirmed high-grade cervical disease. There is currently a diagnostic test accuracy study being undertaken called Validation of Human Papillomavirus Assays and Collection Devices for Self-samples and Urine Samples (VALHUDES) [58], which seeks to examine the clinical sensitivity and specificity of urine, and vaginal self-collected samples, against both a clinician-collected samples and against the histological diagnosis of followed up cases. It is likely that VALHUDES will address a number of other issues surrounding the development of a standardized urine collection and testing protocol [59], although this trial utilises a specific urine collection device, the Colli-Pee (Novosanis NV, Wijnegem, Belgium), which may be cost prohibitive in low- and middle-income countries.

There is a need for further research into the acceptance and feasibility of using self-collection in diverse under- or never-screened populations, but this falls outside the scope of the current review which is more focused on the technical aspects of self-collection.

## 4. Conclusions

This review sought to examine how the current evidence base could be used to support self-collection as a method of increasing cervical screening. There were three key findings relating to what type of test should be used, what type of collection devices should be used and what factors need to be considered moving forwards:

1. Self-collection is an attractive mechanism to increase participation in cervical screening worldwide. The current evidence strongly supports the need for PCR-based HPV assays with internal controls, specifically for both sample adequacy and detection of inhibition.
2. Whilst there does not appear to be major differences in the sensitivity of different collection devices, their cost, acceptability, and scalability as part of population level screening programs need to be considered.
3. This area of clinical testing for HPV as part of organized cervical screening programs will likely be a fast evolving one, with the continued development of new HPV assays, pre-analytic devices, and hopefully, manufacturer-validated claims for the use of self-collection.

**Author Contributions:** Conceptualization, D.H., M.S., and J.M.L.B. Writing—original draft preparation, M.H.T.K., Y.H., T.L.M., J.R., and D.H. Writing—review and editing, D.H., and J.M.L.B. All authors have read and agreed to the published version of the manuscript.

**Funding:** This research received no external funding.

**Conflicts of Interest:** M.S. is the Primary Investigator, J.M.L.B. and D.H. are Investigators, on the Compass Trial which received funding from Roche. D.H. has received funding for conference travel and attendance from Roche, Abbott and Seegene. No authors have received any personal benefit from any of these interactions.

## References


1. Arbyn, M.; Weiderpass, E.; Bruni, L.; de Sanjosé, S.; Saraiya, M.; Ferlay, J.; Bray, F. Estimates of incidence and mortality of cervical cancer in 2018: A worldwide analysis. *Lancet. Glob. Health* **2020**, *8*, e191–e203. [CrossRef]
2. Zucchetto, A.; Ronco, G.; Giorgi Rossi, P.; Zappa, M.; Ferretti, S.; Franzo, A.; Falcini, F.; Visioli, C.B.; Zanetti, R.; Biavati, P.; et al. Screening patterns within organized programs and survival of Italian women with invasive cervical cancer. *Prev. Med.* **2013**, *57*, 220–226. [CrossRef] [PubMed]
3. Ibáñez, R.; Alejo, M.; Combalia, N.; Tarroch, X.; Autonell, J.; Codina, L.; Culubret, M.; Bosch, F.X.; de Sanjosé, S. Underscreened women remain overrepresented in the pool of cervical cancer cases in Spain: A need to rethink the screening interventions. *Biomed. Res. Int* **2015**, *2015*, 605375. [CrossRef] [PubMed]
4. Harder, E.; Juul, K.E.; Jensen, S.M.; Thomsen, L.T.; Frederiksen, K.; Kjaer, S.K. Factors associated with non-participation in cervical cancer screening - a nationwide study of nearly half a million women in Denmark. *Prev. Med.* **2018**, *111*, 94–100. [CrossRef] [PubMed]
5. AIHW. *Cervical Screening in Australia*; Australian Institute of Health and Welfare: Canberra, Australia, 2018.
6. Chorley, A.J.; Marlow, L.A.V.; Forster, A.S.; Haddrell, J.B.; Waller, J. Experiences of cervical screening and barriers to participation in the context of an organised programme: A systematic review and thematic synthesis. *Psychooncology* **2017**, *26*, 161–172. [CrossRef] [PubMed]
7. Saville, M.; Hawkes, D.; McLachlan, E.; Anderson, S.; Arabena, K. Self-collection for under-screened women in a national cervical screening program: Pilot study. *Current Oncology* **2018**, *25*, e27. [CrossRef]
8. Birdsong, G.G.; Davey, D.D. Specimen adequacy. In *The Bethesda System for Reporting Cervical Cytology*, 3rd ed.; Nayar, R., Wilbur, D.C., Eds.; Springer: Berlin/Heidelberg, Germany, 2015.

9. Cancer Council Australia. Cervical Cancer Screening Guidelines Working Party. National Cervical Screening Program: Guidelines for the management of screen-detected abnormalities, screening in specific populations and investigation of abnormal vaginal bleeding. Cancer Council Australia: Sydney, Australia; Available online: https://wiki.cancer.org.au/australia/Guidelines:Cervical_cancer/Screening (accessed on 21 April 2020).

10. Arbyn, M.; Smith, S.B.; Temin, S.; Sultana, F.; Castle, P. Detecting cervical precancer and reaching underscreened women by using HPV testing on self samples: Updated meta-analyses. *BMJ* **2018**, *363*, k4823. [CrossRef]

11. Pedersen, H.N.; Smith, L.W.; Racey, C.S.; Cook, D.; Krajden, M.; van Niekerk, D.; Ogilvie, G.S. Implementation considerations using HPV self-collection to reach women under-screened for cervical cancer in high-income settings. *Curr. Oncol. Toronto Ont.* **2018**, *25*, e4–e7. [CrossRef]

12. Franco, E.L. Self-sampling for cervical cancer screening: Empowering women to lead a paradigm change in cancer control. *Curr. Oncol. (Toronto Ont.)* **2018**, *25*, e1–e3. [CrossRef]

13. Ketelaars, P.J.W.; Bosgraaf, R.P.; Siebers, A.G.; Massuger, L.F.A.G.; van der Linden, J.C.; Wauters, C.A.P.; Rahamat-Langendoen, J.C.; van den Brule, A.J.C.; IntHout, J.; Melchers, W.J.G.; et al. High-risk human papillomavirus detection in self-sampling compared to physician-taken smear in a responder population of the Dutch cervical screening: Results of the Vera study. *Prev. Med.* **2017**, *101*, 96–101. [CrossRef]

14. Tranberg, M.; Jensen, J.S.; Bech, B.H.; Blaakær, J.; Svanholm, H.; Andersen, B. Good concordance of HPV detection between cervico-vaginal self-samples and general practitioner-collected samples using the cobas 4800 HPV DNA test. *BMC Infect. Dis.* **2018**, *18*, 348. [CrossRef] [PubMed]

15. Leinonen, M.K.; Schee, K.; Jonassen, C.M.; Lie, A.K.; Nystrand, C.F.; Rangberg, A.; Furre, I.E.; Johansson, M.J.; Tropé, A.; Sjøborg, K.D.; et al. Safety and acceptability of human papillomavirus testing of self-collected specimens: a methodologic study of the impact of collection devices and HPV assays on sensitivity for cervical cancer and high-grade lesions. *J. Clin. Virol. Off. Publ. Pan Am. Soc. Clin. Virol.* **2018**, *99*, 22–30. [CrossRef] [PubMed]

16. Jentschke, M.; Chen, K.; Arbyn, M.; Hertel, B.; Noskowicz, M.; Soergel, P.; Hillemanns, P. Direct comparison of two vaginal self-sampling devices for the detection of human papillomavirus infections. *J. Clin. Virol. Off. Publ. Pan Am. Soc. Clin. Virol.* **2016**, *82*, 46–50. [CrossRef] [PubMed]

17. Chen, K.; Ouyang, Y.; Hillemanns, P.; Jentschke, M. Excellent analytical and clinical performance of a dry self-sampling device for human papillomavirus detection in an urban Chinese referral population. *J. Obstet. Gynaecol. Res.* **2016**, *42*, 1839–1845. [CrossRef] [PubMed]

18. Ejegod, D.M.; Pedersen, H.; Alzua, G.P.; Pedersen, C.; Bonde, J. Time and temperature dependent analytical stability of dry-collected Evalyn HPV self-sampling brush for cervical cancer screening. *Papillomavirus Res. Amst. Neth.* **2018**, *5*, 192–200. [CrossRef] [PubMed]

19. Toliman, P.J.; Kaldor, J.M.; Badman, S.G.; Phillips, S.; Tan, G.; Brotherton, J.M.L.; Saville, M.; Vallely, A.J.; Tabrizi, S.N. Evaluation of self-collected vaginal specimens for the detection of high-risk human papillomavirus infection and the prediction of high-grade cervical intraepithelial lesions in a high-burden, low-resource setting. *Clin. Microbiol. Infect. Off. Publ. Eur. Soc. Clin. Microbiol. Infect. Dis.* **2019**, *25*, 496–503. [CrossRef]

20. Woo, Y.L. The feasibility and acceptability of self-sampling and HPV testing using Cepheid Xpert(®) HPV in a busy primary care facility. *J. Virus Erad.* **2019**, *5*, 10–11.

21. Brisson, M.; Kim, J.J.; Canfell, K.; Drolet, M.; Gingras, G.; Burger, E.A.; Martin, D.; Simms, K.T.; Bénard, É.; Boily, M.-C.; et al. Impact of HPV vaccination and cervical screening on cervical cancer elimination: A comparative modelling analysis in 78 low-income and lower-middle-income countries. *Lancet Lond. Engl.* **2020**, *395*, 575–590. [CrossRef]

22. WHO. Who Director-General Calls for All Countries to Take Action to Help End the Suffering Caused by Cervical Cancer. Available online: https://www.who.int/reproductivehealth/call-to-action-elimination-cervical-cancer/en/ (accessed on 10 February 2020).

23. WHO. *Draft: Global Strategy Towards Eliminating Cervical Cancer as a Public Health Problem*; WHO: Geneva, Switzerland, 2019.

24. NPAAC. *Requirements for Laboratories Reporting Tests for the National Cervical Screening Program*, 1st ed.; Australian Government Department of Health: Canberra, Australia, 2017.

25.  Saville, M.; Hawkes, D.; Keung, M.H.; Ip, E.; Silvers, J.; Sultana, F.; Malloy, M.; Velentzis, L.; Canfell, K.; Wrede, C.D.; et al. Analytical performance of HPV assays on vaginal self-collected vs practitioner-collected samples: The SCOPE study. *J. Clin. Virol.* **2020**, in press. [CrossRef]

26.  Arbyn, M.; Verdoodt, F.; Snijders, P.J.; Verhoef, V.M.; Suonio, E.; Dillner, L. Accuracy of human papillomavirus testing on self-collected versus clinician-collected samples: a meta-analysis. *Lancet Oncol.* **2014**, *15*, 172–183. [CrossRef]

27.  Chernesky, M.; Jang, D.; Gilchrist, J.; Elit, L.; Lytwyn, A.; Smieja, M.; Dockter, J.; Getman, D.; Reid, J.; Hill, C. Evaluation of a new Aptima specimen collection and transportation kit for high-risk human papillomavirus e6/e7 messenger RNA in cervical and vaginal samples. *Sex. Transm Dis.* **2014**, *41*, 365–368. [CrossRef] [PubMed]

28.  Asciutto, K.C.; Ernstson, A.; Forslund, O.; Borgfeldt, C. Self-sampling with HPV mRNA analyses from vagina and urine compared with cervical samples. *J. Clin. Virol.* **2018**, *101*, 69–73. [CrossRef] [PubMed]

29.  Nieves, L.; Enerson, C.L.; Belinson, S.; Brainard, J.; Chiesa-Vottero, A.; Nagore, N.; Booth, C.; Perez, A.G.; Chavez-Aviles, M.N.; Belinson, J. Primary cervical cancer screening and triage using an mRNA human papillomavirus assay and visual inspection. *Int. J. Gynecol. Cancer Off. J. Int. Gynecol. Cancer Soc.* **2013**, *23*, 513–518. [CrossRef] [PubMed]

30.  Borgfeldt, C.; Forslund, O. Increased HPV detection by the use of a pre-heating step on vaginal self-samples analysed by Aptima HPV assay. *J. Virol. Methods* **2019**, *270*, 18–20. [CrossRef] [PubMed]

31.  Cuschieri, K.; Geraets, D.; Cuzick, J.; Cadman, L.; Moore, C.; Vanden Broeck, D.; Padalko, E.; Quint, W.; Arbyn, M. Performance of a cartridge-based assay for detection of clinically significant human papillomavirus (HPV) infection: Lessons from VALGENT (Validation of HPV genotyping tests). *J. Clin. Microbiol.* **2016**, *54*, 2337–2342. [CrossRef] [PubMed]

32.  Brotherton, J.M.; Hawkes, D.; Sultana, F.; Malloy, M.J.; Machalek, D.A.; Smith, M.A.; Garland, S.M.; Saville, M. Age-specific HPV prevalence among 116,052 women in Australia's renewed cervical screening program: A new tool for monitoring vaccine impact. *Vaccine* **2019**, *37*, 412–416. [CrossRef] [PubMed]

33.  Machalek, D.A.; Roberts, J.M.; Garland, S.M.; Thurloe, J.; Richards, A.; Chambers, I.; Sivertsen, T.; Farnsworth, A. Routine cervical screening by primary HPV testing: Early findings in the renewed national cervical screening program. *Med. J. Aust.* **2019**, *211*, 113–119. [CrossRef]

34.  Vassilakos, P.; Catarino, R.; Bougel, S.; Munoz, M.; Benski, C.; Meyer-Hamme, U.; Jinoro, J.; Heriniainasolo, J.L.; Petignat, P. Use of swabs for dry collection of self-samples to detect human papillomavirus among Malagasy women. *Infect. Agents Cancer* **2016**, *11*, 13. [CrossRef]

35.  Chen, W.; Jeronimo, J.; Zhao, F.-H.; Qiao, Y.-L.; Valdez, M.; Zhang, X.; Kang, L.-N.; Bansil, P.; Paul, P.; Bai, P.; et al. The concordance of HPV DNA detection by hybrid capture 2 and careHPV on clinician- and self-collected specimens. *J. Clin. Virol.* **2014**, *61*, 553–557. [CrossRef]

36.  Obiri-Yeboah, D.; Adu-Sarkodie, Y.; Djigma, F.; Hayfron-Benjamin, A.; Abdul, L.; Simpore, J.; Mayaud, P. Self-collected vaginal sampling for the detection of genital human papillomavirus (HPV) using CareHPV among Ghanaian women. *BMC Women's Health* **2017**, *17*, 1–6. [CrossRef]

37.  Baussano, I.; Tshering, S.; Choden, T.; Lazzarato, F.; Tenet, V.; Plummer, M.; Franceschi, S.; Clifford, G.M.; Tshomo, U. Cervical cancer screening in rural Bhutan with the careHPV test on self-collected samples: An ongoing cross-sectional, population-based study (Reach-Bhutan). *BMJ Open* **2017**, *7*. [CrossRef] [PubMed]

38.  Swanson, M.; Ibrahim, S.; Blat, C.; Oketch, S.; Olwanda, E.; Maloba, M.; Huchko, M.J. Evaluating a community-based cervical cancer screening strategy in western Kenya: A descriptive study. *BMC Womens Health* **2018**, *18*, 116. [CrossRef] [PubMed]

39.  Tin-Oo, C.; Hlaing, H.N.T.; Nandar, C.S.; Aung, T.; Fishbein, D. Why the cost of purchasing the careHPV test in Myanmar was many times greater than that reported in the international literature. *J. Glob. Oncol.* **2018**, *4*, 49s. [CrossRef]

40.  Polman, N.J.; Ebisch, R.M.F.; Heideman, D.A.M.; Melchers, W.J.G.; Bekkers, R.L.M.; Molijn, A.C.; Meijer, C.J.L.M.; Quint, W.G.V.; Snijders, P.J.F.; Massuger, L.F.A.G.; et al. Performance of human papillomavirus testing on self-collected versus clinician-collected samples for the detection of cervical intraepithelial neoplasia of grade 2 or worse: A randomised, paired screen-positive, non-inferiority trial. *Lancet. Oncol.* **2019**, *20*, 229–238. [CrossRef]

41. Van Baars, R.; Bosgraaf, R.P.; ter Harmsel, B.W.A.; Melchers, W.J.G.; Quint, W.G.V.; Bekkers, R.L.M. Dry storage and transport of a cervicovaginal self-sample by use of the Evalyn brush, providing reliable human papillomavirus detection combined with comfort for women. *J. Clin. Microbiol.* **2012**, *50*, 3937–3943. [CrossRef] [PubMed]

42. Leeman, A.; Del Pino, M.; Molijn, A.; Rodriguez, A.; Torné, A.; de Koning, M.; Ordi, J.; van Kemenade, F.; Jenkins, D.; Quint, W. HPV testing in first-void urine provides sensitivity for CIN2+ detection comparable with a smear taken by a clinician or a brush-based self-sample: Cross-sectional data from a triage population. *BJOG Int. J. Obstet. Gynaecol.* **2017**, *124*, 1356–1363. [CrossRef]

43. Geraets, D.T.; van Baars, R.; Alonso, I.; Ordi, J.; Torné, A.; Melchers, W.J.G.; Meijer, C.J.L.M.; Quint, W.G.V. Clinical evaluation of high-risk HPV detection on self-samples using the indicating FTA-elute solid-carrier cartridge. *J. Clin. Virol. Off. Publ. Pan Am. Soc. Clin. Virol.* **2013**, *57*, 125–129. [CrossRef]

44. Dijkstra, M.G.; Heideman, D.A.M.; van Kemenade, F.J.; Hogewoning, K.J.A.; Hesselink, A.T.; Verkuijten, M.C.G.T.; van Baal, W.M.; Boer, G.M.N.-d.; Snijders, P.J.F.; Meijer, C.J.L.M. Brush-based self-sampling in combination with GP5+/6+-PCR-based HRHPV testing: High concordance with physician-taken cervical scrapes for HPV genotyping and detection of high-grade CIN. *J. Clin. Virol. Off. Publ. Pan Am. Soc. Clin. Virol.* **2012**, *54*, 147–151. [CrossRef]

45. Aiko, K.Y.; Yoko, M.; Saito, O.M.; Ryoko, A.; Yasuyo, M.; Mikiko, A.S.; Takeharu, Y.; Fumiki, H.; Etsuko, M. Accuracy of self-collected human papillomavirus samples from Japanese women with abnormal cervical cytology. *J. Obstet. Gynaecol. Res.* **2017**, *43*, 710–717. [CrossRef]

46. Gyllensten, U.; Sanner, K.; Gustavsson, I.; Lindell, M.; Wikstrom, I.; Wilander, E. Short-time repeat high-risk HPV testing by self-sampling for screening of cervical cancer. *Br. J. Cancer* **2011**, *105*, 694–697. [CrossRef]

47. Sanner, K.; Wikstrom, I.; Strand, A.; Lindell, M.; Wilander, E. Self-sampling of the vaginal fluid at home combined with high-risk HPV testing. *Br. J. Cancer* **2009**, *101*, 871–874. [CrossRef] [PubMed]

48. Stenvall, H.; Wikstrom, I.; Wilander, E. High prevalence of oncogenic human papilloma virus in women not attending organized cytological screening. *Acta Dermato Venereol.* **2007**, *87*, 243–245. [PubMed]

49. Wikstrom, I.; Lindell, M.; Sanner, K.; Wilander, E. Self-sampling and HPV testing or ordinary Pap-smear in women not regularly attending screening: A randomised study. *Br. J. Cancer* **2011**, *105*, 337–339. [CrossRef] [PubMed]

50. Aprovix. Qvintip Instructions for Use. 2019, p. 1. Available online: http://www.genoid.net/images/uploads/sy_Qvintip_user_guide_20160531.pdf (accessed on 21 April 2020).

51. Viviano, M.; Willame, A.; Cohen, M.; Benski, A.C.; Catarino, R.; Wuillemin, C.; Tran, P.L.; Petignat, P.; Vassilakos, P. A comparison of cotton and flocked swabs for vaginal self-sample collection. *Int. J. Womens Health* **2018**, *10*, 229–236. [CrossRef]

52. Catarino, R.; Vassilakos, P.; Bilancioni, A.; Vanden Eynde, M.; Meyer-Hamme, U.; Menoud, P.-A.; Guerry, F.; Petignat, P. Randomized comparison of two vaginal self-sampling methods for human papillomavirus detection: Dry swab versus FTA cartridge. *PLoS ONE* **2015**, *10*, e0143644. [CrossRef]

53. Meijer, C.J.; Berkhof, J.; Castle, P.E.; Hesselink, A.T.; Franco, E.L.; Ronco, G.; Arbyn, M.; Bosch, F.X.; Cuzick, J.; Dillner, J.; et al. Guidelines for human papillomavirus DNA test requirements for primary cervical cancer screening in women 30 years and older. *Int. J. Cancer* **2009**, *124*, 516–520. [CrossRef]

54. Arbyn, M.; Depuydt, C.; Benoy, I.; Bogers, J.; Cuschieri, K.; Schmitt, M.; Pawlita, M.; Geraets, D.; Heard, I.; Gheit, T.; et al. VALGENT: A protocol for clinical validation of human papillomavirus assays. *J. Clin. Virol.* **2016**, *76* (Suppl. S1), S14–S21. [CrossRef]

55. Stanczuk, G.; Baxter, G.; Currie, H.; Lawrence, J.; Cuschieri, K.; Wilson, A.; Arbyn, M. Clinical validation of HRHPV testing on vaginal and urine self-samples in primary cervical screening (cross-sectional results from the Papillomavirus Dumfries and Galloway—PAVDAG study). *BMJ Open* **2016**, *6*, e010660. [CrossRef]

56. Asciutto, K.C.; Henningsson, A.J.; Borgfeldt, H.; Darlin, L.; Borgfeldt, C. Vaginal and urine self-sampling compared to cervical sampling for HPV-testing with the cobas 4800 HPV test. *Anticancer. Res.* **2017**, *37*, 4183–4187.

57. Lam, J.U.H.; Rebolj, M.; Ejegod, D.M.; Pedersen, H.; Rygaard, C.; Lynge, E.; Harder, E.; Thomsen, L.T.; Kjaer, S.K.; Bonde, J. Prevalence of human papillomavirus in self-taken samples from screening nonattenders. *J. Clin. Microbiol.* **2017**, *55*, 2913–2923. [CrossRef]

58. Arbyn, M.; Peeters, E.; Benoy, I.; Vanden Broeck, D.; Bogers, J.; De Sutter, P.; Donders, G.; Tjalma, W.; Weyers, S.; Cuschieri, K.; et al. Valhudes: A protocol for validation of human papillomavirus assays and collection devices for HPV testing on self-samples and urine samples. *J. Clin. Virol.* **2018**, *107*, 52–56. [CrossRef] [PubMed]
59. Pattyn, J.; Van Keer, S.; Téblick, L.; Van Damme, P.; Vorsters, A. HPV DNA detection in urine samples of women: 'An efficacious and accurate alternative to cervical samples?'. *Expert Rev. Anti-Infect. Ther.* **2019**, *17*, 755–757. [CrossRef] [PubMed]

*Review*

# Management of HPV-Related Squamous Cell Carcinoma of the Head and Neck: Pitfalls and Caveat

**Francesco Perri [1,*], Francesco Longo [2], Francesco Caponigro [1], Fabio Sandomenico [3], Agostino Guida [2], Giuseppina Della Vittoria Scarpati [4], Alessandro Ottaiano [5], Paolo Muto [6] and Franco Ionna [2]**

[1]   Head and Neck Medical Oncology Unit, Istituto Nazionale Tumori, IRCCS G. Pascale, 80131 Naples, Italy; f.caponigro@istitutotumori.na.it

[2]   Division of Surgical Oncology Maxillo-Facial Unit, Istituto Nazionale Tumori—IRCCS—Fondazione G. Pascale, Via Mariano Semmola, 80131 Naples, Italy; f.longo@istitutotumori.na.it (F.L.); a.guida@istitutotumori.na.it (A.G.); f.ionna@istitutotumori.na.it (F.I.)

[3]   Unit, Istituto Nazionale Tumori—IRCCS—G. Pascale, 80131 Naples, Italy; f.sandomenico@istitutotumori.na.it

[4]   Medical Oncology Unit, Ospedale Pollena, ASL NA3, 80040 Naples, Italy; giuseppina.dellavittoria@gmail.com

[5]   SSD Innovative Therapies for Abdominal Metastases, Department of Abdominal Oncology, INT IRCCS Fondazione G. Pascale, 80131 Naples, Italy; a.ottaiano@istitutotumori.na.it

[6]   Radiation Oncology Unit, Istituto Nazionale Tumori—IRCCS—G. Pascale, 80131 Naples, Italy; p.muto@istitutotumori.na.it

*   Correspondence: f.perri@istitutotumori.na.it; Tel.: +39-081-590-3362

Received: 13 March 2020; Accepted: 10 April 2020; Published: 15 April 2020

**Abstract:** Head and neck squamous cell carcinomas (HNSCCs) are a very heterogeneous group of malignancies arising from the upper aerodigestive tract. They show different clinical behaviors depending on their origin site and genetics. Several data support the existence of at least two genetically different types of HNSCC, one virus-related and the other alcohol and/or tobacco and oral trauma-related, which show both clinical and biological opposite features. In fact, human papillomavirus (HPV)-related HNSCCs, which are mainly located in the oropharynx, are characterized by better prognosis and response to therapies when compared to HPV-negative HNSCCs. Interestingly, virus-related HNSCC has shown a better response to conservative (nonsurgical) treatments and immunotherapy, opening questions about the possibility to perform a pretherapy assessment which could totally guide the treatment strategy. In this review, we summarize molecular differences and similarities between HPV-positive and HPV-negative HNSCC, highlighting their impact on clinical behavior and on therapeutic strategies.

**Keywords:** human papilloma virus; squamous cell carcinoma of the head and neck; translational research; oncogenes; tumor suppressor genes

## 1. Background

Head and neck squamous cell carcinoma (HNSCC) is an heterogeneous group of malignancies comprised of several entities arising from different anatomical subsites, including the oral cavity, hypopharynx, oropharynx, larynx and nasopharynx. They are characterized by different etiologies, genetics and clinical behaviors. HNSCC development is strongly related to tobacco and/or alcohol consumption and/or oral trauma, and/or, in particular in some countries such as Iran and Southeast Asia, betel quid chewing. Nevertheless, in the last 15 years, remarkable changes in HNSCC epidemiology have been observed and a critical increase in the diagnosis of some kinds of HSNCC, e.g., the oropharyngeal carcinoma, have been noted. This feature is probably due to the increasing incidence of human papilloma virus (HPV)-related tumors [1–3]. Several lines of evidence support the existence of

*Cancers* **2020**, *12*, 975; doi:10.3390/cancers12040975         www.mdpi.com/journal/cancers

at least two genetically different types of HNSCC, one virus-related and the other alcohol and/or tobacco and oral trauma-related, characterized by both clinical and biological opposite features [3,4]. Unlike HPV-negative HNSCC, HPV-positive HNSCC often occurs in younger patients with minimal or no tobacco exposure [5,6]. HPV-positive HNSCC, similarly to its HPV-negative counterpart, has a male predominance, with men suffering a three-to-five times higher incidence than women worldwide [7].

HPV-positive HNSCC carries a favorable prognosis if compared to HPV-negative tumors. In fact, five-year survival rates for patients with advanced-stage HPV-positive HNSCC are 75–80%, versus values less than 50% in patients with similarly staged HPV-negative tumors [8,9]. The cause of the aforementioned different behavior is the different chemo- and radiosensitivity shown by the HPV-positive and HPV-negative HNSCCs. In fact, several clinical trials have shown that HPV-positive HNSCC patients have a better response to chemotherapy and radiation therapy than HPV-negative cases [10–12]. The reasons for this different behavior should be searched in the opposite genetic features which characterize the two types of tumors.

In this review, we will analyze the genetics of both HPV-positive and HPV-negative HNSCC, highlighting their impact on the clinical behavior and finally on the therapeutic strategies.

## 2. Genetics of HPV-Positive HNSCC

Carcinogenesis, which is the complex process through which the normal cell is pushed to transform itself into a cancer cell, is very different between HPV-related and non-HPV-related HNSCC. Viral carcinogenesis in HNSCC is partly due to HPV infection, with the oropharynx being the most commonly involved site. HPV-mediated carcinogenesis is driven by a few viral oncoproteins expressed by high-risk HPV genotypes. In particular, E6 and E7 have shown to inactivate p53 and retinoblastoma protein (pRb), respectively, and to impair several metabolic pathways in the infected cells [13–15].

The dominant viral type associated with the development of HNSCC (especially oropharyngeal carcinoma) is HPV16, while HPV18, 31 and 33 are less frequently detected [16,17]. The commonly acknowledged paradigm of HPV carcinogenesis, based on studies conducted upon uterine cervical cancer, highlights the importance of HPV genome integration as a premalignant lesion [18]. Nevertheless, recent acquisitions support the issue that nearly 30% of HPV-positive HNSCC contained only episomal HPV, prompting new theories about the alternative mechanisms of HPV-driven carcinogenesis.

The HPV life cycle is related to the host cell capability of proliferation, since its genome does not encode the polymerase or other enzymes necessary for viral replication; so HPV, once entered in the cell, both by integrating itself in the host DNA remaining in the form of episomes, becomes able to promote cell cycle progression mainly through the above mentioned genes, E6 and E7 [19,20]. In particular, E6 inactivates p53, affecting its capability to activate p21, which, in turns, blocks CDK (Kinase Cyclin Dependent)/Cyclin heterodimers. This latter protein, through phosphorylation of pRB, provokes E2F release. E2F is able to act as an oncogene stimulating the G1/S transition. The final result is a deregulation of the cell cycle [21,22]. On the other hand, E7 allows for retinoblastoma protein (pRB) degradation, which in normal conditions binds and puts off the transcriptional factor E2F. Moreover, pRB degradation is able to remove the inhibitory feedback upon p16 synthesis, thus leading to its hyperproduction. For this reason, in HPV-infected tumor cells, p16 is always upregulated [23,24]. The cell cycle dysregulation caused by HPV infection can lead to accumulation of DNA damage, and thus promotes carcinogenesis (see Figure 1).

The E7 protein is also able to induce methylation of suppressors with a morphogenetic effect on genitalia (SMG-1) gene promoter, provoking its dysfunction. SMG-1 is a tumor suppressor gene encoding for a protein able to arrest the cell cycle in response to DNA damage [25,26].

In addition, HPV DNA often integrates in host DNA, and the integration takes place in specific loci, such as RAD51. This nonrandom integration leads to RAD51 gene dysfunction. RAD51 is an enzyme involved in double-strand DNA break repair, and its function results in impairment in virus-related HNSCC [27].

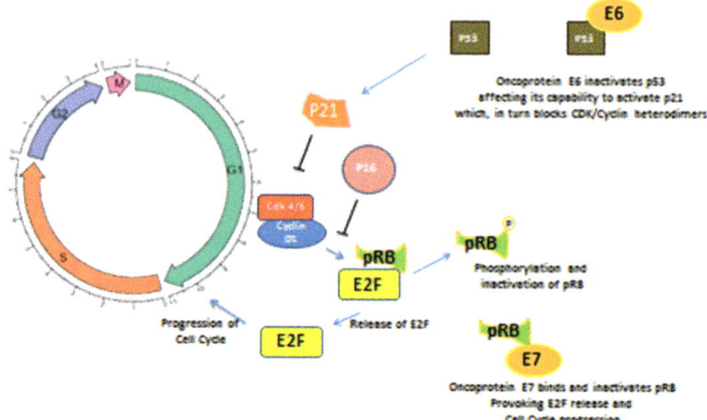



**Figure 1.** Oncoproteins E6 and E7 induce cell cycle progression acting upon cell cycle regulation, in particular during G1/S transition. E6 binds and inactivates P53, affecting its capability to activate P21, which in turn is not able to arrest CyclinD1/Cdk4/6 heterodimer. Oncoprotein E7 directly acts on RB, linking and inactivating it, leading to E2F upregulation and cell cycle progression.E2F: elongation factor 2; pRB: retinoblastoma protein; Cdk4/6: cyclin dependent kinase; p21: protein 21; E6: oncoprotein 6; E7: oncoprotein 7.

Overall, from a genetic point of view, HPV-related tumors present unique features, such as p16 overexpression, CyclinD1 and pRB down-regulation, a low EGFR (Epithelial growth factor receptor) expression with a high proliferating index (Ki-67). Interestingly, the most mutated and disrupted pathway in those tumors is the Akt-related one [28,29].

HPV-related carcinogenesis is characterized by a relatively low number of DNA mutations and chromosomal changes while, on the other hand, by a higher percentage of epigenetic changes. As a matter of fact, virus-related HNSCCs are associated with a significantly lower mutational rate, when compared to non-virus-related HNSCC [30–32]. The lower tumor mutational burden (TMB) typical of HPV-related HNSCC leads to a generation of oligoclonal tumors which are intrinsically more chemo- and radiosensive. Table 1 describes the main genetic differences between HPV-positive and the HPV-negative HSNCC.

**Table 1.** The picture describes the main genetic features characterizing the virus-related, the mutagens-related HSNCC and the HPV positive but not related HSNCC.

| HPV-related HSNCC | Alcohol and Tobacco related HSNCC | HPV-positive (but not related) HSNCC |
|---|---|---|
| -P16 upregulation (not mutated INK-4 gene) | -P16 downregulation (INK-4 mutations) | -P16 upregulation (not mutated INK-4 gene) |
| -CCND1 wild type | -TP53 mutations | -TP53 mutations |
| -TP53 wild type | -CCND1 amplification | -CCND1 amplification |
| -Low number of genic/chromosomal abnormalities | -High number of genic/chromosomal | -High number of genic/chromosomal |
| -Higher rate of PI3Kca mutations | abnormalities | abnormalities |
| -Extensive TSG promoters methylation | -Low immune infiltrate | -Low immune infiltrate |
| -High immune infiltrate | | |

CCND1: Cyclin D1; PI3Kca: the gene encoding for protein 3 Kinase; TSG: tumor suppressor genes; INK-4: INhibitors of CDK4; HPV: human papillomavirus; HSNCC: Head and neck squamous cell carcinoma.

On the other hand, in the HPV-negative (mutagen-related) HSNCCs, the gradual acquisition of mutations involving both "oncogenes" and "tumor suppressor genes", by effect of the mutagens contained in alcohol, tobacco and betel, is critical to cause neoplastic transformation and is reflected by their high mutation load. Thus, the pathogenesis of mutagen-related HSNCC is strongly linked to progressive accumulation of mutations which affect several DNA traits. Interestingly, these DNA changes often concern some important and crucial genes, namely TP53, CCND1, INK-4, EGFR, and NOTCH1. Mutagen-related HSNCC often presents EGFR and CCND1 amplification, INK-4 mutations, inducing p16 down-regulation, and disruption of the NOTCH-1 pathway [4]. The wide number of genic aberrations generates heterogeneous neoplastic populations, which are responsible for chemo- and radioresistance often observed in alcohol- and tobacco-related SCCHN.

In addition, it is very important to discern between HPV-positive and HPV-"related" HNSCC. These two categories of tumors may be very different from each other, due to the fact that in the HPV-related HNSCC, the entire carcinogenesis process is initiated and sustained by HPV, which may be defined as the sole "driver". In this case only, the tumor is characterized by a low TMB, a wild-type p53 status, p16 overexpression and the concomitant wild-type status of the genes INK4 (which encodes for p16), EGFR and CCND1 (the gene encoding for cyclin D1). Nevertheless, there is another category of HPV-positive tumors which harbor p53 mutations, CCND1 amplification and INK4 mutations, too. The latter have a poor prognosis, similarly to the HPV-negative HNSCC.

Weinberger et al. [33] performed a molecular analysis on 80 cases of oropharyngeal carcinomas, analyzing the status of a panel of biomarkers, such as p16, p53 and pRB and correlating it with the clinical outcome. On the basis of the results obtained, they divided HNSCCs into three classes. Class I was characterized by the absence of the HPV, and the contemporaneous presence of p53 mutations and p16 inactivation; these tumors were considered to be mutagen-related and showed poor prognosis. Class II was characterized by HPV positivity, and in concomitance, p53 mutations and p16 inactivation; these were considered to be HPV-positive but not HPV-related, and showed poor prognosis. Finally, class III encompassed all the HPV-related HNSCCs characterized by HPV positivity, wild-type for p16 and p53. Figure 2 describes the classification of oropharyngeal carcinomas suggested by Weinberger et al.

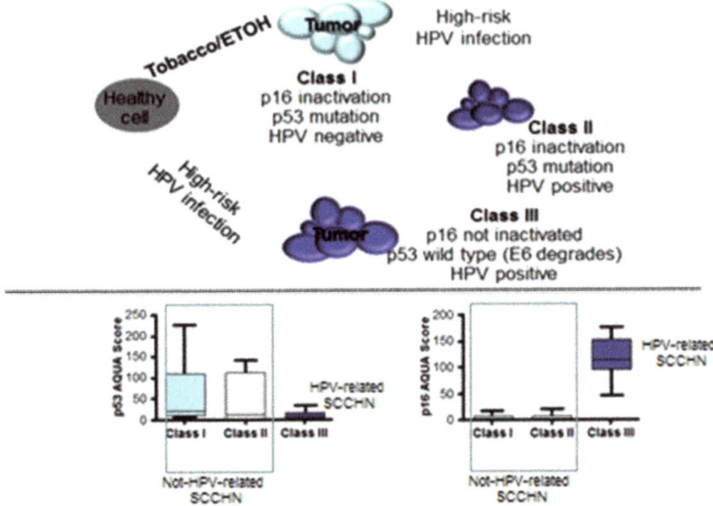


**Figure 2.** Proposed model for head and neck squamous cell carcinomas (HNSCC) carcinogenesis. Not all the human papillomavirus (HPV)-positive HNSCCs are also HPV-related. Both HPV-negative and HPV-positive but not HPV-related HSNCC (Class I and II, respectively) are characterized by INK-4 (the gene encoding for P16) and TP53 mutations which allow for a high expression of P53 and a down-regulation of P16 on the immunohistochemical assay. On the other hand, HPV-related HSNCC (Class III) show reverse features. 33. Source: Weinberger PM et al., J Clin Oncol. **2006**, *24*, 736–747 [33]. ETOH: alcohol; SCCHN: squamous cell carcinoma of the head and neck.

This last classification may be of help in improving the identification of the virus-related HNSCCs and in separating them from their mutagen-related counterpart, aiming to modulate the therapy options.

### 3. HPV-Related Tumors: Implications in Clinical Practice

Due to the high chemo- and radiosensitivity shown by HPV-related HNSCC, some authors have hypothesized the possibility of under-treating affected patients. The rationale for de-intensification of chemoradiotherapy is to reduce the side-effects caused by combination therapy. In fact, radiotherapy is associated with dose-related adverse side effects, from acute toxicities like mucositis and loss of taste to long-term problems including renal dysfunction, severe dysphagia, significant xerostomia, hearing loss, osteoradionecrosis, strong neck muscle fibrosis, accelerated arteriosclerosis and trismus. These toxicities may cause a cascade of events, such as infections, dysphagia, feeding tube necessity and increased hospitalizations, that can markedly affect the quality of life. All the above side effects are strongly boosted up by the addition of chemotherapy. Thus, it is deductive to think that the reduction of the dose of radiation therapy and/or the chemotherapy may reduce the percent of cumulative toxicities.

In 2014, Cmelak et al. presented at ASCO (American Society of clinical Oncology) the preliminary results of phase II trials enrolling HPV-positive HNSCC patients. Patients underwent induction chemotherapy followed by two different regimens of concurrent cetuximab radiotherapy (RT) on the basis of the obtained response. In particular, patients who completely responded to induction chemotherapy were treated with an underpowered RT regimen, consisting in 50 Gy instead of the standard 70 Gy, while concurrent cetuximab and 70 Gy RT was administered in those who showed only a partial response. As a result, a better outcome was obtained by patients treated with underpowered RT [34]. This study paved the way to different trials assessing the de-intensification strategies in HPV-related HNSCC.

Chera BS et al., in a prospective phase II trial, enrolled 44 patients with diagnosis of T0-T3, N0-N2c, M0, p16-positive HNSCC and treated them with 60 Gy of intensity-modulated radiotherapy with concurrent weekly intravenous cisplatin. As a result, 3-year local control, regional control, cause-specific survival, distant metastasis-free survival, and overall survival rates were 100%, 100%, 100%, 100%, and 95%, respectively. The authors concluded that, in HPV-HNSCC patients, a protocol consisting of under-dosed radiation therapy (IMRT) given concomitantly with weekly cisplatin was able to obtain a good preservation of quality of life and an excellent 3-year tumor control and survival [35].

Woody NM et al. treated a cohort of patients with HPV-positive oropharyngeal squamous cell carcinoma with definitive chemoradiotherapy (70–74.4 Gy) to the primary site and, since a postradiation neck dissection was planned, 54 Gy to the involved nodal areas. The authors observed a five-year locoregional control, disease-free survival and overall survival of 96%, 81% and 86%, respectively. The conclusion was that regional lymph node control in HPV-positive oropharyngeal cancer was not compromised by a de-escalated dose of radiotherapy to involved nodes in the setting of concurrent cisplatin-based chemotherapy [36].

These aforementioned trials, although being considered positive due to the good activity shown by the de-intensified treatments, suffered from being only phase II and nonrandomized trials.

Onita et al. [37] performed a retrospective review of patients with p16-positive oropharyngeal carcinomas who underwent, from 2006 to 2016, definitive radiotherapy concurrently with either triweekly cisplatin ($n = 251$) or cetuximab ($n = 40$). The study comprised also patients with stage I disease. Median follow-up was 40 months. On multivariate analysis comparing cisplatin and cetuximab, the 3-year locoregional recurrence (LRR) was 6% vs. 16% ($p = 0.07$); the 3-year distant metastasis rate (DM) was 8% vs. 21% ($p = 0.04$), the 3-year overall recurrence rate (ORR) was 11% vs. 29% ($p = 0.01$), and the 3-year cause-specific survival (CSS) was 94% vs. 79% ($p = 0.06$), respectively. The aforementioned results sharply favored the cisplatin arm. Nevertheless, when a stage-based subgroup analysis was done, the results were interesting; in fact, for stage I-II patients, 3-year LRR, DM, ORR and CSS did not significantly differ. The same parameters were significantly superior in the cisplatin arm, only when the authors considered stage III diseases. The authors concluded that, when given concurrently with radiotherapy, cetuximab and triweekly cisplatin demonstrated comparable efficacy for stage I–II p16-positive oropharyngeal squamous carcinomas (OPSCC). However, cetuximab appeared to be associated with higher rates of treatment failure and cancer-related deaths in stage III disease. Lately,

Gillison ML et al. [38] published the results of a large (987 patients enrolled) prospective phase III trial comparing concurrent cisplatin (at the standard dose of 100 mg for square meter of body surface) and 70 Gy radiation therapy, with the combination of cetuximab and the same radiation therapy regimen. This trial aimed to demonstrate the noninferiority of cetuximab and radiotherapy with respect to the standard of care, namely cisplatin-radiation therapy, in a population of patients affected by locally advanced HPV-related HNSCC. As a result, the experimental combination of cetuximab and radiation therapy did not meet the primary endpoint, showing to be inferior to the standard cisplatin radiotherapy (estimated 5-year overall survival was 77.9% in the cetuximab group versus 84.6% in the cisplatin group). Similar results were obtained also with regard to progression-free survival. The authors concluded that for patients with HPV-positive oropharyngeal carcinoma, radiotherapy plus cetuximab showed inferior overall and progression-free survival if compared to radiotherapy plus cisplatin, so that in those patients, radiotherapy plus cisplatin remained the standard of care (for eligible patients).

Mehanna H et al. performed a very similar trial aiming to demonstrate the noninferiority of the cetuximab-radiotherapy combination, in comparison with cisplatin and radiotherapy (De-ESCALaTE HPV trial). They enrolled 334 patients with locally advanced HPV-positive oropharyngeal carcinoma and randomized them to receive standard 70 Gy radiation therapy associated with cetuximab or cisplatin (100 mg for square meter of body surface). The results were similar to the previously mentioned trial, showing the significant less efficacy of cetuximab if compared with the standard cisplatin (2-year overall survival 97.5% in the cisplatin arm vs. 89.4% in the cetuximab arm, $p = 0.001$) [39].

Recently, Jones et al. published the updated results of the aforementioned phase III De-ESCALaTE HPV trial. Three hundred and thirty-four (334) patients were randomized to cisplatin (166) or cetuximab (168). Two-year overall survival (97.5% vs. 90.0%, HR: 3.268, $p = 0.0251$) and recurrence rates (6.4% vs. 16.0%, HR: 2.67; $p = 0.0024$) favored the cisplatin arm. Furthermore, the results of this phase III large trial highlighted that in HPV-positive patients, the standard association of cisplatin-radiotherapy should not be avoided [40].

On the basis of the conflicting results obtained in clinical trials, de-intensification therapies have not been taken into account in clinical practice, and moreover, it is not yet clear if the de-intensification should involve systemic therapy or radiation therapy. Table 2 shows the main studies exploring the concept of de-intensification of the standard chemoradiotherapy regimen in patients with HPV-positive HSNCC.

Overall, HPV status has a prognostic significance in HSNCC, but it has not yet altered the treatment guidelines. As a matter of fact, the last version of the National Comprehensive Cancer Network (NccN) Guidelines has sharply separated treatment pathways for p16-positive and p16-negative oropharyngeal carcinomas, but the treatment options for p16-positive and p16-negative oropharyngeal carcinomas are almost identical, with the below-mentioned differences only.

HPV-positive oropharyngeal carcinomas staged as T1 N1 M0, which in the previous version of the guidelines were suitable for chemoradiotherapy, should be treated now with upfront surgery or alternatively with radiation alone. T2 N1 M0 tumors (with a single <3 cm lymph node metastasis) may be suitable for chemoradiation, but concomitant chemoradiation is considered to be only a 2B (namely, not strongly supported by evidence) category of choice [41]. The surgical approach that should be chosen is the trans oral robotic surgery (TORS) which in clinical trials is able to guarantee the same efficacy at a price of a significantly minor morbidity [42,43].

Another difference between HPV-positive and HPV-negative oropharyngeal carcinoma takes into account the role of adjuvant chemoradiotherapy for the so-called "high-risk" disease. Adjuvant concurrent chemoradiation represents a category 1 recommendation for all the patients surgically resected and with presence of extranodal extension. Nevertheless, recent data have highlighted that HPV-related carcinomas staged as T1–2 N1 (single <3 cm metastases) M0, should undergo adjuvant radiotherapy alone [44,45]. On these bases, the NCCN panel of experts recommend the omission of chemotherapy in concomitance with adjuvant radiotherapy in patients with T1–2 N1 M0 HPV-related oropharyngeal carcinoma surgically resected with extranodal extension; nevertheless, this recommendation is only a 2B category option.

**Table 2.** Trials exploring the concept of de-intensification of the standard chemoradiotherapy regimen in patients with HPV-positive HSNCC.

| Study | Design | Type of Study | Number of Patients | Setting | Results |
|---|---|---|---|---|---|
| *Eur. J. Cancer* **2020**, *124*, 178–185 (Update De-Escalate trial) [40] | cDDP-RT vs. Cet-RT in HPV-positive oropharyngeal Carcinomas | Phase III randomized trial | 334 | Stage II-IV oropharyngeal carcinoma | cDDP-RT better than Cet-RT in terms of 2-year OS |
| *Lancet* **2019**, *393*, 51–60 [39] | cDDP-RT vs. Cet-RT in HPV-positive oropharyngeal Carcinomas | Phase III randomized trial | 334 | Stage II-IV oropharyngeal carcinoma | cDDP-RT better than Cet-RT in terms of 2-year OS and ORR |
| *Lancet* **2019**, *393*, 40–50 [38] | cDDP-RT vs. Cet-RT in HPV-positive oropharyngeal Carcinomas | Phase III randomized trial | 849 | Stage II-IV oropharyngeal carcinoma | cDDP-RT better than Cet-RT in terms of OS and PFS |
| *Rep. Pract. Oncol. Radiother.* **2018**, *23*, 451–457 [37] | cDDP-RT vs. Cet-RT in HPV-positive oropharyngeal Carcinomas | Retrospective Study | 291 | Stage I-IV oropharyngeal carcinoma | cDDP-RT better than Cet-RT in terms of ORR and CSS |
| *Oral Oncol.* **2016**, *53*, 91–96 [36] | Reduced RT dose (54 vs. 70 Gy upon nodes) plus cisplatin in HPV-positive oropharyngeal Carcinomas | Phase II prospective trial | 50 | Stage II-IV oropharyngeal carcinoma | 5-year LCR, DFS and OS were 96%, 81% and 86% |
| *Cancer* **2018**, *124*, 2347–2354. [35] | Reduced RT dose (60 vs. 70 Gy) plus weekly cisplatin in HPV-positive oropharyngeal Carcinomas | Phase II prospective trial | 44 | Stage II-IV oropharyngeal carcinoma | 3-year LCR, CSS, DMFS and OS were 100%, 100%, 100% and 95% |
| *J. Clin. Oncol.* **2014**, *32*, 5s, (suppl; abstr LBA6006) [34] | Induction Cddp-Pac and Cet followed by Cet + underpowered IMRT (54 Gy) in patients obtaining a CR or a PR | Phase II prospective trial | 90 | Stage II-IV oropharyngeal carcinoma | 70% of CR after IC 2-year PFS and OS were 80 and 94% |

cDDP: cisplatin; RT: radiotherapy; IMRT: intensity modulated radiation therapy; Cet: cetuximab; Pac: paclitaxel; OS: overall survival; PFS: progression-free survival; ORR: overall response rate; LCR: locoregional failure rate; DFS: disease-free survival; CSS: cause-specific survival; DMFS: distant-metastases-free survival; CR: complete response; IC: induction chemotherapy.

## 4. HPV Infection and Immunotherapy

Immunotherapy is a therapeutic strategy aiming to reinforce the host immune system, helping it in reacting against tumor cells. The entire rationale of the immunotherapy has its roots in the existence of the so-called "tumor-associated antigens" (TAAs), namely protein antigens exposed by the tumor cells, able to elicit a strong immune response. Several strategies of immunotherapy are available in clinical practice and others are still being tested, but the most recent immunotherapeutic drugs are those acting in the "checkpoint" phases. The two well-acknowledged checkpoint phases are the "priming phase" (during which the naïve T-lymphocytes mature and became able to attack the tumor cells) and the "effector phase" (during which the matured T-lymphocytes attack and destroy the tumor cells, by recognizing the TAA).

Virus-related and mutagen-related HNSCCs display different genetic and immunologic features. Some data indicate that tobacco and alcohol, provoking several DNA mutations, also alter the tumor immune microenvironment. These last immune microenvironment alterations strongly affect tumor response to immunotherapy, thus leading to lower response rates after immunotherapy [46]. Desrichard et al. demonstrated a significant correlation between a specific "smoking-signature", which characterizes the smoke-related HNSCC, and the entity of the tumor-immune-infiltrate. In particular, they observed that a specific smoke-associated signature (the signature defined by Alexandrov), characterized by a wide number of DNA changes and a very high mutational burden, significantly correlated to a low immune tumor infiltrate. Interestingly, this signature also related to poor response to immunotherapy [47,48]. On the other hand, the HPV-related HNSCC subgroup showed the opposite features, being characterized by a robust CD8 lymphocyte-mediated response and a better response to immunotherapy. The main implications of the aforementioned features are that the HPV-related HNSCC responds better to immunotherapy, and in particular, to checkpoint inhibitors, while the mutagen-related one does not, being characterized by a noninflamed phenotype.

## 5. Future Implications on Therapy

The future therapeutic approaches should start from the statement that HPV-related HNSCC represents an entity which is very different genetically to mutagen-related HNSCC, mainly due to their intrinsic chemo- and radiosensitivity. Consequently, when, in clinical practice, there is a doubt whether to choose surgery or chemoradiotherapy, we can hypothesize, to address HPV-related tumors, that conservative treatment instead of surgery be used, especially if the surgical procedure is burdened with greater compromise of quality of life. On these bases, our effort should be aimed at rapid identification of the HPV-related tumors, in particular those belonging to the class III, as reported by Weinberger et al. [33]. According to more and more data [49–51], the latter seem to be characterized by a genetic signature, namely a pattern of genetic and epigenetic changes, which may sharply distinguish them from the other non-virus-related tumors. HPV-related HNSCC (in particular, oropharyngeal cancers) often shows the following features—P16 overexpression, low EGFR expression, wild-type TP53 and low CyclinD1 expression; in addition they show a lower TMB and an higher number of epigenetic changes, if compared with the mutagen-related counterpart [4,52,53].

Further complicating the matter is the molecular heterogeneity existing within HPV-positive tumors. As a matter of fact, a 2014 study of HNSCC from The Cancer Genome Atlas (TCGA) found that in a group of 35 HPV-positive tumors, 25 had integration of the viral genome while 10 tumors lacked integration [54] Further data have highlighted that nearly 30% of HPV-positive oropharyngeal carcinomas contained only episomal HPV [55]. Oropharyngeal cancers showing integrated versus nonintegrated HPV have differences in somatic gene methylation, gene expression patterns, mRNA processing, and inter- and intrachromosomal rearrangements [56]. In a recent biomolecular analysis of a subgroup of HPV-positive HNSCC, authors identified the presence of deletions or mutations of two proteins that inhibit NF-kB and activate interferon, TNF receptor-associated factor 3 (TRAF3) and cylindromatosis (CYLD) [57]. Furthermore, the presence of these DNA changes was related to the prognosis, with survival improved for patients whose tumors carried defects in either TRAF3 or CYLD. Conversely, the survival of HPV-positive patients without these mutations was similar to that of HPV-negative patients [58].

TRAF3 and CYLD gene deletions or disruptive mutations were identified in 28% of HPV-positive specimens in the initial TCGA HNSCC cohort and it correlated to the absence of HPV gene integration and decreased tobacco exposure [59], leading to the consideration that both DNA damage and the presence of reactive oxygen species (induced by tobacco mutagens) may favor HPV integration.

In conclusion, we can assert that the positivity for HPV (p16 test) is not enough to consider the tumor as HPV-related, and other markers should be taken into account for this scope.

A subgroup analysis carried out in the TAX 324 study, as well as the results of the ECOG 2399 trial, clearly demonstrated that HPV-related HSNCC responded better to induction TPF (docetaxel, cisplatin and 5-FU) followed by chemoradiation, when compared with the p16-negative counterpart [60,61], (Figure 3).

Moreover, Feng et al. [62] demonstrated that wild-type (WT) CCND1 (the gene encoding for CyclinD1) HNSCC displayed a significantly better response to induction chemotherapy compared with tumors showing CCND1 gene amplification (Figure 4). HPV-related HNSCCs often show the wild-type status for CCND1 concomitantly with wild-type status of p16. The authors concluded that the WT status for CCND1 could predict good response to induction chemotherapy and, consequently, we can assume that the presence of HPV-related carcinogenesis may represent a predictive factor of good response to induction chemotherapy, followed by chemoradiation (Figure 4).

The two latter studies have furtherly highlighted the concept that HPV-related tumors not only have a better prognosis but also a better response to conservative treatments when compared with the HPV-negative counterpart.

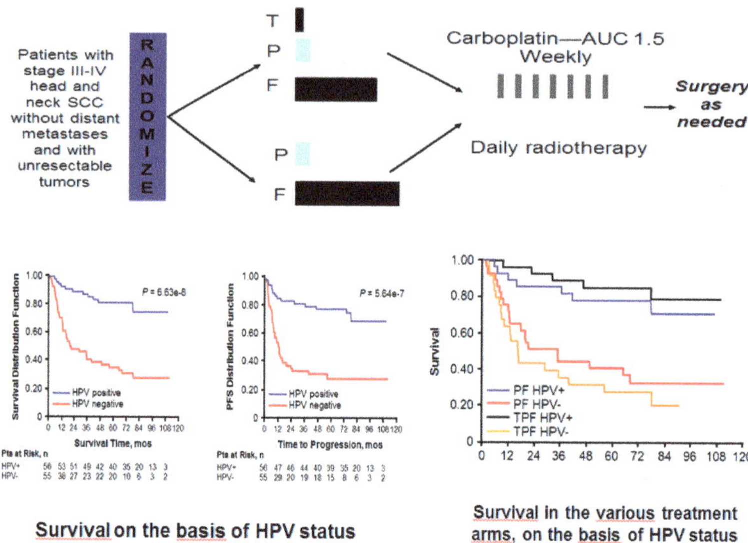



**Figure 3.** The subgroup analysis of Tax342 study revealed that HPV-positive patients treated with both induction TPF and induction PF performed better when compared with those who were HPV-negative, suggesting that HPV positivity could be considered a factor predictive of good response to sequential chemoradiotherapy. Source: Fakhry C et al. J. Natl. Cancer Inst., **2008**, *100*, 261–269 [61]. PF: cisplatin-5Fluorouracil induction chemotherapy; TPF: docetaxel-cisplatin-5Fluorouracil induction chemotherapy.

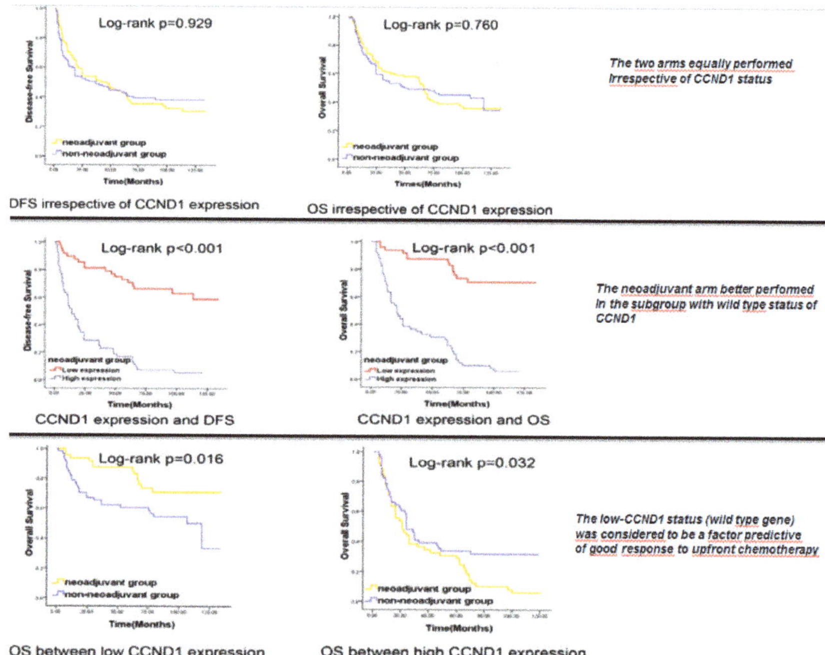



**Figure 4.** The wild type status for CCND1 is sharply linked to the HPV positivity in HSNCC. It is strongly related to the good response to sequential chemoradiotherapy and with the poor response to the upfront surgery. Source: Feng Z et al. PLoS One. **2011**, *6*, e26399 [62]. CCND1: cyclin D1; DFS: disease free survival; OS: overall survival.

HPV presence in the tumor cells could also guide the immunotherapy strategies. Starting from the hypothesis that viral antigens are much more immunogenic than those "self", a number of clinical trials have tested vaccination strategies which selectively target the viral antigens, such as E6 and E7 proteins [63]. Results are encouraging but data are still immature.

Regarding the use of the checkpoint inhibitors, only nivolumab and pembrolizumab are presently approved drugs for the treatment of recurrent/metastatic HSNCC. Both the drugs are indifferently employed in HPV-positive and HPV-negative patients. Nevertheless, it is interesting to highlight that a subgroup analysis in the context of the Keynote 141 trial (the study that led to the approval of nivolumab in clinical practice) has revealed that, among patients with HPV-positive tumors, the median OS was 9.1 months for patients treated with nivolumab versus 4.4 months for those treated with the standard-therapy, confirming the possibility that virus-related HSNCC better responds to checkpoint inhibitors [64]. More data are needed to assert the aforementioned issue, but it seems clear enough that virus-related HSNCC is more suitable to respond to immunotherapy compared with its mutagen-related counterpart.

## 6. Conclusions

HNSCCs are a very heterogeneous group of tumors affecting more than 65,000 patients per year in the United States. Mortality is strongly related to the initial staging, with both advanced and locally-advanced diseases having a poor prognosis. Lately, the knowledge of HNSCC genetics, as well as the translational research in this field, have gained more and more importance in the management of patients. Thus, the discovery of a subgroup of HNSCC, HPV-related HNSCC, particularly different from the others, paved the way to a different approach to HNSCC in clinical practice.

As largely demonstrated by scientific literature, HPV-related tumors are much more radiosensitive and chemosensitive when compared with their mutagen-related counterpart, and this feature can significantly impact on the clinical management of the patients.

Nevertheless, there are at least two problems to face—the importance of sharply distinguishing the HPV-related tumors from the non-HPV-related ones, independently from the presence of the viral DNA in the tumor cells, and the possibility to employ this information in clinical practice.

The identification of the viral protein E6 and E7 may be the best way to identify virus-related HNSCC [15], but this methodology does not take into account some important considerations. As a matter of fact, there is a subgroup of HPV-positive HNSCC that displays both viral DNA and E6/E7 proteins which is characterized by nonviral carcinogenesis. In this last case, some particular DNA and chromosomal changes such as p53, CCND1 and EGFR mutations, as well as a high TMB, are often present. In these cases, the carcinogenesis is due to mutagens from alcohol and tobacco, and the presence of HPV is not relevant, with these tumors having a prognosis comparable with those that are HPV-negative.

Different markers, other than p16, have been taken into account with the aim to best identify the HPV-driven carcinogenesis—TP53, pRB and CCND1, with their expression being very peculiar in HPV-related HNSCC (class III according to Weinberger).

The second and most relevant problem to be solved is the applicability of the aforementioned information in clinical practice. In fact, the last TNM version distinguishes between HPV-related (p16-positive) and non-HPV-related tumors, highlighting the impact that HPV has on the prognosis. Nevertheless, the therapeutic strategies used for HPV-related oropharyngeal cancers are almost the same as in non-HPV-related tumors, with few exceptions.

According to the latest translational acquisitions, we can speculate that in the near future, the HPV-related HNSCC could have different treatments when compared with the mutagen-related tumors. In particular, the locally advanced virus-related HNSCC could be treated with conservative strategies in spite of radical surgery, being very chemo and radiosensitive. On the same bases, HPV-related tumors, which are more suitable to respond to immunotherapy, could benefit from a single drug immunotherapy, such as checkpoint inhibitors; the mutagen-related counterpart, which often have a noninflamed phenotype, necessitates stronger immune modulation with two or more immunotherapeutic drugs.

Further studies on translational research should be designed with the aim to discern therapeutic options on the bases of the genetics of tumors.

**Author Contributions:** Conceptualization, F.P.; F.I.; G.D.V.S.; methodology: F.P., A.O.; software: F.P.; validation: F.P., F.C.; formal analysis: G.D.V.S.; investigation: F.P.; G.D.V.S., F.S.; data curation: A.G., F.P.; writing: F.P., F.I., F.L., A.G.; writing—review and editing: F.P., F.C., A.G.; visualization: P.M., F.P.; supervision: F.P., F.I.; project administration: G.D.V.S.; funding acquisition: N/A. All authors have read and agreed to the published version of the manuscript.

**Funding:** This research received no external funding.

**Conflicts of Interest:** The authors declare no conflict of interest.

## References

1.  Bron, L.; Jandus, C.; Andrejevic-Blant, S.; Speiser, D.E.; Monnier, P.; Romero, P.; Rivals, J.-P. Prognostic value of arginase-II expression and regulatory T-cell infiltration in head and neck squamous cell carcinoma. *Int. J. Cancer* **2012**, *132*, E85–E93. [CrossRef] [PubMed]

2.  Chai, R.C.; Lambie, D.; Verma, M.; Punyadeera, C. Current trends in the etiology and diagnosis of HPV-related head and neck cancers. *Cancer Med.* **2015**, *4*, 596–607. [CrossRef] [PubMed]

3.  Hafkamp, H.C.; Mooren, J.J.; Claessen, S.M.; Klingenberg, B.; Voogd, A.C.; Bot, F.J.; Klussmann, J.P.; Hopman, A.H.; Manni, J.J.; Kremer, B.; et al. P21 Cip1/WAF1 expression is strongly associated with HPV-positive tonsillar carcinoma and a favorable prognosis. *Mod. Pathol.* **2009**, *22*, 686–698. [CrossRef] [PubMed]

4.  Caponigro, F.; Ionna, F.; Scarpati, G.D.V.; Longo, F.; Addeo, R.; Manzo, R.; Muto, P.; Pisconti, S.; Leopaldi, L.; Perri, F. Translational Research: A Future Strategy for Managing Squamous Cell Carcinoma of the Head and Neck? *Anti-Cancer Agents Med. Chem.* **2019**, *18*, 1220–1227. [CrossRef] [PubMed]

5.  Pytynia, K.B.; Dahlstrom, K.; Sturgis, E.M. Epidemiology of HPV-associated oropharyngeal cancer. *Oral Oncol.* **2014**, *50*, 380–386. [CrossRef]

6.  Gillison, M.L. Current topics in the epidemiology of oral cavity and oropharyngeal cancers. *Head Neck* **2007**, *29*, 779–792. [CrossRef]

7.  Chaturvedi, A.K.; Engels, E.A.; Anderson, W.F.; Gillison, M.L. Incidence Trends for Human Papillomavirus–Related and –Unrelated Oral Squamous Cell Carcinomas in the United States. *J. Clin. Oncol.* **2008**, *26*, 612–619. [CrossRef]

8.  Shiboski, C.H.; Schmidt, B.; Jordan, R.C.K. Tongue and tonsil carcinoma. *Cancer* **2005**, *103*, 1843–1849. [CrossRef]

9.  Frisch, M.; Hjalgrim, H.; Jæger, A.B.; Biggar, R.J. Changing patterns of tonsillar squamous cell carcinoma in the United States. *Cancer Causes Control.* **2000**, *11*, 489–495. [CrossRef]

10. D'Souza, G.; Agrawal, Y.; Halpern, J.; Bodison, S.; Gillison, M.L. Oral sexual behaviors associated with prevalent oral human papillomavirus infection. *J. Infect. Dis.* **2009**, *199*, 1263–1269. [CrossRef]

11. Kjaer, S.K.; Chackerian, B.; Brule, A.J.V.D.; Svare, E.I.; Paull, G.; Walbomers, J.M.; Schiller, J.T.; Bock, J.E.; Sherman, M.E.; Lowy, D.R.; et al. High-risk human papillomavirus is sexually transmitted: Evidence from a follow-up study of virgins starting sexual activity (intercourse). *Cancer Epidemiol. Biomark. Prev.* **2001**, *10*, 101–106.

12. Clinician FAQs: CDC Recommendations for HPV Vaccine 2-Dose Schedule|Human Papillomavirus (HPV)|CDC. 2017. Available online: https://www.cdc.gov/hpv/hcp/2-dose/clinician-faq.html (accessed on 24 February 2018).

13. Westra, W.H. Detection of human papillomavirus (HPV) in clinical samples: Evolving methods and strategies for the accurate determination of HPV status of head and neck carcinomas. *Oral Oncol.* **2014**, *50*, 771–779. [CrossRef] [PubMed]

14. Speel, E.J.M. HPV Integration in Head and Neck Squamous Cell Carcinomas: Cause and Consequence. *Recent Results Cancer Res.* **2017**, *206*, 57–72. [PubMed]

15. Bussu, F.; Ragin, C.; Boscolo-Rizzo, P.; Rizzo, D.; Gallus, R.; Delogu, G.; Morbini, P.; Tommasino, M. HPV as a marker for molecular characterization in head and neck oncology: Looking for a standardization of clinical use and of detection method(s) in clinical practice. *Head Neck* **2019**, *41*, 1104–1111. [CrossRef]

16. Agalliu, I.; Gapstur, S.; Chen, Z.; Wang, T.; Anderson, R.L.; Teras, L.; Kreimer, A.R.; Hayes, R.B.; Freedman, N.D.; Burk, R.D. Associations of Oral α-, β-, and γ-Human Papillomavirus Types With Risk of Incident Head and Neck Cancer. *JAMA Oncol.* **2016**, *2*, 599–606. [CrossRef]
17. Snow, A.; Laudadio, J. Human Papillomavirus Detection in Head and Neck Squamous Cell Carcinomas. *Adv. Anat. Pathol.* **2010**, *17*, 394–403. [CrossRef]
18. Muñoz, N.; Castellsagué, X.; De Gonzalez, A.B.; Gissmann, L. Chapter 1: HPV in the etiology of human cancer. *Vaccine* **2006**, *24*, S1–S10. [CrossRef]
19. Chow, L.T. Model systems to study the life cycle of human papillomaviruses and HPV-associated cancers. *Virol. Sin.* **2015**, *30*, 92–100. [CrossRef]
20. Doorbar, J.; Quint, W.; Banks, L.; Bravo, I.G.; Stoler, M.; Broker, T.R.; Stanley, M.A. The Biology and Life-Cycle of Human Papillomaviruses. *Vaccine* **2012**, *30*, F55–F70. [CrossRef]
21. Ganti, K.; Broniarczyk, J.; Manoubi, W.; Massimi, P.; Mittal, S.; Pim, D.; Szalmás, A.; Thatte, J.; Thomas, M.; Tomaić, V.; et al. The Human Papillomavirus E6 PDZ Binding Motif: From Life Cycle to Malignancy. *Viruses* **2015**, *7*, 3530–3551. [CrossRef]
22. Delury, C.P.; Marsh, E.; James, C.D.; Boon, S.S.; Banks, L.; Knight, G.; Roberts, S. The role of protein kinase A regulation of the E6 PDZ-binding domain during the differentiation-dependent life cycle of human papillomavirus type 18. *J. Virol.* **2013**, *87*, 9463–9472. [CrossRef]
23. Songock, W.K.; Kim, S.-M.; Bodily, J. The human papillomavirus E7 oncoprotein as a regulator of transcription. *Virus Res.* **2017**, *231*, 56–75. [CrossRef] [PubMed]
24. Tomaić, V. Functional Roles of E6 and E7 Oncoproteins in HPV-Induced Malignancies at Diverse Anatomical Sites. *Cancers* **2016**, *8*, 95. [CrossRef] [PubMed]
25. Gubanova, E.; Brown, B.; Ivanov, S.; Helleday, T.; Mills, G.B.; Yarbrough, W.G.; Issaeva, N. Downregulation of SMG-1 in HPV-positive head and neck squamous cell carcinoma due to promoter hypermethylation correlates with improved survival. *Clin. Cancer Res.* **2012**, *18*, 1257–1267. [CrossRef] [PubMed]
26. Bol, V.; Gregoire, V. Biological Basis for Increased Sensitivity to Radiation Therapy in HPV-Positive Head and Neck Cancers. *BioMed Res. Int.* **2014**, *2014*, 696028. [CrossRef] [PubMed]
27. Rusan, M.; Li, Y.Y.; Hammerman, P.S. Genomic landscape of human papillomavirus-associated cancers. *Clin. Cancer Res.* **2015**, *21*, 2009–2019. [CrossRef]
28. Dreyer, J.H.; Hauck, F.; Barros, M.; Niedobitek, G. pRb and CyclinD1 Complement p16 as Immunohistochemical Surrogate Markers of HPV Infection in Head and Neck Cancer. *Appl. Immunohistochem. Mol. Morphol.* **2017**, *25*, 366–373. [CrossRef]
29. Beck, T.N.; Georgopoulos, R.; Shagisultanova, E.I.; Sarcu, D.; Handorf, E.A.; Dubyk, C.; Lango, M.N.; Ridge, J.A.; Astsaturov, I.; Serebriiskii, I.G.; et al. EGFR and RB1 as Dual Biomarkers in HPV-Negative Head and Neck Cancer. *Mol. Cancer Ther.* **2016**, *15*, 2486–2497. [CrossRef]
30. Stransky, N.; Egloff, A.M.; Tward, A.D.; Kostic, A.D.; Cibulskis, K.; Sivachenko, A.; Kryukov, G.V.; Lawrence, M.S.; Sougnez, C.; McKenna, A.; et al. The mutational landscape of head and neck squamous cell carcinoma. *Science* **2011**, *333*, 1157–1160. [CrossRef]
31. Tinhofer, I.; Budach, V.; Saki, M.; Konschak, R.; Niehr, F.; Jöhrens, K.; Weichert, W.; Linge, A.; Lohaus, F.; Krause, M. Targeted next-generation sequencing of locally advanced squamous cell carcinomas of the head and neck reveals druggable targets for improving adjuvant chemoradiation. *Eur. J. Cancer* **2016**, *57*, 78–86.
32. Mooren, J.J.; Kremer, B.; Claessen, S.M.; Voogd, A.C.; Bot, F.J.; Klussmann, J.P.; Huebbers, C.; Hopman, A.H.; Ramaekers, F.C.; Speel, E.M. Chromosome stability in tonsillar squamous cell carcinoma is associated with HPV16 integration and indicates a favorable prognosis. *Int. J. Cancer* **2012**, *132*, 1781–1789. [CrossRef] [PubMed]
33. Weinberger, P.; Yu, Z.; Haffty, B.G.; Kowalski, D.; Harigopal, M.; Brandsma, J.; Sasaki, C.; Joe, J.; Camp, R.L.; Rimm, D.L.; et al. Molecular Classification Identifies a Subset of Human Papillomavirus–Associated Oropharyngeal Cancers With Favorable Prognosis. *J. Clin. Oncol.* **2006**, *24*, 736–747. [CrossRef] [PubMed]
34. Cmelak, A.; Li, S.; Marur, S.; Zhao, W.; Westra, W.H.; Chung, C.H.; Gillison, M.L.; Gilbert, J.; Bauman, J.E.; Wagner, L.I.; et al. Reduced-dose IMRT in human papilloma virus (HPV)-associated resectable oropharyngeal squamous carcinomas (OPSCC) after clinical complete response (cCR) to induction chemotherapy (IC). *J. Clin. Oncol.* **2014**, *32*, 5s. [CrossRef]
35. Chera, B.S.; Amdur, R.J.; Tepper, J.; Tan, X.; Weiss, J.M.; Grilley-Olson, J.E.; Hayes, D.N.; Zanation, A.; Hackman, T.G.; Patel, S.; et al. Mature results of a prospective study of deintensified chemoradiotherapy

for low-risk human papillomavirus-associated oropharyngeal squamous cell carcinoma. *Cancer* **2018**, *124*, 2347–2354. [CrossRef] [PubMed]

36. Woody, N.M.; Koyfman, S.A.; Xia, P.; Yu, N.; Shang, Q.; Adelstein, D.J.; Scharpf, J.; Burkey, B.; Nwizu, T.; Saxton, J.; et al. Regional control is preserved after dose de-escalated radiotherapy to involved lymph nodes in HPV positive oropharyngeal cancer. *Oral Oncol.* **2016**, *53*, 91–96. [CrossRef] [PubMed]

37. Bhattasali, O.; Thompson, L.D.R.; Abdalla, I.A.; Chen, J.; Iganej, S. Comparison of high-dose Cisplatin-based chemoradiotherapy and Cetuximab-based bioradiotherapy for p16-positive oropharyngeal squamous cell carcinoma in the context of revised HPV-based staging. *Rep. Pract. Oncol. Radiother.* **2018**, *23*, 451–457. [CrossRef]

38. Gillison, M.L.; Trotti, A.M.; Harris, J.; Eisbruch, A.; Harari, P.M.; Adelstein, D.J.; Jordan, R.C.K.; Zhao, W.; Sturgis, E.M.; Burtness, B.; et al. Radiotherapy plus cetuximab or cisplatin in human papillomavirus-positive oropharyngeal cancer (NRG Oncology RTOG 1016): A randomised, multicentre, non-inferiority trial. *Lancet* **2019**, *393*, 40–50. [CrossRef]

39. Mehanna, H.; Robinson, M.; Hartley, A.; Kong, A.; Foran, B.; Fulton-Lieuw, T.; Dalby, M.; Mistry, P.; Sen, M.; O'Toole, L.; et al. Radiotherapy plus cisplatin or cetuximab in low-risk human papillomavirus-positive oropharyngeal cancer (De-ESCALaTE HPV): An open-label randomised controlled phase 3 trial. *Lancet* **2019**, *393*, 51–60. [CrossRef]

40. Jones, D.A.; Mistry, P.; Dalby, M.; Fulton-Lieuw, T.; Kong, A.; Dunn, J.; Mehanna, H.; Gray, A. Concurrent cisplatin or cetuximab with radiotherapy for HPV-positive oropharyngeal cancer: Medical resource use, costs, and quality-adjusted survival from the De-ESCALaTE HPV trial. *Eur. J. Cancer* **2020**, *124*, 178–185. [CrossRef]

41. Nccn Practice Guidelines in Oncology. Head and Neck Cancers. Version 3/2019. Available online: https://www.nccn.org/store/login/login.aspx?ReturnURL=https://www.nccn.org/professionals/physician_gls/pdf/head-and-neck.pdf (accessed on 28 February 2020).

42. Howard, J.; Masterson, L.; Dwivedi, R.C.; Riffat, F.; Benson, R.; Jefferies, S.; Jani, P.; Tysome, J.R.; Nutting, C. Minimally invasive surgery versus radiotherapy/chemoradiotherapy for small-volume primary oropharyngeal carcinoma. *Cochrane Database Syst. Rev.* **2016**, *12*, CD010963. [CrossRef]

43. Mäkitie, A.A.; Keski-Säntti, H.; Markkanen-Leppänen, M.; Bäck, L.; Koivunen, P.; Ekberg, T.; Sandström, K.; Laurell, G.; Von Beckerath, M.; Nilsson, J.S.; et al. Transoral Robotic Surgery in the Nordic Countries: Current Status and Perspectives. *Front. Oncol.* **2018**, *8*, 289. [CrossRef] [PubMed]

44. Psyrri, A.; Rampias, T.; Vermorken, J.B. The current and future impact of human papillomavirus on treatment of squamous cell carcinoma of the head and neck. *Ann. Oncol.* **2014**, *25*, 2101–2115. [CrossRef] [PubMed]

45. Kofler, B.; Laban, S.; Busch, C.J.; Lôrincz, B.; Knecht, R. New treatment strategies for HPV-positive head and neck cancer. *Eur. Arch. Oto-Rhino-Laryngol.* **2013**, *271*, 1861–1867. [CrossRef] [PubMed]

46. Ferris, R.L.; Blumenschein, G.; Fayette, J.; Guigay, J.; Colevas, A.D.; Licitra, L.; Harrington, K.; Kasper, S.; Vokes, E.E.; Even, C.; et al. Nivolumab for Recurrent Squamous-Cell Carcinoma of the Head and Neck. *N. Engl. J. Med.* **2016**, *375*, 1856–1867. [CrossRef]

47. Alexandrov, L.B.; Initiative, A.P.C.G.; Nik-Zainal, S.; Wedge, D.C.; Aparicio, S.A.J.R.; Behjati, S.; Biankin, A.V.; Bignell, G.R.; Bolli, N.; Borg, A.; et al. Signatures of mutational processes in human cancer. *Nature* **2013**, *500*, 415–421. [CrossRef]

48. Desrichard, A.; Kuo, F.; Chowell, D.; Lee, K.-W.; Riaz, N.; Wong, R.J.; Chan, T.A.; Morris, L.G. Tobacco Smoking-Associated Alterations in the Immune Microenvironment of Squamous Cell Carcinomas. *J. Natl. Cancer Inst.* **2018**, *110*, 1386–1392. [CrossRef] [PubMed]

49. Bali, A.; O'Brien, P.M.; Edwards, L.S.; Sutherland, R.L.; Hacker, N.F.; Henshall, S.M. Cyclin D1, p53, and p21Waf1/Cip1 expression is predictive of poor clinical outcome in serous epithelial ovarian cancer. *Clin. Cancer Res.* **2004**, *10*, 5168–5177. [CrossRef] [PubMed]

50. Li, W.; Thompson, C.H.; Cossart, Y.E.; O'Brien, C.; McNeil, E.; Scolyer, R.A.; Rose, B.R. The expression of key cell cycle markers and presence of human papillomavirus in squamous cell carcinoma of the tonsil. *Head Neck* **2004**, *26*, 1–9. [CrossRef]

51. Queiroz, C.J.D.S.; Nakata, C.M.D.A.G.; Solito, E.; Damazo, A.S. Relationship between HPV and the biomarkers annexin A1 and p53 in oropharyngeal cancer. *Infect. Agents Cancer* **2014**, *9*, 13. [CrossRef]

52. Klussmann, J.P.; Mooren, J.J.; Lehnen, M.; Claessen, S.M.; Stenner, M.; Huebbers, C.U.; Weissenborn, S.J.; Wedemeyer, I.; Preuss, S.F.; Straetmans, J.M.; et al. Genetic signatures of HPV-related and unrelated

oropharyngeal carcinoma and their prognostic implications. *Clin. Cancer Res.* **2009**, *15*, 1779–1786. [CrossRef] [PubMed]

53. Van Kempen, P.M.; Noorlag, R.; Braunius, W.W.; Stegeman, I.; Willems, S.M.; Grolman, W. Differences in methylation profiles between HPV-positive and HPV-negative oropharynx squamous cell carcinoma. *Epigenetics* **2014**, *9*, 194–203. [CrossRef] [PubMed]

54. Network, T.C.G.A.; Network, C.G.A. Comprehensive genomic characterization of head and neck squamous cell carcinomas. *Nature* **2015**, *517*, 576–582. [CrossRef] [PubMed]

55. Parfenov, M.; Pedamallu, C.S.; Gehlenborg, N.; Freeman, S.; Danilova, L.; Bristow, C.A.; Lee, S.; Hadjipanayis, A.G.; Ivanova, E.V.; Wilkerson, M.D.; et al. Characterization of HPV and host genome interactions in primary head and neck cancers. *Proc. Natl. Acad. Sci. USA* **2014**, *111*, 15544–15549. [CrossRef] [PubMed]

56. Galot, R.; van Marcke, C.; Helaers, R.; Mendola, A.; Goebbels, R.M.; Caignet, X.; Ambroise, J.; Wittouck, K.; Vikkula, M.; Limaye, N.; et al. Liquid biopsy for mutational profiling of locoregional recurrent and/or metastatic head and neck squamous cell carcinoma. 2020 Mar 10;104:104631. *Oral Oncol.* **2020**, *104*, 104631. [CrossRef] [PubMed]

57. Hajek, M.; Sewell, A.; Kaech, S.; Burtness, B.; Yarbrough, W.G.; Issaeva, N. TRAF3/CYLD mutations identify a distinct subset of human papillomavirusassociatedhead and neck squamous cell carcinoma. *Cancer* **2017**, *123*, 1778–1790. [CrossRef] [PubMed]

58. Albers, A.E.; Qian, X.; Kaufmann, A.M.; Coordes, A. Meta analysis: HPV and p16pattern determines survival in patients with HNSCC and identifies potentialnew biologic subtype. *Sci. Rep.* **2017**, *7*, 16715. [CrossRef]

59. Pan, C.; Yarbrough, W.G.; Issaeva, N. Advances in biomarkers and treatment strategies for HPV-associated head and neck cancer. *Oncoscience* **2018**, *5*, 140–141.

60. Posner, M.; Lorch, J.H.; Goloubeva, O.; Tan, M.; Schumaker, L.; Sarlis, N.J.; Haddad, R.I.; Cullen, K.J. Survival and human papillomavirus in oropharynx cancer in TAX 324: A subset analysis from an international phase III trial. *Ann. Oncol.* **2011**, *22*, 1071–1077. [CrossRef]

61. Fakhry, C.; Westra, W.H.; Li, S.; Cmelak, A.; Ridge, J.A.; Pinto, H.; Forastiere, A.; Gillison, M.L. Improved Survival of Patients With Human Papillomavirus-Positive Head and Neck Squamous Cell Carcinoma in a Prospective Clinical Trial. *J. Natl. Cancer Inst.* **2008**, *100*, 261–269. [CrossRef] [PubMed]

62. Feng, Z.; Guo, W.; Zhang, C.; Xu, Q.; Zhang, P.; Sun, J.; Zhu, H.; Wang, Z.; Li, J.; Wang, L.; et al. CCND1 as a Predictive Biomarker of Neoadjuvant Chemotherapy in Patients with Locally Advanced Head and Neck Squamous Cell Carcinoma. *PLoS ONE* **2011**, *6*, e26399. [CrossRef]

63. Wang, C.; Dickie, J.; Sutavani, R.V.; Pointer, C.; Thomas, G.J.; Savelyeva, N. Targeting Head and Neck Cancer by Vaccination. *Front. Immunol.* **2018**, *9*, 830. [CrossRef] [PubMed]

64. Gavrielatou, N.; Doumas, S.; Economopoulou, P.; Foukas, P.G.; Psyrri, A. Biomarkers for immunotherapy response in head and neck cancer. *Cancer Treat Rev.* **2020**, *84*, 101977. [CrossRef] [PubMed]

*Article*

# Analysis of HPV-Positive and HPV-Negative Head and Neck Squamous Cell Carcinomas and Paired Normal Mucosae Reveals Cyclin D1 Deregulation and Compensatory Effect of Cyclin D2

Jiří Novotný [1,2,†], Veronika Bandúrová [3,4,†], Hynek Strnad [1], Martin Chovanec [5], Miluše Hradilová [1], Jana Šáchová [1], Martin Šteffl [5], Josipa Grušanović [6], Roman Kodet [7], Václav Pačes [1], Lukáš Lacina [3], Karel Smetana Jr. [3], Jan Plzák [4], Michal Kolář [1,*] and Tomáš Vomastek [6,*]

[1] Laboratory of Genomics and Bioinformatics, Institute of Molecular Genetics of the Czech Academy of Sciences, 142 20 Prague, Czech Republic; jiri.novotny@img.cas.cz (J.N.); strnad@img.cas.cz (H.S.); miluse.hradilova@img.cas.cz (M.H.); jana.sachova@img.cas.cz (J.Š.); vaclav.paces@img.cas.cz (V.P.)

[2] Department of Informatics and Chemistry, Faculty of Chemical Technology, University of Chemistry and Technology, 160 00 Prague, Czech Republic

[3] Institute of Anatomy, 1st Faculty of Medicine, Charles University, 128 00 Prague, Czech Republic; veronika.bandurova@lf1.cuni.cz (V.B.); lukas.lacina@lf1.cuni.cz (L.L.); karel.smetana@lf1.cuni.cz (K.S.J.)

[4] Department of Otorhinolaryngology, Head and Neck Surgery, 1st Faculty of Medicine, Charles University, Faculty Hospital Motol, 150 06 Prague, Czech Republic; jan.plzak@lf1.cuni.cz

[5] Department of Otorhinolaryngology, 3rd Faculty of Medicine, Charles University, 100 00 Prague, Czech Republic; martin.chovanec@lf3.cuni.cz (M.C.); quethak@gmail.com (M.S.)

[6] Institute of Microbiology of the Czech Academy of Sciences, 142 00 Prague, Czech Republic; josipa.grusanovic@biomed.cas.cz

[7] Department of Pathology and Molecular Medicine, Charles University, 2nd Faculty of Medicine and Faculty Hospital Motol, 150 06 Prague, Czech Republic; roman.kodet@lfmotol.cuni.cz

* Correspondence: michal.kolar@img.cas.cz (M.K.); vomy@biomed.cas.cz (T.V.)

† These authors contributed equally to this work.

Received: 24 January 2020; Accepted: 21 March 2020; Published: 26 March 2020

**Abstract:** Aberrant regulation of the cell cycle is a typical feature of all forms of cancer. In head and neck squamous cell carcinoma (HNSCC), it is often associated with the overexpression of cyclin D1 (*CCND1*). However, it remains unclear how *CCND1* expression changes between tumor and normal tissues and whether human papillomavirus (HPV) affects differential *CCND1* expression. Here, we evaluated the expression of D-type cyclins in a cohort of 94 HNSCC patients of which 82 were subjected to whole genome expression profiling of primary tumors and paired normal mucosa. Comparative analysis of paired samples showed that *CCND1* was upregulated in 18% of HNSCC tumors. Counterintuitively, *CCND1* was downregulated in 23% of carcinomas, more frequently in HPV-positive samples. There was no correlation between the change in D-type cyclin expression and patient survival. Intriguingly, among the tumors with downregulated *CCND1*, one-third showed an increase in cyclin D2 (*CCND2*) expression. On the other hand, one-third of tumors with upregulated *CCND1* showed a decrease in *CCND2*. Collectively, we have shown that *CCND1* was frequently downregulated in HNSCC tumors. Furthermore, regardless of the HPV status, our data suggested that a change in *CCND1* expression was alleviated by a compensatory change in *CCND2* expression.

**Keywords:** human papillomavirus; head and neck squamous cell carcinoma; cell cycle; D-type cyclins; *CCND1*; *CCND2*; *CCND3*; patient survival; paired tumor-normal samples; 11q13 amplification

---

## 1. Introduction

Head and neck squamous cell carcinomas (HNSCC) are malignant neoplasms that arise from the mucosal epithelial surface of the upper respiratory and digestive tracts. The worldwide annual incidence of head and neck cancers is more than 550,000 cases, thus making HNSCC the sixth-most common cancer in the world [1]. Unfortunately, the incidence of this tumor type continues to increase.

HNSCC is more prevalent in men than in women, and the majority of HNSCC cases occur in patients over the age of 60 [2]. The most affected subsites are the oral cavity, oropharynx, hypopharynx, nasopharynx and larynx. Early-stage, locally contained disease responds favorably to treatment, presenting very good cure rates of 70–90%. However, advanced HNSCC exhibit aggressive loco-regional invasions, frequent second primary tumors and lymph node metastases, while distant metastases are relatively rare. Despite advances in the treatment of HNSCC using molecularly targeted therapies, the five-year survival rate of loco-regional or recurrent/metastatic diseases has remained at 50–60% [3,4].

The main risk factors for the development of head and neck cancers include tobacco exposure and alcohol consumption, which are associated with more than 70% of all HNSCC cases [5,6], and the infection of high-risk oncogenic types of human papillomavirus (HPV) [7]. Among HPV high-risk types, HPV16 and HPV18 are the most common, accounting for more than 85% of all HPV-positive (HPV(+)) tumors [8,9]. HPV status, combined with the traditional tumor-node-metastasis (TNM) staging system, is considered to be a particularly significant biological prognostic marker and distinguishes two etiologically different subtypes of HNSCC [10]. Individuals with HPV(+) HNSCCs have a considerably better prognosis compared to those who possess HPV-negative (HPV(−)) cancers [11].

In HNSCC, the alterations which lead to functional loss of the tumor suppressor function are much more frequent than oncogene-activating mutations. These changes in both HPV subtypes almost invariably deregulate cell cycle entry and progression and, thus, are detrimental to cell proliferation. In HPV(−) tumors, p53 tumor suppressor is inactivated by genetic mutations, and retinoblastoma protein (pRb) is inactivated by cyclin D1 (*CCND1*) and the CDK4/6 complex. On the other hand, in HPV(+) tumors, viral proteins E6 and E7 inactivate p53 and pRb, thus enabling these tumors to evade cell cycle checkpoints [8,12]. Functional loss of p53, either by mutation or proteasomal degradation, occurs early in HNSCC development in approximately 75% of cases [13,14]. Likewise, another tumor suppressor gene, *CDKN2A*, that encodes cyclin-dependent kinase inhibitor $p16^{INK4a}$, is frequently inactivated by a copy number loss in HPV(−) HNSCC [15,16]. In contrast, $p16^{INK4a}$ overexpression is typical for HPV(+) HNSCC [17] and is used as a surrogate marker of HPV infection [13,16].

A mutation detected in negative regulators of the cell cycle is often paralleled by the increased expression of cyclin D1, a potent oncogene and positive regulator of the cell cycle that is encoded by the *CCND1* gene [15]. *CCND1* is a member of the D-type cyclin family that includes cyclin D2 (*CCND2*) and cyclin D3 (*CCND3*). Cyclins D1, D2 and D3 show roughly 50% of sequence identity at the amino acid level, and all have been shown to activate CDK4/6 and promote G1/S transition [18]. In nontransformed cells, the expression of D-type cyclins is controlled by mitogenic, adhesion and differentiation signals and, thus, integrates extracellular signals with the cell cycle. Overexpression of the D-type cyclins bypasses the requirement for mitogenic stimuli promoting unchecked proliferation and is believed to be an early cause of tumor formation [18–20]. Cyclin D1 protein overexpression has been reported in up to 70% of HNSCC cases [21,22], and gene amplification represents the most prominent mechanism of increased *CCND1* expression [15]. In addition, amplification of *CCND2* in HNSCC has also been reported [23]. Despite the frequent *CCND1* upregulation and its positive role in cell proliferation, the value of *CCND1* as an independent prognostic marker in HNSCC is inconsistent. Several studies have suggested that elevated *CCND1* expression correlates with poor prognosis of some types of HNSCC [21,24,25]; however, other studies have not found a significant correlation [26–29].

In the present study, we examined the expression of D-type cyclins in a new cohort of 94 HNSCC patients with known clinical data. For 82 of these patients, the expression profiles of both the tumor and matching normal mucosa were obtained using DNA microarrays. Analysis of expression changes between tumor and normal tissues revealed that a considerable fraction of tumors downregulates

*CCND1* expression. However, the reduction of cyclin D1 expression did not correlate with a favorable clinical outcome. We also found that tumors with downregulated cyclin D1 expression frequently upregulate the expression of cyclin D2. Taken together, these results suggest a compensatory mechanism where the upregulation of cyclin D2 may be a direct consequence of the loss of related cyclin D1 function.

## 2. Results

### 2.1. Patient Characteristics

Normal (n = 86) and tumor (n = 90) samples were obtained from a total of 94 HNSCC patients from which clinical and pathological data were collected (Table 1 and Table S1). Overall, patients were users of tobacco (81%) and alcohol (59%). HPV was detected in 26 cases (28%). In contrast to previous reports [30–32], there was almost no difference in the average age of HNSCC diagnosis between HPV(−) and HPV(+) subgroups. Primary treatment consisted of surgery alone in 15 (16%) cases, surgery combined with adjuvant radiotherapy in 73 (78%) cases and surgery combined with adjuvant chemoradiotherapy in 6 (6%) cases. Median follow-up length was 110 months for HPV(+) patients and 29 months for HPV(−) patients. HPV infection was associated with significantly longer overall survival (OS) (Figure 1A, left panel). Five-year OS was 77% (CI 62–95%) and 32% (CI 22–48%) in HPV(+) and HPV(−) groups, respectively. Hazard ratio (HR) of HPV(−) vs HPV(+) 2.13–9.15, $p < 0.001$. Disease-free survival (DFS) was also shorter for HPV(−) patients (five-year survival 61%; CI 46–81%) than for HPV(+) patients (five-year survival 85%; CI 72–100%) (Figure S1A). However, the difference between HPV(−) and HPV(+) patients did not reach statistical significance (HR = 2.89, CI 0.95-8.77, $p = 0.06$).

**Table 1.** Summary of clinical data for all patients in the present cohort (n = 94). For four patients, only samples from normal mucosa were available, and thus, the human papillomavirus (HPV) status was not determined. Statistical significance gauges the association between a variable and the HPV subtypes: (n.s.) non-significant, (*) $p < 0.05$ and (***) $p < 0.001$. Percentages are column-wise by default, row-wise if "r" is behind the value. (NA) not available.

| Variable (Stat. Signif.) | All Patients N (%) | HPV(−) N (%) | HPV(+) N (%) |
|---|---|---|---|
| *No. patients* | 94 | - | - |
| *Sample groups* | | | |
| No. samples per normal mucosa | 86 | - | - |
| No. samples per tumor | 90 | 64 (71r) | 26 (29r) |
| *Patients* | | | |
| No. patients with paired samples | 82 | 57 (70r) | 25 (30r) |
| No. patients with paired samples and known follow-up | 76 | 51 (68r) | 25 (32r) |
| *Age at surgery (n.s.)* | | | |
| Median (range) | 60 (26–94) | 59 (26–94) | 62 (41–72) |
| No. age < 40 | 1 | 1 | 0 |
| *Gender (n.s.)* | | | |
| Female | 10 (11) | 6 (9) | 4 (15) |
| Male | 84 (89) | 58 (91) | 22 (85) |
| *Smoking (*)* | | | |
| No | 17 (18) | 7 (11) | 9 (35) |
| Yes | 76 (81) | 56 (88) | 17 (65) |
| NA | 1 (1) | 1 (2) | 0 |
| *Alcohol usage (n.s.)* | | | |
| No | 38 (40) | 22 (34) | 14 (54) |
| yes | 55 (59) | 41 (64) | 12 (46) |
| NA | 1 (1) | 1 (2) | 0 |
| *Tumour site (***)* | | | |
| base of the tongue | 16 (17) | 9 (14) | 7 (27) |
| hypopharynx | 4 (4) | 4 (6) | 0 |
| larynx | 21 (22) | 20 (31) | 0 |
| oral cavity | 21 (22) | 18 (28) | 1 (4) |
| oropharynx part | 9 (10) | 6 (9) | 2 (8) |
| tonsils | 23 (24) | 7 (11) | 16 (62) |
| *Stage (n.s.)* | | | |
| I | 8 (9) | 6 (9) | 2 (8) |
| II | 8 (9) | 5 (8) | 3 (12) |
| III | 21 (22) | 17 (27) | 4 (15) |
| IV | 56 (60) | 35 (55) | 17 (65) |
| NA | 1 (1) | 1 (2) | 0 |
| *Grade (*)* | | | |
| G1 | 19 (20) | 18 (28) | 1 (4) |
| G2 | 46 (49) | 26 (41) | 18 (69) |
| G3 | 28 (30) | 19 (30) | 7 (27) |
| G4 | 1 (1) | 1 (2) | 0 |
| *Months of follow-up (median) (range) (n.s.)* | 47 (0–139) | 29 (0–138) | 110 (16–139) |

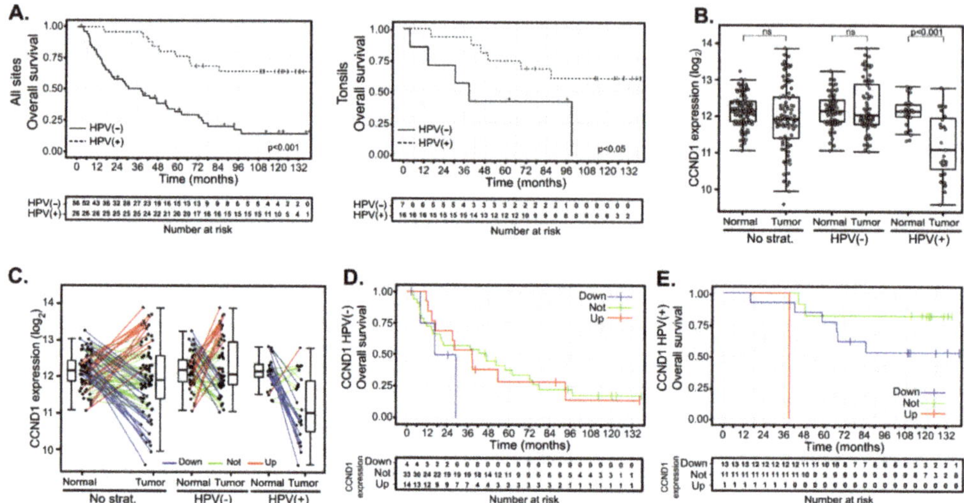

**Figure 1.** *CCND1* gene expression changes between tumor and normal tissues and patient survival in human papillomavirus (HPV) (+) and HPV(−) groups. (**A**) Kaplan-Meier plots of overall survival (OS) for HPV(−) and HPV(+) groups: entire cohort (left panel, n = 82) and tonsillar carcinoma (right panel, n = 23). (**B**) *CCND1* gene expression in a cohort of 94 patients (left panel) and stratified according to the HPV status (middle and right panel). The boxplot displays interquartile range with the whiskers indicating the greatest and smallest observations, excluding outliers. (**C**) *CCND1* deregulation between paired tumor and normal tissues. Paired samples are connected by the lines color-coded according to the change of *CCND1* expression (red—*CCND1* upregulation, fold-changes (FC) > 2; green—no deregulation, 0.5 < FC < 2 and blue—*CCND1* downregulation, FC < 0.5). (**D**) Kaplan–Meier plots for OS stratified according to *CCND1* gene deregulation in the HPV(−) group. (**E**) Kaplan–Meier plot for OS by *CCND1* deregulation in the HPV(+) group. Up—group with *CCND1* upregulated, not—group without *CCND1* deregulation and down—group with *CCND1* downregulated. Tick marks and crosses in (**A,D,E**) indicate right censoring.

We next analyzed whether HPV infection results in better prognosis of patients with the same tumor location. We analyzed tumors from the tonsils and the base of the tongue, as these subsites were significantly represented in respect of both the number and the proportion of HPV(+) tumors. Similar to the entire cohort, HPV infection of tonsil tumors was associated with significantly longer overall survival (Figure 1A, right panel). Five-year OS was 69% (CI 49–96%) and 43% (CI 18–100%) in the HPV(+) and HPV(−) groups, respectively. The hazard ratio of HPV(+) tonsil tumors was 0.3 (CI 0.09–0.99, *p* < 0.05). A similar trend was observed in patients with tumors originating in the base of the tongue, although in this case, statistical significance was not reached (Supplementary Figure S1B). Five-year OS was 71% (CI 24–100%) and 29% (CI 9–92%) in the HPV(+) and HPV(−) groups, respectively. The hazard ratio of the HPV(+) base of the tongue tumors was 0.27 (CI 0.07–1.11, *p* = 0.07).

Whole genome expression profiles were obtained for a cohort of 94 patients (90 tumor samples, 86 normal tissue samples). From this cohort, a group of 82 patients had expression profiles from both tumor and matched normal tissues. Of these 82 patients, 25 and 57 individuals had HPV(+) and HPV(−) tumors, respectively.

### 2.2. Deregulation of Cyclin D1 Expression in HPV(−) and in HPV(+) HNSCC Tumors Does Not Correlate with Patient Outcomes

Analysis of expression profiles from the cohort of 94 HNSCC cases revealed that the mean expression of *CCND1* does not differ between the normal mucosae and carcinomas. However, *CCND1* expression in tumors showed a visible spread toward both increased and decreased values (Figure 1B). Decreased *CCND1* expression was significant mainly in the HPV(+) group. We thus determined the changes in expression of *CCND1* between 82 paired tumor and normal tissues patient by patient. We considered *CCND1* to be deregulated in the tumor if the fold-change of expression between tumor and normal tissues was above two (*CCND1* upregulation) or less than 0.5 (*CCND1* downregulation), respectively. This categorization resulted in the formation of three different clusters: 47 cases (57.3%) in which there was no change in *CCND1* expression and 15 cases (18.3%) where *CCND1* was upregulated as expected in proliferating tumor tissue. However, a substantial number of tumors downregulated *CCND1* (20 cases, 24.4%) (Figure 1C). Further stratification according to HPV status revealed that *CCND1* downregulation was typical for HPV(+) tumors (13 cases, 52.0%), while almost all of the remaining 11 (44.0%) HPV(+) tumors did not deregulate *CCND1* expression. Only in one HPV(+) case (4.0%), *CCND1* was found to be upregulated (Figure 1C). On the other hand, HPV(−) tumors were spread across all three categories: tumors with *CCND1* expression upregulated (14 cases, 24.6%), unchanged (36 cases, 63.2%) or downregulated (7 cases, 12.3%). We examined whether these changes in *CCND1* expression correlated with patient outcomes and clustered 76 patients with known follow-up according to their *CCND1* regulation. Survival analysis revealed that neither OS (Figure 1D, Table 2 and Table S2) nor DFS (Supplementary Figure S1C and Table S2) were affected by *CCND1* upregulation in the HPV(−) group. *CCND1* downregulation was associated with a shorter five-year OS (Figure 1D and Table 2); however, the clinical data for this group was available for four patients only, showing no statistically significant difference (Table 2 and Table S2). In the HPV(+) group, *CCND1* downregulation did not affect OS or DFS significantly (Figure 1E, Supplementary Figure S1D and Table 2). This data suggested that there is no significant association between worse clinical outcome and *CCND1* upregulation and, conversely, between better outcomes in patients with *CCND1* downregulation.

**Table 2.** Survival analysis, overall survival (OS). Patients with paired samples and known survival (n = 76, see Table 1) were clustered according to the deregulation of D-type cyclin expression in matching samples. Thresholds for deregulation were the following: not (0.5 < fold-changes (FC) < 2), up (FC > 2) and down (FC < 0.5). HR—hazard ratio and NA—not available.

| | Five-Year Survival (%) | | Univariate HR (95% CI), P | |
|---|---|---|---|---|
| | HPV(−) | HPV(+) | HPV(−) | HPV(+) |
| ***CCND1*** | | | | |
| not | 38 | 82 | 1 | 1 |
| up | 29 | 0 | 1.04 (0.5–2.17), 0.92 | 46.95 (2.26–973.39), 0.01 |
| down | 0 | 77 | 2.6 (0.86–7.85), 0.09 | 2.88 (0.58–14.3), 0.2 |
| ***CCND2*** | | | | |
| not | 30 | 81 | 1 | 1 |
| up | 28 | 50 | 1.16 (0.59–2.29), 0.67 | 2.04 (0.54–7.69), 0.29 |
| down | 60 | 100 | 0.51 (0.15–1.72), 0.28 | NA |
| ***CCND3*** | | | | |
| not | 32 | 77 | 1 | 1 |
| up | 100 | 67 | NA | 0.9 (0.11–7.22), 0.92 |
| down | 0 | 0 | NA | NA |

### 2.3. Cyclin D1 Upregulation Correlates with Bona Fide Amplification of Its Genomic Locus

The observed low number of tumors with upregulated *CCND1* expression was unexpected, as *CCND1* overexpression was generally reported in up to 70% of patients at the protein level [21,22]. The common mechanism underlying *CCND1* upregulation is its gene locus amplification, with frequency ranging from 17% to 50% of cases [33,34]. While we do not possess direct data on the amplification, we analyzed changes in the expression of genes surrounding the *CCND1* gene as its surrogate marker [15,22]. Locus amplification would lead to upregulation of other genes in the locus, together with the *CCND1* upregulation. Out of 20 genes present in the two-megabase window around the *CCND1* genomic locus, we could analyze changes in the expression of 12 genes, while the expression of the other six genes was not detected by the microarray technology (Figure 2A). In the group of 15 predominantly HPV(−) cases where *CCND1* expression was upregulated in the tumor tissues (FC > 2), we observed upregulation of the genes that were downstream of *CCND1* in 14 (93%) cases (Figure 2B,C). Notably, *ANO1* was upregulated both in the majority of HPV(−) tumors and in HPV(+) tumors, probably reflecting the fact that this protein is a potential driver oncogene [35]. With the exception of *TPCN2*, the genes upstream of *CCND1* were not upregulated along with *CCND1* in HPV(−) tumors. The expression of *MRGPRF* and *MYEOV* were rather downregulated (Figure 2B,C). The downregulation of *MYEOV* may be explained by frequent epigenomic silencing of the gene [36]. In summary, we observed a strong signal for amplification of the *CCND1* genomic loci in HPV(−) tumors. In the HPV(+) group, the co-amplification was observed only in one case, in agreement with previously published data that amplification of the *CCND1* locus is rare in HPV(+) tumors [22].

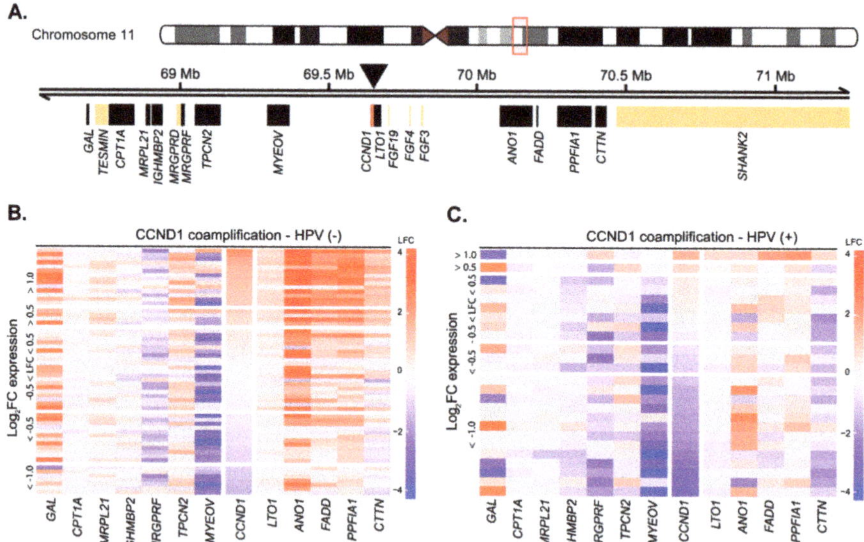

**Figure 2.** *CCND1* gene locus amplification as estimated from expression changes between tumor and normal tissues in HPV(+) and HPV(−) patient groups. (**A**) Schematics of the *CCND1* gene locus spanning ~ 2 Mb indicating position of the surrounding genes. Position of the *CCND1* gene (red) is indicated by the arrowhead. Black rectangles indicate expressed genes, and beige rectangles indicate genes expressed below the detection threshold. Position on chromosome 11 is given in Mb, million base pairs. (**B**) Heatmap of gene expression changes between tumor and normal tissues in the HPV(−) group for expressed genes from the *CCND1* locus. LFC—log₂ fold-change in expression. (**C**) Heatmap of gene expression changes between tumor and normal tissues in the HPV(+) group for expressed genes from the *CCND1* locus.

## 2.4. Deregulation of Cyclin D2 Does Not Correlate with Patient Outcomes

The expression levels of other D-type cyclins, *CCND2* and *CCND3*, were also examined. In contrast to *CCND1*, the median expression of *CCND2* was slightly elevated, yet it was statistically significant in both HPV(−) and HPV(+) tumors (Figure 3A). Comparative analysis of *CCND2* gene expression between paired tumor and normal samples revealed upregulation of *CCND2* expression in 26 cases (31.7%) (Figure 3B). In 49 cases (59.8%), there was no change in *CCND2* expression, and downregulation of *CCND2* was observed in seven cases (8.5%) only. Stratification according to HPV status revealed that both HPV(+) and HPV(−) tumors displayed significant upregulation of *CCND2* (Figure 3B), while downregulation of *CCND2* was observed only in a few HPV(−) tumors and a single HPV(+) tumor. Neither *CCND2* upregulation nor downregulation affected the OS (Figure 3C,D and Table 2) and DFS (Supplementary Figure S1E,F and Table S2).

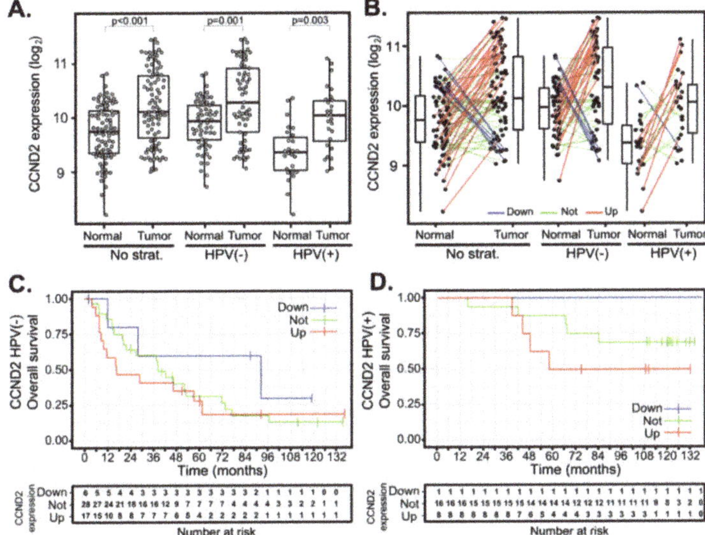

**Figure 3.** *CCND2* gene expression changes between tumor and normal tissues and patient survival in HPV(+) and HPV(−) groups. (**A**) *CCND2* gene expression in a cohort of 94 patients (left panel) and stratified according to the HPV status (middle and right panels). The boxplot displays interquartile range with the whiskers indicating the greatest and smallest observations, excluding outliers. (**B**) *CCND2* deregulation between paired tumor and normal tissues. Paired samples are connected by the lines color-coded according to the change of *CCND2* expression (red—*CCND2* upregulation, FC > 2; green—no deregulation, 0.5 < FC < 2 and blue—*CCND2* downregulation, FC < 0.5). (**C,D**) Kaplan–Meier plots for OS stratified according to *CCND2* gene deregulation in HPV(−) and HPV(+) groups. Up—group with *CCND2* upregulated, not—group without *CCND2* deregulation and down—*CCND2* group with *CCND2* downregulated. Tick marks in (**C,D**) indicate right censoring.

The analysis of *CCND3* mean expressions revealed elevated *CCND3* expression only in HPV(+) tumors (Figure 4A). Expression changes between paired tumor and normal samples also revealed the upregulation of *CCND3* expression in four cases (6.6%), predominantly in HPV(−) tumors (Figure 4B). Downregulation of *CCND3* was not observed. Due to the small number of patients with *CCND3* expression changes, statistical analysis did not yield any significant results (Table 2 and Table S2).

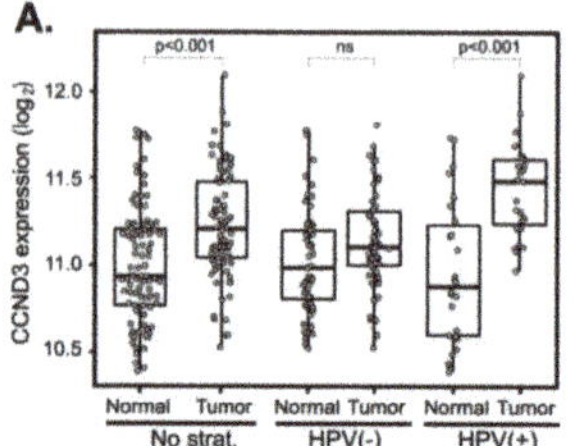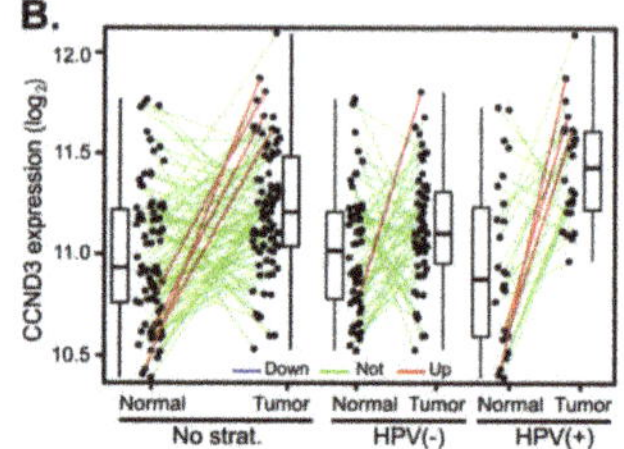

**Figure 4.** *CCND3* gene expression changes between tumor and normal tissues. (**A**) *CCND3* gene expression in a cohort of 94 patients (left panel) and stratified according to the HPV status (middle and right panels). The boxplot displays interquartile range with the whiskers indicating the greatest and smallest observations, excluding outliers. (**B**) *CCND3* deregulation between paired tumor and normal tissues. Paired samples are connected by the lines color-coded according to the change of *CCND3* expression (red—*CCND3* upregulation, FC > 2; green—no deregulation, 0.5 < FC < 2 and blue—*CCND3* downregulation, FC < 0.5).

## 2.5. Cyclin D2 is Often Upregulated in Tumors with Downregulated Cyclin D1 and Vice Versa

The finding that *CCND1* was downregulated in more than a fourth of the cases (Figure 1C) and that these patients did not display a better prognosis (Table 2) suggests the hypothesis that additional cyclins could compensate for the *CCND1* deficiency. We thus examined whether *CCND1* mRNA downregulation is paralleled by upregulation of *CCND2* and/or *CCND3*. Indeed, the comparison revealed that a substantial number of tumors with downregulated *CCND1* upregulated *CCND2* (7 out of 20; 35%) and *CCND3* in one individual case (Figure 5A). Intriguingly, we also found that tumors that upregulated *CCND1* often downregulated *CCND2* expression (5 out of 15; 33.3%) (Figure 5A). This observation suggests that expression of *CCND1* and *CCND2* is coupled.

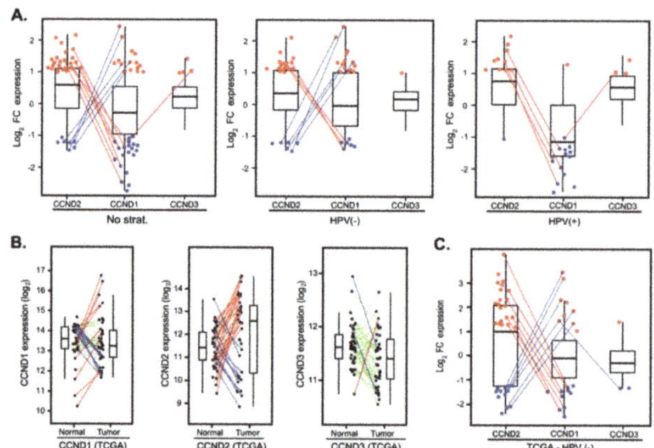

**Figure 5.** Compensatory expression of *CCND2* in tumors with deregulated *CCND1*. (**A**) D-type cyclins expression changes presented as log₂ fold-change between tumor and normal tissues in a cohort of 82 patients (left panel) and stratified according to the HPV status (middle and right panels). For clarity, only tumors with FC > 2 (red dots) and FC < 0.5 (blue dots) are shown. Lines connect the same patients. (**B**) D-type cyclins deregulation between paired tumor and normal tissues in The Cancer Genome Atlas (TCGA) cohort of 39 HPV(−) patients. Paired samples are connected by the lines color-coded according to the change in D-type cyclin expression (red—upregulation, FC > 2; green—no deregulation, 0.5 < FC < 2 and blue—downregulation, FC < 0.5). (**C**) D-type cyclins expression changes presented as log₂ fold-change between paired tumor and normal tissues in the TCGA cohort of 39 HPV(−) patients. For clarity, only tumors with FC > 2 (red dots) and FC < 0.5 (blue dots) are shown. Lines connect the same patients.

### 2.6. Validation in an Independent Cohort

To validate our findings in an independent cohort, we analyzed the expression profiles from a cohort of 490 patients from The Cancer Genome Atlas (TCGA) obtained from cBioPortal (see Materials and Methods for details). In this cohort, there were 42 paired tumor and normal mucosa samples with follow-up clinical data; however, only three of these pairs contained HPV(+) tumors. Hence, we focused on HPV(−) pairs. Analysis of *CCND1*, *CCND2* and *CCND3* expression recapitulated the findings in our dataset. *CCND1* deregulation between tumor and normal tissues clustered patients in three groups (Figure 5B) with no correlation with OS and DFS (Table S3). Substantial fraction of patients had *CCND2* in tumors upregulated (20 cases, 50.0%) or downregulated (12 cases, 30.0%) (Figure 5B). Importantly, we observed again that a significant fraction of the tumors with upregulated *CCND1* had downregulated *CCND2* (6 out of 9; 60.0%) or vice versa (7 out of 10, 70.0%) (Figure 5C).

### 2.7. Immunohistochemical Analysis of Tumor Samples

To evaluate the expression of cyclins at the protein level, we performed immunohistochemical analysis of 10 independent tumor samples (Figure 6). We observed cyclin D1 focal staining with variable intensity in the cytoplasm (10/10) and in the nuclei as well (9/10) (Figure 6A,B). The expression of cyclin D2 was weaker, and it was completely negative in three samples (Figure 6A); the nuclear signal of cyclin D2 was observed only in three cases (3/10) (Figure 6B). Frequently, we observed that the nuclear signal of cyclin D1 was accompanied by low or negative nuclear signals of cyclin D2 (Figure 6C,D). Vice versa, in the regions with high nuclear staining of cyclin D2, we observed that the level of cyclin D1 was low (Figure 6E,F).

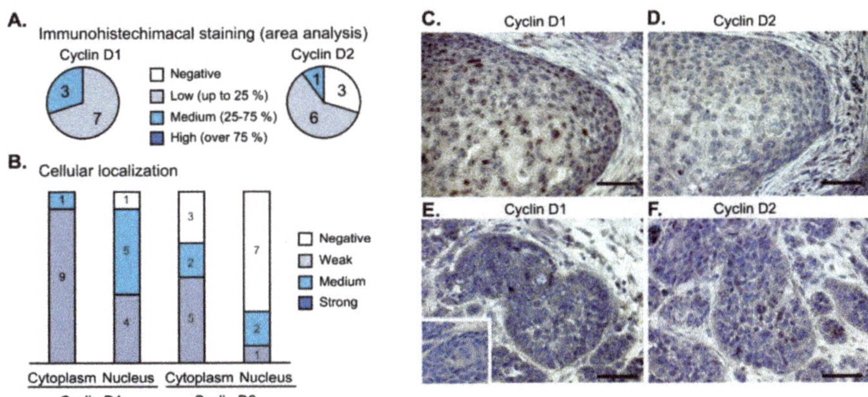


**Figure 6.** Immunohistochemistry of cyclin D1 and cyclin D2 expression in head and neck squamous cell carcinomas (n = 10). (**A**) Low-to-medium intensity staining of cyclin D1 was detected in all tumor samples. The staining of cyclin D2 was negative in three cases. (**B**) Nuclear and cytoplasmic distribution of cyclin D1 and cyclin D2 staining in the tumor samples. (**C,D**) In cases with high nuclear intensity of cyclin D1 (panel C), the nuclear staining of cyclin D2 was weak or negative (panel D). (**E,F**) Vice versa, higher intensity of cyclin D2 (panel F) was observed in tumor buds with a lack of nuclear cyclin D1 (panel E). Inset in panel E shows a negative control. The bar is 0.1 mm.

## 3. Discussion

Amplification of the *CCND1* gene encoding cyclin D1 is one of the most frequent genomic alterations in human cancers [37]. Altered *CCND1* expression has been reported in many different cancers [18,19], including head and neck squamous cell carcinoma [13,38]. In this work, we have focused on the expression patterns of cyclin D1, as well as the other two members of the cyclin D family, in a new cohort of 94 HNSCC patients. This dataset contains expression data from 82 tumor samples and paired normal mucosa and, to our best knowledge, represents the largest collection of paired HNSCC samples. Analysis of expression changes between paired tumor and normal samples revealed that expression of the *CCND1* gene is upregulated mainly in HPV(−) patients, and, unexpectedly, it may also be downregulated, predominantly in HPV(+) patients. Nonetheless, there was no statistically significant correlation between *CCND1* deregulation and the disease outcome. Further analysis showed that *CCND1* upregulation was frequently accompanied by *CCND2* downregulation, and, conversely, its downregulation was found to be often accompanied by upregulation of *CCND2* or, in one case, of *CCND3*. Based on these data, we speculate that upregulation of one D-type cyclin could be compensated by downregulation of another one. Such compensatory effects may explain previously observed poor correlations between *CCND1* expression and patient prognosis [26–28].

The importance of D-type cyclins for tumorigenesis has been demonstrated in mouse models, where the presence of cyclin D proteins is required for both tumor initiation and maintenance, while they seem largely dispensable for normal development [39,40]. It is presumed that the increase in the cyclin D expression is an early oncogenic event that causes tumor formation and continuous uncontrolled proliferation of tumor cells [18,19,41]. Concordantly, with these studies, the analysis of almost 500 HNSCC samples showed amplification of chromosomal region 11q13 containing the *CCND1* gene in 31% of cases [15], although the *CCND1* amplification frequency has been reported to range from 17% to 50% [33,34]. The amplified 11q13 locus harbors several other genes that could promote tumorigenesis. Among them, *ANO1* coding for anoctamin-1 and *CTTN* coding for cortactin are frequently coamplified with *CCND1*. These genes were shown to promote HNSCC progression and to be associated with poor prognosis [22,29,35,42]. In agreement with that, we found that *LTO1*, *ANO1* and *CTTN* are coregulated in the HPV(−) tumors, and they could cooperate with *CCND1* to affect the clinical outcome.

In addition to gene amplification, other mechanisms likely contribute to cyclin D1 protein upregulation, as its overexpression estimated by immunohistochemistry has been documented in up to 70% of HNSCC tumors [21,22]. In our dataset, we observed *CCND1* mRNA upregulation in approximately 20% of cases, and, surprisingly, roughly 25% of tumors showed a decreased expression of the *CCND1* gene when compared to normal mucosa. Since the repression of cyclin D1 promotes cell cycle exit, cellular quiescence and, in some cases, cell differentiation [43] and, in HNSCC tumor models, reduces cell growth and survival [44], these findings indicated that a *CCND1* decrease may have a positive impact on patient outcome. However, our data did not show any statistically significant correlation between *CCND1* expression and disease outcomes.

Why patient outcome does not correlate with the changes in *CCND1* expression remains unknown. According to our data, we speculate that a functional significance of change in *CCND1* expression could be alleviated by the compensatory change in the expression of another D-type cyclin. Compensation among D-type cyclins has been observed in mice, where cyclins D1, D2 and D3 are expressed redundantly in most tissues. Genetic ablation of individual D-type cyclin does not generally affect normal development; however, the simultaneous ablation of cyclins resulted in more severe developmental defects and embryonal lethality. These studies suggested that D-cyclins can compensate for the loss of another family member [45]. A compensatory effect was notably evident in mice engineered to express only one D-type cyclin. In embryos, the sole expressed cyclin became upregulated in most tissues which were otherwise negative for this type of cyclin [46]. Cyclin compensation was also observed in adult mice in the proliferating uterine epithelium, where the cyclin D1 absence can be compensated by cyclin D2 [47], and in B-lymphocytes, where D2 absence is compensated by upregulation of cyclin D3 [48]. All these observations point to a mechanism where perturbation of one cyclin D member may induce the compensatory expression of other family members and correspond to our findings, where one-third of tumors with downregulated cyclin D1 mRNA upregulated the expression of cyclin D2. We also observed the opposite trend, where in one-third of tumors with cyclin D1 upregulation, cyclin D2 was downregulated, further supporting the hypothesis that genetic compensation in response to the perturbation of cyclin D function is a common phenomenon in HNSCC. We have validated our results in an independent TCGA dataset and also by immunohistochemical staining.

The best prognostic factor for HNSCC remains to be HPV infection, as HPV(+) patients display better clinical outcomes. The worse prognosis of HPV(−) patients may be the consequence of different mutational burdens in HPV(+) and HPV(−) HNSCC tumors [38]. It may also reflect the fact that the *CCND1* locus is rarely amplified in HPV(+) tumors [22,49], which is consistent with our findings that a positive correlation of expression of *CCND1* and the genes in the 11q13 locus is observed almost exclusively in HPV(−) tumors. The genes coamplified with *CCND1*, such as *LTO1/ORAOV1*, *ANO1* and *CTTN,* are associated with HNSCC metastasis and recurrence and may therefore account for the worse prognosis of HPV(−) patients [22,29,42,50,51]. Our observations that cyclin D1 expression is downregulated in half of HPV(+) tumors was surprising. Given that cyclin D1 is required for tumor formation induced by HPV proteins E6 and E7 [52], it indicated that cyclin D1 downregulation may also contribute to a better prognosis of HPV(+) patients. Indeed, previous studies have associated low cyclin D1 expression in HPV(+) tumors with improved survival rates when in combination with high $p16^{INK4a}$, low pRb and low p53 [53,54]. However, we have not observed an improved outcome in HPV(+) patients who exhibit low expressions of cyclin D1 mRNA. In addition to the proposed compensatory effect of cyclin D2, it is also possible that the decrease in *CCND1* expression in HPV(+) tumors is a consequence of low pressure on cyclin D expression and lower CDK4/6 activity in general, as the G1 transcriptional repressor complex pRb/E2F is already disrupted by the viral protein E7 [14,38]. It is of note that, in HPV-associated cervical tumors, D-type cyclins and CDK4/6 activity inhibition by $p16^{INK4a}$ promotes tumor cell survival [55]. It is possible to speculate that cyclin D1 downregulation contributes to the survival of HPV(+) HNSCC tumor cells, similarly to cervical cancer.

In conclusion, in our study with independent validation, we have analyzed the mRNA expression level changes between paired tumor and normal samples at the patient level. Availability of paired samples provided us with the opportunity to show a significant variation in D-type cyclin gene expression between individual patients and to further stratify the patients according to the changes in the expression of the D-type cyclins. Although our results do not support a role for a specific D-cyclin in patient prognosis, it is possible to speculate about a compensatory mechanism between D-type cyclin expressions. These results are in agreement with the current opinion that tumors can be addicted to cyclin D-dependent CDK4/6 kinase activity [56] and that CDK4/6–cyclin D complexes represent promising targets in cancer therapy [18,20,41]. In the clinical trials conducted in HNSCC patients with radioresistant recurrent tumors, a combination of cetuximab and palbociclib targeting the EGF receptor and CDK4/6, respectively, indeed show promising results [57,58]. Analysis of cyclin expression changes between patient tumors and normal tissues may lead to the employment of more effective personalized therapies that could enhance the efficiency of currently adopted cancer treatment regimens.

## 4. Materials and Methods

### 4.1. Sample Collection

Sample collection, detection of HPV and microarray analysis were performed as described in Valach et al. [59] and Szabo et al. [60]. In short, normal mucosa and tumor tissue specimens were collected from 94 patients suffering from HNSCC, after their informed consent in full agreement with the local ethical committee and the Declaration of Helsinki. Specimens were chemically protected from RNA degradation and stored at −85 °C. Detection of high-risk oncogenic HPV types (16, 19, 31, 33 and 45) in tumor samples was done by RT-qPCR of the *E6* and *E7* genes.

For immunohistochemical analysis, the tissues were fixed in 10% neutral buffered formalin and routinely processed for paraffin block preparation. Tissue sections (5-µm thick) were deparaffinized in xylene and ethanol baths with decreasing concentrations of ethanol (5 min each). Heat-induced epitope retrieval was performed at pH = 6.0 (citrate buffer). Sections were blocked in hydrogen peroxide blocking reagent and protein block (ab64218 and ab64226, respectively; both Abcam, Cambridge,

UK). Primary antibodies (cyclin D1, clone DCS-6, mouse monoclonal; Dako, Glostrup, Denmark and cyclin D2 (D52F9) rabbit monoclonal #3741; Cell Signaling Technology, Danvers (MA), USA) were diluted 1:100, and sections were incubated at room temperature for 2 hours followed by incubation with HRP-tagged secondary antibody (Histofine simple stain MAX PO MULTI; Nichirei Biosciences, Tokyo, Japan) for 15 minutes. The reaction was developed for 10 minutes using Histofine simple stain AEC solution (Nichirei Biosciences, Tokyo, Japan) and counterstained with Gill's hematoxylin (Sigma Aldrich, Prague, Czech Republic). For the negative control, the primary antibody was replaced by control nonimmune serum (X0910 and X0903; Dako, Glostrup, Denmark). Imaging was performed with a Leica DM2000 microscope using the LAS software package (Leica Microsystems GmbH, Wetzlar, Germany). The annotation of immunohistochemical results was performed using the method described in detail in Protein Atlas version 19.2 (https://www.proteinatlas.org/about/assays+annotation).

### 4.2. Transcription Profiling

Whole genome transcription profiling was performed on HumanWG-6 v3 Expression BeadChip microarrays (Illumina, San Diego, California, USA). Subsequent data analysis was done in R statistical environment [61]: raw data was processed using the limma package [62] of the Bioconductor Project [63]. Background corrected and quantile normalized data were corrected for batch effects using the ComBat function from the R package sva [64], and expression intensities of technical replicates were averaged at the probe level. Detected transcripts were annotated using the provided manifest file (HumanWG-6_V3_0_R2_11282955_A.bgx; Illumina, San Diego, CA, USA). Expression changes between the sample groups (tumor HPV(−), tumor HPV(+) and normal) were detected by the moderated *t*-test within the limma package.

### 4.3. Clinical Data

The following patient data was recorded: TNM stages, tumor grade and localization, presence of keratinization, extracapsular extensions, angiogenesis and perineural spread. Information regarding treatment was also recorded (surgery and adjuvant chemo-/radiotherapy), together with patient characteristics (smoking, alcohol usage, personal and family cancer anamnesis). Vital status was recorded by the clinician/investigator at time of last follow-up. For overall survival, vital status was alive or dead. For disease-free survival, four statuses were recorded: disease free, exitus of other reason, recidive and exitus of the disease. The latter two vital statuses were taken as events in the disease-free survival analysis. Association between the HPV(−) and HPV(+) groups and clinical variables was tested using either a Kruskal-Wallis or Fisher's exact test.

### 4.4. Analysis of CCND1 Coamplification

To verify the co-amplification of the 11q13 region in HPV(−) tumors, we calculated $\log_2$ fold-changes of genes near *CCND1* (one megabase upstream and downstream of the *CCND1* start and end, respectively). The intensity of the probes targeting a particular gene was averaged, and genes with a mean probe $\log_2$-intensity lower than 5 were omitted. We used Gviz [65] and ComplexHeatmap [66] packages of the Bioconductor Project for visualization.

### 4.5. The Cancer Genome Atlas (TCGA) Validation Dataset

To provide an external validation, we used the head and neck squamous cell carcinoma TCGA PanCancer RNA-seq dataset of 490 HNSCC patients [15,67], provided through cBioPortal [68,69]. Read count matrices, clinical and follow-up data were downloaded and imported to the R statistical environment. DESeq2 [70] package was used for sequencing depth and variance-stabilizing normalization. DESeq2 was also used for statistical testing of the mean difference between sample groups (normal, tumor HPV(−) and tumor HPV(+)) using the Wald test.

*4.6. Survival Analysis*

We calculated fold-changes (FC) in expression of the D-type cyclins between tumor and normal mucosa samples for each patient and split the patients in three groups depending on FC: patients with cyclin upregulation (FC > 2), no deregulation (0.5 < FC < 2) and downregulation (FC < 0.5). These groups were used in survival analyses using the R packages survival [71] and survminer [72]. For each D-type cyclin, univariate Cox proportional hazard model, Kaplan–Meier estimator and its associated log-rank test were computed, using patients with no deregulation of cyclin as a reference.

*4.7. Data Availability*

The present dataset is available in MIAME-compliant form from the ArrayExpress database under accession E-MTAB-8588. The dataset used to validate our findings is available from The Cancer Genome Atlas via cBioPortal (https://www.cbioportal.org).

*4.8. Ethics Approval and Consent to Participate*

Ethical approval for patient recruitment, sample collection, clinical follow-up and data analysis based on the Declaration of Helsinki was granted by the Ethics Committee for Multi-Centric Clinical Trials of the University Hospital Motol and 2nd Faculty of Medicine, Charles University in Prague (Approval No EK-890/15).

## 5. Conclusions

In this study, we examined the expression of cyclin D1 and other D-type cyclins in a new cohort of HNSCC patients. The availability of both tumor sample and normal tissue sample from the same individual allowed us to analyze the relative changes in D-type cyclin expression rather than their absolute expression levels. The comparison of the D-type cyclin expression in tumor and healthy tissue from the same patients revealed an unexpected pattern of their expression. A subset of tumors increased the cyclin D1 expression that was specific, with one exception, to HPV-negative tumors. Unexpectedly, we also observed that the cyclin D1 expression was often downregulated in some tumors, with a higher frequency observed in HPV-positive patients. We have not found any direct effect of the changes in D-type cyclin expression on patient prognosis, even after stratification of the patients according to their HPV status. This observation could be a consequence of cyclin D2 upregulation, which frequently compensates cyclin D1 downregulation and vice versa. The compensatory expression among D-type cyclins was also observed in an independent dataset obtained from The Cancer Genome Atlas, further validating our data. We thus propose that absent correlation between the cyclin D1 expression and patient survival in both subtypes of HNSCCs may be a consequence of a compensatory mechanism where the effect of change in the cyclin D1 expression could be alleviated by the reciprocal change in the expression of cyclin D2. These findings highlight the importance of analyses of matched tissues from the same individual, as they can reveal molecular changes associated with the cancer development.

**Supplementary Materials:** The following are available online at http://www.mdpi.com/2072-6694/12/4/792/s1: Figure S1: Disease-free survival analysis of groups stratified according to HPV status, *CCND1* and *CCND2* expression and overall survival of tumors from the base of the tongue stratified according to the HPV status. Table S1: Summary of clinical data for present cohort. Table S2: Survival analyses in the presented dataset. Table S3: Survival analyses in the TCGA dataset. MS Word file supplementary_captions.docx: Captions for supplementary files.

**Author Contributions:** Conceptualization, J.N., H.S., V.P., L.L., K.S.J., J.P., M.K. and T.V.; Data Curation, J.N., H.S. and M.K.; Formal Analysis, J.N., H.S. and M.K.; Funding Acquisition, H.S., V.P., L.L., K.S.J., J.P., M.K. and T.V.; Investigation, V.B., M.C., M.H., J.Š., M.Š., R.K. and L.L.; Methodology, H.S., K.S.J., J.P., M.K. and T.V.; Resources, H.S., K.S.J., J.P., M.K. and T.V.; Software, J.N., H.S. and M.K.; Supervision, M.K. and T.V.; Visualization, J.N., M.K. and T.V.; Writing Original Draft Preparation, J.N., J.Š., J.G., M.K. and T.V. and Writing Review & Editing, J.N., J.Š., J.G., M.K. and T.V. All authors have read and agreed to the published version of the manuscript.

**Funding:** This research was funded by the Ministry of Health of the Czech Republic under grant no. AZV 16-29032A, by the Czech Science Foundation under grant no. 18-11908S, by the Operational Programme Research, Development and Education under the project "Center for Tumor Ecology—Research of the Cancer Microenvironment Supporting Cancer Growth and Spread" (reg. No. CZ.02.1.01/0.0/0.0/16_019/0000785) and by the Research and Development for Innovations Operational Program under project no. CZ.1.05/2.1.00/19.0400 (co-financed by the European Regional Development Fund and the state budget of the Czech Republic).

**Acknowledgments:** Authors are thankful to the service laboratory at Institute of Molecular Genetics of the Czech Academy of Sciences and especially to Martina Krausová and Šárka Kocourková for excellent technical assistance. Some results shown here are in whole or part based upon data generated by the TCGA Research Network: https://www.cancer.gov/tcga. We are deeply grateful to all specimen donors and research groups that have made the data publicly available. The authors thank J.E. Manning and Šárka Takáčová for critical reading of the manuscript and English grammar check.

**Conflicts of Interest:** The authors declare no conflicts of interest. The funders had no role in the design of the study; in the collection, analyses or interpretation of data; in the writing of the manuscript or in the decision to publish the results.

## References

1. Ferlay, J.; Shin, H.R.; Bray, F.; Forman, D.; Mathers, C.; Parkin, D.M. Estimates of worldwide burden of cancer in 2008: GLOBOCAN 2008. *Int. J. Cancer* **2010**, *127*, 2893–2917. [CrossRef] [PubMed]

2. Ferlay, J.; Steliarova-Foucher, E.; Lortet-Tieulent, J.; Rosso, S.; Coebergh, J.W.W.; Comber, H.; Forman, D.; Bray, F. Cancer incidence and mortality patterns in Europe: Estimates for 40 countries in 2012. *Eur. J. Cancer* **2013**, *49*, 1374–1403. [CrossRef] [PubMed]

3. Kalavrezos, N.; Bhandari, R. Current trends and future perspectives in the surgical management of oral cancer. *Oral Oncol.* **2010**, *46*, 429–432. [CrossRef] [PubMed]

4. Fung, C.; Grandis, J.R. Emerging drugs to treat squamous cell carcinomas of the head and neck. *Expert Opin. Emerg. Drugs* **2010**, *15*, 355–373. [CrossRef]

5. Hashibe, M.; Brennan, P.; Chuang, S.C.; Boccia, S.; Castellsague, X.; Chen, C.; Curado, M.P.; Dal Maso, L.; Daudt, A.W.; Fabianova, E.; et al. Interaction between Tobacco and Alcohol Use and the Risk of Head and Neck Cancer: Pooled Analysis in the International Head and Neck Cancer Epidemiology Consortium. *Cancer Epidemiol. Biomark. Prev.* **2009**, *18*, 541–550.

6. Warnakulasuriya, S. Global epidemiology of oral and oropharyngeal cancer. *Oral Oncol.* **2009**, *45*, 309–316. [CrossRef]

7. Gillison, M.L.; Koch, W.M.; Capone, R.B.; Spafford, M.; Westra, W.H.; Wu, L.; Zahurak, M.L.; Daniel, R.W.; Viglione, M.; Symer, D.E.; et al. Evidence for a Causal Association Between Human Papillomavirus and a Subset of Head and Neck Cancers. *J. Natl. Cancer Inst.* **2000**, *92*, 709–720.

8. Tommasino, M. The human papillomavirus family and its role in carcinogenesis. *Semin. Cancer Biol.* **2014**, *26*, 13–21. [CrossRef]

9. Kreimer, A.R.; Clifford, G.M.; Boyle, P.; Franceschi, S. Human papillomavirus types in head and neck squamous cell carcinomas worldwide: A systemic review. *Cancer Epidemiol. Biomark. Prev.* **2005**, *14*, 467–475. [CrossRef]

10. Psyrri, A.; Rampias, T.; Vermorken, J.B. The current and future impact of human papillomavirus on treatment of squamous cell carcinoma of the head and neck. *Ann. Oncol.* **2014**, *25*, 2101–2115. [CrossRef]

11. Dayyani, F.; Etzel, C.J.; Liu, M.; Ho, C.-H.; Lippman, S.M.; Tsao, A.S. Meta-analysis of the impact of human papillomavirus (HPV) on cancer risk and overall survival in head and neck squamous cell carcinomas (HNSCC). *Head Neck Oncol.* **2010**, *2*, 15. [CrossRef] [PubMed]

12. Ragin, C.C.R.; Modugno, F.; Gollin, S.M. The epidemiology and risk factors of head and neck cancer: A focus on human papillomavirus. *J. Dent. Res.* **2007**, *86*, 104–114. [CrossRef] [PubMed]

13. Leemans, C.R.; Braakhuis, B.J.M.; Brakenhoff, R.H. The molecular biology of head and neck cancer. *Nat. Rev. Cancer* **2011**, *11*, 9–22. [CrossRef]

14. Leemans, C.R.; Snijders, P.J.F.; Brakenhoff, R.H. The molecular landscape of head and neck cancer. *Nat. Rev. Cancer* **2018**, *18*, 269–282. [CrossRef]

15. Lawrence, M.S.; Sougnez, C.; Lichtenstein, L.; Cibulskis, K.; Lander, E.; Gabriel, S.B.; Getz, G.; Ally, A.; Balasundaram, M.; Birol, I.; et al. Comprehensive genomic characterization of head and neck squamous cell carcinomas. *Nature* **2015**, *517*, 576–582.

16. Chen, W.S.; Bindra, R.S.; Mo, A.; Hayman, T.; Husain, Z.; Contessa, J.N.; Gaffney, S.G.; Townsend, J.P.; Yu, J.B. CDKN2A Copy Number Loss Is an Independent Prognostic Factor in HPV-Negative Head and Neck Squamous Cell Carcinoma. *Front. Oncol.* **2018**, *8*, 95. [CrossRef]

17. Dok, R.; Nuyts, S. HPV Positive Head and Neck Cancers: Molecular Pathogenesis and Evolving Treatment Strategies. *Cancers* **2016**, *8*, 41. [CrossRef]

18. Musgrove, E.A.; Caldon, C.E.; Barraclough, J.; Stone, A.; Sutherland, R.L. Cyclin D as a therapeutic target in cancer. *Nat. Rev. Cancer* **2011**, *11*, 558–572. [CrossRef]

19. Deshpande, A.; Sicinski, P.; Hinds, P.W. Cyclins and cdks in development and cancer: A perspective. *Oncogene* **2005**, *24*, 2909–2915. [CrossRef]

20. VanArsdale, T.; Boshoff, C.; Arndt, K.T.; Abraham, R.T. Molecular Pathways: Targeting the Cyclin D-CDK4/6 Axis for Cancer Treatment. *Clin. Cancer Res.* **2015**, *21*, 2905–2910. [CrossRef]

21. Bova, R.J.; Quinn, D.I.; Nankervis, J.S.; Cole, I.E.; Sheridan, B.F.; Jensen, M.J.; Morgan, G.J.; Hughes, C.J.; Sutherland, R.L. Cyclin D1 and p16INK4A expression predict reduced survival in carcinoma of the anterior tongue. *Clin. Cancer Res.* **1999**, *5*, 2810–2819. [PubMed]

22. Hermida-Prado, F.; Menéndez, S.; Albornoz-Afanasiev, P.; Granda-Diaz, R.; Álvarez-Teijeiro, S.; Villaronga, M.; Allonca, E.; Alonso-Durán, L.; León, X.; Alemany, L.; et al. Distinctive Expression and Amplification of Genes at 11q13 in Relation to HPV Status with Impact on Survival in Head and Neck Cancer Patients. *J. Clin. Med.* **2018**, *7*, 501. [CrossRef]

23. Li, H.; Wawrose, J.S.; Gooding, W.E.; Garraway, L.A.; Lui, V.W.Y.; Peyser, N.D.; Grandis, J.R. Genomic analysis of head and neck squamous cell carcinoma cell lines and human tumors: A rational approach to preclinical model selection. *Mol. Cancer Res.* **2014**, *12*, 571–582. [CrossRef] [PubMed]

24. Michalides, R.; Van Veelen, N.; Hart, A.; Loftus, B.; Wientjens, E.; Balm, A. Overexpression of Cyclin D1 Correlates with Recurrence in a Group of Forty-seven Operable Squamous Cell Carcinomas of the Head and Neck. *Cancer Res.* **1995**, *55*, 975–978. [PubMed]

25. Mineta, H.; Miura, K.; Takebayashi, S.; Ueda, Y.; Misawa, K.; Harada, H.; Wennerberg, J.; Dictor, M. Cyclin D1 overexpression correlates with poor prognosis in patients with tongue squamous cell carcinoma. *Oral Oncol.* **2000**, *36*, 194–198. [CrossRef]

26. Wong, R.J.; Keel, S.B.; Glynn, R.J.; Varvares, M.A. Histological pattern of mandibular invasion by oral squamous cell carcinoma. *Laryngoscope* **2000**, *110*, 65–72. [CrossRef]

27. Vielba, R.; Bilbao, J.; Ispizua, A.; Zabalza, I.; Alfaro, J.; Rezola, R.; Moreno, E.; Elorriaga, J.; Alonso, I.; Baroja, A.; et al. p53 and cyclin D1 as prognostic factors in squamous cell carcinoma of the larynx. *Laryngoscope* **2003**, *113*, 167–172. [CrossRef]

28. Thomas, G.R.; Nadiminti, H.; Regalado, J. Molecular predictors of clinical outcome in patients with head and neck squamous cell carcinoma. *Int. J. Exp. Pathol.* **2005**, *86*, 347–363. [CrossRef]

29. Rodrigo, J.P.; García-Carracedo, D.; García, L.A.; Menéndez, S.T.; Allonca, E.; González, M.V.; Fresno, M.F.; Suárez, C.; García-Pedrero, J.M. Distinctive clinicopathological associations of amplification of the cortactin gene at 11q13 in head and neck squamous cell carcinomas. *J. Pathol.* **2009**, *217*, 516–523. [CrossRef]

30. Mellin, H.; Friesland, S.; Lewensohn, R.; Dalianis, T.; Munck-Wikland, E. Human papilloma virus (HPV) DNA in tonsillar cancer: Clinical correlates, risk of relapse, and survival. *Int. J. Cancer* **2000**, *89*, 300–304. [CrossRef]

31. Strome, S.E.; Savva, A.; Brissett, A.E.; Gostout, B.S.; Lewis, J.; Clayton, A.C.; McGovern, R.; Weaver, A.L.; Persing, D.; Kasperbauer, J.L. Squamous cell carcinoma of the tonsils: A molecular analysis of HPV associations. *Clin. Cancer Res.* **2002**, *8*, 1093–1100.

32. Chaturvedi, A.K.; Engels, E.A.; Anderson, W.F.; Gillison, M.L. Incidence trends for human papillomavirus-related and -unrelated oral squamous cell carcinomas in the United States. *J. Clin. Oncol.* **2008**, *26*, 612–619. [CrossRef] [PubMed]

33. Schwaederlé, M.; Daniels, G.A.; Piccioni, D.E.; Fanta, P.T.; Schwab, R.B.; Shimabukuro, K.A.; Parker, B.A.; Kurzrock, R. Cyclin alterations in diverse cancers: Outcome and co-amplification network. *Oncotarget* **2015**, *6*, 3033–3042. [CrossRef] [PubMed]

34. Vossen, D.; Verhagen, C.; Van der Heijden, M.; Essers, P.; Bartelink, H.; Verheij, M.; Wessels, L.; Van den Brekel, M.; Vens, C. Genetic Factors Associated with a Poor Outcome in Head and Neck Cancer Patients Receiving Definitive Chemoradiotherapy. *Cancers* **2019**, *11*, 445. [CrossRef]

35. Britschgi, A.; Bill, A.; Brinkhaus, H.; Rothwell, C.; Clay, I.; Duss, S.; Rebhan, M.; Raman, P.; Guy, C.T.; Wetzel, K.; et al. Calcium-activated chloride channel ANO1 promotes breast cancer progression by activating EGFR and CAMK signaling. *Proc. Natl. Acad. Sci. USA* **2013**, *110*, E1026–E1034. [CrossRef]

36. Janssen, J.W.G.; Imoto, I.; Inoue, J.; Shimada, Y.; Ueda, M.; Imamura, M.; Bartram, C.R.; Inazawa, J. MYEOV, a gene at 11q13, is coamplified with CCND1, but epigenetically inactivated in a subset of esophageal squamous cell carcinomas. *J. Hum. Genet.* **2002**, *47*, 460–464. [CrossRef]

37. Beroukhim, R.; Mermel, C.H.; Porter, D.; Wei, G.; Raychaudhuri, S.; Donovan, J.; Barretina, J.; Boehm, J.S.; Dobson, J.; Urashima, M.; et al. The landscape of somatic copy-number alteration across human cancers. *Nature* **2010**, *463*, 899–905. [CrossRef]

38. Faraji, F.; Schubert, A.D.; Kagohara, L.T.; Tan, M.; Xu, Y.; Zaidi, M.; Fortin, J.-P.; Fakhry, C.; Izumchenko, E.; Gaykalova, D.A.; et al. The Genome-Wide Molecular Landscape of HPV-Driven and HPV-Negative Head and Neck Squamous Cell Carcinoma. In *Molecular Determinants of Head and Neck Cancer*; Humana Press: Totowa, NJ, USA, 2018; pp. 293–325. [CrossRef]

39. Yu, Q.; Geng, Y.; Sicinski, P. Specific protection against breast cancers by cyclin D1 ablation. *Nature* **2001**, *411*, 1017–1021. [CrossRef] [PubMed]

40. Choi, Y.J.; Li, X.; Hydbring, P.; Sanda, T.; Stefano, J.; Christie, A.L.; Signoretti, S.; Look, A.T.; Kung, A.L.; Von Boehmer, H.; et al. The requirement for cyclin D function in tumor maintenance. *Cancer Cell* **2012**, *22*, 438–451. [CrossRef] [PubMed]

41. Otto, T.; Sicinski, P. Cell cycle proteins as promising targets in cancer therapy. *Nat. Rev. Cancer* **2017**, *17*, 93–115. [CrossRef] [PubMed]

42. Sugahara, K.; Michikawa, Y.; Ishikawa, K.; Shoji, Y.; Iwakawa, M.; Shibahara, T.; Imai, T. Combination effects of distinct cores in 11q13 amplification region on cervical lymph node metastasis of oral squamous cell carcinoma. *Int. J. Oncol.* **2011**, 761–769.

43. Klein, E.A.; Assoian, R.K. Transcriptional regulation of the cyclin D1 gene at a glance. *J. Cell Sci.* **2008**, *121*, 3853–3857. [CrossRef] [PubMed]

44. Sauter, E.R.; Nesbit, M.; Litwin, S.; Klein-Szanto, A.J.; Cheffetz, S.; Herlyn, M. Antisense cyclin D1 induces apoptosis and tumor shrinkage in human squamous carcinomas. *Cancer Res.* **1999**, *59*, 4876–4881.

45. Sherr, C.J.; Roberts, J.M. Living with or without cyclins and cyclin-dependent kinases. *Genes Dev.* **2004**, *18*, 2699–2711. [CrossRef]

46. Ciemerych, M.A.; Kenney, A.M.; Sicinska, E.; Kalaszczynska, I.; Bronson, R.T.; Rowitch, D.H.; Gardner, H.; Sicinski, P. Development of mice expressing a single D-type cyclin. *Genes Dev.* **2002**, *16*, 3277–3289. [CrossRef] [PubMed]

47. Chen, B.; Pollard, J.W. Cyclin D2 compensates for the loss of cyclin D1 in estrogen-induced mouse uterine epithelial cell proliferation. *Mol. Endocrinol.* **2003**, *17*, 1368–1381. [CrossRef] [PubMed]

48. Lam, E.W.-F.; Glassford, J.; Banerji, L.; Thomas, N.S.B.; Sicinski, P.; Klaus, G.G.B. Cyclin D3 Compensates for Loss of Cyclin D2 in Mouse B-lymphocytes Activated via the Antigen Receptor and CD40. *J. Biol. Chem.* **2000**, *275*, 3479–3484. [CrossRef] [PubMed]

49. Van Kempen, P.M.W.; Noorlag, R.; Braunius, W.W.; Moelans, C.B.; Rifi, W.; Savola, S.; Koole, R.; Grolman, W.; Van Es, R.J.J.; Willems, S.M. Clinical relevance of copy number profiling in oral and oropharyngeal squamous cell carcinoma. *Cancer Med.* **2015**, *4*, 1525–1535. [CrossRef]

50. Clark, E.S.; Brown, B.; Whigham, A.S.; Kochaishvili, A.; Yarbrough, W.G.; Weaver, A.M. Aggressiveness of HNSCC tumors depends on expression levels of cortactin, a gene in the 11q13 amplicon. *Oncogene* **2009**, *28*, 431–444. [CrossRef]

51. Togashi, Y.; Arao, T.; Kato, H.; Matsumoto, K.; Terashima, M.; Hayashi, H.; De Velasco, M.A.; Fujita, Y.; Kimura, H.; Yasuda, T.; et al. Frequent amplification of oraov1 gene in esophageal squamous cell cancer promotes an aggressive phenotype via proline metabolism and ros production. *Oncotarget* **2014**, *5*, 2962–2973. [CrossRef]

52. Al Moustafa, A.-E.; Foulkes, W.D.; Wong, A.; Jallal, H.; Batist, G.; Yu, Q.; Herlyn, M.; Sicinski, P.; Alaoui-Jamali, M.A. Cyclin D1 is essential for neoplastic transformation induced by both E6/E7 and E6/E7/ErbB-2 cooperation in normal cells. *Oncogene* **2004**, *23*, 5252–5256. [CrossRef] [PubMed]

53.  Holzinger, D.; Flechtenmacher, C.; Henfling, N.; Kaden, I.; Grabe, N.; Lahrmann, B.; Schmitt, M.; Hess, J.; Pawlita, M.; Bosch, F.X. Identification of oropharyngeal squamous cell carcinomas with active HPV16 involvement by immunohistochemical analysis of the retinoblastoma protein pathway. *Int. J. Cancer* **2013**, *133*, 1389–1399. [CrossRef] [PubMed]

54.  Plath, M.; Broglie, M.A.; Förbs, D.; Stoeckli, S.J.; Jochum, W. Prognostic significance of cell cycle-associated proteins p16, pRB, cyclin D1 and p53 in resected oropharyngeal carcinoma. *J. Otolaryngol. Head Neck Surg.* **2018**, *47*, 1–9. [CrossRef]

55.  McLaughlin-Drubin, M.E.; Park, D.; Munger, K. Tumor suppressor p16INK4A is necessary for survival of cervical carcinoma cell lines. *Proc. Natl. Acad. Sci. USA* **2013**, *110*, 16175–16180. [CrossRef] [PubMed]

56.  Choi, Y.J.; Anders, L. Signaling through cyclin D-dependent kinases. *Oncogene* **2014**, *33*, 1890–1903. [CrossRef]

57.  Michel, L.; Ley, J.; Wildes, T.M.; Schaffer, A.; Robinson, A.; Chun, S.-E.; Lee, W.; Lewis, J.; Trinkaus, K.; Adkins, D. Phase I trial of palbociclib, a selective cyclin dependent kinase 4/6 inhibitor, in combination with cetuximab in patients with recurrent/metastatic head and neck squamous cell carcinoma. *Oral Oncol.* **2016**, *58*, 41–48. [CrossRef]

58.  Adkins, D.; Ley, J.; Neupane, P.; Worden, F.; Sacco, A.G.; Palka, K.; Grilley-Olson, J.E.; Maggiore, R.; Salama, N.N.; Trinkaus, K.; et al. Palbociclib and cetuximab in platinum-resistant and in cetuximab-resistant human papillomavirus-unrelated head and neck cancer: A multicentre, multigroup, phase 2 trial. *Lancet. Oncol.* **2019**, *20*, 1295–1305. [CrossRef]

59.  Valach, J.; Fík, Z.; Strnad, H.; Chovanec, M.; Plzák, J.; Čada, Z.; Szabo, P.; Šáchová, J.; Hroudová, M.; Urbanová, M.; et al. Smooth muscle actin-expressing stromal fibroblasts in head and neck squamous cell carcinoma: Increased expression of galectin-1 and induction of poor prognosis factors. *Int. J. Cancer* **2012**, *131*, 2499–2508. [CrossRef]

60.  Szabó, P.; Kolář, M.; Dvořánková, B.; Lacina, L.; Štork, J.; Vlček, Č.; Strnad, H.; Tvrdek, M.; Smetana, K. Mouse 3T3 fibroblasts under the influence of fibroblasts isolated from stroma of human basal cell carcinoma acquire properties of multipotent stem cells. *Biol. Cell* **2011**, *103*, 233–248. [CrossRef]

61.  R Core Team R: A Language and Environment for Statistical Computing. Available online: http://www.r-project.org/ (accessed on 22 March 2020).

62.  Ritchie, M.E.; Phipson, B.; Wu, D.; Hu, Y.; Law, C.W.; Shi, W.; Smyth, G.K. Limma powers differential expression analyses for RNA-sequencing and microarray studies. *Nucleic Acids Res.* **2015**, *43*, e47. [CrossRef]

63.  Huber, W.; Carey, V.J.; Gentleman, R.; Anders, S.; Carlson, M.; Carvalho, B.S.; Bravo, H.C.; Davis, S.; Gatto, L. Orchestrating high-throughput genomic analysis with Bioconductor. *Nat. Methods* **2015**, *12*, 115–121. [CrossRef] [PubMed]

64.  Leek, J.T.; Johnson, W.E.; Parker, H.S.; Fertig, E.J.; Jaffe, A.E.; Storey, J.D.; Zhang, Y.; Torres, L.C. sva: Surrogate Variable Analysis. R package version 3.34.0. Available online: http://bioconductor.org/packages/release/bioc/html/sva.html (accessed on 22 March 2020).

65.  Hahne, F.; Ivanek, R. Visualizing Genomic Data Using Gviz and Bioconductor. In *Statistical Genomics*; Mathé, E., Davis, S., Eds.; Humana Press: New York, NY, USA, 2016; Volume 1418, pp. 391–416. ISBN 9780849331664.

66.  Gu, Z.; Eils, R.; Schlesner, M. Complex heatmaps reveal patterns and correlations in multidimensional genomic data. *Bioinformatics* **2016**, *32*, 2847–2849. [CrossRef] [PubMed]

67.  Liu, J.; Lichtenberg, T.; Hoadley, K.A.; Poisson, L.M.; Lazar, A.J.; Cherniack, A.D.; Kovatich, A.J.; Benz, C.C.; Levine, D.A.; Lee, A.V.; et al. An Integrated TCGA Pan-Cancer Clinical Data Resource to Drive High-Quality Survival Outcome Analytics. *Cell* **2018**, *173*, 400–416. [CrossRef]

68.  Cerami, E.; Gao, J.; Dogrusoz, U.; Gross, B.E.; Sumer, S.O.; Aksoy, B.A.; Jacobsen, A.; Byrne, C.J.; Heuer, M.L.; Larsson, E.; et al. The cBio Cancer Genomics Portal: An Open Platform for Exploring Multidimensional Cancer Genomics Data. *Cancer Discov.* **2012**, *2*, 401–404. [CrossRef]

69.  Gao, J.; Aksoy, B.A.; Dogrusoz, U.; Dresdner, G.; Gross, B.; Sumer, S.O.; Sun, Y.; Jacobsen, A.; Sinha, R.; Larsson, E.; et al. Integrative Analysis of Complex Cancer Genomics and Clinical Profiles Using the cBioPortal. *Sci. Signal* **2014**, *6*, 1–34. [CrossRef]

70.  Love, M.I.; Huber, W.; Anders, S. Moderated estimation of fold change and dispersion for RNA-seq data with DESeq2. *Genome Biol.* **2014**, *15*, 1–21. [CrossRef]

71. Therneau, T. A Package for Survival Analysis in S. version 2.38. Available online: https://github.com/therneau/survival (accessed on 22 March 2020).
72. Kassambara, A.; Kosinski, M.; Biecek, P.; Scheipl, F. Survminer: Drawing Survival Curves using "ggplot2.". Available online: https://rpkgs.datanovia.com/survminer/index.html (accessed on 22 March 2020).

*Article*

# Establishment and Molecular Phenotyping of Organoids from the Squamocolumnar Junction Region of the Uterine Cervix

Yoshiaki Maru [1,†], Akira Kawata [2,†], Ayumi Taguchi [2,3,*], Yoshiyuki Ishii [4], Satoshi Baba [2], Mayuyo Mori [2], Takeshi Nagamatsu [2], Katsutoshi Oda [2], Iwao Kukimoto [4], Yutaka Osuga [2], Tomoyuki Fujii [2] and Yoshitaka Hippo [1,*]

[1]  Department of Molecular Carcinogenesis, Chiba Cancer Center Research Institute, Chiba 260-8717, Japan; ymaru@chiba-cc.jp

[2]  Department of Obstetrics and Gynecology, Graduate School of Medicine, The University of Tokyo, Tokyo 113-8655, Japan; akira2612@gmail.com (A.K.); s.baba.91888@gmail.com (S.B.); mayuyo1976@gmail.com (M.M.); tnag-tky@umin.ac.jp (T.N.); katsutoshi-tky@umin.ac.jp (K.O.); yutakaos-tky@umin.ac.jp (Y.O.); fujiit-tky@umin.org (T.F.)

[3]  Department of Gynecology, Tokyo Metropolitan Cancer and Infectious Diseases Center, Komagome Hospital, Tokyo 113-8677, Japan

[4]  Pathogen Genomics Center, National Institute of Infectious Diseases, Tokyo 208-0001, Japan; yishii@nih.go.jp (Y.I.); ikuki@nih.go.jp (I.K.)

*  Correspondence: aytaguchi-tky@umin.ac.jp (A.T.); yhippo@chiba-cc.jp (Y.H.)

†  These authors contributed equally to this work.

Received: 27 February 2020; Accepted: 13 March 2020; Published: 15 March 2020

**Abstract:** The metaplastic epithelium of the transformation zone (TZ) including the squamocolumnar junction (SCJ) of the uterine cervix is a prime target of human papilloma virus (HPV) infection and subsequent cancer development. Due to the lack of adequate in vitro models for SCJ, however, investigations into its physiological roles and vulnerability to carcinogenesis have been limited. By using Matrigel-based three-dimensional culture techniques, we propagated organoids derived from the normal SCJ region, along with metaplastic squamous cells in the TZ. Consisting predominantly of squamous cells, organoids basically exhibited a dense structure. However, at least in some organoids, a small but discrete population of mucin-producing endocervix cells co-existed adjacent to the squamous cell population, virtually recapitulating the configuration of SCJ in a TZ background. In addition, transcriptome analysis confirmed a higher expression level of many SCJ marker genes in organoids, compared to that in the immortalized cervical cell lines of non-SCJ origin. Thus, the obtained organoids appear to mimic cervical SCJ cells and, in particular, metaplastic squamous cells from the TZ, likely providing a novel platform in which HPV-driven cervical cancer development could be investigated.

**Keywords:** organoid; uterine cervix; squamocolumnar junction; human papillomavirus; Matrigel

---

## 1. Introduction

The uterine cervix consists of three distinct epithelial types; tall mucin-secreting columnar cells of the endocervix in a single layer, glycogenated stratified squamous cells in the ectocervix, and a transformation zone (TZ) in between, which results from gradual metaplastic replacement of columnar cells by squamous cells during the reproductive age [1]. Reserve cells, putative stem cells in the squamocolumnar junction (SCJ) region, are implicated in this metaplastic process; thereby, their roles have been intensively investigated [2,3]. Whereas the SCJ originally resides at the boundary of the endocervix and ectocervix, the newly formed SCJ is shifted, alongside the extension of the TZ toward the endocervix, to the region connecting the TZ and endocervix. The SCJ and the TZ have been regarded as the most important cytological and colposcopic landmarks in the clinic, based on the fact that the large majority of uterine cervical cancers (UCC) and high-grade squamous intraepithelial lesions (HSIL) arise at this region [4,5]. Whereas human papillomavirus (HPV) is a major cause of neoplastic changes in the cervix for both squamous cell carcinoma (SCC) and adenocarcinoma [6], the incidence of UCC is significantly higher than that of cancers arising from other genital tract tissues [7]. However, the precise mechanisms underlying the predisposition of the cervix toward HPV-driven carcinogenesis have remained elusive.

Recently, a residual embryonic cell population harboring the capacity to differentiate and the vulnerability to undergo neoplastic transformation was documented in both gastro-esophageal [8] and ecto-endocervical junctions [9]. With regard to the uterine cervix, a small discrete population of cuboidal cells in the SCJ region was histologically identified. By micro-dissection and microarray analysis, over 70 genes were identified as upregulated genes by more than two-fold, compared to adjacent squamous or columnar cell populations. In particular, Cytokeratin7 (KRT7), Anterior gradient protein 2 homolog (AGR2), Cluster differentiation 63 (CD63), Matrix metalloproteinase-7 (MMP7) and Guanine deaminase (GDA) were further demonstrated to specifically mark these cuboidal SCJ cells by immunohistochemistry [9]. Intriguingly, all these five markers remained positive in all HPV-related neoplastic tissues and cervix-derived cancer cell lines, but not in the SCC of other tissues in the lower genital tract [9]. Besides, it was demonstrated that SCJ cells give rise to reserve cells [10] and are specific targets of HPV infection in the cervix [11]. These observations point toward the notion that SCJ cells might be highly vulnerable to, and a major cell of origin for, HPV-driven cervical carcinogenesis [12]. As a resource for in vitro studies investigating the relationship between HPV and UCC, several cell lines have been generated. For example, End1/E6E7 and Ect1/E6E7, which are widely used as normal controls for cervical cells originating from columnar cells and squamous cells in the cervix, respectively, were immortalized by the introduction of HPV-derived oncogenes E6 and E7 [13]. Normal immortal human keratinocytes (NIKS) comprise an undifferentiated keratinocyte cell line derived from neonatal foreskin [14] and has been intensively used for the investigation of biological impacts mediated by the introduction of the HPV genome [15]. However, none of these cell lines are, in fact, derived from a discrete population of the SCJ, limiting detailed analysis that focuses on HPV-driven UCC development from SCJ cells.

Organoid culture is an emerging technique that enables the infinite expansion of normal stem cells in culture [16]. It has been applied to various research fields, including infectious diseases [17], developmental biology [18], and tissue regeneration [19]. By taking advantage of propagating normal stem cells in vitro, we have established murine organoid-based ex vivo carcinogenesis models for the intestine [20], lungs [21], hepatobiliary tract [22], and pancreas [23], by in vitro lentiviral gene transduction [24] or chemical treatment [25], followed by inoculation in the dorsal skin of nude mice. More recently, organoid culture techniques have been further applied to patient-derived tumor samples of diverse organs, which revealed that organoids basically retained the histological features and genetic aberrations of the original tumors [26–28]. However, there was little progress in their applicability to gynecologic tumors until recently [29], when we established an efficient culture method for ovarian and endometrial tumors [30], by modification of our Matrigel bilayer organoid culture (MBOC) protocol [31], which we previously developed for various murine cells. Moreover, for the first time, we established patient-derived organoids of cervical clear cell carcinoma, a rare type of cervical adenocarcinoma [32], further confirming the validity of the modified culture protocol for gynecologic tumors.

In this study, we aimed to propagate normal cervical cells from the SCJ region by applying our modified MBOC protocol. We successfully expanded HPV-negative SCJ organoids, which were proved to retain many features of the SCJ. These organoids would, therefore, likely contribute to gaining mechanistic insights into how SCJ cells could be deregulated for neoplastic changes.

## 2. Results

### 2.1. Propagation of Patient-Derived Organoids from the Cervical SCJ Region

For future elucidation of the mechanisms underlying UCC development, we set out to conduct an organoid culture of SCJ cells with a modified MBOC protocol [30]. We first tested outpatient biopsy samples targeting the SCJ region (Figure S1A), but we did not achieve robust propagation of organoids. Typical failures included cases where organoids stopped proliferating in early passages or dissociated cells predominantly appeared flat, reminiscent of surface squamous cells (Figure S1B). Based on these observations, we reasoned that collecting tiny amounts of tissue by single biopsy might not be ideal to accurately spot the SCJ and to practically obtain stem or progenitor populations in sufficient amounts. To address this issue, we then selected normal uteri that were surgically co-resected with non-cervical gynecologic tumors. Through careful observation of surface texture, the areas for columnar and squamous epithelium were macroscopically estimated. New SCJ was postulated to be located around their borderline zones. Unlike original SCJ, new SCJ could appear broad in width. Consequently, to ensure collection from the new SCJ region, we comprehensively collected epithelial cells from the areas within a sufficient margin from the estimated SCJ region. Such tissue samples were collected in a hospital immediately after surgery, and processed the next morning following overnight transfer to the lab while maintained in a cold media (Figure 1A). As a culture media, we tested the standard culture media supplemented with EGF, R-spondin-1, Noggin, Jagged-1, and Rho-associated, coiled-coil containing protein kinase (ROCK) inhibitor Y27632, which we have confirmed robustly applicable to organoid culture of murine primary cells from various organs and human gynecological neoplasms [20,22,23,30,32], to facilitate future cross-referencing of organoids from various organs.

We tested SCJ samples from four independent patients (Table 1). Dissociated cells were plated onto solidified Matrigel and subjected to 3D culture. Organoids kept proliferating for at least one month over several passages until we ceased the culture. For example, organoids Cx-1 (Figure 1B) derived from Patient #1, initially exhibited a round morphology with dense and cystic features, but grew in an irregular shape with multiple budding or chain-like structures after several passages (Figure 1C). Similar results were obtained for Cx-2 and Cx-3 from Patients #2 and #3, respectively, while some organoids became resistant to enzymatic and physical dissociation, forming dense homogeneous cell aggregates, as seen in Cx-2 (Figure 1C). Since these observed features are not common to the usual organoids from the intestine [33] but from those of the esophagus [34], we speculated that squamous cell differentiation might be prominent in organoids. With regard to Cx-4 from Patient #4, the majority of the cells appeared to be differentiated squamous surface cells, as frequently observed in biopsy samples, and resulted in a gradual decline of an actively proliferating population, suggesting that the SCJ was not properly collected in this case. A total of 31 HPV genotypes proved negative in the three propagated organoids (Table 1). Besides, organoids were feasible for highly efficient lentiviral gene transduction (Figure S2) and tolerated cryopreservation (Figure 1D). These observations strongly suggested that normal SCJ cells were established for studies to reconstitute UCC development ex vivo. In Papanicoloau staining, a routine procedure for cervical smear samples in the clinic, organoids Cx-1 and Cx-3 appeared to exhibit a solid nature with a layer of cuboidal or thin epithelial cells on the surface, and orange-colored flat cells were observed in Cx-2 organoids, both suggestive of the presence of terminally differentiated squamous cells (Figure 1E).

**Table 1.** Summary of clinicopathological features of the patients.

| Patient | Organoid | Age | Parity | Menstrual Cycle | Disease | HPV |
|---|---|---|---|---|---|---|
| #1 | Cx-1 | 53 | 0 | Proliferative phase | Ovarian cancer | Negative |
| #2 | Cx-2 | 33 | 0 | Secretory phase | Ovarian borderline tumor | Negative |
| #3 | Cx-3 | 40 | 0 | Secretory phase | Ovarian cancer | Negative |
| #4 | Cx-4 | 50 | 0 | Not available | Uterine body tumor | Not tested |

Abbreviation: HPV, human papillomavirus

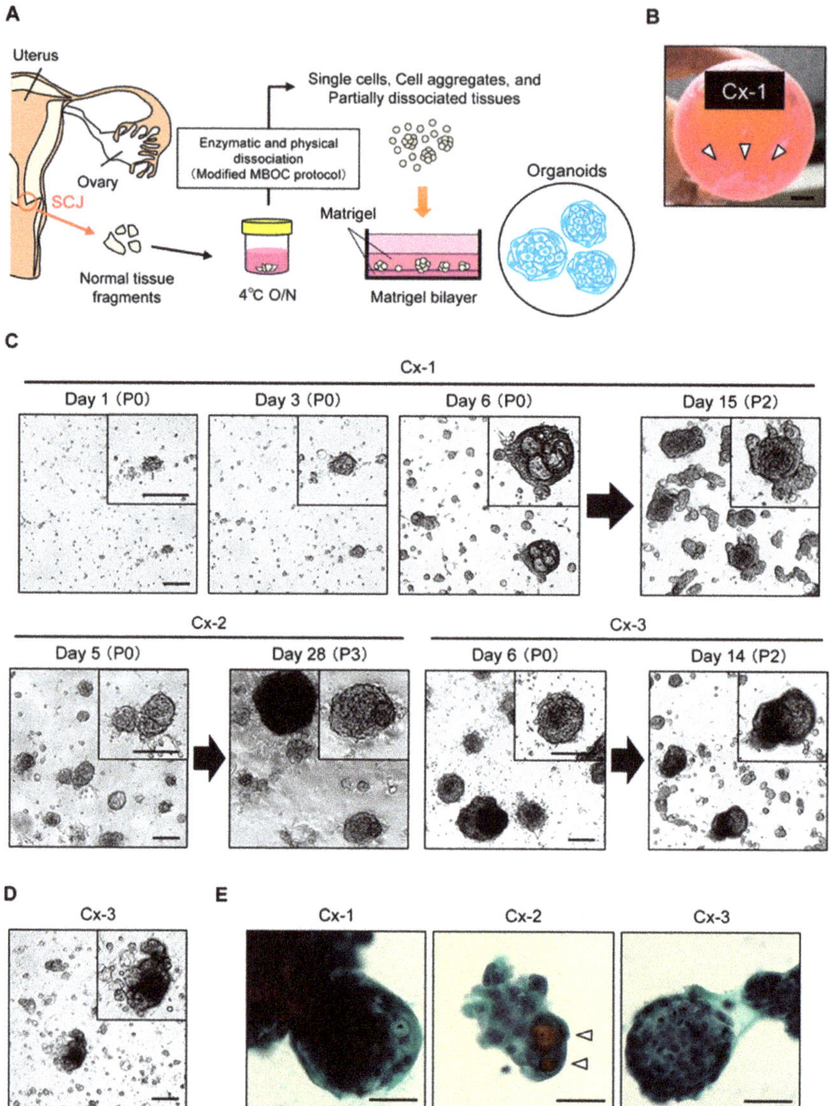

**Figure 1.** Organoid culture of normal cervical cells derived from the squamocolumnar junction (SCJ) region. (**A**) Schematic presentation of procedures for the establishment of SCJ-derived organoids. A modified Matrigel bilayer organoid culture (MBOC) protocol was used (see Materials and Methods). (**B**) Tissue samples collected from the cervical SCJ. A scale bar indicates 10 mm. White arrowheads depict tissue fragments. (**C**) Phase-contrast images of organoids. Upper panel, representative time-lapse images of SCJ-derived organoids (Cx-1) in the bright field (passage P0 and P2) at 1–6 days and 15 days. Lower panel: representative bright field images of the other SCJ-derived organoids (Cx-2 and Cx-3). Note dense homogeneous cell aggregates in the upper left area of the panel for Cx-2 at day 28. Insets show magnified images of organoids. Scale bars indicate 200 μm. (**D**) Propagation of organoids after cryopreservation. A phase contrast image of Cx-3 organoids at day 14 (passage P1) after thawing is shown. (**E**) Papanicolaou staining of SCJ-derived organoids. Open arrowheads show superficial squamous cells. Scale bars indicate 50 μm.

## 2.2. Mutually Exclusive Localization of Ectocervix-Like Cells and Endocervix-Like Cells within Organoids

With thin sections of formalin-fixed and paraffin-embedded (FFPE) samples, we found that organoids basically displayed dense structures (Figure 2A). Although cavity-like structures or cysts were occasionally observed within organoids, they tended to lack the regular lining of cells with apico-basal cell polarity, unlike typical cystic structures seen in columnar cell organoids (Figure 2A). Immunostaining for the squamous cell marker p40, the delta N isoform of p63, revealed that most cells, if not all, were positively stained throughout organoids (Figure 2A). This observation is in line with the notion that collected samples would contain TZ/squamous cells in large quantities. On the other hand, given that the SCJ region resides in the interface between TZ/squamous cells and columnar endocervix cells, the propagated organoids are supposed to contain endocervical cells as well. However, cystic structures reminiscent of columnar epithelial cell features were not particularly evident and those few cystic structures were invariably p40-positive in organoids (Figure 2A). To facilitate identification of cells with endocervical differentiation, we conducted Periodic acid-Schiff (PAS) reaction for staining organoids, anticipating that visualized mucins would serve as a surrogate marker of endocervical differentiation. Whereas intense staining was detected only in a subset of organoids (~10%), its distribution invariably showed a reciprocal pattern to that of p40 (Figure 2B). Indeed, p40-positive cells were circumferentially lined along cystic structures as in Cx-1, concentrated in the middle as in Cx-2 or exclusively along one half of an organoid, as in Cx-3, whereas PAS-positive cells were all p40-negative (Figure 2B). After careful histological examination of many organoids, we concluded that the staining patterns of p40 and PAS were, indeed, mutually exclusive, with no intermediate cells, such as both positive or both negative, for p40 and PAS staining.

The co-existence of two distinct cell populations in a back-to-back manner within single organoids greatly resembles the configuration of the actual SCJ, which prompted us to ask whether previously described cuboidal SCJ cells [9] could also be identified in the organoids. As KRT7 and AGR2 have been shown to specifically mark cuboidal SCJ cells by immunohistochemistry [9], we examined expression of these two proteins in organoids. In all three cases, a pan-cytokeratin antibody targeting epithelial cells diffusely stained organoids, while KRT7 was only focally detected (Figure 2C). Similarly, AGR2 was also focally detected, albeit to a lesser extent in terms of the stain-positive area (Figure 2C). With regard to Cx-2 organoids, serial sections were subjected to PAS staining and immunostaining, which clearly indicated that SCJ marker-positive cells appeared to coincide with PAS-positive cells (Figure 2B,C). These observations suggest that SCJ cells might be present in organoids, but only as a fraction of PAS-positive cells. Given the absence of markers highly specific to each population, however, endocervical cells and SCJ cells were indistinguishable. There were only a few cells that were positive for Ki-67, but their localization did not seem to be correlated with that of SCJ marker- or p40-positive cells (Figure 2C).

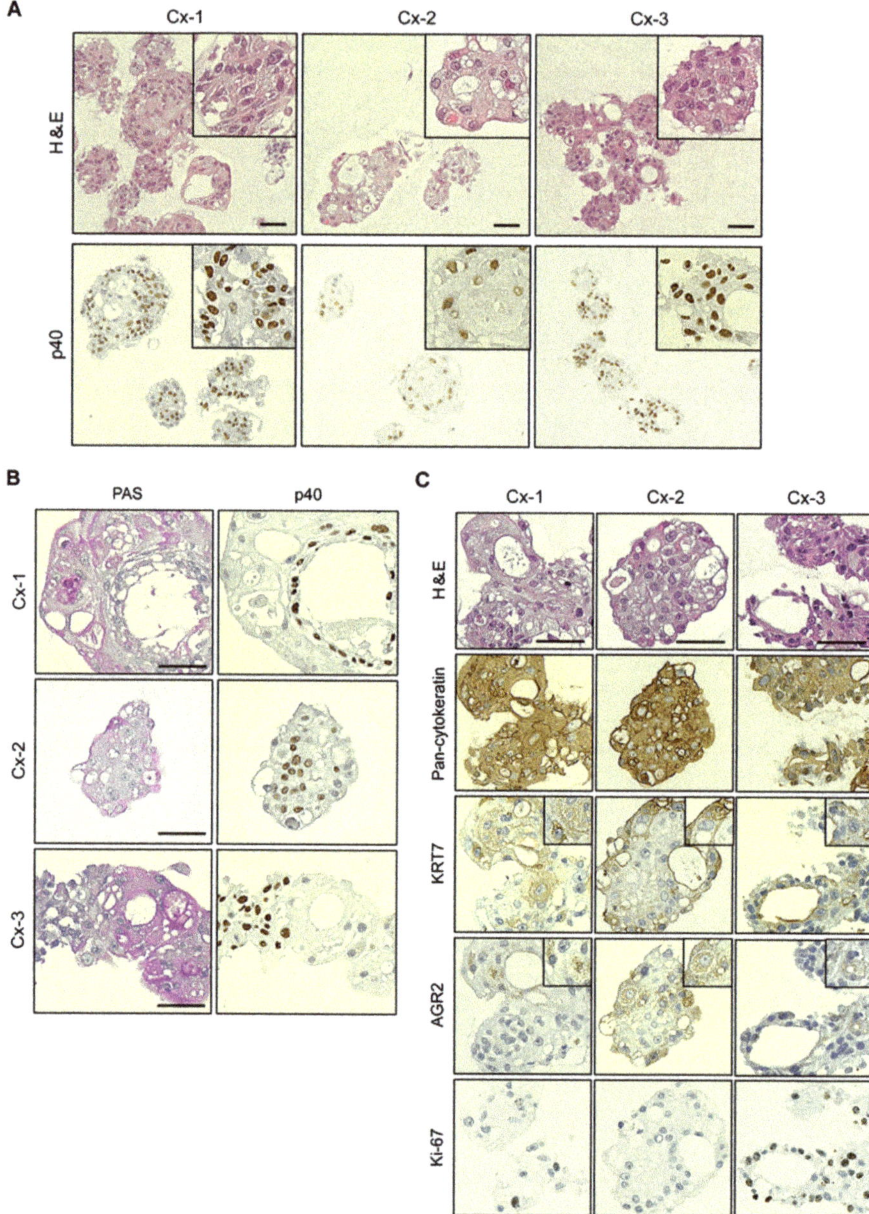

**Figure 2.** Histological characterization of SCJ-derived organoids. (**A**) Histological examination of thin sections. Organoids Cx-1 to Cx-3 were analyzed. Upper panel, hematoxylin and eosin (H&E) staining. Lower panel, immunostaining for the squamous cell marker p40. Insets show magnified images. Scale bars indicate 50 μm. (**B**) Characterization of cell lineages in SCJ-derived organoids. Serial sections of organoids were immunostained for p40 and by Periodic acid-Schiff (PAS) reaction. PAS reaction visualizes mucins produced by endocervix cells. Note that each staining shows a reciprocal pattern. Scale bars indicate 50 μm. (**C**) Expression of SCJ markers in organoids. Serial sections of organoids were histologically analyzed. H&E staining and immunohistochemical staining are shown. Scale bars indicate 50 μm. Insets show magnified images.

### 2.3. Many SCJ Markers Exhibited Higher Levels of Expression in Organoids than in Non-SCJ Cervical Cell Lines

To gain insights into the extent to which the organoids functionally reflect the properties of SCJ cuboidal cells, we next performed microarray analysis of the organoids. Immortalized cervical cell lines Ect1/E6E7 and End1/E6E7 [13] were analyzed as normal references representing ectocervix cells and endocervix cells, respectively. NIKS, undifferentiated keratinocytes [14], were also included as a reference in light of their frequent use in HPV infection experiments. Of 77 genes proposed as SCJ markers [9], 12 genes did not match the corresponding probes in the microarray used in this study (Table S2). Among the remaining 65 genes, 17 genes only had low expression throughout all samples in this study (Table S1). Twenty-eight genes had more than a two-fold expression level in the organoids compared to the three cervical cell lines (Figure 3A). Notably, 13 genes, including MMP7 and AGR2, had more than a 10-fold upregulation in organoids. The expression level was similar and lower in the organoids in 13 and five genes, respectively. GDA, KRT7, and CD63 were not necessarily upregulated compared to non-SCJ cell lines (Figure 3A). These results strongly suggest that the organoids might retain most, if not all, expressions of many SCJ markers, strongly suggesting the superiority of organoids to non-SCJ cervical cell lines as a model of SCJ cells.

For five representative SCJ markers previously well characterized in immunohistochemistry [9], we performed RT-qPCR to validate the microarray data. Whereas we included one primary SCJ tissue sample, all five SCJ markers were expressed in the primary SCJ tissue sample in RT-qPCR analysis, confirming their validity as SCJ marker genes. The expression levels of MMP7 and AGR2 were strikingly higher in all the organoids compared to those in the cell lines (Figure 3B). Although the organoids abundantly expressed KRT7 and GDA, their high-level expression was also observed in End1/E6E7 and Ect1/E6E7, respectively, questioning the specificity of these SCJ markers. Besides, all organoids and cell lines expressed CD63 at similar levels, also negating its specificity for SCJ cells. These results were essentially consistent with the microarray data, suggesting the validity as a transcriptional profile. In addition, these observations also pointed toward the notion that, among 77 genes previously proposed as SCJ markers, a subset of genes, including MMP7 and AGR2, might be SCJ cell markers with the highest specificity and sensitivity that can be expressed, even in an in vitro setting, in a cell-autonomous manner.

Now that we had established that SCJ cells would likely reside in organoids, we asked if reserve cells, putative stem-like cells in the cervix and those implicated in squamous metaplasia [10], could also be detected in organoids. We conducted immunostaining of organoids for reserve cell marker KRT17, to unexpectedly find that it was diffusely expressed in nearly all cells throughout organoids for Cx-1, Cx-2, and Cx-3 (Figure S3A). Moreover, a similar level of KRT17 expression was observed in microarray analysis for all non-SCJ-derived cell lines (Figure S3B). Considering that KRT17 also marks squamous metaplasia and immature types of cells [35], we supposed that the observed high level and ubiquitous expression of KRT17 in vitro might not be informative in spotting reserve cells, but rather might reflect the metaplastic and undifferentiated status of the cell lines and organoids, respectively.

**A**

| Gene | Cell line | | | Organoid | | | Ratio |
|------|-----------|-----|------|----------|------|------|-------|
| | Ect | End | NIKS | Cx-1 | Cx-2 | Cx-3 | (O/C) |
| *MMP7* | 67 | 26 | 18 | 5408 | 4916 | 6423 | 151.38 |
| *LCN2* | 159 | 86 | 34 | 8477 | 15713 | 8738 | 118.33 |
| *TCN1* | 4 | 4 | 28 | 899 | 1161 | 785 | 77.47 |
| *VTCN1* | 1 | 1 | 2 | 42 | 218 | 81 | 71.50 |
| *SPP1* | 3 | 6 | 1 | 181 | 146 | 380 | 70.37 |
| *CFB* | 19 | 26 | 9 | 610 | 2363 | 842 | 69.38 |
| *SLC6A14* | 9 | 4 | 7 | 389 | 584 | 274 | 64.79 |
| *RARRES1* | 7 | 3 | 2 | 155 | 445 | 161 | 61.86 |
| *CXCL5* | 10 | 74 | 6 | 737 | 1919 | 1920 | 51.32 |
| *AGR2* | 19 | 25 | 31 | 1387 | 1322 | 1074 | 50.59 |
| *PROM1* | 1 | 1 | 1 | 13 | 70 | 60 | 37.92 |
| *CFTR* | 21 | 18 | 23 | 111 | 324 | 334 | 12.46 |
| *TMC5* | 4 | 3 | 2 | 20 | 49 | 43 | 11.52 |
| *IL8* | 299 | 896 | 61 | 3409 | 5628 | 3013 | 9.60 |
| *KYNU* | 43 | 509 | 31 | 1874 | 2386 | 1263 | 9.49 |
| *HLA-DRB5* | 3 | 9 | 6 | 24 | 67 | 56 | 8.31 |
| *CFH* | 31 | 63 | 19 | 161 | 206 | 233 | 5.28 |
| *IGFBP3* | 221 | 51 | 28 | 424 | 551 | 435 | 4.70 |
| *LRRN1* | 6 | 5 | 4 | 17 | 31 | 23 | 4.66 |
| *GLRX* | 401 | 185 | 90 | 473 | 1445 | 683 | 3.84 |
| *S100P* | 330 | 602 | 855 | 1388 | 4695 | 694 | 3.79 |
| *SOD2* | 1028 | 660 | 474 | 1704 | 3284 | 2629 | 3.52 |
| *C4orf19* | 76 | 63 | 1 | 55 | 314 | 92 | 3.28 |
| *SAA1* | 1040 | 666 | 618 | 2062 | 2922 | 2407 | 3.18 |
| *SLC26A4* | 14 | 23 | 22 | 34 | 63 | 36 | 2.22 |
| *CLIC4* | 457 | 256 | 653 | 1084 | 917 | 793 | 2.04 |
| *SRD5A3* | 247 | 306 | 210 | 304 | 964 | 282 | 2.03 |
| *TSPAN1* | 713 | 928 | 1053 | 1555 | 1824 | 2026 | 2.01 |
| *RDH10* | 63 | 49 | 90 | 63 | 174 | 49 | 1.41 |
| *FGA* | 30 | 33 | 34 | 41 | 51 | 38 | 1.34 |
| *GDA* | 109 | 10 | 6 | 20 | 73 | 73 | 1.34 |
| *CYP24A1* | 47 | 22 | 4 | 14 | 48 | 34 | 1.32 |
| *STEAP1* | 876 | 419 | 502 | 711 | 643 | 867 | 1.24 |
| *CD63* | 3229 | 6047 | 3541 | 4908 | 4261 | 4128 | 1.04 |
| *PGK1* | 368 | 251 | 334 | 313 | 177 | 222 | 0.75 |
| *CFL2* | 166 | 143 | 175 | 93 | 101 | 128 | 0.66 |
| *LLPH* | 546 | 805 | 542 | 463 | 321 | 391 | 0.62 |
| *KRT7* | 184 | 8197 | 51 | 1480 | 1340 | 2142 | 0.59 |
| *EPS8* | 34 | 66 | 27 | 20 | 22 | 29 | 0.56 |
| *IL6* | 118 | 97 | 1 | 12 | 80 | 29 | 0.56 |
| *RPS27L* | 343 | 398 | 693 | 221 | 318 | 245 | 0.55 |
| *RPL41* | 480 | 648 | 467 | 272 | 245 | 264 | 0.49 |
| *IFI44L* | 33 | 422 | 6 | 38 | 23 | 127 | 0.41 |
| *CAMK2N1* | 642 | 872 | 863 | 341 | 318 | 245 | 0.38 |
| *G0S2* | 2290 | 920 | 4 | 108 | 135 | 122 | 0.11 |
| *IFIT2* | 306 | 736 | 28 | 38 | 26 | 57 | 0.11 |

**B**

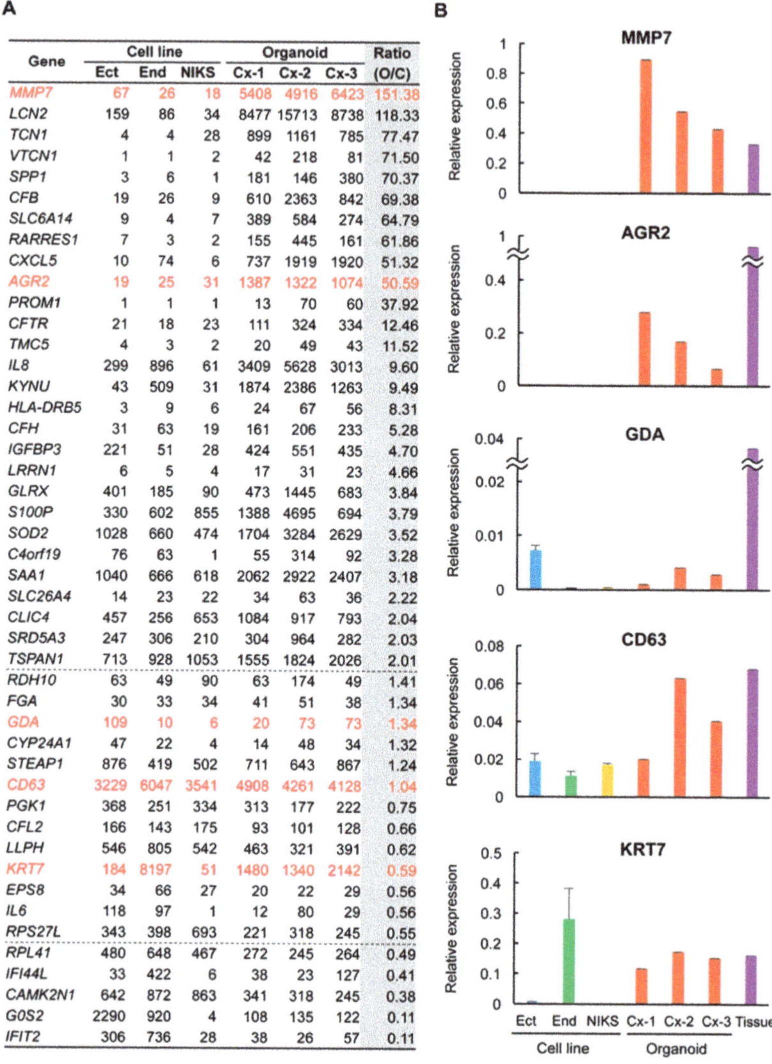


**Figure 3.** Expression profiles of putative SCJ cell markers in SCJ-derived organoids. (**A**) Expression levels of SCJ marker genes. Microarray analysis of non-SCJ-derived immortalized cell lines: End, Ect, normal immortal human keratinocytes (NIKS), three SCJ-derived organoids Cx-1 to Cx-3, and one SCJ tissue sample. Five commonly used SCJ markers are highlighted in red. Ratio of mean signal intensity for cell line and organoids was calculated. Dashed lines indicate two-fold and 0.5-fold changes. (**B**) Validation of microarray data by RT-qPCR. Expression of the five SCJ markers was examined.

## 2.4. Genes Related to Inflammation and Immune Response Were Highly Expressed in Organoids

To further explore common features of organoids in an unbiased manner, we conducted a two-way hierarchical cluster analysis on microarray data. Intriguingly, the organoids displayed highly similar gene expression patterns and segregated as a cluster (Figure 4A), although they were derived from different patients. We focused on three gene clusters, Clusters 1–3, in which gene expression levels were significantly higher in the organoids than in the cell lines. KEGG pathway analysis was conducted based on the genes included in each cluster. Genes related to inflammatory reactions, such as the

IL-17 signaling pathway, TNF signaling pathway, and rheumatoid arthritis pathway, were significantly enriched in Clusters 1 and 2. Among the SCJ cell markers most upregulated in organoids, MMP7, AGR2, LCN2, CXCL5, and CFB were co-segregated in Cluster 1, raising the possibility that transcription of SCJ markers might be commonly regulated as a result of the activation of specific pathways. On the other hand, genes related to ECM-receptor interaction were significantly enriched in Cluster 3 (Figure 4B), which might reflect the presence of Matrigel in the organoid culture.

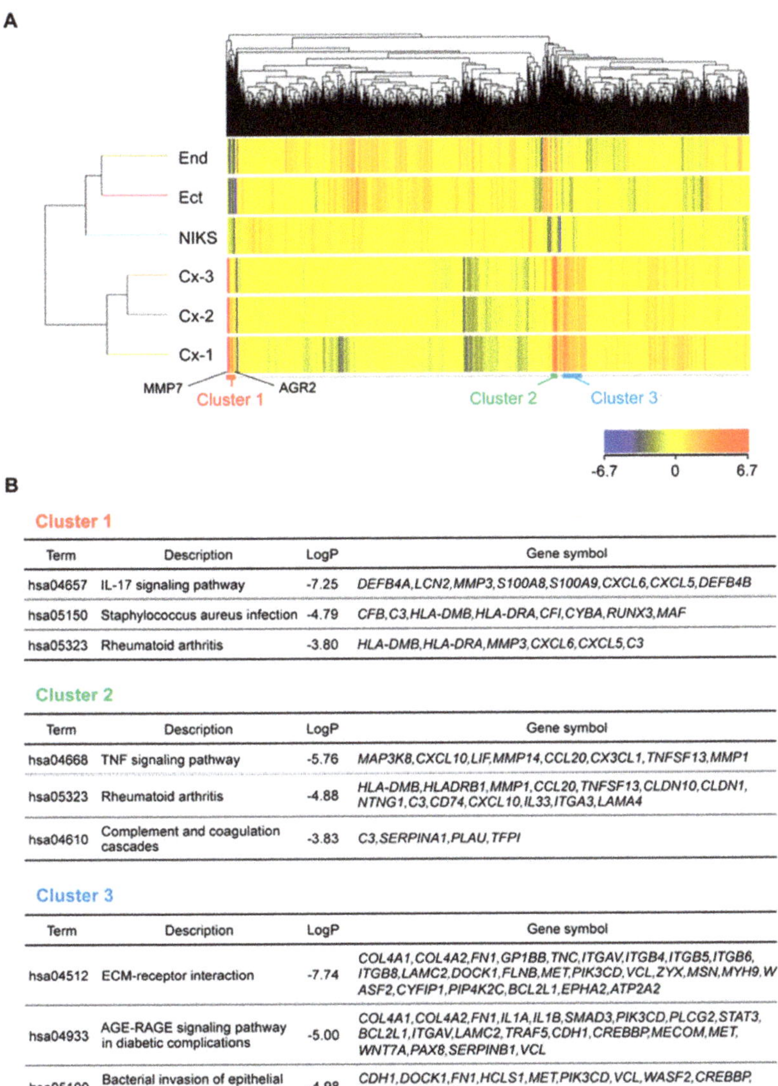


**B**

### Cluster 1

| Term | Description | LogP | Gene symbol |
| --- | --- | --- | --- |
| hsa04657 | IL-17 signaling pathway | -7.25 | DEFB4A,LCN2,MMP3,S100A8,S100A9,CXCL6,CXCL5,DEFB4B |
| hsa05150 | Staphylococcus aureus infection | -4.79 | CFB,C3,HLA-DMB,HLA-DRA,CFI,CYBA,RUNX3,MAF |
| hsa05323 | Rheumatoid arthritis | -3.80 | HLA-DMB,HLA-DRA,MMP3,CXCL6,CXCL5,C3 |

### Cluster 2

| Term | Description | LogP | Gene symbol |
| --- | --- | --- | --- |
| hsa04668 | TNF signaling pathway | -5.76 | MAP3K8,CXCL10,LIF,MMP14,CCL20,CX3CL1,TNFSF13,MMP1 |
| hsa05323 | Rheumatoid arthritis | -4.88 | HLA-DMB,HLADRB1,MMP1,CCL20,TNFSF13,CLDN10,CLDN1,NTNG1,C3,CD74,CXCL10,IL33,ITGA3,LAMA4 |
| hsa04610 | Complement and coagulation cascades | -3.83 | C3,SERPINA1,PLAU,TFPI |

### Cluster 3

| Term | Description | LogP | Gene symbol |
| --- | --- | --- | --- |
| hsa04512 | ECM-receptor interaction | -7.74 | COL4A1,COL4A2,FN1,GP1BB,TNC,ITGAV,ITGB4,ITGB5,ITGB6,ITGB8,LAMC2,DOCK1,FLNB,MET,PIK3CD,VCL,ZYX,MSN,MYH9,WASF2,CYFIP1,PIP4K2C,BCL2L1,EPHA2,ATP2A2 |
| hsa04933 | AGE-RAGE signaling pathway in diabetic complications | -5.00 | COL4A1,COL4A2,FN1,IL1A,IL1B,SMAD3,PIK3CD,PLCG2,STAT3,BCL2L1,ITGAV,LAMC2,TRAF5,CDH1,CREBBP,MECOM,MET,WNT7A,PAX8,SERPINB1,VCL |
| hsa05100 | Bacterial invasion of epithelial cells | -4.98 | CDH1,DOCK1,FN1,HCLS1,MET,PIK3CD,VCL,WASF2,CREBBP,SMAD3,FBXW11 |

**Figure 4.** Transcriptome analysis of SCJ-derived organoids. (**A**) Heat map for two-way clustering analysis. SCJ-derived immortalized cell lines: End, Ect, and NIKS, and three SCJ-derived organoids Cx-1 to Cx-3 were analyzed. Note that MMP7 and AGR2 were included in Cluster 1. (**B**) Clusters of specifically upregulated genes in SCJ-derived organoids.

## 3. Discussion

No SCJ-derived cells in culture have been documented to date, despite their relevance in HPV infection to the cervix and subsequent carcinogenesis. In this study, we demonstrated that normal SCJ samples from patients could give rise to organoids that are robustly propagated and exhibit many features of cuboidal SCJ cells. Intriguingly, they specifically expressed many SCJ markers and comprised dual cell lineages, with many squamous cells resembling TZ. Moreover, unlike widely used cervical cell lines, long-term culture was feasible in this study, even without immortalization by introduction of HPV-derived E6 and E7 oncogenes. These organoids are therefore likely the first in vitro model for normal SCJ cells under physiological conditions. Whereas 77 genes were originally reported to be specifically upregulated in SCJ cells [9], we confirmed through transcriptome analysis that only a subset of genes in fact showed higher levels of expression in organoids compared to those in non-SCJ cell lines. Notably, of five representative SCJ markers in immunohistochemistry, only two (MMP7 and AGR2) were specifically upregulated in organoids. We, therefore, assume that these 77 genes need to be curated by thorough comparison with genes specifically upregulated in SCJ-derived organoids, as revealed in this study. In microarray and qPCR analyses, KRT7 and GDA were highly expressed in End1 and Ect1, respectively. These observations suggest that KRT7 and GDA might be broadly expressed in cells with endocervical and ectocervical differentiation, respectively, and that these immortalized non-SCJ-derived cell lines might likely retain the features of their original tissues. In this regard, SCJ markers superior to the five molecules currently used might well be identified based on this study.

We found that the obtained organoids were mostly solid and showed a significant bias toward squamous differentiation. Considering that the SCJ region harbors reserve cells that drive squamous metaplasia, we initially assumed that reserve cells would be present in SCJ-derived organoids and yield metaplastic squamous cells, leading to the reconstitution of a TZ equivalent in organoids. Naturally, we aimed to detect reserve cells in organoids by immunostaining for a reserve cell marker KRT17. However, nearly all the cells in organoids were positively stained for KRT17, making it impossible to pinpoint reserve cells. As KRT17 is also highly expressed in squamous metaplasia, these observations might rather reflect the predominant presence of metaplastic cells, mimicking the TZ in organoids. Future development of reserve cell-specific markers will clarify KRT17's localization in organoids. Another feature of the SCJ region is that it accommodates two distinct cell populations with differentiation into ectocervix/squamous cells and endocervix/columnar cells. We adopted PAS to readily identify mucin-secreting columnar cells in organoids with mostly squamous cells. Whereas PAS staining is generally used for detection of glycogen in the surface and intermediate layer of squamous epithelium, or mucin in the glandular epithelium, PAS-positive cells in this study were confined to a p40-negative non-squamous cell population within organoids. Hence, it is likely that PAS-positive cells in the organoid would represent an endocervix cell-like population. Since PAS-positive cells largely overlapped with a population positively stained for SCJ markers, SCJ cells in organoids might be present as PAS-positive cells, which could, in turn, give rise to reserve-like cells. Given the lack of highly specific markers, it is currently difficult to distinguish between endocervix-like cells and SCJ cells solely by histological approaches.

The co-existence of a dual cell population of distinct lineages in a single organoid strongly suggests that SCJ cells gave rise to both populations. However, there is a possibility that a similar observation can be alternatively achieved. One is that squamous cells and endocervix cells have their own progenitor cells, which can propagate on their own, but which coincidentally aggregate during subculture to form single organoids. Given that PAS-positive cells were detected only in a subset of organoids, the presence of organoids genuinely consisting of squamous cells cannot be completely ruled out. Nonetheless, since PAS was only focally stained in any PAS-positive organoids, we speculate that the seeming lack of PAS-positive cells across organoids in many cases might be simply attributable to examination on cross-sections. Whole organoid-basis analysis, such as flow cytometry, might be able to address this issue, given the future development of highly specific markers that distinguish endocervix cells from SCJ cells. We also assume that the presence of endocervix cell-only organoids would be unlikely, because there were no such organoids in any cross-section. The other possibility is that differentiated squamous cells and endocervix cells derived from the original SCJ cells coincidentally attached to each other to form hybrid organoids as we observed. We presume that this is also unlikely, because we learned that differentiated squamous cells in the clinical specimens were extremely difficult to proliferate in organoid culture conditions. Collectively, it is likely that the new SCJ, along with the TZ, was functionally reconstituted in organoids, although it remains to be confirmed by further investigations, such as single cell transcriptome analysis. It should be also noted that, as a limitation of this study, SCJ-derived organoids were characterized only in terms of expression markers and histological features. Functional evaluation of organoids and optimization of culture conditions are to be pursued, in order to further solidify the authenticity of the organoids. By making these efforts, the mechanisms of how estrogens could induce development of TZ, for example, might be clarified in a future study.

Previous HPV infection studies have mainly utilized NIKS as a host and provided enormous knowledge on HPV-dependent transformation of keratinocytes and, thereby, SCC development. Although the implications from these studies are still valid in terms of the relationship between HPV and keratinocytes, it is unclear whether the results could be extrapolated to SCJ-originating UCC development, as it is highly vulnerable to HPV infection and subsequent carcinogenesis, unlike keratinocytes. Now that we have established SCJ-derived organoids, investigations on cervical carcinogenesis by using more common and susceptible cells of origin would pave the way to gaining more detailed insights into HPV-driven UCC development. It is also tempting to speculate that high expression level of genes related to inflammation immune response might underlie the high affinity of HPV to SCJ cells. Synchronous transcriptional regulation across organoids from different patients suggests that there might be a common upstream regulator for SCJ cells, which could also serve as a specific stem cell marker. By using these organoids, exploration of such factors will be warranted.

## 4. Materials and Methods

### 4.1. Patient Information

Normal cervical tissues were obtained from four patients who underwent hysterectomy in the University of Tokyo Hospital. None of the four patients had any malignant findings on the cervix. They had regular menstrual cycles and no episode of hormonal treatment for at least 6 months before hysterectomy. Menstrual cycle was determined by histological examination. HPV status was determined by genotyping of organoids. The Ethics Committee of The University of Tokyo Hospital (approval number 12017) and Chiba Cancer Center (approval number H30-216) approved all experimental procedures in this study. Written informed consent was obtained from all the patients.

### 4.2. HPV Genotyping

HPV genotyping assays were performed by PCR with PGMY primers followed by reverse line blot hybridization, as previously described [36]. This assay can detect 31 HPV genotypes, including HPV 6, 11, 16, 18, 26, 31, 33, 34, 35, 39, 40, 42, 44, 45, 51, 52, 53, 54, 55, 56, 57, 58, 59, 66, 68, 69, 70, 73, 82, 83 and 84.

### 4.3. Isolation and Organoid Culture of SCJ Cells

A median section in the anterior wall of the resected uterus, from the endocervical canal to the fundus, was obtained immediately after surgical resection. Through careful observation of the surface texture of the uterine mucosa, the areas for columnar and squamous epithelium were roughly estimated. The SCJ region was postulated to be located around their borderline zone. SCJ samples were obtained with scissors, from almost the entire area around the postulated SCJ region, approximately 2 cm width and 2–3 mm depth, except for the median part of the posterior wall, so as not to interfere with pathological diagnosis. Such samples were collected in a hospital immediately after surgery and preserved overnight at 4 °C in Advanced DMEM/F12 (Thermo Fisher Scientific, Waltham, MA, USA). The next morning, SCJ samples were further dissociated into cell aggregates or single cells by enzymatic digestion with 2 μ/mL dispase II, 1 mg/mL collagenase P (Roche Diagnostics K.K., Tokyo, Japan) and Accumax (Innovative Cell Technologies, San Diego, CA, USA). Primary organoid culture was conducted according to the modified MBOC protocol, as previously described [30]. Briefly, resuspended cells were plated on solidified Matrigel (BD Biosciences, Franklin Lakes, NJ, USA). The following morning, viable cells attached onto Matrigel were covered with Matrigel and overlaid with media to start the organoid culture. Organoid culture media was advanced DMEM/F12 (Thermo Fisher Scientific) supplemented with 50 ng/mL human EGF (Peprotech, Rocky Hill, NJ, USA), 250 ng/mL R-spondin1 (R&D, Minneapolis, MN, USA), 100 ng/mL Noggin (Peprotech), 10 μM Y27632 (Wako, Osaka, Japan), 1 μM Jagged-1 (AnaSpec, Fremont, CA, USA), L-glutamine solution (Wako), penicillin/streptomycin (Sigma-Aldrich, St. Louis, MO, USA), and amphotericin B suspension (Wako). pCDH-CMV-MCS-EF1-copGFP (System Biosciences, Mountain View, CA, USA) was introduced into organoids as a green fluorescent protein (GFP)-expressing vector, as previously described [24].

### 4.4. Pathological Analysis

Following de-polymerization of Matrigel with Cell Recovery Solution (BD Biosciences, Franklin Lakes, NJ, USA), organoids were collected at day 14 (P2) for Cx-1, d21 (P2) for Cx-2, and d10 (P1) for Cx-3, followed by resuspension in iPGell (GenoStaff, Tokyo, Japan). The iPGell-embedded organoids were fixed in 10–15% buffered neutral formalin, dehydrated and embedded in paraffin. FFPE samples were sectioned at 3 μm thickness and stained with hematoxylin and eosin (H&E). Periodic acid-Schiff (PAS) reaction was conducted to visualize mucin production. Dako Autostainer Link48 (Agilent, Santa Clara, CA, USA) was used for automatic immunohistochemical (IHC) staining with the following primary antibodies; KRT7 (clone OV-TL 12/30, Thermo Fisher, 1:100), AGR2 (clone D9V2F, Cell Signaling Technology, 1:800), pan-cytokeratin (clone AE1/AE3, Abcam, 1:40), Ki-67 (clone MIB-1, Dako, ready-to-use). The following primary antibodies were used for manual staining: p40 (clone BC28, Abcam, 1:40), Cytokeratin 17 (clone E-4, Santa Cruz, 1:250). The reactions were visualized with the Dako REAL EnVision Detection System (DAKO, Glostrup, Denmark) using diaminobenzidine chromogen as the substrate. For cytology, organoids were immediately fixed in 95% ethyl alcohol and subjected to Papanicolaou staining.

### 4.5. Cell Lines

Ect1/E6E7 (ATCC CRL-2614$^{TM}$, Manassas, VA, USA) and End1/E6E7 (ATCC CRL-2615$^{TM}$) were maintained as previously described [13]. Normal immortal human keratinocytes (NIKS, ATCC CRL-12191$^{TM}$) were cultured in F medium in the presence of mitomycin C-treated NIH3T3 cells, as previously described [14].

### 4.6. Microarray Analysis

Total RNA of cell lines and organoids was extracted using an RNeasy Mini Kit (QIAGEN, Hilden, Germany), at day 14 (P2) for Cx-1, d25 (P3) for Cx-2, and d10 (P1) for Cx-3. These RNA samples were subjected to the TORAY 3D-gene analysis service using the 3D-Gene Human Oligo chip 25K (TORAY, Tokyo, Japan). Total RNA was amplified and labeled with Cy5, then hybridized with a 3D-Gene chip. Signals were detected on a 3D-Gene scanner (TORAY) and normalized according to a global normalization method in which the median value of the detected signal intensities was adjusted to 25. With regard to SCJ markers, genes without corresponding probes and with an average signal intensity lower than 20 are listed in Table S2. Differentially expressed genes between three cell lines and organoids were extracted from the microarray data. Genes with normalized signal intensities less than 20 in more than three samples were excluded from the analysis. GeneSpring GX (Ver.14.9.1, Agilent Technologies, Santa Clara, CA, USA) was used to generate a heat map. Metascape [37] was used to identify enriched KEGG pathways of the focused clusters. A $p$-value $< 0.05$ was considered statistically significant. Microarray data were deposited to GEO accession: GSE138554.

*4.7. RT-qPCR*

Extracted RNA was reverse transcribed using ReverTra Ace® qPCR RT Master Mix (TOYOBO, Osaka, Japan) according to the manufacturer's instructions. To assess mRNA expression of GAPDH, KRT7, AGR2, MMP7, CD63 and GDA, qRT–PCR was performed using a Light Cycler 480 (Roche Diagnostics, Mannheim, Germany). Relative expression to GAPDH is shown. PCR reactions were conducted in triplicate and means ± SD are shown. Primer sequences and annealing temperatures are listed in Table S2.

## 5. Conclusions

In conclusion, normal cervical organoids were established, which exhibited features of SCJ cuboidal cells and TZ metaplastic cells. To the best of our knowledge, this is the first demonstration of stably propagating primary SCJ cells. As an in vitro model for cell of origin in UCC development, these organoids would likely become a useful resource relevant for the elucidation of multi-step processes of HPV-dependent cervical carcinogenesis.

**Supplementary Materials:** The following are available online at http://www.mdpi.com/2072-6694/12/3/694/s1, Figure S1: A failed case of organoid culture of biopsy samples, Figure S2: Highly efficient gene transduction of SCJ-derived organoids, Figure S3: KRT17 expression in organoids, Table S1: SCJ marker genes whose expression was not verified in organoids, Table S2: Primers for RT-qPCR.

**Author Contributions:** Y.M. and A.K. conducted the experiments. Y.I. and I.K. performed a detection test for HPV. S.B. analyzed the microarray data. M.M. collected clinical samples. T.N., K.O., Y.O., and T.F. assisted in experimental design and data interpretation. A.T. and Y.H. designed the study and wrote the manuscript. All authors have read and agreed to the published version of the manuscript.

**Funding:** This work was supported by Grants-in-Aid for Japanese Initiative for Progress of Research on Infectious Disease for global Epidemic (J-PRIDE) from the Japan Agency for Medical Research and Development (AMED). (Grant Number: 19fm0208013h0003 and 19fm0208013h0203).

**Acknowledgments:** The authors are grateful to Kei Kawana for helpful discussion and mentoring.

**Conflicts of Interest:** The authors declare no conflict of interest.

## References

1.  Reich, O.; Regauer, S.; McCluggage, W.G.; Bergeron, C.; Redman, C. Defining the Cervical Transformation Zone and Squamocolumnar Junction: Can We Reach a Common Colposcopic and Histologic Definition? *Int. J. Gynecol. Pathol.* **2017**, *36*, 517–522. [CrossRef] [PubMed]

2.  Mukonoweshuro, P.; Oriowolo, A.; Smith, M. Audit of the histological definition of cervical transformation zone. *J. Clin. Pathol.* **2005**, *58*, 671. [PubMed]

3.  Delvenne, P.; Herman, L.; Kholod, N.; Caberg, J.H.; Herfs, M.; Boniver, J.; Jacobs, N.; Hubert, P. Role of hormone cofactors in the human papillomavirus-induced carcinogenesis of the uterine cervix. *Mol. Cell Endocrinol.* **2007**, *264*, 1–5. [CrossRef] [PubMed]

4.  Marsh, M. Original site of cervical carcinoma; topographical relationship of carcinoma of the cervix to the external os and to the squamocolumnar junction. *Obstet. Gynecol.* **1956**, *7*, 444–452. [CrossRef]

5.  Richart, R.M. Cervical intraepithelial neoplasia. *Pathol. Annu.* **1973**, *8*, 301–328.

6.  Li, N.; Franceschi, S.; Howell-Jones, R.; Snijders, P.J.; Clifford, G.M. Human papillomavirus type distribution in 30,848 invasive cervical cancers worldwide: Variation by geographical region, histological type and year of publication. *Int. J. Cancer* **2011**, *128*, 927–935. [CrossRef]

7.  Kreimer, A.R.; Pierce Campbell, C.M.; Lin, H.Y.; Fulp, W.; Papenfuss, M.R.; Abrahamsen, M.; Hildesheim, A.; Villa, L.L.; Salmeron, J.J.; Lazcano-Ponce, E.; et al. Incidence and clearance of oral human papillomavirus infection in men: The HIM cohort study. *Lancet* **2013**, *382*, 877–887. [CrossRef]

8.  Wang, X.; Ouyang, H.; Yamamoto, Y.; Kumar, P.A.; Wei, T.S.; Dagher, R.; Vincent, M.; Lu, X.; Bellizzi, A.M.; Ho, K.Y.; et al. Residual embryonic cells as precursors of a Barrett's-like metaplasia. *Cell* **2011**, *145*, 1023–1035. [CrossRef]

9.  Herfs, M.; Yamamoto, Y.; Laury, A.; Wang, X.; Nucci, M.R.; McLaughlin-Drubin, M.E.; Munger, K.; Feldman, S.; McKeon, F.D.; Xian, W.; et al. A discrete population of squamocolumnar junction cells implicated in the pathogenesis of cervical cancer. *Proc. Natl. Acad. Sci. USA* **2012**, *109*, 10516–10521. [CrossRef]

10. Herfs, M.; Vargas, S.O.; Yamamoto, Y.; Howitt, B.E.; Nucci, M.R.; Hornick, J.L.; McKeon, F.D.; Xian, W.; Crum, C.P. A novel blueprint for 'top down' differentiation defines the cervical squamocolumnar junction during development, reproductive life, and neoplasia. *J. Pathol.* **2013**, *229*, 460–468. [CrossRef]

11. Mirkovic, J.; Howitt, B.E.; Roncarati, P.; Demoulin, S.; Suarez-Carmona, M.; Hubert, P.; McKeon, F.D.; Xian, W.; Li, A.; Delvenne, P.; et al. Carcinogenic HPV infection in the cervical squamo-columnar junction. *J. Pathol.* **2015**, *236*, 265–271. [CrossRef] [PubMed]

12. Doorbar, J.; Griffin, H. Refining our understanding of cervical neoplasia and its cellular origins. *Papillomavirus Res.* **2019**, *7*, 176–179. [CrossRef] [PubMed]

13. Fichorova, R.N.; Rheinwald, J.G.; Anderson, D.J. Generation of papillomavirus-immortalized cell lines from normal human ectocervical, endocervical, and vaginal epithelium that maintain expression of tissue-specific differentiation proteins. *Biol. Reprod.* **1997**, *57*, 847–855. [CrossRef] [PubMed]

14. Allen-Hoffmann, B.L.; Schlosser, S.J.; Ivarie, C.A.; Sattler, C.A.; Meisner, L.F.; O'Connor, S.L. Normal growth and differentiation in a spontaneously immortalized near-diploid human keratinocyte cell line, NIKS. *J. Investig. Dermatol.* **2000**, *114*, 444–455. [CrossRef]

15. Nakahara, T.; Peh, W.L.; Doorbar, J.; Lee, D.; Lambert, P.F. Human papillomavirus type 16 E1circumflexE4 contributes to multiple facets of the papillomavirus life cycle. *J. Virol.* **2005**, *79*, 13150–13165. [CrossRef]

16. Sato, T.; Vries, R.G.; Snippert, H.J.; van de Wetering, M.; Barker, N.; Stange, D.E.; van Es, J.H.; Abo, A.; Kujala, P.; Peters, P.J.; et al. Single Lgr5 stem cells build crypt-villus structures in vitro without a mesenchymal niche. *Nature* **2009**, *459*, 262–265. [CrossRef]

17. Bartfeld, S.; Bayram, T.; van de Wetering, M.; Huch, M.; Begthel, H.; Kujala, P.; Vries, R.; Peters, P.J.; Clevers, H. In vitro expansion of human gastric epithelial stem cells and their responses to bacterial infection. *Gastroenterology* **2015**, *148*, 126–136. [CrossRef]

18. Chen, Y.W.; Huang, S.X.; de Carvalho, A.; Ho, S.H.; Islam, M.N.; Volpi, S.; Notarangelo, L.D.; Ciancanelli, M.; Casanova, J.L.; Bhattacharya, J.; et al. A three-dimensional model of human lung development and disease from pluripotent stem cells. *Nat. Cell Biol.* **2017**, *19*, 542–549. [CrossRef]

19. Schumacher, M.A.; Aihara, E.; Feng, R.; Engevik, A.; Shroyer, N.F.; Ottemann, K.M.; Worrell, R.T.; Montrose, M.H.; Shivdasani, R.A.; Zavros, Y. The use of murine-derived fundic organoids in studies of gastric physiology. *J. Physiol.* **2015**, *593*, 1809–1827. [CrossRef]

20. Onuma, K.; Ochiai, M.; Orihashi, K.; Takahashi, M.; Imai, T.; Nakagama, H.; Hippo, Y. Genetic reconstitution of tumorigenesis in primary intestinal cells. *Proc. Natl. Acad. Sci. USA* **2013**, *110*, 11127–11132. [CrossRef]

21. Sato, T.; Morita, M.; Tanaka, R.; Inoue, Y.; Nomura, M.; Sakamoto, Y.; Miura, K.; Ito, S.; Sato, I.; Tanaka, N.; et al. Ex vivo model of non-small cell lung cancer using mouse lung epithelial cells. *Oncol. Lett.* **2017**, *14*, 6863–6868. [CrossRef] [PubMed]

22. Ochiai, M.; Yoshihara, Y.; Maru, Y.; Tetsuya, M.; Izumiya, M.; Imai, T.; Hippo, Y. Kras-driven heterotopic tumor development from hepatobiliary organoids. *Carcinogenesis* **2019**. [CrossRef] [PubMed]

23. Matsuura, T.; Maru, Y.; Izumiya, M.; Hoshi, D.; Kato, S.; Ochiai, M.; Hori, M.; Yamamoto, S.; Tatsuno, K.; Imai, T.; et al. Organoid-based ex vivo reconstitution of Kras-driven pancreatic ductal carcinogenesis. *Carcinogenesis* **2019**. [CrossRef] [PubMed]

24. Maru, Y.; Orihashi, K.; Hippo, Y. Lentivirus-Based Stable Gene Delivery into Intestinal Organoids. *Methods Mol. Biol.* **2016**, *1422*, 13–21. [CrossRef] [PubMed]

25. Naruse, M.; Masui, R.; Ochiai, M.; Maru, Y.; Hippo, Y.; Imai, T. An organoid-based carcinogenesis model induced by in vitro chemical treatment. *Carcinogenesis* **2020**. [CrossRef] [PubMed]

26. Boj, S.F.; Hwang, C.I.; Baker, L.A.; Chio, I.I.C.; Engle, D.D.; Corbo, V.; Jager, M.; Ponz-Sarvise, M.; Tiriac, H.; Spector, M.S.; et al. Organoid models of human and mouse ductal pancreatic cancer. *Cell* **2015**, *160*, 324–338. [CrossRef] [PubMed]

27. Li, X.; Francies, H.E.; Secrier, M.; Perner, J.; Miremadi, A.; Galeano-Dalmau, N.; Barendt, W.J.; Letchford, L.; Leyden, G.M.; Goffin, E.K.; et al. Organoid cultures recapitulate esophageal adenocarcinoma heterogeneity providing a model for clonality studies and precision therapeutics. *Nat. Commun.* **2018**, *9*, 2983. [CrossRef]
28. Yan, H.H.N.; Siu, H.C.; Law, S.; Ho, S.L.; Yue, S.S.K.; Tsui, W.Y.; Chan, D.; Chan, A.S.; Ma, S.; Lam, K.O.; et al. A Comprehensive Human Gastric Cancer Organoid Biobank Captures Tumor Subtype Heterogeneity and Enables Therapeutic Screening. *Cell Stem Cell* **2018**, *23*, 882–897. [CrossRef]
29. Maru, Y.; Hippo, Y. Current Status of Patient-Derived Ovarian Cancer Models. *Cells* **2019**, *8*, 505. [CrossRef]
30. Maru, Y.; Tanaka, N.; Itami, M.; Hippo, Y. Efficient use of patient-derived organoids as a preclinical model for gynecologic tumors. *Gynecol. Oncol.* **2019**, *154*, 189–198. [CrossRef]
31. Maru, Y.; Onuma, K.; Ochiai, M.; Imai, T.; Hippo, Y. Shortcuts to intestinal carcinogenesis by genetic engineering in organoids. *Cancer Sci.* **2019**, *110*, 858–866. [CrossRef] [PubMed]
32. Maru, Y.; Tanaka, N.; Ebisawa, K.; Odaka, A.; Sugiyama, T.; Itami, M.; Hippo, Y. Establishment and characterization of patient-derived organoids from a young patient with cervical clear cell carcinoma. *Cancer Sci.* **2019**, *110*, 2992–3005. [CrossRef] [PubMed]
33. Sato, T.; Stange, D.E.; Ferrante, M.; Vries, R.G.; Van Es, J.H.; Van den Brink, S.; Van Houdt, W.J.; Pronk, A.; Van Gorp, J.; Siersema, P.D.; et al. Long-term expansion of epithelial organoids from human colon, adenoma, adenocarcinoma, and Barrett's epithelium. *Gastroenterology* **2011**, *141*, 1762–1772. [CrossRef] [PubMed]
34. Whelan, K.A.; Muir, A.B.; Nakagawa, H. Esophageal 3D Culture Systems as Modeling Tools in Esophageal Epithelial Pathobiology and Personalized Medicine. *Cell Mol. Gastroenterol. Hepatol.* **2018**, *5*, 461–478. [CrossRef]
35. Escobar-Hoyos, L.F.; Yang, J.; Zhu, J.; Cavallo, J.A.; Zhai, H.; Burke, S.; Koller, A.; Chen, E.I.; Shroyer, K.R. Keratin 17 in premalignant and malignant squamous lesions of the cervix: Proteomic discovery and immunohistochemical validation as a diagnostic and prognostic biomarker. *Mod. Pathol.* **2014**, *27*, 621–630. [CrossRef]
36. Azuma, Y.; Kusumoto-Matsuo, R.; Takeuchi, F.; Uenoyama, A.; Kondo, K.; Tsunoda, H.; Nagasaka, K.; Kawana, K.; Morisada, T.; Iwata, T.; et al. Human papillomavirus genotype distribution in cervical intraepithelial neoplasia grade 2/3 and invasive cervical cancer in Japanese women. *Jpn. J. Clin. Oncol.* **2014**, *44*, 910–917. [CrossRef]
37. Zhou, Y.; Zhou, B.; Pache, L.; Chang, M.; Khodabakhshi, A.H.; Tanaseichuk, O.; Benner, C.; Chanda, S.K. Metascape provides a biologist-oriented resource for the analysis of systems-level datasets. *Nat. Commun.* **2019**, *10*, e1523. [CrossRef]

*Review*

# The Interplay between Antiviral Signalling and Carcinogenesis in Human Papillomavirus Infections

Ana Rita Ferreira [†], Ana Catarina Ramalho [†], Mariana Marques and Daniela Ribeiro *

Institute of Biomedicine—iBiMED & Department of Medical Sciences, University of Aveiro, 3810-198 Aveiro, Portugal; arferreira@ua.pt (A.R.F.); ana.catarina.ramalho@ua.pt (A.C.R.); mar.marques@ua.pt (M.M.)
* Correspondence: daniela.ribeiro@ua.pt; Tel.: +351-234-247 014; Fax: +351-234-372-587
† These authors contributed equally to this paper.

Received: 14 February 2020; Accepted: 6 March 2020; Published: 10 March 2020

**Abstract:** Human papillomaviruses (HPV) are the causative agents of the most common sexually transmitted infection worldwide. While infection is generally asymptomatic and can be cleared by the host immune system, when persistence occurs, HPV can become a risk factor for malignant transformation. Progression to cancer is actually an unintended consequence of the complex HPV life cycle. Different antiviral defence mechanisms recognize HPV early in infection, leading to the activation of the innate immune response. However, the virus has evolved several specific strategies to efficiently evade the antiviral immune signalling. Here, we review and discuss the interplay between HPV and the host cell innate immunity. We further highlight the evasion strategies developed by different HPV to escape this cellular response and focus on the correlation with HPV-induced persistence and tumorigenesis.

**Keywords:** human papillomavirus; innate immunity; cancer; intracellular antiviral response; immune evasion

---

## 1. Introduction

Human papillomaviruses (HPV) are the main causative agents of cervical cancer and represent the most common sexually transmitted infection worldwide [1,2]. To this date, over 200 HPV types have already been identified [2]. HPV infections are usually asymptomatic and cleared by the immune system within 12 months [3]. However, HPV-infected immunocompromised individuals are susceptible to the development of HPV-associated carcinomas and, depending on the HPV type, infection can be a major risk factor for malignant progression [1]. High-risk HPV (HR-HPV) types are associated with the development of several other carcinomas, such as anal, vulvovaginal and penile, head and neck cancers [1,3,4].

The innate immune system is an early defence mechanism triggered upon detection of pathogens, such as viruses [5]. The efficient activation of the immune response is the key between viral clearance and viral persistence. Upon infection, the recognition of essential viral components, named pathogen-associated molecular patterns (PAMPs), by the cellular pattern-recognition receptors (PRRs) leads to the activation of the innate immune response and, ultimately, the adaptive immune response [6].

In general, cellular PRRs can detect either viral RNA or DNA, and they can be either associated with membranes or localize freely in the cytosol [7,8]. These different classes of PRRs use common pathways to convey their signals, ultimately culminating in the expression of pro-inflammatory cytokines, such as type I interferons (IFNs), and IFN-stimulated genes (ISGs), restricting infection establishment and spreading [9]. This is accomplished by triggering the activation of downstream signalling pathways, namely the IFN-regulatory factors (IRFs) pathways, the janus kinase/signal transducers and activators of transcription (JAK-STAT) pathway and the nuclear factor-κB (NF-κB) signalling pathway [6,10].

Understanding the biology of HPV infection was only possible in the last few decades with the advent of molecular cloning and the development of organotypic cultures, allowing not only the study of individual viral genes but also the analysis of viral infections and their progression. Nevertheless, there is still a gap in the knowledge concerning the interplay between innate immune evasion and cancer progression during HPV infection. Here, we review and clarify these different evasion mechanisms and discuss their correlation with cancer progression during infection.

## 2. Human Papillomavirus Biology

HPV can infect cutaneous epithelial cells or mucosal tissues and, depending on their tropism, HPV are categorized either as cutaneous or mucosal. Additionally, HPV can be further divided into two categories: Low-risk HPV (LR-HPV), which cause benign warts, and HR-HPV with oncogenic potential (Table 1) [1,2,4,11].

**Table 1.** Human papillomavirus (HPV) types and associated lesions. Represented in bold are the most frequent high-risk HPV. HPV types that are not yet fully established as high-risk are represented between brackets.

| Group | Type of Lesions | HPV Types |
| --- | --- | --- |
| High-risk | Intraepithelial neoplasia and cervical cancer | **16, 18**, 31, 33, 35, 39, 45, 51, 52, 56, 68, 73, 82 (26, 53, 66) |
| Low-risk | Intraepithelial neoplasia or genital warts | 6, 11, 40, 12, 43, 44, 53, 54, 61, 72, 73, 81 |

HPV particles are composed by a non-enveloped icosahedral capsid that shields the viral genome. The HPV genome consists of a single circular double-stranded DNA (dsDNA) molecule associated with host-derived histones, and, typically, encodes seven to eight open reading frames (ORFs) (Figure 1).

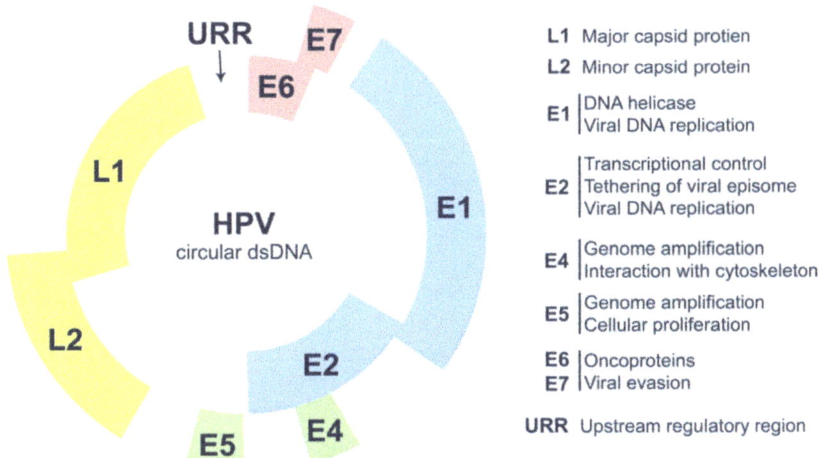


**Figure 1.** HPV genome organization and functions of the main viral proteins.

The genome's main structural and organizational features are shared, with some variations, among different HPV types [12,13]. The genome is divided into three functional regions: Early (E), late (L) and the upstream regulatory region (URR) (Figure 1). The E region is composed of six ORFs that codify the proteins E1, E2, E6, E7, E4 and E5. Additionally, it has been reported that some HPV can also express E3, E8 and an 'E5 like protein' also known as E10 [13,14]. The L region encodes for the structural proteins L1 and L2. The URR (previously known as long control region or LCR) corresponds to a non-coding segment containing the cis elements essential for viral replication and transcription by the host machinery [2,13]. E1, E2, L1 and L2 have well-conserved sequences, while the remaining genes display a greater variability, resulting in the differences observed during infections with the different HPV types [15].

## 2.1. HPV Life Cycle

HPV infects the basal stem cells of the stratified epithelia (which are the only epithelial cells capable of undergoing cell division) of the skin, oral cavity and anogenital track [1,2]. As the basal layer of the epithelium is protected by several layers of differentiated cells, HPV reaches this area via micro wounds in the tissue (Figure 2a) [16,17].

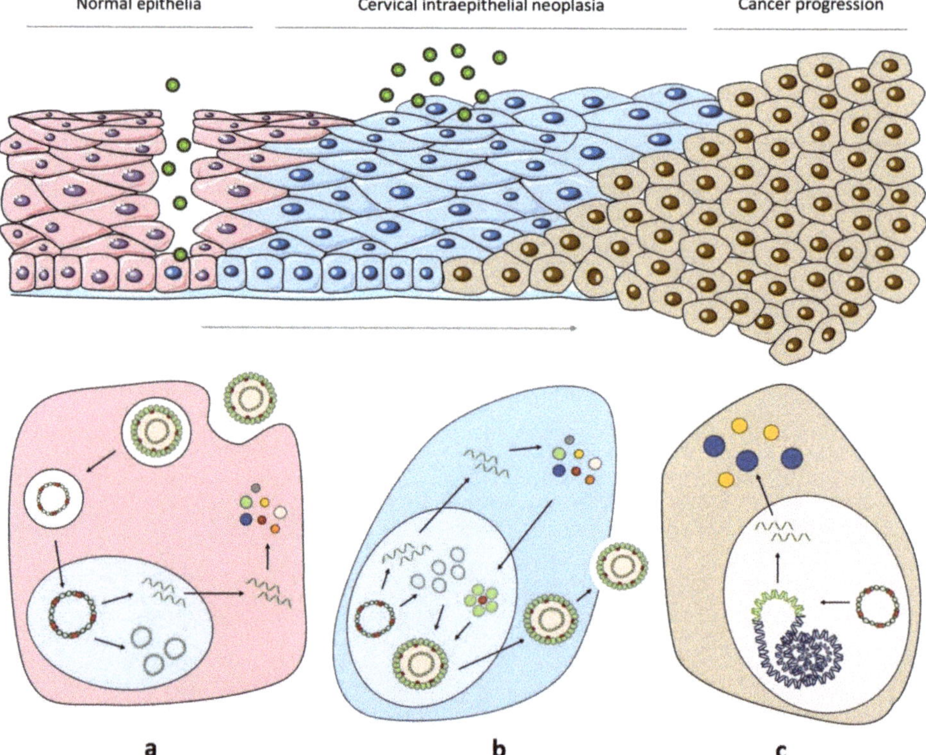


**Figure 2.** HPV life cycle and progression to cancer. (**a**) HPV reaches the basal cells of the epithelium through tissue abrasions. Upon recognition, endocytosis of the virion occurs, and HPV is transported through the endosomal pathway. At endosomes, L2 mediates viral egressing and HPV vesicles are transported along microtubules to the nucleus, where early transcription is initiated, with a quick but transient expression of the early proteins and through the recruitment of the cellular DNA replication machinery. Afterwards HPV can enter the latency phase. (**b**) Once basal cells start to differentiate, they migrate towards the surface of the tissue. Here, structural proteins are expressed, allowing virion assembly and release, which occurs alongside with tissue desquamation. (**c**) Viral persistence in basal cells can result in HPV genome integration, which promotes cancer progression.

Upon recognition of L1 at the cell surface, the viral capsid undergoes structural modifications, required for endocytosis of the virion [18–20]. Then, while HPV travels along the endosomal pathway, L1 dissociates from the viral genome and L2 mediates viral egressing from the endosomes, guiding HPV vesicles along microtubules [21–23]. When the nuclear envelope is disrupted during mitosis, L2-viral DNA complexes are delivered into the nucleus [17,24,25]. Upon nuclear entry, viral early transcription is initiated with the expression of the early proteins E1 and E2, essential factors for viral DNA replication, as they recruit the cellular DNA replication machinery [26,27]. Additionally, E2 is also critical for transcription, since it has splicing-regulating activities that control the processing of viral pre-mRNA [28–30]. E1, E2, E6 and E7 are the most early proteins and are found in the basal layers of the epithelium, while E5 and E4 only start to be detected at the suprabasal layers [11].

Three phases of replication in the viral life cycle have been identified: Initial amplification, maintenance replication and vegetative amplification (reviewed in [11,31,32]). Initially, replication starts at the URR, causing a quick but transient increase of viral genome copies (Figure 2a). Afterwards, viral DNA is stably maintained in low copy numbers during basal cells division. This is accomplished through the establishment of stable episomes at specific regions of the nucleus. E2 is then responsible for tethering the viral genome to host chromatin, thus guarantying its successful partition in equal amounts upon cell division [32]. HPV infection can persist in the latent form for years to decades until a switch from genome maintenance to vegetative viral replication occurs, allowing virion production (Figure 2b) [2,11].

In non-infected epithelia, the daughter cells from the proliferative basal layer lose contact with the basement membrane and stop dividing, initiating the process of terminal differentiation and ceasing their proliferative capacity [11,17]. However, as HPV is highly dependent on the host-cell machinery for viral genome replication and translation, it has evolved to carry out its replication cycle in concert with epithelial differentiation, alongside viral gene expression. HPV-infected cells retain their differentiation capacity and are capable of moving into the upper epithelial layers [11]. The molecular mechanisms behind this cell reprogramming are not yet well understood, but viral-encoded oncogenes seem to have a central role in this process. Moreover, vegetative amplification, besides being associated to an increase in HPV genome copy numbers, is also followed by the expression of the structural proteins L1 and L2 [2,11]. In the infected cells from the upper epithelium layers, virion assembly starts at the nucleus, where the capsid and genome packaging occur. Since HPV is a non-lytic virus, virions are only released when the infected cells reach the granular surface layers of the epithelium (Figure 2b) [11].

## 2.2. From HPV Infection to Malignant Transformation

Progression to cancer is a rare event in HPV infection. Indeed, this is an unwanted event for the virus, since infected cells that suffer transformation do not produce any virions. The mechanisms behind transformation of infected cells are not yet clearly understood, but it is known to be highly dependent on HR-HPV E6 and E7 oncogenes [33]. These proteins present a broad-spectrum functionality and modify different pathways, mainly associated to cell growth, differentiation and host genome stability (reviewed in [34]). It is thought that the trigger for cell transformation starts with the integration of viral episomes into the host genome (Figure 2c). In the reported cases, the viral genome is partially integrated occurring the loss of E2 ORF, which codifies the transcription repressor of the oncoproteins E6 and E7 genes [33,35,36]. It was also suggested that integration may occur as an indirect effect of episomes tethering, close to chromosomal regions that accumulate DNA break repair factors, which are required for viral genome replication [37].

Without the E2 repressor function, E6 and E7 start to be highly expressed during progression from cervical intraepithelial neoplasia (CIN) in HR-HPV infected tissues, contributing to a malignant evolution to invasive cancer [38–40]. Additionally, the integrated HPV genome seems to recruit DNA repair/recombination systems, which induce alterations in the host cell genome and eventually affect several genes, including cellular oncogenes that aid the transformation initiated by HPV oncoproteins.

It was even suggested that de novo infections by new HPVs in cells that already present integrated HR-HPV genome can occur, potentiating the genomic instability [31,41].

The most important functions of E6 and E7 for carcinogenesis are the impairment of p53 and pRb tumour suppressor's pathways, respectively [42]. Independently, E6 also stimulates the telomerase, which prevents senescence by stabilizing telomere length at the chromosomes' ends [34,42]. E6 also contains a binding motif for PDZ (PSD-95/DLG/ZO-1) proteins, which induces the degradation or subcellular localization alterations of numerous cellular proteins. Both of these functions are assigned exclusively to HR-HPV E6 proteins [34].

## 3. Activation of the Antiviral Immune Signalling

The epithelial cells that constitute the cutaneous and mucosal tissues are the first line of defence of the innate immune system, and constitutively express low levels of IFNs and cytokines. Additionally, in the case of mucosal epithelia, cells produce mucin that prevents viral attachment and penetration [43]. HPV is able to bypass this protection through abrasions in the tissue [17].

While viral infections are generally extremely immunogenic, the efficient HPV life cycle grants protection against epithelia defences. The hallmarks of HPV infection are the slow replication cycle and the virus capability to maintain low levels of viral proteins expression and secretion. Additionally, since HPV virions are only released at the epithelium surface without inducing cell lysis, no inflammatory response or viremia occur [1,12]. Thus, HPV is able to delay viral detection and elimination by the innate immune system. Nonetheless, to further ensure its ability to surpass immune surveillance, HPV has developed numerous strategies to evade and manipulate the cellular antiviral defences, by interfering with the function of host antiviral proteins or by inhibiting their expression (Figure 3).

Upon HPV infection, viral DNA is released into the cytosol and can be sensed by the PRRs IFN-inducible protein 16 (IFI16) [44,45], absent in melanoma 2 (AIM2) [46], and the toll-like receptors 4 (TLR4) [47] and TLR9 [48]. It has been reported that the activation of the cytosolic dsDNA sensors IFI16 and AIM2, induces the formation of the inflammasome complex that is required for processing and release of the pro-inflammatory cytokines interleukin-1β (IL-1β) and IL-18. Moreover, it was demonstrated that the ISG IFI16 ultimately restricts HPV replication by inducing epigenetic alterations on the viral genome [45].

Regarding the TLRs signalling, it has been shown that TLR4 is able to recognize the association of HPV11 with the heparin sulfate [49] or the glycosaminoglycans at the cell surface [49,50]. Additionally, it has been shown that TLR9 is able to recognize and be stimulated by the CpG motifs present in the HPV16 E6 gene sequence [48], during viral capsid disassembling in the endosome [20]. Induction of TLRs signalling pathways ultimately leads to the expression of IFNs and pro-inflammatory cytokines [51]. In fact, an association between the clearance of initial HPV infection in young women and higher expression levels of TLR3, TLR7, TLR8 and TLR9 has been reported [52,53]. Nevertheless, the signalling behind HPV genome recognition by TLRs remains to be elucidated.

It was reported that several cytokines, namely IL-1α, IL-4 IL-13, transforming growth factor β (TGF-β), tumor necrosis factor α (TNF-α), IFN-α and IFN-β, act as inhibitors of HPV16 URR activity, suppressing early gene transcription [54]. Furthermore, type I IFNs, TNF-α and TGF-β restrain the growth of non-infected and HPV-infected keratinocytes, while this suppression tends to cease in oncogenic cells [55,56]. Additionally, epithelial cells of cutaneous and mucosal tissues produce a single specific type I IFN, IFN-κ [57], which has been shown to have antiviral functions during HPV16 and HPV31 infection [58,59].

In addition to cytokines and ISGs, it was found that apolipoprotein B mRNA-editing catalytic polypeptide 3 (APOBEC3) proteins, an IFN-inducible antiviral family, promote hypermutations in the HPV genome [60], and reduces HPV infectivity [61]. Recently, the Myb-related transcription factor partner of profilin (MYPOP) was also shown to have antiviral activity against HPV by repressing the URR function [62]. Another important class of molecules that impede HPV infection are human α-defensins, immune system components that have broad antimicrobial effects [63]. Different mechanisms of inhibition have been proposed: For example, the α-defensin 5 (HD5) was reported to inhibit the furin cleavage of HPV L2, as well as to impede the dissociation of L1 from L2, which is essential for the entry process of HPV [64,65]. Moreover, it has been shown that human neutrophil peptides 1-4 (HNP1-4) and HD5 block escape from endocytic vesicles instead of inhibiting the binding or internalization in multiple serotypes of HPV infection [63,66]. Furthermore, defensins have been proposed to recruit immune cells, thus contributing to the activation of adaptive immunity [67].

These findings demonstrate that several immune effectors mount a response to HPV infection, eventually leading to viral clearance.

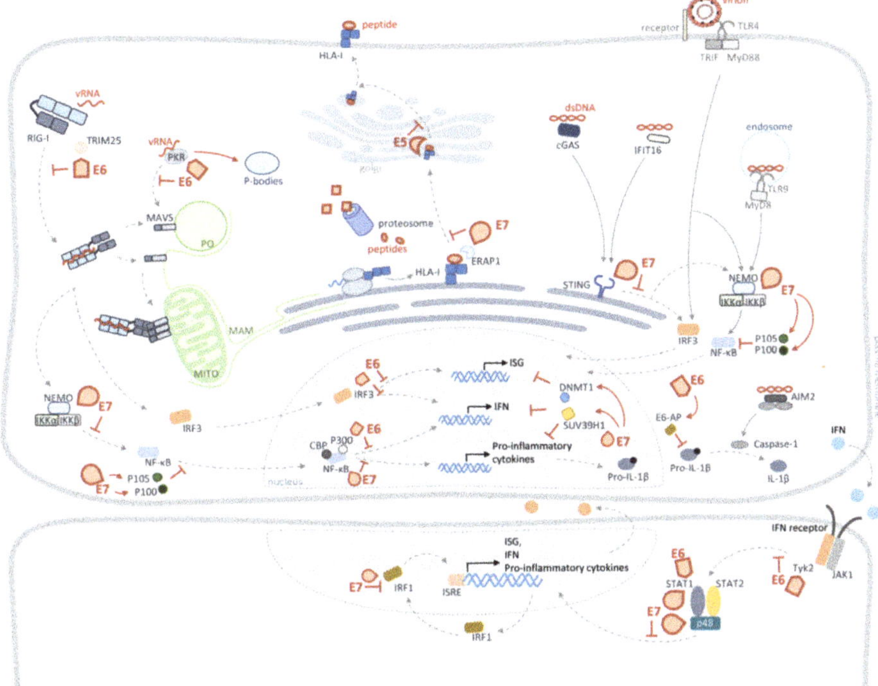


**Figure 3.** Evasion of the cellular innate immunity response by HPV proteins. HPV E2, E5, E6 and E7 target several steps of the pattern-recognition receptors (PRRs) signalling, downregulating the expression of interferons (IFNs), pro-inflammatory cytokines and IFN-stimulated genes (ISGs), and consequently inhibiting the cellular antiviral response, as well as antigen presentation at cell surface.

## 4. Cellular Innate Immunity Evasion by HPV

During its life cycle, HPV produces a small number of proteins, which therefore have multifunctional roles during infection. Curiously, HR-HPV oncoproteins E6 and E7 are the ones that interfere the most with the cellular innate immunity, alongside with minor roles of E5 and E2 (summarized in Figure 3 and Table 2).

**Table 2.** HPV proteins' functions on the evasion of the cellular antiviral response.

| Viral Protein | HPV Type | Effects on Innate Immunity |
|---|---|---|
| E2 | HPV16 HPV18 | impairs transcription of the stimulator of IFN genes (STING) and IFN-κ, and their downstream antiviral genes [68] |
| E5 | HPV16 | inhibits the human leukocyte antigen class I (HLA-I) transport, decreasing its surface expression [69–71] |
| E6 | HPV16 | induces degradation of pro-IL-β through the ubiquitin ligase E6-AP [72] |
| | HPV16 | induces the tripartite motif-containing protein 25 (TRIM25) degradation suppressing the retinoic acid-inducible gene-I (RIG-I)-mediated expression of IFN-β, chemokines, and ISGs [73] |
| | HPV16 | binds to IRF3, inhibiting its transcription activities [74] |
| | HPV16 | reduces transcription activity of CBP/p300, and consequentially NF-κB promoter activity [75,76] |
| | HPV18 | binds to the tyrosine-protein kinase (TYK2), inhibiting the downstream signalling [77] |
| | HPV16 HPV31 | re-localizes the protein kinase R (PKR) to P-bodies, impeding the downstream signalling [78] |
| | HPV31 HPV16 HPV18 | inhibits STAT1 binding and ISGs transcription [79,80] |
| E7 | HPV16 | induces the overexpression of endoplasmic reticulum aminopeptidase 1 (ERAP1) decreases epitopes presentation [81] |
| | HPV16 HPV11 HPV18 | binds to IRF1, inhibiting its promoter activity through histone deacetylases (HDACs) [82–84] |
| | HPV16 HPV18 | binds to the NF-κB kinase (IKK) complex, impairing NF-κB signalling [75] |
| | HPV16 | impairs DNA binding activity of NF-κB, through impairment of p65 subunit functions [82,85] |
| | HPV16 | binds to p48, inhibiting the IFN-stimulated gene factor 3 (ISGF3)-mediated gene expression [86] |
| | HPV16 | binds to the DNA methyltransferase (DNMT1), impairing antiviral gene transcription through epigenetic modification [87] |
| | HPV18 | binds to STING, impairing IFNs and pro-inflammatory cytokines expression [88] |
| | HPV18 | induces of the H3K9 methyltransferase (SUV39H1) transcription, which promotes epigenetic silencing of PRRs [89] |
| | HPV16 HPV38 | inhibits of TLR9 promoter region through the formation of a inhibitory complex and induction of HDACs [90,91] |
| E6 E7 | HPV16 | competes with the transcription promoters of interleukin-18 (IL-18), impairing its expression [85] |
| E6/E7 | HPV16 | inhibit of STAT1 binding to DNA and antiviral genes transcription [92] |
| E6/E7 | HPV16 | induce the overexpression of p100 and p105 and their re-localization, which inhibit the transcriptional activity of NF-κB [93] |
| E6/E7 | HPV38 | decrease MHC-I (human HLA-I) expression through downregulation of STAT1 expression [94] |

Besides interfering with the cellular antiviral signalling pathways, HPV also impairs the antigen processing machinery, impeding T-cell recognition of infected cells. HPV directly impedes the generation of cytotoxic T-lymphocyte (CTL) epitopes through different mechanisms mediated by E7 and E5 [69–71,81].

HPV16 E7 also increases the expression of ERAP1, an immunopeptidase essential for epitope editing [81], which leads to a reduction of CD8+ T cell responses. It has furthermore been shown that the attenuation of ERAP1 induces CTL-mediated HPV-infected cell death [81]. Similarly, HPV16 E5 also impairs the transport of HLA-I to the cell surface [69–71]. In the last few years, new data on HPV epigenetic control has been emerging. Lo Cigno et al. showed that HPV E7 upregulates the H3K9 methyltransferase SUV39H1, which, through alterations in the chromatin structure, promotes epigenetic inhibition of nucleic acid sensors, such as RIG-I, cyclic GMP-AMP synthase (cGAS) or even STING [89].

### 4.1. HPV Targets Pattern Recognition Receptors

Several reports have shown different HPV strategies to directly impair PRRs signalling. It has been demonstrated that HPV16 and HPV18 E7 bind to STING, inhibiting the cGAS-STING signalling pathway [88]. STING is the adaptor protein that mediates the immune signalling upon recognition of viral DNA by a numerous set of cytosolic receptors, leading to the expression of type I IFNs and pro-inflammatory cytokines [95–99]. The role of STING in HPV infection recognition is still unclear and further studies should be performed.

Interestingly, it has been shown that HPV16 E6, but not E7, forms a complex with TRIM25 and its regulator ubiquitin carboxyl-terminal hydrolase 15 (USP15), inducing TRIM25 degradation [73]. TRIM25 is an E3 ubiquitin ligase essential for RIG-I activation, allowing the induction of its downstream signalling and consequential expression of IFNs and ISGs [100–102]. Whereas RIG-I only recognizes dsRNA, different studies have shown the activation and evasion of this signalling pathway by DNA viruses [103–111]. Thus, the role of RIG-I signalling in HPV sensing needs to be further investigated, but its targeting by E6 suggests that it plays a critical role on the HPV immune signalling.

### 4.2. HPV Targets Interferon Regulatory Factors Signalling

Several PRRs signalling pathways converge to activate the transcription factor IRF3, responsible for IFNs expression. As expected, HPV also targets this transcription factor to inhibit its translocation to the nucleus. It has been shown that HPV16 E6 binds to IRF3, although the same was not observed for HPV6 or HPV18 [74]. The IFN-κ induced IRF1 is another target of E7 from HPV16, HPV 18 and HPV11, resulting in the inactivation of its promoter activity [82–84,94]. It was proposed that IRF1 targeting prevents the correct binding to the IFN-β promoter region, in a mechanism that involves HDACs, leading to reduced IFN-β production [83]. Thus, IRFs targeting by HPV proteins leads to the impairment of IFN-α [79], IFN-β [74,83] and IFN-κ expression [112].

HPV16 was also reported to upregulate the ubiquitin carboxyl-terminal hydrolase L1 (UCHL1), to indirectly impair IRF signalling. This protein inhibits the K63 poly-ubiquitination of TNF receptor-associated factor 3 (TRAF3), suppressing IRFs activation [113].

### 4.3. HPV Targets NF-κBs Signalling

Another transcription promoter activated by PRRs signalling is the NF-κB. NF-κB signalling is a tightly regulated pathway that culminates on the regulation of several genes associated with immune and stress responses, as well as apoptosis, proliferation, differentiation and development (reviewed in [114,115]). HPV has evolved mechanisms to abrogate the immune and inflammatory response promoted by NF-κB signalling through E6 and E7.

When inactivated, NF-κB is complexed with its repressors and, upon activation, their degradation is induced in order to allow translocation of NF-κB to the nucleus [114,115]. It has been reported that HPV E7 associates with the inhibitor of IKK complex, impairing the release of NF-κB [75]. Furthermore, HPV16 E6 reduces the transcriptional activity of NF-κB, by interacting with its coactivators CREB binding protein (CBP) and p300 [75,76]. Interestingly, HPV6 E6 was also reported to bind the same coactivators, although less efficiently, and with a smaller inhibitory effect [76]. E7 was also reported to decrease the DNA binding activity of NF-κB and to reduce nuclear translocation and acetylation of the p65 subunit of NF-κB [82,116,117]. Furthermore, E7 was shown to bind to P/CAF, impairing its interaction with p65 [83], and to target the transcriptional coactivator p300 [85]. However, it has been demonstrated that the E2-dependent transcription requires CBP/p300. Thus, E7 interferes with the regulation of E2 transcriptional activity by associating with p300 [118]. It has also been reported that HPV16 E6 and E7 proteins induce the overexpression and modulate the subcellular localization of p105 and p100, NF-κB precursors [93].

As previously mentioned, HPV16 induces the overexpression of UCHL1 [113]. Binding of UCHL1 to TRAF6 leads to the degradation of NF-κB essential modulator (NEMO), which in turn results in the suppression of p65 phosphorylation, blocking the canonical NF-κB signalling [113]. UCHL1 also targets IκBα by attenuating its ubiquitination, preventing the release of NF-κB [119].

NF-κB signalling is a crucial mediator of inflammatory responses and regulates the expression of different interleukins [120]. Cellular inflammatory responses are also targets of HPV E6 and E7 proteins. It has been shown that E6 inhibits IL-18, a pro-inflammatory cytokine, by binding to its receptor [121,122]. E7 was also reported to bind and impair IL-18 receptor signalling [122]. Moreover, HPV16 E6 binds to the ubiquitin ligase E6-AP, inducing the degradation of pro-IL-1β in a proteasome-dependent manner, impairing IL-1β processing and secretion [72]. HPV also represses NF-κB-mediated transcription of AIM2, through overexpression of sirtuin 1 (SIRT1) [123].

While these studies propose an anti-inflammatory role for HPV E6 and E7, there is still some contradictions associated with the HPV effect over NF-κB signalling, since in vivo data suggests that HPV promotes chronic inflammation, which correlates with HPV-induced carcinogenesis [124,125].

*4.4. HPV Targets JAK/STAT Signalling*

E6, from HPV16 and HPV18 but not HPV11, directly interacts with TYK2 impairing binding with the transmembrane IFN-α/β receptor (IFNAR) and the consequent activation of the downstream signalling [77]. Additionally, HPV16 E7 targets p48 (also known as IRF9) impairing the translocation to the nucleus of ISGF3, a heterodimer formed by STAT1-STAT2-p48, and consequential activation of antiviral genes expression [86,126,127].

Microarray analysis of HPV31 or HPV16 infected-keratinocytes showed a decrease in STAT1 transcription promoted by E6 and E7 [79,80,92]. Like STAT1, several other ISGs were shown to be downregulated upon HPV31, HPV18 and HPV16 infection [79,80]. PKR (an IFN-inducible protein that recognizes dsRNA, activating IFNs expression, and shuts-down host transcription) was shown to have its transcription impaired during HPV infection, and to be re-localized from the cytosol to P-bodies by E6 [78,128]. The same downregulation on transcription induced by E6 and E7 has been reported in different studies for several PRRs, such as TLR3, TLR9 and RIG-I [48,80,90,129,130]. The mechanism proposed for E7-mediated downregulation of TLR9 transcription was through histone modifications [90,91]. HPV16 and HPV18 E2 were even associated with the decrease in STING and IFN-κ transcription [68]. Moreover, this was also observed in clinical samples of low-grade CIN [68].

HPV E7 also impairs antiviral genes transcription through the induction of host DNA methylation by DNMT1, a transcription repressor [87]. Transcription of the chemokine CXCL14, essential for leukocyte attraction to the infection site, is affected by this process. HPV E6 was also associated to DNA methylation of the IFN-κ gene [112].

Curiously, IFNs have been used as therapy in clinical cases with HPV lesions, and non-responsiveness to IFNs was associated with higher levels of E7 protein [131].

## 5. Cellular Innate Immunity and Cancer Progression in HPV Infection

As previously mentioned, malignant transformation is an unwanted consequence for HPV, as it implies a lack of production of new virions. This event is rather a consequence of the unspecific targeting of HPV E6 and E7, whose sequences get integrated into the host genome while losing their viral transcription regulator. Besides modifying cell growth, differentiation and genome stability processes, these proteins alter specific cellular antiviral response mechanisms that have been associated with cancer progression and have a critical role in inflammatory processes and tumorigenesis (summarized in Figure 4).

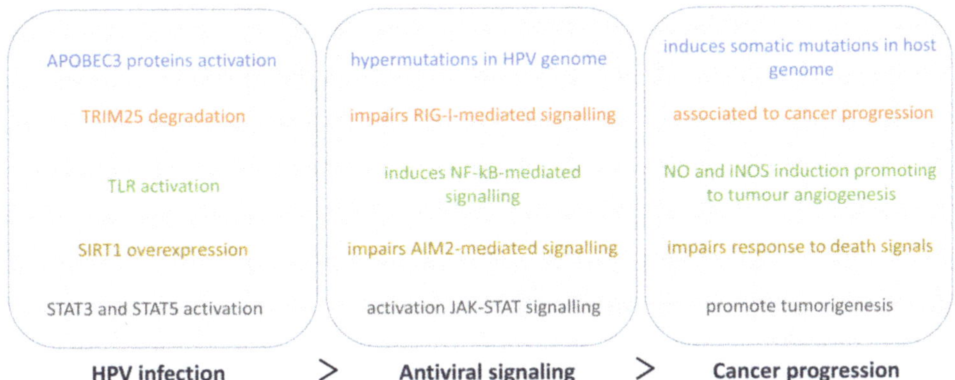


**Figure 4.** HPV evasion of the antiviral signalling and corresponding impact on carcinogenesis.

APOBEC3 proteins exert their antiviral function through the induction of hypermutations in the HPV genome [60,61]. They may furthermore contribute to cancer mutagenesis by inducing somatic mutations in host genome [61]. Similarly, SIRT1 overexpression induced by HPV impairs AIM2-mediated immunity, and this inhibition allows HPV-infected cervical cancer cells to escape death and continue their growth. Moreover, SIRT1 expression in HPV-infected cervical cancers was associated with a poor clinical outcome [123].

Additionally, as TRIM25 has been associated to cancer–related pathways, its targeting by E6 may also regulate other functions of this protein that should be addressed in the context of HPV-associated cancer progression [132].

The influence of HPV on the inflammatory response is still controversial. While most studies using cell lines show an inhibition of the NF-κB pathway, in CIN and cervical cancer, HPV seems to induce the expression of inflammatory cytokines, which correlates with cancer progression [124,125]. A recent study has shown an increase of nitric oxide (NO) and inducible NO synthase (iNOS), which was suggested to be mediated by TLR-induced NF-κB signalling, in cervical samples from HR-HPV-infected patients [133]. Since NO is a critical component of the tumour microenvironment and promotes tumour angiogenesis, as well as tumour cell invasion and metastasis, it has been studied as a possible target for cancer therapy [134].

Contrary to the mentioned components of JAK/STAT, which are inhibited during viral infection, HPV also modulate STAT3 and STAT5, respectively, by increasing their activity. STAT3 stimulation leads to cell cycle progression and cell survival, suggesting its importance in the life cycle of HPV18 [135–139]. Likewise, STAT5 activation induces genome amplification in differentiating cells, through the exploitation of the DNA damage response [140]. Both STAT3 and STAT5 have been extensively studied in the context of tumorigenesis [141,142]. Moreover, it has been suggested that STAT3 expression correlates with increased severity of HPV lesions, being a possible target for therapy [139].

## 6. Conclusions

The antiviral defence mechanisms that recognize HPV early in infection are still poorly elucidated and most of what is known was inferred from HPV proteins' overexpression studies. Nonetheless, several reports have demonstrated that different HPV efficiently evade the cellular antiviral signalling pathways using diverse strategies throughout their life cycle.

The oncogenes E6 and E7 are the viral proteins most involved in immune evasion, targeting almost all cellular innate immune pathways in a synergetic manner. However, most of the evasion mechanisms reported for HPV have been observed in vitro, and whether these results can be translated to the clinic remains unknown. The effectiveness of these studies is highly impaired by the fact that the interplay between HPV and their host cells changes during the different cell differentiation stages and disease progression. Nonetheless, it has been shown that E6 and E7 levels are inversely correlated to IFN treatment response and, more importantly, as discussed above, many of these evasion strategies directly correlate to the development of HPV-induced tumorigenesis (Figure 5).

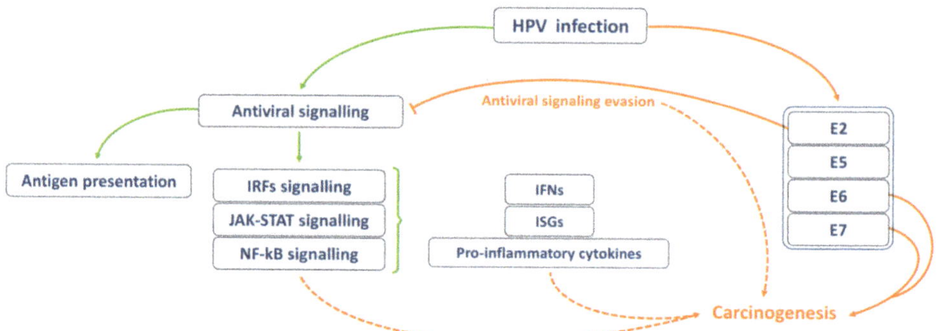


**Figure 5.** Overview of the interplay between HPV and the cellular antiviral signalling and its impact on carcinogenesis. Orange arrows represent HPV-mediated events, while green arrows represent host defence processes.

Furthermore, although prophylactic vaccines are effective in averting infection of the most medically relevant HR-HPV, they do not exert any effect on existing infections [143]. Hence, it is essential to further analyse and understand the mechanisms behind HPV evasion of the cellular innate immunity and their correlation to HPV-induced persistence and tumorigenesis. These studies may reveal essential to the discovery of new cellular targets for the development of novel antiviral and anticancer therapies.

**Funding:** This work was supported by the Portuguese Foundation for Science and Technology (FCT): PTDC/BIA-CEL/31378/2017 (POCI-01-0145-FEDER-031378), CEECIND/03747/2017, SFRH/BD/137851/2018 and UIDB/04501/2020, under the scope of the Operational Program "Competitiveness and internationalization", in its FEDER/FNR component.

**Conflicts of Interest:** The authors declare no conflict of interest.

## References


1.  Burd, E.M.; Dean, C.L. Human papillomavirus. *Microbiol. Spectr.* **2016**, *4*. [CrossRef]
2.  Harden, M.E.; Munger, K. Human papillomavirus molecular biology. *Mutat. Res.* **2017**, *772*, 3–12. [CrossRef]
3.  De Sanjosé, S.; Brotons, M.; Pavón, M.A. The natural history of human papillomavirus infection. *Best Pract. Res. Clin. Obstet. Gynaecol.* **2018**, *47*, 2–13. [CrossRef]
4.  IARC Working Group on the Evaluation of Carcinogenic Risk to Humans. *Human Papillomaviruses*; International Agency for Research on Cancer: Lyon, France, 2007; Volume 90, ISBN 978-92-832-1290-4.
5.  Medzhitov, R.; Janeway, C.A. Innate immunity: The virtues of a nonclonal system of recognition. *Cell* **1997**, *91*, 295–298. [CrossRef]
6.  Fensterl, V.; Chattopadhyay, S.; Sen, G.C. No love lost between viruses and interferons. *Annu. Rev. Virol.* **2015**, *2*, 549–572. [CrossRef]
7.  Janeway, C.A.; Medzhitov, R. Innate immune recognition. *Annu. Rev. Immunol.* **2002**, *20*, 197–216. [CrossRef] [PubMed]

8. Kawai, T.; Akira, S. The role of pattern-recognition receptors in innate immunity: Update on toll-like receptors. *Nat. Immunol.* **2010**, *11*, 373–384. [CrossRef] [PubMed]

9. Takeuchi, O.; Akira, S. Pattern recognition receptors and inflammation. *Cell* **2010**, *140*, 805–820. [CrossRef] [PubMed]

10. Schneider, W.M.; Chevillotte, D.; Rice, C.M. Interferon-stimulated genes: A complex web of host defenses. *Annu. Rev. Immunol.* **2014**, *32*, 513–545. [CrossRef]

11. Graham, S.V. The human papillomavirus replication cycle, and its links to cancer progression: A comprehensive review. *Clin. Sci.* **2017**, *131*, 2201–2221. [CrossRef]

12. Woodworth, C.D. HPV innate immunity. *Front. Biosci.* **2002**, *7*. [CrossRef]

13. Harari, A.; Chen, Z.; Burk, R.D. HPV genomics: Past, present and future. *Curr. Probl. Dermatol.* **2014**, *45*, 1–18. [CrossRef] [PubMed]

14. Van Doorslaer, K.; Li, Z.; Xirasagar, S.; Maes, P.; Kaminsky, D.; Liou, D.; Sun, Q.; Kaur, R.; Huyen, Y.; Mcbride, A.A. The papillomavirus episteme: A major update to the papillomavirus sequence database. *Nucleic Acids Res.* **2017**, *45*, 499–506. [CrossRef]

15. García-Vallvé, S.; Alonso, A.; Bravo, I.G. Papillomaviruses: Different genes have different histories. *Trends Microbiol.* **2005**, *13*, 514–521. [CrossRef]

16. Egawa, K. Do human papillomaviruses target epidermal stem cells? *Dermatology* **2003**, *207*, 251–254. [CrossRef] [PubMed]

17. Pyeon, D.; Pearce, S.M.; Lank, S.M.; Ahlquist, P.; Lambert, P.F. Establishment of human papillomavirus infection requires cell cycle progression. *PLoS Pathog.* **2009**, *5*, e1000318. [CrossRef] [PubMed]

18. Bousarghin, L.; Touzé, A.; Sizaret, P.-Y.; Coursaget, P. Human papillomavirus types 16, 31, and 58 use different endocytosis pathways to enter cells. *J. Virol.* **2003**, *77*, 3846–3850. [CrossRef]

19. Schelhaas, M.; Shah, B.; Holzer, M.; Blattmann, P.; Kühling, L.; Day, P.M.; Schiller, J.T.; Helenius, A. Entry of human papillomavirus type 16 by actin-dependent, clathrin-and lipid raft-independent endocytosis. *PLoS Pathog.* **2012**, *8*, e1002657. [CrossRef]

20. Richards, R.M.; Lowy, D.R.; Schiller, J.T.; Day, P.M. Cleavage of the papillomavirus minor capsid protein, L2, at a furin consensus site is necessary for infection. *Proc. Natl. Acad. Sci. USA* **2006**, *103*, 1522–1527. [CrossRef]

21. Kämper, N.; Day, P.M.; Nowak, T.; Selinka, H.-C.; Florin, L.; Bolscher, J.; Hilbig, L.; Schiller, J.T.; Sapp, M. A membrane-destabilizing peptide in capsid protein L2 is required for egress of papillomavirus genomes from endosomes. *J. Virol.* **2006**, *80*, 759–768. [CrossRef]

22. Bienkowska-Haba, M.; Williams, C.; Kim, S.M.; Garcea, R.L.; Sapp, M. Cyclophilins facilitate dissociation of the human papillomavirus type 16 capsid protein L1 from the L2/DNA complex following virus entry. *J. Virol.* **2012**, *86*, 9875–9887. [CrossRef] [PubMed]

23. DiGiuseppe, S.; Bienkowska-Haba, M.; Hilbig, L.; Sapp, M. The nuclear retention signal of HPV16 L2 protein is essential for incoming viral genome to transverse the trans-Golgi network. *Virology* **2014**, *458–459*, 93–105. [CrossRef] [PubMed]

24. Aydin, I.; Weber, S.; Snijder, B.; Samperio Ventayol, P.; Kühbacher, A.; Becker, M.; Day, P.M.; Schiller, J.T.; Kann, M.; Pelkmans, L.; et al. Large scale RNAi reveals the requirement of nuclear envelope breakdown for nuclear import of human papillomaviruses. *PLoS Pathog.* **2014**, *10*, e1004162. [CrossRef] [PubMed]

25. DiGiuseppe, S.; Bienkowska-Haba, M.; Sapp, M. Human papillomavirus entry: Hiding in a bubble. *J. Virol.* **2016**, *90*, 8032–8035. [CrossRef] [PubMed]

26. Wallace, N.A.; Galloway, D.A. Manipulation of cellular DNA damage repair machinery facilitates propagation of human papillomaviruses. *Semin. Cancer Biol.* **2014**, *26*, 30–42. [CrossRef]

27. Wang, X.; Helfer, C.M.; Pancholi, N.; Bradner, J.E.; You, J. Recruitment of Brd4 to the human papillomavirus type 16 DNA replication complex is essential for replication of viral DNA. *J. Virol.* **2013**, *87*, 3871–3884. [CrossRef]

28. Lai, M.C.; Teh, B.H.; Tarn, W.Y. A human papillomavirus E2 transcriptional activator: The interactions with cellular splicing factors and potential function in pre-mRNA processing. *J. Biol. Chem.* **1999**, *274*, 11832–11841. [CrossRef]

29. Mole, S.; Milligan, S.G.; Graham, S.V. Human papillomavirus type 16 E2 protein transcriptionally activates the promoter of a key cellular splicing factor, SF2/ASF. *J. Virol.* **2009**, *83*, 357–367. [CrossRef]

30. Graham, S.V.; Faizo, A.A.A.; Ali Faizo, A.A. Control of human papillomavirus gene expression by alternative splicing. *Virus Res.* **2017**, *231*, 83–95. [CrossRef]

31. Kadaja, M.; Silla, T.; Ustav, E.; Ustav, M. Papillomavirus DNA replication—From initiation to genomic instability. *Virology* **2009**, *384*, 360–368. [CrossRef]

32. McBride, A.A. Replication and partitioning of papillomavirus genomes. *Adv. Virus Res.* **2008**, *72*, 155–205. [CrossRef] [PubMed]

33. Moody, C.A.; Laimins, L.A. Human papillomavirus oncoproteins: Pathways to transformation. *Nat. Rev. Cancer* **2010**, *10*, 550–560. [CrossRef] [PubMed]

34. Hoppe-Seyler, K.; Bossler, F.; Braun, J.A.; Herrmann, A.L.; Hoppe-Seyler, F. The HPV E6/E7 oncogenes: Key factors for viral carcinogenesis and therapeutic targets. *Trends Microbiol.* **2018**, *26*, 158–168. [CrossRef] [PubMed]

35. Ziegert, C.; Wentzensen, N.; Vinokurova, S.; Kisseljov, F.; Einenkel, J.; Hoeckel, M.; Von Knebel Doeberitz, M. A comprehensive analysis of HPV integration loci in anogenital lesions combining transcript and genome-based amplification techniques. *Oncogene* **2003**, *22*, 3977–3984. [CrossRef] [PubMed]

36. Peter, M.; Rosty, C.; Couturier, J.; Radvanyi, F.; Teshima, H.; Sastre-Garau, X. MYC activation associated with the integration of HPV DNA at the MYC locus in genital tumors. *Oncogene* **2006**, *25*, 5985–5993. [CrossRef] [PubMed]

37. Jang, M.K.; Shen, K.; McBride, A.A. Papillomavirus genomes associate with BRD4 to replicate at fragile sites in the host genome. *PLoS Pathog.* **2014**, *10*. [CrossRef] [PubMed]

38. Gammoh, N.; Isaacson, E.; Tomaić, V.; Jackson, D.J.; Doorbar, J.; Banks, L. Inhibition of HPV-16 E7 oncogenic activity by HPV-16 E2. *Oncogene* **2009**, *28*, 2299–2304. [CrossRef]

39. Baldwin, A.; Huh, K.-W.; Munger, K. Human papillomavirus E7 oncoprotein dysregulates steroid receptor coactivator 1 localization and function. *J. Virol.* **2006**, *80*, 6669–6677. [CrossRef]

40. Yeo-Teh, N.S.L.; Ito, Y.; Jha, S. High-Risk human papillomaviral oncogenes E6 and E7 target key cellular pathways to achieve oncogenesis. *Int. J. Mol. Sci.* **2018**, *19*, 1706. [CrossRef]

41. Kadaja, M.; Sumerina, A.; Verst, T.; Ojarand, M.; Ustav, E.; Ustav, M. Genomic instability of the host cell induced by the human papillomavirus replication machinery. *EMBO J.* **2007**, *26*, 2180–2191. [CrossRef]

42. Narisawa-Saito, M.; Kiyono, T. Basic mechanisms of high-risk human papillomavirus-induced carcinogenesis: Roles of E6 and E7 proteins. *Cancer Sci.* **2007**, *98*, 1505–1511. [CrossRef] [PubMed]

43. Medzhitov, R. Recognition of microorganisms and activation of the immune response. *Nature* **2007**, *449*, 819–826. [CrossRef] [PubMed]

44. Cai, H.H.; Yan, L.; Liu, N.; Xu, M.; Cai, H.H. IFI16 promotes cervical cancer progression by upregulating PD-L1 in immunomicroenvironment through STING-TBK1-NF-kB pathway. *Biomed. Pharmacother.* **2020**, *123*, 109790. [CrossRef] [PubMed]

45. Lo Cigno, I.; De Andrea, M.; Borgogna, C.; Albertini, S.; Landini, M.M.; Peretti, A.; Johnson, K.E.; Chandran, B.; Landolfo, S.; Gariglio, M. The nuclear DNA sensor IFI16 acts as a restriction factor for human papillomavirus replication through epigenetic modifications of the viral promoters. *J. Virol.* **2015**, *89*, 7506–7520. [CrossRef] [PubMed]

46. Reinholz, M.; Kawakami, Y.; Salzer, S.; Kreuter, A.; Dombrowski, Y.; Koglin, S.; Kresse, S.; Ruzicka, T.; Schauber, J. HPV16 activates the AIM2 inflammasome in keratinocytes. *Arch. Dermatol. Res.* **2013**, *305*, 723–732. [CrossRef] [PubMed]

47. Bahramabadi, R.; Dabiri, S.; Iranpour, M.; Kazemi Arababadi, M. TLR4: An important molecule participating in either anti-human papillomavirus immune responses or development of its related cancers. *Viral Immunol.* **2019**, *32*, 417–423. [CrossRef]

48. Hasan, U.A.; Bates, E.; Takeshita, F.; Biliato, A.; Accardi, R.; Bouvard, V.; Mansour, M.; Vincent, I.; Gissmann, L.; Iftner, T.; et al. TLR9 expression and function is abolished by the cervical cancer-associated human papillomavirus type 16. *J. Immunol.* **2007**, *178*, 3186–3197. [CrossRef]

49. Joyce, J.G.; Tung, J.-S.; Przysiecki, C.T.; Cook, J.C.; Lehman, E.D.; Sands, J.A.; Jansen, K.U.; Keller, P.M. The L1 major capsid protein of human papillomavirus type 11 recombinant virus-like particles interacts with heparin and cell-surface glycosaminoglycans on human keratinocytes. *J. Biol. Chem.* **1999**, *274*, 5810–5822. [CrossRef]

50. Yan, M.; Peng, J.; Jabbar, I.A.; Liu, X.; Filgueira, L.; Frazer, I.H.; Thomas, R. Activation of dendritic cells by human papillomavirus-like particles through TLR4 and NF-κB-mediated signalling, moderated by TGF-β. *Immunol. Cell Biol.* **2005**, *83*, 83–91. [CrossRef]

51. Kawasaki, T.; Kawai, T. Toll-Like receptor signaling pathways. *Front. Immunol.* **2014**, *5*, 461. [CrossRef]

52. Daud, I.I.; Scott, M.E.; Ma, Y.; Shiboski, S.; Farhat, S.; Moscicki, A.B. Association between toll-like receptor expression and human papillomavirus type 16 persistence. *Int. J. Cancer* **2011**, *128*, 879–886. [CrossRef] [PubMed]

53. Scott, M.E.; Ma, Y.; Farhat, S.; Moscicki, A.B. Expression of nucleic acid-sensing toll-like receptors predicts HPV16 clearance associated with an E6-directed cell-mediated response. *Int. J. Cancer* **2015**, *136*, 2402–2408. [CrossRef] [PubMed]

54. Lembo, D.; Donalisio, M.; De Andrea, M.; Cornaglia, M.; Scutera, S.; Musso, T.; Landolfo, S. A cell-based high-throughput assay for screening inhibitors of human papillomavirus-16 long control region activity. *FASEB J.* **2006**, *20*, 148–150. [CrossRef] [PubMed]

55. Scott, M.; Nakagawa, M.; Moscicki, A.B. Cell-Mediated immune response to human papillomavirus infection. *Clin. Diagn. Lab. Immunol.* **2001**, *8*, 209–220. [CrossRef] [PubMed]

56. Kowli, S.; Velidandla, R.; Creek, K.E.; Pirisi, L. TGF-β regulation of gene expression at early and late stages of HPV16-mediated transformation of human keratinocytes. *Virology* **2013**, *447*, 63–73. [CrossRef]

57. Lafleur, D.W.; Nardelli, B.; Tsareva, T.; Mather, D.; Feng, P.; Semenuk, M.; Taylor, K.; Buergin, M.; Chinchilla, D.; Roshke, V.; et al. Interferon-κ, a novel type I Interferon expressed in human keratinocytes. *J. Biol. Chem.* **2001**, *276*, 39765–39771. [CrossRef]

58. Woodby, B.L.; Songock, W.K.; Scott, M.L.; Raikhy, G.; Bodily, J.M. Induction of Interferon kappa in human papillomavirus 16 infection by transforming growth factor beta-induced promoter demethylation. *J. Virol.* **2018**, *92*, 1–17. [CrossRef]

59. Habiger, C.; Jäger, G.; Walter, M.; Iftner, T.; Stubenrauch, F. Interferon kappa inhibits human papillomavirus 31 transcription by inducing Sp100 proteins. *J. Virol.* **2016**, *90*, 694–704. [CrossRef]

60. Vartanian, J.P.; Guétard, D.; Henry, M.; Wain-Hobson, S. Evidence for editing of human papillomavirus DNA by APOBEC3 in benign and precancerous lesions. *Science* **2008**, *320*, 230–233. [CrossRef]

61. Warren, C.J.; Westrich, J.A.; Van Doorslaer, K.; Pyeon, D. Roles of APOBEC3A and APOBEC3B in human papillomavirus infection and disease progression. *Viruses* **2017**, *9*, 233. [CrossRef]

62. Wüstenhagen, E.; Boukhallouk, F.; Negwer, I.; Rajalingam, K.; Stubenrauch, F.; Florin, L. The Myb-related protein MYPOP is a novel intrinsic host restriction factor of oncogenic human papillomaviruses. *Oncogene* **2018**, *37*, 6275–6284. [CrossRef]

63. Wilson, S.S.; Wiens, M.E.; Smith, J.G. Antiviral mechanisms of human defensins. *J. Mol. Biol.* **2013**, *425*, 4965–4980. [CrossRef] [PubMed]

64. Gulati, N.M.; Miyagi, M.; Wiens, M.E.; Smith, J.G.; Stewart, P. α-Defensin HD5 stabilizes human papillomavirus 16 capsid/core interactions. *Pathog. Immun.* **2019**, *4*, 196–234. [CrossRef] [PubMed]

65. Wiens, M.E.; Smith, J.G. Alpha-Defensin HD5 inhibits furin cleavage of human papillomavirus 16 L2 to block infection. *J. Virol.* **2015**, *89*, 2866–2874. [CrossRef] [PubMed]

66. Buck, C.B.; Day, P.M.; Thompson, C.D.; Lubkowski, J.; Lu, W.; Lowy, D.R.; Schiller, J.T. Human α-defensins block papillomavirus infection. *Proc. Natl. Acad. Sci. USA* **2006**, *103*, 1516–1521. [CrossRef] [PubMed]

67. Hubert, P.; Herman, L.; Maillard, C.; Caberg, J.H.; Nikkels, A.; Pierard, G.; Foidart, J.M.; Noel, A.; Boniver, J.; Delvenne, P. Defensins induce the recruitment of dendritic cells in cervical human papillomavirus-associated (pre)neoplastic lesions formed In Vitro and transplanted In Vivo. *FASEB J.* **2007**, *21*, 2765–2775. [CrossRef] [PubMed]

68. Sunthamala, N.; Thierry, F.; Teissier, S.; Pientong, C.; Kongyingyoes, B.; Tangsiriwatthana, T.; Sangkomkamhang, U.; Ekalaksananan, T. E2 proteins of high risk human papillomaviruses down-modulate STING and IFN-κ transcription in keratinocytes. *PLoS ONE* **2014**, *9*, e91473. [CrossRef] [PubMed]

69. Ashrafi, G.H.; Haghshenas, M.R.; Marchetti, B.; O'Brien, P.M.; Campo, M.S. E5 protein of human papillomavirus type 16 selectively downregulates surface HLA class I. *Int. J. Cancer* **2005**, *113*, 276–283. [CrossRef]

70. Ashrafi, G.H.; Haghshenas, M.; Marchetti, B.; Campo, M.S. E5 protein of human papillomavirus 16 downregulates HLA class I and interacts with the heavy chain via its first hydrophobic domain. *Int. J. Cancer* **2006**, *119*, 2105–2112. [CrossRef]

71. Gruener, M.; Bravo, I.G.; Momburg, F.; Alonso, A.; Tomakidi, P. The E5 protein of the human papillomavirus type 16 down-regulates HLA-I surface expression in calnexin-expressing but not in calnexin-deficient cells. *Virol. J.* **2007**, *4*, 116. [CrossRef]

72. Niebler, M.; Qian, X.; Höfler, D.; Kogosov, V.; Kaewprag, J.; Kaufmann, A.M.; Ly, R.; Böhmer, G.; Zawatzky, R.; Rösl, F.; et al. Post-Translational control of IL-1β via the human papillomavirus type 16 E6 oncoprotein: A novel mechanism of innate immune escape mediated by the E3-ubiquitin ligase E6-AP and p53. *PLoS Pathog.* **2013**, *9*, e1003536. [CrossRef] [PubMed]

73. Chiang, C.; Pauli, E.-K.; Biryukov, J.; Feister, K.F.; Meng, M.; White, E.A.; Münger, K.; Howley, P.M.; Meyers, C.; Gack, M.U. The human papillomavirus E6 oncoprotein targets USP15 and TRIM25 to suppress RIG-I-mediated innate immune signaling. *J. Virol.* **2018**, *92*, e01737-17. [CrossRef]

74. Ronco, L.V.; Karpova, A.Y.; Vidal, M.; Howley, P.M. Human papillomavirus 16 E6 oncoprotein binds to interferon regulatory factor-3 and inhibits its transcriptional activity. *Genes Dev.* **1998**, *12*, 2061–2072. [CrossRef] [PubMed]

75. Spitkovsky, D.; Hehner, S.P.; Hofmann, T.G.; Möller, A.; Schmitz, M.L. The human papillomavirus oncoprotein E7 attenuates NF-kappa B activation by targeting the ikappa B kinase complex. *J. Biol. Chem.* **2002**, *277*, 25576–25582. [CrossRef] [PubMed]

76. Patel, D.; Huang, S.M.; Baglia, L.A.; McCance, D.J. The E6 protein of human papillomavirus type 16 binds to and inhibits co-activation by CBP and p300. *EMBO J.* **1999**, *18*, 5061–5072. [CrossRef]

77. Li, S.; Labrecque, S.; Gauzzi, M.C.; Cuddihy, A.R.; Wong, A.H.T.; Pellegrini, S.; Matlashewski, G.J.; Koromilas, A.E. The human papilloma virus (HPV)-18 E6 oncoprotein physically associates with Tyk2 and impairs Jak-STAT activation by interferon-α. *Oncogene* **1999**, *18*, 5727–5737. [CrossRef]

78. Hebner, C.M.; Wilson, R.; Rader, J.; Bidder, M.; Laimins, L.A. Human papillomaviruses target the double-stranded RNA protein kinase pathway. *J. Gen. Virol.* **2006**, *87*, 3183–3193. [CrossRef]

79. Chang, Y.E.; Laimins, L.A. Microarray analysis identifies interferon-inducible genes and Stat-1 as major transcriptional targets of human papillomavirus type 31. *J. Virol.* **2000**, *74*, 4174–4182. [CrossRef]

80. Reiser, J.; Hurst, J.; Voges, M.; Krauss, P.; Münch, P.; Iftner, T.; Stubenrauch, F. High-Risk human papillomaviruses repress constitutive kappa interferon transcription via E6 to prevent pathogen recognition receptor and antiviral-gene expression. *J. Virol.* **2011**, *85*, 11372–11380. [CrossRef]

81. Steinbach, A.; Winter, J.; Reuschenbach, M.; Blatnik, R.; Klevenz, A.; Bertrand, M.; Hoppe, S.; Von Knebel Doeberitz, M.; Grabowska, A.K.; Riemer, A.B. ERAP1 overexpression in HPV-induced malignancies: A possible novel immune evasion mechanism. *Oncoimmunology* **2017**, *6*, e1336594. [CrossRef]

82. Perea, S.E.; Massimi, P.; Banks, L. Human papillomavirus type 16 E7 impairs the activation of the interferon regulatory factor-1. *Int. J. Mol. Med.* **2000**, *5*, 661–666. [CrossRef] [PubMed]

83. Park, J.S.; Kim, E.J.; Kwon, H.J.; Hwang, E.S.; Namkoong, S.E.; Um, S.J. Inactivation of interferon regulatory factor-1 tumor suppressor protein by HPV E7 oncoprotein. *J. Biol. Chem.* **2000**, *275*, 6764–6769. [CrossRef] [PubMed]

84. Um, S.J.; Rhyu, J.W.; Kim, E.J.; Jeon, K.C.; Hwang, E.S.; Park, J.S. Abrogation of IRF-1 response by high-risk HPV E7 protein In Vivo. *Cancer Lett.* **2002**, *179*, 205–212. [CrossRef]

85. Huang, S.-M.; McCance, D.J. Down regulation of the interleukin-8 promoter by human papillomavirus type 16 E6 and E7 through effects on CREB binding protein/p300 and P/CAF. *J. Virol.* **2002**, *76*, 8710–8721. [CrossRef]

86. Barnard, P.; McMillan, N.A.J. The human papillomavirus E7 oncoprotein abrogates signaling mediated by interferon-α. *Virology* **1999**, *259*, 305–313. [CrossRef]

87. Burgers, W.A.; Blanchon, L.; Pradhan, S.; De Launoit, Y.; Kouzarides, T.; Fuks, F. Viral oncoproteins target the DNA methyltransferases. *Oncogene* **2007**, *26*, 1650–1655. [CrossRef]

88. Lau, A.; Gray, E.E.; Brunette, R.L.; Stetson, D.B. DNA tumor virus oncogenes antagonize the cGAS-STING DNA-sensing pathway. *Science* **2015**, *350*, 568–571. [CrossRef]

89. Lo Cigno, I.; Calati, F.; Borgogna, C.; Zevini, A.; Albertini, S.; Martuscelli, L.; De Andrea, M.; Hiscott, J.; Landolfo, S.; Gariglio, M. HPV E7 oncoprotein subverts host innate immunity via SUV39H1-mediated epigenetic silencing of immune sensor genes. *J. Virol.* **2019**, *94*, e01812–e018119. [CrossRef]

90. Hasan, U.A.; Zannetti, C.; Parroche, P.; Goutagny, N.; Malfroy, M.; Roblot, G.; Carreira, C.; Hussain, I.; Müller, M.; Taylor-Papadimitriou, J.; et al. The human papillomavirus type 16 E7 oncoprotein induces a transcriptional repressor complex on the toll-like receptor 9 promoter. *J. Exp. Med.* **2013**, *210*, 1369–1387. [CrossRef]

91. Pacini, L.; Savini, C.; Ghittoni, R.; Saidj, D.; Lamartine, J.; Hasan, U.A.; Accardi, R.; Tommasino, M. Downregulation of toll-like receptor 9 expression by beta human papillomavirus 38 and implications for cell cycle control. *J. Virol.* **2015**, *89*, 11396–11405. [CrossRef]

92. Nees, M.; Geoghegan, J.M.; Hyman, T.; Frank, S.; Miller, L.; Woodworth, C.D. Papillomavirus type 16 oncogenes downregulate expression of interferon-responsive genes and upregulate proliferation-associated and NF-κB-responsive genes in cervical keratinocytes. *J. Virol.* **2001**, *75*, 4283–4296. [CrossRef] [PubMed]

93. Havard, L.; Rahmouni, S.; Boniver, J.; Delvenne, P. High levels of p105 (NFKB1) and p100 (NFKB2) proteins in HPV16-transformed keratinocytes: Role of E6 and E7 oncoproteins. *Virology* **2005**, *331*, 357–366. [CrossRef] [PubMed]

94. Cordano, P.; Gillan, V.; Bratlie, S.; Bouvard, V.; Banks, L.; Tommasino, M.; Campo, M.S. The E6E7 oncoproteins of cutaneous human papillomavirus type 38 interfere with the interferon pathway. *Virology* **2008**, *377*, 408–418. [CrossRef] [PubMed]

95. Ishikawa, H.; Barber, G.N. STING is an endoplasmic reticulum adaptor that facilitates innate immune signalling. *Nature* **2008**, *455*, 674–678. [CrossRef] [PubMed]

96. Jin, L.; Waterman, P.M.; Jonscher, K.R.; Short, C.M.; Reisdorph, N.A.; Cambier, J.C. MPYS, a novel membrane tetraspanner, is associated with major histocompatibility complex class II and mediates transduction of apoptotic signals. *Mol. Cell. Biol.* **2008**, *28*, 5014–5026. [CrossRef]

97. Sun, W.; Li, Y.; Chen, L.; Chen, H.; You, F.; Zhou, X.; Zhou, Y.; Zhai, Z.; Chen, D.; Jiang, Z. ERIS, an endoplasmic reticulum IFN stimulator, activates innate immune signaling through dimerization. *Proc. Natl. Acad. Sci. USA* **2009**, *106*, 8653–8658. [CrossRef]

98. Zhong, B.; Yang, Y.; Li, S.; Wang, Y.-Y.; Li, Y.; Diao, F.; Lei, C.; He, X.; Zhang, L.; Tien, P.; et al. The adaptor protein MITA links virus-sensing receptors to IRF3 transcription factor activation. *Immunity* **2008**, *29*, 538–550. [CrossRef]

99. Paludan, S.R.; Bowie, A.G. Immune sensing of DNA. *Immunity* **2013**, *38*, 870–880. [CrossRef]

100. Gack, M.U.; Shin, Y.C.; Joo, C.-H.; Urano, T.; Liang, C.; Sun, L.; Takeuchi, O.; Akira, S.; Chen, Z.; Inoue, S.; et al. TRIM25 RING-finger E3 ubiquitin ligase is essential for RIG-I-mediated antiviral activity. *Nature* **2007**, *446*, 916–920. [CrossRef]

101. Jiang, X.; Kinch, L.N.; Brautigam, C.A.; Chen, X.; Du, F.; Grishin, N.V.; Chen, Z.J. Ubiquitin-Induced oligomerization of the RNA sensors RIG-I and MDA5 activates antiviral innate immune response. *Immunity* **2012**, *36*, 959–973. [CrossRef]

102. Yoneyama, M.; Onomoto, K.; Jogi, M.; Akaboshi, T.; Fujita, T. Viral RNA detection by RIG-I-like receptors. *Curr. Opin. Immunol.* **2015**, *32*, 48–53. [CrossRef] [PubMed]

103. Marques, M.; Ferreira, A.R.; Ribeiro, D. The interplay between human cytomegalovirus and pathogen recognition receptor signaling. *Viruses* **2018**, *10*, 514. [CrossRef] [PubMed]

104. Choi, M.K.; Wang, Z.; Ban, T.; Yanai, H.; Lu, Y.; Koshiba, R.; Nakaima, Y.; Hangai, S.; Savitsky, D.; Nakasato, M.; et al. A selective contribution of the RIG-I-like receptor pathway to type I interferon responses activated by cytosolic DNA. *Proc. Natl. Acad. Sci. USA* **2009**, *106*, 17870–17875. [CrossRef] [PubMed]

105. Melchjorsen, J.; Rintahaka, J.; Søby, S.; Horan, K.A.; Poltajainen, A.; Østergaard, L.; Paludan, S.R.; Matikainen, S.; Soby, S.; Horan, K.A.; et al. Early innate recognition of herpes simplex virus in human primary macrophages is mediated via the MDA5/MAVS-dependent and MDA5/MAVS/RNA polymerase III-independent pathways. *J. Virol.* **2010**, *84*, 11350–11358. [CrossRef] [PubMed]

106. Samanta, M.; Iwakiri, D.; Kanda, T.; Imaizumi, T.; Takada, K. EB virus-encoded RNAs are recognized by RIG-I and activate signaling to induce type I IFN. *EMBO J.* **2006**, *25*, 4207–4214. [CrossRef] [PubMed]

107. Chiu, Y.-H.; MacMillian, J.B.; Chen, Z.J. RNA polymerase III detects cytosolic DNA and induces type I interferons through the RIG-I pathway. *Cell* **2009**, *138*, 576–591. [CrossRef]

108. Ablasser, A.; Bauernfeind, F.; Hartmann, G.; Latz, E.; Fitzgerald, K.A.; Hornung, V. RIG-I-dependent sensing of poly(dA:dT) through the induction of an RNA polymerase III-transcribed RNA intermediate. *Nat. Immunol.* **2009**, *10*, 1065–1072. [CrossRef]

109. Chiang, J.J.; Sparrer, K.M.J.; Van Gent, M.; Lässig, C.; Huang, T.; Osterrieder, N.; Hopfner, K.P.; Gack, M.U. Viral unmasking of cellular 5S rRNA pseudogene transcripts induces RIG-I-mediated immunity article. *Nat. Immunol.* **2018**, *19*, 53–62. [CrossRef]

110. Magalhães, A.C.; Ferreira, A.R.; Gomes, S.; Vieira, M.; Gouveia, A.; Valença, I.; Islinger, M.; Nascimento, R.; Schrader, M.; Kagan, J.C.; et al. Peroxisomes are platforms for cytomegalovirus' evasion from the cellular immune response. *Sci. Rep.* **2016**, *6*, 26028. [CrossRef]

111. Ferreira, A.R.; Marques, M.; Ribeiro, D. Peroxisomes and innate immunity: Antiviral response and beyond. *Int. J. Mol. Sci.* **2019**, *20*, 3795. [CrossRef]

112. Rincon-Orozco, B.; Halec, G.; Rosenberger, S.; Muschik, D.; Nindl, I.; Bachmann, A.; Ritter, T.M.; Dondog, B.; Ly, R.; Bosch, F.X.; et al. Epigenetic silencing of interferon-κ in human papillomavirus type 16-positive cells. *Cancer Res.* **2009**, *69*, 8718–8725. [CrossRef] [PubMed]

113. Karim, R.; Tummers, B.; Meyers, C.; Biryukov, J.L.; Alam, S.; Backendorf, C.; Jha, V.; Offringa, R.; van Ommen, G.J.B.; Melief, C.J.M.; et al. Human papillomavirus (HPV) upregulates the cellular deubiquitinase UCHL1 to suppress the keratinocyte's innate immune response. *PLoS Pathog.* **2013**, *9*, e1003384. [CrossRef] [PubMed]

114. Oeckinghaus, A.; Ghosh, S. The NF-κB family of transcription factors and its regulation. *Cold Spring Harb. Perspect. Biol.* **2009**, *1*, a000034. [CrossRef] [PubMed]

115. Oeckinghaus, A.; Hayden, M.S.; Ghosh, S. Crosstalk in N-κB signaling pathways. *Nat. Immunol.* **2011**, *12*, 695–708. [CrossRef]

116. Richards, K.H.; Wasson, C.W.; Watherston, O.; Doble, R.; Blair, G.E.; Wittmann, M.; Macdonald, A.; Eric Blair, G.; Wittmann, M.; Macdonald, A. The human papillomavirus (HPV) E7 protein antagonises an Imiquimod-induced inflammatory pathway in primary human keratinocytes. *Sci. Rep.* **2015**, *5*, 12922. [CrossRef]

117. Sugiura, A.; Mattie, S.; Prudent, J.; McBride, H.M. Newly born peroxisomes are a hybrid of mitochondrial and ER-derived pre-peroxisomes. *Nature* **2017**, *542*, 251–254. [CrossRef]

118. Bernat, A.; Avvakumov, N.; Mymryk, J.S.; Banks, L. Interaction between the HPV E7 oncoprotein and the transcriptional coactivator p300. *Oncogene* **2003**, *22*, 7871–7881. [CrossRef]

119. Takami, Y.; Nakagami, H.; Morishita, R.; Katsuya, T.; Cui, T.X.; Ichikawa, T.; Saito, Y.; Hayashi, H.; Kikuchi, Y.; Nishikawa, T.; et al. Ubiquitin carboxyl-terminal hydrolase L1, a novel deubiquitinating enzyme in the vasculature, attenuates NF-κB activation. *Arterioscler. Thromb. Vasc. Biol.* **2007**, *27*, 2184–2190. [CrossRef]

120. Liu, T.; Zhang, L.; Joo, D.; Sun, S.C. NF-κB signaling in inflammation. *Signal. Transduct. Target. Ther.* **2017**, *2*, e17023. [CrossRef]

121. Cho, Y.S.; Kang, J.W.; Cho, M.C.; Cho, C.W.; Lee, S.J.; Choe, Y.K.; Kim, Y.M.; Choi, I.P.; Park, S.N.; Kim, S.H.; et al. Down modulation of IL-18 expression by human papillomavirus type 16 E6 oncogene via binding to IL-18. *FEBS Lett.* **2001**, *501*, 139–145. [CrossRef]

122. Lee, S.-J.; Cho, Y.-S.; Cho, M.-C.; Shim, J.-H.; Lee, K.-A.; Ko, K.-K.; Choe, Y.K.; Park, S.-N.; Hoshino, T.; Kim, S.; et al. Both E6 and E7 oncoproteins of human papillomavirus 16 inhibit IL-18-induced IFN-γ production in human peripheral blood mononuclear and NK cells. *J. Immunol.* **2001**, *167*, 497–504. [CrossRef] [PubMed]

123. So, D.; Shin, H.W.; Kim, J.; Lee, M.; Myeong, J.; Chun, Y.S.; Park, J.W. Cervical cancer is addicted to SIRT1 disarming the AIM2 antiviral defense. *Oncogene* **2018**, *37*, 5191–5204. [CrossRef] [PubMed]

124. Senba, M.; Buziba, N.; Mori, N.; Fujita, S.; Morimoto, K.; Wada, A.; Toriyama, K. Human papillomavirus infection induces NF-κB activation in cervical cancer: A comparison with penile cancer. *Oncol. Lett.* **2011**, *2*, 65–68. [CrossRef] [PubMed]

125. Hemmat, N.; Bannazadeh Baghi, H. Association of human papillomavirus infection and inflammation in cervical cancer. *Pathog. Dis.* **2019**, *77*, ftz048. [CrossRef]

126. Barnard, P.; Payne, E.; McMillan, N.A.J. The human papillomavirus E7 protein is able to inhibit the antiviral and anti-growth functions of interferon-α. *Virology* **2000**, *277*, 411–419. [CrossRef]

127. Antonsson, A.; Payne, E.; Hengst, K.; McMillan, N.A.J. The human papillomavirus type 16 E7 protein binds human interferon regulatory factor-9 via a novel PEST domain required for transformation. *J. Interf. Cytokine Res.* **2006**, *26*, 455–461. [CrossRef]

128. Kazemi, S.; Papadopoulou, S.; Li, S.; Su, Q.; Wang, S.; Yoshimura, A.; Matlashewski, G.; Dever, T.E.; Koromilas, A.E. Control of subunit of eukaryotic translation initiation factor 2 (eIF2) phosphorylation by the human papillomavirus type 18 E6 oncoprotein: Implications for eIF2-dependent gene expression and cell death. *Mol. Cell. Biol.* **2004**, *24*, 3415–3429. [CrossRef]

129. Karim, R.; Meyers, C.; Backendorf, C.; Ludigs, K.; Offringa, R.; van Ommen, G.J.B.; Melief, C.J.M.; van der Burg, S.H.; Boer, J.M. Human papillomavirus deregulates the response of a cellular network comprising of chemotactic and proinflammatory genes. *PLoS ONE* **2011**, *6*, e17848. [CrossRef]

130. Morale, M.G.; da Silva Abjaude, W.; Silva, A.M.; Villa, L.L.; Boccardo, E. HPV-Transformed cells exhibit altered HMGB1-TLR4/MyD88-SARM1 signaling axis. *Sci. Rep.* **2018**, *8*, 3476. [CrossRef]

131. Arany, I.; Goel, A.; Tyring, S.K. Interferon response depends on viral transcription in human papillomavirus-containing lesions. *Anticancer Res.* **1995**, *15*, 2865–2869.

132. Martín-Vicente, M.; Medrano, L.M.; Resino, S.; García-Sastre, A.; Martínez, I. TRIM25 in the regulation of the antiviral innate immunity. *Front. Immunol.* **2017**, *8*, 1187. [CrossRef] [PubMed]

133. Li, J.; Rao, H.; Jin, C.; Liu, J. Involvement of the toll-like receptor/nitric oxide signaling pathway in the pathogenesis of cervical cancer caused by high-risk human papillomavirus infection. *Biomed. Res. Int.* **2017**, *2017*, 7830262. [CrossRef] [PubMed]

134. Salimian Rizi, B.; Achreja, A.; Nagrath, D. Nitric oxide: The forgotten child of tumor metabolism. *Trends Cancer* **2017**, *3*, 659–672. [CrossRef] [PubMed]

135. Morgan, E.L.; Wasson, C.W.; Hanson, L.; Kealy, D.; Pentland, I.; McGuire, V.; Scarpini, C.; Coleman, N.; Arthur, J.S.C.; Parish, J.L.; et al. STAT3 activation by E6 is essential for the differentiation-dependent HPV18 life cycle. *PLoS Pathog.* **2018**, *14*, e1006975. [CrossRef] [PubMed]

136. Shishodia, G.; Verma, G.; Srivastava, Y.; Mehrotra, R.; Das, B.C.; Bharti, A.C. Deregulation of microRNAs Let-7a and miR-21 mediate aberrant STAT3 signaling during human papillomavirus-induced cervical carcinogenesis: Role of E6 oncoprotein. *BMC Cancer* **2014**, *14*, 996. [CrossRef]

137. Li, Y.X.; Zhang, L.; Simayi, D.; Zhang, N.; Tao, L.; Yang, L.; Zhao, J.; Chen, Y.Z.; Li, F.; Zhang, W.J. Human papillomavirus infection correlates with inflammatory stat3 signaling activity and IL-17 level in patients with colorectal cancer. *PLoS ONE* **2015**, *10*, e0118391. [CrossRef]

138. De Andrea, M.; Rittà, M.; Landini, M.M.; Borgogna, C.; Mondini, M.; Kern, F.; Ehrenreiter, K.; Baccarini, M.; Marcuzzi, G.P.; Smola, S.; et al. Keratinocyte-specific stat3 heterozygosity impairs development of skin tumors in human papillomavirus 8 transgenic mice. *Cancer Res.* **2010**, *70*, 7938–7948. [CrossRef]

139. Shukla, S.; Shishodia, G.; Mahata, S.; Hedau, S.; Pandey, A.; Bhambhani, S.; Batra, S.; Basir, S.F.; Das, B.C.; Bharti, A.C. Aberrant expression and constitutive activation of STAT3 in cervical carcinogenesis: Implications in high-risk human papillomavirus infection. *Mol. Cancer* **2010**, *9*, 282. [CrossRef]

140. Hong, S.; Laimins, L.A. The JAK-STAT transcriptional regulator, STAT-5, activates the ATM DNA damage pathway to induce HPV 31 genome amplification upon epithelial differentiation. *PLoS Pathog.* **2013**, *9*, e1003295. [CrossRef]

141. Lu, R.; Zhang, Y.G.; Sun, J. STAT3 activation in infection and infection-associated cancer. *Mol. Cell. Endocrinol.* **2017**, *451*, 80–87. [CrossRef]

142. Yu, H.; Pardoll, D.; Jove, R. STATs in cancer inflammation and immunity: A leading role for STAT3. *Nat. Rev. Cancer* **2009**, *9*, 798–809. [CrossRef] [PubMed]

143. Yang, A.; Farmer, E.; Wu, T.C.; Hung, C.F. Perspectives for therapeutic HPV vaccine development. *J. Biomed. Sci.* **2016**, *23*, 75. [CrossRef] [PubMed]

*Article*

# Hypoxia-Induced Centrosome Amplification Underlies Aggressive Disease Course in HPV-Negative Oropharyngeal Squamous Cell Carcinomas

Karuna Mittal [1,†], Da Hoon Choi [1,†], Guanhao Wei [1], Jaspreet Kaur [1], Sergey Klimov [1], Komal Arora [1], Christopher C. Griffith [2], Mukesh Kumar [1], Precious Imhansi-Jacob [1], Brian D. Melton [1], Sonal Bhimji-Pattni [2], Remus M. Osan [1], Padmashree Rida [3], Paweł Golusinski [4,5] and Ritu Aneja [1,*]

[1] Department of Biology, Georgia State University, Atlanta, GA 30303, USA; kkaruna.goel@gmail.com (K.M.); dchoi@msm.edu (D.H.C.); gwei1@student.gsu.edu (G.W.); jkaur2@student.gsu.edu (J.K.); sklimov1@student.gsu.edu (S.K.); karora@gsu.edu (K.A.); mkumar8@gsu.edu (M.K.); pimhansijacob1@student.gsu.edu (P.I.-J.); meltonb@janelia.hhmi.org (B.D.M.); rosan@gsu.edu (R.M.O.)
[2] Emory Hospital Midtown, Atlanta, GA 30308, USA; griffic8@ccf.org (C.C.G.); sonal.pattni1@gmail.com (S.B.-P.)
[3] Novazoi Theranostics, Rancho Palos Verdes, CA 90725, USA; cgp_rida@yahoo.com
[4] Department of Otolaryngology and Maxillofacial Surgery, University of Zielona Gora, 65-417 Zielona Gora, Poland; pgolusinski@uz.zgora.pl
[5] Department of Biology and Environmental Sciences, Poznan University of Medical Sciences, 61-701 Poznan, Poland
* Correspondence: raneja@gsu.edu; Tel.: +1-404-413-5417; Fax: +1-404-413-5301
† These authors contributed equally to this work.

Received: 11 December 2019; Accepted: 17 February 2020; Published: 24 February 2020

**Abstract:** Human papillomavirus-negative (HPV-neg) oropharyngeal squamous cell carcinomas (OPSCCs) are associated with poorer overall survival (OS) compared with HPV-positive (HPV-pos) OPSCCs. The major obstacle in improving outcomes of HPV-neg patients is the lack of robust biomarkers and therapeutic targets. Herein, we investigated the role of centrosome amplification (CA) as a prognostic biomarker in HPV-neg OPSCCs. A quantitative evaluation of CA in clinical specimens of OPSCC revealed that (a) HPV-neg OPSCCs exhibit higher CA compared with HPV-pos OPSCCs, and (b) CA was associated with poor OS, even after adjusting for potentially confounding clinicopathologic variables. Contrastingly, CA was higher in HPV-pos cultured cell lines compared to HPV-neg ones. This divergence in CA phenotypes between clinical specimens and cultured cells can therefore be attributed to an inaccurate recapitulation of the in vivo tumor microenvironment in the cultured cell lines, namely a hypoxic environment. The exposure of HPV-neg OPSCC cultured cells to hypoxia or stabilizing HIF-1$\alpha$ genetically increased CA. Both the 26-gene hypoxia signature as well as the overexpression of HIF-1$\alpha$ positively correlated with increased CA in HPV-neg OPSCCs. In addition, we showed that HIF-1$\alpha$ upregulation is associated with the downregulation of miR-34a, increase in CA and expression of cyclin- D1. Our findings demonstrate that the evaluation of CA may aid in therapeutic decision-making, and CA can serve as a promising therapeutic target for HPV-neg OPSCC patients.

**Keywords:** hypoxia; human papillomavirus; oropharyngeal squamous cell carcinoma; centrosome amplification; miRNA

---

## 1. Introduction

Head and neck squamous cell carcinoma (HNSCC) is the sixth most common cancer worldwide [1]. Although the incidence and mortality rates for HNSCC are declining globally, there has been a gradual increase in the incidence rate of oropharyngeal squamous cell carcinomas (OPSCCs) in recent years [2]. OPSCC is a type of HNSCC, which includes cancer of the tonsils, base of the tongue, back of the roof of the mouth and the side and back walls of the throat. A major factor underlying the increased incidence rate of OPSCCs is human papillomavirus (HPV) infection [3]. Studies have shown that HPV-positive (HPV-pos) OPSCC patients respond better to treatment compared to patients with HPV-negative (HPV-neg) OPSCC. A randomized trial comparing accelerated-fractionation with standard-fractionation radiotherapy, in combination with cisplatin therapy showed better rates of three-year overall survival (82.4%) in HPV-pos OPSCC patients compared to their HPV-neg counterparts (57.1%) [4]. In the absence of a good therapeutic target, conventional chemotherapy is unfortunately still the mainstay of treatment for HPV-neg OPSCC patients, whose prognosis remains dismal.

Treatment decisions for OPSCC patients are made based on disease stage and tumor location, without taking into account the tumor's HPV status [5]. Recent studies have suggested that HPV-pos and HPV-neg OPSCCs are biologically unique entities [4], with different tumor biologies and characteristics. Thus, a new stage categorization has been introduced for HPV-pos OPSCC in the 8th edition TNM [T describes the size of the tumor and any spread of cancer into nearby tissue; N describes spread of cancer to nearby lymph nodes; and M describes metastasis (spread of cancer to other parts of the body)] classification system [6], wherein p16 staining serves as a surrogate for HPV status. Studies have indicated that HNSCC is a heterogeneous disease with higher chromosomal instability (CIN) in HPV-neg (50% higher mutational load) compared to HPV-pos HNSCCs [7]. The inactivation of tumor suppressor genes such as TP53 and CDKNA2 and the oncogenic activation of the *CCND1* gene have been shown to be crucial for pathogenesis and disease progression in HPV-neg HNSCCs [8]. By contrast, the inactivation of the tumor suppressor genes and activation of oncoproteins in HPV-pos tumors have been linked to the viral E6 and E7 oncoproteins [9]. A key feature of E6 and E7 oncoproteins is that they both converge to induce centrosome amplification (CA) [10], a hallmark of cancer. Amplified centrosomes are well-recognized to drive CIN, which fuels tumorigenesis, tumor progression, drug resistance and, as a result, poor prognosis [11]. CA can be numerical (increase in the number of centrosomes) as well as structural (increase in size/volume of centrosomes) and can arise in multiple ways, including the failure of the cell to undergo cytokinesis, inappropriate replication of centrosomes, and de novo generation of centrosomes [12,13].

Past genomic analyses of HNSCCs have described CIN to be a more prominent feature in HPV-neg tumors than in HPV-pos tumors [14]. Drivers of CA are distinct in HPV-pos and HPV-neg HNSCCs; CA in the former subtype is driven by viral oncoproteins, and that in the latter is driven by the misexpression of host cell-encoded proteins. Aurora A kinase and PLK1 are major factors contributing to CIN in HPV-neg HNSCCs [15]. Both Aurora A and PLK1 promote CIN by deregulating the spindle assembly checkpoint, resulting in chromosome mis-segregation and the amplification of centrosomes [16]. Therefore, we reasoned that CA may be a readily quantifiable prognostic biomarker and a potentially druggable target for HPV-neg tumors.

Another critical factor underlying poorer prognosis in patients with HPV-neg cancers is tumor hypoxia. Tumor hypoxia has long been associated with poor responses to radiotherapy and chemotherapy [17]. A recent study has shown that HPV-neg oropharyngeal tumors display higher tumor hypoxia [18]. Additionally, reduced partial pressure of oxygen inside tumors plays a significant role in overexpression of Aurora-A/STK15 [19], and this overexpression results in CA, CIN, and aneuploidy. The hypoxia-mediated overexpression of PLK4 has been well-documented to induce CA [20]. Recently, we have shown that hypoxic tumor microenvironment can induce CA via the stabilization of the transcriptional factor HIF-1$\alpha$ in breast cancer, facilitating an aggressive disease course [21]. Thus, there is mounting evidence that hypoxia-associated CA may underlie the aggressive disease course and treatment resistance of HPV-neg OPSCCs.

No studies to date have reported quantitation of centrosomal aberrations in OPSCCs with inherently different HPV status. Herein, we performed a thorough quantitative analysis of centrosomal aberrations in OPSCC tumors to compare differences in incidence and severity of CA between HPV-pos and HPV-neg OPSCCs. Interestingly, we found that HPV-neg OPSCCs exhibited significantly higher CA when compared with HPV-pos OPSCCs, and CA was associated with the poor overall survival in HPV-neg OPSCCs. Furthermore, we found that HPV-neg tumors show a higher expression of HIF-1$\alpha$ than HPV-pos OPSCCs, and there was a strong association between CA and HIF-1$\alpha$ expression in HPV-neg OPSCCs. In addition, we found that HPV-neg tumors exhibited a higher expression of the CA-associated protein, cyclin D1. Our study also showed that HIF-1$\alpha$ upregulation is associated with the downregulation of miR-34a, an increase in CA and the expression of cyclin- D1. Taken together, these results shed new light on the drivers of tumor biology in HPV-neg tumors and emphasize the role of CA as new prognostic marker and actionable target to improve outcomes in HPV-neg OPSCC patients.

## 2. Results

### 2.1. Discordant Relationship between CA in Cultured Cells versus Patient Samples for HPV-Pos and HPV-Neg OPSCC

Previous studies from our group have shown that higher levels of CA are associated with poor prognosis and aggressive disease course in multiple malignancies, including breast cancer, pancreatic cancer, and serous ovarian adenocarcinoma [22–24]. We are the first to perform a rigorous quantitative study to compare CA in HPV-pos and HPV-neg OPSCCs. Given that the HPV-neg OPSCCs exhibit a higher expression of DNA damage response genes and are associated with a more aggressive disease course and poorer overall survival than HPV-pos OPSCCs, we postulated that HPV-neg tumors might exhibit significant CA. In this study, centrosomes in resection samples from 98 OPSCC patients (n = 47 HPV-pos and n = 51 HPV-neg samples) were visualized by immunofluorescence confocal imaging (details in supplementary information). Centrosome numbers and volumes were evaluated to generate a cumulative percentage of structurally and numerically amplified centrosomes for each sample (patient cohort details are shown in Table 1). Wilcoxon distribution scores revealed that HPV-neg OPSCCs exhibit significantly higher CA (numerical and structural) compared to HPV-pos OPSCCs ($p = 0.034$; Figure 1A,B). In agreement with the view that CA drives tumor progression, our data (Figure S1A) showed that, among HPV-neg tumors, higher CA was associated with a higher disease stage ($p = 0.1013$). Moreover, HPV-neg tumors displayed higher CA compared to stage-matched (stage III and stage IV; $p = 0.1225$ and $p = 0.0551$, respectively) HPV-pos tumors (Figure S1B,C).

However, when cultured cells from HPV-pos and HPV-neg OPSCCs were stained for CA, we observed that HPV-pos OPSCC cells (SCC090) exhibit higher CA ($p < 0.05$) relative to the HPV-neg OPSCC cells (FaDu and SCC-25) (Figure 1C,D). Our findings from RT-PCR also indicate significantly higher levels of centrosomal (pericentrin, $\gamma$-tubulin and centrin-1) and CA-associated (PLK4) mRNA expression in HPV-pos OPSCC cells compared to levels observed in HPV-neg OPSCC cells (Figure 1E). Taken together, these findings clearly demonstrate a high prevalence of CA in HPV-neg OPSCC tissue sections from patient samples, but not in cultured cell lines, and suggest that differences between the in vivo tumor microenvironment and in vitro culture conditions are at least partly responsible for this discrepant observation.

**Table 1.** Descriptive statistics of clinicopathological characteristics and treatment for OPSCC patients in the clinical samples analyzed for centrosome amplification phenotypes.

| Baseline Characteristics | HPV-Neg | HPV-Pos | $p$ Value |
|---|---|---|---|
| **Gender, n (%)** | | | |
| Female | 11 (21.57) | 19 (40.43) | 0.0430 |
| Male | 40 (78.43) | 28 (59.57) | |
| **Tumor Grade, n (%)** | | | |
| 1 | 5 (9.80) | 1 (2.13) | |
| 2 | 22 (43.14) | 27 (57.45) | 0.2957 |
| 3 | 18 (32.29) | 13 (27.66) | |
| N/A | 6 (11.76) | 6 (12.77) | |
| **Tumor Stage, n (%)** | | | |
| II | 1 (1.96) | 0 (0.00) | |
| III | 15 (29.41) | 9 (19.15) | 0.1933 |
| IV | 28 (54.90) | 35 (74.47) | |
| N/A | 7 (13.73) | 3 (6.38) | |
| **Smoking, n (%)** | | | |
| 0 | 6 (11.76) | 5 (10.64) | |
| 1 | 3 (5.58) | 29 (61.70) | <0.0001 |
| 2 | 4 (7.84) | 6 (12.77) | |
| 3 | 38 (74.51) | 7 (14.89) | |
| **Alcohol, n (%)** | | | |
| No | 42 (82.35) | 46 (97.87) | |
| Yes | 8 (15.69) | 1 (2.13) | 0.0393 |
| N/A | 1 (1.96) | 0 (0.00) | |
| **CA, n (%)** | | | |
| High | 34 (66.67) | 26 (55.32) | 0.2494 |
| Low | 17 (33.33) | 21 (44.58) | |
| **Chemotherapy, n (%)** | | | |
| None | 32 (62.75) | 18 (38.30) | |
| Concomitant | 12 (23.53) | 26 (55.32) | 0.0070 |
| Neoadjuvant | 2 (3.92) | 0 (0.00) | |
| Adjunctive | 2 (3.92) | 0 (0.00) | |
| N/A | 3 (5.88) | 3 (6.39) | |
| **Radiotherapy, n (%)** | | | |
| None | 5 (9.80) | 2 (4.26) | |
| Primary | 11 (21.57) | 14 (29.79) | 0.1828 |
| Adjuvant | 26 (50.98) | 29 (61.70) | |
| Palliative | 6 (11.76) | 1 (2.13) | |
| N/A | 3 (5.88) | 1 (2.13) | |

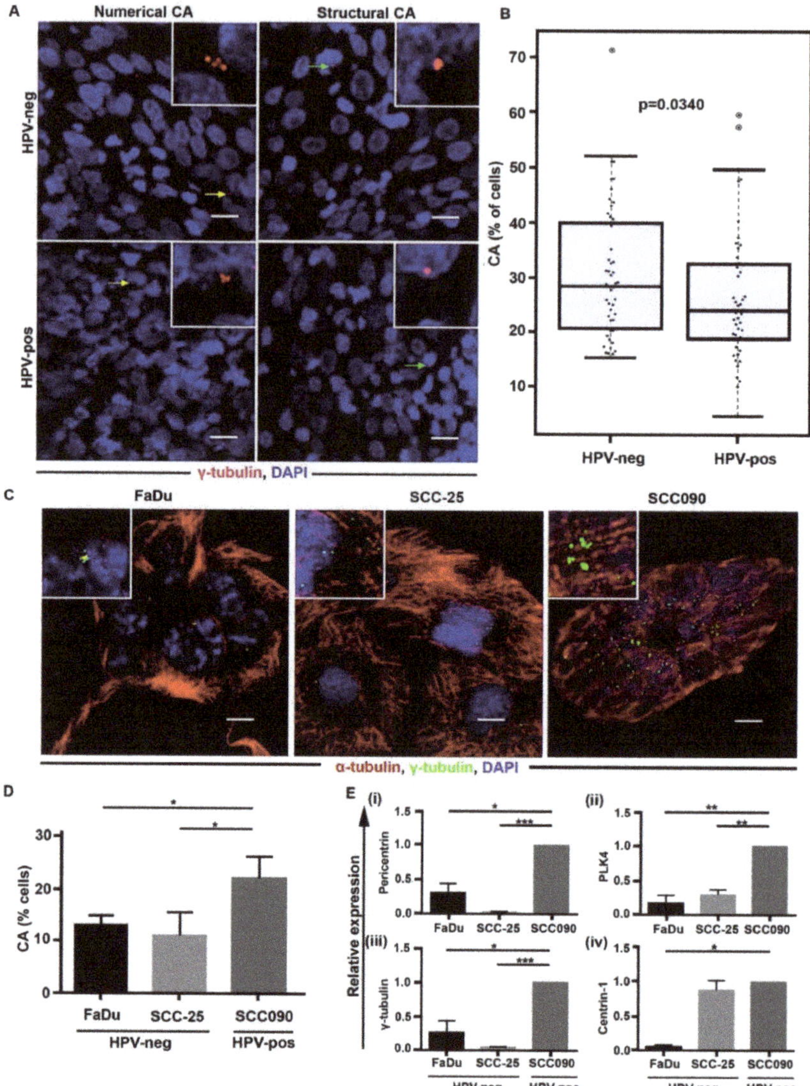

**Figure 1.** HPV-neg OPSCC specimens exhibit higher CA when compared with the HPV-pos OPSCCs. (**A**) Confocal immunomicrographs showing numerical and structural CA in HPV-pos and HPV-neg OPSCC sections. OPSCC tissue sections were immunostained for centrosomes (γ-tubulin, red) and counterstained with Hoechst (blue). Scale bar (white), 20 μm. (**B**) Percentage distribution of cells with CA (structural and numerical) in HPV-neg (n = 51) and HPV-pos (n = 47) patients ($p < 0.0340$). (**C**) Confocal immunomicrographs showing numerical CA in HPV-pos and HPV-neg tumor cells. OPSCC cells were immunostained for centrosomes (γ-tubulin, green), microtubules (α- tubulin, red) and counterstained with Hoechst (blue). Scale bar (white), 20 μm. (**D**) Percentage distribution of cells with CA (structural and numerical) in HPV-neg and HPV-pos cells ($p < 0.05$). (**E**) qRT-PCR analysis of mRNAs for γ-tubulin, pericentrin, centrin-1, and PLK4 in FaDu, SCC25, and SCC090 cells. Data were normalized by the amount of GAPDH mRNA, expressed relative to the corresponding value for all the cells and are means ± SD from triplicate data. * $p \leq 0.05$, ** $p \leq 0.01$, *** $p \leq 0.001$, **** $p \leq 0.0001$.

## 2.2. CA Is Associated with Poor Overall Survival in HPV-Neg OPSCCs

Consistent with previous studies, our data show that patients with HPV-neg OPSCCs had poorer overall survival when compared to patients with HPV-pos OPSCCs (HR = 4.332; $p$ = 0.005) (Figure 2A). When HPV-pos and HPV-neg patients were stratified into low- and high-CA (threshold used was the one that minimized log-rank $p$-value), we found that high-CA HPV-neg OPSCCs were associated with a poorer overall survival relative to high-CA HPV-pos OPSCCs (HR = 9.848; $p$ = 0.007) (Figure 2B). Furthermore, within HPV-neg OPSCCs, the high-CA subgroup was associated with worse overall survival relative to the subgroup with low-CA HPV-neg OPSCCs (Figure 2C). This association of overall survival with CA was significant (HR = 7.3; $p$ = 0.02) in our multivariable analysis when potentially confounding factors like smoking, alcohol consumption, grade and tumor stage were accounted for (Table 2). In univariate and multivariate analyses, only CA remained significantly associated with OS (Table 2). In contrast, no significant differences were found between the overall survival of low-CA HPV-pos and HPV-neg subgroups. Taken together, HPV-neg OPSCCs exhibit higher CA when compared with HPV-pos OPSCCs and higher CA in the HPV-neg OPSCCs is associated with poorer overall survival.

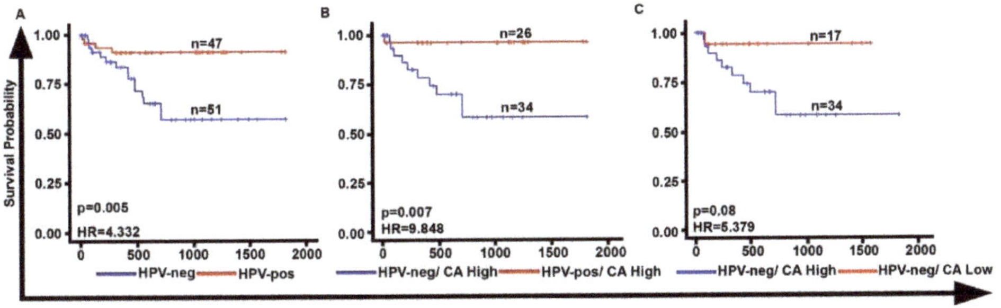


**Figure 2.** CA is associated with poor overall survival in HPV-neg OPSCCs. (**A**) Kaplan Meier survival curves representing the survival probabilities of HPV-neg (n = 51) and HPV-pos (n = 47) OPSCC patients (HR = 4.629; $p$ = 0.0062), (**B**) Kaplan Meier survival curves representing the survival probabilities of high-CA HPV-neg (n = 34) and HPV-pos (n = 26) OPSCC patients (HR = 5.07; $p$ = 0.003), and (**C**) Kaplan Meier survival curves representing the survival probabilities of high-CA HPV-neg (n = 34) and low CA HPV-neg (n = 17) OPSCC patients (HR = 5.379, $p$ = 0.08).

**Table 2.** Univariate and multivariate Cox proportional regression analysis for overall survival in HPV-neg OPSCCs, comparing the influence of common clinicopathological variables relative to the percentage of CA on patients' overall survival.

| Variable | | Univariate Analysis | | | Multivariate Analysis | | |
|---|---|---|---|---|---|---|---|
| | | $p$-Value | Hazard Ratio | 95% Hazard Ratio Confidence Limits | $p$-Value | Hazard Ratio | 95% Hazard Ratio Confidence Limits |
| CA | High vs. low | 0.05 | 5.38 | 1.69 | 42.17 | 0.02 | 7.38 | 1.32 | 41.28 |
| Gender | Female | 0.35 | 0.37 | 0.05 | 2.90 | 0.61 | 1.64 | 0.23 | 11.49 |
| Age at diagnosis | High vs. low | 0.37 | 0.96 | 0.86 | 1.06 | 0.27 | 1.09 | 0.93 | 1.29 |
| Overall stage | IV vs. other | 0.77 | 0.84 | 0.27 | 2.64 | 0.49 | 1.87 | 0.32 | 11.04 |
| Alcohol abuse | Yes vs. no | 0.46 | 1.79 | 0.38 | 8.37 | 0.18 | 7.12 | 0.39 | 129.29 |
| Radiotherapy | Yes | 0.81 | 1.29 | 0.16 | 10.20 | 0.99 | 0.00 | 0.00 | - |
| Smoking | No vs. yes | 0.95 | 0.96 | 0.26 | 3.57 | 0.85 | 1.16 | 0.24 | 5.69 |
| Tumor size | >vs. ≤4 | 0.35 | 1.86 | 0.50 | 6.88 | 0.56 | 0.58 | 0.09 | 3.62 |
| Nodal metastasis | Present vs. absent | 0.94 | 0.92 | 0.11 | 7.85 | 0.17 | 0.29 | 0.05 | 1.71 |

To strengthen our clinical findings, we performed an in silico analysis of the publicly available the cancer genome atlas (TCGA) microarray data of HNSCC patients (Table S1) [25]. Herein, we evaluated the expression levels of seven CA-related genes and the associations between this signature and overall survival in these patients. A cumulative score (CA7) was generated by adding the log-transformed values, normalized gene expression for CCND1, NEK2, PIN1, TUBG1, PLK1, BIRC5 and AURKA. First, we evaluated the CA7 score in all (n = 521) HNSCCs, regardless of subtype and HPV status. The patients were then stratified into high- and low-CA subgroups using the optimal CA7 score cut-off point (based on the log-rank test). Our findings demonstrate that (Figure S2A) high CA7 score HNSCCs (n = 420) were associated with poorer survival (HR = 1.497; $p$ = 0.0389) when compared with the low CA7 HNSCCs (n = 101). Interestingly, among OPSCC patients (n = 80; HPV-neg = 26 and HPV-pos = 54) we found that a high CA7 score was associated with poor overall survival (Figure S2B), regardless of HPV status (HR = 11.369; $p$ < 0.0001).

CA7 score was not able to stratify HPV-neg HNSCCs into high- and low-risk subgroups significantly. Since HPV-pos and HPV-neg tumors are distinct disease entities, we designed a subtype-specific weighted gene expression signature based on the appropriately weighted expression of the CA7 genes in each subgroup. To develop this signature, the expression for each CA7 gene was split into high-versus low-expression subgroups through the optimization of the log-rank test statistic, and then the hazard ratio parameter estimate for the high expression group was determined. High-expression gene groups that had a negative impact on survival had a positive parameter estimate, while genes that correlated positively with good prognosis had a negative parameter estimate. The total weighted sum for each patient was generated by adding the parameter estimates for each gene which had above threshold expression (if they were in the low expression group, they were given a 0 for that gene weight). The cutoff between high- and low-weighted CA7 scores was optimized using log-rank statistics. Notably, this new model was able to stratify HPV-neg HNSCCs into high- and low-risk groups with higher significance (HR = 1.867; $p$ < 0.001) (Figure S2C,D). Among HPV-neg OPSCCs, the high CA7 group (n = 6) showed a strong trend towards poorer overall survival (HR = 2.242; $p$ = 0.113) when compared to the low CA7 group (n = 20) (data not shown), but owing to the small sample size, this difference did not attain achieve statistical significance.

*2.3. Hypoxia Enhances CA in HPV-Neg OPSCCs via HIF-1α*

Our laboratory has previously identified that hypoxia induces CA in breast tumors via HIF-1α. Given that the discordance in CA is much higher between patient samples and cultured cell lines in HPV-neg OPSCCs compared to HPV-pos ones, we hypothesized that this discrepancy may be a result of the divergent tumor cell micro-environments under in vivo and in vitro conditions. Hypoxia, which is inadequately reflected in in vitro cell cultures, may underlie this observed divergence in CA. To test this hypothesis, we exposed HPV-neg (SCC-25 and FaDu) and HPV-pos (SCC90) OPSCC cell lines to hypoxia for 48 h using a hypoxic chamber flushed with a 1% $O_2$ gas mixture. The presence of hypoxia was confirmed by the upregulation of HIF-1α through Western blot (Figure 3C and Figure S8 and Table S3). We observed that following hypoxia treatment, HPV-neg OPSCC cells (SCC-25 and FaDu) exhibited significantly higher CA ($p$ < 0.05) compared to their normoxic controls, whereas no significant difference in CA levels was observed in HPV-pos OPSCC cells (SCC90) (Figure 3A,B). It is well-recognized that hypoxia mediates its function through the regulation of hypoxia-regulated genes by the transcription factor hypoxia-inducible factor-1 (HIF-1). The functional HIF-1 heterodimer (composed of alpha and beta subunits) is stabilized under hypoxic conditions and binds to hypoxia response elements (HREs) in target gene promoters. To confirm that the increase in CA under hypoxia was due to HIF-1α, cells cultured under normoxic conditions were transfected with GFP-tagged degradation-resistant HIF-1α (Figures S3C and S9 and Table S3). These transfected SCC25 and FaDu (HPV-neg) cell lines displayed higher CA (FaDu HIF-1α OE ~24%, SCC-25 HIF-1α OE ~22%) under normoxic conditions than vector controls (FaDu HIF-1α CV 12%, SCC-25 HIF-1α CV 9%) (Figure S3A,B) and the increase in CA was further confirmed with RT- PCR (Figure S3D). Meanwhile, no significant difference was observed between the control (control vector- CV) and HIF-1α overexpressing (overexpression- OE) HPV-pos OPSCC cells (SCC90).

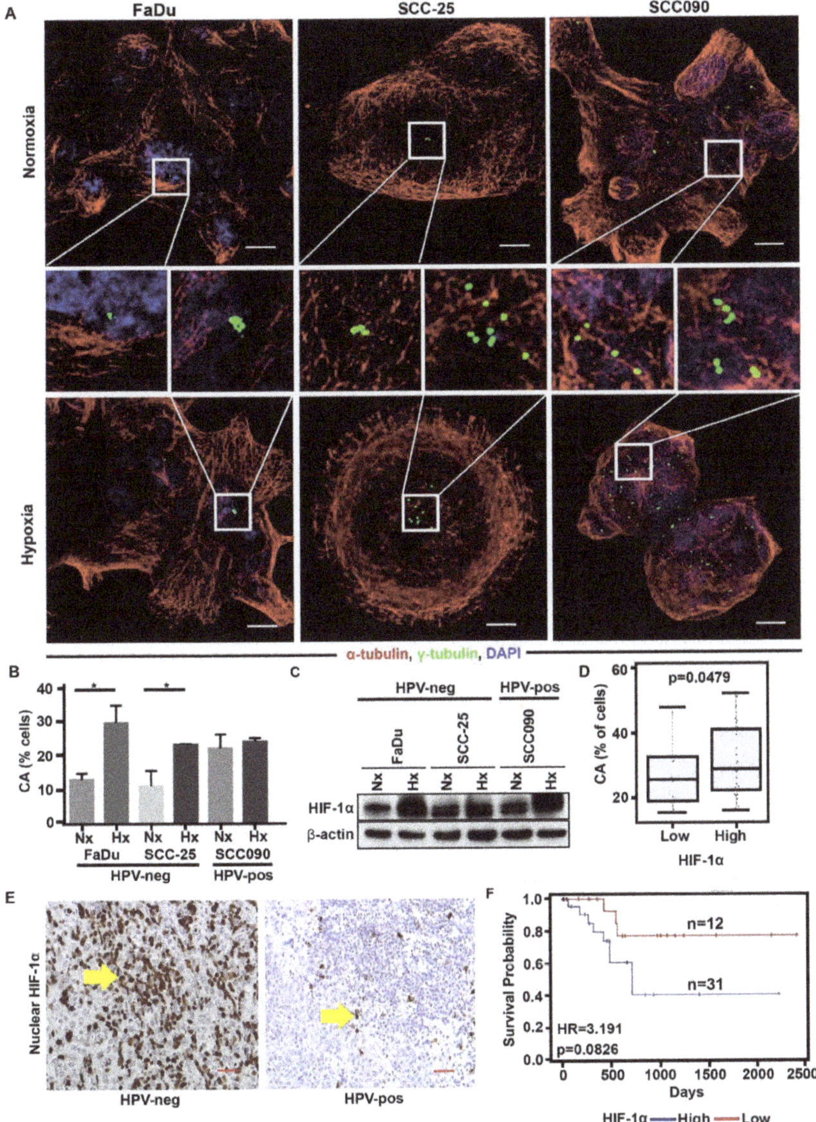



**Figure 3.** HPV-neg OPSCCs show a strong association between CA and HIF-1α expression. (**A**) Confocal immunomicrographs showing numerical CA in HPV-pos and HPV-neg tumor cells cultured in normoxic and hypoxic conditions. OPSCC cells were immunostained for centrosomes (γ-tubulin, green), microtubules (α- tubulin, red) and counterstained with Hoechst (blue). Scale bar (white), 20 μm. (**B**) Percentage distribution of cells with CA (structural and numerical) in HPV-neg and HPV-pos cells in normoxic (Nx) and hypoxic (Hx) conditions. (**C**) Immunoblots of HIF-1α in FaDu, SCC-25, and SCC090 OPSCC cells cultured in normoxic (Nx) and hypoxic (Hx) conditions. (**D**) Box plot depicting the distribution of CA in HIF-1α-high (n = 30) and -low (n = 12) HPV-neg tumors (*p* = 0.0479). (**E**) Representative immunohistochemical micrographs of HPV-pos and HPV-neg OPSCC tumors stained for HIF-1α. (**F**) Kaplan Meier survival analysis representing overall survival in HPV-neg OPSCCs stratified based on HIF-1α scores (HR = 3.191; *p* = 0.0826). * *p* < 0.05.

Next, to corroborate our findings in the clinical samples, we asked whether the high-CA HPV-neg OPSCC patient samples also showed higher HIF-1α expression compared to the low-CA ones. For this, serial sections of the 87 OPSCC samples (used to quantify CA in Figure 1—due to the loss of tissue, eleven cases were not included in the HIF-1α analysis) were immunohistochemically stained for HIF-1α. The nuclear HIF-1α weighted index (WI) was calculated as indicated in Materials and Methods. The patients were stratified into high- and low- HIF-1α expressing subgroups using the optimal HIF-1α expression cut-point (based on the log-rank test). In the whole cohort, the expression of HIF-1α was positively correlated with the CA (R = 0.319; $p$ = 0.03). Figure S4 shows that OPSCCs with high HIF-1α showed higher CA compared with low HIF-1α OPSCCs. Figure 3D depicts that HPV-neg OPSCCs with high HIF-1α (n = 30) showed higher CA ($p$ = 0.0479) compared with low HIF-1α HPV-neg OPSCCs (n = 12). The survival analysis (Figure 3E,F) demonstrated that the high HIF-1α expressing HPV-neg OPSCCs were associated with poorer overall survival (HR = 3.191; $p$ = 0.0826) than the low HIF-1α HPV-neg OPSCCs.

We further evaluated the differences in the expression of hypoxia-associated genes in HPV-neg and HPV-pos HNSCCs. A 26-gene hypoxic signature was probed in the same dataset (TCGA) used in Results Section 2 [26]. Interestingly, our results indicate significantly higher expression ($p$ = 3.77 × 10$^{-7}$) of the 26 hypoxia-associated genes *(ALDOA, ANGPTL4, ANLN, BNC1, C20ORF20, CA9, CDKN3, COL456, DCBLD1, ENO1, FAM83B, FOSL1, GNAL1, HIG2, KCTD11, KR717, LDHA, MPRS17, P4HA1, PGAM1, PGK1, SDC1, SLC16A1, SLAC2A1, TPI1, VEGFA)* and HIF-1α in HPV-neg (n = 422) head and neck tumors compared with those that were HPV-pos (n = 97) (Figure S5A). More so, in the analysis of the subset consisting of only OPSCCs among the cohort, we found similar results, wherein HPV-neg OPSCCs (n = 26) showed significantly ($p$ < 0.001) higher expression of the hypoxia gene signature compared with the HPV-pos OPSCCs (n = 54) (Figure S5B). In addition, the hypoxia score was able to stratify the OPSCCs into high and low-risk groups. The high-hypoxia group was associated with significantly poorer overall survival when compared with the low-hypoxia group (HR = 3.297; $p$ = 0.0127). Notably, among the HPV-neg OPSCCS, high-hypoxia HPV-neg OPSCCs exhibited poorer overall survival (HR = 2.197; $p$ = 0.205) relative to the low-hypoxia HPV-neg OPSCCs (Figure S6). We also observed a positive correlation between the CA7 and the hypoxia 26 gene scores in HPV-neg OPSCCs (R = 0.34760; $p$ = 0.0819).

Collectively, the findings from our clinical data and TCGA analysis confirm that HIF-1α and CA are positively correlated in OPSCCs with higher significance in the HPV-neg tumors. These results suggest that the CA observed in HPV-neg OPSCCs may be hypoxia-induced and may underlie their poorer prognoses relative to the HPV-pos ones.

### 2.4. HIF-1α Upregulation Is Associated with Downregulation of miR-34a, Increase in CA and Expression of Cyclin D1

Having established the relationship between hypoxia and CA in HPV-neg OPSCCs, we sought to delineate the possible mechanism through which HIF-1α induces CA in OPSCC. Studies have also shown that hypoxia and HIF-1α regulate a panel of miRNAs [27]. miRNAs regulate the expression of genes involved in many vital events related to angiogenesis, tumorigenesis and even CA in multiple malignancies, including head and neck cancer [28]. Through our in silico analysis of the publicly-available TCGA miRNA-seq data, we analyzed the differential expression of the top 19 CA-associated miRNAs (list and expression reported in Table S2) in HPV-neg vs. HPV-pos OPSCCs. Among these 19, 12 miRNAs were upregulated in HPV-pos and seven were upregulated in the HPV-neg OPSCCs. The significant overexpression of miR-34a in HPV-pos tumors compared to HPV-neg ones ($p$ = 0.000248) was particularly interesting. CCND1 mRNA is a known target of miR-34a, which has been shown to downregulate cyclin D1 expression [29]. In line with this, we observed *CCND1* gene expression levels to be significantly downregulated in HPV-pos tumors ($p$ = 9.88 × 10$^{-9}$), with a negative correlation between miR-34a and *CCND1* (Figure S7) expression levels. Furthermore, it has been reported that HIF-1$\alpha$ represses the expression of miR-34a in p53 deficient cancer cells [30]. Intriguingly, we observed a significantly higher expression of HIF-1$\alpha$ in HPV-neg HNSCCs in the TCGA dataset.

Based on these in silico findings, we hypothesized that hypoxia may induce the expression of CA-associated genes through the regulation of miR-34a in HPV-neg OPSCCs. To test this hypothesis, we performed quantitative PCR to evaluate the levels of miRNA-34a in HPV-pos and HPV-neg cells, both in control vector- and HIF-1$\alpha$-overexpressing cells. Interestingly, in line with published studies, we observed that in all the cells, regardless of HPV status, HIF-1$\alpha$ overexpression negatively correlated with the levels of miRNA-34a, and this effect was more pronounced in HPV-neg cells (Figure 4A(i,ii)) than in HPV-pos cells. This decrease in miRNA-34a inversely correlated with the expression of *CCND1* which encodes cyclin D1 (which otherwise was not significantly different in the OPSCC cells) (Figure 4Bi). More so, cyclin D1 expression only increased in HPV-neg OPSCC cells (Figure 4Bii). Collectively, we observed that in HPV-neg OPSCC tumors, CCND1 expression is associated with HIF-1$\alpha$ upregulation and the downregulation of miR-34a, and an increase in CA.

To further bolster our in silico and in vitro findings, we examined the relationship of HIF-1$\alpha$ and cyclin D1 in clinical samples. To this end, we immunohistochemically stained the serial sections of the 87 OPSCC samples used in Results Sections 1 and 2. Nuclear cyclin D1 WI was calculated as indicated in Materials and Methods. We found that cyclin D1 expression was significantly ($p = 0.0001$) higher in the HPV-neg ($n = 43$) OPSCCs when compared with the HPV-pos ($n = 44$) OPSCCs (Figure 4A,B) and high cyclin D1 expression was associated with poorer overall survival in OPSCCs (HR = 3.409; $p = 0.0646$) (Figure 4C). These findings are in line with the previous studies that showed a low expression of Cyclin D1 in HPV-pos HNSCCs (p16 inhibits cyclinD1-CDK4/6 (cyclin-dependent kinase 4/6) complexes). Also, cyclin D1 further stratified with HPV-neg OPSCCs in high- and low- risk groups (HR = 3.62; $p = 0.0152$) (Figure 4D). We also observed a strong positive correlation between HIF-1$\alpha$ and cyclin D1 scores in HPV-neg OPSCCs (Spearman's rho $\varrho = 0.642$, $p < 0.001$). Several studies have confirmed the overexpression of cyclin D1 in promoting CA, aneuploidy, and tumorigenesis [31]. In line with this, we also observed that the percentage of CA in HPV-neg OPSCC samples was positively correlated with cyclin D1 expression (Spearman's rho $\varrho = 0.637$, $p < 0.001$. Thus, these findings substantiate the paradigm that hypoxia induces CA in HPV-neg OPSCCs, at least in part, by the overexpression of cyclin D1.

Finally, we wanted to gain insight into the most informative biomarker for clinical decision making. We thus asked which biomarker (CA, HIF-1$\alpha$, or cyclin D1) was best able to stratify HPV-neg OPSCCs into high- and low-risk groups. First, we performed a multivariate analysis, and noted that only CA showed a significant association with poor overall survival when other confounding factors like stage, therapy, gender, alcohol consumption, as well as expression levels of HIF-1$\alpha$ and cyclin D1, were taken in account (HR = 4.43; $p = 0.062$) (Table 3A). Next, to measure the performance of the prognostic models, we used a measure of model fit, 2 Log Likelihood (−2LogL) (the model that minimized the −2LogL was considered superior). The results from this statistical test indicate that CA is the best-fit model (Table 3B). Similarly, when we performed the same test for our in silico findings, the weighted CA7 score better stratified the HPV-neg HNSCCs in high- and low-risk groups than the hypoxia score (Table 3C). Thus, collectively, these findings suggest that CA can serve as a clinically-informative phenotypic biomarker for the identification of high-risk HPV-neg OPSCC patients and can potentially also serve as a novel therapeutic target for these patients.

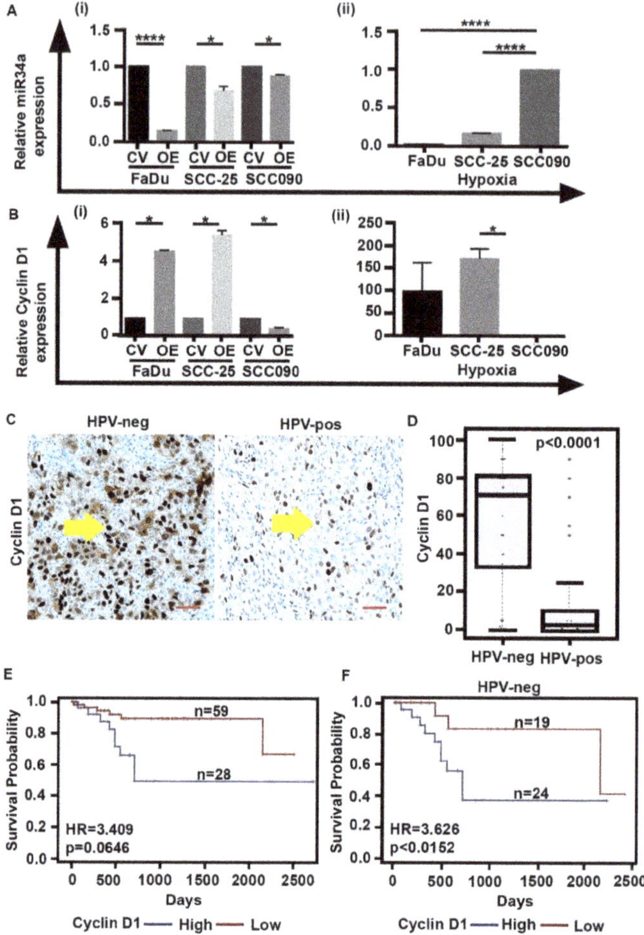



**Figure 4.** Hypoxia induces CA by upregulating the expression of cyclin D1 via downregulating miR-34a in HPV-neg OPSCCs. qRT-PCR analysis of (**Ai**) miR-34a in FaDu, SCC25, and SCC090 OPSCC cells cultured in normoxic and hypoxic conditions (normalized to normoxic conditions for each individual cell line), (**Aii**) miR-34a in FaDu, SCC25, and SCC090 OPSCC cells grown under hypoxic conditions compared to the expression of miR-34a in FaDu, SCC25, and SCC090 OPSCC cells grown under normoxic conditions, (**Bi**) Cyclin D1 in FaDu, SCC25, and SCC090 OPSCC cells cultured in normoxic and hypoxic conditions (normalized to normoxic conditions for each individual cell line), (**Bii**) Cyclin D1 in FaDu, SCC25, and SCC090 OPSCC cells cultured in hypoxic conditions. Data were normalized by the amount of Ctrl_miRTC_1 and beta-actin mRNA, for miR34a and *CCND1*, respectively, expressed relative to the corresponding value for all the cells and are means ± SD from triplicate data. (**C**) Immunohistochemical micrographs of HPV-pos and HPV-neg OPSCC tumors labeled with nuclear cyclin D1. Scale bar (red), 50 μm (**D**) Box plot representing the distribution of cyclin D1 WI in HPV-pos (n = 44) and HPV-neg (n = 43) tumors ($p < 0.0001$). (**E**) Kaplan Meier survival curves representing the overall survival of cyclin D1 high (n = 28) and low (n = 59) groups in HPV-neg and HPV-pos OPSCC patients (HR = 3.409; $p = 0.0646$). (**F**) Kaplan Meier survival curve representing the overall survival of cyclin D1-high (n = 24) and -low (n = 19) groups in HPV-neg OPSCC patients (HR = 3.626; $p = 0.0152$). * $p \leq 0.05$, **** $p \leq 0.0001$.

**Table 3.** Multivariate analysis for HPV-neg OPSCCs comparing the high- and low-CA group. (**A**) −2log L model fit test for clinical samples. (**B**) and (**C**) −2log L model fit test for in silico TCGA dataset.

| A | | | |
|---|---|---|---|
| **Variables** | | **Multivariate Analysis** | |
| | | *p*-**Value** | **Hazard Ratio** |
| **CA** | High vs. low | 0.063 | 4.433 |
| **Gender** | Male | 0.939 | 1.070 |
| **Age at diagnosis** | High vs. low | 0.227 | 1.076 |
| **Overall stage** | IV | 0.852 | 1.163 |
| **Alcohol abuse** | Yes vs. no | 0.166 | 0.160 |
| **Radiotherapy** | Yes | 0.994 | 0.000 |
| **HIF-1$\alpha$** | Yes | 0.994 | 0.991 |
| **Cyclin D1** | Yes | 0.287 | 3.343 |

| B | | | |
|---|---|---|---|
| **Variable** | **Cyclin D1** | **HIF-1$\alpha$** | **CA** |
| −**2 Log L** | 74.741 | 76.340 | 74.547 |

| C | |
|---|---|
| **Variable** | −**2 Log L** |
| **Null** | 1912.9 |
| **Weighted Index of CA7** | 1897.6 |
| **Basic Sum CA7** | 1911.0 |
| **Hyp26 Score** | 1911.9 |
| **CCNDI** | 1907.1 |

## 3. Discussion

CA, a key driver of CIN and an early driver of intratumoral heterogeneity [32], is associated with tumorigenesis and tumor progression in multiple cancers, including that of the head and neck. Previous studies in surgically resected HNSCCs have shown an association between higher CA and local recurrence, with CA being a better predictor of recurrence than other commonly used parameters such as T stage [33,34]; furthermore, higher CA has been linked with poor overall survival in HNSCCs [35]. Despite this, no rigorous comparison of CA in OPSCCs that differ in their HPV status has been performed to date. Neither has the prognostic potential of CA in OPSCCs been investigated. The generation of CA in HPV-pos HNSCCs is known in greater detail and is related to HPV infection, jump-started by the viral oncoproteins (E6 and E7). For a long time, the comparatively lower CA observed in in vitro HPV-neg HNSCC cell lines was presumed to be due to the absence of the drivers E6 and E7. In fact, the HPV-neg HNSCCs display a CA driving mechanism distinct from that of HPV-pos HNSCCs, which lacks a detailed description. To shed light on this subject, we performed rigorous quantitation of CA (structural and numerical centrosomal aberrations) in a large cohort of HPV-neg and HPV-pos OPSCC tumor samples and explored the role of hypoxia, a hitherto overlooked driver of CA, in the generation of CA in HPV-neg tumors.

The findings from this study uncover that HPV-neg OPSCCs exhibit higher CA than HPV-pos OPSCCs, and that CA was associated with poorer overall survival in HPV-neg OPSCCs, even when all the other confounding factors were controlled for. In addition, higher CA was associated with poorer overall survival in OPSCCs, regardless of HPV status. These findings were corroborated by our in silico analysis of CA-associated genes in the large, well-annotated TCGA microarray dataset (Figure 2). Since CA can be induced by perturbations in the expression of many different genes, we focused on a set of seven genes commonly associated with centrosome structure/biogenesis and whose dysregulation is known to induce CA. Impressively, this "CA7" gene signature was significantly prognostic in HNSCCs as well as in OPSCCs, and was able to stratify HPV-neg HNSCCs into high and low-risk groups.

Multiple factors are responsible for differences in the biology of HPV-neg and HPV-pos OPSCCs, the prime among which is a varied tumor microenvironment [36]. Hypoxia is a classic feature of the tumor microenvironment that makes tumors more resistant to treatments and is associated with poor prognosis across a variety of cancers. In accordance with this, we observed higher HIF-1α expression in the HPV-neg OPSCCs [37]. Also, high HIF-1α expression strongly correlated with high CA in HPV-neg OPSCCs. Previously, however, separate studies have reported conflicting results regarding HIF-1α expression in HPV-pos versus HPV-neg tumors [38,39]. Other endogenous hypoxia markers such as CA IX have also not yielded definitive results [40]. Scrutinizing a panel of hypoxia markers rather than relying on individual ones can provide a far more comprehensive picture of the hypoxic environment in HNSCCs as well as the intratumoral heterogeneity and cellular responses it drives. Therefore, in this study, a previously established 26-gene hypoxia signature with addition of HIF-1α gene expression was used in in silico analyses of the publicly-available TCGA dataset, revealing higher expression of hypoxia-associated genes in HPV-neg OPSCCs, which correlated with poorer overall survival within this subset. The observation that hypoxia and the resulting HIF-1α activation can induce CA provides a compelling explanation for the higher CA observed in our HPV-neg OPSCC clinical samples, and consequently the poorer survival, compared with HPV-pos OPSCCs.

The analysis of miRNA expression in OPSCC has revealed a possible mechanistic link between hypoxia and CA. In this study, we have newly identified miR-34a as a probable player in driving CA in OPSCC. The dysregulation of miR-34a has been observed in different types of cancers. For example, by targeting *CCND1*, miR-34a controls the expression of cyclin D1, a protein whose upregulation triggers CA. Since hypoxia through HIF-1α has been shown to repress the expression of miR-34a [41], a downregulation of miR-34a in HPV-neg tumors leads to cyclin D1 overexpression and CA. Our in vitro findings are commensurate with this observation, wherein clinical samples of HPV-neg OPSCCs expressed higher levels of cyclin D1 and higher CA, and was associated with poor overall survival in these patients. Therefore, our findings indicate a HIF-1α-mediated downregulation of miR-34a in HPV-neg tumors which drives its distinct tumor biology and establishes a causative link between hypoxia and CA, which co-occur in many solid tumors.

Studies have shown that EGFR inhibitors such as cetuximab are effective in treating HPV-neg HNSCCs [42]. However, only a modest effect on survival has been shown when cetuximab was co-administered with conventional chemotherapy. Therefore, new molecular targets are required to improve survival in HPV-neg HNSCCs. Our study has uncovered CA as an objectively evaluable and actionable phenotypic biomarker and has yielded novel insights into potential therapeutic targets for HPV-neg OPSCCs.

## 4. Materials and Methods

### 4.1. Clinical Tissue Samples

Formalin-fixed paraffin-embedded OPSCC tissue microarrays (TMA) of tonsil, base of tongue, and soft palate tumors were procured from the Greater Poland Cancer Centre, Poznan in Poland. Patients were diagnosed between the years 2007 and 2014, and based on the location of tumor (tonsil, base of tongue and soft palate) the patients were selected for the TMA construction. Clinicopathologic characteristics of the patients are provided in Table 1. The Institutional Review Board of Greater Poland Cancer Centre, Poznan, PL approved all aspects of the study (The IRB code: 412/18). The methods were carried out in accordance with the approved guidelines stipulated in the Material Transfer Agreements and Data User Agreements between Greater Poland Cancer Centre, Poznan, PL, and Georgia State University. Informed consent was obtained from all subjects.

### 4.2. HPV Analysis

The detection of high-risk HPV was performed using GP5+/GP6+ HPV DNA PCR with enzyme-immunoassay (EIA). For the genotyping of the viral DNA, the Luminex platform was used for bead-based array. The EIA detected 14 HPV types: 16, 18, 31, 33, 35, 39, 45, 51, 52, 56, 58, 59, 66, 68. β-globin PCR was used to test for sample quality post-DNA extraction [43].

### 4.3. Immunohistofluorescence Staining and Quantitation

The TMA slides were deparaffinized as described previously [21]. Antigen retrieval was performed by heating in a pressure cooker with citrate buffer (pH 6.0). The tissues were blocked with 5% BSA in PBS solution for 30 min. After blocking, primary antibody incubation with γ-tubulin (Sigma, St. Louis, MO, USA) at a dilution of 1:1000 was performed for 1 h at room temperature. The tissues were then washed 3× with PBS after which incubation with secondary antibody (Alexa-488 anti-mouse) was performed at room temperature for 1 h at a dilution of 1:2000. After 3× washes in PBS, coverslips were mounted with Prolong-Gold Antifade with DAPI (Invitrogen, Waltham, MA, USA). Tissue sections were imaged using the Zeiss LSM 700 confocal microscope (Ziess, Oberkochen, Germany). Details on imaging and analysis are included in Supplementary Materials.

### 4.4. Immunohistochemistry of Cyclin D1 and HIF-1α and Scoring

The initial steps from deparaffinization to antigen retrieval were performed as described for immunofluorescence. The tissues were then blocked with Ultravision Protein Block (ThermoFisher, Waltham, MA, USA) for 30 min followed by hydrogen peroxide block with Ultravision Hydrogen Peroxide Block (ThermoFisher) for 10 min. The tissues were then incubated with antibodies directed against HIF-1α (Abcam, Cambridge, MA, USA) or cyclin D1 (ThermoFisher) at a dilution of 1:1000 for 1 h. After 3× washes in TBST, the slides were subjected to secondary antibody incubation using anti-Rabbit HRP (Biocare, Pacheco, CA, USA) for 1 h. Enzymatic detection was performed using DAB Chromogen Kit (Biocare). Nuclear HIF-1α and cyclin D1 staining were categorized as 0 = none, 1 = low, 2 = moderate, and 3 = high. The percentage of positive cells, defined as cells with a staining intensity of 1+, quantitated from a total of around 500 cells, was determined. The weighted index (WI) for each sample was calculated as the product of the percentage of cell positivity and staining intensity.

### 4.5. Cell Culture and Hypoxia Treatment

FaDu, SCC25 (HPV-negative) and SCC90 (HPV-positive) cells were purchased from American Type Culture Collection, Manassas, VA. FaDu cells were maintained in EMEM medium supplemented with 10% FBS and 1% penicillin/streptomycin. SCC25 cells were maintained in 1:1 mixture of DMEM and Ham's F12 medium supplemented with 400 ng/mL hydrocortisone, 10% FBS and 1% penicillin/streptomycin. SCC 90 cells were maintained in EMEM supplemented with 2mM L-glutamine, 10% FBS, and 1% penicillin/streptomycin. All the cell lines were cultured under standard culture conditions—37 °C and 5% $CO_2$. For hypoxia induction, the cells were placed in a hypoxic modulated incubator chamber which was flushed with 5% $CO_2$ and a gas mixture containing 1% $CO_2$ and 94% $N_2$ gas mixture at 20 L/min for 7–10 min every 3–6 h. HIF-1α was genetically overexpressed by transfecting cells with GFP-tagged degradation resistant HIF-1α. HA-HIF-1α P402A/P564A-pcDNA3 was a generous gift from Dr. Willian Kaelin (Addgene plasmid # 18955). Cells at a confluency of ~70% were transfected using Lipofectamine LTX according to the manufacturer's instructions.

### 4.6. RNA Extraction and Quantitative Real Time PCR

The total RNA from the cell lines after transfection and hypoxia induction was extracted using TRIzol reagent (Takara Bio, Inc., Otsu, Japan). Genomic DNA contamination was eliminated from RNA preparation by digesting with RNase-free DNase (Qiagen). cDNA was prepared using iScript™ cDNA Synthesis Kit (Bio-Rad). All PCR reactions were performed using the fluorescent SYBR Green methodology. Quantitative RT-PCR (qRT-PCR) was run in Quant Studio 3 Real-Time PCR system (Thermo Fisher Scientific, Inc.) with SYBR-Green PCR Master Mix (ThermoFisher Scientific) following the manufacturer's protocol. The relative quantification was calculated by the $2^{-\Delta\Delta Ct}$ method. The expression of all mRNAs was normalized with respect to GAPDH or beta-actin.

### 4.7. miRNA Extraction and Quantitative Real Time PCR

The total RNA was isolated from the cell lines after transfection and hypoxia induction using miRNeasy Micro Kit (Qiagen) per the manufacturer's protocol. Genomic DNA contamination was eliminated by digesting the RNA with RNase-free DNase (Qiagen). cDNA was prepared using the miScript II RT Kit (Qiagen). qRT-PCR was performed using the miScript SYBR green PCR kit, and specific miRNA primer assays (Qiagen) following the manufacturer's protocol. qRT-PCR was run in Quant Studio 3 Real-Time PCR system (Thermo Fisher Scientific, Inc.) and relative expression was quantitated using the $2^{-\Delta\Delta Ct}$ method. The expression of mir-34 A was normalized to Ctrl_miRTC_1.

### 4.8. Immunocytofluorescence Staining

HPV-pos and HPV-neg cells were grown on poly-L-lysine-coated cover glasses. Following hypoxia treatment, the cells were fixed with ice cold methanol for 10 min. After sequentially blocking in PBS with 5% BSA and 10% goat serum for 30 min, the cells were incubated with primary antibodies directed against $\alpha$-tubulin and $\gamma$-tubulin (1:1000 dilution) prepared in 1% BSA-PBS (with 0.1% Triton X-100) at 37 °C for 35 min. Following primary antibody incubation, the cells were washed five times with PBS before adding the secondary antibody. The cells were incubated with Alexa 555-and 488-conjugated antibodies (Invitrogen) at 37 °C for 35 min, after which the cells were washed with PBS for 10 min. For DNA visualization, the cells were stained with Heochst33342 (Invitrogen) and then mounted onto glass slides with ProLong Gold Antifade Reagent. Fluorescent images were captured using the Zeiss LSM 700 confocal microscope (Oberkochen, Germany) and processed with Zen software (Oberkochen, Germany).

### 4.9. Statistical Analyses

For clinical data, as well as our in silico data analysis, patients' overall survival was used as the endpoint for survival analysis. A log-rank test was applied to determine the significance of survival differences between subgroups. The cut-off points that we found for CA and HIF1-$\alpha$ were those which maximized survival differences between high- and low-risk subgroups. The range of CA value was 4.85–71.31 and 23 was used as the cut-off point as it resulted in the minimization of the log-rank p-value. The test of group mean differences shown in box-whisker plots is based on the Mann–Whitney U test. In cases with more than two groups, the differences were evaluated by the Kruskal–Wallis test. Statistical analyses were performed using SAS software 9.4 (SAS Institute Inc., Cary, NC). Details of the survival model used in our in silico analysis are included in the Supplementary Materials.

## 5. Conclusions

This is the first report to substantiate the previously unrecognized role of HIF-1$\alpha$-induced CA in HPV-neg OPSCCs, revealing a molecular pathway that may be responsible for the CIN, intratumoral heterogeneity, and poor prognosis associated with these tumors. The prognostic potential of CA is especially resounding within HPV-neg OPSCCs, facilitating the enhanced identification of higher risk patients, influencing future treatment strategies, and providing a platform for the discovery of effective molecular targets.

**Supplementary Materials:** The following are available online at http://www.mdpi.com/2072-6694/12/2/517/s1, Figure S1: Analysis of CA in OPSCC clinical samples. Figure S2: Upregulation of CA7 genes is associated with poor overall survival in HNSCCs. Figure S3: (A) Confocal immunomicrographs showing numerical CA in HPV-pos and HPV-neg OPSCC cells transfected with empty vector or degradation-resistant HIF-1$\alpha$. OPSCC cells were immunostained for centrosomes ($\gamma$-tubulin, green), microtubules ($\alpha$- tubulin, red) and counterstained with Hoechst (blue). Scale bar (white), 20 µm. (B) Quantitation of centrosome aberrations per microscopic examination for HPV-pos and HPV-neg OPSCC cells transfected with empty vector or degradation-resistant HIF-1$\alpha$. (C) Immunoblots of HIF-1$\alpha$ in FaDu, SCC-25, and SCC090 OPSCC cells transfected with degradation-resistant HIF-1$\alpha$ and control vector. (D) qRT- PCR analysis of mRNAs for $\gamma$-tubulin, pericentrin, centrin-1, and PLK4 in FaDu, SCC-25, and SCC090 OPSCC cells transfected with empty vector or degradation-resistant HIF-1$\alpha$. Data were normalized by the amount of beta-actin mRNA, expressed relative to the corresponding value for all the cells and are means ± SD from triplicate data. Figure S4: Box plot depicting the distribution of CA in HIF-1α-high (n = 40) and -low (n = 47) OPSCCs (*p* = 0.0007). Figure S5: Comparison of 26-gene hypoxia gene signature in HPV-pos versus HPV-neg HNSCC tumors. Figure S6: (A) and (B) Kaplan Meier survival curves representing the survival probabilities of high and low hypoxia score OPSCC patients (HR = 3.297; *p* = 0.0127) and the survival probabilities of high and low hypoxia score hypoxia HPV-neg OPSCC patients (HR = 2.197; *p* = 0.2050), respectively. Figure S7: *CCND1* gene expression is negatively correlated with the miR-34a expression in HPV-pos HNSCCs. Figure S8: Whole blot showing all the bands with all molecular weight markers are shown for Figure 3C. Figure S9: Whole blot showing all the bands with all molecular weight markers are shown for Figure S3C. Table S1: Descriptive statistics of clinicopathological characteristics for HNSCC patients in the cohort (TCGA) used for in silico analysis of the prognostic value of CA7 signature. Table S2: List of CA associated miRNAs sorted by logFC values. Table S3: Densitometry values relative to loading control β-actin calculated using Image-J for immunoblot assays provided in main manuscript and supplementary data.

**Author Contributions:** K.M. and D.H.C. carried out the major experiments, analyzed and interpreted the data, and wrote the manuscript. G.W. performed all the statistical analyses and S.K. performed in silico data analysis. J.K. and K.A. carried out the RT-PCR experiments, analyzed and interpreted the data. C.C.G. helped in scoring of tissues. P.I.-J. carried out the immunoblot assays. B.D.M. carried out immunofluorescence staining and imaging. S.B.-P. carried out the immunohistochemical staining. P.G. contributed tissue microarray and associated clinical data and revised the manuscript. R.M.O. helped with statistical analysis and M.K. revised the manuscript and contributed to discussion. P.R. and R.A. conceived and designed the study and critically revised the manuscript. All authors have read and agreed to the published version of the manuscript.

**Funding:** This study was supported by a grant to RA from the National Cancer Institute (U01 CA179671).

**Conflicts of Interest:** The authors declare that there are no conflicts of interest.

## References


1. Vigneswaran, N.; Williams, M.D. Epidemiologic trends in head and neck cancer and aids in diagnosis. *Oral Maxillofac. Surg. Clin. N. Am.* **2014**, *26*, 123–141. [CrossRef] [PubMed]

2. van Monsjou, H.S.; Balm, A.J.; van den Brekel, M.M.; Wreesmann, V.B. Oropharyngeal squamous cell carcinoma: A unique disease on the rise? *Oral Oncol.* **2010**, *46*, 780–785. [CrossRef] [PubMed]

3. Viens, L.J.; Henley, S.J.; Watson, M.; Markowitz, L.E.; Thomas, C.C.; Thompson, T.D.; Razzaghi, H.; Saraiya, M. Human Papillomavirus-Associated Cancers—United States, 2008–2012. *MMWR Morb. Mortal. Wkly. Rep.* **2016**, *65*, 661–666. [CrossRef] [PubMed]

4. Ang, K.K.; Harris, J.; Wheeler, R.; Weber, R.; Rosenthal, D.I.; Nguyen-Tan, P.F.; Westra, W.H.; Chung, C.H.; Jordan, R.C.; Lu, C.; et al. Human papillomavirus and survival of patients with oropharyngeal cancer. *N. Engl. J. Med.* **2010**, *363*, 24–35. [CrossRef]

5. Lowy, D.R.; Munger, K. Prognostic implications of HPV in oropharyngeal cancer. *N. Engl. J. Med.* **2010**, *363*, 82–84. [CrossRef]

6. Huang, S.H.; O'Sullivan, B. Overview of the 8th Edition TNM Classification for Head and Neck Cancer. *Curr. Treat. Options Oncol.* **2017**, *18*, 40. [CrossRef]

7. Stransky, N.; Egloff, A.M.; Tward, A.D.; Kostic, A.D.; Cibulskis, K.; Sivachenko, A.; Kryukov, G.V.; Lawrence, M.S.; Sougnez, C.; McKenna, A.; et al. The mutational landscape of head and neck squamous cell carcinoma. *Science* **2011**, *333*, 1157–1160. [CrossRef]

8. Rothenberg, S.M.; Ellisen, L.W. The molecular pathogenesis of head and neck squamous cell carcinoma. *J. Clin. Investig.* **2012**, *122*, 1951–1957. [CrossRef]

9. Munger, K.; Scheffner, M.; Huibregtse, J.M.; Howley, P.M. Interactions of HPV E6 and E7 oncoproteins with tumour suppressor gene products. *Cancer Surv.* **1992**, *12*, 197–217.

10. Duensing, S.; Munger, K. Human papillomaviruses and centrosome duplication errors: Modeling the origins of genomic instability. *Oncogene* **2002**, *21*, 6241–6248. [CrossRef]

11. Chan, J.Y. A clinical overview of centrosome amplification in human cancers. *Int. J. Biol. Sci.* **2011**, *7*, 1122–1144. [CrossRef] [PubMed]

12. D'Assoro, A.B.; Lingle, W.L.; Salisbury, J.L. Centrosome amplification and the development of cancer. *Oncogene* **2002**, *21*, 6146–6153. [CrossRef] [PubMed]

13. Ogden, A.; Rida, P.C.; Aneja, R. Heading off with the herd: How cancer cells might maneuver supernumerary centrosomes for directional migration. *Cancer Metastasis Rev.* **2013**, *32*, 269–287. [CrossRef] [PubMed]

14. Leemans, C.R.; Braakhuis, B.J.; Brakenhoff, R.H. The molecular biology of head and neck cancer. *Nat. Rev. Cancer* **2011**, *11*, 9–22. [CrossRef] [PubMed]

15. Broustas, C.G.; Lieberman, H.B. DNA damage response genes and the development of cancer metastasis. *Radiat. Res.* **2014**, *181*, 111–130. [CrossRef] [PubMed]

16. van Vugt, M.A.; Medema, R.H. Getting in and out of mitosis with Polo-like kinase-1. *Oncogene* **2005**, *24*, 2844–2859. [CrossRef]

17. Nurwidya, F.; Takahashi, F.; Minakata, K.; Murakami, A.; Takahashi, K. From tumor hypoxia to cancer progression: The implications of hypoxia-inducible factor-1 expression in cancers. *Anat. Cell Biol.* **2012**, *45*, 73–78. [CrossRef]

18. Hanns, E.; Job, S.; Coliat, P.; Wasylyk, C.; Ramolu, L.; Pencreach, E.; Suarez-Carmona, M.; Herfs, M.; Ledrappier, S.; Macabre, C.; et al. Human Papillomavirus-related tumours of the oropharynx display a lower tumour hypoxia signature. *Oral Oncol.* **2015**, *51*, 848–856. [CrossRef]

19. Klein, A.; Flugel, D.; Kietzmann, T. Transcriptional regulation of serine/threonine kinase-15 (STK15) expression by hypoxia and HIF-1. *Mol. Biol. Cell* **2008**, *19*, 3667–3675. [CrossRef]

20. Rosario, C.; Swallow, C.J. Abstract 4045: Hypoxia induces Plk4 expression and promotes immortalization of proliferating cells. *Cancer Res.* **2013**, 4045. [CrossRef]

21. Mittal, K.; Choi, D.H.; Ogden, A.; Donthamsetty, S.; Melton, B.D.; Gupta, M.V.; Pannu, V.; Cantuaria, G.; Varambally, S.; Reid, M.D.; et al. Amplified centrosomes and mitotic index display poor concordance between patient tumors and cultured cancer cells. *Sci. Rep. UK* **2017**, *7*, 43984. [CrossRef] [PubMed]

22. Mittal, K.; Choi, D.H.; Klimov, S.; Pawar, S.; Kaur, R.; Mitra, A.K.; Gupta, M.V.; Sams, R.; Cantuaria, G.; Rida, P.C.G.; et al. A centrosome clustering protein, KIFC1, predicts aggressive disease course in serous ovarian adenocarcinomas. *J. Ovarian Res.* **2016**, *9*, 17. [CrossRef] [PubMed]

23. Mittal, K.; Ogden, A.; Reid, M.D.; Rida, P.C.G.; Varambally, S.; Aneja, R. Amplified centrosomes may underlie aggressive disease course in pancreatic ductal adenocarcinoma. *Cell Cycle* **2015**, *14*, 2798–2809. [CrossRef] [PubMed]

24. Pannu, V.; Mittal, K.; Cantuaria, G.; Reid, M.D.; Li, X.; Donthamsetty, S.; McBride, M.; Klimov, S.; Osan, R.; Gupta, M.V.; et al. Rampant centrosome amplification underlies more aggressive disease course of triple negative breast cancers. *Oncotarget* **2015**, *6*, 10487–10497. [CrossRef] [PubMed]

25. Head, T.R. TCGA Releases Head and Neck Cancer Data. *Cancer Discov.* **2015**, *5*, 340–341.

26. Eustace, A.; Mani, N.; Span, P.N.; Irlam, J.J.; Taylor, J.; Betts, G.N.; Denley, H.; Miller, C.J.; Homer, J.J.; Rojas, A.M. A 26-gene hypoxia signature predicts benefit from hypoxia-modifying therapy in laryngeal cancer but not bladder cancer. *Clin. Cancer Res.* **2013**, *19*, 4879–4888. [CrossRef]

27. Chan, Y.C.; Khanna, S.; Roy, S.; Sen, C.K. miR-200b targets Ets-1 and is down-regulated by hypoxia to induce angiogenic response of endothelial cells. *J. Biol. Chem.* **2011**, *286*, 2047–2056. [CrossRef]

28. Shivdasani, R.A. MicroRNAs: Regulators of gene expression and cell differentiation. *Blood* **2006**, *108*, 3646–3653. [CrossRef]

29. Sun, F.; Fu, H.; Liu, Q.; Tie, Y.; Zhu, J.; Xing, R.; Sun, Z.; Zheng, X. Downregulation of CCND1 and CDK6 by miR-34a induces cell cycle arrest. *FEBS Lett.* **2008**, *582*, 1564–1568. [CrossRef]

30. Li, H.; Rokavec, M.; Jiang, L.; Horst, D.; Hermeking, H. Antagonistic Effects of p53 and HIF1A on microRNA-34a Regulation of PPP1R11 and STAT3 and Hypoxia-induced Epithelial to Mesenchymal Transition in Colorectal Cancer Cells. *Gastroenterology* **2017**, *153*, 505–520. [CrossRef]

31. Nelsen, C.J.; Kuriyama, R.; Hirsch, B.; Negron, V.C.; Lingle, W.L.; Goggin, M.M.; Stanley, M.W.; Albrecht, J.H. Short term cyclin D1 overexpression induces centrosome amplification, mitotic spindle abnormalities, and aneuploidy. *J. Biol. Chem.* **2005**, *280*, 768–776. [CrossRef] [PubMed]

32. McBride, M.; Rida, P.C.; Aneja, R. Turning the headlights on novel cancer biomarkers: Inspection of mechanics underlying intratumor heterogeneity. *Mol. Asp. Med.* **2015**, *45*, 3–13. [CrossRef] [PubMed]

33. Gustafson, L.M.; Gleich, L.L.; Fukasawa, K.; Chadwell, J.; Miller, M.A.; Stambrook, P.J.; Gluckman, J.L. Centrosome hyperamplification in head and neck squamous cell carcinoma: A potential phenotypic marker of tumor aggressiveness. *Laryngoscope* **2000**, *110*, 1798–1801. [CrossRef]

34. Syed, M.I.; Syed, S.; Minty, F.; Harrower, S.; Singh, J.; Chin, A.; McLellan, D.R.; Parkinson, E.K.; Clark, L.J. Gamma tubulin: A promising indicator of recurrence in squamous cell carcinoma of the larynx. *Otolaryngol. Head Neck Surg.* **2009**, *140*, 498–504. [CrossRef] [PubMed]

35. Reiter, R.; Gais, P.; Steuer-Vogt, M.K.; Boulesteix, A.L.; Deutschle, T.; Hampel, R.; Wagenpfeil, S.; Rauser, S.; Walch, A.; Bink, K.; et al. Centrosome abnormalities in head and neck squamous cell carcinoma (HNSCC). *Acta Otolaryngol.* **2009**, *129*, 205–213. [CrossRef] [PubMed]

36. Bonilla-Velez, J.; Mroz, E.A.; Hammon, R.J.; Rocco, J.W. Impact of human papillomavirus on oropharyngeal cancer biology and response to therapy: Implications for treatment. *Otolaryngol. Clin. N. Am.* **2013**, *46*, 521–543. [CrossRef]

37. Bristow, R.G.; Hill, R.P. Hypoxia and metabolism. Hypoxia, DNA repair and genetic instability. *Nat. Rev. Cancer* **2008**, *8*, 180–192. [CrossRef]

38. Knuth, J.; Sharma, S.J.; Wurdemann, N.; Holler, C.; Garvalov, B.K.; Acker, T.; Wittekindt, C.; Wagner, S.; Klussmann, J.P. Hypoxia-inducible factor-1alpha activation in HPV-positive head and neck squamous cell carcinoma cell lines. *Oncotarget* **2017**, *8*, 89681–89691. [CrossRef]

39. Jung, Y.S.; Najy, A.J.; Huang, W.; Sethi, S.; Snyder, M.; Sakr, W.; Dyson, G.; Huttemann, M.; Lee, I.; Ali-Fehmi, R.; et al. HPV-associated differential regulation of tumor metabolism in oropharyngeal head and neck cancer. *Oncotarget* **2017**, *8*, 51530–51541. [CrossRef]

40. Brockton, N.; Dort, J.; Lau, H.; Hao, D.; Brar, S.; Klimowicz, A.; Petrillo, S.; Diaz, R.; Doll, C.; Magliocco, A. High stromal carbonic anhydrase IX expression is associated with decreased survival in P16-negative head-and-neck tumors. *Int. J. Radiat. Oncol. Biol. Phys.* **2011**, *80*, 249–257. [CrossRef]

41. Du, R.; Sun, W.; Xia, L.; Zhao, A.; Yu, Y.; Zhao, L.; Wang, H.; Huang, C.; Sun, S. Hypoxia-induced down-regulation of microRNA-34a promotes EMT by targeting the Notch signaling pathway in tubular epithelial cells. *PLoS ONE* **2012**, *7*, e30771. [CrossRef] [PubMed]

42. Michel, L.; Ley, J.; Wildes, T.M.; Schaffer, A.; Robinson, A.; Chun, S.E.; Lee, W.; Lewis, J., Jr.; Trinkaus, K.; Adkins, D. Phase I trial of palbociclib, a selective cyclin dependent kinase 4/6 inhibitor, in combination with cetuximab in patients with recurrent/metastatic head and neck squamous cell carcinoma. *Oral Oncol.* **2016**, *58*, 41–48. [CrossRef] [PubMed]

43. Rietbergen, M.M.; Leemans, C.R.; Bloemena, E.; Heideman, D.A.; Braakhuis, B.J.; Hesselink, A.T.; Witte, B.I.; Baatenburg de Jong, R.J.; Meijer, C.J.; Snijders, P.J.; et al. Increasing prevalence rates of HPV attributable oropharyngeal squamous cell carcinomas in the Netherlands as assessed by a validated test algorithm. *Int. J. Cancer.* **2013**, *132*, 1565–1567. [CrossRef] [PubMed]

*Article*

# Multistate Markov Model to Predict the Prognosis of High-Risk Human Papillomavirus-Related Cervical Lesions

Ayumi Taguchi [1,2], Konan Hara [3,4], Jun Tomio [5], Kei Kawana [6,*], Tomoki Tanaka [1], Satoshi Baba [1], Akira Kawata [1], Satoko Eguchi [1], Tetsushi Tsuruga [1], Mayuyo Mori [1], Katsuyuki Adachi [1], Takeshi Nagamatsu [1], Katsutoshi Oda [1], Toshiharu Yasugi [1,2], Yutaka Osuga [1] and Tomoyuki Fujii [1]

[1] Department of Obstetrics and Gynecology, Graduate School of Medicine, The University of Tokyo, Tokyo 113-8655, Japan; aytaguchi-tky@umin.ac.jp (A.T.); tomotanaka-tky@umin.ac.jp (T.T.); BABAS-GYN@h.u-tokyo.ac.jp (S.B.); kawataa-gyn@h.u-tokyo.ac.jp (A.K.); satokojolly@yahoo.co.jp (S.E.); tsuruga-tky@umin.ac.jp (T.T.); mayumori-tky@umin.ac.jp (M.M.); kadachi-gyn@umin.ac.jp (K.A.); tnag-tky@umin.ac.jp (T.N.); katsutoshi-tky@umin.ac.jp (K.O.); yasugi-tky@cick.jp (T.Y.); yutakaos-tky@umin.ac.jp (Y.O.); fujiit-tky@umin.org (T.F.)
[2] Gynecology Division, Tokyo Metropolitan Cancer and Infectious Diseases Center, Komagome Hospital, Tokyo 113-8677, Japan
[3] Graduate School of Economics, The University of Tokyo, Tokyo 113-0033, Japan; hara.konan@e.u-tokyo.ac.jp
[4] Hematology Division, Tokyo Metropolitan Cancer and Infectious Diseases Center, Komagome Hospital, Tokyo 113-8677, Japan
[5] Department of Public Health, Graduate School of Medicine, The University of Tokyo, Tokyo 113-0033, Japan; juntomio@m.u-tokyo.ac.jp
[6] Department of Obstetrics and Gynecology, School of Medicine, Nihon University, Tokyo 173-8610, Japan
* Correspondence: kkawana-tky@umin.org; Tel.: +81-3-3972-8111

Received: 19 November 2019; Accepted: 20 January 2020; Published: 22 January 2020

**Abstract:** Cervical intraepithelial neoplasia (CIN) has a natural history of bidirectional transition between different states. Therefore, conventional statistical models assuming a unidirectional disease progression may oversimplify CIN fate. We applied a continuous-time multistate Markov model to predict this CIN fate by addressing the probability of transitions between multiple states according to the genotypes of high-risk human papillomavirus (HPV). This retrospective cohort comprised 6022 observations in 737 patients (195 normal, 259 CIN1, and 283 CIN2 patients at the time of entry in the cohort). Patients were followed up or treated at the University of Tokyo Hospital between 2008 and 2015. Our model captured the prevalence trend satisfactory, particularly for up to two years. The estimated probabilities for 2-year transition to CIN3 or more were the highest in HPV 16-positive patients (13%, 30%, and 42% from normal, CIN1, and CIN2, respectively) compared with those in the other genotype-positive patients (3.1–9.6%, 7.6–16%, and 21–32% from normal, CIN1, and CIN2, respectively). Approximately 40% of HPV 52- or 58-related CINs remained at CIN1 and CIN2. The Markov model highlights the differences in transition and progression patterns between high-risk HPV-related CINs. HPV genotype-based management may be desirable for patients with cervical lesions.

**Keywords:** cervical intraepithelial neoplasia; high-risk human papillomavirus; multistate Markov model; retrospective cohort study; survival analysis

---

## 1. Introduction

Cervical cancer is the second most commonly diagnosed type of cancer, and the third leading cause of cancer-related death among females in developed countries [1]. In 2012, an estimated 527,600 new cases of cervical cancer were diagnosed and 265,700 related deaths were reported worldwide [1]. In Japan, 33,114 women were newly diagnosed with cervical cancer in 2013 [2], and the mortality rate was 4.1 per 100,000 person-years [3]. Infection with human papillomaviruses (HPVs) is the main cause of cervical cancer development [4–6]. The types of HPV are categorized on the basis of their carcinogenesis. The International Agency for Research on Cancer divided the HPV genotypes into the following groups according to their carcinogenesis: the highly carcinogenic Group 1 (HPVs 16, 18, 31, 33, 35, 39, 45, 51, 52, 56, 58, and 59); the probably carcinogenic Group 2A (HPV 68); and the possibly carcinogenic Group 2B (HPVs 26, 30, 34, 53, 66, 67, 69, 70, 73, 82, 85, and 97) [7]. The high-risk HPV (hrHPV)-derived oncogenes E6 and E7 are necessary for malignant conversion. E6 and E7 inactivate the p53 tumor suppressor, and suppress the expression of retinoblastoma proteins, resulting in resistance to apoptosis and the promotion of cell proliferation [6]. The stabilized expression of E6 and E7 is the critical step in the progression of cervical cancer [8,9].

Worldwide, patients with a high-grade squamous intraepithelial lesion (HSIL) undergo surgical interventions (e.g., conization and loop electrosurgical excision or laser vaporization), regardless of the HPV genotype [10,11]. In Japan, surgical intervention to patients with CIN2 was not common, because approximately half of these patients regress within two years [12]. A prospective cohort study involving patients with cervical intraepithelial neoplasia (CIN) has demonstrated that, without treatment, approximately 20% of CIN2 patients with HPVs 16, 18, 31, 33, 35, 45, 52 or 58 progress to CIN3 within 5 years [12]. On the basis of the available evidence, excision strategies can be considered in patients with CIN2 as a risk-reducing treatment when the lesion is caused by these hrHPVs [13]. However, surgical excision in patients with HSIL occasionally leads to poor obstetric outcome, including preterm birth, low-birth-weight infants, and cesarean delivery, due to cervical incompetence after surgery [14,15]. Therefore, surgical treatment is unfavorable for patients who desire to be pregnant when their risk for progression to CIN3 or cancer is low.

Among the hrHPVs, HPVs 16 and 18 are the most frequently observed genotypes in patients with cervical cancer, and approximately 70% of cervical cancers are HPV 16 or 18 positive [4,16]. Several studies, including prospective cohort studies, demonstrated that the risk of HPV 16-positive patients to develop CIN2 or CIN3 lesions is higher than that reported in patients positive for other hrHPVs [12,17–19]. Matsumoto et al. revealed that seven hrHPV types, including HPV 16, show a high rate of progression of CIN1–2 to CIN3 compared with the other hrHPVs [12]. In addition, several reports demonstrated that patterns of persistent infection or persistent cervical lesions may differ according to the HPV types [20–22]. Collectively, we consider that the natural history of HPV-infected CIN lesions differs depending on the HPV genotypes; i.e., some HPV genotype-related lesions are likely to progress to cervical cancer, some are likely to regress, and others are likely to result in persistent disease.

The Cox proportional hazards model is most frequently used to predict the risk of transition from one state to another. However, the natural history of CIN shows a bidirectional transition between different states, and patients intricately move between a series of states. For example, CIN2 can progress to CIN3 or regress to CIN1 during the follow-up period. A model capable of estimating the risks of transitions from and to multiple states is more suitable for assessing the fate of CIN patients compared with the traditional Cox proportional hazards model, accounting only for the transition between two states. A solution is the use of multistate Markov models that enable investigators to estimate the transitions between multiple states [23–25]. Use of this approach has recently been adopted in various clinical settings [26–34].

In the present study, we retrospectively analyzed a cohort of patients with HPV genotype-confirmed CIN to investigate the relationship between HPV genotypes and clinicopathological features in CIN/cervical cancer. We applied a continuous-time multistate Markov model to independently estimate the prognosis of cervical lesions for designated HPV categories.

## 2. Results

### 2.1. Patients

Among the 1417 patients, 737 patients were enrolled with the following diagnoses at the time of entry: normal, 195 patients; CIN1, 259 patients; and CIN2, 283 patients. The median age was 37.6 years (interquartile range [IQR]: 31.8–44.5 years), the median number of visits was seven (IQR: 4–12), and the median duration of the follow-up was 3.03 years (IQR: 1.21–4.91 years). Table 1 displays the summary statistics for the combination of the defined HPV categories and diagnoses at the time of entry. In these patients, HPVs 16, 52, and 58 were the three most frequently observed HPV types. Because HPV 18 is the second most frequently observed HPV genotype in cervical cancer [35], in this study, we focused on HPVs 16, 52, 58, and 18.

**Table 1.** Sample size and summary statistics for the combination of patients in the six HPV categories and diagnoses at the time of entry.

| Diagnosis at the Time of Entry | | HPV 16 | HPV 18 | HPV 52 | HPV 58 | Other hrHPVs | No hrHPVs | All |
|---|---|---|---|---|---|---|---|---|
| Normal | N | 13 | 11 | 17 | 13 | 30 | 122 | 195 |
| | Age at the time of entry (years), mean (SD) | 42.4 (13.8) | 39.5 (15.0) | 36.7 (10.0) | 41.3 (16.8) | 42.5 (16.9) | 41.2 (10.5) | 41.3 (12.1) |
| | Number of visits, mean (SD) | 5.1 (4.8) | 7.2 (3.4) | 7.5 (5.7) | 6.8 (4.6) | 7.0 (4.8) | 6.5 (3.9) | 6.8 (4.3) |
| | Follow-up interval (years), mean (SD) | 0.49 (0.37) | 0.46 (0.27) | 0.50 (0.35) | 0.49 (0.42) | 0.47 (0.27) | 0.53 (0.40) | 0.51 (0.37) |
| | Follow-up period (years), mean (SD) | 2.1 (2.5) | 2.9 (1.6) | 3.7 (3.1) | 2.9 (2.3) | 2.9 (2.2) | 3.1 (2.1) | 3.1 (2.2) |
| CIN1 | N | 23 | 11 | 38 | 24 | 79 | 111 | 259 |
| | Age at the time of entry (years), mean (SD) | 34.6 (8.2) | 33.0 (10.0) | 36.6 (8.6) | 36.1 (7.7) | 34.5 (7.2) | 39.0 (10.3) | 36.9 (9.2) |
| | Number of visits, mean (SD) | 7.6 (5.3) | 7.6 (4.7) | 10.7 (5.9) | 10.5 (6.4) | 9.0 (4.9) | 9.3 (5.2) | 9.3 (5.2) |
| | Follow-up interval (years), mean (SD) | 0.42 (0.43) | 0.51 (0.53) | 0.38 (0.18) | 0.45 (0.42) | 0.39 (0.22) | 0.42 (0.31) | 0.41 (0.29) |
| | Follow-up period (years), mean (SD) | 3.2 (2.6) | 3.2 (2.2) | 4.0 (2.4) | 3.9 (2.6) | 3.4 (2.2) | 3.9 (2.3) | 3.7 (2.3) |
| CIN2 | N | 67 | 15 | 65 | 57 | 67 | 54 | 283 |
| | Age at the time of entry (years), mean (SD) | 37.6 (7.9) | 42.0 (5.7) | 41.2 (8.2) | 39.7 (8.5) | 37.9 (8.5) | 36.6 (9.2) | 39.1 (8.4) |
| | Number of visits, mean (SD) | 6.8 (5.4) | 7.0 (4.4) | 7.2 (5.4) | 8.9 (5.8) | 8.6 (5.6) | 8.6 (5.8) | 8.0 (5.5) |
| | Follow-up interval (years), mean (SD) | 0.35 (0.15) | 0.34 (0.14) | 0.38 (0.24) | 0.38 (0.24) | 0.39 (0.33) | 0.38 (0.33) | 0.37 (0.26) |
| | Follow-up period (years), mean (SD) | 2.2 (2.2) | 2.1 (1.6) | 2.5 (2.2) | 3.5 (2.6) | 3.2 (2.3) | 3.3 (2.5) | 2.9 (2.3) |

CIN, cervical intraepithelial neoplasia; HPV, human papillomavirus; hrHPV, high-risk human papillomavirus; SD, standard deviation. Cytological and histological results were combined to classify the results into the following diagnoses: normal, CIN1, and CIN2. HPVs 16, 18, 31, 33, 35, 39, 45, 51, 52, 56, 58, 59, and 68 were classified as hrHPVs. Of these, HPV 16, 18, 52, and 58 were categorized separately. hrHPVs other than HPVs 16, 18, 52, and 58 were classified as "other hrHPVs." Patients who were not infected with any hrHPVs were referred to as "no hrHPVs" patients. In cases of observed coinfection with different HPV genotypes, it was possible to include the same patient in the summary statistics of multiple HPV categories.

Table 2 displays a summary of the transitions from each diagnosis of cervical epithelial lesions for the six HPV categories. Of the total 6022 transitions, 702, 270, 1008, and 852 transitions were observed for HPVs 16, 18, 52, and 58, respectively. The remaining 1501 and 2300 transitions were observed for other hrHPVs and no hrHPVs, respectively. For normal patients, 75–90% remained at the same state (e.g., 84.4% of HPV 16-positive normal patients at a visit were also diagnosed as normal at the subsequent visit). For CIN1 patients, 29–55% regressed to normal, 34–44% remained at the same state, and the rest progressed to CIN2 or CIN3. For CIN2 patients, >50% remained at the same state and the probability of progression to CIN3 was HPV type dependent: 15.6%, 10.3%, 11.1%, 7.7%, and 8.5% for HPVs 16, 18, 52, 58, and other hrHPVs, respectively. HPV 16-positive patients tended to shift to more severe states compared with those who were positive for the other HPV genotypes (e.g., 37% of those with CIN1 shifted to more severe states at the subsequent visit).

**Table 2.** Summary of the transitions from each diagnosis of cervical epithelial lesions for the six human papillomavirus (HPV) categories.

| Diagnosis at $(t-1)^{th}$ Visit | HPV Category | Diagnosis at $t^{th}$ Visit | | | | |
|---|---|---|---|---|---|---|
| | | Normal | CIN1 | CIN2 | CIN3 | Cancer |
| Normal | HPV 16 | 206 (84.4) | 13 (5.3) | 21 (8.6) | 4 (1.6) | 0 (0.0) |
| | HPV 18 | 89 (81.6) | 12 (11.0) | 8 (7.3) | 0 (0.0) | 0 (0.0) |
| | HPV 52 | 277 (75.8) | 54 (14.7) | 32 (8.7) | 2 (0.5) | 0 (0.0) |
| | HPV 58 | 230 (80.4) | 33 (11.5) | 21 (7.3) | 2 (0.6) | 0 (0.0) |
| | Other hrHPVs | 611 (86.1) | 72 (10.1) | 23 (3.2) | 3 (0.4) | 0 (0.0) |
| | No hrHPVs | 1289 (90.2) | 109 (7.6) | 26 (1.8) | 3 (0.2) | 1 (0.0) |
| CIN1 | HPV 16 | 29 (28.9) | 34 (34.0) | 35 (35.0) | 2 (2.0) | 0 (0.0) |
| | HPV 18 | 18 (38.2) | 19 (40.4) | 8 (17.0) | 2 (4.2) | 0 (0.0) |
| | HPV 52 | 80 (35.0) | 90 (39.4) | 53 (23.2) | 5 (2.1) | 0 (0.0) |
| | HPV 58 | 51 (31.6) | 68 (42.2) | 40 (24.8) | 2 (1.2) | 0 (0.0) |
| | Other hrHPVs | 132 (40.7) | 143 (44.1) | 45 (13.8) | 4 (1.2) | 0 (0.0) |
| | No hrHPVs | 203 (54.5) | 132 (35.4) | 34 (9.1) | 3 (0.8) | 0 (0.0) |
| CIN2 | HPV 16 | 31 (12.1) | 37 (14.4) | 147 (57.4) | 40 (15.6) | 1 (0.3) |
| | HPV 18 | 10 (12.9) | 8 (10.3) | 51 (66.2) | 8 (10.3) | 0 (0.0) |
| | HPV 52 | 41 (13.8) | 53 (17.9) | 168 (56.9) | 33 (11.1) | 0 (0.0) |
| | HPV 58 | 32 (10.2) | 45 (14.4) | 210 (67.5) | 24 (7.7) | 0 (0.0) |
| | Other hrHPVs | 49 (16.7) | 52 (17.8) | 166 (56.8) | 25 (8.5) | 0 (0.0) |
| | No hrHPVs | 58 (27.2) | 31 (14.5) | 114 (53.5) | 10 (4.6) | 0 (0.0) |

CIN, cervical intraepithelial neoplasia; HPV, human papillomavirus; hrHPV, high-risk human papillomavirus. Values are the number (percentage) of transitions observed from prior diagnosis to current diagnosis for each HPV category. Cytological and histological results were combined to classify the results into the following diagnoses: normal, CIN1, CIN2, CIN3, and cervical cancer. HPVs 16, 18, 31, 33, 35, 39, 45, 51, 52, 56, 58, 59, and 68 were classified as hrHPVs. Of these, HPV 16, 18, 52, and 58 were categorized separately. hrHPVs other than HPVs 16, 18, 52, and 58 were classified as "other hrHPVs." Patients who were not infected with any hrHPVs were referred to as "no hrHPVs" patients. In cases of observed coinfection with different HPV genotypes, it was possible to include the same patient in multiple HPV categories.

### 2.2. Prognosis of HPV-Infected Cervical Lesions Estimated Using the Markov Model

Table 3 shows the predicted probabilities of transitions from states to states within two years for the six HPV categories, as estimated by the Markov model. Regarding HPV 16-positive patients, 13%, 30%, and 42% of normal, CIN1, and CIN2 patients, respectively, progressed to CIN3/cancer. In contrast, 43% and 34% of CIN1 and CIN2 patients, respectively, regressed to the normal state. The fates of HPV 18-infected cervical lesions were similar to those of HPV 16-infected cervical lesions. However, exceptions to these were the probabilities for progression to CIN3/cancer: from the normal state, 7.6%; from CIN1, 15%; and from CIN2, 32%, respectively. The fates of HPV 52/58-positive cervical lesions differed considerably versus those of HPV 16/18-positive cervical lesions: one-third to two-fifths of HPV 52/58-positive patients were eventually diagnosed with CIN1 or CIN2 regardless of the initial state, and their probability of progression from CIN2 to CIN3/cancer was approximately 25%. Among the hrHPV-positive patients, those with other hrHPVs were most likely to regress to the normal state and least likely to progress to CIN3/cancer: probability of regression to normal state, 66% and 52% from CIN1 and CIN2, respectively, and probability of progression to CIN3/cancer, 3.4%, 8.4%, and 22% from normal, CIN1, and CIN2, respectively. The probability of regression to the normal state of no hrHPVs patients was higher than that of hrHPV-positive patients: 81% and 71% for CIN1 and CIN2, respectively. The Cox model yielded the progression probability from CIN1 to CIN2 or more severe lesions to be 50%, 9%, 53%, 50%, 25%, and 13% for HPVs 16, 18, 52, 58 other hrHPVs, and no hrHPVs, respectively (Supplementary Table S1).

**Table 3.** Predicted 2-year transition probabilities from states to states and their 95% confidence intervals (CIs) for the six HPV categories.

| Current State | HPV Category | State after 2 Years | | | |
|---|---|---|---|---|---|
| | | Normal | CIN1 | CIN2 | CIN3/Cancer |
| Normal | HPV 16 | 0.598 (0.506–0.684) | 0.099 (0.074–0.128) | 0.169 (0.127–0.215) | 0.132 (0.090–0.183) |
| | HPV 18 | 0.610 (0.479–0.719) | 0.156 (0.109–0.215) | 0.156 (0.093–0.230) | 0.076 (0.033–0.148) |
| | HPV 52 | 0.533 (0.474–0.593) | 0.189 (0.162–0.219) | 0.180 (0.146–0.216) | 0.096 (0.070–0.130) |
| | HPV 58 | 0.559 (0.484–0.627) | 0.171 (0.140–0.205) | 0.206 (0.162–0.255) | 0.062 (0.041–0.089) |
| | Other hrHPVs | 0.723 (0.676–0.766) | 0.155 (0.132–0.182) | 0.085 (0.066–0.108) | 0.034 (0.023–0.050) |
| | No hrHPVs | 0.838 (0.814–0.861) | 0.105 (0.090–0.121) | 0.042 (0.032–0.054) | 0.012 (0.007–0.020) |
| CIN1 | HPV 16 | 0.434 (0.349–0.512) | 0.089 (0.067–0.115) | 0.175 (0.134–0.223) | 0.300 (0.225–0.378) |
| | HPV 18 | 0.535 (0.396–0.652) | 0.146 (0.100–0.207) | 0.172 (0.102–0.257) | 0.146 (0.069–0.267) |
| | HPV 52 | 0.473 (0.413–0.529) | 0.178 (0.152–0.208) | 0.183 (0.150–0.221) | 0.164 (0.122–0.219) |
| | HPV 58 | 0.469 (0.399–0.535) | 0.165 (0.135–0.197) | 0.239 (0.192–0.291) | 0.126 (0.084–0.181) |
| | Other hrHPVs | 0.656 (0.606–0.702) | 0.156 (0.133–0.181) | 0.102 (0.079–0.128) | 0.084 (0.058–0.119) |
| | No hrHPVs | 0.808 (0.781–0.835) | 0.107 (0.091–0.123) | 0.049 (0.038–0.065) | 0.034 (0.021–0.054) |
| CIN2 | HPV 16 | 0.335 (0.266–0.404) | 0.079 (0.059–0.101) | 0.165 (0.121–0.218) | 0.418 (0.330–0.512) |
| | HPV 18 | 0.373 (0.245–0.501) | 0.119 (0.074–0.178) | 0.186 (0.099–0.302) | 0.320 (0.178–0.507) |
| | HPV 52 | 0.381 (0.324–0.434) | 0.156 (0.129–0.184) | 0.175 (0.138–0.216) | 0.286 (0.220–0.367) |
| | HPV 58 | 0.356 (0.291–0.419) | 0.150 (0.122–0.181) | 0.260 (0.209–0.319) | 0.232 (0.167–0.307) |
| | Other hrHPVs | 0.518 (0.453–0.571) | 0.146 (0.122–0.169) | 0.117 (0.089–0.148) | 0.218 (0.159–0.299) |
| | No hrHPVs | 0.706 (0.643–0.749) | 0.106 (0.090–0.123) | 0.063 (0.045–0.089) | 0.124 (0.079–0.191) |

CIN, cervical intraepithelial neoplasia; HPV, human papillomavirus; hrHPV, high-risk human papillomavirus. Values are the predicted probabilities (95% confidence intervals) of transitions from the current state to the state after two years for each HPV category. Cytological and histological results were combined to classify the results into the following diagnoses: normal, CIN1, CIN2, CIN3, and cervical cancer. HPVs 16, 18, 31, 33, 35, 39, 45, 51, 52, 56, 58, 59, and 68 were classified as hrHPVs. Of these, HPV 16, 18, 52, and 58 were categorized separately. hrHPVs other than HPVs 16, 18, 52, and 58 were classified as "other hrHPVs." Patients who were not infected with any hrHPVs were referred to as "no hrHPVs" patients. In cases of observed coinfection with different HPV genotypes, it was possible to include the same patient in multiple HPV categories. We used the continuous-time multistate Markov model to estimate the prognosis of each patient. We defined four states: normal (state 1), cervical intraepithelial neoplasia 1 (CIN1, state 2), CIN2 (state 3), and CIN3/cancer (state 4). Arrows in Figure 2 specify possible transitions between the states defined in our model; all transitions between adjacent states, except the backward transition from CIN3/cancer to CIN2, were allowed. CIN3/cancer was the absorbing state. We truncated observations after the diagnosis of CIN3 or cancer.

Figure 1 demonstrates the observed and simulated prevalence transition of each state for the HPV categories. Overall, our model captured the trend of the prevalence, especially up to two years. However, as time progressed, the overestimation (underestimation) of the prevalence of the normal (CIN3/cancer) state expanded.

Supplementary Tables S2 and S3 display the summary statistics and the summary of the transitions from the dataset of the alternative model. Supplementary Table S4 and Supplementary Figure S1 show the results of all sensitivity analyses using the dataset and categories of the alternative model. These data indicate that the results of the main model were fairly robust when we altered the study population and categories.

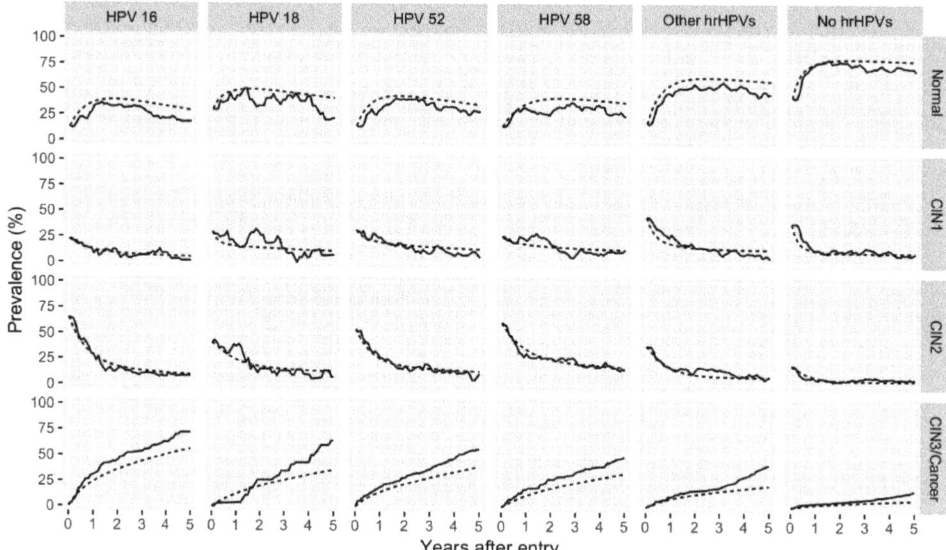

**Figure 1.** Observed and simulated prevalence transition of each state for the human papillomavirus (HPV) categories. The figure demonstrates the observed (solid line) and simulated (dotted line) prevalence transition of each state for the six HPV categories. Cytological and histological results were combined to classify the results into the states of the model. HPVs 16, 18, 31, 33, 35, 39, 45, 51, 52, 56, 58, 59, and 68 were classified as high-risk HPVs (hrHPVs). Of these, HPV 16, 18, 52, and 58 were separately categorized. hrHPVs other than HPVs 16, 18, 52, and 58 were classified as "other hrHPVs." Patients who were not infected with any hrHPVs were referred to as "no hrHPVs" patients. In cases of observed coinfection with different HPV genotypes, it was possible to include the same patient in multiple HPV categories.

## 3. Discussion

This study is the first to estimate the parameters of the continuous-time multistate Markov model from one data set to shed light on the prognosis of cervical lesions based on the infected HPV types. The results indicated that the prognosis of hrHPV-infected cervical lesions differs according to the types of HPV and the grades of CIN.

The Markov model revealed that the rate of progression to CIN3 from any condition was highest in HPV 16-positive patients, followed by HPV 18-positive patients. These results are consistent with those reported in previous studies. In addition, our Markov model also revealed that HPV 52/58-positive patients tend to remain between CIN1 and CIN2. The Markov model clarifies the characteristics of HPV 52- and HPV 58-related lesions, as it demonstrates the probabilities of transitioning from state to state. Contrary to HPV 16-positive lesions, which highly progress to CIN3/cancer, HPV 52- and 58-related lesions were characterized by their stability between CIN1 and CIN2. In this study, we estimated the 2-year prognosis of patients with CIN using two types of models. The main model reflects transitions based upon objective results, and it is not always in line with clinical practice, because some patients with CIN2 undergo treatment without a final diagnosis of CIN3. On the other hand, the alternative model well reflects the trajectory of clinical practice. However, treatment is occasionally affected by the physicians' decision and patients' choice; therefore, the alternative model may lack objectivity. Admitting that our models both have advantages and disadvantages, we consider the results for the 2-year prognosis of CIN patients to be reasonably robust, as both models essentially yielded identical results.

The continuous-time multistate Markov model is more capable of reflecting the features of HPV-infected cervical lesions than the frequently used Cox proportional hazards model. Most of the previous studies have focused on the risk of progression to CIN3 or cervical cancer according to the types of HPV or grades of CIN, implicitly assuming a unidirectional disease progression. Considering the bidirectional nature of CIN, the Markov model has the potential to more accurately estimate the fates of HPV-infected cervical lesions than the standard Cox proportional hazards model. Indeed, the Markov model and the Cox model led to quite different predicted 2-year transition probabilities from CIN1 to CIN2 or more severe lesions. Notably, we could not detect higher disease progression risk in HPV 16-positive patients than in HPV 52/58-positive patients in the Cox model. The Markov model can estimate the transition between multiple states besides the probability of the customarily used endpoint, e.g., CIN3 and cervical cancer. The use of this model highlighted the difference between the natural history of cervical lesions infected with the focused four HPV types. We conclude that the model may improve our understanding of the natural history of cervical lesions and aid in the management of hrHPV-related lesions.

The Markov model was suitable for HPV-infected cervical lesions because the virus–host interaction (virus vs. host immune system) leads to bidirectional transitions between states. Activated host immune responses contribute to CIN regression, whereas immunosuppressive conditions promote CIN progression [36–38]. Likewise, this model can be applied to virus-induced persistent conditions, such as hepatitis B virus- or hepatitis C virus-infected liver conditions, and human immunodeficiency virus-related conditions. Indeed, previous studies applied the model to hepatitis C virus-related liver diseases with three states (i.e., none or mild fibrosis, moderate fibrosis and cirrhosis) to quantify differences in the prognosis among cohorts [34].

In this study, we could not fully estimate the transitions of HPV 18-positive patients with the present Markov model due to the limited sample size. HPV 18 is the second most frequently observed HPV genotype in patients with cervical cancer, whereas it is not frequently detected in those with precancer lesions [35]. In addition, the detection rate of HPV 18 is higher in adenocarcinoma than that in squamous cell carcinoma [39]. Further accumulation of cases may enable us to model the complicated transitions of HPV 18-positive patients.

This study has several limitations. First, we performed the HPV genotyping test once for each patient. Therefore, the change to negative for the HPV test (latent infection or clearance) was not taken into account. Previous reports demonstrate that patients with HPV clearance tend to regress compared with those with persistent HPV infection [40,41]. Patients with the transient HPV infection and persistent infection might take different courses.

Second, in this study, we did not evaluate the effect of concurrent multiple HPV infections. In previous reports, coinfections with multiple $\alpha 9$ species were associated with an increased risk of CIN2 or more severe lesions compared with single infections [42,43]. Because there are many different possible combinations of concurrent infection, the numbers of patients for each possible combination was expected to be too small for an analysis of the effect of concurrent infection. Further analysis with a larger sample size is warranted to unveil the fates of cervical lesions with multiple HPV infections.

Third, the evaluation of model fitness uncovered that the simulated prevalence of the normal state (CIN3/cancer) was overstated (understated) compared with the observed prevalence. These discrepancies imply a violation of the Markov property, i.e., the distribution of the next states depends only upon the present state. Misclassification and omitted covariates, such as age, can be a potential source of the violation. The application of a hidden Markov model that can incorporate the possibility of misclassification through the information from the previous or more distant visits may be an intriguing direction for further research.

Fourth, we have not assessed the validity of our model with the use of a test set separate from the training set, which is the dataset used for parameter estimation. Instead, we prioritized improving the parameter estimation accuracy by using all data for parameter estimation. This is currently a common practice in the literature [26–34]. If a sufficient sample size can be acquired, the external validity can be assessed to some degree using a test set, which will be an issue for future research.

Lastly, this study was based on data obtained from a single institute with some censoring. Especially, there is a selection bias; for example, patients enrolled in this study previously had abnormal cytology. In addition, some patients with relatively severe CIN2 were censored because of the treatment according to the physician's decision. However, by conducting the sensitivity analysis, we confirmed that the impact on the results was limited. Nevertheless, as we have not assessed the external validity of our model in the general population, care must be taken when extrapolating the present results to the entire population, unless similar results are yielded from other institutes.

## 4. Materials and Methods

### 4.1. Study Design and Patients

This study was performed in accordance with the Declaration of Helsinki. Approval was obtained from the internal institutional review board (Approval number: 1390-1, G10082-7, and G0637-6) for this study.

This study was a retrospective cohort analysis of data extracted from the electronic health records of a teaching hospital in Japan. Since 2008, human papillomavirus (HPV) genotyping has been performed in the University of Tokyo Hospital (Tokyo, Japan) for outpatients. The patients who had abnormal cytology in the population-based screening, or whose abnormal cytology was found in the outpatient visits at the Obstetrics and Gynecology Department of the University of Tokyo Hospital or other hospitals, were enrolled. At the first visit, we confirmed the diagnosis by performing punch biopsy under colposcopic examination (based on the histological diagnosis). As follow-up screening, we routinely performed combination examinations of cytology and colposcopy because these examinations are non-invasive. We performed histological diagnosis only when the disease progressed. We reviewed the electronic health records of 1417 patients for whom genotyping was performed between October 1, 2008, and March 31, 2015, to construct a retrospective cohort. Patients were included in the study if they were (i) diagnosed with normal cervical lesion, CIN1, or CIN2 and (ii) observed for at least two visits during the study period. Patients were excluded if they had HPV 6- or HPV 11-single-positive lesions with only the diagnosis of condyloma during their follow-up period, or had only glandular lesions. Patients were followed up until they were diagnosed with cervical cancer, underwent treatment, transferred to another hospital, or until the date March 31, 2018, whichever came first. In our practical management, most patients with CIN2 were followed up without any treatment. Patients diagnosed with CIN3 underwent surgical intervention, such as conization and loop electrosurgical excision or laser vaporization.

### 4.2. Variables

The date of birth was extracted for each patient. Patient age was defined as their age at the time of entry. Follow-up interval was defined as the length of time between two consecutive visits.

Patient cytological and histological results and the date were recorded for each visit. Cytological and histological results were combined to classify the results into any of the following four diagnoses: normal, CIN1, CIN2, and CIN3/cancer. The investigators (experts in gynecologic oncology) convened and determined the criteria for pathological diagnosis as follows: (1) CIN1–2 was classified into CIN2; (2) CIN2–3 and carcinoma in situ were classified into CIN3; (3) uncertain diagnoses (e.g., atypical squamous cells of uncertain significance, atypical squamous cells that cannot exclude HSIL, and dysplasia without grading) were excluded from the study due to concerns regarding diagnostic reliability; and (4) in the presence of histological and cytological examinations, the most severe classification was adopted as the final diagnosis.

The results of the HPV genotyping in cervical samples collected using swabs were recorded. Genotyping was performed once for each patient; thus, the HPV type assigned to a patient did not change over time. It was permitted to assign multiple genotypes to a single patient. On the basis of the classification of the International Agency for Research on Cancer, we defined HPVs classified in Group 1 or Group 2A (HPVs 16, 18, 31, 33, 35, 39, 45, 51, 52, 56, 58, 59, and 68) as "high-risk HPVs (hrHPVs)". Of these, HPVs 16, 18, 52, and 58 were separately categorized. hrHPVs other than HPVs 16, 18, 52, and 58 were classified as "other hrHPVs." Patients who were not infected with any hrHPVs were referred to as "no hrHPVs" patients. Patients without HPV infection were also categorized in this group.

### 4.3. DNA Extraction and HPV Genotyping

Cervical samples were tested for HPV DNA using PGMY line-blot hybridization, as previously described [44]. DNA was extracted from cervical samples using the DNeasy Blood Mini Kit (Qiagen, Crawley, UK). HPV genotyping was performed using the PGMY-CHUV assay method. Briefly, standard polymerase chain reaction (PCR) was performed using the PGMY09/11 L1 consensus primer set and the human leukocyte antigen-DQ primer set, as previously described [44]. Reverse blotting hybridization was performed. Heat-denatured PCR amplicons were hybridized to probes specific for 32 HPV genotypes and the human leukocyte antigen-DQ references.

### 4.4. Continuous-Time Multistate Markov Model

We used the continuous-time multistate Markov model to estimate the prognosis of each patient with HPV-infected cervical lesions [23–25]. We defined the following four states: normal (state 1), CIN1 (state 2), CIN2 (state 3), and CIN3/cancer (state 4) (Figure 2). The arrows in Figure 2 specify possible transitions between the states defined in our model; all transitions between adjacent states, except the backward transition from CIN3/cancer to CIN2, were allowed. CIN3/cancer was the absorbing state.

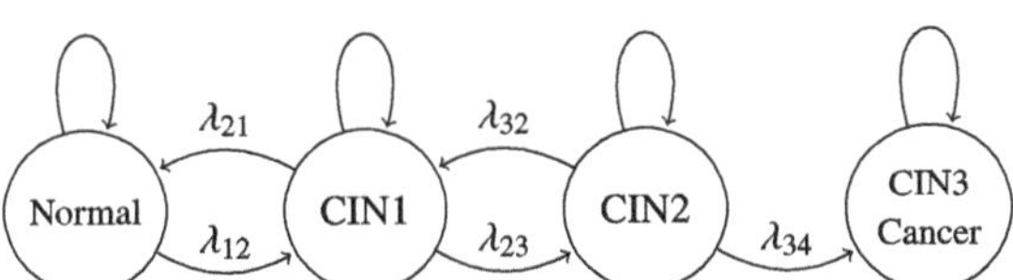


**Figure 2.** Markov model for the disease progression and regression of cervical epithelial lesions. The figure displays the schema of the Markov model for the model specified in this study. We defined four states: normal (state 1), cervical intraepithelial neoplasia 1 (CIN1, state 2), CIN2 (state 3), and CIN3/cancer (state 4). The arrows in the figure specify possible transitions between these states; all transitions between adjacent states, except the backward transition from CIN3/cancer to CIN2, were allowed. CIN3/cancer was the absorbing state. Each transition parameter $\lambda$ indicates the transition intensity; i.e., $\lambda_{ij}$ is interpreted as an "instantaneous risk" of transition from state i to j.

Each transition parameter $\lambda$ in Figure 2 indicates the transition intensity; i.e., $\lambda_{ij}$ is interpreted as an "instantaneous risk" of transition from state i to j [24]. Identification of transition parameters in the continuous-time multistate Markov model requires observation of the corresponding transitions.

### 4.5. Dataset Construction

Initially, an unbalanced panel data with the unit of observation being one visit was constructed on the basis of the chart review. We truncated observations after the diagnosis of CIN3 or cancer to make the data used in the estimation compatible with the model in Figure 2. Additionally, we excluded samples that left only one observation after the truncation to estimate the model, as these samples do not contribute to the estimation. Figure 3 shows the transitions of typical samples from the baseline dataset, which illustrates that the patterns of disease progression or regression varied among patients.

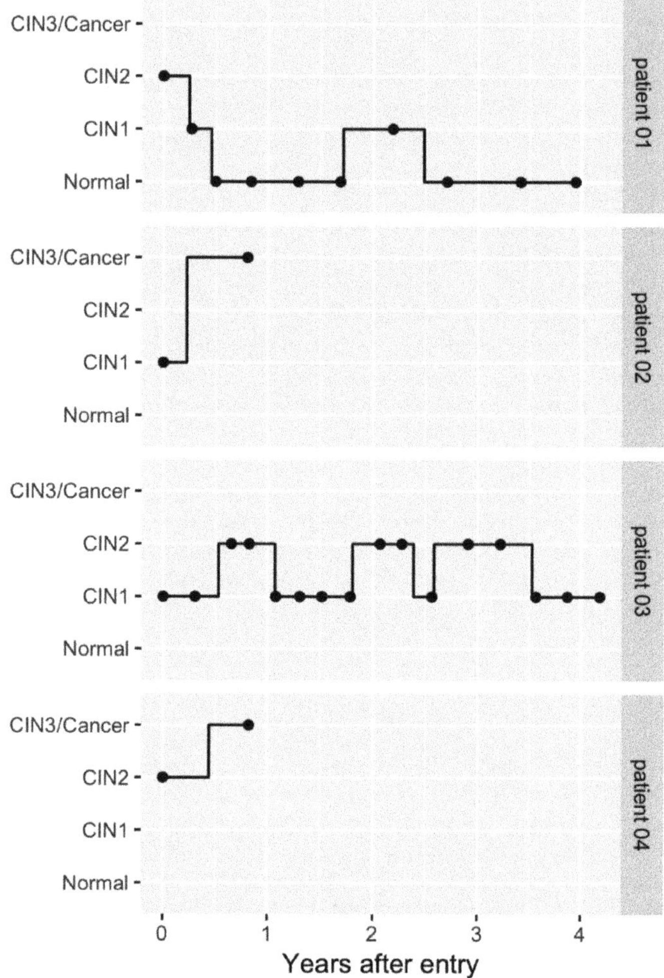

**Figure 3.** Possible transition paths for the selected patients. The figure shows possible transition paths for four selected patients (patient 01–patient 04). Cytological and histological results were combined to classify the results into the states of the model. We truncated observations after the diagnosis of cervical intraepithelial neoplasia 3 or cancer. The filled circles indicate actual observations or visits. The solid line is a possible transition path during the follow-up period. The possible transition paths were randomly selected on the basis of the observed states.

*4.6. Statistical Analysis*

Summary statistics are reported for the combination of defined HPV categories (HPV 16, HPV 18, HPV 52, HPV 58, other hrHPVs, and no hrHPVs) and diagnoses at the time of entry (normal, CIN1, and CIN2). In cases of observed coinfection with different HPV genotypes, it was possible to include the same patient in the summary statistics of multiple HPV categories.

A maximum likelihood estimation was performed to estimate the parameters using the *msm* package in R [45]. Intuitively, the likelihood function consists of the sum of the probabilities of all possible paths given the observations, and each observed transition contributes to it. Parameters for the defined HPV categories were independently estimated. In cases of coinfection, the patients could contribute to the parameter estimation of several HPV categories. According to the estimated parameters, we simulated the probabilities of transitioning from state to state after two years. This 2-year transition probability was selected for two reasons. First, since the typical follow-up duration of this study was <5 years (upper quartile value of the follow-up duration was 4.9 years), a prediction beyond this period was unwarranted. Second, a 2-year prediction is considered a good benchmark for the prognosis of hrHPV-related cervical lesions. Notably, ≥50% of these lesions regress to the normal state, and approximately 10% of these lesions progress to CIN3 within two years [12]. The Cox proportional hazards model was also used to predict the 2-year transition probability from CIN1 to CIN2 or more severe lesions for each HPV category. To this end, we restricted the patients to those who were diagnosed with CIN1 at the time of entry and followed them until they were diagnosed as CIN2 or more severe lesions or at the end of their observation period.

Additionally, we simulated the change in the prevalence of each state over the five years for the HPV categories. To make it easier to compare the observed and the simulated prevalence, the initial prevalence of each state was set to be the same as the data used for parameter estimation. The fit of the model was assessed on the basis of the comparison of the observed and the simulated prevalence.

### 4.7. Sensitivity Analysis

Some patients with CIN2 underwent treatment without a final diagnosis of CIN3. Some patients with CIN3 were observed without any treatment if the lesion immediately regressed to CIN2 or a lower state. Therefore, we specified an alternative model that encompasses these situations with the following four states: normal (state 1), CIN1 (state 2), CIN2/CIN3 (state 3), and treatment (state 4) (Supplementary Figure S2). The arrows in Supplementary Figure S2 specify possible transitions between these states; all transitions between adjacent states, except the backward transition from treatment to CIN2/CIN3, were allowed. Treatment was the absorbing state. We truncated observations after the diagnosis of treatment intervention in this alternative model, and excluded samples that left only one observation after the truncation to estimate the model. Typical sampling situations from this alternative dataset are illustrated in Supplementary Figure S3.

## 5. Conclusions

In this study, we applied a continuous-time multistate Markov model to predict the prognosis of patients with HPV-infected cervical lesions. We demonstrated that the Markov model is a promising analytical method considering the bidirectional nature of these lesions. The study revealed that the natural history of hrHPV-related cervical lesions differed among the focused four HPV types: HPV 16-positive lesions were likely to be upgraded to CIN states in a step-by-step manner; HPV 52/58-positive lesions were likely to be maintained between CIN1 and CIN2; lesions positive for the other hrHPVs were most likely to regress to the normal state and least likely to progress to CIN3/cancer. On the basis of the present findings, HPV genotype-based management may be desirable for patients with cervical lesions.

**Supplementary Materials:** The following are available online at http://www.mdpi.com/2072-6694/12/2/270/s1, Supplementary Materials: Figure S1: Observed and simulated prevalence transition of each state for the human papillomavirus (HPV) categories in the alternative model. Figure S2: Markov model for the disease progression and regression of cervical epithelial lesions of the alternative model. Figure S3: Possible transition paths of the alternative model for the selected patients. Table S1: Predicted 2-year transition probabilities from CIN1 to CIN2 or more severe lesions and their 95% confidence intervals for the six HPV categories derived from the Cox regression analysis. Table S2: Sample size and summary statistics for the combination of patients in the six HPV categories and diagnoses at the time of entry in the dataset for the alternative model. Table S3: Summary of the transitions from each diagnosis of cervical epithelial lesions for the six HPV categories in the dataset for the alternative model. Table S4: Predicted 2-year transition probabilities from states to states and their 95% confidence intervals for the six HPV categories in the alternative model.

**Author Contributions:** Conceptualization, A.T., K.H., and K.K.; methodology, A.T. and K.H.; data collection and data curation, A.T., K.H., T.T. (Tomoki Tanaka), S.B., A.K., S.E., M.M., T.T. (Tetsushi Tsuruga), and K.A.; formal analysis, K.H.; writing—original draft preparation, A.T. and K.H.; writing—review and editing, J.T., T.N., K.O., and K.K.; supervision, T.N., K.O., T.Y., J.T., K.K, Y.O., and T.F.; funding acquisition, A.T. and K.K. All authors have read and agreed to the published version of the manuscript.

**Funding:** This work was supported by Grants-in-Aid for Practical Research for Innovative Cancer Control (KK, Grant Number: 15656298) and J-PRIDE (AT, Grant number: 19fm0208013h0003) from the Japan Agency for Medical Research and Development (AMED).

**Acknowledgments:** We thank Terufumi Yokoyama for the excellent technical support.

**Conflicts of Interest:** The authors declare no conflict of interest.

## References

1.  Torre, L.A.; Bray, F.; Siegel, R.L.; Ferlay, J.; Lortet-Tieulent, J.; Jemal, A. Global cancer statistics, 2012. *Ca Cancer J. Clin.* **2015**, *65*, 87–108. [CrossRef]

2.  National Cancer Center. National Estimates of Cancer Incidence based on Cancer Registries in Japan (1975–2013). Available online: https://ganjoho.jp/en/professional/statistics/table_download.html (accessed on 1 January 2019).

3.  National Cancer Center. Cancer mortality from Vital Statistics in Japan (1958–2016). Available online: https://ganjoho.jp/en/professional/statistics/table_download.html (accessed on 1 January 2019).

4.  Lowy, D.R.; Schiller, J.T. Reducing HPV-Associated Cancer Globally. *Cancer Prev. Res.* **2012**, *5*, 18–23. [CrossRef] [PubMed]

5.  Viens, L.J.; Henley, S.J.; Watson, M.; Markowitz, L.E.; Thomas, C.C.; Thompson, T.D.; Razzaghi, H.; Saraiya, M. Human Papillomavirus–Associated Cancers—United States, 2008–2012. *Mmwr. Morb. Mortal. Wkly. Rep.* **2016**, *65*, 661–666. [CrossRef] [PubMed]

6.  Yim, E.; Park, J. The Role of HPV E6 and E7 Oncoproteins in HPV-associated Cervical Carcinogenesis. *Cancer Res. Treat.* **2005**, *37*, 319. [CrossRef] [PubMed]

7.  Bzhalava, D.; Guan, P.; Franceschi, S.; Dillner, J.; Clifford, G. A systematic review of the prevalence of mucosal and cutaneous human papillomavirus types. *Virology* **2013**, *445*, 224–231. [CrossRef]

8.  Jeon, S.; Allen-Hoffmann, B.L.; Lambert, P.F. Integration of human papillomavirus type 16 into the human genome correlates with a selective growth advantage of cells. *J. Virol.* **1995**, *69*, 2989–2997. [CrossRef]

9.  McBride, A.A.; Warburton, A. The role of integration in oncogenic progression of HPV-associated cancers. *Plos Pathog.* **2017**, *13*, e1006211. [CrossRef]

10. Wright, T.C.; Massad, L.S.; Dunton, C.J.; Spitzer, M.; Wilkinson, E.J.; Solomon, D. 2006 consensus guidelines for the management of women with cervical intraepithelial neoplasia or adenocarcinoma in situ. *Am. J. Obs. Gynecol.* **2007**, *197*, 340–345. [CrossRef]

11. Wright, T.C.; Massad, L.S.; Dunton, C.J.; Spitzer, M.; Wilkinson, E.J.; Solomon, D. 2006 consensus guidelines for the management of women with abnormal cervical cancer screening tests. *Am. J. Obs. Gynecol.* **2007**, *197*, 346–355. [CrossRef]

12. Matsumoto, K.; Oki, A.; Furuta, R.; Maeda, H.; Yasugi, T.; Takatsuka, N.; Mitsuhashi, A.; Fujii, T.; Hirai, Y.; Iwasaka, T.; et al. Predicting the progression of cervical precursor lesions by human papillomavirus genotyping: A prospective cohort study. *Int. J. Cancer* **2011**, *128*, 2898–2910. [CrossRef]

13. Japan Society of Obstetrics and Gynecology; Japan Association of Obstetricians and Gynecologists (Eds.) *Guideline for Gynecological Practice 2017*; Japan Society of Obstetrics and Gynecology: Tokyo, Japan, 2017; pp. 53–57. (in Japanese)

14. Kyrgiou, M.; Koliopoulos, G.; Martin-Hirsch, P.; Arbyn, M.; Prendiville, W.; Paraskevaidis, E. Obstetric outcomes after conservative treatment for intraepithelial or early invasive cervical lesions: Systematic review and meta-analysis. *Lancet* **2006**, *367*, 489–498. [CrossRef]

15. Kyrgiou, M.; Athanasiou, A.; Paraskevaidi, M.; Mitra, A.; Kalliala, I.; Martin-Hirsch, P.; Arbyn, M.; Bennett, P.; Paraskevaidis, E. Adverse obstetric outcomes after local treatment for cervical preinvasive and early invasive disease according to cone depth: Systematic review and meta-analysis. *BMJ* **2016**, *71*, i3633. [CrossRef] [PubMed]

16. Miura, S.; Matsumoto, K.; Oki, A.; Satoh, T.; Tsunoda, H.; Yasugi, T.; Taketani, Y.; Yoshikawa, H. Do we need a different strategy for HPV screening and vaccination in East Asia? *Int. J. Cancer* **2006**, *119*, 2713–3715. [CrossRef] [PubMed]

17. Adebamowo, S.N.; Olawande, O.; Famooto, A.; Dareng, E.O.; Offiong, R.; Adebamowo, C.A. Persistent Low-Risk and High-Risk Human Papillomavirus Infections of the Uterine Cervix in HIV-Negative and HIV-Positive Women. *Front. Public Heal.* **2017**, *5*, 1–11. [CrossRef]

18. Wright, T.C.; Stoler, M.H.; Behrens, C.M.; Sharma, A.; Zhang, G.; Wright, T.L. Primary cervical cancer screening with human papillomavirus: End of study results from the ATHENA study using HPV as the first-line screening test. *Gynecol. Oncol.* **2015**, *136*, 189–197. [CrossRef]

19. Naucler, P.; Ryd, W.; Törnberg, S.; Strand, A.; Wadell, G.; Hansson, B.G.; Rylander, E.; Dillner, J. HPV type-specific risks of high-grade CIN during 4 years of follow-up: A population-based prospective study. *Br. J. Cancer* **2007**, *97*, 129–132. [CrossRef]

20. Rositch, A.F.; Koshiol, J.; Hudgens, M.G.; Razzaghi, H.; Backes, D.M.; Pimenta, J.M.; Franco, E.L.; Poole, C.; Smith, J.S. Patterns of persistent genital human papillomavirus infection among women worldwide: A literature review and meta-analysis. *Int. J. Cancer* **2013**, *133*, 1271–1285. [CrossRef]

21. Rodriguez, A.C.; Schiffman, M.; Herrero, R.; Wacholder, S.; Hildesheim, A.; Castle, P.E.; Solomon, D.; Burk, R. Rapid Clearance of Human Papillomavirus and Implications for Clinical Focus on Persistent Infections. *Jnci. J. Natl. Cancer Inst.* **2008**, *100*, 513–517. [CrossRef]

22. Schiffman, M.; Herrero, R.; DeSalle, R.; Hildesheim, A.; Wacholder, S.; Cecilia Rodriguez, A.; Bratti, M.C.; Sherman, M.E.; Morales, J.; Guillen, D.; et al. The carcinogenicity of human papillomavirus types reflects viral evolution. *Virology* **2005**, *337*, 76–84. [CrossRef]

23. Cox, D.R.; Miller, H.D. *The Theory of Stochastic Processes*, 1st ed.; Routledge: London, NY, USA; Chapman & Hall/CRC: London, UK, 1965.

24. Kalbfleisch, J.D.; Lawless, J.F. The Analysis of Panel Data under a Markov Assumption. *J. Am. Stat. Assoc.* **1985**, *80*, 863–871. [CrossRef]

25. Kay, R. A Markov Model for Analysing Cancer Markers and Disease States in Survival Studies. *Biometrics* **1986**, *42*, 855–865. [CrossRef] [PubMed]

26. Buchman, A.S.; Leurgans, S.E.; Yu, L.; Wilson, R.S.; Lim, A.S.; James, B.D.; Shulman, J.M.; Bennett, D.A. Incident parkinsonism in older adults without Parkinson disease. *Neurology* **2016**, *87*, 1036–1044. [CrossRef] [PubMed]

27. Buter, T.C.; van den Hout, A.; Matthews, F.E.; Larsen, J.P.; Brayne, C.; Aarsland, D. Dementia and survival in Parkinson disease: A 12-year population study. *Neurology* **2008**, *70*, 1017–1022. [CrossRef] [PubMed]

28. Dancourt, V.; Quantin, C.; Abrahamowicz, M.; Binquet, C.; Alioum, A.; Faivre, J. Modeling recurrence in colorectal cancer. *J. Clin. Epidemiol.* **2004**, *57*, 243–251. [CrossRef]

29. Jack, C.R.; Therneau, T.M.; Wiste, H.J.; Weigand, S.D.; Knopman, D.S.; Lowe, V.J.; Mielke, M.M.; Vemuri, P.; Roberts, R.O.; Machulda, M.M.; et al. Transition rates between amyloid and neurodegeneration biomarker states and to dementia: A population-based, longitudinal cohort study. *Lancet Neurol.* **2016**, *15*, 56–64. [CrossRef]

30. Pan, S.-L.; Lien, I.-N.; Yen, M.-F.; Lee, T.-K.; Chen, T.H.-H. Dynamic Aspect of Functional Recovery after Stroke Using a Multistate Model. *Arch. Phys. Med. Rehabil.* **2008**, *89*, 1054–1060. [CrossRef]

31. Price, M.J.; Ades, A.E.; De Angelis, D.; Welton, N.J.; Macleod, J.; Soldan, K.; Simms, I.; Turner, K.; Horner, P.J. Risk of Pelvic Inflammatory Disease Following Chlamydia trachomatis Infection: Analysis of Prospective Studies With a Multistate Model. *Am. J. Epidemiol.* **2013**, *178*, 484–492. [CrossRef]

32. Sharples, L.D.; Jackson, C.H.; Parameshwar, J.; Wallwork, J.; Large, S.R. Diagnostic accuracy of coronary angiography and risk factors for post-heart-transplant cardiac allograft vasculopathy. *Transplantation* **2003**, *76*, 679–682. [CrossRef]

33. Skogvoll, E.; Eftestøl, T.; Gundersen, K.; Kvaløy, J.T.; Kramer-Johansen, J.; Olasveengen, T.M.; Steen, P.A. Dynamics and state transitions during resuscitation in out-of-hospital cardiac arrest. *Resuscitation* **2008**, *78*, 30–37. [CrossRef]

34. Sweeting, M.J.; De Angelis, D.; Neal, K.R.; Ramsay, M.E.; Irving, W.L.; Wright, M.; Brant, L.; Harris, H.E. Estimated progression rates in three United Kingdom hepatitis C cohorts differed according to method of recruitment. *J. Clin. Epidemiol.* **2006**, *59*, 144–152. [CrossRef]

35. Bulk, S.; Berkhof, J.; Rozendaal, L.; Fransen Daalmeijer, N.C.; Gök, M.; de Schipper, F.A.; van Kemenade, F.J.; Snijders, P.J.; Meijer, C.J. The contribution of HPV18 to cervical cancer is underestimated using high-grade CIN as a measure of screening efficiency. *Br. J. Cancer* **2007**, *96*, 1234–1236. [CrossRef] [PubMed]

36. De Gruijl, T.D.; Bontkes, H.J.; Walboomers, J.M.M.; Stukart, M.J.; Doekhie, F.S.; Remmink, A.J.; Helmerhorst, T.J.M.; Verheijen, R.H.M.; Duggan-Keen, M.F.; Stern, P.L.; et al. Differential T helper cell responses to human papillomavirus type 16 E7 related to viral clearance or persistence in patients with cervical neoplasia: A longitudinal study. *Cancer Res.* **1998**, *58*, 1700–1706. [PubMed]

37. Ahdieh, L.; Munoz, A.; Vlahov, D.; Trimble, C.L.; Timpson, L.A.; Shah, K. Cervical Neoplasia and Repeated Positivity of Human Papillomavirus Infection In Human Immunodeficiency Virus-seropositive and -seronegative Women. *Am. J. Epidemiol.* **2000**, *151*, 1148–1157. [CrossRef]

38. Brennan, D.C.; Aguado, J.M.; Potena, L.; Jardine, A.G.; Legendre, C.; Säemann, M.D.; Mueller, N.J.; Merville, P.; Emery, V.; Nashan, B. Effect of maintenance immunosuppressive drugs on virus pathobiology: Evidence and potential mechanisms. *Rev. Med. Virol.* **2013**, *23*, 97–125. [CrossRef]

39. De Sanjose, S.; Quint, W.G.V.; Alemany, L.; Geraets, D.T.; Klaustermeier, J.E.; Lloveras, B.; Tous, S.; Felix, A.; Bravo, L.E.; Shin, H.-R.; et al. Human papillomavirus genotype attribution in invasive cervical cancer: A retrospective cross-sectional worldwide study. *Lancet Oncol.* **2010**, *11*, 1048–1056. [CrossRef]

40. Brown, D.R.; Shew, M.L.; Qadadri, B.; Neptune, N.; Vargas, M.; Tu, W.; Juliar, B.E.; Breen, T.E.; Fortenberry, J.D. A longitudinal study of genital human papillomavirus infection in a cohort of closely followed adolescent women. *J. Infect. Dis.* **2005**, *191*, 182–192. [CrossRef]

41. Cho, H.W.; So, K.A.; Lee, J.K.; Hong, J.H. Type-specific persistence or regression of human papillomavirus genotypes in women with cervical intraepithelial neoplasia 1: A prospective cohort study. *Obstet. Gynecol. Sci.* **2015**, *58*, 40–45. [CrossRef]

42. Wentzensen, N.; Schiffman, M.; Dunn, T.; Zuna, R.E.; Gold, M.A.; Allen, R.A.; Zhang, R.; Sherman, M.E.; Wacholder, S.; Walker, J.; et al. Multiple human papillomavirus genotype infections in cervical cancer progression in the study to understand cervical cancer early endpoints and determinants. *Int. J. Cancer* **2009**, *125*, 2151–2158. [CrossRef]

43. Chaturvedi, A.K.; Katki, H.A.; Hildesheim, A.; Rodríguez, A.C.; Quint, W.; Schiffman, M.; Van Doorn, L.-J.; Porras, C.; Wacholder, S.; Gonzalez, P.; et al. Human Papillomavirus Infection with Multiple Types: Pattern of Coinfection and Risk of Cervical Disease. *J. Infect. Dis.* **2011**, *203*, 910–920. [CrossRef]

44. Kojima, S.; Kawana, K.; Tomio, K.; Yamashita, A.; Taguchi, A.; Miura, S.; Adachi, K.; Nagamatsu, T.; Nagasaka, K.; Matsumoto, Y.; et al. The Prevalence of Cervical Regulatory T Cells in HPV-Related Cervical Intraepithelial Neoplasia (CIN) Correlates Inversely with Spontaneous Regression of CIN. *Am. J. Reprod Immunol.* **2013**, *69*, 134–141. [CrossRef]

45. Jackson, C.H. Multi-State Models for Panel Data: The msm Package for R. *J. Stat. Softw.* **2011**, *38*, 1–29. [CrossRef]

*Article*

# Survival-Associated Metabolic Genes in Human Papillomavirus-Positive Head and Neck Cancers


Martin A. Prusinkiewicz [1,†], Steven F. Gameiro [1,†], Farhad Ghasemi [2], Mackenzie J. Dodge [1], Peter Y. F. Zeng [3], Hanna Maekebay [1], John W. Barrett [3], Anthony C. Nichols [3,4,5] and Joe S. Mymryk [1,3,4,5,*]

[1]  Department of Microbiology and Immunology, The University of Western Ontario, London, ON N6A 3K7, Canada; mprusink@uwo.ca (M.A.P.); sgameiro@uwo.ca (S.F.G.); mdodge@uwo.ca (M.J.D.); hmaekeba@uwo.ca (H.M.)

[2]  Department of Surgery, The University of Western Ontario, London, ON N6A 3K7, Canada; fghasemi2019@meds.uwo.ca

[3]  Department of Otolaryngology, Head & Neck Surgery, The University of Western Ontario, London, ON N6A 3K7, Canada; yzeng2023@meds.uwo.ca (P.Y.F.Z.); john.barrett@lhsc.on.ca (J.W.B.); anthony.nichols@lhsc.on.ca (A.C.N.)

[4]  Department of Oncology, The University of Western Ontario, London, ON N6A 3K7, Canada

[5]  London Regional Cancer Program, Lawson Health Research Institute, London, ON N6C 2R5, Canada

*  Correspondence: jmymryk@uwo.ca; Tel.: +1-519-685-8600 (ext. 53012)

†  Contributed equally to this project.



Received: 29 November 2019; Accepted: 16 January 2020; Published: 20 January 2020



**Abstract:** Human papillomavirus (HPV) causes an increasing number of head and neck squamous cell carcinomas (HNSCCs). Altered metabolism contributes to patient prognosis, but the impact of HPV status on HNSCC metabolism remains relatively uncharacterized. We hypothesize that metabolism-related gene expression differences unique to HPV-positive HNSCC influences patient survival. The Cancer Genome Atlas RNA-seq data from primary HNSCC patient samples were categorized as 73 HPV-positive, 442 HPV-negative, and 43 normal-adjacent control tissues. We analyzed 229 metabolic genes and identified numerous differentially expressed genes between HPV-positive and negative HNSCC patients. HPV-positive carcinomas exhibited lower expression levels of genes involved in glycolysis and higher levels of genes involved in the tricarboxylic acid cycle, oxidative phosphorylation, and β-oxidation than the HPV-negative carcinomas. Importantly, reduced expression of the metabolism-related genes *SDHC, COX7A1, COX16, COX17, ELOVL6, GOT2*, and *SLC16A2* were correlated with improved patient survival only in the HPV-positive group. This work suggests that specific transcriptional alterations in metabolic genes may serve as predictive biomarkers of patient outcome and identifies potential targets for novel therapeutic intervention in HPV-positive head and neck cancers.



**Keywords:** human papillomavirus; head and neck cancer; cancer metabolism; glycolysis; cellular respiration; TCGA


## 1. Introduction

As of 2018, head and neck squamous cell carcinomas (HNSCC), namely cancers of the oral cavity, oropharynx, nasopharynx, larynx, and hypopharynx, had the 8th highest combined incidence rate and the 5th highest 5-year prevalence as interpreted from GLOBOCAN data [1,2]. This translates to 834,860 new head and neck cancers per year and 2,164,271 active head and neck cancers within the past five years worldwide [1,2]. Recent incidence rates of some oropharyngeal cancers, such as those of the tonsils and base of the tongue, have been rapidly increasing due to high-risk human papillomavirus

(HPV) infection [3]. Infection by specific high-risk HPVs, such as HPV16, was only recognized as a contributing factor for oropharyngeal cancer by the International Agency for Research on Cancer in 2003 [4]. However, the number of oropharyngeal cancers caused by HPV has risen at epidemic rates over the last decades in many parts of the world [5], while the number of HNSCCs caused by exposure to mutagens from excessive smoking and drinking has been decreasing [6].

HPV-positive (HPV+) HNSCCs are distinct from their HPV-negative (HPV-) counterparts from a molecular perspective, with characteristic genetic, epigenetic, and protein expression profiles [7–9]. In addition, patient outcomes are generally far more favorable for HPV+ than HPV- HNSCC [10]. The underlying molecular reasons for this difference are not entirely clear. However, approximately 10% of all HPV+ HNSCC patients still succumb to their disease [11]. Identification of prognostic markers predicting favorable survival outcomes in patients could allow for treatment deintensification, thereby avoiding potential lifelong complications from unnecessarily aggressive treatments. Alternatively, identification of cellular pathways contributing to poor prognosis could lead to the development of new effective therapies for those not responding to the current standard of care.

Altered metabolism is a cancer hallmark that was recognized decades ago with the discovery of the Warburg effect, also known as aerobic glycolysis [12]. Aerobic glycolysis involves an upregulation of glycolysis despite the presence of ample oxygen for efficient cellular respiration. Many tumours exhibit this metabolic phenotype, as it provides rapid energy and an ample supply of precursors for macromolecule biosynthesis. Tumours can also rely on cellular respiration, often via glutaminolysis, which is the breakdown of glutamine into intermediates of the tricarboxylic acid (TCA) cycle. [13]. TCA intermediates can be funneled off for macromolecule biosynthesis. Many viruses are known to extensively modulate cellular metabolic processes to facilitate infection [14]. These changes can include similar tumour-associated metabolic changes as described above. Infection with HPV has been shown to phenocopy cancer-like metabolic changes that are maintained in HPV+ HNSCC [15]. Examination of how the metabolic phenotype differs between HPV+ and HPV- HNSCC could lead to the identification of targetable metabolic changes and potential new treatment options that are specific for either HPV+ HNSCCs or HPV- HNSCCs. Admittedly, cancer metabolism is complex as it impinges on a variety of other cellular processes and can vary across an individual tumour [16]. In addition, tumours can be highly adaptive to metabolic perturbations [16]. Identifying multiple metabolic targets that are specific to a cancer type from a large dataset that contains information from an extensive number of tumours, such as The Cancer Genome Atlas (TCGA), is an ideal step towards selecting or generating useful anti-metabolic cancer therapeutics.

In this study, we used RNA-seq data from over 500 HNSCC primary tumour samples from the TCGA to comprehensively compare the expression of genes across key cellular metabolic pathways between HPV+, HPV-, and normal-adjacent control tissues. Expression of a number of metabolic genes were significantly altered in HPV+ versus HPV- HNSCC. Specifically, genes involved in glycolysis were expressed at lower levels in HPV+ compared to HPV- HNSCC. In contrast, genes involved in the TCA cycle, oxidative phosphorylation, and β-oxidation exhibited higher expression in HPV+ samples compared to their HPV- counterparts. Importantly, we identified that low expression of multiple metabolic genes—*SDHC*, *COX7A1*, *COX16*, *COX17*, *ELOVL6*, *GOT2*, and *SLC16A2*—correlated with markedly improved patient survival in HPV+, but not HPV- HNSCC. The products of these genes could potentially be exploited as targets for therapeutic intervention. Furthermore, the low expression of these seven genes appears useful in predicting improved overall survival in HPV+ HNSCC and could serve as biomarkers of patient outcome.

## 2. Results

### 2.1. Expression of Pathway-specific Metabolic Genes Were Altered Between HPV+, HPV-, and Normal Control Samples from The TCGA HNSC Cohort

In order to identify differences in metabolic gene expression between HPV+ and HPV- HNSCC, we analyzed the TCGA Illumina HiSeq RNA expression dataset from the HNSC cohort for expression of 229 metabolic genes in central metabolic pathways (Supplementary Table S1). This clinical cohort is comprised of 73 HPV+, 442 HPV-, and 43 normal control samples with available RNA-seq data. Significant differences were seen in a subset of genes in each pathway between HPV+ and HPV- HNSCC. In addition, significant changes were observed between either HPV+ HNSCC or HPV- HNSCC and normal control tissue. To simplify interpretation of these differences, we plotted the fraction of genes in each pathway that were significantly different for each pairwise comparison (Figure 1).

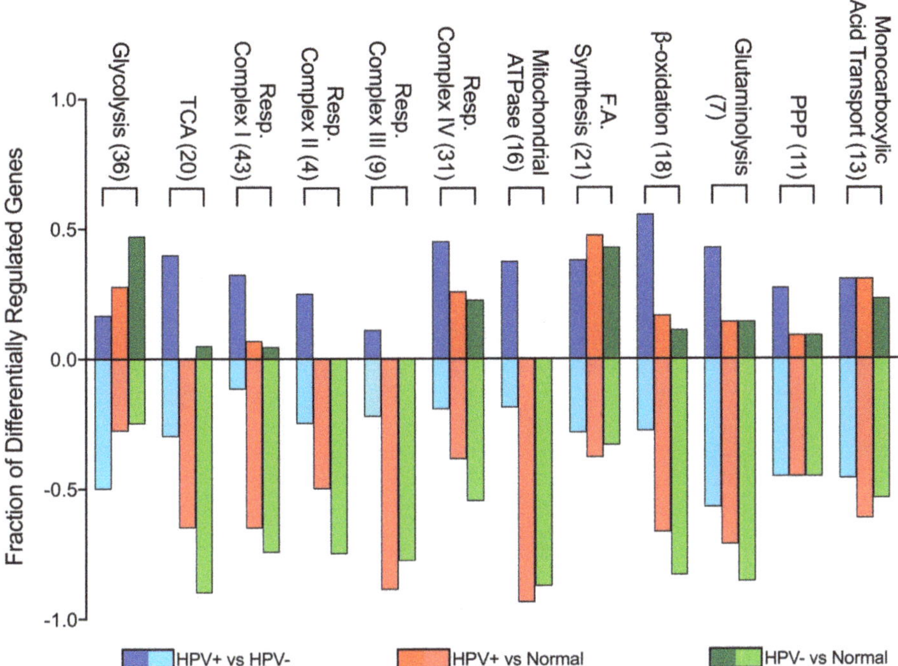


**Figure 1.** Genes differentially expressed between HPV-positive (HPV+) head and neck squamous cell carcinomas (HNSCC), HPV-negative (HPV-) HNSCC, and normal control tissues by metabolic pathway. The y-axis reflects the proportion of genes that are up or downregulated in a given pathway comparison. For example, the positive fraction of genes reflects the proportion of genes upregulated in the first group (e.g., HPV+) when compared to the second group (e.g., HPV-). The negative fraction of genes reflects the proportion of genes downregulated in the same comparison. Blue = HPV+ tissue vs HPV- tissue comparison; Red = HPV+ tissue vs normal control tissue comparison; Green = HPV- tissue vs normal control tissue comparison. Numbers in brackets denote total number of genes analyzed from each pathway. Abbreviations: TCA, tricarboxylic acid cycle; Resp., respiratory; F.A., fatty acid; PPP, pentose phosphate pathway.

This analysis illustrates that glycolytic genes in HPV- HNSCC tissue are more upregulated in comparison to normal tissue than in HPV+ HNSCC when compared to normal tissue. This is particularly evident between HPV+ HNSCC and HPV- HNSCC tissues, since most glycolytic genes are downregulated in HPV+ HNSCC when compared to HPV- HNSCC. However, genes involved in the TCA cycle, oxidative phosphorylation, and β-oxidation were downregulated in both HPV+ and HPV- HNSCC in comparison to normal control tissue. This means that, despite differences in metabolic gene expression between HPV+ and HPV- HNSCC, the metabolism of both types of tumours could resemble the Warburg effect, with lower cellular respiration rates compared to normal control tissue. When compared to one another, HPV+ HNSCC had generally higher expression of these genes than HPV- HNSCC. Other metabolic pathways appeared to have a more similar split between upregulated and downregulated genes in all comparisons, suggesting that the presence of HPV may have minimal impact on transcriptionally mediated changes in these pathways. Overall, it appears that HPV+ HNSCC may be more reliant on cellular respiration than HPV- HNSCC.

*2.2. Low Expression of Genes Encoding Multiple Components of the Mitochondrial Electron Transport Chain Are Associated with Improved Patient Survival in HPV+ HNSCC*

Tumour-associated metabolic alterations have functional consequences that impact disease progression, response to therapy, and patient survival [13]. We dichotomized the expression data for each of the 229 metabolic genes by median expression and calculated the impact of high versus low expression on overall patient survival for HPV+ HNSCC patients, as well as HPV- HNSCC patients (Supplementary Table S2). We identified seven genes that were significantly associated with patient survival in HPV+, but not HPV- HNSCC patients. These genes were *SDHC*, part of the mitochondrial respiratory complex II; *COX7A1*, *COX16*, and *COX17*, all part of the mitochondrial respiratory complex IV; *ELOVL6*, involved in fatty acid elongation; *GOT2*, involved in amino acid metabolism; and *SLC16A2* (also known as *MCT8*), which encodes a thyroid hormone transporter. We performed a pathway enrichment analysis of our seven significant genes utilizing a PANTHER overrepresentation test which indicated that 5 of these 7 genes were significantly associated with the mitochondria ($p = 5.10 \times 10^{-5}$; FDR $= 1.46 \times 10^{-2}$). These genes were *SDHC*, *COX7A1*, *COX16*, *COX17*, and *GOT2*. Detailed studies of the relative expression of each of these genes in HPV+, HPV- and normal control tissue, as well as their association with overall patient survival are presented below.

Previous studies indicate that HPV+ HNSCC is more reliant on oxidative phosphorylation as an energy source than HPV- HNSCC [15]. Oxidative phosphorylation requires electron transport via mitochondrial cellular respiratory complexes I-IV [17]. Reduced expression of *SDHC*, which encodes a component of the mitochondrial cellular respiration complex II, correlated with increased HPV+ HNSCC patient survival (Figure 2).

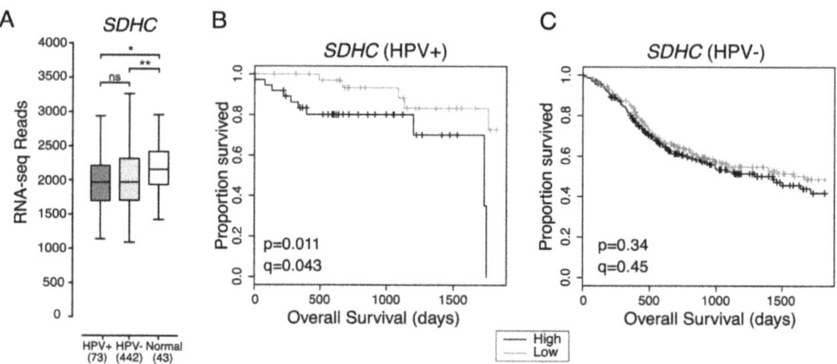

**Figure 2.** Low expression of *SDHC* is associated with favorable survival outcomes in HPV+ HNSCC. (**A**) Transcript levels of *SDHC* across all HNSCC tissues samples and normal control tissues. Bracketed numbers refer to the sample size of each group. Overall five-year survival outcomes in (**B**) HPV+ HNSCC and (**C**) HPV- HNSCC patients dichotomized by *SDHC* expression. $p$ = Two-sided log-rank test, $q$ = Benjamini-Hochberg FDR method. Gray = low transcript expression, Black = high transcript expression. * $p \leq 0.05$, ** $p \leq 0.01$, ns (not significant).

Overall expression of *SDHC* was not significantly different between HPV+ and HPV- HNSCC, and both types of HNSCC expressed lower levels of *SDHC* than normal control tissue (Figure 2A). HPV+ HNSCC patients with tumours exhibiting low *SDHC* expression had better overall five-year survival outcomes than HPV+ HNSCC patients with tumours exhibiting high *SDHC* expression ($p = 0.011$, FDR $= 0.043$) (Figure 2B). However, *SDHC* expression was not correlated with improved patient survival in HPV- HNSCC ($p = 0.34$, FDR $= 0.45$) (Figure 2C).

Consistent with the possibility that HPV+ HNSCCs are more reliant on cellular respiration than HPV- HNSCCs, three genes encoding components of the mitochondrial cellular respiration complex IV were also correlated with HPV+ HNSCC patient survival (Figure 3). These genes were *COX7A1* and *COX17*, which encode structural components of complex IV, and *COX16*, whose product is involved in complex IV assembly.

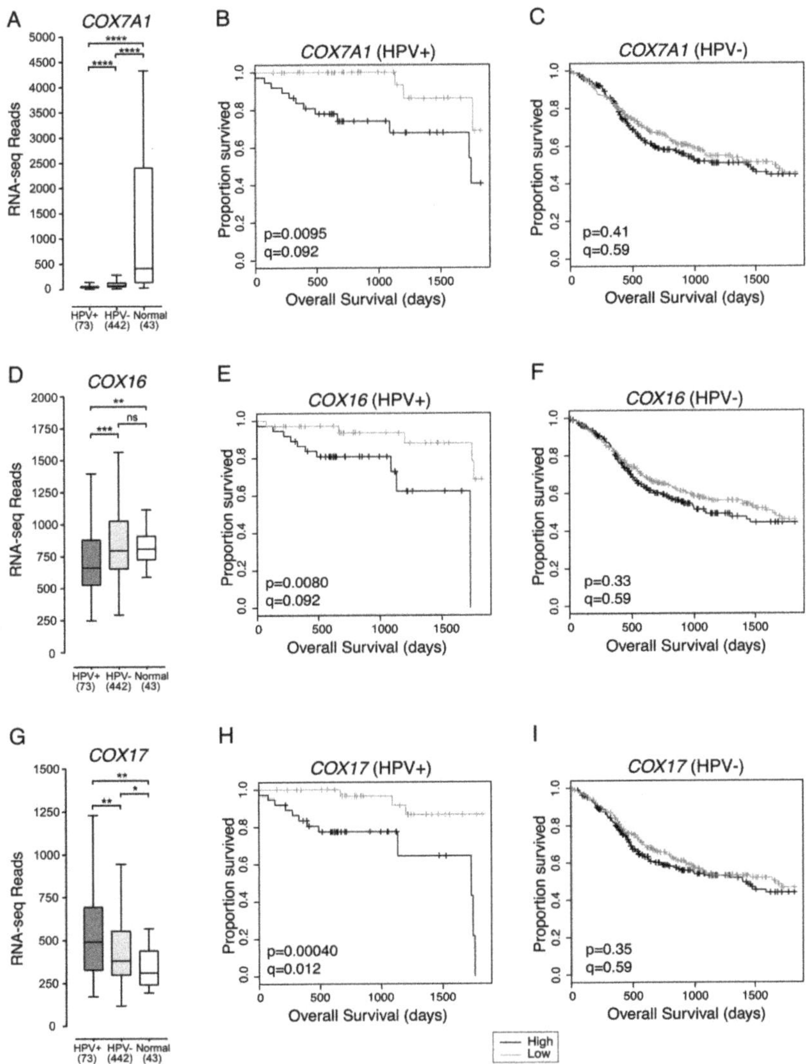



**Figure 3.** Low expression of three mitochondrial respiration complex IV genes in HPV+ HNSCC is associated with improved survival. Expression of (**A**) *COX7A1*, (**D**) *COX16*, and (**G**) *COX17* in HNSCC tissues samples and normal control tissues. Overall 5-year survival outcomes in HPV+ HNSCC patients and HPV- HNSCC dichotomized by median (**B,C**) *COX7A1* expression, (**E,F**) *COX16* expression, and (**H,I**) *COX17* expression. $p$ = Two-sided log-rank test, $q$ = Benjamini-Hochberg FDR method. Gray = low transcript expression, Black = high transcript expression. * $p \leq 0.05$, ** $p \leq 0.01$, *** $p \leq 0.001$, **** $p \leq 0.0001$, ns (not significant).

*COX7A1* expression was significantly lower in both HPV+ and HPV- HNSCC samples compared to normal control tissues (Figure 3A). In addition, HPV+ HNSCC had significantly lower expression of *COX7A1* than HPV- HNSCC (Figure 3A). Overall survival of patients with HPV+ (Figure 3B) or HPV- HNSCC (Figure 3C) were dichotomized based on median *COX7A1* expression. We found that low expression of *COX7A1* was correlated with favourable survival outcomes in patients with HPV+ ($p$ = 0.0095, FDR = 0.092) (Figure 3B), but not HPV- HNSCC ($p$ = 0.41, FDR = 0.59) (Figure 3C).

Compared to normal control tissue, *COX16* expression was lower in HPV+, but not in HPV- HNSCC (Figure 3D). *COX16* expression was also significantly lower in HPV+ compared to HPV- HNSCC (Figure 3D). HPV+ and HPV- HNSCC samples were dichotomized based on median *COX16* expression. Low levels of *COX16* expression were correlated with improved survival in patients with HPV+ ($p$ = 0.0080, FDR = 0.092) (Figure 3E), but not in patients with HPV- HNSCC ($p$ = 0.33, FDR = 0.59) (Figure 3F).

In contrast to *COX7A1* and *COX16*, *COX17* expression was higher in HPV+ than HPV- HNSCC (Figure 3G). Expression of *COX17* across HPV+ and HPV- HNSCC samples was significantly higher than in normal control tissues (Figure 3G). However, the same association between low expression of *COX17* and better overall 5-year patient survival was observed for patients with HPV+ HNSCC ($p$ = 0.00040, FDR = 0.012) (Figure 3H), but not patients with HPV- HNSCC ($p$ = 0.35, FDR = 0.59) (Figure 3I).

### 2.3. Low Expression of ELOVL6, Involved in Fatty Acid Synthesis, Is Associated with Better Overall Survival in Patients with HPV+ HNSCC

*ELOVL6* expression in HPV+ HNSCC was not significantly different from HPV- HNSCC. However, both HPV+ and HPV- HNSCC had significantly lower overall levels of *ELOVL6* expression than normal control tissues (Figure 4A). HPV+ HNSCC samples were dichotomized based on median *ELOVL6* expression. HPV+ patients with tumours expressing low levels of *ELOVL6* had significantly better five-year overall survival than patients with tumours expressing high levels of *ELOVL6* ($p$ = 0.0040, $q$ = 0.084) (Figure 4B). Reduced *ELOVL6* expression was not correlated with altered survival in HPV- HNSCC patients ($p$ = 0.34, $q$ = 0.83) (Figure 4C).

### 2.4. Low Expression of GOT2, Involved in Amino Acid Metabolism, Is Associated with Better HPV+ HNSCC Patient Survival

*GOT2* expression was significantly lower in HPV+ than HPV- HNSCC or normal control tissues and significantly lower in HPV- HNSCC than normal control tissues (Figure 4D). The high normalized RNA-seq read levels for *GOT2* suggest that it is abundantly expressed in normal head and neck tissues. HPV+ HNSCC samples were dichotomized based on median *GOT2* expression. Low expression of *GOT2* was associated with better five-year overall survival outcomes in patients with HPV+ HNSCC ($p$ = 0.012, $q$ = 0.086; Figure 4E). Although low *GOT2* expression appeared to be significantly correlated with favourable patient survival in HPV- HNSCC ($p$ = 0.029; Figure 4F), it lost its significance after correcting for FDR ($q$ = 0.20).

### 2.5. Low Expression of SLC16A2, a Thyroid Hormone Transporter, in HPV+ HNSCC Is Associated with Better Overall Survival

Increased expression of the monocarboxylic acid transporter family member, *SLC16A1* (*MCT1*) was recently reported to be associated with poor survival outcomes in HNSCC [15]. Although, expression of *SLC16A1* was significantly higher than normal head and neck tissues for both HPV+ and HPV- HNSCC, and expression of *SLC16A1* was higher in HPV- HNSCC than HPV+ HNSCC, the impact of differential expression of *SLC16A1* on overall survival in either HPV+ or HPV- HNSCC was not significant ($p$ > 0.05). Of the various family members, only expression of *SLC16A2* (*MCT8*), whose main function is thyroid hormone transport, appeared to be associated with altered overall survival (Figure 4G–I).

Expression of *SLC16A2* was significantly lower in HPV+ HNSCC when compared to either HPV-HNSCC or normal control tissues (Figure 4G). HPV+ HNSCC samples were dichotomized based on median *SLC16A2* expression. Again, low expression of *SLC16A2* was associated with improved patient survival in HPV+ HNSCC patients ($p = 0.0036$, FDR = 0.047) (Figure 4H), but not in patients with HPV-HNSCC ($p = 0.26$, FDR = 0.48) (Figure 4I).

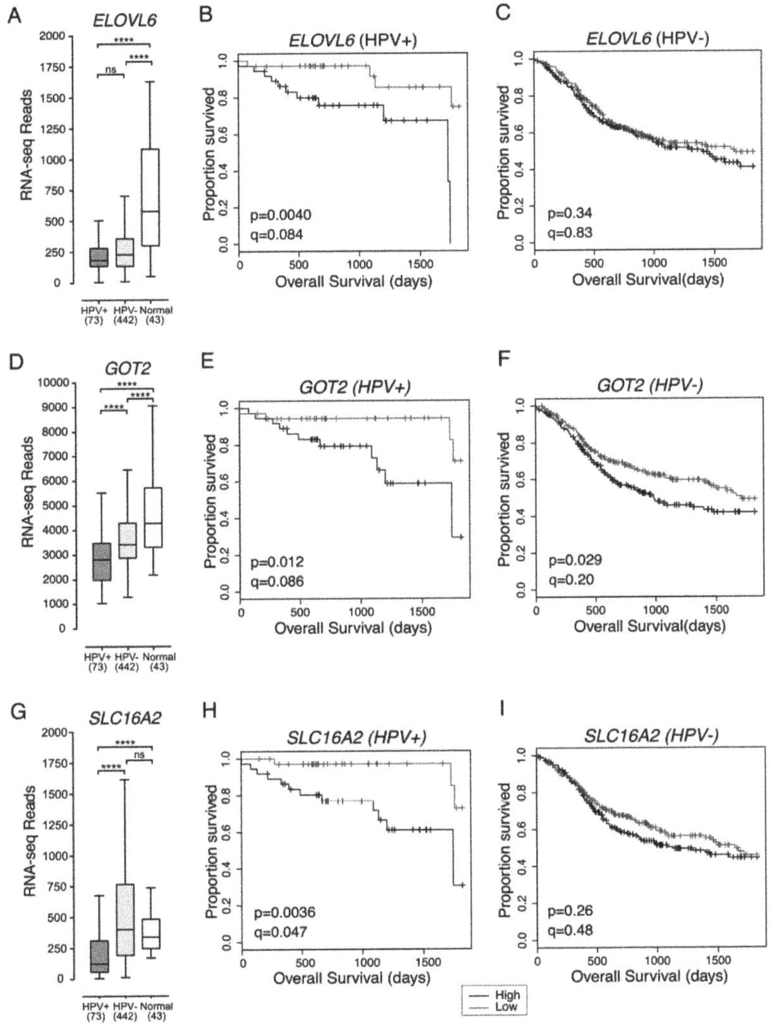

**Figure 4.** Low expression of *ELOVL6*, *GOT2* and *SLC16A2* is associated with improved patient survival in HPV+ HNSCC. Expression of (**A**) *ELOVL6*, (**D**) *GOT2*, and (**G**) *SLC16A2* in HPV+, HPV- HNSCC tissue samples and normal control tissues. Overall five-year survival outcomes in HPV+ HNSCC and HPV- HNSCC patients dichotomized by median (**B,C**) *ELOVL6* expression, (**E,F**) *GOT2* expression, and (**H,I**) *SLC16A2* expression. $p$ = Two-sided log-rank test and $q$ = Benjamini-Hochberg FDR method. Gray = low transcript expression, Black = high transcript expression. **** $p \leq 0.0001$, ns (not significant).

### 2.6. COX16, COX17, and SLC16A2 Are Independently Correlated with Favourable Survival Outcomes in HPV+ HNSCC

To determine the extent that each of the HPV+ HNSCC survival-associated genes could influence patient outcomes, we generated a hazard ratio (HR) for each gene and a variety of clinical variables by univariate analysis (Table 1). Each HR describes the relative increase in risk of death for the first variable x vs y [18]. As expected, the HR for each metabolic gene was significantly below 1, indicating a greatly reduced risk of death. In contrast, a comparison of the oral cavity vs the oropharynx subsites for HPV+ HNSCC generated a hazard ratio of 2.82, indicating that HPV+ HNSCC in the oral cavity is associated with a 2.82x increased risk of mortality compared to oropharynx.

**Table 1.** Univariate and multivariate analysis of the association of clinical variables and expression of metabolic genes with overall survival in HPV+ HNSCC.

| Variables | | Univariate | | Multivariate | |
|---|---|---|---|---|---|
| | | HR (95% CI) | P value | HR (95% CI) | P value |
| Sex | Male vs Female | 0.81 (0.18–3.62) | 0.78 | | |
| Age | per every additional year | 0.99 (0.94–1.04) | 0.74 | | |
| Subsite | Oral cavity vs Oropharynx | 2.82 (1.02–7.80) | **0.045** | 13.59 (2.67–69.11) | **0.002** |
| | Larynx vs Oropharynx | $1.52 \times 10^{-8}$ (0–Inf) | 1.00 | $3.67 \times 10^{-10}$ (0–Inf) | 1.00 |
| | Hypopharynx vs Oropharynx | $1.57 \times 10^{-8}$ (0–Inf) | 1.00 | $1.10 \times 10^{-8}$ (0–Inf) | 1.00 |
| T Stage | T3–T4 vs T1–T2 | 1.03 (0.36–2.91) | 0.96 | | |
| N Stage | N2b–N3 vs N0–N2a | 0.41 (0.14–1.19) | 0.10 | | |
| Overall Stage | IV vs I–III | 0.76 (0.26–2.24) | 0.62 | | |
| HPV Type | 33, 35, 56 vs 16 | 3.33 (1.14–9.78) | **0.028** | 16.88 (3.24–87.88) | **0.0008** |
| COX16 | Low vs High Expression | 0.19 (0.05–0.72) | **0.015** | 0.059 (0.009–0.39) | **0.003** |
| COX17 | Low vs High Expression | 0.13 (0.03–0.47) | **0.002** | 0.03 (0.003–0.35) | **0.005** |
| COX7A1 | Low vs High Expression | 0.22 (0.06–0.77) | **0.018** | 0.25 (0.04–1.56) | 0.14 |
| ELOVL6 | Low vs High Expression | 0.17 (0.04–0.65) | **0.009** | | |
| GOT2 | Low vs High Expression | 0.24 (0.07–0.79) | **0.019** | | |
| SDHC | Low vs High Expression | 0.23 (0.07–0.77) | **0.018** | | |
| SLC16A2 | Low vs High Expression | 0.17 (0.04–0.63) | **0.008** | 0.07 (0.01–0.37) | **0.002** |

As the contribution of these genes to overall survival might not be independent of one another, we also analyzed the relationship between survival and gene expression for all survival-associated metabolic genes and clinical variables concurrently by multivariate analysis (Table 1). The hazard ratios for COX16, COX17, and SLC16A2 remained significant, indicating that low expression of each of these genes is a significant, and potentially independent, contributor to overall survival. COX7A1 had a minor contribution to survival in this model. The multivariate model also included subsite (oral cavity vs. oropharynx) and HPV type as significant contributing factors to survival as previously reported in the literature [19–22].

To test whether concurrent low expression of these genes had an additive effect on survival, we stratified the HPV+ HNSCCs into groups, based on high expression or low expression for a combination of any two of these seven survival-associated genes. When survival of HPV+ HNSCC patients with low expression of both *COX16* and *COX17* in their tumours was compared to survival of patients with high expression of both genes, survival was significantly greater in the *COX16* and *COX17* double low expression group ($p = 0.0015$) (Figure 5A). In patients with low expression of both *COX16* and *SLC16A2*, survival was virtually 100% until approximately 4.75 years and significantly better ($p = 0.0021$) than samples expressing high levels of both genes (Figure 5B). In patients with low expression of both *COX17* and *SLC16A2* (Figure 5C), survival was 100%, which was significantly higher than patients expressing high levels of both *COX17* and *SLC16A2* ($p = 0.00027$; Figure 5C).

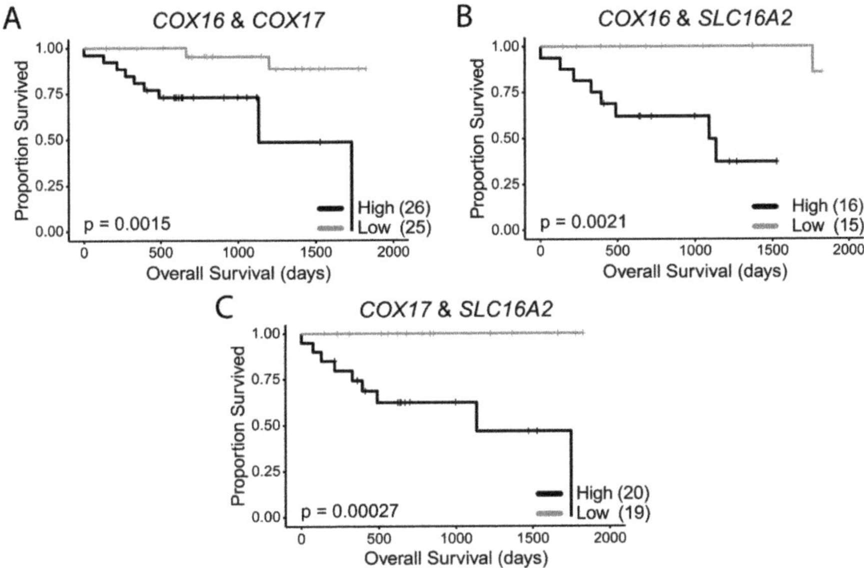

**Figure 5.** HPV+ HNSCC patient survival stratified by double low or double high expression of survival-associated metabolic genes. (**A**) *COX16* and *COX17*, (**B**) *COX16* and *SLC16A2*, (**C**) *COX17* and *SLC16A2*. Comparisons made with a two-sided log-rank test. Gray = low transcript expression, Black = high transcript expression. Bracketed value indicates number of HPV+ HNSCCs with double high or double low expression for genes of interest.

Supplementary Figure S1 shows that there was no significant correlation between the expression of *SLC16A2* and *COX16* or *SLC16A2* and *COX17*. This provides further evidence that the improved survival associated with low expression for each of these genes may occur independently of one another. Additionally, the multivariate analysis indicates that *COX16* and *COX17* independently contribute to survival despite the correlation between these two genes (Supplementary Figure S1).

## 3. Discussion

Our analysis identified many changes in expression of metabolism-associated genes between HPV+ and HPV- HNSCC when compared to normal control tissues. The expression of seven genes was predictive of survival for HPV+ HNSCC patients. In each case, reduced expression correlated with improved survival, suggesting that reduced tumour cell metabolism is prognostically favorable. None of these genes were associated with altered survival in HPV- HNSCC, reinforcing the concept that HPV+ and HPV- HNSCC are distinct tumour entities [7–9,23]. This is not unexpected, as E6 from HPV16 and HPV18 can increase the expression of mitochondrial cellular respiration genes in a head and neck cancer cell line [24], which matches our observation of increased expression of these genes in HPV+ HNSCCs when compared to HPV- HNSCCs. In addition, E6 and E7 may also be responsible for perturbing glycolysis in HPV+ cervical cancer cells [25,26], which could explain our observation of increased expression of glycolytic genes in HPV+ HNSCC as compared to normal control tissues. Interestingly, most of the survival-associated genes we identified in our study can be inhibited by small molecule inhibitors as outlined below.

As shown in our results, limiting *SDHC* may serve as a unique target in virally transformed HPV+ HNSCCs. *SDHC* encodes part of mitochondrial respiratory complex II, for which a few selective inhibitors exist. α-tocopheryl succinate (α-TOS) is a vitamin E analogue, which has selective growth inhibitory properties for some human cancer cells [27]. α-TOS can induce apoptosis by increasing the levels of reactive oxidative species (ROS), triggering stress response pathways [27]. α-TOS is also effective at inhibiting tumour growth [28] in in vivo xenograft mouse models. Interestingly, α-TOS inhibited growth of several HNSCC cell lines in vitro and in vivo [29]. Given that all of these experiments were done using HPV- HNSCC, α-TOS may be even more toxic to HPV+ HNSCCs based on the correlation between *SDHC* expression and HPV+ HNSCC survival outcomes we observed.

Specific small molecule inhibitors for both the *COX7A1* and *COX16* gene products have not yet been identified. However, as both are part of mitochondrial respiratory complex IV, it is possible that the complex IV inhibitors ADDA 5 [30] and tetrathiomolybdate [31] could prove useful to phenocopy any metabolic effects associated with low gene expression, promoting enhanced survival in HPV+ HNSCC. It is also important to note that COX16 is an inhibitor of p53 activity, which means that non-mitochondrial functions of COX16 should not be discounted [32]. E6 has been shown to inhibit expression of COX16 [33], which may contribute to the lower levels observed in HPV+ versus HPV- HNSCC (Figure 3D). MitoBloCK-6 is an inhibitor of mitochondrial respiratory complex IV that specifically targets the COX17 protein [34]. Whether inhibition of cellular respiration is less effective in HPV- HNSCCs because they are already more oxidatively stressed than HPV+ HNSCCs [35], perhaps as a result of being less adapted to utilize cellular respiration, is an open question.

Two inhibitors of ELOVL6 were able to reduce the fatty acid composition of hepatocytes and the liver in a murine model of obesity [36,37], but the effects of these compounds on cancer cells have not been explored. Another potential druggable target to influence ELOVL6 expression is ATP citrate lyase (ACLY), which has a wide variety of inhibitors [38]. Expression of ELOVL6 has been shown to decrease concurrently with ACLY inhibition [39]. Whether ELOVL6 inhibition would reduce the growth of HPV+ HNSCC cell lines remains to be examined.

In our study, we found that low expression of *GOT2* was associated with statistically significant survival in both groups (p < 0.05). However, only expression of *GOT2* in our HPV+ HNSCC group met the FDR cut-off of $q = 0.1$. This means that while *GOT2* may be important for survival outcomes in both HPV+ and HPV- HNSCC, it is likely that it has a more substantial contribution to patient survival in HPV+ HNSCC. In breast cancer, sensitivity to a nucleotide synthesis inhibitor, methotrexate, has been linked to high *GOT2* expression [40]. This is likely due to the function of GOT2 in providing aspartate for nucleotide biosynthesis [40]. It is possible that HPV+ HNSCCs, or potentially any HNSCCs expressing high levels of *GOT2*, may be sensitive to methotrexate, but this remains to be explored.

*SLC16A2* encodes a plasma membrane T3/T4 transporter. Once inside the cell, T3/T4 can bind nuclear and mitochondrial-localized thyroid hormone receptors, which are key regulators of mitochondrial biogenesis [41]. As HPV+ HNSCC may be more reliant on cellular respiration than HPV- HNSCC, it is possible that inhibiting SLC16A2-mediated thyroid hormone transport across the plasma membrane could preferentially inhibit ATP generation in HPV+ HNSCCs. Some tyrosine kinase inhibitors (TKIs), such as sunitinib, imatinib, dasatinib, and bosutinib, may inhibit SLC16A2 [42]. These TKIs are already employed to treat a wide variety of cancers and are being evaluated for the treatment of HNSCC [43]. As such, they may be especially suitable for the treatment of HPV+ HNSCCs expressing *SLC16A2* at high levels. We extracted data from our previous study of 27 HNSCC cell lines (6 HPV+ and 21 HPV- HNSCC cell lines) examining the effects of a variety of agents on cell growth and proliferation, including the TKI inhibitors mentioned above [44]. Of the TKIs tested, dasatinib was selectively cytotoxic to HPV+, but not HPV- HNSCC cell lines (Supplementary Figure S2), suggesting that it could be used as a treatment for HPV+ HNSCC that expresses high levels of *SLC16A2*. A flavonoid, silychristin, also inhibits SLC16A2 [45], but has not been studied in cancer models. Other antithyroid hormones used to treat hyperthyroidism, such as carbimazole, methimazole, and potassium perchlorate, could preferentially inhibit HPV+ HNSCC by mimicking inhibition of the SLC16A2 transporter. In one study, methimazole was used to experimentally treat patients with end-stage solid tumours and resulted in improved survival [46]. High levels of thyroid hormone can promote proliferation of some cancers [47], and thyroid hormone mimetics which function as antagonists, such as tetraiodothyroacetic acid (tetrac) and triiodothyroacetic acid (triac) appear to exhibit an antiproliferative effect on breast cancer [47] and T cell lymphomas [48].

While our univariate analysis of all seven genes confirmed that their expression was significantly correlated with survival, our multivariate analysis indicated that *COX16*, *COX17*, and *SLC16A2* were independently associated with overall survival. This suggested that the collective changes in the expression of these three genes could be a powerful predictor of clinical outcome. This was supported by our observation that the overall survival for the group of patients exhibiting simultaneously low expression of any two of these genes was far better than those simultaneously expressing high levels. Thus, expression of these genes may have prognostic utility. Furthermore, *COX16*, *COX17*, and *SLC16A2* may represent attractive therapeutic targets for HPV+ HNSCC that warrant further exploration, especially considering that treatment with specific inhibitors may phenocopy the effects of low expression of these metabolic genes.

It is important to be cognizant of the limitations to this kind of study. One limitation to this study was the lack of protein data in the TCGA to corroborate our HPV+ HNSCC mRNA expression data. This is an important consideration, as levels of protein expression do not necessarily mirror mRNA expression [49]. As with any high throughput dataset, batch effects that result from processing could be reflected in the data [50]. In addition, the RNA-seq data contained within the TCGA reflects average mRNA expression within the whole tumour and does not identify expression differences between the various tumour cells as could be obtained from single-cell RNA sequencing platforms [51,52]. However, the bulk of the tissue that was sequenced is of tumour origin [52]. Also, it is important to be aware that the TCGA contains HNSCC data from a single, albeit high quality, cohort and it would be useful to validate these genes of interest in other cohorts in the future. Finally, the concept of biomarker identification itself has its own caveats. Specifically, all of the survival-associated genes we identified in this study are only correlated with survival, which does not equal causation [53]. In addition, as with all studies of this type, there exists the possibility that these correlations are due to chance or occur as the result of another confounder [53]. However, despite these limitations, the seven metabolic genes identified in this study provide an interesting starting point for considering the metabolic differences between HPV+ and HPV- HNSCC as new prognostic markers or potential targets for therapy.

## 4. Materials and Methods

### 4.1. Data Collection

Level 3 RNA-seq by Expectation Maximization normalized Illumina HiSeq RNA expression data (build 2016012800) for the TCGA HNSC cohort, was downloaded from the Broad Genome Data Analysis Centers Firehose server (https://gdac.broadinstitute.org/). Patient survival data for the TCGA HNSC cohort, as reported by the Pan-Cancer Atlas [54], was downloaded from: https://www.cell.com/cms/10.1016/j.cell.2018.02.052/attachment/f4eb6b31-8957-4817-a41f-e46fd2a1d9c3/mmc1.xlsx.

### 4.2. RNA Expression Comparisons

RNA-seq by Expectation Maximization normalized expression data for the TCGA HNSC cohort was extracted and analyzed as described previously [23]. Briefly, primary patient samples with known HPV status were manually grouped as HPV+, HPV-, or normal-adjacent control tissue based on other previously published datasets [21,35]. In these datasets, HPV status was assigned by aligning RNA-seq reads for the HPV oncogenes expressed in the tumours to the different high risk HPV types. This resulted in 73 HPV+, 442 HPV-, and 43 normal-adjacent control samples with data available for gene expression comparisons. Note that all HPV+ samples had high risk HPV as follows: HPV16 (61 samples), HPV33 (8 samples), HPV35 (3 samples), and HPV56 (1 sample). The five-number summary, mean, and pairwise statistical tests were calculated using R (version 3.4.0) for all 229 metabolic genes analyzed (see Supplementary Table S1). These 229 genes were manually selected (Supplementary Table S1) as they are involved in eight central cellular metabolic pathways or processes, of which cellular respiration was then separated into its respective complexes. The number of genes examined from each pathway was as follows: glycolysis, 36 genes (adapted from GO:0061621 and [55]); TCA cycle, 20 genes (adapted from GO:0006099 and [56]); mitochondrial respiratory complex I, 43 genes (adapted from GO:0045333, GO:0046043 and [57]); mitochondrial respiratory complex II, 4 genes (adapted from GO:0045333, GO:0046043 and [58]); mitochondrial respiratory complex III, 9 genes (adapted from GO:0045333, GO:0046043 and [59,60]); mitochondrial respiratory complex IV, 31 genes (adapted from GO:0045333, GO:0046043 and [61]); mitochondrial ATPase, 16 genes (adapted from GO:0046043 and [62]); fatty acid synthesis, 21 genes (adapted from GO:0019368, GO:0046949, GO: 0006629 and [63–66]); β-oxidation, 18 genes (adapted from GO:0003995, GO:0003985, GO: 0004300 and [67]); glutaminolysis, 7 genes (adapted from GO:0004069, GO:0004352, GO: 0004359, GO: 0004021 and [68]); pentose phosphate pathway, 11 genes (GO:0006098 and [69]); monocarboxylic acid transport (MCT) family, 13 genes (adapted from GO:0008028 and [15]). Boxplot comparisons of gene expression were made with GraphPad Prism v7.0 (Graphpad Software, Inc., San Diego, California, USA). For the boxplots, center lines show the medians, box limits indicate the 25th and 75th percentiles, and whiskers extend 1.5 times the interquartile range from the 25th and 75th percentiles. P-values were assigned using a two-tailed non-parametric Mann-Whitney U test using Graphpad Prism. Bivariate analysis for selected genes was performed through R (version 3.4.0) using the Spearman rank correlation coefficient. The proportion of genes in each pathway that were up or downregulated among each comparison was represented in a bar graph and was calculated as follows: number of genes upregulated (or downregulated) in a comparison (e.g., HPV+ HNSCC vs HPV- HNSCC) divided by the total number of genes in that pathway. The proportion of downregulated genes were represented as a negative value.

### 4.3. Survival Analysis

Five-year overall survival outcomes were compared in both HPV+ and HPV- subsets of HNSCC patients dichotomized by median expression for all metabolic genes listed in Supplementary Table S1. Log-rank statistical p-values were calculated for each Cox survival model. The derived log-rank p-values for all tested genes (listed in Supplementary Table S2) were assessed for significance after correcting for false discovery rate (FDR) using the Benjamini–Hochberg method, and an FDR threshold of 0.1 was set for significance. Univariate analysis was performed through R (version 3.4.0) based on a Cox Proportional Hazard Model using the survival package (version 2.41-3). Stepwise bidirectional multivariate analysis was then carried out with clinical variables (sex, age, subsite, T stage, N stage, Overall stage, and HPV type), and *SDHC, COX7A1, COX16, COX17, ELOVL6, GOT2,* and *SLC16A2* expression—low expression of these 7 genes were found to be statistically correlated with improved survival after univariate analysis. The p-values derived from the Wald test on survival coefficients were reported for investigated variables. Furthermore, a second set of survival outcomes were determined to compare HPV+ tumours expressing low levels of each combination of genes that were significantly correlated with improved survival after multivariate analysis—*COX16, COX17,* and *SLC16A2*.

### 4.4. Gene Enrichment Analysis

We performed a gene enrichment analysis on our seven survival-associated genes using the Go Enrichment Analysis feature on http://geneontology.org [70,71]. This analysis is powered by PANTHER14.1 (PANTHER Overrepresentation Test) using the "GO cellular component complete" annotation data set with a Fisher's exact test followed by a calculation of false discovery rate (cutoff = FDR P < 0.05) to determine statistical significance [72].

### 4.5. Analysis of Differential Cell Line Sensitivity to Tyrosine Kinase Inhibitors Based on HPV Status

B-scores (mean ± SEM) reflecting drug activity were extracted from a previously conducted high throughput drug screen using 27 HNSCC cell lines [44]. The average B-scores for the indicated tyrosine kinase inhibitors (TKIs) was calculated for the 6 HPV+ and 21 HPV- HNSCC lines and plotted (Supplementary Figure S2).

## 5. Conclusions

In summary, our analysis of HNSCC TCGA data stratified by HPV status indicated that the metabolic profile of HPV+ and HPV- HNSCC are strikingly different. HPV- HNSCCs may utilize glycolysis to a greater extent than HPV+ HNSCCs, while HPV+ HNSCCs may be more reliant on the TCA cycle, cellular respiration, and β-oxidation than HPV- HNSCCs. Despite this difference, both types of HNSCCs likely exhibit far less cellular respiration than normal head and neck tissues, consistent with a cancer-associated Warburg phenotype [73]. Importantly, expression of genes involved in mitochondrial complex II and mitochondrial complex IV were associated with survival for HPV+ HNSCC patients. Namely, low expression of *SDHC, COX7A1, COX16,* or *COX17* was associated with better survival outcomes. Low expression of *ELOVL6*, involved in fatty acid elongation; *GOT2*, involved in amino acid metabolism; and *SLC16A2*, involved in thyroid hormone transport, were also all associated with better survival outcomes in HPV+ HNSCC patients. However, of these genes, only *COX16, COX17* and *SLC16A2* were independently correlated with survival outcomes according to our multivariate analysis. Importantly, *COX16, COX17,* and *SLC16A2* were associated with near 100% survival in all patients with low expression of any two of these genes. The products of these genes may represent useful new therapeutic targets for HPV+ HNSCC, as inhibition of their functions could phenocopy the metabolism of those tumours with low levels of metabolic gene expression, leading to improved survival in HPV+ HNSCC patients.

**Supplementary Materials:** The following are available online at http://www.mdpi.com/2072-6694/12/1/253/s1, Figure S1: Correlation plots of independently significant genes identified by multivariate analysis, Figure S2: Tyrosine kinase inhibitor activity in 27 HNSCC cell lines based on HPV status, Table S1: Metabolic pathway gene expression, Table S2: Survival curve statistics.

**Author Contributions:** Writing—Original Draft: M.A.P. and J.S.M.; Writing—Review and Editing: M.A.P., J.S.M., S.F.G., F.G., M.J.D., P.Y.F.Z., H.M., J.W.B., and A.C.N.; Conceptualization: J.S.M., M.A.P., S.F.G., J.W.B. and A.C.N.; Software: S.F.G. and F.G.; Formal Analysis: M.A.P., S.F.G., F.G., M.J.D., P.Y.F.Z., H.M. and J.S.M.; Supervision: J.S.M., J.W.B. and A.C.N. Visualization: M.A.P. and S.F.G. All authors have read and agreed to the published version of the manuscript.

**Funding:** This work was supported by grants provided by the Canadian Institutes of Health Research to J.S.M./A.C.N. (MOP#148689) and A.C.N. (MOP#340674). A.C.N. was supported by the Wolfe Surgical Research Professorship in the Biology of Head and Neck Cancers Fund. S.F.G. was supported from a Cancer Research and Technology Transfer studentship. M.A.P. was supported by a studentship from the Natural Sciences and Engineering Research Council of Canada and from the Western University Schulich School of Medicine and Dentistry.

**Conflicts of Interest:** The authors declare no conflict of interest.

## References

1. Bray, F.; Ferlay, J.; Soerjomataram, I.; Siegel, R.L.; Torre, L.A.; Jemal, A. Global cancer statistics 2018: GLOBOCAN estimates of incidence and mortality worldwide for 36 cancers in 185 countries. *CA Cancer J. Clin.* **2018**, *68*, 394–424. [CrossRef]

2. Ferlay, J.; Colombet, M.; Soerjomataram, I.; Mathers, C.; Parkin, D.M.; Piñeros, M.; Znaor, A.; Bray, F. Estimating the global cancer incidence and mortality in 2018: GLOBOCAN sources and methods. *Int. J. Cancer* **2018**, *144*, 1941–1953. [CrossRef]

3. De Martel, C.; Plummer, M.; Vignat, J.; Franceschi, S. Worldwide burden of cancer attributable to HPV by site, country and HPV type. *Int. J. Cancer* **2017**, *141*, 664–670. [CrossRef]

4. Herrero, R.; Castellsagué, X.; Pawlita, M.; Lissowska, J.; Kee, F.; Balaram, P.; Rajkumar, T.; Sridhar, H.; Rose, B.; Pintos, J.; et al. Human papillomavirus and oral cancer: The International Agency for Research on Cancer multicenter study. *J. Natl. Cancer Inst.* **2003**, *95*, 1772–1783. [CrossRef]

5. Michmerhuizen, N.L.; Birkeland, A.C.; Bradford, C.R.; Brenner, J.C. Genetic determinants in head and neck squamous cell carcinoma and their influence on global personalized medicine. *Genes Cancer* **2016**, *7*, 182–200.

6.  Chaturvedi, A.K.; Engels, E.A.; Pfeiffer, R.M.; Hernandez, B.Y.; Xiao, W.; Kim, E.; Jiang, B.; Goodman, M.T.; Sibug-Saber, M.; Cozen, W.; et al. Human papillomavirus and rising oropharyngeal cancer incidence in the United States. *J. Clin. Oncol.* **2011**, *29*, 4294–4301. [CrossRef]

7.  Seiwert, T.Y.; Zuo, Z.; Keck, M.K.; Khattri, A.; Pedamallu, C.S.; Stricker, T.; Brown, C.; Pugh, T.J.; Stojanov, P.; Cho, J.; et al. Integrative and comparative genomic analysis of HPV-positive and HPV-negative head and neck squamous cell carcinomas. *Clin. Cancer Res.* **2015**, *21*, 632–641. [CrossRef]

8.  Worsham, M.J.; Chen, K.M.; Ghanem, T.; Stephen, J.K.; Divine, G. Epigenetic modulation of signal transduction pathways in HPV-associated HNSCC. *Otolaryngol. Head Neck Surg.* **2013**, *149*, 409–416. [CrossRef]

9.  Gameiro, S.F.; Ghasemi, F.; Barrett, J.W.; Koropatnick, J.; Nichols, A.C.; Mymryk, J.S.; Maleki Vareki, S. Treatment-naïve HPV+ head and neck cancers display a T-cell-inflamed phenotype distinct from their HPV-counterparts that has implications for immunotherapy. *Oncoimmunology* **2018**, *7*, 1–14. [CrossRef]

10. Fakhry, C.; Westra, W.H.; Li, S.; Cmelak, A.; Ridge, J.A.; Pinto, H.; Forastiere, A.; Gillison, M.L. Improved survival of patients with human papillomavirus-positive head and neck squamous cell carcinoma in a prospective clinical trial. *J. Natl. Cancer Inst.* **2008**, *100*, 261–269. [CrossRef]

11. Weller, M.A.; Ward, M.C.; Berriochoa, C.; Reddy, C.A.; Trosman, S.; Greskovich, J.F.; Nwizu, T.I.; Burkey, B.B.; Adelstein, D.J.; Koyfman, S.A. Predictors of distant metastasis in human papillomavirus–associated oropharyngeal cancer. *Head Neck* **2017**, *39*, 940–946. [CrossRef]

12. Lunt, S.Y.; Vander Heiden, M.G. Aerobic Glycolysis: Meeting the Metabolic Requirements of Cell Proliferation. *Annu. Rev. Cell Dev. Biol.* **2011**, *27*, 441–464. [CrossRef]

13. Vander Heiden, M.G.; DeBerardinis, R.J. Understanding the Intersections between Metabolism and Cancer Biology. *Cell* **2017**, *168*, 657–669. [CrossRef]

14. Goodwin, C.M.; Xu, S.; Munger, J. Stealing the Keys to the Kitchen: Viral Manipulation of the Host Cell Metabolic Network. *Trends Microbiol.* **2015**, *23*, 789–798. [CrossRef]

15. Fleming, J.C.; Woo, J.; Moutasim, K.; Mellone, M.; Frampton, S.J.; Mead, A.; Ahmed, W.; Wood, O.; Robinson, H.; Ward, M.; et al. HPV, tumour metabolism and novel target identification in head and neck squamous cell carcinoma. *Br. J. Cancer* **2019**, *120*, 356–367. [CrossRef]

16. Montrose, D.C.; Galluzzi, L. Drugging cancer metabolism: Expectations vs. reality. *Int. Rev. Cell Mol. Biol.* **2019**, *347*, 1–26.

17. Enríquez, J.A. Supramolecular Organization of Respiratory Complexes. *Annu. Rev. Physiol.* **2016**, *78*, 533–561. [CrossRef]

18. Spruance, S.L.; Reid, J.E.; Grace, M.; Samore, M. Hazard ratio in clinical trials. *Antimicrob. Agents Chemother.* **2004**, *48*, 2787–2792. [CrossRef]

19. Goodman, M.T.; Saraiya, M.; Thompson, T.D.; Steinau, M.; Hernandez, B.Y.; Lynch, C.F.; Lyu, C.W.; Wilkinson, E.J.; Tucker, T.; Copeland, G.; et al. Human papillomavirus genotype and oropharynx cancer survival in the United States of America. *Eur. J. Cancer* **2015**, *51*, 2759–2767. [CrossRef]

20. O'Rorke, M.A.; Ellison, M.V.; Murray, L.J.; Moran, M.; James, J.; Anderson, L.A. Human papillomavirus related head and neck cancer survival: A systematic review and meta-analysis. *Oral Oncol.* **2012**, *48*, 1191–1201. [CrossRef]

21. Bratman, S.V.; Bruce, J.P.; O'Sullivan, B.; Pugh, T.J.; Xu, W.; Yip, K.W.; Liu, F.F. Human papillomavirus genotype association with survival in head and neck squamous cell carcinoma. *JAMA Oncol.* **2016**, *2*, 823–826. [CrossRef]

22. Chatfield-Reed, K.; Gui, S.; O'Neill, W.Q.; Teknos, T.N.; Pan, Q. HPV33+ HNSCC is associated with poor prognosis and has unique genomic and immunologic landscapes. *Oral Oncol.* **2020**, *100*, 104488. [CrossRef]

23. Gameiro, S.F.; Kolendowski, B.; Zhang, A.; Barrett, J.W.; Nichols, A.C.; Torchia, J.; Mymryk, J.S. Human papillomavirus dysregulates the cellular apparatus controlling the methylation status of H3K27 in different human cancers to consistently alter gene expression regardless of tissue of origin. *Oncotarget* **2017**, *8*, 72564–72576. [CrossRef]

24. Cruz-Gregorio, A.; Aranda-Rivera, A.K.; Aparicio-Trejo, O.E.; Coronado-Martínez, I.; Pedraza-Chaverri, J.; Lizano, M. E6 oncoproteins from high-risk human papillomavirus induce mitochondrial metabolism in a head and neck squamous cell carcinoma model. *Biomolecules* **2019**, *9*, 351. [CrossRef]

25. Ma, D.; Huang, Y.; Song, S. Inhibiting the HPV16 oncogene-mediated glycolysis sensitizes human cervical carcinoma cells to 5-fluorouracil. *Onco Targets Ther.* **2019**, *12*, 6711–6720. [CrossRef]

26. Guo, Y.; Meng, X.; Ma, J.; Zheng, Y.; Wang, Q.; Wang, Y.; Shang, H. Human papillomavirus 16 E6 contributes HIF-1α induced warburg effect by attenuating the VHL-HIF-1α interaction. *Int. J. Mol. Sci.* **2014**, *15*, 7974–7986. [CrossRef]

27. Neuzil, J.; Dyason, J.C.; Freeman, R.; Dong, L.F.; Prochazka, L.; Wang, X.F.; Scheffler, I.; Ralph, S.J. Mitocans as anti-cancer agents targeting mitochondria: Lessons from studies with vitamin e analogues, inhibitors of complex II. *J. Bioenerg. Biomembr.* **2007**, *39*, 65–72. [CrossRef]

28. Kanai, K.; Kikuchi, E.; Mikami, S.; Suzuki, E.; Uchida, Y.; Kodaira, K.; Miyajima, A.; Ohigashi, T.; Nakashima, J.; Oya, M. Vitamin E succinate induced apoptosis and enhanced chemosensitivity to paclitaxel in human bladder cancer cells in vitro and in vivo. *Cancer Sci.* **2010**, *101*, 216–223. [CrossRef]

29. Gu, X.; Song, X.; Dong, Y.; Cai, H.; Walters, E.; Zhang, R.; Pang, X.; Xie, T.; Guo, Y.; Sridhar, R.; et al. Vitamin E Succinate Induces Ceramide-Mediated Apoptosis in Head and Neck Squamous Cell Carcinoma In vitro and In vivo. *Clin. Cancer Res.* **2008**, *14*, 1840–1848. [CrossRef]

30. Oliva, C.R.; Markert, T.; Ross, L.J.; White, E.L.; Rasmussen, L.; Zhang, W.; Everts, M.; Moellering, D.R.; Bailey, S.M.; Suto, M.J.; et al. Identification of small molecule inhibitors of human cytochrome c oxidase that target chemoresistant glioma cells. *J. Biol. Chem.* **2016**, *291*, 24188–24199. [CrossRef]

31. Kim, K.K.; Abelman, S.; Yano, N.; Ribeiro, J.R.; Singh, R.K.; Tipping, M.; Moore, R.G. Tetrathiomolybdate inhibits mitochondrial complex IV and mediates degradation of hypoxia-inducible factor-1α in cancer cells. *Sci. Rep.* **2015**, *5*, 14296. [CrossRef] [PubMed]

32. Siebring-van Olst, E.; Blijlevens, M.; de Menezes, R.X.; van der Meulen-Muileman, I.H.; Smit, E.F.; van Beusechem, V.W. A genome-wide siRNA screen for regulators of tumor suppressor p53 activity in human non-small cell lung cancer cells identifies components of the RNA splicing machinery as targets for anticancer treatment. *Mol. Oncol.* **2017**, *11*, 534–551. [CrossRef]

33. Butz, K.; Whitaker, N.; Denk, C.; Ullmann, A.; Geisen, C.; Hoppe-Seyler, F. Induction of the p53-target gene GADD45 in HPV-positive cancer cells. *Oncogene* **1999**, *18*, 2381–2386. [CrossRef] [PubMed]

34. Dabir, D.V.; Hasson, S.A.; Setoguchi, K.; Johnson, M.E.; Wongkongkathep, P.; Douglas, C.J.; Zimmerman, J.; Damoiseaux, R.; Teitell, M.A.; Koehler, C.M. A Small Molecule Inhibitor of Redox-Regulated Protein Translocation into Mitochondria. *Dev. Cell* **2013**, *25*, 81–92. [CrossRef]

35. The Cancer Genome Atlas Network Comprehensive genomic characterization of head and neck squamous cell carcinomas. *Nature* **2015**, *517*, 576–582. [CrossRef]

36. Shimamura, K.; Kitazawa, H.; Miyamoto, Y.; Kanesaka, M.; Nagumo, A.; Yoshimoto, R.; Aragane, K.; Morita, N.; Ohe, T.; Takahashi, T.; et al. 5,5-Dimethyl-3-(5-methyl-3-oxo-2-phenyl-2,3-dihydro-1H-pyrazol-4-yl)-1-phenyl-3-(trifluoromethyl)-3,5,6,7-tetrahydro-1H-indole-2,4-dione, a Potent Inhibitor for Mammalian Elongase of Long-Chain Fatty Acids Family 6. *J. Pharmacol. Exp. Ther.* **2009**, *330*, 249–256. [CrossRef]

37. Shimamura, K.; Nagumo, A.; Miyamoto, Y.; Kitazawa, H.; Kanesaka, M.; Yoshimoto, R.; Aragane, K.; Morita, N.; Ohe, T.; Takahashi, T.; et al. Discovery and characterization of a novel potent, selective and orally active inhibitor for mammalian ELOVL6. *Eur. J. Pharmacol.* **2010**, *630*, 34–41. [CrossRef]

38. Granchi, C. ATP citrate lyase (ACLY) inhibitors: An anti-cancer strategy at the crossroads of glucose and lipid metabolism. *Eur. J. Med. Chem.* **2018**, *157*, 1276–1291. [CrossRef]

39. Migita, T.; Okabe, S.; Ikeda, K.; Igarashi, S.; Sugawara, S.; Tomida, A.; Soga, T.; Taguchi, R.; Seimiya, H. Inhibition of ATP citrate lyase induces triglyceride accumulation with altered fatty acid composition in cancer cells. *Int. J. Cancer* **2014**, *135*, 37–47. [CrossRef]

40. Hong, R.; Zhang, W.; Xia, X.; Zhang, K.; Wang, Y.; Wu, M.; Fan, J.; Li, J.; Xia, W.; Xu, F.; et al. Preventing BRCA1/ZBRK1 repressor complex binding to the GOT2 promoter results in accelerated aspartate biosynthesis and promotion of cell proliferation. *Mol. Oncol.* **2019**, *13*, 959–977. [CrossRef]

41. Weitzel, J.M.; Alexander Iwen, K. Coordination of mitochondrial biogenesis by thyroid hormone. *Mol. Cell. Endocrinol.* **2011**, *342*, 1–7. [CrossRef]

42. Braun, D.; Kim, T.D.; Le Coutre, P.; Köhrle, J.; Hershman, J.M.; Schweizer, U. Tyrosine kinase inhibitors noncompetitively inhibit MCT8-mediated iodothyronine transport. *J. Clin. Endocrinol. Metab.* **2012**, *97*, 100–105. [CrossRef]

43. Schmitz, S.; Ang, K.K.; Vermorken, J.; Haddad, R.; Suarez, C.; Wolf, G.T.; Hamoir, M.; Machiels, J.-P. Targeted therapies for squamous cell carcinoma of the head and neck: Current knowledge and future directions. *Cancer Treat. Rev.* **2014**, *40*, 390–404. [CrossRef]

44. Ghasemi, F.; Black, M.; Sun, R.X.; Vizeacoumar, F.; Pinto, N.; Ruicci, K.M.; Yoo, J.; Fung, K.; MacNeil, D.; Palma, D.A.; et al. High-throughput testing in head and neck squamous cell carcinoma identifies agents with preferential activity in human papillomavirus-positive or negative cell lines. *Oncotarget* **2018**, *9*, 26064–26071. [CrossRef]

45. Johannes, J.; Jayarama-Naidu, R.; Meyer, F.; Wirth, E.K.; Schweizer, U.; Schomburg, L.; Köhrle, J.; Renko, K. Silychristin, a flavonolignan derived from the milk thistle, is a potent inhibitor of the thyroid hormone transporter MCT8. *Endocrinology* **2016**, *157*, 1694–1701. [CrossRef]

46. Hercbergs, A.; Johnson, R.E.; Ashur-Fabian, O.; Garfield, D.H.; Davis, P.J. Medically Induced Euthyroid Hypothyroxinemia May Extend Survival in Compassionate Need Cancer Patients: An Observational Study. *Oncologist* **2015**, *20*, 72–76. [CrossRef]

47. Hercbergs, A.; Mousa, S.A.; Leinung, M.; Lin, H.Y.; Davis, P.J. Thyroid Hormone in the Clinic and Breast Cancer. *Horm. Cancer* **2018**, *9*, 139–143. [CrossRef]

48. Cayrol, F.; Sterle, H.A.; Díaz Flaqué, M.C.; Barreiro Arcos, M.L.; Cremaschi, G.A. Non-genomic Actions of Thyroid Hormones Regulate the Growth and Angiogenesis of T Cell Lymphomas. *Front. Endocrinol. (Lausanne)* **2019**, *10*, 63. [CrossRef]

49. Fortelny, N.; Overall, C.M.; Pavlidis, P.; Cohen Freue, G.V. Can we predict protein from mRNA levels? *Nature* **2017**, *547*, E19–E23. [CrossRef]

50. Leek, J.T.; Scharpf, R.B.; Bravo, H.C.; Simcha, D.; Langmead, B.; Johnson, W.E.; Geman, D.; Baggerly, K.; Irizarry, R.A. Tackling the widespread and critical impact of batch effects in high-throughput data. *Nat. Rev. Genet.* **2010**, *11*, 733–739. [CrossRef]

51. Hwang, B.; Lee, J.H.; Bang, D. Single-cell RNA sequencing technologies and bioinformatics pipelines. *Exp. Mol. Med.* **2018**, *50*, 96. [CrossRef] [PubMed]

52. Aran, D.; Sirota, M.; Butte, A.J. Systematic pan-cancer analysis of tumour purity. *Nat. Commun.* **2015**, *6*, 8971. [CrossRef] [PubMed]

53. Kaelin, W.G. Common pitfalls in preclinical cancer target validation. *Nat. Rev. Cancer* **2017**, *17*, 441–450. [CrossRef]

54. The Cancer Genome Atlas Research Network; Chang, K.; Creighton, C.J.; Davis, C.; Donehower, L.; Drummond, J.; Wheeler, D.; Ally, A.; Balasundaram, M.; Birol, I.; et al. The Cancer Genome Atlas Pan-Cancer analysis project. *Nat. Genet.* **2013**, *45*, 1113–1120. [CrossRef]

55. Bolaños, J.P.; Almeida, A.; Moncada, S. Glycolysis: A bioenergetic or a survival pathway? *Trends Biochem. Sci.* **2010**, *35*, 145–149. [CrossRef]

56. Chen, J.Q.; Russo, J. Dysregulation of glucose transport, glycolysis, TCA cycle and glutaminolysis by oncogenes and tumor suppressors in cancer cells. *Biochim. Biophys. Acta Rev. Cancer* **2012**, *1826*, 370–384. [CrossRef]

57. Sharma, L.K.; Lu, J.; Bai, Y. Mitochondrial respiratory complex I: Structure, function and implication in human diseases. *Curr. Med. Chem.* **2009**, *16*, 1266–1277. [CrossRef]

58. Kluckova, K.; Bezawork-Geleta, A.; Rohlena, J.; Dong, L.; Neuzil, J. Mitochondrial complex II, a novel target for anti-cancer agents. *Biochim. Biophys. Acta Bioenerg.* **2013**, *1827*, 552–564. [CrossRef]

59. Owens, K.M.; Kulawiec, M.; Desouki, M.M.; Vanniarajan, A.; Singh, K.K. Impaired OXPHOS complex III in breast cancer. *PLoS ONE* **2011**, *6*, e23846. [CrossRef]

60. Guo, R.; Gu, J.; Wu, M.; Yang, M. Amazing structure of respirasome: Unveiling the secrets of cell respiration. *Protein Cell* **2016**, *7*, 854–865. [CrossRef]

61. Mansilla, N.; Racca, S.; Gras, D.E.; Gonzalez, D.H.; Welchen, E. The complexity of mitochondrial complex iv: An update of cytochrome c oxidase biogenesis in plants. *Int. J. Mol. Sci.* **2018**, *19*, 662. [CrossRef] [PubMed]

62. Jonckheere, A.I.; Smeitink, J.A.M.; Rodenburg, R.J.T. Mitochondrial ATP synthase: Architecture, function and pathology. *J. Inherit. Metab. Dis.* **2012**, *35*, 211–225. [CrossRef] [PubMed]

63. Kuo, C.Y.; Ann, D.K. When fats commit crimes: Fatty acid metabolism, cancer stemness and therapeutic resistance. *Cancer Commun.* **2018**, *38*, 47. [CrossRef] [PubMed]

64. Currie, E.; Schulze, A.; Zechner, R.; Walther, T.C.; Farese, R.V. Cellular fatty acid metabolism and cancer. *Cell Metab.* **2013**, *18*, 153–161. [CrossRef]

65. Koundouros, N.; Poulogiannis, G. Reprogramming of fatty acid metabolism in cancer. *Br. J. Cancer* **2019**, *122*, 4–22. [CrossRef]

66. Guillou, H.; Zadravec, D.; Martin, P.G.P.; Jacobsson, A. The key roles of elongases and desaturases in mammalian fatty acid metabolism: Insights from transgenic mice. *Prog. Lipid Res.* **2010**, *49*, 186–199. [CrossRef]

67. Houten, S.M.; Wanders, R.J.A. A general introduction to the biochemistry of mitochondrial fatty acid β-oxidation. *J. Inherit. Metab. Dis.* **2010**, *33*, 469–477. [CrossRef]

68. Jin, L.; Alesi, G.N.; Kang, S. Glutaminolysis as a target for cancer therapy. *Oncogene* **2016**, *35*, 3619–3625. [CrossRef]

69. Patra, K.C.; Hay, N. The pentose phosphate pathway and cancer. *Trends Biochem. Sci.* **2014**, *39*, 347–354. [CrossRef]

70. Ashburner, M.; Ball, C.A.; Blake, J.A.; Botstein, D.; Butler, H.; Cherry, J.M.; Davis, A.P.; Dolinski, K.; Dwight, S.S.; Eppig, J.T.; et al. Gene Ontology: Tool for the unification of biology. *Nat. Genet.* **2000**, *25*, 25–29. [CrossRef]

71. The Gene Ontology Consortium The Gene Ontology Resource: 20 years and still GOing strong. *Nucleic Acids Res.* **2018**, *47*, D330–D338.

72. Mi, H.; Muruganujan, A.; Ebert, D.; Huang, X.; Thomas, P.D. PANTHER version 14: More genomes, a new PANTHER GO-slim and improvements in enrichment analysis tools. *Nucleic Acids Res.* **2019**, *47*, D419–D426. [CrossRef] [PubMed]

73. Pavlova, N.N.; Thompson, C.B. The Emerging Hallmarks of Cancer Metabolism. *Cell Metab.* **2016**, *23*, 27–47. [CrossRef] [PubMed]

*Article*

# High Level Expression of MHC-II in HPV+ Head and Neck Cancers Suggests that Tumor Epithelial Cells Serve an Important Role as Accessory Antigen Presenting Cells


Steven F. Gameiro [1], Farhad Ghasemi [2], John W. Barrett [2], Anthony C. Nichols [2,3,4] and Joe S. Mymryk [1,2,3,4,*]

[1]  Department of Microbiology and Immunology, The University of Western Ontario, London, ON N6A 3K7, Canada
[2]  Department of Otolaryngology, Head & Neck Surgery, The University of Western Ontario, London, ON N6A 3K7, Canada
[3]  Department of Oncology, The University of Western Ontario, London, ON N6A 3K7, Canada
[4]  London Regional Cancer Program, Lawson Health Research Institute, London, ON N6C 2R5, Canada
*  Correspondence: jmymryk@uwo.ca; Tel.: +1-519-685-8600 (ext. 53012)


Received: 23 July 2019; Accepted: 5 August 2019; Published: 7 August 2019


**Abstract:** High-risk human papillomaviruses (HPVs) are responsible for a subset of head and neck squamous cell carcinomas (HNSCC). Expression of class II major histocompatibility complex (MHC-II) is associated with antigen presenting cells (APCs). During inflammation, epithelial cells can be induced to express MHC-II and function as accessory APCs. Utilizing RNA-seq data from over 500 HNSCC patients from The Cancer Genome Atlas, we determined the impact of HPV-status on the expression of MHC-II genes and related genes involved in their regulation, antigen presentation, and T-cell co-stimulation. Expression of virtually all MHC-II genes was significantly upregulated in HPV+ carcinomas compared to HPV− or normal control tissue. Similarly, genes that encode products involved in antigen presentation were also significantly upregulated in the HPV+ cohort. In addition, the expression of *CIITA* and *RFX5*—regulators of MHC-II—were significantly upregulated in HPV+ tumors. This coordinated upregulation of MHC-II genes was correlated with higher intratumoral levels of interferon-gamma in HPV+ carcinomas. Furthermore, genes that encode various co-stimulatory molecules involved in T-cell activation and survival were also significantly upregulated in HPV+ tumors. Collectively, these results suggest a previously unappreciated role for epithelial cells in antigen presentation that functionally contributes to the highly immunogenic tumor microenvironment observed in HPV+ HNSCC.

**Keywords:** human papillomavirus; MHC-II; major histocompatibility complex; antigen presentation; head and neck carcinoma; co-stimulatory molecules; survival signals; T-cell


## 1. Introduction

High-risk human papillomaviruses (HPVs) are small, non-enveloped, double-stranded DNA viruses that are responsible for an estimated 5% of all human cancers [1,2]. These biological carcinogens are the causative agents of virtually all cervical cancers and a subset of head and neck squamous cell carcinomas (HNSCC) [1]. HNSCC are a heterogenous group of malignancies caused by multiple distinct etiologies. Infection with high-risk HPVs is responsible for approximately 85,000 of the 600,000 global annual cases of HNSCC, making it the second most common cause of HPV-induced cancers [3–5]. HPV-positive (HPV+) tumors are distinct from their HPV-negative (HPV−) counterparts from a molecular perspective, with distinct genetic, epigenetic, and protein expression profiles [6–11].

Interestingly, patients with HPV+ HNSCC tumors have markedly better clinical outcomes compared to those with HPV− tumors, leading to the recognition of HPV+ HNSCC as unique clinical entities [12–14].

We and others have noted significant differences in the immune landscape between the tumor microenvironments of HPV+ and HPV− HNSCC [7,9,15–19]. Specifically, HPV+ tumors express higher levels of class I major histocompatibility complex (MHC-I) compared to their HPV− counterparts, which could be a consequence of the higher intratumoral levels of interferon-gamma (IFNγ) observed in the tumors of the HPV+ cohort [9]. Utilizing an immunogenomic approach, we have also shown that HPV+ HNSCC tumors exhibited a strong Th1 response characterized by increased infiltration with multiple types of T-cells—CD4+, CD8+, and regulatory T-cells—and expression of their effector molecules [7]. In addition, HPV+ HNSCC also expressed higher levels of *CD39* and multiple T-cell exhaustion markers including *LAG3*, *PD1*, *TIGIT*, and *TIM3* compared to HPV− HNSCC. Importantly, patients with higher expression of these exhaustion markers—indicative of a T-cell-inflamed tumor—exhibited markedly improved survival in HPV+, but not HPV−, HNSCC [7].

In order for an effective T-cell-specific anti-tumor response to occur, a tumor associated antigen must be presented in either the context of MHC-I or class II MHC (MHC-II) [20]. This process is dependent on the initial acquisition of specific antigenic peptides by surveilling antigen presenting cells (APCs). APCs present these exogenous peptides on their cell surface in the context of major histocompatibility complex-II (MHC-II) to activate cognate CD4+ helper T-cells in an antigen-specific fashion [21]. This crosslinking of T-cell receptor (TCR) with its cognate antigen-MHC-II complex is the initial step in the activation of T-cells. The next step is the crosslinking of co-stimulatory molecules between T-cell and APCs that will provide the appropriate signals to initiate T-cell proliferation and survival [22]. Once activated, CD4+ T-cells then stimulate the proliferation of CD8+ cytotoxic T-cells (CTLs) that recognize and respond to the initial antigenic peptide. CTLs subsequently target tumor cells for lysis based on presentation of the cognate endogenously derived antigenic peptides on the tumor cell surface in the context of MHC-I [23,24].

Although the ability of HPV to suppress MHC-I expression in cell culture systems is well known [25,26], this is not likely the case in actual human tumors. Indeed, we previously reported that HPV+ head and neck cancers express higher levels of MHC-I than HPV− tumors [9]. Thus, HPV+ tumor cells may be more effective at displaying endogenously derived viral or neo-antigenic peptides, making them more easily targeted for CTL lysis. Expression of MHC-II molecules is typically restricted to professional APCs, such as dendritic cells (DCs), macrophages, and B-cells [27]. However, epithelial cells can be stimulated by proinflammatory cytokines—specifically IFNγ—to express MHC-II and function as accessory APCs to stimulate T-cell responses [28–30]. As mentioned above, MHC-II proteins play a key role in presenting exogenously-derived peptide antigens that ultimately lead to an effective CTL response, and it is likely that the induced tumor-specific MHC-II expression on epithelial cells may accentuate this process. Indeed, the role of tumor cell derived MHC-II in anti-tumor immunity has become increasingly appreciated [31], with accumulating evidence suggesting that tumor-specific MHC-II expression is correlated with favorable outcomes in melanoma, classic Hodgkin's lymphoma, breast cancer, and oropharyngeal cancers [31–35].

In this study, we used RNA-seq data from over 500 human head and neck tumors from The Cancer Genome Atlas (TCGA) to determine if the presence of oncogenic HPV is associated with altered expression of all classical MHC-II genes and other key genes involved in their regulation, antigen presentation, and T-cell co-stimulation. We found that expression of virtually all classical MHC-II genes was significantly upregulated in HPV+ tumors compared to their HPV− counterparts or normal control tissue. Similarly, genes that encode products that are fundamental to proper antigen loading and presentation were also significantly upregulated in HPV+ tumors. Importantly, the relative level of expression of these inducible MHC-II genes was far beyond those genes associated with professional APCs. Furthermore, the expression of *class II major histocompatibility complex transactivator* (*CIITA*) and *regulatory factor X5* (*RFX5*)—essential master regulators of the MHC-II transcriptional control system—were significantly upregulated in the HPV+ cohort. This coordinated upregulation of the mRNA levels of genes involved in the MHC-II antigen presentation pathway and their regulation were correlated with the higher intratumoral levels of IFNγ observed in HPV+ carcinomas. In addition, genes that encode various T-cell co-stimulatory molecules involved in T-cell activation and survival were found to be significantly upregulated in HPV+ tumors compared to HPV− tumors and normal control tissue. Taken together, these results suggest a previously unappreciated role for epithelial cells in antigen presentation that functionally contributes to the highly immunogenic tumor microenvironment observed in HPV+ HNSCC. This further illustrates the profound differences in the immune landscape between the tumor microenvironments of HPV+ and HPV− HNSCC and may in part contribute to the superior clinical outcomes that are associated with HPV+ HNSCC.

## 2. Results

### 2.1. Classical MHC Class II α- and β-Chain Genes are Expressed at Higher Levels in HPV+ Head and Neck Carcinomas

Constitutive expression of classical MHC-II molecules is typically restricted to professional antigen presenting cells—DCs, macrophages, and B-cells [27]. However, in non-immune cells that lack constitutive expression, such as those of the epithelia, their expression can be induced through exposure to proinflammatory cytokines [28,29,36]. In humans, the genes that encode the three classical polymorphic MHC-II molecules HLA-DP, HLA-DQ, and HLA-DR are expressed as α- and β-chains that form heterodimers on the cell surface [25]. We began by analyzing the Illumina HiSeq RNA expression data from the TCGA HNSC cohort for expression of *HLA-DPA1*, *-DPB1*, *-DQA1*, *-DQA2*, *-DQB1*, *-DQB2*, *-DRA*, *-DRB1*, *-DRB5*, and *-DRB6* genes, encoding the various α- and β-chains for all three classical isotypes (Figure 1). Uniformly, all HPV+ patient samples expressed significantly higher levels of mRNA for all 10 MHC-II genes analyzed compared to HPV− patient samples and virtually all normal control tissues—with the exception of *HLA-DQB2* (HPV+ versus Normal). As the majority of the HPV+ samples are from the oropharynx subsite, we repeated this analysis with HPV+ and HPV− samples that occur only in the oropharynx. Similarly, to the analysis including all subsites, HPV+ oropharyngeal tumors expressed significantly higher levels of MHC-II α- and β-chain genes compared to their HPV− oropharyngeal tumor counterparts (Supplementary Materials Figure S1). Collectively, these results indicate that HPV+ head and neck tumors express high levels of the mRNAs encoding the α- and β-chain heterodimers of the classical MHC-II molecules versus HPV− tumors or normal control tissues. It is noteworthy that based on the normalized read levels, all of these genes are expressed at levels several orders of magnitude above any markers of professional APCs, such as *CD19* (B-cells) [37], *CCL13* (DCs) [38], and *CD84* (macrophages) [39] (Supplementary Materials Figure S2A). However, these normalized read levels are comparable to that of an established epithelial cell marker, E-cadherin (*CDH1*) [40] (Supplementary Materials Figure S2B). Thus, it is likely that these genes are being expressed by epithelial cells within the actual tumor.

### 2.2. Genes Encoding Key Components of the MHC-II Antigen Presentation Pathway are Expressed at Higher Levels in HPV+ Head and Neck Carcinomas

In the endoplasmic reticulum (ER), newly synthesized MHC-II α- and β-chains form a trimeric complex with a non-polymorphic protein called the invariant chain (Ii), which is encoded by the *HLA-DR antigens-associated invariant chain* or *Cluster of Differentiation 74* (CD74) gene [41]. This association with the Ii chain prevents premature peptide loading and also dictates the trafficking of the Ii-MHC-II complex to the endosomal-lysosomal antigen-processing compartments, which contain the antigenic peptides [21,42]. Once in this compartment, Ii is proteolytically cleaved, leaving only a small fragment in the peptide-binding groove called the class II-associated invariant chain peptide (CLIP). Similarly, to the genes encoding the classical MHC-II α- and β-chains, *CD74* was found to be significantly upregulated in HPV+ HNSCC compared to their HPV− counterparts or normal control tissues (Figure 2).

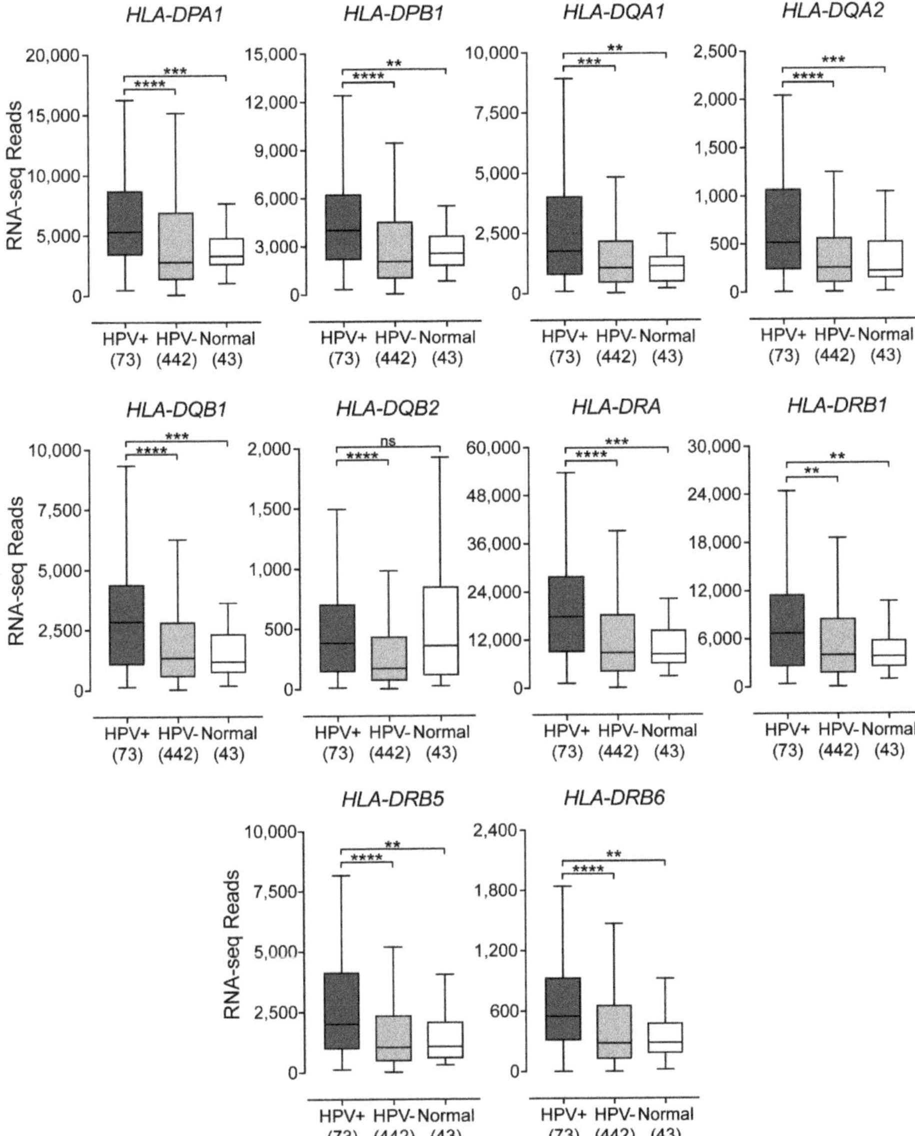

**Figure 1.** Expression of classical class II major histocompatibility complex (MHC-II) α- and β-chain genes in head and neck squamous cell carcinomas stratified by high-risk human papillomaviruses (HPV)-status. RNA-Seq by Expectation Maximization (RSEM)-normalized RNA-seq data for the indicated MHC-II genes was extracted from The Cancer Genome Atlas (TCGA) database for the head and neck squamous cell carcinoma (HNSC) cohort for HPV+, HPV−, and normal control tissues. Numbers in brackets refer to the number of samples included in each analysis. Statistical analysis was performed using a two-tailed non-parametric Mann–Whitney U test. ** $p \leq 0.01$, *** $p \leq 0.001$, **** $p \leq 0.0001$, ns—not significant.

In order for antigenic peptide-binding to occur, CLIP must be removed from the peptide-binding groove [21,41]. The enzymatic removal of CLIP is mediated by the MHC class II-like heterodimer,

HLA-DM. After HLA-DM-mediated removal of CLIP, the class II molecules can now bind lysosomally generated antigenic peptides [43,44]. The binding of antigenic peptides is influenced by another MHC class II-like heterodimer, HLA-DO, which regulates the MHC-II peptide repertoire by modulating the activity of HLA-DM [45,46]. The α- and β-chains of these dimeric class II-like molecules are encoded by the *HLA-DMA, HLA-DMB, HLA-DOA,* and *HLA-DOB* genes. Like the classical MHC-II genes, all four genes encoding the class II-like MHC molecules are similarly upregulated in HPV+ HNSCC versus HPV− tumors or normal control tissues (Figure 2).

When repeated only considering the oropharynx subsite, HPV+ oropharyngeal tumors expressed significantly higher levels of the invariant chain and MHC-II-like genes compared to their HPV− oropharyngeal tumor counterparts (Supplementary Materials Figure S1). Taken together, the upregulation of the genes encoding the MHC-II invariant chain and class II like genes, suggests that key components of the MHC-II antigen presentation pathway are transcribed in HPV+ HNSCC at levels that are significantly higher than observed in HPV− tumors or normal control tissues. Furthermore, these genes are also expressed at very high levels, with the exception of *HLA-DOB*, that are indicative of being expressed by epithelial cells within the actual tumor.

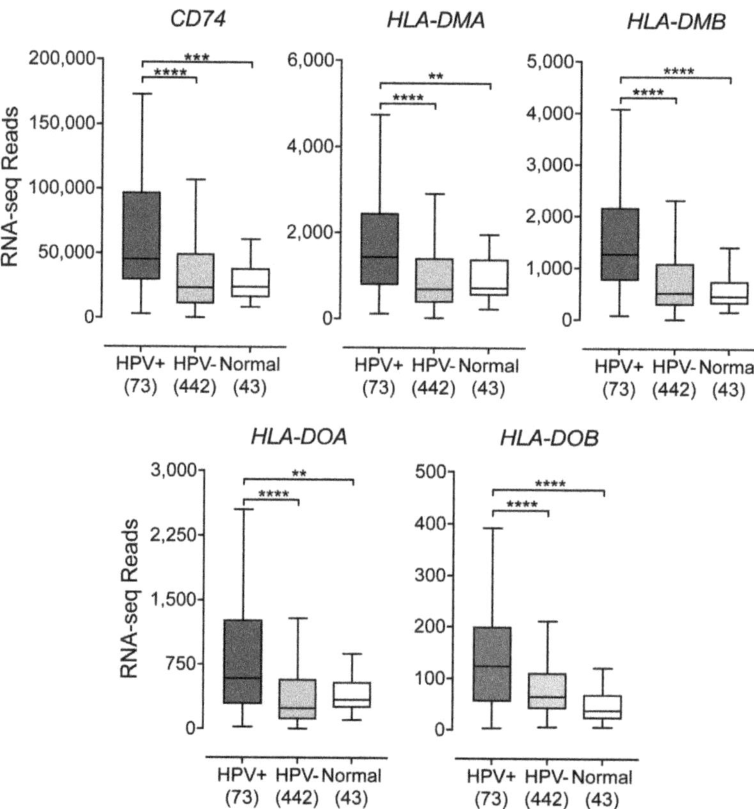

**Figure 2.** Expression of the invariant chain and MHC class II-like genes in head and neck carcinomas stratified by HPV-status. RSEM normalized RNA-seq data for the indicated genes involved in MHC-II-dependent antigen processing and presentation was extracted from the TCGA database for the HNSC cohort for HPV+, HPV−, and normal control tissues. Numbers in brackets refer to the number of samples included in each analysis. Statistical analysis was performed using a two-tailed non-parametric Mann–Whitney U test. ** $p \leq 0.01$, *** $p \leq 0.001$, **** $p \leq 0.0001$, ns—not significant.

### 2.3. Impact of HPV-Status on the Expression of Transcriptional Regulators of MHC-II Gene Expression

Transcriptional control of MHC-II genes and the related genes that encode key components of the MHC-II antigen presentation pathway is among one of the best understood systems in mammals. It is a complex transcriptional system with a unique method of regulation that is completely dependent on the master transcriptional regulator CIITA [29,47]. In agreement with the high levels of MHC-II genes and related genes, analysis of the TCGA data reveals significantly higher levels of *CIITA* in HPV+ samples compared to HPV− or normal control tissues (Figure 3). In addition, *RFX5*—another important transcriptional regulator of MHC-II genes [29]—was similarly expressed at significantly higher levels in HPV+ samples with respect to HPV− tumors or normal control tissues (Figure 3). Again, these differences were also observed when only considering the expression of these genes in the oropharynx (Supplementary Materials Figure S1).

Expression of MHC-II genes and related genes are restricted to APCs, however non-hematopoietic cells such as fibroblasts, endothelial cells, and epithelial cells can be stimulated by IFNγ to express MHC-II molecules and related proteins involved in the antigen presentation pathway [28,29,36]. We analyzed the expression level of the IFNγ gene (*IFNG*) (Figure 3). As expected, the relative level of *IFNG* expression was similar in magnitude to that of other leukocyte specific genes (Supplementary Materials Figure S2A). However, *IFNG* was expressed at significantly higher levels in HPV+ tumors compared to its HPV− counterpart or normal control tissues (Figure 3). In addition, when only considering the oropharynx, the expression of the IFNγ gene was significantly higher in HPV+ samples compared to HPV− oropharyngeal tumors (Supplementary Materials Figure S1).

To further illustrate the IFNγ-specific coordinated upregulation of MHC-II genes and related genes that encode products essential for antigen processing and presentation, we generated a correlation matrix for both HPV+ (Figure 4: upper triangle) and HPV− samples (Figure 4: lower triangle). As expected, regardless of HPV-status, we found that expression of all MHC-II antigen presentation-specific genes were statistically correlated in a pairwise fashion in each patient sample (see also Supplementary MaterialsTable S1).

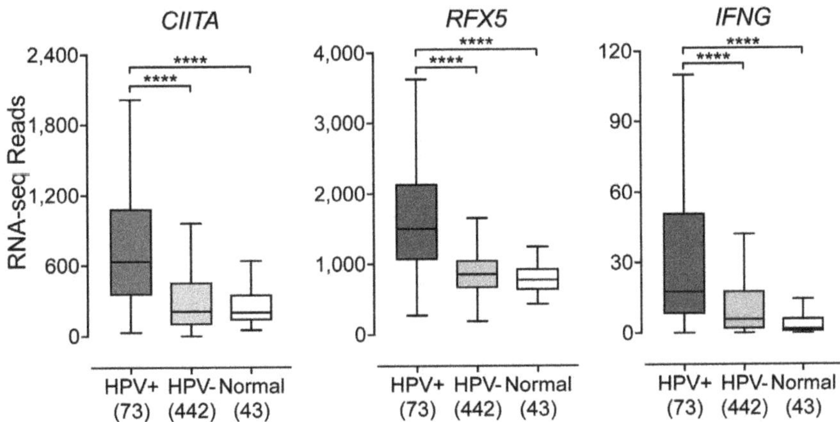


**Figure 3.** Expression of *class II major histocompatibility complex transactivator (CIITA), regulatory factor X5 (RFX5), and interferon-gamma (IFNG)* mRNA in head and neck carcinomas stratified by HPV-status. RSEM normalized RNA-seq data for the *CIITA*, *RFX5*, and *IFNG* genes were extracted from the TCGA database for the HNSC cohort for HPV+, HPV−, and normal control tissues. Numbers in brackets refer to the number of samples included in each analysis. Statistical analysis was performed using a two-tailed non-parametric Mann–Whitney U test. ** $p \leq 0.01$, *** $p \leq 0.001$, **** $p \leq 0.0001$, ns—not significant.

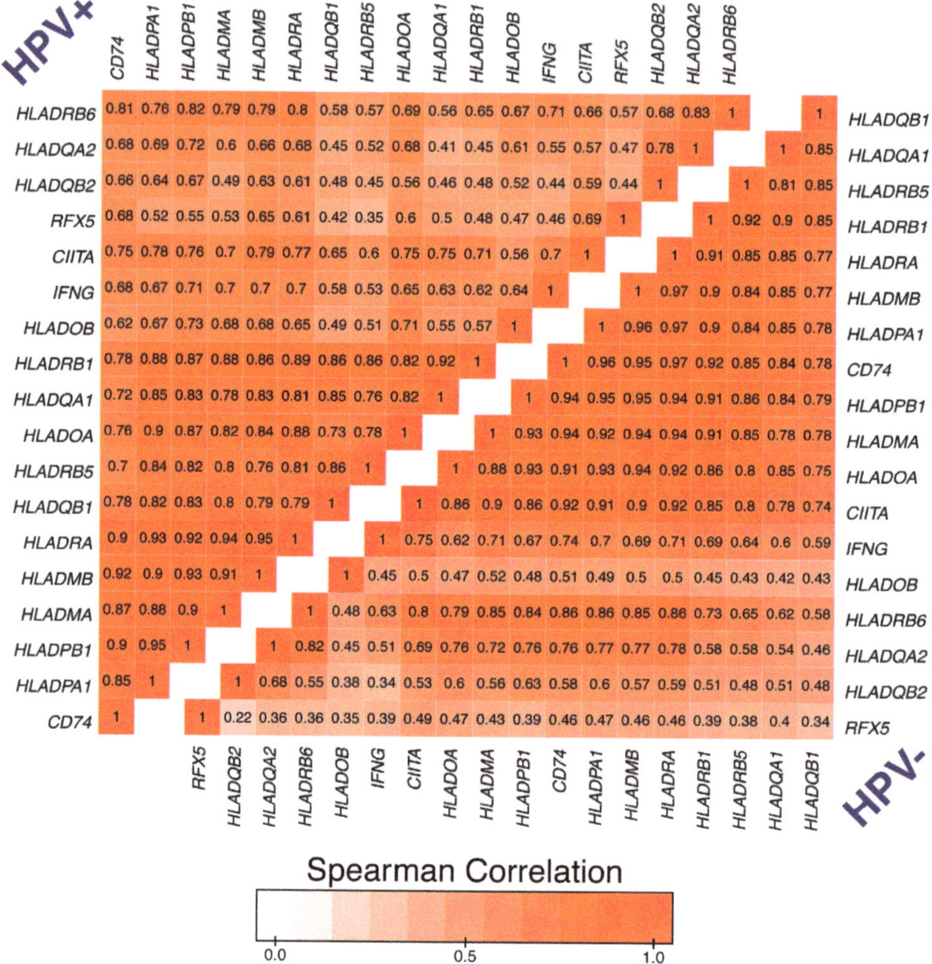


**Figure 4.** Correlation matrix of genes involved in the MHC-II antigen presentation pathway in head and neck carcinomas stratified by HPV-status. RSEM normalized RNA-seq data for the genes listed above were extracted from the TCGA database for the HNSC cohort for HPV+ (upper triangle) and HPV− samples (lower triangle). Pairwise spearman correlation was performed followed by hierarchical clustering to group based on correlation. Number in boxes indicate Spearman's rank correlation coefficient of analyzed gene pairs.

Thus, the upregulated expression of *CIITA*, *RFX5*, and subsequent expression of all the classical MHC-II genes and related genes required for antigen loading and presentation observed in HPV+ head and neck carcinomas are likely a consequence of IFNγ exposure. Furthermore, the correlation matrix illustrates the unique simultaneous coordination of the MHC-II transcriptional control system that has been shown to be dictated by the master transcriptional regulator CIITA [29,47].

*2.4. Impact of HPV-Status on the Expression of T-Cell Co-Stimulatory Molecules in HPV-Positive Head and Neck Carcinomas*

Co-stimulation of T-cells occurs through the interaction of its constitutively expressed CD28 receptor with either CD80 or CD86 on APCs [22]. Utilizing RNA expression data for the levels of each of these co-stimulatory molecules, we found higher levels of *CD28* in HPV+ tumors compared to HPV− or normal control tissues (Figure 5). While the levels of both *CD80* and *CD86* in HPV+ tumors were not significantly different compared to their HPV− counterparts, they were significantly higher compared to normal control tissues (Figure 5). In addition, we found that the mRNA levels of *CD152*, which encodes for CTLA-4, a marker of T-cell activation [48], was significantly upregulated in HPV+ tumors compared to HPV− and normal control tissues (Figure 5). Again, these differences were also observed when only considering the expression of these genes in the oropharynx (Supplementary Materials Figure S1). These results suggest that, like the MHC-II genes and the genes involved in antigen loading and presentation, co-stimulatory molecules are similarly present at higher levels in HPV+ tumors and this is correlated with a higher level of T-cell activation.

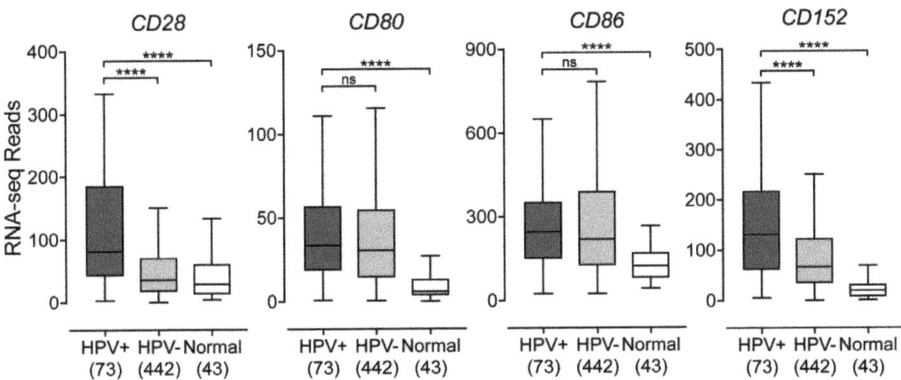


**Figure 5.** Expression of genes that encode for T-cell co-stimulatory molecules in head and neck carcinomas stratified by HPV-status. RSEM normalized RNA-seq data for genes that encode T-cell specific co-stimulatory molecules was extracted from the TCGA database for the HNSC cohort for HPV+, HPV−, and normal control tissues. Statistical analysis was performed using a two-tailed non-parametric Mann–Whitney U test. Numbers in brackets refer to the number of samples included in each analysis. **** $p \leq 0.0001$, ns—not significant.

### 2.5. Impact of HPV-Status on the Expression of Inducible T-cell Survival Signal Molecules in HPV-Positive Head and Neck Carcinomas

Utilizing the RNA-seq HNSC dataset from the TCGA, we looked at genes that encode for inducible, T-cell activation-dependent, survival signal molecules and their respective ligands [22,49]. We found that *CD137* (4-1BB, TNFRSF9) was significantly upregulated in HPV+ tumors compared to HPV− or normal control tissues (Figure 6). However, its ligand *TNFSF9* (CD137L, 4-1BBL) was found to be significantly downregulated in HPV+ tumors compared to HPV− and not significantly different compared to normal control tissues (Figure 6). Next, we looked at the genes that encode for the inducible T-cell co-stimulator (*ICOS*) and its ligand *ICOSLG* and found that both were significantly upregulated in HPV+ tumors compared to HPV−, but only *ICOS* was significantly upregulated in comparison to normal control tissues (Figure 6). Finally, we looked at OX40 (*TNFRSF4*, CD134) and its ligand OX40L (*TNFSF4*, CD252) and found that both genes were significantly upregulated in HPV+ tumors compared to their HPV− counterparts and normal control tissues (Figure 6). Again, these differences were also observed when only considering the expression of these genes in the oropharynx with the exception of *TNFSF4* (Supplementary Materials Figure S1). Taken together, these results suggest that T-cells are activated and proliferating within the HPV+ tumor microenvironment via the observation of an increase in expression of genes that encode for survival signal molecules that are only induced following TCR-mediated antigen-specific T-cell activation and/or CD28 co-stimulation [22,49].

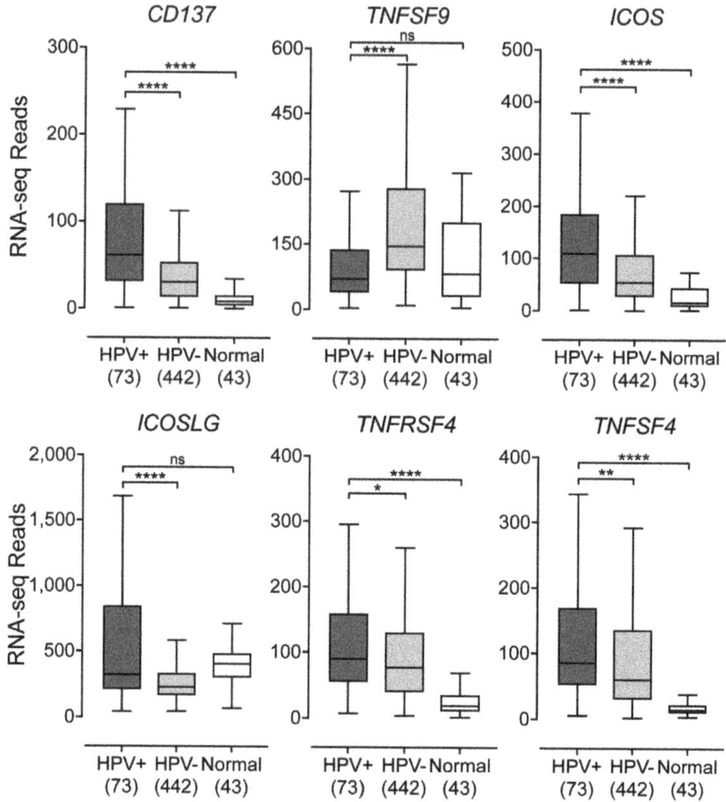

**Figure 6.** Expression of inducible T-cell survival signal molecules in head and neck carcinomas stratified by HPV-status. RSEM normalized RNA-seq data for inducible genes that encode for T-cell survival molecules was extracted from the TCGA database for the HNSC cohort for HPV+, HPV−, and normal control tissues. Statistical analysis was performed using a two-tailed non-parametric Mann–Whitney U test. Numbers in brackets refer to the number of samples included in each analysis. * $p \leq 0.05$, ** $p \leq 0.01$, *** $p \leq 0.001$, **** $p \leq 0.0001$, ns—not significant.

## 3. Discussion

Expression of MHC-II is typically associated with professional APCs, which are considered essential for the initiation of the adaptive immune response. They function by sampling their local environment via phagocytosis, acquiring particles and processing them internally in order to present them on their cell surface to CD4+ T-cells in the context of an antigen-MHC-II complex [21]. The crosslinking of the CD4+ TCR and antigen-MHC-II complex initiates the T-cell activation protocol that, in conjunction with co-stimulatory signals, can ultimately lead to an effective adaptive immune response against a threat of internal or external origin, such as cancerous cells or bacteria and viruses, respectively [22].

In non-hematopoietic cells, such as epithelia, MHC-II expression can be induced through exposure to the pro-inflammatory cytokine IFNγ [28,29,36]. This induction of MHC-II on epithelial cells bestows on them the ability to act as accessory APCs, and this can accentuate the presentation of antigens to CD4+ T-cells [30]. However, the ability of epithelial cells to function as accessory APCs is generally underappreciated, with most existing information related to the gastrointestinal and respiratory tracts [50]. Interestingly, cancerous tissues can retain tumor-specific MHC-II expression, and this has the potential to increase recognition of a tumor by the immune system [31]. Indeed, tumor specific

MHC-II expression has been associated with superior prognosis and/or improved response to immune checkpoint inhibitor therapy in several human cancers, as well as increased tumor rejection in murine models [31–35]. The importance of the immune response for successful resolution of cancers cannot be understated. Indeed, mouse models have shown that neither chemotherapy nor radiation treatment functioned effectively in the absence of a functional immune system, at least in HPV+ HNSCC [51].

While MHC-II expression has been reported in head and neck tumors [35], existing studies have been limited to individual classical isotypes, such as HLA-DR, based on limitations in available antibodies [52]. Cell culture models based on established head and neck cancer lines have also demonstrated MHC-II expression, often in response to IFNγ, or transfection with CIITA—master regulator of MHC-II transcription [53,54]. However, no existing studies have comprehensively assessed the transcriptional status of the entire MHC-II antigen presentation system in head and neck cancers. In this study, our goal was to determine if MHC-II components were widely expressed in human head and neck cancers and whether expression was influenced by HPV-status.

Using data from over 500 primary human head and neck tumors, we provide evidence that HPV+ head and neck carcinomas display high mRNA levels for virtually all MHC-II genes, including the classical and non-classical α- and β-chains, the invariant γ chain, as well as factors required for MHC-II loading and trafficking (Figures 1 and 2). Increased expression of these genes was observed whether the comparison included all head and neck cancer subsites or was restricted to just the oropharynx, where the majority of HPV+ head and neck cancers arise (Supplementary Materials Figure S1). These results are in good agreement with the concurrent detection of high levels of expression of *CIITA* and *RFX5*, important global regulators of MHC-II transcription [29,47], in HPV+ tumors. HPV+ head and neck carcinomas express significantly higher levels of these genes as compared to normal control tissues, and these levels are generally higher than in HPV− carcinomas (Figure 3). This likely reflects the T-cell inflamed nature of HPV+ cancers [7], and specifically the higher levels of IFNγ expressed in these tumors (Figure 3). This IFNγ-dependent coordinated upregulation of MHC-II genes and related genes involved in antigen processing and presentation was further illustrated in Figure 4, where we observed a strong global correlation with all genes analyzed in the MHC-II transcriptional control system.

After generation and programming in the thymus, CD8+ and CD4+ T-cells circulate in the body until they encounter their specific antigen presented on either class I or class II MHC molecules, respectively [23,24,55]. This interaction between TCR and antigen-loaded MHC complex represents signal 1, which triggers the activation of T-cells. However, in order for the activated T-cell to fully respond to the presented threat, and not enter a state of unresponsiveness, it requires a secondary signal through co-stimulatory molecules [22]. We found higher levels of *CD28* in HPV+ tumors compared to their HPV− counterparts and normal control tissues (Figure 5). The binding of CD28 with either CD80 or CD86 leads to clonal expansion of the T-cell pool that is specific to the recognized antigen [22,49]. In order to attenuate this response, the aforementioned interaction leads to the induction of the co-inhibitory molecule CTLA-4, which is encoded by the gene *CD152*. This co-inhibitory molecule will compete with CD28 for binding to either CD80 or CD86 to attenuate the T-cell response [22,48]. We found that *CD152* was significantly upregulated in the HPV+ cohort (Figure 5). Collectively, this data indirectly illustrates the higher number of infiltrating T-cells within the tumor microenvironments of HPV+ HNSCC through the identification of significantly higher levels of the constitutively expressed T-cell-specific *CD28* marker. In addition, it provides good evidence that the interaction of co-stimulatory molecules in HPV+ HNSCC is effective, given that we detected significantly higher levels of expression of *CD152* mRNA, which encodes for the inducible co-inhibitory molecule CTLA-4.

In order for proliferating T-cells to persist and survive after antigen-recognition and subsequent stimulation with co-stimulatory molecules, they require survival signals that are delivered through the cross-linking of various molecules [22,49]. Interestingly, unlike other co-stimulatory molecules—such as CD28—that can be found constitutively expressed on T-cells, the molecules that convey these survival signals are encoded by genes that are only expressed following TCR-mediated antigen-specific

T-cell activation and/or CD28 co-stimulation [22,49]. We found that all inducible T-cell survival genes analyzed, with the exception of *TNFSF9*, were expressed at significantly higher levels in HPV+ tumors compared to HPV− and normal control tissues (Figure 6). *TNFSF9* was significantly downregulated in HPV+ tumors compared to its HPV− counterpart (Figure 6). This downregulation of *TNFSF9* is indicative of a mechanistic response to excessive CD137-mediated signaling [56–58]. This data indirectly illustrates that the HPV+ tumor microenvironment contains increased levels of activated and proliferating T-cells via the observation of an increase in expression of genes that encode for survival signal molecules that are induced following TCR-mediated antigen-specific T-cell activation and/or CD28 co-stimulation. These results agree well with previous reports by our group and others that HPV+ tumors contain more T-cells [7,16–19], but goes further in that it indirectly provides evidence of productive MHC-II-dependent tumor-antigen recognition.

## 4. Materials and Methods

### 4.1. TCGA RNA-Seq Boxplot Comparisons

Level 3 RNA-Seq by Expectation Maximization (RSEM)-normalized Illumina HiSeq RNA expression data for the TCGA head and neck cancer (HNSC) cohort was downloaded from the Broad Genome Data Analysis Centers Firehose server (https://gdac.broadinstitute.org/). The normalized, gene level Firehose dataset was utilized for all the genes analyzed. RSEM-normalized RNA-Seq data was extracted into Microsoft Excel and the HPV-status was determined based on published datasets [59–62]. For all genes analyzed in this study, patient samples from primary tumors with known HPV-status were grouped as HPV+, HPV−, or normal control samples. Patient samples with undetermined HPV-status or samples from secondary metastatic lesions were omitted from our calculations. This resulted in 73 HPV+, 442 HPV−, and 43 normal control samples with RNA-Seq data available for gene expression analysis. Oropharynx-only gene reanalysis was performed by utilizing patient samples that were isolated from the tissues of the oropharynx (tonsils, base of tongue, or oropharynx) for both HPV+ and HPV− samples. This resulted in 53 HPV+ and 26 HPV− samples with RNA-Seq data available for gene expression analysis. Graphpad Prism v7.0 (Graphpad Software, Inc., San Diego, California, USA) was used to generate boxplot comparisons of gene expression between the indicated HPV+, HPV−, and normal control samples as well as the oropharynx-only reanalysis between HPV+ and HPV− samples. For each boxplot, the center line indicates the median, the lower and upper box limits represent Q1 (25th percentile) and Q3 (75th percentile), respectively, and the whiskers extend 1.5 times the interquartile range (IQR) from Q1 (lower whisker) and Q3 (upper whisker). An unpaired, two-tailed non-parametric Mann–Whitney U test was utilized to assign p-values. G*Power Software version 3.1.9.2 [63] was used to perform post-hoc power calculations, with effect size selected as 0.8 and $\alpha = 0.05$. All boxplot comparisons achieved a power value > 0.8. Figures were assembled into final form using Adobe Illustrator CS6.

### 4.2. Correlation Matrix

Level 3 RSEM normalized RNA-seq data for the genes listed above were extracted from the TCGA database and processed as detailed in Materials and Methods 4.1. For the HPV+ (upper triangle) and HPV− samples (lower triangle), pairwise spearman correlation was performed for each gene involved in the MHC-II transcriptional control system. Hierarchical clustering was utilized to group genes based on strength of correlation. Correlations and clustering were performed using R statistical environment (version 3.4.0; https://cran.r-project.org/bin/macosx/) utilizing packages ggplot2 and reshape2. Correlation matrix figure was assembled into final form using Adobe Illustrator.

## 5. Conclusions

HPV+ HNSCC tumors are remarkably different from their HPV− counterparts in that they express high levels of all components of the MHC-II antigen presentation apparatus. Importantly, while some of this signal can be attributed to professional APCs, the extremely high relative level of expression supports a model whereby the T-cell inflamed environment in HPV+ HNSCC induces a functionally effective MHC-II presentation system based on tumor epithelial cells. As MHC-II-dependent antigen presentation is critical for CD4+ help in CD8+ T-cell responses, which are essential for the control and clearance of cancerous cells, it is likely that the expression of non-self-derived viral antigens or tumor derived neoantigens, combined with intact MHC-II presentation and appropriate co-stimulation contributes to the markedly better patient outcomes for HPV+ versus HPV− head and neck carcinomas. As immune checkpoint inhibition therapy in other cancers has been reported to be most effective for tumors with high MHC-II levels, suggesting that stratification based on MHC-II levels may help predict those likely to respond to checkpoint inhibition therapy.

**Supplementary Materials:** The following are available online at http://www.mdpi.com/2072-6694/11/8/1129/s1, Figure S1: Gene expression reanalysis in the oropharynx. Figure S2: Gene expression comparisons of (A) APC and (B) Epithelia markers. Table S1: MHC-II genes pairwise correlation, Spearman coefficients, and P-values.

**Author Contributions:** Conceptualization, S.F.G., J.W.B., A.C.N. and J.S.M.; Data curation, S.F.G. and F.G.; Formal analysis, S.F.G., F.G. and J.S.M.; Funding acquisition, A.C.N. and J.S.M.; Methodology, S.F.G., F.G. and J.S.M.; Supervision, J.S.M.; Writing—original draft, S.F.G. and J.S.M.; Writing—review and editing, S.F.G. and J.S.M.

**Funding:** This work was supported from grants provided by the Canadian Institutes of Health Research to JSM and ACN (MOP#142491) and to ACN (MOP#340674). ACN was supported by the Wolfe Surgical Research Professorship in the Biology of Head and Neck Cancers Fund. SFG was supported in part from a Cancer Research and Technology Transfer studentship.

**Acknowledgments:** We thank Scott Bratman, James Pipas, and Paul Cantalupo for providing information regarding HPV status of various TCGA samples.

**Conflicts of Interest:** The authors declare no conflict of interest.

## References

1. Doorbar, J.; Egawa, N.; Griffin, H.; Kranjec, C.; Murakami, I. Human Papillomavirus Molecular Biology and Disease Association. *Rev. Med. Virol.* **2015**, *25* (Suppl. 1), 2–23. [CrossRef] [PubMed]

2. Forman, D.; de Martel, C.; Lacey, C.J.; Soerjomataram, I.; Lortet-Tieulent, J.; Bruni, L.; Vignat, J.; Ferlay, J.; Bray, F.; Franceschi, S.; et al. Global Burden of Human Papillomavirus and Related Diseases. *Vaccine* **2012**, *30* (Suppl. 5), F12–F23. [CrossRef] [PubMed]

3. Gillison, M.L.; Koch, W.M.; Capone, R.B.; Spafford, M.; Westra, W.H.; Wu, L.; Zahurak, M.L.; Daniel, R.W.; Viglione, M.; Symer, D.E.; et al. Evidence for a Causal Association between Human Papillomavirus and a Subset of Head and Neck Cancers. *J. Natl. Cancer Inst.* **2000**, *92*, 709–720. [CrossRef] [PubMed]

4. Syrjänen, S. Human Papillomavirus (HPV) in Head and Neck Cancer. *J. Clin. Virol.* **2005**, *32* (Suppl. 1), S59–S66.

5. Gillison, M.L.; Castellsagué, X.; Chaturvedi, A.; Goodman, M.T.; Snijders, P.; Tommasino, M.; Arbyn, M.; Franceschi, S. Eurogin Roadmap: Comparative Epidemiology of HPV Infection and Associated Cancers of the Head and Neck and Cervix. *Int. J. Cancer* **2013**, *134*, 497–507. [CrossRef] [PubMed]

6. Sepiashvili, L.; Bruce, J.P.; Huang, S.H.; O'Sullivan, B.; Liu, F.-F.; Kislinger, T. Novel Insights into Head and Neck Cancer using Next-generation "Omic" Technologies. *Cancer Res.* **2015**, *75*, 480–486. [CrossRef] [PubMed]

7. Gameiro, S.F.; Ghasemi, F.; Barrett, J.W.; Koropatnick, J.; Nichols, A.C.; Mymryk, J.S.; Maleki Vareki, S. Treatment-naïve HPV+ Head and Neck Cancers Display a T-cell-inflamed Phenotype Distinct from Their HPV− Counterparts that Has Implications for Immunotherapy. *Oncoimmunology* **2018**, *7*, e1498439. [CrossRef] [PubMed]

8. Gameiro, S.F.; Kolendowski, B.; Zhang, A.; Barrett, J.W.; Nichols, A.C.; Torchia, J.; Mymryk, J.S. Human Papillomavirus Dysregulates the Cellular Apparatus Controlling the Methylation Status of H3K27 in Different Human Cancers to Consistently Alter Gene Expression Regardless of Tissue of Origin. *Oncotarget* **2017**, *8*, 72564–72576. [CrossRef]

9. Gameiro, S.F.; Zhang, A.; Ghasemi, F.; Barrett, J.W.; Nichols, A.C.; Mymryk, J.S. Analysis of Class I Major Histocompatibility Complex Gene Transcription in Human Tumors Caused by Human Papillomavirus Infection. *Viruses* **2017**, *9*, 252. [CrossRef]

10. Seiwert, T.Y.; Zuo, Z.; Keck, M.K.; Khattri, A.; Pedamallu, C.S.; Stricker, T.; Brown, C.; Pugh, T.J.; Stojanov, P.; Cho, J.; et al. Integrative and Comparative Genomic Analysis of HPV-positive and HPV-negative Head and Neck Squamous Cell Carcinomas. *Clin. Cancer Res.* **2015**, *21*, 632–641. [CrossRef]

11. Worsham, M.J.; Chen, K.M.; Ghanem, T.; Stephen, J.K.; Divine, G. Epigenetic Modulation of Signal Transduction Pathways in HPV-Associated HNSCC. *Otolaryngol. Head Neck Surg.* **2013**, *149*, 409–416. [CrossRef] [PubMed]

12. Fakhry, C.; Westra, W.H.; Li, S.; Cmelak, A.; Ridge, J.A.; Pinto, H.; Forastiere, A.; Gillison, M.L. Improved Survival of Patients with Human Papillomavirus-positive Head and Neck Squamous Cell Carcinoma in a Prospective Clinical Trial. *J. Natl. Cancer Inst.* **2008**, *100*, 261–269. [CrossRef] [PubMed]

13. Psyrri, A.; Rampias, T.; Vermorken, J.B. The Current and Future Impact of Human Papillomavirus on Treatment of Squamous Cell Carcinoma of the Head and Neck. *Ann. Oncol.* **2014**, *25*, 2101–2115. [CrossRef] [PubMed]

14. Ang, K.K.; Harris, J.; Wheeler, R.; Weber, R.; Rosenthal, D.I.; Nguyen-Tân, P.F.; Westra, W.H.; Chung, C.H.; Jordan, R.C.; Lu, C.; et al. Human Papillomavirus and Survival of Patients with Oropharyngeal Cancer. *N. Engl. J. Med.* **2010**, *363*, 24–35. [CrossRef] [PubMed]

15. Baruah, P.; Bullenkamp, J.; Wilson, P.O.G.; Lee, M.; Kaski, J.C.; Dumitriu, I.E. TLR9 Mediated Tumor-Stroma Interactions in Human Papilloma Virus (HPV)-positive Head and Neck Squamous Cell Carcinoma Up-regulate PD-L1 and PD-L2. *Front. Immunol.* **2019**, *10*, 561. [CrossRef] [PubMed]

16. Nguyen, N.; Bellile, E.; Thomas, D.; McHugh, J.; Rozek, L.; Virani, S.; Peterson, L.; Carey, T.E.; Walline, H.; Moyer, J.; et al. Head and Neck SPORE Program Investigators Tumor Infiltrating Lymphocytes and Survival in Patients with Head and Neck Squamous Cell Carcinoma. *Head Neck* **2016**, *38*, 1074–1084. [CrossRef] [PubMed]

17. Solomon, B.; Young, R.J.; Bressel, M.; Urban, D.; Hendry, S.; Thai, A.; Angel, C.; Haddad, A.; Kowanetz, M.; Fua, T.; et al. Prognostic Significance of PD-L1+ and CD8+ Immune Cells in HPV+ Oropharyngeal Squamous Cell Carcinoma. *Cancer Immunol. Res.* **2018**, *6*, 295–304. [CrossRef] [PubMed]

18. Mandal, R.; Şenbabaoğlu, Y.; Desrichard, A.; Havel, J.J.; Dalin, M.G.; Riaz, N.; Lee, K.-W.; Ganly, I.; Hakimi, A.A.; Chan, T.A.; et al. The Head and Neck Cancer Immune Landscape and Its Immunotherapeutic Implications. *JCI Insight* **2016**, *1*, e89829. [CrossRef] [PubMed]

19. Chen, X.; Yan, B.; Lou, H.; Shen, Z.; Tong, F.; Zhai, A.; Wei, L.; Zhang, F. Immunological Network Analysis in HPV Associated Head and Neck Squamous Cancer and Implications for Disease Prognosis. *Mol. Immunol.* **2018**, *96*, 28–36. [CrossRef] [PubMed]

20. Rock, K.L.; Reits, E.; Neefjes, J. Present Yourself! By MHC Class I and MHC Class II Molecules. *Trends Immunol.* **2016**, *37*, 724–737. [CrossRef] [PubMed]

21. Roche, P.A.; Furuta, K. The Ins and Outs of MHC Class II-mediated Antigen Processing and Presentation. *Nat. Rev. Immunol.* **2015**, *15*, 203–216. [CrossRef] [PubMed]

22. Chen, L.; Flies, D.B. Molecular Mechanisms of T cell Co-Stimulation and Co-Inhibition. *Nat. Rev. Immunol.* **2013**, *13*, 227–242. [CrossRef] [PubMed]

23. Zhang, N.; Bevan, M.J. CD8+ T Cells: Foot Soldiers of the Immune System. *Immunity* **2011**, *35*, 161–168. [CrossRef] [PubMed]

24. Luckheeram, R.V.; Zhou, R.; Verma, A.D.; Xia, B. CD4+ T Cells: Differentiation and Functions. *Clin. Dev. Immunol.* **2012**, *2012*, 925135-12. [CrossRef] [PubMed]

25. Heller, C.; Weisser, T.; Mueller-Schickert, A.; Rufer, E.; Hoh, A.; Leonhardt, R.M.; Knittler, M.R. Identification of Key Amino Acid Residues that Determine the Ability of High Risk HPV16-E7 to Dysregulate Major Histocompatibility Complex Class I Expression. *J. Biol. Chem.* **2011**, *286*, 10983–10997. [CrossRef]

26. Ferris, R.L. Immunology and Immunotherapy of Head and Neck Cancer. *J. Clin. Oncol.* **2015**, *33*, 3293–3304. [CrossRef] [PubMed]

27. Van den Elsen, P.J.; Holling, T.M.; Kuipers, H.F.; van der Stoep, N. Transcriptional Regulation of Antigen Presentation. *Curr. Opin. Immunol.* **2004**, *16*, 67–75. [CrossRef]

28. Collins, T.; Korman, A.J.; Wake, C.T.; Boss, J.M.; Kappes, D.J.; Fiers, W.; Ault, K.A.; Gimbrone, M.A.; Strominger, J.L.; Pober, J.S. Immune Interferon Activates Multiple Class II Major Histocompatibility Complex Genes and the Associated Invariant Chain Gene in Human Endothelial Cells and Dermal Fibroblasts. *Proc. Natl. Acad. Sci. USA* **1984**, *81*, 4917–4921. [CrossRef]

29. Boss, J.M.; Jensen, P.E. Transcriptional Regulation of the MHC Class II Antigen Presentation Pathway. *Curr. Opin. Immunol.* **2003**, *15*, 105–111. [CrossRef]
30. Kim, B.S.; Miyagawa, F.; Cho, Y.-H.; Bennett, C.L.; Clausen, B.E.; Katz, S.I. Keratinocytes Function as Accessory Cells for Presentation of Endogenous Antigen Expressed in the Epidermis. *J. Investig. Dermatol.* **2009**, *129*, 2805–2817. [CrossRef]
31. Axelrod, M.L.; Cook, R.S.; Johnson, D.B.; Balko, J.M. Biological Consequences of MHC-II Expression by Tumor Cells in Cancer. *Clin. Cancer Res.* **2019**, *25*, 2392–2402. [CrossRef] [PubMed]
32. Johnson, D.B.; Estrada, M.V.; Salgado, R.; Sanchez, V.; Doxie, D.B.; Opalenik, S.R.; Vilgelm, A.E.; Feld, E.; Johnson, A.S.; Greenplate, A.R.; et al. Melanoma-specific MHC-II Expression Represents a Tumour-autonomous Phenotype and Predicts Response to Anti-PD-1/PD-L1 Therapy. *Nature Commun.* **2016**, *7*, 10582. [CrossRef] [PubMed]
33. Roemer, M.G.M.; Redd, R.A.; Cader, F.Z.; Pak, C.J.; Abdelrahman, S.; Ouyang, J.; Sasse, S.; Younes, A.; Fanale, M.; Santoro, A.; et al. Major Histocompatibility Complex Class II and Programmed Death Ligand 1 Expression Predict Outcome after Programmed Death 1 Blockade in Classic Hodgkin Lymphoma. *JCO* **2018**, *36*, 942–950. [CrossRef] [PubMed]
34. Forero, A.; LI, Y.; Chen, D.; Grizzle, W.E.; Updike, K.L.; Merz, N.D.; Downs-Kelly, E.; Burwell, T.C.; Vaklavas, C.; Buchsbaum, D.J.; et al. Expression of the MHC Class II Pathway in Triple-Negative Breast Cancer Tumor Cells Is Associated with a Good Prognosis and Infiltrating Lymphocytes. *Cancer Immunol. Res.* **2016**, *4*, 390–399. [CrossRef] [PubMed]
35. Cioni, B.; Jordanova, E.S.; Hooijberg, E.; Linden, R.; van der Linden, R.; de Menezes, R.X.; Menezes, R.X.; Tan, K.; Willems, S.; Elbers, J.B.W.; et al. HLA Class II Expression on Tumor Cells and Low Numbers of Tumor-associated Macrophages Predict Clinical Outcome in Oropharyngeal Cancer. *Head Neck* **2018**, *26*, 123–478. [CrossRef] [PubMed]
36. Steimle, V.; Siegrist, C.A.; Mottet, A.; Lisowska-Grospierre, B.; Mach, B. Regulation of MHC Class II Expression by Interferon-gamma Mediated by the Transactivator Gene CIITA. *Science* **1994**, *265*, 106–109. [CrossRef] [PubMed]
37. Wang, K.; Wei, G.; Liu, D. CD19: A Biomarker for B cell Development, Lymphoma Diagnosis and Therapy. *Exp. Hematol. Oncol.* **2012**, *1*, 36. [CrossRef] [PubMed]
38. Danaher, P.; Warren, S.; Dennis, L.; D'Amico, L.; White, A.; Disis, M.L.; Geller, M.A.; Odunsi, K.; Beechem, J.; Fling, S.P. Gene Expression Markers of Tumor Infiltrating Leukocytes. *J. Immunother. Cancer* **2017**, *5*, 18. [CrossRef]
39. Zaiss, M.; Hirtreiter, C.; Rehli, M.; Rehm, A.; Kunz-Schughart, L.A.; Andreesen, R.; Hennemann, B. CD84 Expression on Human Hematopoietic Progenitor Cells. *Exp. Hematol.* **2003**, *31*, 798–805. [CrossRef]
40. Gall, T.M.H.; Frampton, A.E. Gene of the Month: E-cadherin (CDH1). *J. Clin. Pathol.* **2013**, *66*, 928–932. [CrossRef]
41. Cresswell, P. Invariant Chain Structure and MHC Class II Function. *Cell* **1996**, *84*, 505–507. [CrossRef]
42. Roche, P.A.; Cresswell, P. Invariant Chain Association with HLA-DR Molecules Inhibits Immunogenic Peptide Binding. *Nature* **1990**, *345*, 615–618. [CrossRef] [PubMed]
43. Roche, P.A. HLA-DM: An in Vivo Facilitator of MHC Class II Peptide Loading. *Immunity* **1995**, *3*, 259–262. [CrossRef]
44. Busch, R.; Rinderknecht, C.H.; Roh, S.; Lee, A.W.; Harding, J.J.; Burster, T.; Hornell, T.M.C.; Mellins, E.D. Achieving Stability Through Editing and Chaperoning: Regulation of MHC Class II Peptide Binding and Expression. *Immunol. Rev.* **2005**, *207*, 242–260. [CrossRef] [PubMed]
45. Poluektov, Y.O.; Kim, A.; Sadegh-Nasseri, S. HLA-DO and Its Role in MHC Class II Antigen Presentation. *Front. Immunol.* **2013**, *4*, 260. [CrossRef] [PubMed]
46. Denzin, L.K.; Fallas, J.L.; Prendes, M.; Yi, W. Right place, right time, right peptide: DO Keeps DM Focused. *Immunol. Rev.* **2005**, *207*, 279–292. [CrossRef] [PubMed]
47. Van den Elsen, P.J. Expression Regulation of Major Histocompatibility Complex Class I and Class II Encoding Genes. *Front. Immunol.* **2011**, *2*, 48. [CrossRef]
48. Rudd, C.E.; Taylor, A.; Schneider, H. CD28 and CTLA-4 Coreceptor Expression and Signal Transduction. *Immunol. Rev.* **2009**, *229*, 12–26. [CrossRef]
49. Beier, K.C.; Kallinich, T.; Hamelmann, E. Master Switches of T-cell Activation and Differentiation. *Eur. Respir. J.* **2007**, *29*, 804–812. [CrossRef]

50. Wosen, J.E.; Mukhopadhyay, D.; Macaubas, C.; Mellins, E.D. Epithelial MHC Class II Expression and Its Role in Antigen Presentation in the Gastrointestinal and Respiratory Tracts. *Front. Immunol.* **2018**, *9*, 221. [CrossRef]

51. Spanos, W.C.; Nowicki, P.; Lee, D.W.; Hoover, A.; Hostager, B.; Gupta, A.; Anderson, M.E.; Lee, J.H. Immune Response During Therapy with Cisplatin or Radiation for Human Papillomavirus-related Head and Neck Cancer. *Arch. Otolaryngol. Head Neck Surg.* **2009**, *135*, 1137–1146. [CrossRef] [PubMed]

52. Johnson, D.B.; Nixon, M.J.; Wang, Y.; Wang, D.Y.; Castellanos, E.; Estrada, M.V.; Ericsson-Gonzalez, P.I.; Cote, C.H.; Salgado, R.; Sanchez, V.; et al. Tumor-specific MHC-II Expression Drives a Unique Pattern of Resistance to Immunotherapy via LAG-3/FCRL6 Engagement. *JCI Insight* **2018**, *3*, 5250. [CrossRef] [PubMed]

53. Meissner, M.; Whiteside, T.L.; Kaufmann, R.; Seliger, B. CIITA Versus IFN-γ Induced MHC Class II Expression in Head and Neck Cancer Cells. *Arch. Dermatol. Res.* **2008**, *301*, 189–193. [CrossRef] [PubMed]

54. Arosarena, O.A.; Baranwal, S.; Strome, S.; Wolf, G.T.; Krauss, J.C.; Bradford, C.R.; Carey, T.E. Expression of Major Histocompatibility Complex Antigens in Squamous Cell Carcinomas of the Head and Neck: Effects of Interferon Gene Transfer. *Otolaryngol. Head Neck Surg.* **2016**, *120*, 665–671. [CrossRef] [PubMed]

55. Pennock, N.D.; White, J.T.; Cross, E.W.; Cheney, E.E.; Tamburini, B.A.; Kedl, R.M. T Cell Responses: Naïve to Memory and Everything in Between. *Adv. Physiol. Educ.* **2013**, *37*, 273–283. [CrossRef] [PubMed]

56. Eun, S.-Y.; Lee, S.-W.; Xu, Y.; Croft, M. 4-1BB Ligand Signaling to T Cells Limits T cell Activation. *J. Immunol.* **2014**, *194*, 134–141. [CrossRef] [PubMed]

57. Ho, W.T.; Pang, W.L.; Chong, S.M.; Castella, A.; Al-Salam, S.; Tan, T.E.; Moh, M.C.; Koh, L.K.; Gan, S.U.; Cheng, C.K.; et al. Expression of CD137 on Hodgkin and Reed-Sternberg Cells Inhibits T-cell Activation by Eliminating CD137 Ligand Expression. *Cancer Res.* **2013**, *73*, 652–661. [CrossRef]

58. Kwon, B. Is CD137 Ligand (CD137L) Signaling a Fine Tuner of Immune Responses? *Immune Netw.* **2015**, *15*, 121–124. [CrossRef]

59. The Cancer Genome Atlas Research Network. Integrated Genomic and Molecular Characterization of Cervical Cancer. *Nature* **2017**, *543*, 378–384. [CrossRef]

60. The Cancer Genome Atlas Research Network. Comprehensive Genomic Characterization of Head and Neck Squamous Cell Carcinomas. *Nature* **2015**, *517*, 576–582. [CrossRef]

61. Bratman, S.V.; Bruce, J.P.; O'Sullivan, B.; Pugh, T.J.; Xu, W.; Yip, K.W.; Liu, F.-F. Human Papillomavirus Genotype Association with Survival in Head and Neck Squamous Cell Carcinoma. *JAMA Oncol.* **2016**, *2*, 823–826. [CrossRef]

62. Banister, C.E.; Liu, C.; Pirisi, L.; Creek, K.E.; Buckhaults, P.J. Identification and Characterization of HPV-Independent Cervical Cancers. *Oncotarget* **2017**, *8*, 13375–13386. [CrossRef]

63. Faul, F.; Erdfelder, E.; Lang, A.-G.; Buchner, A. G*Power 3: A Flexible Statistical Power Analysis Program for the Social, Behavioral, and Biomedical Sciences. *Behav. Res. Methods* **2007**, *39*, 175–191. [CrossRef]

*Article*

# The Antiviral Agent Cidofovir Induces DNA Damage and Mitotic Catastrophe in HPV-Positive and -Negative Head and Neck Squamous Cell Carcinomas In Vitro


Femke Verhees [1,*], Dion Legemaate [2], Imke Demers [2], Robin Jacobs [2], Wisse Evert Haakma [2], Mat Rousch [2], Bernd Kremer [1] and Ernst Jan Speel [2]

[1] Department of Otorhinolaryngology, Head and Neck Surgery, GROW-school for Oncology and Development Biology, Maastricht University Medical Centre, PO Box 5800, 6202 AZ Maastricht, The Netherlands

[2] Department of Pathology, GROW-school for Oncology and Development Biology, Maastricht University Medical Centre, PO Box 5800, 6202 AZ Maastricht, The Netherlands

* Correspondence: femke.verhees@mumc.nl


Received: 29 May 2019; Accepted: 26 June 2019; Published: 30 June 2019


**Abstract:** Cidofovir (CDV) is an antiviral agent with antiproliferative properties. The aim of our study was to investigate the efficacy of CDV in HPV-positive and -negative head and neck squamous cell carcinoma (HNSCC) cell lines and whether it is caused by a difference in response to DNA damage. Upon CDV treatment of HNSCC and normal oral keratinocyte cell lines, we carried out MTT analysis (cell viability), flow cytometry (cell cycle analysis), (immuno) fluorescence and western blotting (DNA double strand breaks, DNA damage response, apoptosis and mitotic catastrophe). The growth of the cell lines was inhibited by CDV treatment and resulted in $\gamma$-H2AX accumulation and upregulation of DNA repair proteins. CDV did not activate apoptosis but induced S- and G2/M phase arrest. Phospho-Aurora Kinase immunostaining showed a decrease in the amount of mitoses but an increase in aberrant mitoses suggesting mitotic catastrophe. In conclusion, CDV inhibits cell growth in HPV-positive and -negative HNSCC cell lines and was more profound in the HPV-positive cell lines. CDV treated cells show accumulation of DNA DSBs and DNA damage response activation, but apoptosis does not seem to occur. Rather our data indicate the occurrence of mitotic catastrophe.

**Keywords:** human papillomavirus; head and neck cancer; double-stranded DNA breaks; DNA repair; cell line; cyclin B1; Aurora Kinase A


---

## 1. Introduction

Each year ~600,000 people worldwide are diagnosed with head and neck squamous cell carcinoma (HNSCC), making HNSCC the sixth most common cancer in the world [1]. Important risk factors for the development of HNSCC are alcohol consumption and/or smoking as well as high-risk human papillomavirus (HPV) infections. HPV-positive HNSCC is considered to be a distinct clinical and molecular entity in comparison to HPV-negative HNSCC [2]. The mortality rates have hardly decreased over the last decades and the five-year survival rate still ranges between 40–50%, even though improvements in detection and treatment have been achieved [3]. The HPV status of the tumor possesses powerful prognostic value, where HPV-positive patients have a more favorable prognosis [4,5]. There is an urgent need for new agents that can be integrated into or replace current treatment regimens to improve outcome and quality of life of HNSCC patients.

Cidofovir (CDV) is an acyclic nucleoside phosphonate which targets DNA viruses that encode for their own DNA polymerase, because the active diphosphate metabolite (CDVpp) has a higher affinity for viral DNA polymerase compared to cellular DNA polymerase. CDVpp competitively inhibits the incorporation of deoxycytidine triphosphate (dCTP) into viral DNA by viral DNA polymerase, which results in reduction in the rate of viral DNA synthesis [6,7]. Currently, CDV is approved by the Food and Drug Administration for intravenous administration in the therapy of cytomegalovirus retinitis in AIDS patients [8,9]. CDV is also used off-label for the treatment of infections caused by other DNA viruses, including papilloma- and polyomaviruses. In earlier studies, CDV has shown to have anti-proliferative properties against HPV-positive cervical carcinoma and HPV-negative transformed cell lines [10]. CDV has also been reported to be effective in a number of HPV-negative malignancies in vivo, such as glioblastoma and nasopharyngeal carcinoma [11,12]. The effects of CDV on HPV-positive induced benign and malignant proliferations should be linked to the antiproliferative effects of the compound as HPV uses the host DNA polymerase for replication [10,13]. Today, the molecular mechanisms underlying the effectivity of CDV are not completely understood. One hypothesis is that the selectivity of CDV for HPV-transformed cells is based on differences in replication rate, CDV incorporation into the cellular DNA, and in response to DNA damage caused by CDV [14]. The aim of our study was to investigate the in vitro efficacy of CDV in HPV-positive and -negative HNSCC cell lines and the normal oral keratinocyte (NOK) cell line, which is immortalized by the activation of hTERT [15], and whether this efficacy is caused by a difference in response to DNA damage.

## 2. Results

### 2.1. Effect of CDV Treatment on the Cell Viability of HNSCC and Uterine Cervical Carcinoma (UCC) Cell Lines

To determine the cell viability in the presence of CDV, all cell lines were cultured for 3, 6 and 9 days with increasing concentrations of CDV. CDV inhibited cell growth in the HPV-positive and -negative HNSCC-, the HPV-positive UCC- and the NOK cell lines as determined by the MTT assay. The anti-proliferative activity of CDV increased over time from day 3 to day 9 in all the cell lines tested. There was only a significant difference between the $IC_{50}$ of the HPV-positive HNSCC and UCC cell lines versus the HPV-negative HNSCC cell lines after 6 days of treatment ($p = 0.0102$). The $IC_{50}$ values of day 6 and 9 varied considerably between the different cell lines (Figure 1). We used the $IC_{50}$ of day 9 for further experiments.

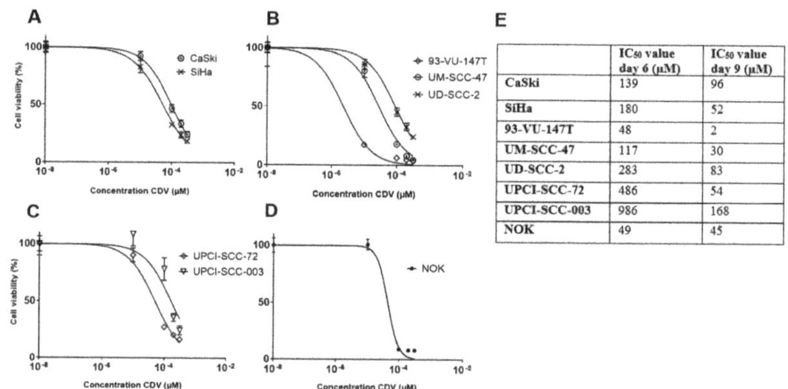

|  | $IC_{50}$ value day 6 ($\mu$M) | $IC_{50}$ value day 9 ($\mu$M) |
|---|---|---|
| CaSki | 139 | 96 |
| SiHa | 180 | 52 |
| 93-VU-147T | 48 | 2 |
| UM-SCC-47 | 117 | 30 |
| UD-SCC-2 | 283 | 83 |
| UPCI-SCC-72 | 486 | 54 |
| UPCI-SCC-003 | 986 | 168 |
| NOK | 49 | 45 |

**Figure 1.** Effect of CDV on cell viability. The viability of the used cell lines was assessed using an MTT assay. The $IC_{50}$ value is the drug dose that causes 50% growth inhibition. Showing the results of 9 days CDV treatment: (**A**) HPV-positive UCC cell lines, (**B**) HPV-positive HNSCC cell lines, (**C**) HPV-negative HNSCC cell lines, (**D**) NOK cell line, (**E**) Overview of $IC_{50}$ values after 6 and 9 days of treatment. The experiments were performed in triplicate.

## 2.2. CDV Treatment Results in DNA Damage

The HPV-positive cell lines 93-VU-147T and UM-SCC-47, HPV-negative cell line UPCI-SCC-72 and NOK were used to investigate DNA damage induction by CDV. The occurrence of DNA damage induction in the cell lines was confirmed by irradiation of 93-VU-147T, as there was an increase of γ-H2AX in the irradiated cells compared to the non-irradiated cells after both 4 and 24 h (Figure S1).

All four cell lines were treated for 3 and 6 days with CDV and processed for γ-H2AX immunofluorescence. Figure 2A illustrates representative nuclei of the untreated and treated cells of 93-VU-147T. γ-H2AX was visible after 3 days of CDV treatment and increased further after 6 days (Figure 2B). The increased expression of phospho-H2AX (p-H2AX) in CDV treated cells was also seen in western blot analyses (Figure 2C). Similar results were observed for UM-SCC-47 and UPCI-SCC-72. NOK showed in the control and treated cells accumulation of DNA damage. There was more upregulation of γ-H2AX in the cell lines with the highest anti-proliferative effects (93-VU-147T and UM-SCC-47), compared to the cell line with the lowest anti-proliferative effect (UPCI-SCC-72).

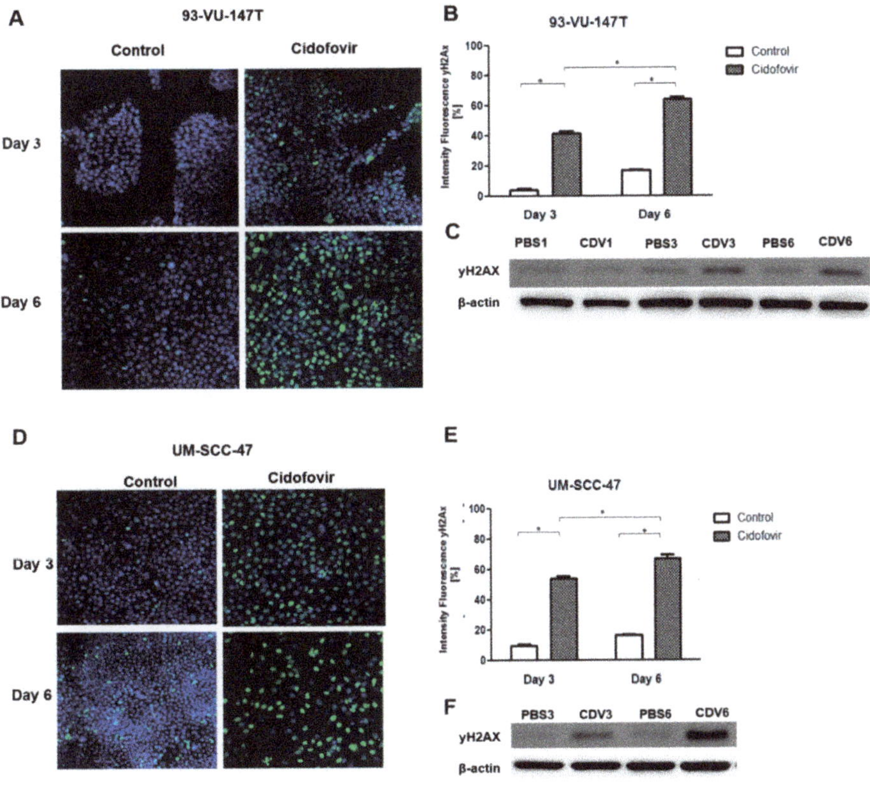


**Figure 2.** *Cont.*

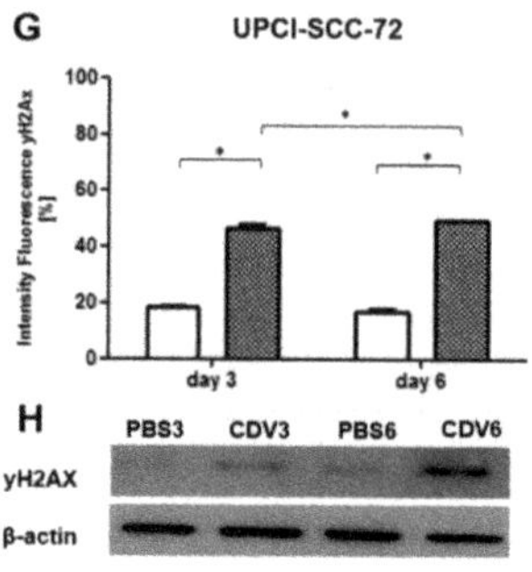
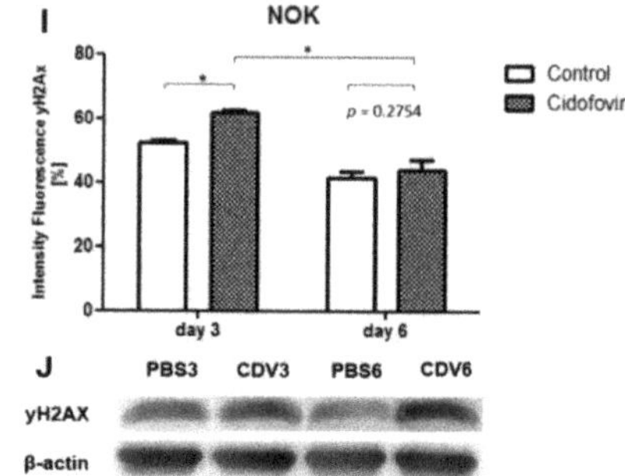



**Figure 2.** DNA damage induced by CDV as detected by γ-H2AX analysis. Cells were treated with CDV or PBS (control) and after 3 and 6 days immunostaining of γ-H2AX was performed. (**A**) DNA-damage is accumulated in the treated 93-VU-147T cells. Nuclei are stained with Hoechst in blue, DSBs are shown by γ-H2AX in green. (**B**) Quantification of γ-H2AX positive cells after 3 and 6 days CDV treatment. (**C**) Cell lysates of 93-VU-147T were examined by western blotting with p-H2AX after 3 and 6 days. β-actin was used as loading control. (**D**) DNA damage is accumulated in treated UM-SCC-47 cells. (**E,F**) Quantification of y-H2AX positive cells after 3 and 6 days CDV treatment and western blotting analysis of p-H2AX for UM-SCC-47, (**G,H**) UPCI-SCC-72, (**I,J**) and NOK. Statistical significance was indicated as follows: $p < 0.05$ (*). The experiments were performed in triplicate.

## 2.3. Activation of DNA Damage Response by CDV

Since increased γ-H2AX expression upon CDV treatment suggests accumulation of DNA double strand breaks (DNA DSBs), the DNA damage response pathway was investigated at protein level. In response to DNA damage, cells normally activate the DNA damage response pathway, which causes G1/S arrest via the p53 pathway and G2/M arrest via checkpoint kinases Chk1 and Chk2. We performed both western blotting of DNA damage response proteins and p53 mutation analysis on the cell lines. In 93-VU-147T, starting from day 3 a strongly increased expression of the phosphorylated checkpoint kinases Chk1 (p-Chk1) and Chk2 (p-Chk2), phosphorylated BRCA1 (p-BRCA1) and a moderately increased expression of phosphorylated p53 at ser15 (ser15p53) was observed upon CDV treatment compared to the control. In addition, cdc2 was phosphorylated at Tyr15 (p-cdc2), which is one of the two inhibition sites for the activation of the cdc2-cyclin B complex. P53 and p21 were upregulated in the treated and untreated cells (Figure 3A). This may be explained by presence of both wild type and mutant TP53 (L275R; allelic frequency (AF) 51%) in this cell line. In UM-SCC-47 the upregulation of the pathway appeared at day 6. In this cell line, there is only an upregulation of p53 and p21 in the CDV treated cells (Figure 3B). This cell line proved to harbor wild type TP53, which is down regulated by HPV oncoprotein E6. In the two HPV-positive cell lines, there was still a significant amount of DNA damage visible in the treated cells after 6 days. Analysis of UPCI-SCC-72 and NOK showed lower expression levels of the DNA damage response proteins in comparison to UM-SCC-47 and 93-VU-147T. UPCI-SCC-72 showed an upregulation of p-Chk1, p-Chk2 and ser15p53 after 6 days. p53, p-BRCA1 and p-cdc2 were detected at similar levels in the treated and untreated cells, and p21 showed lower expression levels in CDV treated cells (Figure 3C). This cell line harbors a pathogenic TP53 mutation (H179N; AF 100%), which is in agreement with earlier observations [16]. NOK showed upregulation of p-Chk1, p-Chk2, ser15p53 and p-cdc2. p53 and p-BRCA1 were detected at similar levels in the treated and untreated cells, and p21 showed reduced expression in CDV treated cells (Figure 3D). This cell line has both wild type and mutant TP53 (R213Ter; AF 39%).

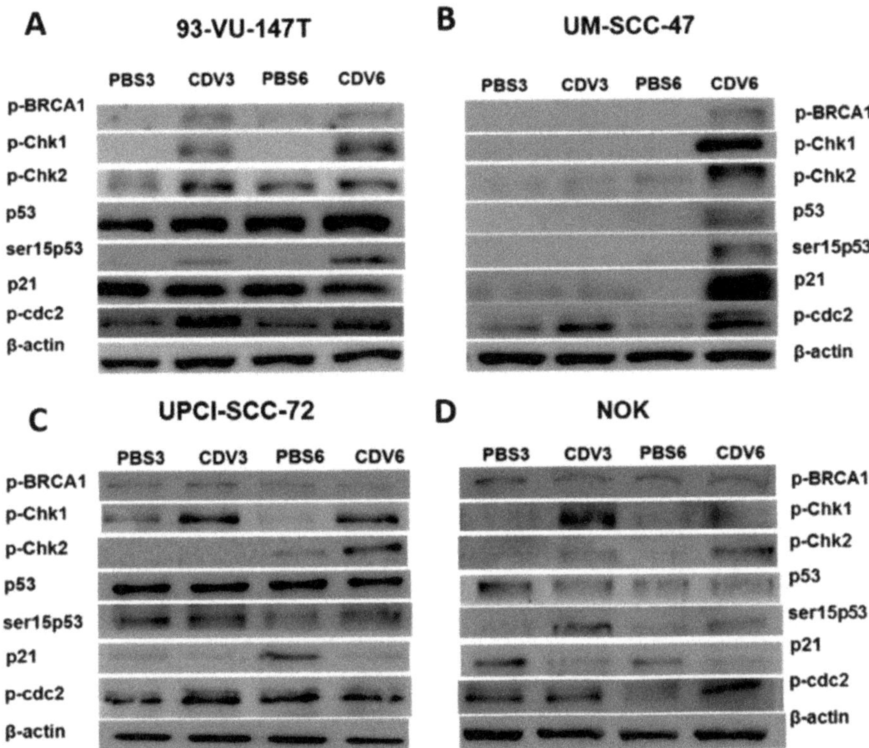

**Figure 3.** Expression levels of proteins involved in the DNA damage response pathway by western blot analysis of whole protein extracts. The cells were treated for 3 and 6 days with the IC$_{50}$ value of CDV or control (PBS). β-actin was used as loading control. For the cell lines (**A**) 93-VU-147T and (**B**) UM-SCC-47 protein extracts of 10µg were used, where for (**C**) UPCI-SCC-72 and (**D**) NOK protein extracts of 30µg were used. The experiments were performed in triplicate.

## 2.4. CDV Treatment Results in Mitotic Catastrophe

A consequence of the activation of the DNA damage response pathway may be cell cycle arrest followed by apoptosis. For this purpose, we first analyzed the cell cycle distribution by Flow Cytometry analysis after 3 and 6 days of CDV treatment. In the four cell lines there was a decrease of cells in the G1 phase and an increase of cells in the S-phase compared to the control. Furthermore, in the UM-SCC-47, UPCI-SCC-72 and NOK also after 6 days an increase in cells in the G2/M phase was observed. These results indicate that under CDV treatment cells accumulate in S- and G2/M-phase (Figure 4).

This was further confirmed by cyclin B1 immunostaining in CDV treated cell lines, showing an increase in intensity as well as the number of cyclin B1 positive cells after 6 days of CDV treatment (Figure 5). The most significant increase of cells in the G2/M phase after 6 days was seen for UM-SCC-47 and NOK. These cell lines showed also the most significant increase in cyclin B1 intensity after 6 days treatment.

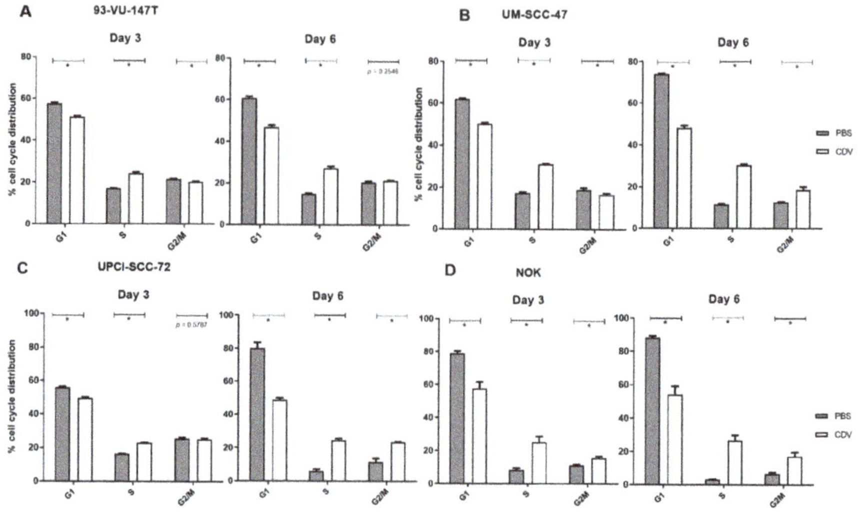

**Figure 4.** Cell cycle distribution of the HNSCC cell lines and NOK treated for 3 and 6 days with CDV or not treated (PBS). (**A**) 93-VU-147T, (**B**) UM-SCC-47, (**C**) UPCI-SCC-72, (**D**) NOK. Statistical significance was indicated as follows: $p < 0.05$ (*). The experiments were performed in triplicate.

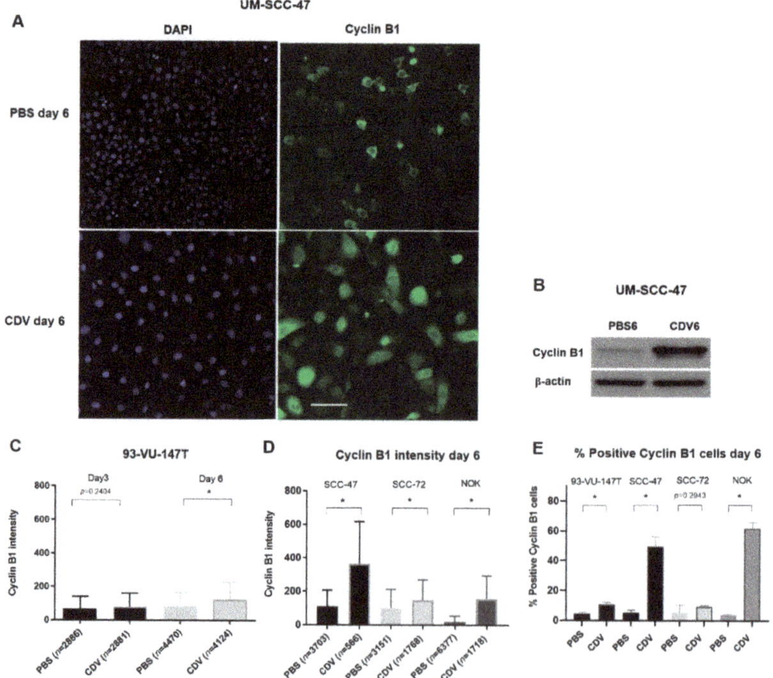

**Figure 5.** Upregulation of cyclin B1 expression in the nucleus after treatment of cell lines with CDV. The cells were treated for 3 and 6 days with the $IC_{50}$ value of CDV followed by cyclin B1 immunofluorescence staining. Nuclei were considered positive if the intensity was higher than the average intensity plus two times standard deviation of the negative control. (**A**) Representative images of cyclin B1 immunofluorescence (right side) of the HPV-positive UM-SCC-47 cell line after 6 days CDV

treatment vs. PBS control, left side showing blue nuclear DAPI staining. (**B**) Cell lysates of UM-SCC-47 were examined by western blotting of cyclin B1 after 6 days. β-actin was used as loading control. (**C**) cyclin B1 intensity of 93-VU-147T after 3 and 6 days of treatment. (**D**) cyclin B1 intensity of UM-SCC-47, UPCI-SCC-72 and NOK after 6 days of treatment. (**E**) % positive cyclin B1 cells of 93-VU-147T, UM-SCC-47, UPCI-SCC-72 and NOK after 6 days treatment. $n$ = number of analyzed cells. Statistical significance was indicated as follows: $p < 0.05$ (*). The experiments were performed in triplicate. Scale bar of (**A**): 100 μm.

In order to assess if cells go into apoptosis under CDV treatment, we performed an Annexin-V assay. First, all cell lines were treated with 1 μM Staurosporine for 1 day, a known inducer of apoptosis. In the three HNSCC cell lines there was a strong increase of apoptotic cells observed, whereas only a slight increase was observed in the NOK cell line. In contrast, after CDV treatment there was no increase in apoptotic cells observed in the HNSCC cell lines, except for the 93-VU-147T, showing a significant increase of apoptotic cells after CDV treatment, but this was an increase of 2.7%. The NOK cell line showed a strong increase in apoptotic cells. Taken together, CDV induced apoptosis in the NOK cell line, but not in the HNSCC cell lines (Figure 6).

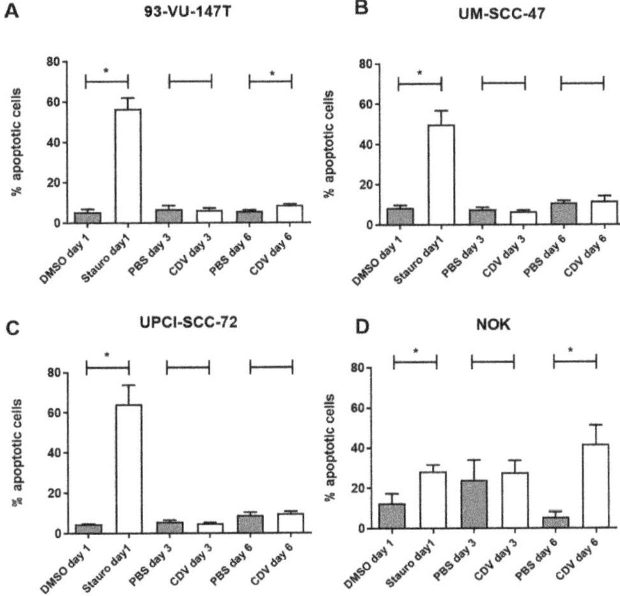

**Figure 6.** Effect of CDV treatment on induction of apoptosis. Cells were either treated for 1 day with 1μM Staurosporine, a known inducer of apoptosis or for 3 and 6 days with CDV, followed by analysis of Annexin V staining. Results are shown for (**A**) 93-VU-147T, (**B**) UM-SCC-47, (**C**) UPCI-SCC-72 and (**D**) NOK. Statistical significance was indicated as follows: $p < 0.05$ (*). The experiments were performed in triplicate.

Cyclin B1 accumulation in the nucleus indicates that a part of the cells enter mitosis and with an inactive apoptosis machinery, this may lead to mitotic catastrophe. To visualize this process, we used immunofluorescence detection of phospho-Aurora Kinase, which is detected at the centrosomes along mitotic spindle microtubules and plays a role in the mitotic chromatid segregation. The first observation in these experiments were an increase in cell nuclei size after CDV treatment in comparison with the control cells (Figure S2). CDV treated cells showed a decrease in number of mitotic figures and an increase in cells in mitotic catastrophe (Figure 7). NOK showed a slight increase in mitoses after treatment with CDV instead of a decrease, but also an increase in mitotic catastrophe. Because so far, the cell lines were treated with CDV concentrations resulting in equal toxicity ($IC_{50}$ value), we also wanted to investigate if mitotic catastrophes could explain the differences in sensitivity. Indeed, Figure 7I shows that more mitotic catastrophes were observed with increasing sensitivity for CDV.

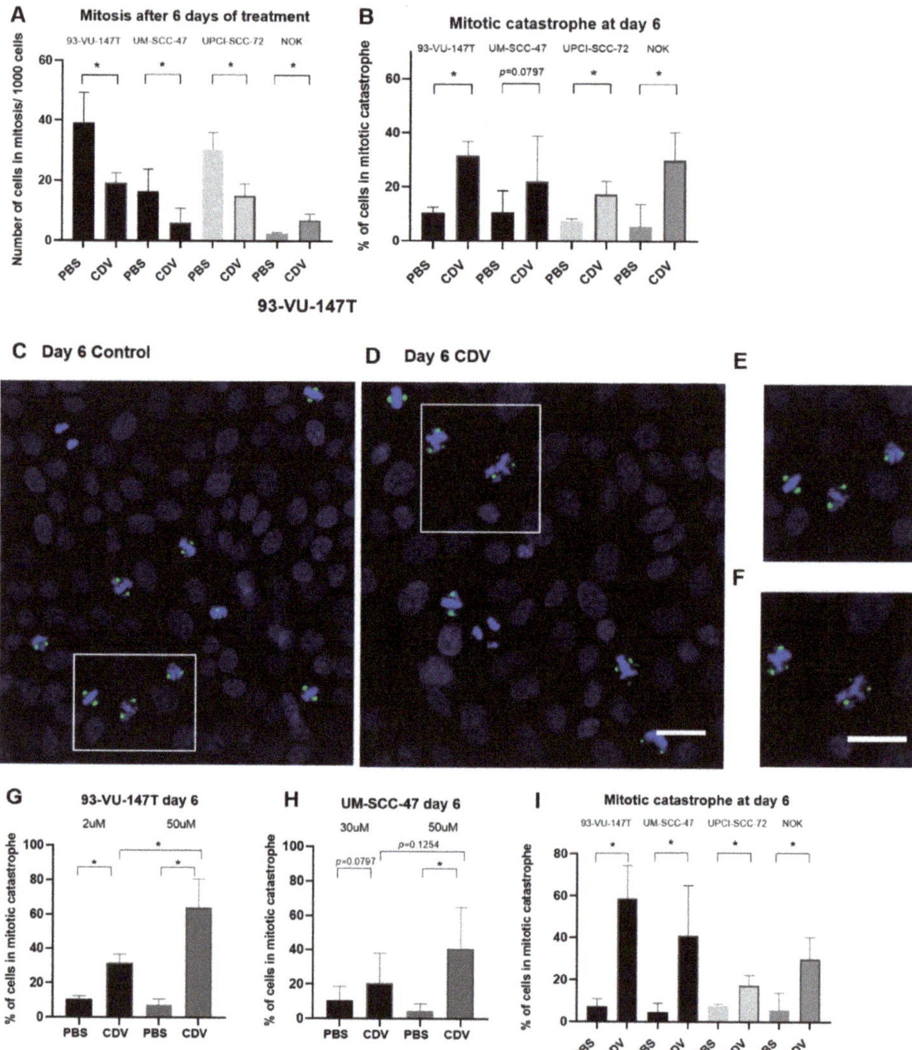

**Figure 7.** Induction of mitosis and mitotic catastrophe after treatment with CDV. The cells were treated with CDV or PBS for 3 and 6 days after which immunostaining of phospho-Aurora Kinase was performed. The cells were treated with an equal toxicity ($IC_{50}$) and with the same CDV concentration (50 μM). (**A**) The number of cells in mitosis (2 centrosomes) per 1000 counted cells and (**B**) percentage of cells in mitosis undergoing mitotic catastrophe when treated with PBS or CDV ($IC_{50}$). (**C**) Representative nuclei of 93-VU-147T untreated and (**D**) treated with CDV for 6 days. (**E**) Magnification of a normal mitotic figures and (**F**) 2 nuclei in mitotic catastrophe with multiple spindles visible (**G**) 93-VU-147T and (**H**) UM-SCC-47 cell line treated with $IC_{50}$ vs. 50 μM. (**I**) Percentage of control and treated cells in mitotic catastrophe when treated with 50 μM. Statistical significance was indicated as follows: $p < 0.05$ (*). The experiments were performed in triplicate. Scale bar of (**C–F**): 50 μm.

## 3. Discussion

The antiproliferative effects of CDV were studied in three HPV-positive, two HPV-negative HNSCC cell lines, two HPV-positive UCC cell lines and the immortalized NOK cell line. In all the cell lines the cell growth was inhibited by CDV with differences in response between the cell lines. Treatment with CDV caused DNA damage by means of DNA DSBs and as a result the DNA damage response pathway became activated. There was an accumulation of cells in the S- and G2/M phase and with an inappropriate apoptosis machinery, the cells appeared to undergo mitotic catastrophe.

CDV targets DNA viruses that encode for their own DNA polymerase. In addition, CDV has been shown to have antiproliferative properties against HPV-positive and HPV-negative malignancies in vitro and vivo [10–12]. The molecular mechanism underlying the efficacy of CDV is not completely understood, as HPV uses the host DNA polymerase for replication [10,13]. The aim of our study was to investigate the efficacy of CDV in HPV-positive and -negative HNSCC cell lines in vitro and whether this efficacy is caused by a difference in response to DNA damage. Our results show that CDV inhibits the cell growth of all the HPV-positive and -negative HNSCC, the UCC cell lines and the NOK cell line, and is more effective in the HPV-positive cell lines than in the HPV-negative cell lines after 6 days. Treatment with CDV caused DNA damage by means of DNA DSB's. There was more DNA damage visible in the two HPV-positive cell lines showing the strongest inhibition as compared to the HPV-negative cell line showing much less inhibition by CDV. The $IC_{50}$ values of the cell lines SiHa, CaSki, UM-SCC-47 and UD-SCC-2 were in accordance to those found by Mertens et al. [17]. They reported that CDV incorporation into DNA caused DNA damage, but there was no correlation between the occurrence of DNA damage and the anti-proliferative effects of CDV.

In order to further investigate the mechanism of action of CDV, we examined the activation of the DNA damage response pathway, the cell cycle and the induction of apoptosis. After treatment with CDV, the DNA damage response pathway became activated by means of phosphorylation of the DNA repair proteins (BRCA-1, Chk-1, Chk-2 and p53) in the two HPV-positive HNSCC cell lines. This effect was seen to a lesser extent in the HPV-negative cell line and NOK cell line. In the HPV-positive cell lines only a slight upregulation of phosphorylated p53 would be expected, because of inactivation by E6, which in turn is not influenced by CDV [14,18]. This was observed in UM-SCC-47. The higher expression of p53 in 93-VU-147T might be the consequence of a TP53 mutation in one allele.

We found a S-phase arrest after 3 and 6 days CDV treatment and after 6 days there was also a G2/M arrest visible. The expression of cyclin B1 in the nucleus after treatment with CDV was also increased after 6 days. Additionally, the phosphorylation of cdc-2 on Tyr15 increased, also suggesting G2/M arrest. However, there was still a significant amount of DNA damage visible in the treated cells after 6 days, which implies that DNA repair does not occur efficiently in the HPV-positive cell lines. Similar results were found in HPV-positive UCC cells (SiHa, HeLa) by De Schutter et al. [14]. They found that these tumor cells lacked appropriate cell cycle regulation and DNA repair as did the immortalized keratinocyte cell line (HaCaT). Earlier studies have also indicated that an impaired DNA damage repair is responsible for the elevated radiosensitivity of HPV-positive tumor cells [19,20]. An explanation for this observation might be that the expression of HPV E6 and E7 in cells hinder the homologous recombination pathway through the mislocalization of Rad51 away from the DSBs through a yet unknown mechanism [21].

We noted that CDV treatment did not lead to an increase in Annexin-V staining. Abdulkarim et al. also did not detect apoptosis after CDV treatment in HPV-positive UCC and HNSCC cells and proposed cell cycle arrest to occur [22]. These results are in agreement with studies inducing DNA damage by radiotherapy in HNSCC cell lines, which also showed no occurrence of apoptosis [19,23].

Immunofluorescence of phospho-Aurora Kinase revealed nuclei increased in size and the presence of multiple centrosomes in CDV treated cells. Combined with the suggested G2/M arrest, this finding indicates the development of mitotic catastrophe being the predominant cause leading to cell death. Indeed, more mitotic catastrophes were observed with increasing sensitivity for CDV. Radiation as well as various antitumor drugs have been described to induce mitotic catastrophe [24–26]. Progression from G2- to M-phase is driven by the activation of the cyclin B1/cdc2 complex. Aberrant mitotic entry before the completion of DNA replication can cause mitotic catastrophe and is associated with multinuclear enlarged cells and multipolar spindles [27]. Upregulation of cyclin B1 and prolonged activation of cyclin B1/cdc2 complex are typical features of mitotic catastrophe [28].

In contrast to the HNSCC cell lines that do not show an evident increase in apoptosis due to DNA damage caused by CDV, already substantial apoptosis was detectable at baseline in the NOK cell line which increased under CDV treatment. Assuming that NOK cells contain a least one wild-type allele of TP53, one would expect less DNA damage at baseline and induction of apoptosis under CDV treatment because of functional p53. An alternative explanation of the observed results could be that this cell line is polyclonal, with subclones having homozygous wild-type TP53 or homozygous mutated TP53. This would explain the baseline DNA damage (in the mutated p53 cells) and detection of apoptosis under CDV treatment (occurring in the wild-type p53 cells). Hence, the question is whether or not the NOK cell line is a good normal keratinocyte control. Rather, the observed features, including the presence of a TP53 mutation, more resemble features seen in the HNSCC cell lines. The fact that normal keratinocytes cell lines that are not immortalized do not show DNA damage after CDV treatment, as has been reported by Mertens et al., further underscores this suggestion [17].

In conclusion, we found that CDV inhibits the cell growth of HPV-positive and -negative HNSCC cell lines, and was more profound in HPV-positive cell lines. CDV treated cells showed accumulation of DNA DSBs and DNA damage activation, but apoptosis did not seem to occur. Rather our data indicate the occurrence of mitotic catastrophe.

## 4. Materials and Methods

### 4.1. Cell Lines and Culture Conditions

Three HPV16-positive head and neck squamous cell carcinoma (HNSCC) cell lines: UD-SCC-2 (from Thomas Hoffmann, University of Ulm, Germany), 93-VU-147T (Johan. P. De Winter, VU Medical Center, the Netherlands), and UM-SCC-47 (Thomas E. Carey, University of Michigan, Ann Arbor, MI, USA) were used. Two HPV16-negative HNSCC cell lines: UPCI-SCC-72 and UPCI-SCC-003 (both from Susanne M. Collins, University of Pittsburgh, Pittsburgh, PA, USA) were used. Two HPV16-positive uterine cervical carcinoma cell lines, SiHa and CaSki, were purchased from the American Type Culture Collection (ATCC). The normal oral keratinocyte (NOK) cell line (Karl Munger, Tufts University Medical School, Boston, MA, USA), which is immortalized by activation of h-TERT [15] is a cell line prepared from gingival tissues obtained from oral surgeries [29] as described previously [30].

Cells were cultured at 37 °C in a humidified atmosphere with 5% $CO_2$. All HNSCC cell lines used in this study were cultured in Dulbecco's modified Eagle medium (DMEM) containing 10% fetal calf serum (FCS). CaSki was cultured in Roswell Park Memorial Institute (RPMI) with 10% FCS. SiHa was cultured in Minimum Essential Medium (MEM) with 10% FCS, supplemented with L-glutamine and non-essential amino acids. The NOK cell line was cultured in keratinocyte serum-free medium (KSFM) supplemented with (2.6 μg/mL) bovine pituitary extract (BPE) and (0.16 ng/mL) recombinant epidermal growth factor (rEGF). All the cell lines were regularly tested and found to be mycoplasma-free. All cell lines were confirmed to have unique genotypes, as tested using the ProfilerPlus assay [18]. The presence of HPV DNA was detected by PCR using the consensus primer set GP51/61 [31].

### 4.2. In Vitro Cell Proliferation Assay

Cells were seeded in 96-well flat bottom plates at densities that allowed exponential growth for the duration of the experiment. They were placed in the cell culture incubator overnight at 37 °C allowing the cells to attach, after which they were treated with concentrations of Cidofovir (Vistide, Gilead Sciences Inc, Foster City, CA USA) of 10, 100, 200 and 300 μM or PBS (control). At indicated time points post-treatment (3, 6 and 9 days), the MTT ((3-(4,5-dimethylthiazol-2-yl)-2,5-diphenyl tetrazolium bromide) assay (Sigma-Aldrich, Saint Louis, MO, USA) was performed as previously described [32]. The experiments were performed in triplicate.

### 4.3. Irradiation

The cells were irradiated at room temperature with 4 Gray (Gy). After 4 and 24 h of incubation the irradiated cells and the no irradiated control cells were fixed with methanol for 15 min at −20 °C and analyzed for γ-H2AX expression by immunofluorescence (see below).

### 4.4. Cell Cycle Analysis

Cells were seeded in T25 culture flasks and placed in the cell culture incubator at 37 °C and allowed to attach overnight. Culture medium was added containing CDV ($IC_{50}$) or PBS. After 3 and 6 days, cells were washed with PBS and trypsinized to form a cell pellet. Ice-cold 70% ethanol was added to the cell pellet while vortexing, assuring fixation of the cells and minimizing cell clumping. Cells in 70% ethanol were stored at −20 °C for a minimal duration of 30 min. Cells were washed with PBS and resuspended in 0.5 mL propidium iodide(PI)/RNAse staining solution (100 μg/mL PI and 1 mg/mL RNAse in PBS). Cells were incubated on ice for 30 min and analyzed by flow cytometry using a FACScanto II (BD Biosciences, San Jose, CA, USA). Data analysis was performed using FACSdiva software (BD Biosciences). The different cell cycle regions were set to those defined by the untreated control cells for each cell line individually.

### 4.5. Apoptosis Assay

As a positive control for apoptosis, the cells were treated with 1 μM Staurosporine (Sigma-Aldrich). For the Annexin-V assay cells were seeded in 96-wells plates and allowed to attach overnight at 37 °C. Cells were treated with CDV ($IC_{50}$) or PBS for 3 and 6 days. Cells were stained with Hoechst 33,342 (200 μg/mL, Sigma-Aldrich) in culture medium for 15 min at 37 °C. Cells were washed with Annexin-V binding buffer (10 mM HEPES, 140 mM NaCl, 5 mM $CaCl_2$ in PBS) and stained with Annexin-V-FITC (2.5 μg/mL in Annexin-V binding buffer) for 15 min at 37 °C. Staining intensities of cells were measured in High-Content Imaging. Data was acquired using a BDpathway855 High-Content Bioimager (BD Biosciences). Digitalization and segmentation of acquired data was done with Attovision software (BD Biosciences). Processed data was evaluated by DIVAsoftware (BD Biosciences).

### 4.6. Immunofluorescence Staining of γ-H2AX, Cyclin B1 and Phospho-Aurora Kinase A/B/C

Cells were grown in 96-well plates (γ-H2AX) or on coverslips (cyclin B1 and phospho-Aurora Kinase A/B/C) and allowed to attach overnight at 37 °C. Culture medium containing CDV ($IC_{50}$) or PBS was added, and cells were incubated at 37 °C. After 3 and 6 days, cells were washed with PBS followed by fixation in CytoRich Red for 20 min at RT (γ-H2AX) or methanol for 15 min at −20 °C (cyclin B1 and phospho-Aurora Kinase A/B/C). After washing with PBS, the cells were permeabilized with 0.1% Triton in TBS/T (0.1% Tween20 in TBS) for 20 min and then blocked with 5% bovine serum albumin (BSA) in TBS/T for 30 min at RT. Cells were incubated with the primary antibody (Table S1) diluted in blocking buffer overnight at 4 °C. After washing with TBS/T, the cells were incubated with a fluorescent-labeled secondary antibody directed against the primary antibody (Table S1).

For the quantification of γ-H2AX expression after CDV treatment, cells were stained with (200 μg/mL) Hoechst 33,342 for 10 min at 37 °C. Staining intensities of cells were measured in High-Content Imaging. Data was acquired using a BDpathway855 High-Content Bioimager (BD Biosciences). Digitalization and segmentation of acquired data was done with Attovision software (BD Biosciences). Processed data was evaluated by DIVAsoftware (BD Biosciences).

For cyclin B1, phospho-Aurora Kinase A/B/C, and for γ-H2AX expression in the radiotherapy experiment, nuclear morphology was visualized with 4′6-diadomidino-2-phenylindole (DAPI). Cell images were obtained using a Leica DM5000B microscope (Leica Microsystems, Wetzlar, Germany) with filters for DAPI and fluorescein and Leica Qwin Software (Leica Microsystems). For further analysis of cyclin B1 and phospho-Aurora Kinase A/B/C, Cell Profiler image analysis software (Carpenter Lab, Cambridge, CA, USA) was used [33].

For cyclin B1 and $\gamma$-H2AX analysis, the 'IdentifyPrimaryObjects' module has been run on the DAPI image to identify the cell nuclei and 'MeasureObjectSizeShape' to determine the nucleus diameter. This was followed by the 'MeasureObjectIntensity' to measure the antibody intensity inside the nuclei. The intensity in each nucleus was normalized to the fluorescence background intensity measured in a cell-free area of the image. Nuclei were considered positive if the intensity was higher than the average intensity plus two times standard deviation of the negative control. Phospho-Aurora Kinase A/B/C was analyzed using the 'IdentifyPrimaryObjects' and 'MeasureObjectSizeShape' module. Mitosis and mitotic catastrophes were counted manually.

### 4.7. Western Blot

Cells treated with CDV or PBS were lysed by incubation with RIPA buffer (Cell Signaling, Danvers, MA, USA) containing Protease/Phosphatase Inhibitor Cocktail for 5 min on ice, followed by brief sonication. After centrifugation, the pellet was discarded and the protein extracts were quantified using the Pierce BCA Protein Assay Kit (Thermo Fisher Scientific, Waltham, CA, USA) as per manufacturers' instructions. Equal amounts of the extracts (10 μg for UM-SCC-47 and 93-VU-147T versus 30 μg for UPCI-SCC-72 and NOK) were separated on 8–12% SDS-PAGE and electrotransferred to nitrocellulose membranes according to the manufacturers' instructions using Mini-Protean Tetra System (Bio-Rad, Hercules, CA, USA). Membranes were blocked with non-fat dry milk (NFDM) and incubated with primary antibodies diluted in blocking buffer (5% NFDM or BSA diluted in TBS). For detection, secondary antibodies labeled with Horseradish Peroxidase (HRP) (Dako Agilent, Santa Clara, CA, USA and Cellsignaling) were incubated on membranes during 1 h at RT. Bands were visualized with enhanced chemiluminescence (SuperSignal West Dura Extended Duration Substrate, Thermo Scientific) on the Image reader LAS-3000 (Fuji Film, Minato, Japan).

### 4.8. P53 Mutation Analysis

DNA was extracted using Maxwell FFPE LEV Automated DNA Extraction Kit (Promega Corporation, Madison, WI, USA). DNA concentration was measured using the QuantiFluor dsDNA Dye System (Promega Corporation) [34]. DNA was examined using single molecule molecular inversion probes (smMIP) analysis, as previously described [35]. A smMIP-based library preparation was used to target coding sequences of the TP53 gene; NN_000546 exon 2-11.

### 4.9. Statistical Analysis

GraphPad Prism (version 6, San Diego, CA, USA) was used to conduct all statistical analyses. All results were expressed as the mean ± standard error of the mean. Independent experiments were analyzed by an unpaired Student's $t$-test. Levels of $p < 0.05$ were considered to be of statistical significance.

## 5. Conclusions

CDV inhibits the cell growth of HPV-positive and -negative HNSCC cell lines and was more profound in the HPV-positive cell lines. CDV treated cells showed accumulation of DNA DSBs and DNA damage response activation, but apoptosis did not occur. Instead, our data indicate the occurrence of mitotic catastrophe.

**Supplementary Materials:** The following are available online at http://www.mdpi.com/2072-6694/11/7/919/s1, Table S1: Primary and secondary antibodies used for western blotting and immunofluorescence. Figure S1: The occurrence of DNA-damage in 93-VU-147T treated with irradiation in vitro, Figure S2: Effect of CDV treatment on the cell nucleus diameter.

**Author Contributions:** Conceptualization, F.V. and E.J.S.; Methodology, F.V. and E.J.S.; Validation, F.V., D.L., W.E.H., R.J. and M.R.; Formal analysis, F.V. and D.L.; Investigation, F.V., D.L., W.E.H., R.J., I.D. and M.R.; Resources,

E.J.S., B.K.; Data curation, F.V.; Writing—original draft preparation, F.V.; Writing—review and editing, E.J.S., D.L., W.E.H., R.J., M.R. and B.K. Visualization, F.V., D.L., R.J., W.E.H., I.D. and M.R.; Supervision, E.J.S., B.K.; Project administration, F.V., E.J.S.; Writing—review & editing, I.D.

**Funding:** This research received no external funding

**Acknowledgments:** We would like to thank M. Henfling, P. Delvoux, I. Cornet, G. Kok and P. Wolber for their technical assistance and K. Rouschop for assistance with the control radiotherapy experiments. We would like to thank A. Hoeben for helpful discussion.

**Conflicts of Interest:** Ernst-Jan Speel has received research grants from Pfizer and Novartis. The other authors disclose no potential conflicts of interest.

# References

1. Ferlay, J.; Shin, H.R.; Bray, F.; Forman, D.; Mathers, C.; Parkin, D.M. Estimates of worldwide burden of cancer in 2008: GLOBOCAN 2008. *Int. J. Cancer* **2010**, *127*, 2893–2917. [CrossRef] [PubMed]
2. Stransky, N.; Egloff, A.M.; Tward, A.D.; Kostic, A.D.; Cibulskis, K.; Sivachenko, A.; Kryukov, G.V.; Lawrence, M.S.; Sougnez, C.; McKenna, A.; et al. The mutational landscape of head and neck squamous cell carcinoma. *Science* **2011**, *333*, 1157–1160. [CrossRef] [PubMed]
3. Jemal, A.; Siegel, R.; Ward, E.; Hao, Y.; Xu, J.; Murray, T.; Thun, M.J. Cancer statistics, 2008. *CA Cancer J. Clin.* **2008**, *58*, 71–96. [CrossRef] [PubMed]
4. Ang, K.K.; Harris, J.; Wheeler, R.; Weber, R.; Rosenthal, D.I.; Nguyen-Tan, P.F.; Westra, W.H.; Chung, C.H.; Jordan, R.C.; Lu, C.; et al. Human papillomavirus and survival of patients with oropharyngeal cancer. *N. Engl. J. Med.* **2010**, *363*, 24–35. [CrossRef] [PubMed]
5. Olthof, N.C.; Straetmans, J.M.; Snoeck, R.; Ramaekers, F.C.; Kremer, B.; Speel, E.J. Next-generation treatment strategies for human papillomavirus-related head and neck squamous cell carcinoma: Where do we go? *Rev. Med. Virol.* **2012**, *22*, 88–105. [CrossRef] [PubMed]
6. De Clercq, E.; Holy, A. Acyclic nucleoside phosphonates: A key class of antiviral drugs. *Nat. Rev. Drug Discov.* **2005**, *4*, 928–940. [CrossRef] [PubMed]
7. Lassen, P.; Eriksen, J.G.; Krogdahl, A.; Therkildsen, M.H.; Ulhoi, B.P.; Overgaard, M.; Specht, L.; Andersen, E.; Johansen, J.; Andersen, L.J.; et al. The influence of HPV-associated p16-expression on accelerated fractionated radiotherapy in head and neck cancer: Evaluation of the randomised DAHANCA 6&7 trial. *Radiother. Oncol.* **2011**, *100*, 49–55. [CrossRef]
8. Plosker, G.L.; Noble, S. Cidofovir: A review of its use in cytomegalovirus retinitis in patients with AIDS. *Drugs* **1999**, *58*, 325–345. [CrossRef]
9. Donne, A.J.; Rothera, M.P.; Homer, J.J. Scientific and clinical aspects of the use of cidofovir in recurrent respiratory papillomatosis. *Int. J. Pediatr. Otorhinolaryngol.* **2008**, *72*, 939–944. [CrossRef]
10. Andrei, G.; Snoeck, R.; Piette, J.; Delvenne, P.; De Clercq, E. Antiproliferative effects of acyclic nucleoside phosphonates on human papillomavirus (HPV)-harboring cell lines compared with HPV-negative cell lines. *Oncol. Res.* **1998**, *10*, 523–531.
11. Hadaczek, P.; Ozawa, T.; Soroceanu, L.; Yoshida, Y.; Matlaf, L.; Singer, E.; Fiallos, E.; James, C.D.; Cobbs, C.S. Cidofovir: A novel antitumor agent for glioblastoma. *Clin. Cancer Res.* **2013**, *19*, 6473–6483. [CrossRef] [PubMed]
12. Murono, S.; Raab-Traub, N.; Pagano, J.S. Prevention and inhibition of nasopharyngeal carcinoma growth by antiviral phosphonated nucleoside analogs. *Cancer Res.* **2001**, *61*, 7875–7877. [PubMed]
13. Van Cutsem, E.; Snoeck, R.; Van Ranst, M.; Fiten, P.; Opdenakker, G.; Geboes, K.; Janssens, J.; Rutgeerts, P.; Vantrappen, G.; de Clercq, E.; et al. Successful treatment of a squamous papilloma of the hypopharynx-esophagus by local injections of (S)-1-(3-hydroxy-2-phosphonylmethoxypropyl)cytosine. *J. Med. Virol.* **1995**, *45*, 230–235. [CrossRef] [PubMed]
14. De Schutter, T.; Andrei, G.; Topalis, D.; Naesens, L.; Snoeck, R. Cidofovir selectivity is based on the different response of normal and cancer cells to DNA damage. *BMC Med. Genom.* **2013**, *6*, 18. [CrossRef] [PubMed]
15. Piboonniyom, S.O.; Duensing, S.; Swilling, N.W.; Hasskarl, J.; Hinds, P.W.; Munger, K. Abrogation of the retinoblastoma tumor suppressor checkpoint during keratinocyte immortalization is not sufficient for induction of centrosome-mediated genomic instability. *Cancer Res.* **2003**, *63*, 476–483. [PubMed]

16. Lin, C.J.; Grandis, J.R.; Carey, T.E.; Gollin, S.M.; Whiteside, T.L.; Koch, W.M.; Ferris, R.L.; Lai, S.Y. Head and neck squamous cell carcinoma cell lines: Established models and rationale for selection. *Head Neck* **2007**, *29*, 163–188. [CrossRef] [PubMed]

17. Mertens, B.; Nogueira, T.; Stranska, R.; Naesens, L.; Andrei, G.; Snoeck, R. Cidofovir is active against human papillomavirus positive and negative head and neck and cervical tumor cells by causing DNA damage as one of its working mechanisms. *Oncotarget* **2016**, *7*, 47302–47318. [CrossRef] [PubMed]

18. Olthof, N.C.; Huebbers, C.U.; Kolligs, J.; Henfling, M.; Ramaekers, F.C.; Cornet, I.; van Lent-Albrechts, J.A.; Stegmann, A.P.; Silling, S.; Wieland, U.; et al. Viral load, gene expression and mapping of viral integration sites in HPV16-associated HNSCC cell lines. *Int. J. Cancer* **2015**, *136*, E207–E218. [CrossRef] [PubMed]

19. Rieckmann, T.; Tribius, S.; Grob, T.J.; Meyer, F.; Busch, C.J.; Petersen, C.; Dikomey, E.; Kriegs, M. HNSCC cell lines positive for HPV and p16 possess higher cellular radiosensitivity due to an impaired DSB repair capacity. *Radiother. Oncol.* **2013**, *107*, 242–246. [CrossRef] [PubMed]

20. Park, J.W.; Nickel, K.P.; Torres, A.D.; Lee, D.; Lambert, P.F.; Kimple, R.J. Human papillomavirus type 16 E7 oncoprotein causes a delay in repair of DNA damage. *Radiother. Oncol.* **2014**, *113*, 337–344. [CrossRef]

21. Wallace, N.A.; Khanal, S.; Robinson, K.L.; Wendel, S.O.; Messer, J.J.; Galloway, D.A. High-Risk Alphapapillomavirus Oncogenes Impair the Homologous Recombination Pathway. *J. Virol.* **2017**, *91*. [CrossRef] [PubMed]

22. Abdulkarim, B.; Sabri, S.; Deutsch, E.; Chagraoui, H.; Maggiorella, L.; Thierry, J.; Eschwege, F.; Vainchenker, W.; Chouaib, S.; Bourhis, J. Antiviral agent Cidofovir restores p53 function and enhances the radiosensitivity in HPV-associated cancers. *Oncogene* **2002**, *21*, 2334–2346. [CrossRef] [PubMed]

23. Kimple, R.J.; Smith, M.A.; Blitzer, G.C.; Torres, A.D.; Martin, J.A.; Yang, R.Z.; Peet, C.R.; Lorenz, L.D.; Nickel, K.P.; Klingelhutz, A.J.; et al. Enhanced radiation sensitivity in HPV-positive head and neck cancer. *Cancer Res.* **2013**, *73*, 4791–4800. [CrossRef] [PubMed]

24. Eriksson, D.; Lofroth, P.O.; Johansson, L.; Riklund, K.A.; Stigbrand, T. Cell cycle disturbances and mitotic catastrophes in HeLa Hep2 cells following 2.5 to 10 Gy of ionizing radiation. *Clin. Cancer Res.* **2007**, *13*, 5501s–5508s. [CrossRef] [PubMed]

25. Strauss, S.J.; Higginbottom, K.; Juliger, S.; Maharaj, L.; Allen, P.; Schenkein, D.; Lister, T.A.; Joel, S.P. The proteasome inhibitor bortezomib acts independently of p53 and induces cell death via apoptosis and mitotic catastrophe in B-cell lymphoma cell lines. *Cancer Res.* **2007**, *67*, 2783–2790. [CrossRef] [PubMed]

26. Chen, C.A.; Chen, C.C.; Shen, C.C.; Chang, H.H.; Chen, Y.J. Moscatilin induces apoptosis and mitotic catastrophe in human esophageal cancer cells. *J. Med. Food* **2013**, *16*, 869–877. [CrossRef] [PubMed]

27. Liu, W.T.; Chen, C.; Lu, I.C.; Kuo, S.C.; Lee, K.H.; Chen, T.L.; Song, T.S.; Lu, Y.L.; Gean, P.W.; Hour, M.J. MJ-66 induces malignant glioma cells G2/M phase arrest and mitotic catastrophe through regulation of cyclin B1/Cdk1 complex. *Neuropharmacology* **2014**, *86*, 219–227. [CrossRef] [PubMed]

28. Zajac, M.; Moneo, M.V.; Carnero, A.; Benitez, J.; Martinez-Delgado, B. Mitotic catastrophe cell death induced by heat shock protein 90 inhibitor in BRCA1-deficient breast cancer cell lines. *Mol. Cancer Ther.* **2008**, *7*, 2358–2366. [CrossRef]

29. Krisanaprakornkit, S.; Weinberg, A.; Perez, C.N.; Dale, B.A. Expression of the peptide antibiotic human beta-defensin 1 in cultured gingival epithelial cells and gingival tissue. *Infect. Immun.* **1998**, *66*, 4222–4228.

30. Piboonniyom, S.O.; Timmermann, S.; Hinds, P.; Munger, K. Aberrations in the MTS1 tumor suppressor locus in oral squamous cell carcinoma lines preferentially affect the INK4A gene and result in increased cdk6 activity. *Oral Oncol.* **2002**, *38*, 179–186. [CrossRef]

31. de Roda Husman, A.M.; Walboomers, J.M.; van den Brule, A.J.; Meijer, C.J.; Snijders, P.J. The use of general primers GP5 and GP6 elongated at their 3′ ends with adjacent highly conserved sequences improves human papillomavirus detection by PCR. *J. Gen. Virol.* **1995**, *76*(Pt. 4), 1057–1062. [CrossRef]

32. Mosmann, T. Rapid colorimetric assay for cellular growth and survival: Application to proliferation and cytotoxicity assays. *J. Immunol. Methods* **1983**, *65*, 55–63. [CrossRef]

33. Carpenter, A.E.; Jones, T.R.; Lamprecht, M.R.; Clarke, C.; Kang, I.H.; Friman, O.; Guertin, D.A.; Chang, J.H.; Lindquist, R.A.; Moffat, J.; et al. CellProfiler: Image analysis software for identifying and quantifying cell phenotypes. *Genome Biol.* **2006**, *7*, R100. [CrossRef] [PubMed]

34. Derks, J.L.; Leblay, N.; Thunnissen, E.; van Suylen, R.J.; den Bakker, M.; Groen, H.J.M.; Smit, E.F.; Damhuis, R.; van den Broek, E.C.; Charbrier, A.; et al. Molecular Subtypes of Pulmonary Large-cell Neuroendocrine Carcinoma Predict Chemotherapy Treatment Outcome. *Clin. Cancer Res.* **2018**, *24*, 33–42. [CrossRef] [PubMed]
35. Eijkelenboom, A.; Kamping, E.J.; Kastner-van Raaij, A.W.; Hendriks-Cornelissen, S.J.; Neveling, K.; Kuiper, R.P.; Hoischen, A.; Nelen, M.R.; Ligtenberg, M.J.; Tops, B.B. Reliable Next-Generation Sequencing of Formalin-Fixed, Paraffin-Embedded Tissue Using Single Molecule Tags. *J. Mol. Diagn.* **2016**, *18*, 851–863. [CrossRef] [PubMed]

*Article*

# Prevalence of Human Papillomavirus (HPV) Infection and the Association with Survival in Saudi Patients with Head and Neck Squamous Cell Carcinoma

Ghazi Alsbeih [1,*], Najla Al-Harbi [1], Sara Bin Judia [1], Wejdan Al-Qahtani [1,2], Hatim Khoja [3], Medhat El-Sebaie [4] and Asma Tulbah [3]

[1] Department of Biomedical Physics, King Faisal Specialist Hospital and Research Centre, Riyadh 11211, Saudi Arabia; nharbi@kfshrc.edu.sa (N.A.-H.); sbinjudia55@kfshrc.edu.sa (S.B.J.); walqahtani1@ksu.edu.sa (W.A.-Q.)

[2] Department of Histology, King Saud University, Riyadh 11451, Saudi Arabia

[3] Department of Pathology & Laboratory Medicine, King Faisal Specialist Hospital and Research Centre, Riyadh 11211, Saudi Arabia; hkhoja@kfshrc.edu.sa (H.K.); tulbah@kfshrc.edu.sa (A.T.)

[4] Department of Radiation Oncology, King Faisal Specialist Hospital and Research Centre, Riyadh 11211, Saudi Arabia; melsebaie@gmail.com

* Correspondence: galsbeih@kfshrc.edu.sa; Tel.: +966-11-442-7891; Fax: +966-11-442-4777

Received: 11 May 2019; Accepted: 5 June 2019; Published: 13 June 2019

**Abstract:** Head and neck squamous cell carcinoma (HNSCC) shows wide disparities, association with human papillomavirus (HPV) infection, and prognosis. We aimed at determining HPV prevalence, and its prognostic association with overall survival (OS) in Saudi HNSCC patients. The study included 285 oropharyngeal and oral-cavity HNSCC patients. HPV was detected using HPV Linear-Array and RealLine HPV-HCR. In addition, p16INK4a (p16) protein overexpression was evaluated in 50 representative cases. Oropharyngeal cancers were infrequent (10%) compared to oral-cavity cancers (90%) with no gender differences. Overall, HPV-DNA was positive in 10 HNSCC cases (3.5%), mostly oropharyngeal (21%). However, p16 expression was positive in 21 cases of the 50 studied (42%) and showed significantly higher OS ($p = 0.02$). Kaplan–Meier univariate analysis showed significant associations between patients' OS and age ($p < 0.001$), smoking ($p = 0.02$), and tumor stage ($p < 0.001$). A Cox proportional hazard multivariate analysis confirmed the significant associations with age, tumor stage, and also treatment ($p < 0.01$). In conclusion, HPV-DNA prevalence was significantly lower in our HNSCC patients than worldwide 32–36% estimates ($p \leq 0.001$). Although infrequent, oropharyngeal cancer increased over years and showed 21% HPV-DNA positivity, which is close to the worldwide 36–46% estimates ($p = 0.16$). Besides age, smoking, tumor stage, and treatment, HPV/p16 status was an important determinant of patients' survival. The HPV and/or p16 positivity patients had a better OS than HPV/p16 double-negative patients ($p = 0.05$). Thus, HPV/p16 status helps improve prognosis by distinguishing between the more favorable p16/HPV positive and the less favorable double-negative tumors.

**Keywords:** head and neck cancer; squamous cell carcinoma; human papillomavirus (HPV); oral cavity tumors; oropharyngeal cancer; p16Ink4a biomarker; p16-immunostaining; prognosis; overall survival

---

## 1. Introduction

Head and neck (H&N) cancer is the 9th most frequent malignancy worldwide, accounting for around 5% of all new cases and exhibits wide demographic variations [1,2]. The global incidence is estimated at 600,000 cases per year, with evidence indicating rising trends especially in young adults [3]. Squamous cell carcinoma (SCC) is the most common type of Head and Neck (H&N) cancers [4], and accounts for more than 90%. Lifestyle and several anecdotal risk factors are suspected to contribute to the development of various H&N cancers, including smoking, chewing (smokeless) tobacco and other products, alcohol consumption, dietary factors, chemical irritants, and poor oral hygiene [5]. Recently, however, infection with human papillomavirus (HPV) was recognized as being an important determinant and independent of other risk factors for H&N squamous cell carcinoma (HNSCC) [6–8]. While infection with the Epstein–Barr virus (EBV) is a known risk factor for nasopharyngeal carcinoma, HPV is mostly linked to a subset of HNSCC, particularly oropharyngeal cancer [9,10].

Syrjanen et al. [11] were historically, the first to evoke such an association between HNSCC and HPV based on histopathological observations followed by confirmation of HPV DNA presence in oral lesions [12]. These results gained momentum when it was later observed that the incidence of oropharyngeal squamous cell carcinomas in young patients (<50 years old), particularly in the tonsils and the base of tongue, increased significantly even though most patients are not regular tobacco or alcohol users [9,13]. This observation holds true despite the initial decline in H&N cancer incidence in North America consequent to active antismoking campaigns [9,14,15]. Large meta-analysis studies have estimated that 32% of HNSCC are associated with HPV, with higher rates in oropharyngeal than oral cancers [16]. Globally, HPV16 was considerably the most common subtype, accounting for 82% of all HPV positive cases, followed by HPV18 and ensued by a minority of other sporadic genotypes.

The more prominent turning point is that HNSCC HPV-positive cancers appear to form a distinct tumor entity from smoking- and alcohol-related counterparts with distinguished epidemiology, genetics, characteristic histopathology, therapeutic response, and predictive clinical outcome to chemo-radiation treatment [17–19]. Their noticeable molecular characteristics include p16Ink4a (p16) overexpression, modulation of PI3K/AKT and Wnt pathways, and lack of inactivating p53 mutations [20]. Furthermore, the observed HPV-associated overexpression of p16 protein in HNSCC has been largely considered as a surrogate marker diagnostic for HPV infection and also prognostic for a more favorable treatment outcome [21]. It was incorporated in the recent release of TNM-8, leading to marked changes in the classification of these malignant tumors [17,22]. Thus, the detection of HPV infection and histopathological determination of p16 protein expression in tumor samples are expected to gain importance in clinical settings and marks a major shift in managing HNSCC cancer patients.

The epidemiology of HPV infection is known to have wide variations in human populations, remnant of socioeconomic, ethnic, and genetic predisposing factors [23,24]. According to the Saudi Cancer Registry, H&N cancers, excluding nasopharynx, forms about 4% of all malignancies in this country [25]. If a third of those tumors are HPV-driven, then the projected burden of HPV, along with cervix, uterine, and other anogenital malignancies, would represent, in both genders, approximately 3% of all cancers in Saudi Arabia [26]. This is a significant medical issue for a health authority, particularly for the cost-effectiveness analysis of implementing a nationwide HPV vaccine in order to render these HPV-mediated tumors preventable. This is in addition to introducing personalized treatment modalities to boost cure rate and reduce patients' morbidity and mortality. However, actual data about the implication of HPV infection in HNSCC in Saudi cancer patients is completely lacking. Therefore, the main aims of this retrospective exploratory study were to determine the prevalence of HPV infection and its oncogenic genotypes, and the association with patients' overall survival (OS). The correlation with p16 protein expression was also studied in a subset of these tumors, to assess the prognostic values of HPV status and p16 protein positivity.

## 2. Results

### 2.1. Patients and Clinical Data

The characteristics of the 285 H&N cancer patients included in the study are summarized in Table 1. The age of patients at diagnosis of HNSCC ranged between 22 and 90 years (median = 57 years). The incidence showed a Gaussian distribution that increased with age to reach a peak at 59-year-old, and then decreased gradually (Figure 1A). There were 120 females and 165 males with no noticeable difference by gender in the distribution of cancer patients by age at diagnosis. Although the median age of females (60-year, range 23–90) was slightly higher than that of males (57-year, range 22–90), there was no statistical difference ($p = 0.09$; two-tailed Mann–Whitney Rank Sum test). By anatomical sites, 28 patients (10%) had oropharyngeal while 257 (90%) had oral cavity cancers. Interestingly, the number of oropharyngeal cancer cases increased with time: 8 cases were diagnosed in 2002–2008 (142 patients) compared to 20 cases in 2009–2016 (143 patients). The distribution of sub-anatomical sites of oral and oropharyngeal cancers by 5-year age groups is illustrated in Figure 1B. The stage of the tumors varied from T1N0M0 to T4N2cM0 with 63% of patients having early stage (T1–2) compared to 37% with advanced (T3–4) tumors. Patients followed mainly standardized curative treatment according to the stage of the tumor as described above. The length of patients' follow-up extends to 15 years (mean = 4.36 years; standard deviation = 3.88) after diagnosis. There were seven ambiguous cases with locally advanced invasion, without evidence of distal metastatic cancer, who displayed an overall short mean survival of about six months.

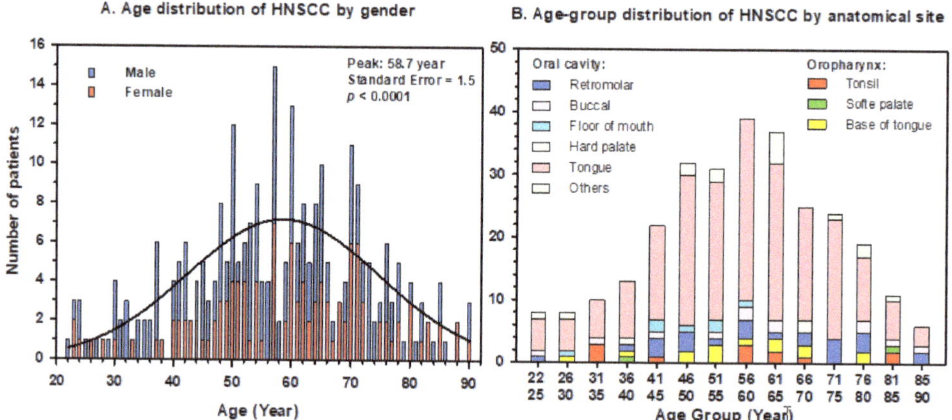



**Figure 1.** Incidence of head and neck squamous cell carcinoma in 285 Saudi cancer patients. (**A**) Age-distribution by gender of patients. Clustering analysis indicates a peak of maximum occurrence at the age of 58.7 years old. (**B**) Distribution by 5-year age groups of oropharyngeal and oral cavity tumors by sub-anatomical sites.

**Table 1.** Descriptive statistics of 285 head and neck squamous cell carcinoma patients with results of HPV and p16[INK4a] tests.

| Cancer Site (ICD Code) | Number of Cases | Age Median (Range) | Gender | Smoking * | Tumor Stage | Treatment | HPV | p16 |
|---|---|---|---|---|---|---|---|---|
| Oropharynx: | | | | | | | | |
| Tonsil (C09) | 12 | 60 (31–83) | M: 9<br>F: 3 | Yes: 4<br>No: 5<br>N/A: 3 | T1–2: 6 (N0: 2, N+: 4, M0: 6)<br>T3–4: 6 (N0: 5, N+: 1, M0: 6) | S: 1, S+RT: 5<br>CRT: 4, S+RT: 2 | P: 3 (HPV16)<br>N: 9 | P: 4<br>N: 8<br>N/A: 0 |
| Soft palate (C05.1) | 2 | 61 (37–85) | M: 2<br>F: 0 | Yes: 1<br>No: 0<br>N/A: 1 | T1–2: 0 (N0: 0, N+: 0, M0: 0)<br>T3–4: 2 (N0: 2, N+: 0, M0:2) | S+CRT: 2 | P: 0<br>N: 2 | P: 0<br>N: 2<br>N/A: 0 |
| Base of tongue C01 | 14 | 54 (27–78) | M: 7<br>F: 7 | Yes: 7<br>No: 5<br>N/A: 2 | T1–2: 8 (N0: 3, N+: 5, M0: 8)<br>T3–4: 6 (N0: 0, N+: 6, M0: 6) | S: 2, S+RT: 5, S+CRT: 1<br>S+RT: 1, CRT: 3, S+CRT: 2 | P: 3 (HPV16)<br>N: 11 | P: 7<br>N: 7<br>N/A: 0 |
| Oral Cavity: | | | | | | | | |
| Retromolar C06.2 | 24 | 62 (24–90) | M: 16<br>F: 8 | Yes: 11<br>No: 8<br>N/A: 5 | T1–2: 11 (N0: 6, N+: 5, M0: 11)<br>T3–4: 13 (N0: 1, N+: 12, M0: 13) | S: 3, S+RT: 5, CRT: 1, S+CRT: 2<br>S+RT: 1, CRT: 1, S+CRT: 11 | P: 2 (HPV16)<br>N: 22 | P: 5<br>N: 5<br>N/A: 14 |
| Tongue C02 | 198 | 57 (22–90) | M: 113<br>F: 85 | Yes: 82<br>No: 45<br>N/A: 71 | T1–2: 134 (N0: 84, N+: 51, M0: 134)<br>T3–4: 64 (N0: 24, N+: 39, M0: 64) | S: 62, S+RT: 56, S+CRT: 16<br>CRT: 1, S+RT: 13, S+CRT: 50 | P: 1 (HPV33)<br>N: 197 | P: 4<br>N: 6<br>N/A: 188 |
| Buccal (C06) | 19 | 62 (24–90) | M: 9<br>F: 10 | Yes: 7<br>No: 3<br>N/A: 9 | T1–2: 11 (N0: 6, N+: 5, M0: 11)<br>T3–4: 8 (N0: 4, N+: 4, M0: 8) | S: 4, S+RT: 7<br>S+RT: 3, S+CRT: 5 | P: 1 (HPV16)<br>N: 18 | P: 1<br>N: 1<br>N/A: 17 |
| Floor of mouth (C04) | 13 | 49 (25–82) | M: 7<br>F: 6 | Yes: 4<br>No: 5<br>N/A: 4 | T1–2: 8 (N0: 6, N+: 2, M0: 8)<br>T3–4: 5 (N0:2, N+: 3, M0: 5) | S: 5, S+RT: 3<br>S+RT: 1, S+CRT: 4 | P: 0<br>N: 13 | N/A: 13 |
| Hard palate (C05.0) | 3 | 66 (37–69) | M: 2<br>F: 1 | Yes: 1<br>No: 0<br>N/A: 2 | T1–2: 1 (N0: 1, N+: 0, M0: 1)<br>T3–4: 2 (N0: 2, N+: 0, M0: 2) | S+RT: 1<br>S+RT: 1, S+CRT: 1 | P: 0<br>N: 3 | P: 0<br>N: 0<br>N/A: 3 |
| All cases | 285 | 57 (22–90) | M: 165<br>F: 120 | Yes: 117<br>No: 71<br>N/A: 97 | T1–2: 179 (N0: 107, N+: 72, M0: 179)<br>T3–4: 106 (N0: 41, N+: 65, M0: 106) | S: 77, CRT: 1, S+RT: 82, S+CRT: 19<br>CRT: 9, S+RT: 22, S+CRT: 75 | P: 10 (3.5%)<br>N: 275 (96.5%) | P: 21 (42%) **<br>N: 29 (58%) **<br>N/A: 238 |

* The smoking category also includes chewing tobacco mixture (Shamma). M: Male. F: Female. T1–2: Tumor size (T1 or T2). T3–4: Tumor size (T3 or T4). N: Lymph nodes. M: Metastasis (note that all M+ = 0). S: Surgery. S+RT: Surgery + radiotherapy. CRT: Chemo-radiotherapy. S+CRT: Surgery + chemo-radiotherapy. P: Positive. N: Negative. N/A: Not Available.
** Percentage out of 50 cases tested.

Patients' characteristics were significantly associated with OS for groups of age, separated by the median of 57 years/old ($p < 0.001$), and smoking ($p = 0.02$), while gender ($p = 0.28$) had no effect (Figure 2). In addition, patients' OS declined significantly ($p < 0.001$) from T1 to T4 (Figure 3A). Although oropharyngeal cancer displayed a slightly improved OS compared to oral cavity patients, the difference was not statistically significant ($p = 0.14$; Figure 3B). Alcohol dependence or abuse was reported in only 8% of the patients, meanwhile years of daily tobacco smoking was common in this cohort (62%) in both genders, comprising 28% who were Shamma (a chewing tobacco mixture) users.

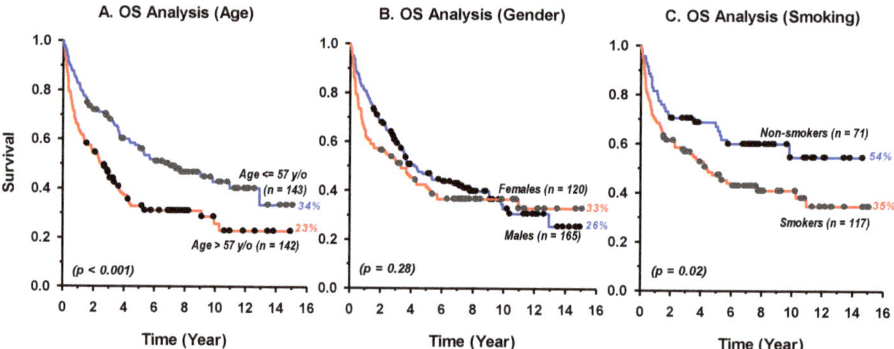


**Figure 2.** Kaplan–Meier Log-Rank overall survival (OS) analysis by patients' characteristics of groups of age (**A**), gender (**B**), and smoking status (**C**) for 285 patients with head and neck squamous cell carcinomas.

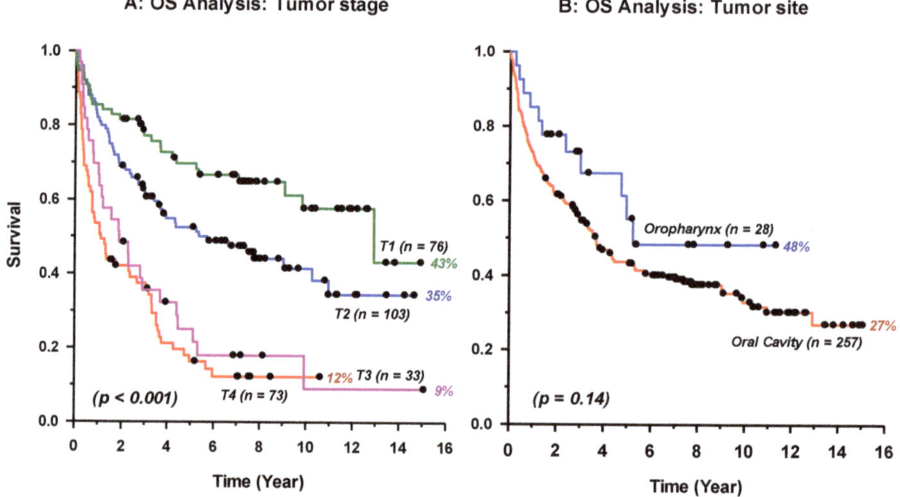


**Figure 3.** Kaplan–Meier Log-Rank overall survival (OS) analysis by tumor stage (**A**) and anatomical tumor site (**B**) of 285 patients with head and neck squamous cell carcinomas. The *p*-value in (**A**) represents the overall significance level. Al pairwise comparisons were statistically significant ($p \leq 0.03$) except T4 vs. T3 ($p = 0.32$).

### 2.2. Detection of HPV Infection and Genotyping

The Linear Array HPV Genotyping Test was first used to detect and genotype HPV infection. Results indicated that only 10 patients (3.5%) were HPV positive while 275 specimens (96.5%) proved to be negative after at least two separate tests and an independent concordant confirmation using the RealLine HPV HCR Genotype (Table 1). By HPV genotype, nine cases were HPV16 and one case was HPV33. These were detected in three females and seven males with a median age of 57 years (range 32–78). A Mann–Whitney Rank Sum test showed no significant difference in the median age between HPV-positive and HPV-negative patients ($p = 0.65$). By anatomical site, 21% (6/28) of the oropharyngeal and 2% (4/257) of the oral cavity cancers were positive for HPV infection. Most frequent HPV-positive cases were recorded from the tonsils (3/12), the base of the tongue (3/14), the retromolar (2/24), followed by the buccal (1/19) and the tongue (1/198). Thus, the highest HPV-DNA positivity was in oropharyngeal cancers (21%), which is not statistically different from the worldwide 36–46% estimates ($p = 0.16$). Overall, survival analyses showed a trend toward better OS for HPV-positive (67% survival) compared to HPV-negative (27% survival) patients but that did not reach statistical significance ($p = 0.12$), most probably due to the small number of HPV-positive cases (Figure 4A).

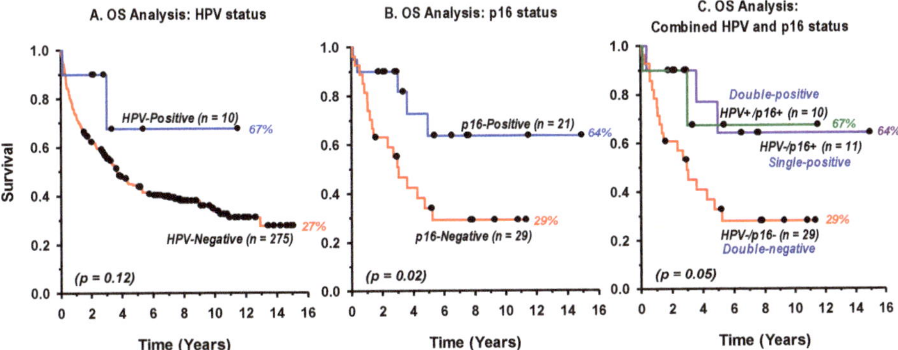


**Figure 4.** Kaplan–Meier Log-Rank overall survival (OS) analysis by HPV (**A**) status of head and neck squamous cell carcinomas in 285 cancer patients, p16INK4a (**B**), and the combination of HPV and p16INK4a (**C**) status in a subset of 50 cases. The *p*-value in (**C**) represents the overall significance level. There was no significant difference between single-positive and double-positive cases ($p = 0.85$).

## 2.3. p16Ink4a Protein Immunohistochemical (IHC) Staining

In view of the small number of HPV-positive HNSCC patients found in this cohort, and the limited amount of pathological materials available, a subset of 50 representative specimens were processed for p16 protein IHC staining. These included all the 28 oropharyngeal cases, which are known as highly suspicious for HPV infection. In addition, 22 cases of oral cavity subsites were processed comprising 10 retromolar, 10 tongue and 2 buccal for which at least one cancer was positive for HPV. Thus, the 10 HPV-positive tumors along with 40 HPV-negative cases were included. Examples of p16 protein IHC strong (positive) and weak (negative) staining is given in Figure 5. In total, p16 was positive in 42% (21 tumors) of the 50 tested cases. Interestingly, p16 was positive in all the 10 HPV-positive tumors (double-positive for HPV and p16-over-expression) in addition to 11 HPV-negative cases (single-positive for p16 overexpression) while the remaining 29 samples were double-negative. The p16 positivity was 39% in the 28 oropharyngeal cases and 45% in the 22 cases studied of oral cavity cancers. There were no significant differences in patients' age or male to female ratios between p16 positive and negative cases ($p > 0.05$). A survival analysis showed a statistically significant ($p = 0.02$) better OS for p16-positive (64% survival) compared to p16-negative (29% survival) patients (Figure 4B). In addition, the survival analysis of the combined HPV/p16 status in the 50 cases studied (Figure 4C), showed an overall significant difference in OS ($p = 0.05$), whereby HPV and/or p16 positive patients displayed better survival (64–67% survival) compared to HPV/p16 double-positive patients (29% survival). However, there was no significant difference between double-positive and single positive patients ($p = 0.85$).

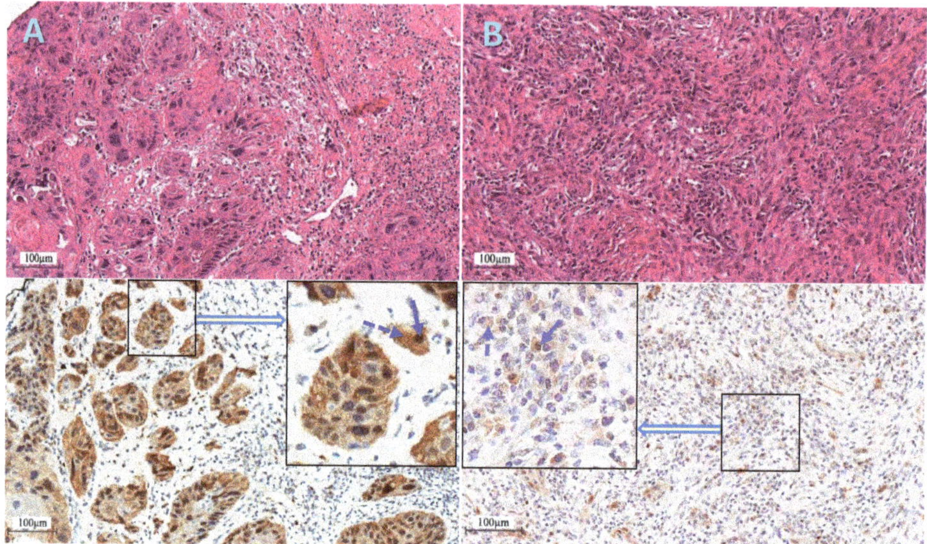

**Figure 5.** Examples of pathological tissue sections of head and neck squamous cell carcinomas stained with hematoxylin and eosin (**upper panels**) and the corresponding immunohistochemical staining for p16Ink4a (**lower panels**) showing nuclear (solid arrows) and cytoplasmic staining (dashed arrows). Sample (**A**) shows strong staining (usually involving >70% of the tumor cells) scored as p16-positive as compared to sample (**B**) with weak patchy staining, arbitrated as p16-negative.

## 3. Discussion

H&N cancer is an important health issue worldwide [1,3,4]. The identification of HPV infection as an independent risk factor, particularly in the oropharynx, with favorable prognosis for treatment response and survival spurred out research to stratify patients to deliver more personalized treatment [6–8]. However, HPV-associated HNSCC cancers are known to display wide epidemiological variation between populations [24]. To the best of our knowledge, this is the first study on the association between HNSCC and HPV in Saudi cancer patients. We have systematically reviewed 1633 medical records spanning more than one decade of H&N cancer patients admitted at our tertiary care hospital. We have first targeted oropharyngeal and secondly oral cavity cancers as they are potentially the most associated with HPV [27]. Following a review of the pathological samples, only 285 cases were available for the study.

Patients' characteristic data showed that the incidence of oropharyngeal and oral cavity HNSCC increased with age from 22 years to reach a peak at 59 years, then decreased to 90 years old (Figure 1A). There were no obvious differences in the incidence by gender or cancer sites. Females composed 42% of the patients compared to 58% of males in this cohort. This relatively high incidence in females is rather unusual for HNSCC; however, it confirms a previous study in the country [28]. Although the exact reason is still unknown, it might be related to the increased habits of females consuming tobacco products (including Shamma) as 55% were smokers (68% in males). Oropharyngeal tumors, however, were infrequent and formed only 10% (28/285) of the patients while 90% (275/285) had oral cavity cancers (Table 1; Figure 1B). This 10% ratio is significantly lower ($p \leq 0.001$) than the projected 32% computed from the estimated number of incident cases worldwide [1], suggesting lower incidence of these types of cancers in our population. Although this is a single institution study, the low rate is representative of the country because the King Faisal Specialist Hospital and Research Centre (KFSHRC) is the primary tertiary care referral hospital, which captures more than 50% of cancer patients in the Kingdom. The low incidence of oropharyngeal cancers found in this study is in agreement with the national registry with an age-standardized rate of 0.07 (Cancer Today, Globocan 2018 statistics on oropharyngeal cancer in Saudi Arabia available at: http://gco.iarc.fr/today/home). Nevertheless, our data indicate an increase in the number of oropharyngeal cancer cases over time as it doubled in the last decade in the country. The subsites of H&N cancers that are most frequently associated with HPV infection are the tonsils, the soft palate, the base of the tongue (oropharynx) and the retromolar (oral cavity). These together formed 18% (52/285) of the cases in this cohort. The remaining 233 patients had other various oral cavity tumors including mostly tongue (Table 1).

As expected, tumor stage had independently major impact on patients' OS which significantly ($p < 0.001$) decreased gradually from T1 to T4 (Figure 3A). This statement incorporates the standardized patient treatment that depended mainly on tumor stage. Interestingly, our results showed that age and tobacco consumption affect prognosis, as statistically significant better survival was observed for younger-age ($p < 0.001$) and non-smokers ($p = 0.02$), but not for patients' gender (Figure 2). As for HPV oncogenic risk factor, only 10 samples were positive for HPV-DNA out of the 285 patients (Table 1). This indicates that only 3.5% (10/285) of HNSCC are infected with HPV in our cancer patients. In an early global systematic meta-analysis that comprised 60 eligible studies and included 5046 cases, the overall HPV prevalence in HNSCC was 25.9% with significantly higher presence in oropharyngeal (35.6%) than in oral (23.5%) and in laryngeal (24%) cancers [29]. In a more recent meta-analysis, HPV-DNA was detected in 32% (3837 out of 12,163 cases) of HNSCC, with again a higher prevalence in oropharyngeal (46%) than oral (24%) or laryngeal and hypopharyngeal (22%) cancers [16]. The 3.5% prevalence of HPV in HNSCC in our patients is significantly lower than the estimated 32–36% worldwide ($p \leq 0.001$, one sample z-test).

In agreement with published data, the highest prevalence of HPV was observed in oropharyngeal cancers (21%), mostly in tonsils (3/12) followed by the base of the tongue (3/14). Although this 21% prevalence remains below the compiled worldwide estimates of 36–46%, it was not statistically different ($p = 0.16$), indicating similar pathogenic association. As for oral cavity, only 2% were HPV-positive, mainly recorded from the retromolar (2 cases), the tongue, and the buccal cavity (one case each). This low prevalence is in agreement with a recent study by Vidal Loustau et al. [30], but again much lower than the worldwide estimate of 23.5% stated above. Overall, these results imply that the prevalence of HPV-driven HNSCC in our population is very low. The reasons for this low rate is unknown, but could be related to the predominance of other risk factors, such as various tobacco products consumption, which is as high as 62% of patients, socio-cultural differences, or the presence of relative protective variants of genetic predisposing factors as has been shown previously for cervical cancer patients [31]. Most infections (90%) were with HPV16 (9/10) followed by 10% with HPV33 (1/10) genotypes. These results are in line with other studies even though HPV33 was much less commonly observed. In fact, the latter was a case of SCC of the tongue (Table 1). The patient was a young male who had bone marrow transplant for leukemia two decades ago. It is probable that his relatively compromised immune response resulted in a persistent HPV infection with this rare HPV33 genotype leading to this neoplasia [32].

One of the most significant advancements in H&N oncology of the precedent decade is the demonstration that cancer patients with HPV-mediated HNSCC, particularly in the oropharynx, have p16 protein expression and are associated with significantly improved treatment outcomes expressed as higher rate of patients' survival, compared to HPV-negative patients [6,33]. Furthermore, these observations have laid the foundations for exploratory clinical trials examining the impact of proposed "treatment deintensification" for patients with HPV-driven cancers [34,35]. The rationale is to improve treatment outcome, by reducing side effects without compromising tumor control. Our results for p16 protein expression in 50 HNSCC cases showed that 42% (21/50) were positive for p16 over-expression (Figure 5, sample A), including all the 10 HPV-positive tumors (double-positive). Interestingly, 11 p16-positive cases were HPV-negative (single-positive). It is known that the clinical relevance of HPV-DNA positivity is a matter of debate, because it is likely to represent both transcriptionally active (RNA+) and inactive (RNA−) HPV genomes. Therefore, detection of HPV-RNA by in situ hybridization is considered the gold standard for clinically relevant, HPV transcriptionally active lesions. However, the availability of this RNA methodology and concern for lower sensitivity compared to the affordable HPV-DNA polymerase chain reaction (PCR) led to the evaluation of IHC-p16 protein as a surrogate marker for the presence of active HPV in tumor cells [15]. Therefore, the Union for International Cancer Control (UICC) 8th edition defines HPV-mediated oropharyngeal cancer by use of p16 immunohistochemistry [36]. With this argument in mind, it is probable that the p16-positive samples in the 50 cases studied represent active HPV infection, as many of them could not be picked by the HPV-DNA PCR-based techniques. In such an arguable case, it is acknowledged that the incidence of HPV-related HNSCC would be higher (up to 42%) in our patients. However, a larger study with more patients is needed to confirm this assumption, particularly that active HPV-RNA is considered rare in non-oropharyngeal tumors [37]. The alternative view could be that it is possible that p16 positivity is not exclusively related to HPV infection, which would debate its use as a surrogate marker for the presence of HPV in all HNSCC [38]. Indeed, discrepancies in the p16/HPV-positivity have been observed and it is questionable if all HPV-positive and/or p16-positive tested cancers are HPV-driven. It is possible that sometimes HPV is an innocent bystander and p16 is independently positive [39]. This highlights the importance of identifying robust fingerprints of HPV-driven carcinogenesis to improve the estimate of HPV-attributable HNSCC and to predict the effectiveness of implementing preventive HPV vaccination and therapeutic interventions.

The relationship between HPV and OS after the treatment showed a clear trend toward a longer survival of HPV-positive patients (Figure 4A) as described elsewhere [40]. However, in our study a survival analysis did not reach statistical significance ($p = 0.12$) due to, most probably, the small number of HPV-positive cases (10/285). In addition, those patients had mainly T1–2 tumor stages with basically a favorable survival prognosis. Nonetheless, in a subset of 50 cases, a statistically significant ($p = 0.02$) better survival was observed for p16-positive compared to p16-negative patients (Figure 4B). Although tumor stages were distributed more evenly in this group (T1–2 = 58%, T3–4 = 42%), there was slight preponderance (32%) of early stages in the p16-positive patients and vice versa. Furthermore, tumor stage as well as age remained significantly associated with OS in multivariate analysis (Table 2). In addition, the treatment offered to patients showed statistically significant ($p = 0.006$) association with OS. It also shows that surgery, which mainly underlies early stage tumors, result in higher survival compared to any other combined treatment. In other word, this result essentially captures that of the tumor stage since the treatment was stage standardized with some subtle adaptation to each individual case. In addition, a trend toward association with OS was apparent for HPV/p16 and smoking status but they did not reach statistical significance in the multivariate analysis. Nevertheless, our results are in overall agreement with published data with the overwhelming belief that p16-positive HNSCC have improved locoregional tumor control and survival with conventional therapy [21,27,41]. Potential future refinement could be brought about by including the copy number variation of the CDKN2A gene that encodes p16ink4a [42], and involving other related prognostic biomarkers such as epidermal growth factor receptor (EGFR), and key transcription factors as molecular signature of HPV presence [19,43], especially for de-escalation of radiotherapy combined with anti-EGFR receptor treatment [44].

While the study points out toward the need of systematic testing of p16 overexpression, results obtained in a subset of patients, the results are also in line with a recent study evaluating the 8th TNM classification that integrates p16 status (as independent or surrogate markers for HPV infection) in oropharyngeal cancer [17,45]. The study included 1204 patients where 32% were p16-positive which is close to the 42% observed in our study despite the limited number of cases processed for p16 expression. Importantly, the authors found that 12% of p16 positive cases were negative for HPV-DNA. This HPV-negative subgroup had distinct features and a poorer OS. Therefore, we have analyzed the OS with the various combination of HPV and p16 status in a subset of 50 cases with sufficient pathological materials (Figure 4C). Interestingly, the Kaplan–Meier Log-Rank survival analysis showed a significant difference ($p = 0.05$) where HPV/p16-positive cases showed substantially better OS than double-negative patients. Although double-positive cases showed slightly better survival than single-positive patients, the difference was not statistically significant, most probably due to the small number of patients who tested positive for HPV and/or p16. Nonetheless, taken together, these results highlight the importance of performing independent HPV and p16 testing when predicting individual patient's prognosis [39,46]. These results are in line with a recent study on oropharyngeal cancer in four Catalonian hospitals where double positivity for HPV-DNA/p16 showed the strongest diagnostic biomarker accuracy and prognostic value [47]. The findings may have major impact in clinical practice, in particular when selecting cases for deintensified treatment regimens.

**Table 2.** Multivariate analysis using a Cox proportional hazard model to test the influence of various risk factors on overall survival in 285 patients with HNSCC.

| Risk Factors | Categories | HR | 95%CI | *p* Value |
|---|---|---|---|---|
| Age | Younger * | 0.57 | 0.38–0.87 | 0.009 |
| Gender | Females | 1.01 | 0.66–1.55 | 0.963 |
| Smoking | Non-smokers | 0.77 | 0.48–1.22 | 0.258 |
| Tumor site | Oropharynx | 0.71 | 0.31–1.63 | 0.422 |
| Tumor stage | Early (T1–2) | 0.53 | 0.33–0.83 | 0.005 |
| Treatment | Surgery ** | 0.40 | 0.20–0.77 | 0.006 |
| HPV/p16 status | Positive | 0.38 | 0.11–1.28 | 0.118 |

HR: Hazard Ratio. CI: Confidence Interval. * Younger denote patients whose age is ≤ the median age of 57 years/old. ** Surgery vs. any combined treatment.

## 4. Materials and Methods

### 4.1. Ethical Considerations

The study was carried out using archival pathologic materials of H&N cancer obtained during routine diagnostic procedures. The samples were anonymized and processed with no patients' identifiable characters. The study was reviewed by the institutional review board and approved by the Research Centre Ethics Committee at the King Faisal Specialist Hospital and Research Centre (KFSHRC) under the number RAC#2130 025.

### 4.2. Clinical Specimens

Medical records of 1633 H&N cancer patients diagnosed between 2002 and 2016 at the KFSHRC tertiary care hospital were screened. The main eligibility criteria were adult patients with squamous cell carcinoma in anatomical location potentially associated with HPV infection. Following the exclusion of palliative cases and cancer sites that had not been proven to be HPV-driven (for instance nasopharynx, salivary glands, and trachea), only 330 patients with oropharyngeal and oral cavity tumors remained for possible inclusion. After the examination of the histopathological slides, 285 patients' samples were included due to the limited amount of pathological blocks available for this study. Patients' treatment with curative intent followed timely standard clinical guidelines that depends on primary tumor location and extension [48]. Briefly, early stage (I–II) oral cavity tumors were treated with conservative surgery (S) and/or external radiotherapy (RT) 66–70 Gy in 33–35 fractions. Locally advanced stages (III–IV) were treated with surgery including reconstruction plus postoperative radiotherapy 60–66 Gy in 30–33 fractions. Patients found at surgery to have high-risk features were treated with post-operative chemoradiotherapy (S + CRT) 66 Gy in 33 fractions with 3 weekly cisplatin 100 mg/m$^2$. Patients having resectable tumors with poor prognosis were treated with combined concomitant CRT 66–70 Gy in 33–35 fractions with 3 weekly cisplatin 100 mg/m$^2$. A combined concomitant CRT was also the standard treatment in oropharyngeal and non-resectable oral cavity cancer patients. Cetuximab was used for patients who were not fit for cisplatin chemotherapy. Radiotherapy modalities included 3D conformal that was gradually replaced with intensity-modulated radiation therapy (IMRT) in 2006, and also RapidArc in 2010 and TomoTherapy in 2012. Although some HPV-related histopathologic features were available for few cases, the treatment followed the same guideline for all patients with no difference between positive and negative HPV.

### 4.3. DNA Extraction

Formalin-fixed, paraffin-embedded (FFPE) tissues proven to contain tumor sections of the 285 patients were obtained from the pathology department's archive. For each case, 3–6 sections of 10 μm thickness were taken from the block for the extraction of DNA using the QIAamp DNA FFPE tissue kit (Qiagen, Dusseldorf, Germany), using the manufacturer's recommended instructions. Briefly, the FFPE sections were deparaffinized using xylene followed by ethanol to extract residual xylene. The specimens are covered with ATL lysis buffer with 20 μL proteinase K (20 mg/mL, Roche, Mannheim, Germany) and incubated at 56 °C and 90 °C for 1 h each. Then, 2 μL 100 mg/mL DNase-free RNase A (Qiagen) was added, mixed and incubated at room temperature for 2 min. After the lysis and heating, followed by binding and washing steps, DNA was eluted in 50 μL of ATE buffer and quantified using a NanoDrop 2000c Spectrophotometer (Thermo Fisher Scientific, Waltham, MA, USA).

### 4.4. HPV Detection and Genotyping

Two different methods were consecutively used to detect and genotype HPV infection in all the H&N samples along with HPV negative (HTB-31) and HPV-16 positive (HTB-35) external controls:

1   The Linear Array HPV Genotyping Test (LA HPV GT; Roche Diagnostics, Mannheim, Germany). This PCR-based test detects and genotypes the 37 most common anogenital HPVs (13 high-risk: 16, 18, 31, 33, 35, 39, 45, 51, 52, 56, 58, 59, 68, and 24 low-risk: 6, 11, 26, 40, 42, 53, 54, 55, 61, 62, 64, 66, 67, 69, 70, 71, 72, 73 (MM9), 81, 82 (MM4), 83 (MM7), 84 (MM8), 89 (CP6108) and IS39). Procedures followed the manufacture's instruction described in detail previously [31,49]. Briefly, the methodology involves the PCR amplification of the target DNA, the hybridization of the amplified DNA segments to oligonucleotide probes immobilized on strips of membranes, and finally, the colorimetric detection of the hybridized products using the Linear Array Detection Kit. The adequacy of samples is determined by the β-globin gene as an internal control. HPV positive reactions show visible blue bands localized on the strip. The HPV genotype is determined using the HPV reference guide provided in the kit. Results were deemed negative when no HPV band was detected after at least 2 independent tests with confirmed adequacy of samples.

2   RealLine HPV HCR Genotype Fla-Format (Bioron, Diagnostics GmbH, Ludwigshafen, Germany). This Real-Time PCR test allows the differential determination of the 12 most frequent high-risk HPV-DNA genotypes, 16, 18, 31, 33, 35, 39, 45, 51, 52, 56, 58 and 59, isolated from clinical specimens. It is based on the detection of the unquenched fluorescence produced by a specific reporter molecule that intensifies as PCR reaction cycles increased. The reporter molecule is a fluorophore-quencher dual-labeled DNA-probe designed to bind exclusively to the HPV-DNA target region. Fluorescent signal increases as a result of the cleavage of the probe by Taq DNA-polymerase exonuclease activity, which separates the fluorescent dye from the quencher during the repeated cycles of hybridization and amplification. The threshold cycle value (Ct) is defined as the cycle number at which the generated fluorescence crosses a set threshold within the reaction where the signal increases significantly above the background fluorescence of the procedure. Ct depends on the initial quantity of the HPV-DNA template present. A positive HPV control is run with the samples and an internal control (IC) detecting the content of human DNA (β-actin) is used to validate the quality of sampling and improve the reliability of results by preventing generation of false negatives which can be caused by the possible loss of a DNA template during sample preparation.

3   Procedures followed the manufacture's recommended methodology. Briefly, to analyze each sample for the detection of the 12 HPV-DNA genotypes, 4 tubes containing Master Mix (MM1, MM2, MM3, MM4) in 0.2 mL 96-well plates were used. The amplification is carried out on the CFX96 Touch Real-Time PCR Detection System (Bio-Rad, Hercules, CA, USA) using the recommended cycling program. The sample is flagged as positive (i.e., containing HPV-DNA) when the Ct value via the

fluorescent dyes, FAM, HEX, and ROX channels, for this sample (in any of MM 1–4 tubes) is less than or equal to 35 for HPV types 31, 33, 35, 39, 45, 51, 52, 56, 58, 59, or is less than or equal to 40 for HPV types 16 and 18. The HPV genotype is determined using a reference table provided by the manufacturer, which correlates each MM with an individual dye channel to one of the 12 specific high-risk HPV types.

### 4.5. Immunohistochemical (IHC) Staining of p16 Protein Expression

Procedures examining the expression of p16 protein were carried out using a Bond-III Automated IHC/ISH Stainer (Leica Biosystem, Wetzlar, Germany) according to manufacturer's instruction and reagents. Briefly, where available, 4 μm FFPE sections were mounted on glass microscope slides coated with Poly-L-Lysine. They were deparaffinized using Bond Dewax Solution (Leica Biosystem), rehydrated, and washed with Bond Wash Solution. The slides were incubated with Bond Epitope Retrieval Solution and heated at 100 °C for 20 min, washed, and Peroxide Wash Solution applied for 5 min. The p16 primary antibody (mouse monoclonal Anti-p16INK4a (E6H4), Ventana, Tucson, AZ, USA) was added on the slides for 15 min, followed by the anti-mouse secondary antibody (Post Primary Rabbit anti mouse IgG, ProClin, Leica Biosystem) for 8 min and the Bond Polymer Refine Detection solutions with intermittent washing. Slides were counterstained with Hematoxylin, and then dehydrated and mounted with DPX by using a Tissue-Tek film coverslipper (Sakura Finetek, Tokyo, Japan). Negative controls were obtained by excluding the primary antibody. Scoring of p16 IHC cytoplasmic and nucleic staining were evaluated by an experienced pathologist, based on defined characteristics whereby p16 was scored as positive if it was strong and diffuse (>70% of tumor cells), and negative if absent, weak, or focal [50].

### 4.6. Statistical Analysis

A one sample z-test was used to detect differences in proportions when the referenced proportion was deemed constant. The non-parametric Mann–Whitney Rank Sum test was used to assess differences between groups. A univariate Kaplan–Meier Log-Rank survival analysis was used to evaluate the relationship between various risk factors and overall survival (OS) represented by the length of patients' follow-up. A multivariate Cox proportional hazards model was used to test the effects of multiple covariates on patients' OS. All statistical tests conducted were two-sided. A $p$-value < 0.05 was considered statistically significant. Statistical analyses were done using the SigmaPlot platform (Version 12.5, SPSS Science, San Jose, CA, USA), and MedCalc, Ostend, Belgium (https://www.medcalc.org/calc/test_one_proportion.php).

## 5. Conclusions

This study indicates an overall low prevalence of HPV infection in our HNSCC patients. Although oropharyngeal cancer cases were infrequent, they increased over years and 21% were associated with HPV infection. Age, smoking, tumor stage, and treatment had important effect on survival. Although all HPV-positive cases were p16-positive (double-positive), the p16 positivity is not exclusive and could be positive in HPV-DNA negative tumors. HPV and/or p16 positivity had better prognosis of survival than HPV and/or p16 negative patients. An important clinical application is in the stratification of patients according to HPV and p16 status. These tests could improve survival predictions by distinguishing between the more favorable HPV-positive/p16-positive group, and the less favorable double-negative HPV/p16 group of HNSCC patients who have the worst prognosis.

**Author Contributions:** Conceptualization, G.A., M.E.-S. and A.T.; methodology and investigation, N.A.-H., S.B.J., W.A.-Q. and H.K.; software, G.A.; validation, G.A., M.E.-S. and H.K.; formal analysis, G.A.; M.E.-S.; resources, A.T.; data curation, G.A., N.A.-H. and S.B.J.; writing—original draft preparation, G.A.; writing—review and editing, all authors; visualization, supervision, project administration and funding acquisition, G.A.

**Funding:** This research was funded by The National Science, Technology and Innovation Plan (NSTIP-KACST), grant number 12-MED2945-20 (RAC# 2130 025).

**Acknowledgments:** We would like to thank Nasser Alrajhi, Khurram Shehzad, Hadeel Almanea, Sherifa Shaker, Shoaib Ahmed, Khaled Al-Hadyan, Sara Elewisy, Aisha Al-Qarni, Jamelah Almusallam, Nawaf Alenazi, and Maha Alrashdi for their help, Mohamed Shoukri for statistical analysis guidance, and Belal Moftah for his support.

**Conflicts of Interest:** The authors declare no conflict of interest.

## References


1. Ferlay, J.; Soerjomataram, I.; Dikshit, R.; Eser, S.; Mathers, C.; Rebelo, M.; Parkin, D.M.; Forman, D.; Bray, F. Cancer incidence and mortality worldwide: Sources, methods and major patterns in GLOBOCAN 2012. *Int. J. Cancer* **2015**, *136*, E359–E386. [CrossRef] [PubMed]

2. Guerrero-Preston, R.; Lawson, F.; Rodriguez-Torres, S.; Noordhuis, M.G.; Pirini, F.; Manuel, L.; Valle, B.L.; Hadar, T.; Rivera, B.; Folawiyo, O.; et al. JAK3 Variant, Immune Signatures, DNA Methylation, and Social Determinants Linked to Survival Racial Disparities in Head and Neck Cancer Patients. *Cancer Prev. Res.* **2019**, *12*, 255–270. [CrossRef] [PubMed]

3. Bray, F.; Colombet, M.; Mery, L.; Piñeros, M.; Znaor, A.; Zanetti, R.; Ferlay, J. *Cancer Incidence in Five Continents (2008-2012), electronic version*; International Agency for Research on Cancer: Lyon, France, 2017; Volume XI.

4. Cruz-Gregorio, A.; Martinez-Ramirez, I.; Pedraza-Chaverri, J.; Lizano, M. Reprogramming of Energy Metabolism in Response to Radiotherapy in Head and Neck Squamous Cell Carcinoma. *Cancers* **2019**, *11*, 182. [CrossRef] [PubMed]

5. Al-Jaber, A.; Al-Nasser, L.; El-Metwally, A. Epidemiology of oral cancer in Arab countries. *Saudi Med. J.* **2016**, *37*, 249–255. [CrossRef] [PubMed]

6. Isayeva, T.; Li, Y.; Maswahu, D.; Brandwein-Gensler, M. Human papillomavirus in non-oropharyngeal head and neck cancers: A systematic literature review. *Head. Neck. Pathol.* **2012**, *6* (Suppl. 1), S104–S120. [CrossRef]

7. Termine, N.; Panzarella, V.; Falaschini, S.; Russo, A.; Matranga, D.; Lo Muzio, L.; Campisi, G. HPV in oral squamous cell carcinoma vs. head and neck squamous cell carcinoma biopsies: A meta-analysis (1988–2007). *Ann. Oncol.* **2008**, *19*, 1681–1690. [CrossRef]

8. Serrano, B.; Brotons, M.; Bosch, F.X.; Bruni, L. Epidemiology and burden of HPV-related disease. *Best Pract. Res. Clin. Obstet. Gynaecol.* **2018**, *47*, 14–26. [CrossRef]

9. Marur, S.; D'Souza, G.; Westra, W.H.; Forastiere, A.A. HPV-associated head and neck cancer: A virus-related cancer epidemic. *Lancet Oncol.* **2010**, *11*, 781–789. [CrossRef]

10. Ngan, H.L.; Wang, L.; Lo, K.W.; Lui, V.W.Y. Genomic Landscapes of EBV-Associated Nasopharyngeal Carcinoma vs. HPV-Associated Head and Neck Cancer. *Cancers* **2018**, *10*, 210. [CrossRef]

11. Syrjanen, K.; Syrjanen, S.; Lamberg, M.; Pyrhonen, S.; Nuutinen, J. Morphological and immunohistochemical evidence suggesting human papillomavirus (HPV) involvement in oral squamous cell carcinogenesis. *Int. J. Oral Surg.* **1983**, *12*, 418–424. [CrossRef]

12. Loning, T.; Ikenberg, H.; Becker, J.; Gissmann, L.; Hoepfer, I.; zur Hausen, H. Analysis of oral papillomas, leukoplakias, and invasive carcinomas for human papillomavirus type related DNA. *J. Investig. Dermatol.* **1985**, *84*, 417–420. [CrossRef] [PubMed]

13. Gillison, M.L.; Chaturvedi, A.K.; Anderson, W.F.; Fakhry, C. Epidemiology of Human Papillomavirus-Positive Head and Neck Squamous Cell Carcinoma. *J. Clin. Oncol.* **2015**, *33*, 3235–3242. [CrossRef] [PubMed]

14. Sturgis, E.M.; Cinciripini, P.M. Trends in head and neck cancer incidence in relation to smoking prevalence: An emerging epidemic of human papillomavirus-associated cancers? *Cancer* **2007**, *110*, 1429–1435. [CrossRef] [PubMed]

15. Vigneswaran, N.; Williams, M.D. Epidemiologic trends in head and neck cancer and aids in diagnosis. *Oral Maxillofac. Surg. Clin. N. Am.* **2014**, *26*, 123–141. [CrossRef] [PubMed]

16. Ndiaye, C.; Mena, M.; Alemany, L.; Arbyn, M.; Castellsague, X.; Laporte, L.; Bosch, F.X.; de Sanjose, S.; Trottier, H. HPV DNA, E6/E7 mRNA, and p16INK4a detection in head and neck cancers: A systematic review and meta-analysis. *Lancet Oncol.* **2014**, *15*, 1319–1331. [CrossRef]

17. Nauta, I.H.; Rietbergen, M.M.; van Bokhoven, A.; Bloemena, E.; Witte, B.I.; Heideman, D.A.M.; Baatenburg de Jong, R.J.; Brakenhoff, R.H.; Leemans, C.R. Evaluation of the 8th TNM classification on p16-positive oropharyngeal squamous cell carcinomas in the Netherlands, and the importance of additional HPV DNA-testing. *Ann. Oncol.* **2018**. [CrossRef] [PubMed]

18. Fung, N.; Faraji, F.; Kang, H.; Fakhry, C. The role of human papillomavirus on the prognosis and treatment of oropharyngeal carcinoma. *Cancer Metastasis Rev.* **2017**, *36*, 449–461. [CrossRef]

19. Dok, R.; Nuyts, S. HPV Positive Head and Neck Cancers: Molecular Pathogenesis and Evolving Treatment Strategies. *Cancers* **2016**, *8*, 41. [CrossRef]

20. Wagner, S.; Sharma, S.J.; Wuerdemann, N.; Knuth, J.; Reder, H.; Wittekindt, C.; Klussmann, J.P. Human Papillomavirus-Related Head and Neck Cancer. *Oncol. Res. Treat.* **2017**, *40*, 334–340. [CrossRef]

21. Lassen, P.; Eriksen, J.G.; Hamilton-Dutoit, S.; Tramm, T.; Alsner, J.; Overgaard, J. Effect of HPV-associated p16INK4A expression on response to radiotherapy and survival in squamous cell carcinoma of the head and neck. *J. Clin. Oncol.* **2009**, *27*, 1992–1998. [CrossRef]

22. Deschuymer, S.; Dok, R.; Laenen, A.; Hauben, E.; Nuyts, S. Patient Selection in Human Papillomavirus Related Oropharyngeal Cancer: The Added Value of Prognostic Models in the New TNM 8th Edition Era. *Front. Oncol.* **2018**, *8*, 273. [CrossRef] [PubMed]

23. Alsbeih, G. Exploring the Causes of the Low Incidence of Cervical Cancer in Western Asia. *Asian Pac. J. Cancer Prev.* **2018**, *19*, 1425–1429. [CrossRef] [PubMed]

24. Mehanna, H.; Franklin, N.; Compton, N.; Robinson, M.; Powell, N.; Biswas-Baldwin, N.; Paleri, V.; Hartley, A.; Fresco, L.; Al-Booz, H.; et al. Geographic variation in human papillomavirus-related oropharyngeal cancer: Data from 4 multinational randomized trials. *Head Neck* **2016**, *38* (Suppl. 1), E1863–E1869. [CrossRef]

25. Al-Shahrani, Z.S.; Al-Rawaji, A.I.; Al-Madouj, A.N.; Hayder, M.S. *Saudi Cancer Registry: Cancer Incidence Report Saudi Arabia 2014*; Saudi Health Council: Riyadh, Saudi Arabia, 2017; pp. 1–82.

26. Alsbeih, G. HPV Infection in Cervical and Other Cancers in Saudi Arabia: Implication for Prevention and Vaccination. *Front. Oncol.* **2014**, *4*, 65. [CrossRef] [PubMed]

27. Wang, F.; Zhang, H.; Xue, Y.; Wen, J.; Zhou, J.; Yang, X.; Wei, J. A systematic investigation of the association between HPV and the clinicopathological parameters and prognosis of oral and oropharyngeal squamous cell carcinomas. *Cancer Med.* **2017**, *6*, 910–917. [CrossRef] [PubMed]

28. Hesham, A.; Syed, K.; Jamal, B.; Alqahtani, A.; Alfaqih, A.; Alshehry, H.; Hameed, M.; Mustafa, A. Incidence, clinical presentation, and demographic factors associated with oral cancer patients in the southern region of Saudi Arabia: A 10-year retrospective study. *J. Int. Oral Health* **2017**, *9*, 105–109. [CrossRef]

29. Kreimer, A.R.; Clifford, G.M.; Boyle, P.; Franceschi, S. Human papillomavirus types in head and neck squamous cell carcinomas worldwide: A systematic review. *Cancer Epidemiol. Prev. Biomarkers* **2005**, *14*, 467–475. [CrossRef]

30. Vidal Loustau, A.C.; Dulguerov, N.; Curvoisier, D.; McKee, T.; Lombardi, T. Low prevalence of HPV-induced oral squamous cell carcinoma in Geneva, Switzerland. *Oral Dis.* **2019**. [CrossRef]

31. Alsbeih, G.A.; Al-Harbi, N.M.; Bin Judia, S.S.; Khoja, H.A.; Shoukri, M.M.; Tulbah, A.M. Reduced rate of human papillomavirus infection and genetic overtransmission of TP53 72C polymorphic variant lower cervical cancer incidence. *Cancer* **2017**, *123*, 2459–2466. [CrossRef]

32. Santegoets, L.A.; van Seters, M.; Heijmans-Antonissen, C.; Kleinjan, A.; van Beurden, M.; Ewing, P.C.; Kuhne, L.C.; Beckmann, I.; Burger, C.W.; Helmerhorst, T.J.; et al. Reduced local immunity in HPV-related VIN: Expression of chemokines and involvement of immunocompetent cells. *Int. J. Cancer* **2008**, *123*, 616–622. [CrossRef]

33. Weinberger, P.M.; Yu, Z.; Haffty, B.G.; Kowalski, D.; Harigopal, M.; Brandsma, J.; Sasaki, C.; Joe, J.; Camp, R.L.; Rimm, D.L.; et al. Molecular classification identifies a subset of human papillomavirus—Associated oropharyngeal cancers with favorable prognosis. *J. Clin. Oncol.* **2006**, *24*, 736–747. [CrossRef] [PubMed]

34. Ang, K.K.; Harris, J.; Wheeler, R.; Weber, R.; Rosenthal, D.I.; Nguyen-Tan, P.F.; Westra, W.H.; Chung, C.H.; Jordan, R.C.; Lu, C.; et al. Human papillomavirus and survival of patients with oropharyngeal cancer. *N. Engl. J. Med.* **2010**, *363*, 24–35. [CrossRef] [PubMed]

35. Mehra, R.; Ang, K.K.; Burtness, B. Management of human papillomavirus-positive and human papillomavirus-negative head and neck cancer. *Semin. Radiat. Oncol.* **2012**, *22*, 194–197. [CrossRef] [PubMed]

36. Bussu, F.; Ragin, C.; Boscolo-Rizzo, P.; Rizzo, D.; Gallus, R.; Delogu, G.; Morbini, P.; Tommasino, M. HPV as a marker for molecular characterization in head and neck oncology: Looking for a standardization of clinical use and of detection method(s) in clinical practice. *Head Neck* **2019**, *41*, 1104–1111. [CrossRef] [PubMed]

37. Combes, J.D.; Franceschi, S. Role of human papillomavirus in non-oropharyngeal head and neck cancers. *Oral Oncol.* **2014**, *50*, 370–379. [CrossRef] [PubMed]

38. Zafereo, M.E.; Xu, L.; Dahlstrom, K.R.; Viamonte, C.A.; El-Naggar, A.K.; Wei, Q.; Li, G.; Sturgis, E.M. Squamous cell carcinoma of the oral cavity often overexpresses p16 but is rarely driven by human papillomavirus. *Oral Oncol.* **2016**, *56*, 47–53. [CrossRef] [PubMed]

39. Albers, A.E.; Qian, X.; Kaufmann, A.M.; Coordes, A. Meta analysis: HPV and p16 pattern determines survival in patients with HNSCC and identifies potential new biologic subtype. *Sci. Rep.* **2017**, *7*, 16715. [CrossRef] [PubMed]

40. Lop, J.; Garcia, J.; Lopez, M.; Taberna, M.; Mena, M.; Alemany, L.; Quer, M.; Leon, X. Competing mortality in oropharyngeal carcinoma according to human papillomavirus status. *Head Neck* **2019**, *41*, 1328–1334. [CrossRef]

41. Lassen, P.; Primdahl, H.; Johansen, J.; Kristensen, C.A.; Andersen, E.; Andersen, L.J.; Evensen, J.F.; Eriksen, J.G.; Overgaard, J.; Danish, H.; et al. Impact of HPV-associated p16-expression on radiotherapy outcome in advanced oropharynx and non-oropharynx cancer. *Radiother. Oncol.* **2014**, *113*, 310–316. [CrossRef] [PubMed]

42. Chen, W.S.; Bindra, R.S.; Mo, A.; Hayman, T.; Husain, Z.; Contessa, J.N.; Gaffney, S.G.; Townsend, J.P.; Yu, J.B. CDKN2A Copy Number Loss Is an Independent Prognostic Factor in HPV-Negative Head and Neck Squamous Cell Carcinoma. *Front. Oncol* **2018**, *8*, 95. [CrossRef] [PubMed]

43. Verma, G.; Vishnoi, K.; Tyagi, A.; Jadli, M.; Singh, T.; Goel, A.; Sharma, A.; Agarwal, K.; Prasad, S.C.; Pandey, D.; et al. Characterization of key transcription factors as molecular signatures of HPV-positive and HPV-negative oral cancers. *Cancer Med.* **2017**, *6*, 591–604. [CrossRef]

44. Taberna, M.; Torres, M.; Alejo, M.; Mena, M.; Tous, S.; Marquez, S.; Pavon, M.A.; Leon, X.; Garcia, J.; Guix, M.; et al. The Use of HPV16-E5, EGFR, and pEGFR as Prognostic Biomarkers for Oropharyngeal Cancer Patients. *Front. Oncol.* **2018**, *8*, 589. [CrossRef] [PubMed]

45. De Felice, F.; Polimeni, A.; Valentini, V.; Brugnoletti, O.; Cassoni, A.; Greco, A.; de Vincentiis, M.; Tombolini, V. Radiotherapy Controversies and Prospective in Head and Neck Cancer: A Literature-Based Critical Review. *Neoplasia* **2018**, *20*, 227–232. [CrossRef] [PubMed]

46. Lechner, M.; Chakravarthy, A.R.; Walter, V.; Masterson, L.; Feber, A.; Jay, A.; Weinberger, P.M.; McIndoe, R.A.; Forde, C.T.; Chester, K.; et al. Frequent HPV-independent p16/INK4A overexpression in head and neck cancer. *Oral Oncol.* **2018**, *83*, 32–37. [CrossRef] [PubMed]

47. Mena, M.; Taberna, M.; Tous, S.; Marquez, S.; Clavero, O.; Quiros, B.; Lloveras, B.; Alejo, M.; Leon, X.; Quer, M.; et al. Double positivity for HPV-DNA/p16(ink4a) is the biomarker with strongest diagnostic accuracy and prognostic value for human papillomavirus related oropharyngeal cancer patients. *Oral Oncol.* **2018**, *78*, 137–144. [CrossRef]

48. Gregoire, V.; Lefebvre, J.L.; Licitra, L.; Felip, E.; EHNS–ESMO–ESTRO guidelines working group. Squamous cell carcinoma of the head and neck: EHNS-ESMO-ESTRO Clinical Practice Guidelines for diagnosis, treatment and follow-up. *Ann. Oncol.* **2010**, *21* (Suppl. 5), v184–v186. [CrossRef] [PubMed]

49. Alsbeih, G.; Ahmed, R.; Al-Harbi, N.; Venturina, L.A.; Tulbah, A.; Balaraj, K. Prevalence and genotypes' distribution of human papillomavirus in invasive cervical cancer in Saudi Arabia. *Gynecol. Oncol.* **2011**, *121*, 522–526. [CrossRef]

50. Barasch, S.; Mohindra, P.; Hennrick, K.; Hartig, G.K.; Harari, P.M.; Yang, D.T. Assessing p16 Status of Oropharyngeal Squamous Cell Carcinoma by Combined Assessment of the Number of Cells Stained and the Confluence of p16 Staining: A Validation by Clinical Outcomes. *Am. J. Surg. Pathol.* **2016**, *40*, 1261–1269. [CrossRef]

MDPI AG
Grosspeteranlage 5
4052 Basel
Switzerland
Tel.: +41 61 683 77 34

*Cancers* Editorial Office
E-mail: cancers@mdpi.com
www.mdpi.com/journal/cancers